MEDICAL PHYSICS MONOGRAPH NO. 13

MULTIPLE REGRESSION ANALYSIS: APPLICATIONS IN THE HEALTH SCIENCES

Edited by

Donald E. Herbert, Ph.D.
Head, Division of Medical Physics & Statistics
Department of Radiology
University of South Alabama, College of Medicine
Mobile, Alabama

and

Raymond H. Myers, Ph.D.
Director of Statistics Consulting Center
Virginia Polytechnic Institute & State University
Blacksburg, Virginia

Proceedings of First Midyear Topical Symposium, "Multiple Regression Analysis: Applications in the Health Sciences," 12–16 March 1984 in Mobile, Alabama, sponsored by the American Association of Physicists in Medicine. The National Institutes of Health participated in the support of this meeting under Grant No. 1 R13 CA37017-01 from the National Cancer Institute.

Published for the
American Association of Physicists in Medicine
by the American Institute of Physics

Further copies of this report may be obtained from:

American Institute of Physics
Publications Sales Dept.
335 East 45 Street
New York, New York 10017

Library of Congress Catalog Card Number: 86-70083
International Standard Book Number: 0-88318-490-7
International Standard Serial Number: 0163–1802

Copyright © 1986 by the American Association
of Physicists in Medicine

All rights reserved. No part of this publication may be
reproduced, stored in a retrieval system, or transmitted
in any form or by any means (electronic, mechanical,
photocopying, recording, or otherwise) without the
prior written permission of the publisher.

Published by the American Institute of Physics, Inc.
335 East 45 Street, New York, NY 10017

Printed in the United States of America

PREFACE

Whatever scientific progress may be, we must account for it by examining the nature of the scientific group, discovering what it values, what it tolerates, what it disdains.

Thomas Kuhn 1970a

The AAPM First Midyear Topical Symposium, "Multiple Regression Analysis: Applications in the Health Sciences," was organized as a first response to the well-documented criticisms that "The standard of statistics in the medical journals is poor" and the recognition that the misuse of statistics in medical science involves ethical and social as well as "bottom-line" and scientific issues (Altman 1982).[a] Multiple Regression Analysis was chosen because several other studies have disclosed that it is the statistical methodology most widely used in the sciences. The Symposium was held 12–16 March 1984 in Mobile, Alabama. This monograph constitutes the Proceedings of the Symposium.

Until now there has been no foregathering of statisticians and physicists by which the latter could be made familiar with the recent, attractive, interesting, and important developments that have occurred in regression methods. However, the enormous success that has been achieved by Professor Raymond Myers in his expositions of regression methods to other groups of scientists and engineers in three to five day "single-shot" courses given over the past several years provided both inspiration and examplar for the present Symposium. The Symposium presents the physicist with detailed and lucid expositions of both standard and state-of-the-art insights and methods of Multiple Regression Analysis by leading statisticians, all of whom have played major roles in the innovation and development of these methods, etc.

The aim of the Symposium was threefold. The first two objectives may be said to have been that of strengthening two of the norms of the scientific institution: 1) to expand the allocation of competent response to the challenge of constructing general linear models from experimental and non-experimental observations and 2) to further organize, inform, and refine the capacity for responsible skepticism and disbelief. Both are achieved by increasing the size and sophistication of the statistics armamentarium—methods *and* insights—of the medical physicist.[b] In this way he or she will be able to provide more effective, imaginative and timely solutions to the statistical aspects of many of the problems that arise both in the course of their own work and in that of their colleagues, the physicians and administrators. This specific augmentation will enlarge their capacity for such work to a level that is more commensurate with their responsibility as a scientist—a problem-oriented, rather than a discipline-limited, professional.

With their increased knowledge, understanding, and insight will come an increased capacity to discriminate between "good" and "bad" statistics, in particular, an increased sensitivity to the presence of errors in statistics in the recent and current medical literature, and the ethical and social, as well as the scientific consequences thereof, will develop. The latter sensitivity is a nontrivial one. Anyone who thoughtfully examines much of the current medical literature may well conclude (to borrow one of Kuhn's locutions) that "... though the fields' practitioners [are] scientists, the net result [is] something less than science." (Kuhn 1970b) But it is, of course, the case that if the physicist makes use of, or refers to, or cites any report in which the results or recommendations derive from a statistical analysis of data, it is his or her *professional responsibility* to be able to competently evaluate the analysis and hence the nature and degree of the *validity* of these results and recommendations that must circumscribe their generalization and exploitation by "reasonable and prudent men."

The third objective was to make the physicist more aware of the intrinsic "scientific" aspect of many of the problems in statistics and of the elegance of much of the methodology.

M. J. R. Healy (1978) has raised the issue of whether statistics is a *science* or a *technology* and after some equivocation appears to opt for the latter. The question does not seem to be altogether moot. R. E. Gomory (1983), in a paper which addressed such topics as the effect of cultural factors on technology transfer, also contrasted science with technology. Gomory pointed out that, with respect to science, technology is different. However, perhaps the most important thing about technology is that it tends to be very complex," and "Simple things [like science?] are hard to keep secret, and complex things like technology are hard to give away." (This seems an accurate description of the experience of many who have taught statistics to first year medical students—or even to residents.) But with respect to the present Symposium and its aims, another remark of Gomory's seems especially appropriate: "... in technology you never run out of ideas, just out of time," for here it is definitely the case that the lectures presented by the statisticians, who are the principal speakers of the Symposium, are not at all exhaustive of their respective "technology." There remains an enormous amount of useful methods and insights which could only be hinted at in the time alloted to each lecture. Therefore, the reader is strongly urged to continue his or her studies in Multiple Regression Analysis from the many excellent references cited in each lecture.

The members of the Symposium Committee were Gary Barnes, Ph.D., Donald Frey, Ph.D., Donald Herbert, Ph.D. (Chairman), Robert Shalek, Ph.D., Larry Simpson, Ph.D., and Jon Trueblood, Ph.D.

The Symposium was organized and conducted under the aegis of the AAPM (President Edward Sternick, Ph.D.) and of the SEAAPM (President Edward Chaney, Ph.D.). It was supported by Grant No. 1 R13 CA37017-01 from the National Cancer Institute.

The organization and conduct of the Symposium were further assisted by many individuals. Notable among these are A. E. Robinson, M.D., Chairman of Radiology and B. G. Brogdon, M.D., Assistant Dean for Medical Education at the College of Medicine, University of South Alabama, and A. R. Smith, Ph.D. and Francis Mahoney, Ph.D. of the National Cancer Institute.

The moderators of the Symposium were Jerry Allison, Ph.D. and Casimir Eubig, Ph.D. of the Medical College of Georgia, Michael Yester, Ph.D. of the University of Alabama at Birmingham, John Lefante, Ph.D., Robert Mee, Ph.D., Alvin Rainosek, Ph.D., and Arvind Shah, Ph.D. of the University of South Alabama, Dept. of Statistics.

The special contributions of Alvin Rainosek, Ph.D., Head, Division of Statistics, University of South Alabama, to the format, organization and substance of the Symposium must be acknowledged.

The major contributions of Marie Woodall, B.S., Executive Secretary for the Symposium and Editorial Assistant for the Proceedings, must be acknowledged.

We also wish to acknowledge the contributions of the following sponsors: ADAC Labs, Atomic Energy of Canada Ltd., BMDP Statistical Software, DuPont Corp., Gammex Inc., S&H X-Ray, SPSS Inc., Varian Medical Services, and Victoreen Nuclear Assn.

D.E.H.

[a] "When a paper containing incorrect results (not necessarily through statistical mistakes) is published there may be serious consequences, although surprisingly this does not seem to be generally appreciated.

(i) Subjects used in the research have been put at risk or inconvenienced for no benefit.

(ii) Other resources have been diverted from more worthwhile use.

(iii) Other patients may subsequently receive an inferior treatment either as a direct consequence of the findings of the study or possibly by delaying the introduction of a better treatment.

(iv) Other scientists' research may be affected.

(v) If the results go unchallenged the researcher(s) involved may use the same substandard statistical methods again in subsequent work and others may copy them.

All of the above points make the misuse of statistics very much an ethical as well as a scientific issue." (Altman 1982).

But, on the other hand, perhaps, "Not to worry!", the seriousness of the consequences may be somewhat abated by an idiosyncracy of the peer group: "...scientists have a strong urge to write papers but only a relatively mild one to read them." (Price 1963). It has been (reliably) estimated that much less than 1% of the peer-reviewed scientific literature is ever read by the peer-group that it services—cited maybe, but *not* read: "For every person who reads the whole of a text of a scientific paper, twenty read through the summary and 500 read the title and stop there. Most papers have their title read and no more."

[b] For "If the only tool you have is a hammer, you tend to treat everything as if it were a nail." (A. Maslow 1966).

REFERENCES

ALTMAN, D.G. (1982) Statistics in Medical Journals. *Statistics in Medicine.* 1: 59-71.
GOMORY, R.E. (1983) Technology Development. *Science.* 220: 576-80.
HEALY, M.J.R. (1978) Is Statistics a Science? *J. Royal Statis. Assoc. A.* 141: (3) 385-93.
KUHN, T. (1970a) Logic of Discovery of Psychology of Research? in *Criticism and the Growth of Knowledge.* I. Lakatos & A. Musgrave, eds., Cambridge Univ. Press, Cambridge.
KUHN, T. (1970b) *The Structure of Scientific Revolutions.* 2nd ed. Int. Encycl. of Unified Science. Vols. I & II. Univ. of Chicago Press. Chicago.
MASLOW, A. (1966) *The Psychology of Science: A Reconnaissance.* Harper and Row, N.Y.
PRICE, D.J. (1963) *Little Science, Big Science.* Columbia University Press, N.Y.

TABLE OF CONTENTS

Welcome Addresses 1
    Arvin E. Robinson, M.D.
    Edward S. Sternick, Ph.D.

Introduction 2
    Donald E. Herbert, Ph.D.

Introduction to Regression Methods. I and II
"Regression Analysis—Basic and Current Frontiers" 4
    Raymond H. Myers, Ph.D.

An Introduction to Regression Diagnostics 17
    Roy E. Welsch, Ph.D.

Biased Estimation and Robust Regression 34
    Douglas C. Montgomery, Ph.D.

Bayesian Regression and Sensitivity Analysis 58
    Edward E. Leamer, Ph.D.

Statistical Graphics and Personal Computer Software for
Regression Methods. "Statistical Graphics on Smaller Computers:
The Data Analyst's New Tools" 75
    Thomas J. Boardman, Ph.D.

Regression Methods for Binomial and Poisson Distributed Data 84
    Edward L. Frome, Ph.D.

Regression Methods in Survival Analysis 124
    Edmund A. Gehan, Ph.D.

Regression Methods for Laboratory and Clinical Experiments.
"Animal Experiments and Clinical Trials" 132
    David A. Schoenfeld, Ph.D.

Overview of Discriminant Analysis 146
    Peter A. Lachenbruch, Ph.D.

Estimate of Radionuclide Intakes From Repetitive Bioassay
Measurements 162
    Kenneth W. Skrable, Ph.D., G. E. Chabot, Ph.D.,
    and C. S. French, Ph.D.

Regression Methods in Clinical Radiobiology 192
    Howard D. Thames, Ph.D., Susan L. Tucker, Ph.D.,
    Shelley L. Rasmussen, Ph.D., and Jerry W. McLarty, Ph.D.

**Clinical Dose-Response Models. I. Regression Diagnostics
and Biased Estimation** 208
    Donald E. Herbert, Ph.D.

**Clinical Radiocarcinogenesis. Applications of Regression
Diagnostics and Bayesian Methods to Poisson Regression Models** 307
    Donald E. Herbert, Ph.D.

**Statistical Analysis of ROC Data in Evaluating Diagnostic
Performance. I and II** 365
    Charles E. Metz, Ph.D.

**Clinical Dose-Response Models. II. Probit and Logit Models
of Experimental and Non-Experimental Data** 385
    Donald E. Herbert, Ph.D.

**Sampled Mean Radiation Treatment Design: Physical and
Biophysical Parameters** 454
    Larry D. Simpson, Ph.D.

**Clinical Diagnostic Models. Discriminant Analysis of *In Vitro*
NMR Measurements on Normal and Malignant Tissues** 487
    Donald E. Herbert, Ph.D.

Non-Linear Regression Analysis in Nuclear Medicine 546
    Michael D. Harpen, Ph.D.

Regression Models and Actuarial Curves. I and II 557
    Richard S. Cox, Ph.D.

Legal Questions in Statistics 579
    Robert J. Shalek, Ph.D.

Summary of Regression Methods 585
    Raymond H. Myers, Ph.D.

**Regression Methods for Problems of "Identification" in
Health Sciences** 588
    Gary T. Barnes, Ph.D.

The Use of Statistics by Medical Physicists 591
    G. Donald Frey, Ph.D.

Regression Methods for Problems of "Yield" in Health Sciences
    Jon H. Trueblood, Ph.D. 592
    Larry D. Simpson, Ph.D. 596

Impressions and Perspectives 597
    Robert J. Shalek, Ph.D.

Closing Comments 598
    Edward L. Chaney, Ph.D.

SYMPOSIUM COMMITTEE

Gary T. Barnes, Ph.D.
University of Alabama
   at Birmingham
Birmingham, AL

Robert J. Shalek, Ph.D.
University of Texas Systems
   Cancer Center
M. D. Anderson Hospital
Houston, TX

Edward L. Chaney, Ph.D.
North Carolina Memorial Hospital
Chapel Hill, NC
(ex officio)

Larry D. Simpson, Ph.D.
University of Rochester
College of Medicine
Rochester, NY

G. Donald Frey, Ph.D.
Medical University of
   South Carolina
Charleston, SC

Edward S. Sternick, Ph.D.
Tufts-New England
   Medical Center
Boston, MA
(ex officio)

Donald E. Herbert, Ph.D.
University of South Alabama
College of Medicine
Mobile, AL

Jon H. Trueblood, Ph.D.
Medical College of Georgia
Augusta, GA

ORGANIZATIONAL CONTRIBUTORS

National Cancer Institute
AAPM Continuing Education Committee
Office of Continuing Medical Education of the
College of Medicine, University of South Alabama
Southeast Chapter of the AAPM

CONBRIBUTORS

GARY T. BARNES, University of Alabama at Birmingham, Dept. of Radiology, Birmingham, AL

THOMAS J. BOARDMAN, Colorado State University, Dept. of Statistics, Fort Collins, CO

EDWARD L. CHANEY, University of North Carolina, School of Medicine, Dept. of Radiation Oncology, Chapel Hill, NC

RICHARD S. COX, Stanford University Medical Center, Dept. of Radiology, Stanford, CA

G. DONALD FREY, Medical University of South Carolina, Dept. of Radiology, Charleston, SC

EDWARD L. FROME, Oak Ridge National Laboratory, Mathematics and Statistics Research Dept., Oak Ridge, TN

EDMUND A. GEHAN, University of Texas Systems Cancer Center, M. D. Anderson Hospital, Dept. of Biomathematics, Houston, TX

MICHAEL D. HARPEN, University of South Alabama, College of Medicine, Dept. of Radiology, Mobile, AL

DONALD E. HERBERT, University of South Alabama, College of Medicine, Dept. of Radiology, Mobile, AL

PETER A. LACHENBRUCH, University of Iowa, Dept. of Preventive Medicine, Iowa City, IA

EDWARD E. LEAMER, University of California at Los Angeles, Dept. of Economics, Los Angeles, CA

CHARLES E. METZ, University of Chicago, Dept. of Radiology, Chicago, IL

DOUGLAS C. MONTGOMERY, Georgia Institute of Technology, School of Industrial and Systems Engineering, Atlanta, GA

RAYMOND H. MYERS, Virginia Polytechnic Institute and State University, Dept. of Statistics, Blacksburg, VA

ARVIN E. ROBINSON, University of South Alabama, College of Medicine, Dept. of Radiology, Mobile, AL

DAVID A. SCHOENFELD, Sidney Farber Cancer Institute, Dept. of Bio-Statistics and Epidemiology, Boston, MA

ROBERT J. SHALEK, University of Texas Systems Cancer Center, M. D. Anderson Hospital, Dept. of Physics, Houston, TX

LARRY D. SIMPSON, University of Rochester Cancer Center, Div. of Radiation Oncology, Rochester, NY

KENNETH W. SKRABLE, University of Lowell, Dept. of Pure and Applied Physics, Lowell, MA

EDWARD S. STERNICK, Tufts-New England Medical Center, Div. of Medical Physics, Boston, MA

HOWARD D. THAMES, University of Texas Systems Cancer Center, M. D. Anderson Hospital, Dept. of Biomathematics, Houston, TX

JON H. TRUEBLOOD, Medical College of Georgia, Dept. of Radiology, Augusta, GA

ROY E. WELSCH, Massachusetts Institute of Technology, Sloan School of Management, Dept. of Economics and Management Sciences, Cambridge, MA

WELCOMING ADDRESSES

It is a pleasure for me to open this program on "Multiple Regression Analysis: Applications in the Health Sciences". It is most essential that physicists, statisticians and physicians have the opportunity to meet together on such important topics in order to provide a better understanding of statistical analysis for research and investigative techniques. Diagnostic radiologists and radiotherapists in clinical practice are dependent on physics support in the analysis of data and mathematical design of experimental protocols. Although many of the fine details discussed in this setting are beyond my capabilities, I know that the outcome of your discussions will directly benefit academic radiology.

I would like to recognize the efforts of Donald Herbert in organizing this program and the support he has been given by the American Association of Physicists in Medicine and the AAPM Symposium Committee, AAPM Continuing Education Committee and the Southeast Chapter of the AAPM. It is a pleasure for me to welcome you to Mobile and the University of South Alabama.

> Arvin E. Robinson, M.D., Professor and Chairman,
> Dept. of Radiology, University of South Alabama,
> College of Medicine, Mobile, Alabama 36688

John Naisbitt, the author of Megatrends, indicates that we are now an "information society" in which scientific and technical information is increasing at a rate of 13 per cent per year with a doubling time of 5.5 years. Within a short time, it is predicted that the rate may increase to 40 per cent because of an increasing population of scientists and the explosive growth of computer-based information systems. This, in turn, will decrease the information doubling time to about 1.5 years. Naisbitt points out that "we are drowning in information but starved for knowledge".

One of the important purposes for which the American Association of Physicists in Medicine is organized is to disseminate scientific and technical knowledge in medical physics and related fields in a manner designed to bring some order to the "chaos of information pollution". Data that otherwise might be of little use will thus be given value. The forum provided at this First Midyear Topical Symposium on "Multiple Regression Analysis: Applications in the Health Sciences" serves that avowed purpose extremely well. Multiple regression techniques have proved to be excellent analytical tools, useful in identifying both the dynamics and causal factors of complex systems in a number of health care disciplines in which medical physicists are expected to take scientific leadership. The organizers of this symposium have shown courage and wisdom in creating a comprehensive, well-structured approach to a most important and wide-ranging topic that heretofore has been given insufficient attention by the medical physics community. Their efforts are reflected in an outstanding faculty presenting a carefully developed program that will enrich both the participants at the meeting and the readers of the published Proceedings.

> Edward S. Sternick, Ph.D., President, American Association of Physicists in Medicine 1984, Tufts-New England Medical Center
> Dept. of Therapeutic Radiology, Boston, Massachusetts 02111

INTRODUCTION

"All models are wrong but some are useful." (G.E.P. Box 1979)

The medical physicist may (or should) often find himself in the circumstances described by Baker and Nelder (1978): "Given a set of sample data we can now describe, in terms of a [model], a theoretical population to which the data are believed to belong. Two questions arise:
a) Do the data support the belief that the proposed [model] is a reasonable description of the population?
b) Since the parameters of the [model] are unknown what values do the data suggest for them?"
In providing himself, or his clients, with responsible and useful answers to these questions - as well as other inferences from his model - the physicist must recognize that, "A regression [model] is constructed using prior knowledge, data, ... and a fitting (estimation) process of some form. It is important to know when the resulting [model estimates] depends heavily on a small part of the prior knowledge, on a small part of the data, or on the exact choice of model or fitting process." (Welsch 1984)
G.E.P. Box (1984) has remarked that there are two kinds of inference required in the construction of useful models: "One kind of inference that may be called <u>criticism</u> involves the <u>contrasting</u> of what might be expected if the assumptions A of some tentative model of interest were true with the data y_d that actually occur. This is conveniently symbolized by subtraction: $y_d - A$. The other kind of inference, which may be called <u>estimation</u>, involves the <u>combination</u> of observed data y_d with the assumptions A of some model tentatively assumed to be true. This process is conveniently symbolized by <u>addition</u>: $y_d + A$." Box further points out that the methods, e.g., goodness of fit, regression diagnostics, etc., for model criticism can be best motivated and justified by Sampling Theory whereas, the methods, e.g., least squares, maximum likelihood, mixed, robust and ridge estimation, for model estimation are better motivated and justified by Bayes' theorem.
Box (1979) and more recently Cook and Weissberg (1982) have described the construction of linear models of sample data (obtained by either an experiment or a non-experiment) as proceeding by <u>iteration</u> of the basic cycle:

                      Estimation
Assumptions ⎯⎯⎯⎯⎯⎯⎯⎯⎯⎯⎯ Consequences
                      Criticism

This sequence is terminated when a model has been constructed so that the sample of data is mapped into a "white noise" sequence by the model (Box 1979).[a]
Leamer's recent (and somewhat iconoclastic) remarks anent inference are also (characteristically) illuminating, "The 'facts' used for statistical inference about [the parameter vector] θ are first the data, symbolized by x, second, a conditional probability density known as a sampling distribution, $f(x|\theta)$, and third, explicitly for a Bayesian and implicitly for 'all others' a marginal or prior probability density function $f(\theta)$ ..." But, "<u>What is a fact? A fact is merely an opinion held by all</u>, or at least <u>held by a set of people you regard to be a close approximation to all</u>." (Leamer 1983)[b]
Lectures <u>1</u> - <u>10</u> will present the views of nine world-class

statisticians on these and other aspects of inferences on general linear models. (McCullagh and Nelder 1983)

Box (1984) has further remarked that the faculties for regression analysis seem to be "hard-wired" into us: "In most people the left half of the cerebral cortex is concerned primarily with language and logical deduction, which plays a major role in estimation, while the right half is concerned primarily with images, patterns and inductive processes which play a major role in criticism." It is usually the case that heritable faculties have considerable "survival value" for their possessors. This is indeed the case for the regression faculties (firmware?) of the medical physicist as the Lectures 11 to 23 and the further remarks of Drs. Barnes, Chaney, Frey, Robinson, Shalek, Simpson, Sternick and Trueblood will disclose.

Lectures 13, 14, 17 and 19 may be read as instances of secondary analysis: "Primary analysis is the original analysis of data ... It is what one typically imagines as the application of statistical methods. Secondary analysis is the reanalysis of data for the purpose of answering the original ... question with better statistical techniques or answering new questions with old data." (Glass 1981) The, "... benefits of a secondary analysis include the verification and refinement of original findings and the refutation of them." (Hedrick 1985). Thus, a characteristic feature of this enterprise is the identification and evaluation of the effects of study weaknesses on study findings. In the published studies cited in these four lectures secondary analysis leads to the surprising conclusion that the data in each appears to have more value for heuristics than it has as evidence on the subject matter issues that motivated its initial collection.

D.E.H.

a Recall Popper's characterization of "Science" as, "... an alternating series of speculative conjectures and empirical refutations." (Popper 1968).
b "Reality - What a concept!" I. Hacking

REFERENCES

BAKER, R.J. and NELDER, J.A. (1978) The GLIM System. Release 3: Generalized Linear Interactive Modelling. Distributed by Numerical Algorithms Group. Oxford, England.

BOX, G.E.P. (1979) Robustness in the Strategy of Scientific Model Building in Robustness In Statistics. R.L. Launer & G.N. Wilkinson, eds., Academic Press. N.Y.

BOX, G.E.P. (1984) The importance of Practice in the Development of Statistics. Technometrics. 26(1): 1-8.

COOK, R.D. and WEISBERG, S. (1982) Residuals and Influence in Regression. Chapman & Hall. N.Y.

GLASS, G.V., MCGAW, B. and SMITH, M.L. (1981) Meta-Analysis In Social Research. Sage Publications. Beverly Hills, CA.

HEDRICK, T. E. (1985) Justification for and Obstacles to Data Sharing In Sharing Research Data. S.E. Fienberg, M.E. Martin and M.L. Straf, eds. National Academy Press. Washington, D.C. pp. 123-147.

LEAMER, E.E. (1983) Let's Take the Con out of Econometrics. Amer. Econ. Review. 73: 31-43.

McCULLAGH, P. and NELDER, J.A. (1983) Generalized Linear Models. Chapman and Hall, N.Y.

POPPER, K. (1968) Logic of Scientific Discovery. Harper and Row, N.Y.

WELSCH, R.E. (1984) An Introduction to Regression Diagnostics. See this publication Lecture 3.

REGRESSION ANALYSIS - BASICS AND CURRENT FRONTIERS

Raymond H. Myers
Virginia Polytechnic Institute and State University
Blacksburg, Virginia 24061

ABSTRACT

In this paper we highlight and illustrate the basics of least squares regression analysis. Performance criteria such as R^2 and error mean square are discussed and used in examples. Properties of least squares estimators are emphasized and assumptions are discussed.

Sequential methods for variable screening are described and recommendations are given regarding their use. In addition, the role of t-tests in model selection is discussed. Prediction intervals on a new observation are given and interpretation is highlighted.

Some modern performance criteria that quantify model prediction performance are discussed. The role of cross validation is presented in some detail and data splitting with the use of the PRESS statistic is discussed. A real-life example is given for purposes of illustration. The concepts associated with bias-variance tradeoff and model under and overspecification is outlined in detail. The C_p statistic, which is based on this tradeoff, is offered and illustrated.

Multicollinearity, the condition whereby linear dependencies among the regression variables make it difficult to determine the proper model, is discussed in detail. Modern techniques of residual analysis, outlier detection, etc., are illustrated. In addition, robust regression procedures that are resistant to violation of standard assumptions are briefly outlined.

1. THE METHOD OF LEAST SQUARES

The technique of multiple linear regression is used to build models in many subject matter areas. No statistical technique receives more attention by statistical researchers and subject matter scientists. Its applications are common in the physical and biological sciences, engineering, the social and behavioral sciences, and others. Its use involves building models that are used for
    (i)   explanatory purposes
    (ii) variable screening
    (iii) control
    (iv) prediction.
If the models are linear, they are often empirical in nature. Theory of statistical inference allows the user to estimate, test hypotheses, and discriminate among models. With these procedures the user is able to draw quantitative conclusions concerning the system from which the data is taken. The theory behind least squares regression has been known by research workers for many decades. However, in the last two decades there has been focus on new advances. In this paper we review in some detail the standard and fundamental procedures in linear regression. In addition we underscore some of the recent developments and point out the role they play in modern regression model building.

Suppose experimental data is collected on m variables x_1, x_2, ..., x_m, and interest is focused on how these influence some measured response y. Suppose these n data points are denoted by

$$\begin{array}{cccc} \underline{y} & \underline{x}_1 & \underline{x}_2 & \cdots & \underline{x}_m \\ y_1 & x_{11} & x_{21} & \cdots & x_{m1} \\ y_2 & x_{12} & x_{22} & \cdots & x_{m2} \\ \vdots & \vdots & \vdots & & \vdots \\ y_n & x_{1n} & x_{2n} & \cdots & x_{m,n} \end{array}$$

where $n \geq m$.

The experimenter postulates a model of the type

$$y_i = \beta_0 + \sum_{j=1}^{m} \beta_j x_{ji} + \varepsilon_i \ . \tag{1}$$

Here, the x_{ji} are assumed measured without error and the ε's are to be estimated. The ε_i are random disturbance terms or model errors. In all the technical detail that follows, the model can be written in general linear model form

$$\underline{y} = X\underline{\beta} + \underline{\varepsilon}$$

where $\underline{y}' = [y_1, y_2, \ldots, y_n]$ and

$$X = \begin{bmatrix} 1 & x_{11} & x_{21} & \cdots & x_{m1} \\ 1 & x_{12} & x_{22} & \cdots & x_{m2} \\ \vdots & \vdots & \vdots & & \vdots \\ 1 & x_{1,n} & x_{2,n} & \cdots & x_{m,n} \end{bmatrix} \qquad \underline{\beta} = \begin{bmatrix} \beta_0 \\ \beta_1 \\ \beta_2 \\ \vdots \\ \beta_m \end{bmatrix} \ .$$

The goal in the least squares procedure is to determine the estimate vector $\hat{\beta}$ for which the sum of squares of deviations is minimized. That is, one requires the minimization of

$$SS_E = (\underline{y} - X\underline{\hat{\beta}})'(\underline{y} - X\underline{\hat{\beta}}) \ . \tag{2}$$

An alternative way of writing the criterion is to say that the regression coefficients $\hat{\beta}_0, \hat{\beta}_1, \ldots, \hat{\beta}_m$ and the fitted regression function $\hat{y} = \hat{\beta}_0 + \sum_{j=1}^{m} \hat{\beta}_j x_j$ is determined for which the residual sum of squares

$$SS_E = \sum_{i=1}^{n} (y_i - \hat{y}_i)^2$$

is minimized. It is well known that the $\underline{\hat{\beta}}$ for which SS_E is minimized is given by

$$\underline{\hat{\beta}} = (X'X)^{-1} X'\underline{y} \ . \tag{3}$$

The least squares estimators given by equation (3) are linear functions of the y-observations. It is important for the data analyst to be aware of what assumptions are being made in order that certain properties of the least squares estimators be achieved. It is assumed that the ε_i are uncorrelated with mean zero and constant variance σ^2. Under these assumptions we know that the estimators are unbiased with variance-covariance matrix given by

$$\text{Var}(\hat{\underset{\sim}{\beta}}) = \sigma^2(\underset{\sim}{X}'\underset{\sim}{X})^{-1} . \qquad (4)$$

In addition, the least squares estimators are b.ℓ.u.e., that is, they are best (minimum variance) of all linear unbiased estimators. In addition, if the further assumption is made that the model errors are normal, the estimators become best (minimum variance) of all unbiased estimators.

2. MODEL PERFORMANCE CRITERIA

Any commercial computer package that does regression analysis computes basic performance criteria to accompany the fitted regression equation. These criteria are designed to assess the quality of fit of the regression function as well as determine how successful the analyst will be in reaching one of the four goals given in 1, namely explanation, variable screening, control, or prediction. The first criterion that one often observes is the well-known coefficient of determination, R^2, given by

$$R^2 = \sum_{i=1}^{n} (\hat{y}_i - \bar{y})^2 / \sum_{i=1}^{n} (y_i - \bar{y})^2$$

The R^2 statistic is designed to indicate to the user the proportion of variation (in the y's) explained by the model. An inexperienced user might view R^2 as a ratio of "sample variance explained" to "sample variance observed". Clearly, an $R^2 = 0.95$ implies that the model fit explains 95% of the observed variation in the response. The interpretation of R^2 as a proportion of variance explained stems from a well-known identity, namely

$$\sum_{i=1}^{n} (y_i - \bar{y})^2 = \sum_{i=1}^{n} (\hat{y}_i - \bar{y})^2 + \sum_{i=1}^{n} (y_i - \hat{y}_i)^2 . \qquad (5)$$

This is a partitioning of the total sum of squares into meaningful elements. $\sum_{i=1}^{n} (y_i - \bar{y})^2$, the "variation explained" is commonly called the regression sum of squares, while $\sum_{i=1}^{n} (y_i - \hat{y}_i)^2$ the "unexplained variation" is the well-known residual sum of squares. Much of fundamental formal statistical inference discussed later will evolve from this identity. Clearly

$$R^2 = SS_{Reg}/SS_{Total}.$$

One very clear danger is associated with the use of R^2 as a performance criterion. If the residual degrees of freedom = $n - (m+1)$ is small in number, R^2 can be artificially high, only indicating a near "force fit" of the model and not giving an indication of how reasonable the model is or how well it predicts y.

One performance criterion that is often used as a relative measure of importance of a fitted model is the error mean square

$$s^2 = \sum_{i=1}^{n} (y_i - \hat{y}_i)^2 / (n-(m+1)). \qquad (6)$$

This statistic is an unbiased estimate of σ^2, the error variance, assuming the model is properly specified. Other criteria that are a standard part of the computer printout include t-statistics on the model coefficients, an overall F-statistic, and sequential and partial F-statistics.

The t-statistics are designed to be informational regarding the strength of the individual variables in the model. The t-statistics for the jth regression coefficient is given by

$$t_j = \hat{\beta}_j / s_{\hat{\beta}_j}$$

where $s_{\hat{\beta}_j}$ is the standard error of the coefficient $\hat{\beta}_j$. The standard errors are retrieved from the estimated variances as the diagonal elements of $s^2(X'X)^{-1}$. Formally, the t-values are designed to be used to test

$$H_0: \beta_j = 0$$
$$H_1: \beta_j \neq 0$$

and thus one may view them as criteria for variable screening. However, there are difficulties associated with their use in this regard in many types of data sets. In data sets where strong linear association (multicollinearity) exists among the regressor variables, the t-statistics lose their value. This results from the fact that the role of the t-statistic on a specific coefficient is to assess the importance of the variable in the model with all others. In the presence of multicollinearity the strength of a variable will depend on what variables are in the model with it.

Partial F-tests are standard components of regression analysis. Again, one F-test is associated with each regressor variable. Each is a ratio

$$F = \frac{R(x_j | x_1, x_2, \ldots, x_{j-1}, x_{j+1}, \ldots, x_m)}{s^2}$$

where $R(x_j | x_1, x_2, \ldots, x_{j-1}, x_{j+1}, \ldots, x_m)$ denotes the increase in SS_{Reg}

when x_j is placed in the model in the presence of the other variables. Thus the F-test determines whether or not the regression attributed to x_j is significant. The information gained from partial F-tests and t-tests are identical. In fact, the partial F on a specific regressor variable is the square of the t-statistic.

Confidence intervals on the mean response can be used to quantify the performance of the regression equation for prediction. Given a data set and a fitted regression, the confidence interval on $E(y|x=x_0)$ is given by

$$\hat{y}(\underline{x}_0) \pm t_{\alpha/2, n-(m+1)} \; s\sqrt{\underline{x}_0'(\underline{X}'\underline{X})^{-1}\underline{x}_0} \quad .$$

Here $\underline{x}_0' = [1, x_{1,0}, x_{2,0}, \ldots, x_{m,0}]$ is a point of interest in the regressor space. The term $s\sqrt{\underline{x}_0'(\underline{X}'\underline{X})^{-1}\underline{x}_0}$ is the very important "standard error of prediction" at $\underline{x}=\underline{x}_0$. It should be noted that the interval given above describes a $100(1-\alpha)\%$ confidence interval on the mean response at $\underline{x}=\underline{x}_0$. Another very fundamental component of standard computer printout in regression analysis is the prediction interval on a new observation at $\underline{x}=\underline{x}_0$. This interval obviously has a different interpretation and use is made of it in cases where an individual observation is critical. This interval is given by

$$\hat{y}(\underline{x}_0) \pm t_{\alpha/2, n-(m+1)} \; s\sqrt{1+\underline{x}_0'(\underline{X}'\underline{X})^{-1}\underline{x}_0} \quad .$$

In both intervals, $t_{\alpha/2, n-(m+1)}$ is the upper $\alpha/2$ percentage point of Student's t-distribution. In many standard commercial packages, these intervals are given for all $\underline{x}=\underline{x}_i$, the locations of the input data. However, there is usually a mechanism that allows for the computation of the interval at any location in the regressor variables.

An experiment[1] was conducted to gain some preliminary insight into the effect of three quantitative factors on the capability of a particular coal cleansing operation. A polymer is being used to clean the coal and the amount of suspended solids (m.g./liter) in the "overflow" solution (following a cleansing operation) is a measure of the efficiency of the operation. The factors that influence the suspended solids are x_1, percent solids in the input solution, x_2, PH of the tank that holds the solution, and x_3, flow rate of the cleansing polymer, mℓ/minute. All three factors were controlled in the experimental process. The data are as follows: (The order of experimentation was random.)

Experiment	x_1	x_2	x_3	y
1	1.5	6.0	1315	243
2	1.5	6.0	1315	261
3	1.5	9.0	1890	244
4	1.5	9.0	1890	285
5	2.0	7.5	1575	202
6	2.0	7.5	1575	180
7	2.0	7.5	1575	183
8	2.0	7.5	1575	207
9	2.5	9.0	1315	216
10	2.5	9.0	1315	160
11	2.5	6.0	1890	104
12	2.5	6.0	1890	110

A regression equation was constructed with the result given by

$$\hat{y} = 397.087 - 110.750 x_1 + 15.5833 x_2 - 0.058 x_3 .$$

Some of the performance criteria we discussed follows.

Sum of Squares	Root MSE	R-Square
$31156.024 = \Sigma(\hat{y}-\bar{y})^2$	20.87730	0.8993
$3486.892 = \Sigma(y-\hat{y})^2$		
$34642.917 = \Sigma(y-\bar{y})^2$		

	Parameter Estimate	Standard Error	t for H_0: Parameter=0
$b_0 =$	397.087	62.75676	6.327
$b_1 =$	-110.750	14.76250	-7.502
$b_2 =$	15.583	4.92083	3.167
$b_3 =$	-0.058	0.02564	-2.274

Obs	Actual	Predicted Value	Residual	Confidence Intervals Lower 95% Mean	Upper 95% Mean
1	243.00	247.808	-4.808	215.508	280.108
2	261.00	247.808	13.192	215.508	280.108
3	244.00	261.040	-17.040	228.174	293.906
4	285.00	261.040	23.960	228.174	293.906
5	202.00	200.652	1.348	186.712	214.592
6	180.00	200.652	-20.652	186.712	214.592
7	183.00	200.652	-17.652	186.712	214.592
8	207.00	200.652	6.348	186.712	214.592
9	216.00	183.808	32.192	151.508	216.108
10	160.00	183.808	-23.808	151.508	216.108
11	104.00	103.540	0.460	70.674	136.406
12	110.00	103.540	6.460	70.674	136.406

3. SEQUENTIAL ALGORITHMS FOR SUBSET SELECTION

In the early 1960's as the use of the high speed computer for statistical analysis reached a state of acceleration, many computer algorithms (See Efroymson (1962)) were developed for doing efficient and reasonably effective sequential searching for the "best" subset of regressor variables when one is faced with a data set containing many candidate variables. Many of these techniques are still used today and thus a presentation of their utility and pitfalls is given here. For a lengthy discussion one should consult Draper and Smith (1981). Most of the techniques are based on a sequential movement in which regressor variables are added or deleted one at a time based on how the change impacted the regression (or residual) sum of squares. There are many procedures, the primary ones being forward selection, backward elimination, and stepwise regression. The latter received a large amount of usage during the late 1960's and 70's among subject matter data analysts.

Forward selection involves sequential model building in which the first variable that enters is that which produces the largest regression sum of squares (i.e., largest R^2) as a single variable. At each stage a variable is entered that produces the largest increase in the regression sum of squares. The stopping rule involves terminating when the next candidate for entry is insignificant as the basis of an F-test.

Backward elimination involves the same criteria as forward selection but the direction is switched. One begins with the full model and eliminates one at a time with termination coming when the candidate for elimination is statistically significant on the basis of a partial F.

Stepwise regression proceeds like forward selection but at each stage partial F-tests are performed on each variable currently in the model. Thus a variable that entered in an early stage may be deleted at later stages. As a result, simultaneous entry and deletion of variables occurs. Termination occurs when no additional variable can be entered on the basis of a significance test and all variables currently in the model are significant.

These sequential procedures were developed in an era in which it was computationally prohibitive to compute and observe the analysis for all possible subset regressions, the number of regressions being 2^m. Their use is more limited now. It should be pointed out that there is no assurance that the "best" regression will result. Generally the algorithms compute regressions for a small subset of all the possibilities. In addition, the results are generally very much dependent on the significance level used in the methodology. This can be particularly disconcerting because, strictly speaking, the F-distribution is not the proper statistical distribution to use. The suggestion is that the F-distribution be used as a guide. In most cases (certainly for $m < 10$) the "shortcut" procedures for arriving at a good model are not necessary because the speed of modern computers allows for the generation of some information on all possible 2^m models.

4. CRITERIA FOR BEST SUBSET MODEL

There is an abundance of criteria that are suggested in the statistics literature for selection of the best of a set of candidate models. For an excellent review of these see Hocking (1974). This is particularly important since there is sufficient computing power available to obtain information from all candidate models. Many of the traditional criteria (R^2, SS_E, s^2, etc.) reflect the capability of the model as a predictor, i.e., determine the quality of

$$\hat{y}(x_0) = x_0'\hat{\beta} .$$

Arriving at the proper model is quite often a matter of "compromise" between underfitting and overfitting. An underspecified model has biased coefficients and a biased prediction. An overspecified model is one that contains unimportant regressor variables, i.e., variables that have marginal or no contribution. Overspecification produces a model with inflated variances of the coefficients. These variances can become particularly excessive when multicollinearity is severe. In many cases it is appropriate to use some measure of quality of $\hat{y}(\underline{x}_0)$ to determine the appropriate compromise.

One useful criterion for evaluation of candidate models is the C_p statistic. See Mallows (1966). The statistic which contains both a bias and variance component is given by

$$C_p = p + (s^2 - \hat{\sigma}^2)(n-p)/\hat{\sigma}^2 \qquad (7)$$

where $p = m+1$, the number of estimated parameters. Note that the C_p statistic is a function of two important quantities, s^2, the error mean square, and p, the size of the model in terms of the number of parameters to be estimated. The C_p is actually an estimate of the "total error" summed over all the data points given by

$$\Gamma_{(p)} = \sum_{i=1}^{n} \{\text{Var } \hat{y}(\underline{x}_i) + [E \, \hat{y}(\underline{x}_i) - E \, y(\underline{x}_i)]^2\}/\hat{\sigma}^2.$$

One can note that $\Gamma_{(p)}$ is the accumulation of the bias and error component over the location of the data points. One disadvantage with the procedure is that an external estimate of σ^2 is required.

A second criterion for evaluating the prediction capability of a candidate model is the PRESS statistic. The PRESS statistic is based on a very imaginative and interesting approach. The philosophy is that a model should be evaluated by allowing it to predict data that was not used in the fitting process. Thus the model takes part in a cross-validation process and prediction errors or PRESS residuals of the type $y_i - \hat{y}_{i,-i}$ are generated. Here $\hat{y}_{i,-i}$ represents the predicted response at the data point $\underline{x}_i$ for a prediction in which the ith data point is not used in building the regression. From this computation, y_i and $\hat{y}_{i,-i}$ are independent and thus the assessment of quality of the candidate model is based on predicting independent observations. The criterion for model selection is given by the Prediction Sum of Squares

$$\text{PRESS} = \sum_{i=1}^{n} (y_i - \hat{y}_{i,-i})^2 .$$

The models with small PRESS are favored in the selection process.

The computation of the PRESS residuals and hence the PRESS statistic would appear to be prohibitive. If one were to mechanically remove data points one at a time and compute regressions each time, the computation of the PRESS statistic for, say, 2^m candidate models would involve $2^m \cdot n$ regressions. However, elimination of points is not necessary. The ith PRESS residual is a simple function of the ordinary residual and the element $\underline{x}_i'(\underline{X}'\underline{X})^{-1}\underline{x}_i$. In fact,

$$y_i - \hat{y}_{i,-i} = (y_i - \hat{y}_i)/(1 - \underline{x}_i'(\underline{X}'\underline{X})^{-1}\underline{x}_i') . \tag{8}$$

The role of the term $\underline{x}_i'(\underline{X}'\underline{X})^{-1}\underline{x}_i$ should be emphasized. It is the variance of the predicted value, $\hat{y}(\underline{x}_i)$, at $\underline{x} = \underline{x}_i$. In addition it is the ith main diagonal element of the matrix

$$H = \underline{X}(\underline{X}'\underline{X})^{-1}\underline{X}' .$$

H is commonly called the HAT matrix. The diagonals are bounded according to the inequality

$$0 < h_{ii} < 1 .$$

Data points that are remote from the center of the data are characterized by h_{ii} values close to 1.0. Thus the obvious result is illustrated, namely that the PRESS residual will be considerably larger than the ordinary residual when the point is a remote point. Clearly these remote points are potentially "high influence" data points.

Consider the same coal cleansing data discussed in the previous section. The following table shows the statistics PRESS, C_p, s^2, and R^2 for all possible models. From this table it seems clear that the full model, i.e., the model containing all three regressor variables is most appropriate.

Table 1. Criteria for All Possible Regressions for Coal Data

Obs	Model	MSE	PRESS	RSQUARE	C_p
1	$x_1 x_2 x_3$	435.86	10062.7	0.899348	4.0000
2	$x_1 x_2$	637.84	12154.0	0.834290	7.1708
3	x_1	1011.18	15842.4	0.708114	15.1995
4	$x_1 x_3$	873.11	16801.4	0.773171	12.0287
5	constant	3149.36	41227.9	-0.000000	69.4815
6	x_2	3027.18	48009.0	0.126177	61.4528
7	x_3	3238.91	51981.0	0.065057	66.3106
8	$x_2 x_3$	3113.11	62146.2	0.191234	58.2819

5. MULTICOLLINEARITY

Since the early 1970's much attention has been dedicated to alternatives to the least squares procedure in the case of data sets with strong multicollinearity. As we indicated earlier, multicollinearity is a result of near linear dependencies in the regressor variables displayed in the data set. In other words, multicollinearity is present when there exists a set of constants $a_0, a_1, \ldots, a_m$ for which

$$a_0 + \sum_{j=1}^{m} a_j x_j \cong 0 \ .$$

The near dependency causes the determinant of the matrix X'X to be near zero and hence relatively large values on the diagonals of $(X'X)^{-1}$, resulting in large variances of regression coefficients. This produces the following difficulties
  (i) Unstable and poorly estimated coefficients; some coefficients take the wrong sign.
  (ii) Questionable prediction; prediction is particularly poor when extrapolation is done.
  (iii) Unreliable information from statistical tests. The power of the t-tests, for example, can be very low.

Difficulties associated with multicollinearity evolve around the conditioning of X'X. One essentially is dealing with a problem of dimension less than m, even though the overspecified model contains m regressors. Much of the modern work in regression analysis deals with proper diagnosis of multicollinearity as well as methods of combating it. The earliest and best known procedure for handling multicollinearity is ridge regression. While we do not choose to give details here, the ridge estimator is given by $\hat{\beta}_R$ where the latter is the solution to

$$(X'X + kI)\hat{\beta}_R = X'y$$

where X'X can be taken to be in correlation form and k is a positive constant.

6. STUDY OF RESIDUALS - OUTLIER DETECTION

The errors in fit of the regression, i.e., the residuals, $y_i - \hat{y}_i$, have for many years been the source of various types of analyses. The residuals are a natural component of the analysis since they are the empirical counterpart to the ε_i, the model errors. Much of the analysis that centers around them involves investigating the palatability of assumptions, normality, independence, and homogeneous variance. There are plotting routines that aid in determining if the homogeneous variance assumption is violated. Analytical techniques are available that give evidence regarding the normality assumption. In addition, plots are suggested that may determine whether or not the model is underspecified. See Draper and Smith (1981).

Residuals are the pivotal ingredient for any analytical procedure for determining if there are suspect data points, i.e., outliers in the regression data set. Basically, large residuals in magnitude are evidence of errors in fit that are more than one would expect due to mere chance. One difficulty that one faces in trying to determine which residuals are "significantly larger in magnitude than zero" is that the residuals are not independent identically distributed random variables as the ε_i are assumed to be. In fact, the variances are generally different and can vary a great deal. As in the case of prediction variance, the variance of a residual depends on how remote the data point is from the data center. Recall that the diagonal of the HAT matrix $h_{ii} = x_i'(X'X)^{-1}x_i$ represents the distance x_i is from the center. It turns out

that
$$\text{Var}(y_i - \hat{y}_i) = \sigma^2(1 - h_{ii}) .$$

As a result, this variance must be taken into account in any effort on the part of the analyst at determining if a large residual is an indication of an outlier. As a result a standardized residual of the type

$$(y_i - \hat{y}_i)/(\sigma\sqrt{1-h_{ii}})$$

is a reasonable type of statistic for diagnosing outliers. Clearly, in a practical situation σ must be replaced by the estimate s, giving the set of studentized residuals

$$r_i = y_i - \hat{y}_i/(s\sqrt{1-h_{ii}})$$

For a test on a potential outlier for an individual data point, the statistic r_i can be used in an approximate t-test. Actually, an exact t-statistic can be produced if s is replaced by s_{-i}, the estimate of σ computed without the use of the ith point.

Several regression computer packages include regression diagnostics involving residuals with studentized residuals being one of several major features.

Consider the coal cleansing data once again. There was concern that the results from data point 9 were erroneous. Conditions were not held constant as was necessary. As a result, the engineers involved felt as if the project might benefit from deletion of the information. However, the analysis was initially conducted with the data point not removed in order to gain some insight into whether or not the analysis supports the conjecture regarding the data point. The following table shows the response, fitted response, residuals, HAT diagonal values, and studentized residuals associated with each of the twelve observations. Note that the residual associated with the 9th point is 32.192 mg/ℓ. This is the largest residual in the data set. The studentized residual computation is designed to determine if the residual differs significantly from zero. The results give t = 2.8695 for data point 9, significant at the 0.05 level. None of the other studentized residuals approach that of data point 9. The results would seem to confirm the suspicions regarding this observation. Thus deletion of observation 9 may be considered as a reasonable approach in the construction of the regression.

Table 2. Results for Coal Cleansing Example

y_i	$\hat{y}_i$	$y_i - \hat{y}_i$	h_{ii}	r_i
243	247.808	-4.808	0.4501	-0.2923
261	247.808	13.192	0.4501	0.8359
244	261.040	-17.040	0.4660	-1.1372
285	261.040	23.960	0.4660	1.7665
202	200.652	1.348	0.0838	0.0631
180	200.652	-20.562	0.0838	-1.0385
183	200.652	-17.652	0.0838	-0.8698
207	200.652	6.348	0.0838	0.2990
{216	183.808	32.192	0.4501	2.8695}
160	183.808	-23.808	0.4501	-1.7141
104	103.540	0.460	0.4660	0.0282
110	103.540	6.460	0.4660	0.4006

Another approach to building regression in the face of suspect data points is the use of robust regression. Robust regression involves a deviation from ordinary least squares. Outliers, or data points with disproportionately large residuals, are not allowed to exert as much influence on the results as they might in the case of least squares. Procedures are available that are based on the minimization of $\sum_{i=1}^{n} g(y_i - \hat{y}_i)$, where $g(\cdot)$ is a function other than the residual sum of squares. The impact of the residuals on the regression results in the case of least squares is illustrated by the complexion of

$$\frac{\partial}{\partial \hat{\beta}} \sum_{i=1}^{n} (y_i - x_i' \hat{\beta})^2 = 0$$

which implies that $\hat{\beta}$ is the solution to

$$\sum_{i=1}^{n} (y_i - \hat{y}_i) x_i = 0 .$$

The above equation indicates that the size of the residual has an important impact on the result. The linear contribution of the residual is a direct result of the quadratic objective function. The function $\sum_{i=1}^{n} (y_i - \hat{y}_i) x_i$ is called an influence function for the least squares estimator. Robust regression provides the estimator $\hat{\beta}$ for which

$$\sum_{i=1}^{n} \psi(y_i - \hat{y}_i) x_i = 0$$

where $\psi(\cdot)$ is a prechosen function of the residuals. The influence function $\psi(\cdot)$ is chosen in such a way that large residuals are not allowed to exert strong influence on the result. The resulting estimation procedure is called the M-estimator. See Montgomery and Peck

(1982) for details. One example of a robust influence function is Huber's (1973) influence function given by

$$\psi(y_i - \hat{y}_i) = y_i - \hat{y}_i \quad |y_i - \hat{y}_i| < c\sigma$$
$$= c\sigma \quad (y_i - \hat{y}_i) > c\sigma$$
$$= -c\sigma \quad (y_i - \hat{y}_i) < -c\sigma$$

The computations involved can be negotiated with fairly straightforward algorithms that are available in some software packages.

FOOTNOTES

[1] Data was generated by Mining Engineering Department and analyzed by the Statistical Consulting Center, Department of Statistics, Virginia Polytechnic Institute and State University, Blacksburg, Virginia.

REFERENCES

ALLEN, DAVID (1974), "Prediction Sum of Squares," Technical Report No. 23, University of Kentucky, Lexington, Kentucky.

BELLSLEY, D.A., KUH, and WELSCH, ROY (1980), *Regression Diagnostics*, New York: John Wiley.

DRAPER, NORMAN and SMITH, H. (1981), *Applied Regression Analysis*, 2nd edition, New York: John Wiley.

EFROYMSON, M. A. (1962), "Multiple Regression Analysis," in *Mathematical Methods for Digital Computers*, New York: John Wiley.

GUNST, R.F. and MASON, R.L. (1980), *Regression Analysis and Its Applications*, New York: Marcel Dekker, Inc.

HOCKING, R. R. (1974), "Misspecification in Regression," *American Statistician*, 28, 39-40.

_____ (1976), "The Analysis and Selection of Variables in Linear Regression," *Biometrics*, 32, 1-51.

HUBER, P.J. (1973), "Robust Regression: Asymptotics, Conjectures, and Monte Carlo," *Annals of Statistics*, 1, 799-821.

MALLOWS, C.L. (1966), "Some Comments on C_p," *Technometrics*, 15, 661-675.

MONTGOMERY, DOUGLAS C. and PECK, ELIZABETH A. (1982), *Introduction to Linear Regression Analysis*, New York: John Wiley.

SEBER, G.A.F. (1977), *Linear Regression Analysis*, New York: John Wiley.

WILLIAM, E. J. (1959), *Regression Analysis*, New York: John Wiley.

AN INTRODUCTION TO REGRESSION DIAGNOSTICS

Roy E. Welsch
Massachusetts Institute of Technology, Cambridge, MA 02139

ABSTRACT

A regression is constructed using prior knowledge, data, models, and a fitting (estimation) process of some form. It is important to know when the resulting regression depends heavily on a small part of the prior knowledge, on a small part of the data, or on the exact choice of model or fitting process. We want models that are sensitive to the issues of interest, but perform well (i.e. are less sensitive) when assumptions are violated. In this paper we present an overview of the theory, application and computation of regression diagnostics, especially those related to the analysis of influential data.

1. INTRODUCTION

Our basic goal in this paper is to learn if our regression is heavily influenced by small subsets of data. A traditional starting point is to look for outliers which may be viewed as observations that appear to be surprising to the investigator or observations that are not a realization from some target distribution. It is essential that all data used in regression models be examined for outliers. The first step is to look at the response and explanatory variables separately to get a feeling for outliers. At this time transformations of these variables might be considered. A heavily skewed data series may appear to have lots of outliers. A logarithmic transformation may make the outliers appear much more like the rest of the data. Such transformations are, of course, tentative and need to be considered in light of prior knowledge and subsequent results.

Such a univariate examination does not help us find outliers relative to a particular model and fitting process. In fact some of the univariate outliers may not look so discrepant in the context of a multivariate model. Conversely, and more commonly, multivariate outliers will arise which cannot be seen in a univariate analysis.

A particularly useful way to detect outliers in the context of a model is to look for overly influential subsets of data. Subsets of data are regarded as influential if their deletion results in substantial changes to important features of an anlysis.

Our discussion of regression diagnostics starts with preliminary steps that are necessary before the decision is made to use a least-squares linear regression model. Then we introduce the idea of adjusted variables and partial regression plots. After a brief discussion of collinearity diagnostics, we define leverage and several different kinds of residuals. We then go on to measure influence and develop plots to summarize influential data diagnostics. We conclude with a brief treatment of diagnostics for generalized linear models and comments about some areas of research in diagnostic methods.

2. PRELIMINARY STEPS

There are many reasons for performing a regression analysis. Two of the most common are:

(a) Fitting an equation or model to data
(b) Attempting to describe local averages of y about values of x

$$E(y|x) = g(x). \tag{1}$$

Both of these involve the response data, y, and the regressors x_1, x_2, etc. All too often, the data and (a) and (b) are combined into

$$y = \underset{\sim}{X}\beta + \varepsilon \tag{2}$$

where $\underset{\sim}{X}$ is an n×p matrix of regressors, possibly including the constant carrier, y is n×1, β is p×1 and ε is n×1. The estimated coefficients, b, are then obtained by a fitting process (usually least-squares) without a great deal of thought.

An important first step is to look at the variables y, and x_1, x_2, etc. separately. These data should be <u>explored</u> using histograms, stem-and-leaf plots, boxplots, etc. (Velleman and Hoaglin 1981) and granularity (clumps, holes), outliers, and asymmetry ought to be noticed. Outliers need to be tagged, and possible transformations considered. Of course, some outliers may not be so prominent when we consider the multivariate nature of the data. Transformations considered now may also be unnecessary later, but asymmetries, outliers, and large changes of magnitude are clues that some variables may be in the wrong units. In short, take a hard look at the raw data. Do nothing if you wish, but set up a list of things to check as you go further.

The response variable, y, is often supposed to be a random variable with some probability distribution. If it has only two values, looks Poisson, etc. do not try ordinary least-squares regression. You will get stupid results. Consider other models such as the generalized linear models discussed by McCullagh and Nelder (1983). The probability plotting techniques discussed by Chambers et al (1983) are useful for checking these assumptions.

When a thorough univariate analysis has been done, it is time to consider the bivariate (an eventually multivariate) nature of the data. Plots of y vs x_j are always worth making, especially for considering transformations to straighten the plot and rough correlations, but can be misleading if used to develop precise models because of the effects of other regressors. Bivariate plots of the regressors are also useful for finding holes, outliers etc. but the number of plots increases rapidly with p. However, these provide the first clues to the fact that information in the "design" or factor space may be spotty, clumpy, have holes or be sparse. Unfortunately, we cannot do this well in higher dimensions -- at least not yet. A number of people are working on this (Chambers et al, 1983, Chapter 5).

Usually a tentative model or equation comes with the data. While doing our exploring we should see if the model is sensible. Possible modifications should be noted for later consideration. It is always tempting to make g(x) linear in some proposed coeffieicnts (a so-called linear model). However, g(x) may be quite different--clues to this effect should be noted for they may require different approaches. The model could be nonlinear in the parameters or perhaps a non-parameteric approach is needed (Friedman and Stuetzle, 1981). The diagnostics to be discussed below assume that, at least tentatively, the model in (2) is considered reasonable. Diagnostics may open our eyes to further problems, but they cannot take the place of a good preliminary look at the data, structural model (g(x) or more specifically, $\underset{\sim}{X}\beta$) and stochastic model (distribution of y or, in some cases, ε). When diagnostics

point to changes, we should use these preliminary procedures after
making changes and again apply diagnostics to see if we have improved
our analysis.

3. PARTIAL REGRESSIONS AND PLOTS

Most of us are used to thinking about the least-squares estimates as

$$\hat{\beta} = (X'X)^{-1}X'y. \tag{3}$$

However, it is often more instructive to think about the estimated
coefficients in a different way. Denote the residuals found when y is
fit by all but the jth regressor by

$$y_{\cdot 12\ldots(j-1)(j+1)\ldots p} = y_{\cdot [j]}. \tag{4}$$

Thus $y_{\cdot [j]}$ is the vector of least-squares residuals obtained by
regressing y on all regressors except the jth and is often called the
<u>adjusted response variable</u>. Similarly let $x_{j \cdot [j]}$ denote the residuals
obtained by regressing x_j on all of the remaining regressors. These
are called the <u>adjusted regressors</u>. It is not hard to show (Mosteller
and Tukey, 1977, page 344) that

$$b_j = \frac{\sum_{i=1}^{n} x_{ij \cdot [j]} y_{i \cdot [j]}}{\sum_{k=1}^{n} x_{kj \cdot [j]}^2} \tag{5}$$

where $x_{ij \cdot [j]}$ is the ith element of the vector $x_{j \cdot [j]}$ and $y_{i \cdot [j]}$ is the
i^{th} element of the vector $y_{\cdot [j]}$. This formula should be compared to
that for simple linear regression through the origin. A great deal of
information about b_j can be obtained by plotting $y_{\cdot [j]}$ against $x_{j \cdot [j]}$
for each j. These are called <u>partial regression plots</u> (or, in some
cases, adjusted variable plots.) Useful references are Belsley, Kuh,
and Welsch (1980) and Chambers et al (1983). Both of these contain
interesting examples.

Some properties of these plots are:
(a) The least-squares linear fit to the plotted data has
    slope = b_j and intercept = 0 (when j is not the intercept
    variable).
(b) The residuals from the least-squares linear fit are the
    final multiple regression residuals, $y-Xb$.
(c) It is relatively easy to see how individual data values
    influence the estimation of b_j.
(d) Often some information about nonlinearity, heteroscedasticity
    and unusual patterns can be obtained.

An example of a partial regression plot is given in Figure 1. The
simple linear regression line is included and some interesting points
have been marked.

Until recently, partial regression plots were thought to be hard
to obtain. Velleman and Welsch (1981) show that this is not the case.

Let
$$\underset{\sim}{C}' = (X'X)^{-1}X' \qquad (6)$$

then
$$\underset{\sim}{b} = \underset{\sim}{C}'\underset{\sim}{y} \qquad (7)$$

and
$$b_j = \sum_{i=1}^{n} c_{ij} y_i . \qquad (8)$$

Using the normal equations and (5), we can show that the jth element of $\underset{\sim}{\beta}$ is given by

$$b_j = \sum_{i=1}^{n} x_{ij} \cdot [j] y_i / \sum_{k=1}^{n} x_{kj}^2 \cdot [j] . \qquad (9)$$

The uniqueness of the least-squares estimates implies that

$$c_{ij} = x_{ij} \cdot [j] / \sum_{k=1}^{n} x_{kj}^2 \cdot [j] \qquad (10)$$

or equivalently

$$x_{ij} \cdot [j] = c_{ij} / \sum_{k=1}^{n} c_{kj}^2 \qquad (11)$$

Furthermore, Mosteller and Tukey (1977) have shown that

$$\underset{\sim}{y} \cdot [j] = \underset{\sim}{e} + b_j \underset{\sim}{x}_j \cdot [j] \qquad (12)$$

where
$$\underset{\sim}{e} = y - Xb. \qquad (13)$$

Therefore $\underset{\sim}{\beta}$, $\underset{\sim}{C}$ and $\underset{\sim}{e}$ are all we need to get partial regression plots. A well-organized regression program can obtain $\underset{\sim}{C}$ very easily. There is no excuse for not making these plots a part of every regression analysis. They are an essential diagnostic tool.

4. COLLINEARITY

Before going further it is advisable to get some feeling for collinearity. A quick way is to note from (9) that

$$\text{var}(b_j) = \frac{\sigma^2}{\sum_{i=1}^{n} x_{ij}^2 \cdot [j]} . \qquad (14)$$

If the sum in the denominator is small compared to say $\sum_{i=1}^{n} x_{ij}^2$, then x_j has been well fit by the other regressors. Since we are used to centering our data we often compute the squared multiple correlation of x_j on the other regressors,

$$R_j^2 = 1 - \sum_{i=1}^{n} x_{ij}^2 \cdot [j] / \sum_{i=1}^{n} (x_{ij} - \bar{x}_j)^2 \qquad (15)$$

where $\bar{x}_j$ is the jth variable mean. Note that if we are interested in using models with an explicit intercept rather than centered data, the denominator of R_j^2 should not be centered. One statistic often proposed as a measure of collinearity is the <u>variance inflation factor</u>, VIF, found from

$$VIF_j = \frac{1}{1-R_j^2} . \qquad (16)$$

If we just had a simple linear regression on x_j,

$$y = c + d x_j + \varepsilon,$$

then

$$var(d) = \sigma^2 / \sum_{i=1}^{n} (x_{ij} - \bar{x}_j)^2 .$$

Thus

$$var(b_j) = var(d) \cdot VIF_j, \qquad (17)$$

and we see that VIF measures the variance inflation due to the presence of additional regressors.

The drawbacks of the above approach (VIF_j large) are that it does not tell us which regressors are involved or how they are involved in the collinear relation with x_j. A very useful way to get this information is described in Chapter 3 of Belsley, Kuh, and Welsch (1980).

If collinearity appears to be a problem, it is wise to reduce it as much as possible before doing influential data diagnostics. When the diagnostics that follow suggest altering or setting aside data it is essential that the new model be rechecked for collinearity.

5. LEVERAGE

One of the preliminary steps for regression analysis that we have discussed is the making of bivariate plots of the regressors. Generally the eye will notice outlying points in these plots. However, it is hard to make higher dimensional plots and a variety of tools exist to overcome this problem. One of the easiest to use measures a type of distance from $\bar{x}$, the row vector of regressor means, to each observation x_i:

$$(x_i - \bar{x})'(\hat{X}'\hat{X})^{-1}(x_i - \bar{x}) \ . \tag{18}$$

Here $\hat{X}$ is the X matrix without the intercept column and with column means subtracted off of the remaining columns. Similarly $(x_i - \bar{x})$ omits the intercept element. Belsley, Kuh and Welsch (1980) show that

$$h_{ii} - \frac{1}{n} = (x_i - \bar{x})'(\hat{X}'\hat{X})^{-1}(x_i - \bar{x})^T \tag{19}$$

where

$$h_{ii} = x_i'(X'X)^{-1}x_i \tag{20}$$

is the diagonal element of the projection or **hat matrix**,

$$H = X(X'X)^{-1}X' \ , \tag{21}$$

so called because

$$\hat{y} = Hy. \tag{22}$$

Most modern regression programs now compute h_{ii} and I would not use one that failed to do so.

Often we will want to hypothesize that x_i is possibly erroneous or "strange". Then we may wish to measure the distance from x_i to the rest of the data. One useful way to do this is to compute.

$$d_i = (x_i - \bar{x}_{-i})'(\hat{X}_{-i}'\hat{X}_{-i})^{-1}(x_i - \bar{x}_{-i})' \tag{23}$$

where $\bar{x}_{-i}$ and X_{-i} are obtained by assuming that the ith observation did not exist.

Since d_i is related to Mahalanobis distance, it is not hard to show (Belsley, Kuh, and Welsch, 1980) that

$$d_i = \left[\frac{h_{ii} - \frac{1}{n}}{1 - h_{ii}}\right] \frac{n}{n-1} \ . \tag{24}$$

Both h_{ii} and $h_{ii}/(1-h_{ii})$ will prove to be useful in what follows.

Hoaglin and Welsch (1978) discuss many properties of h_{ii}. In particular, $0 \le h_{ii} \le 1$ (or $1/n \le h_{ii} \le 1$ when an intercept is present) and $\sum_{i=1}^{n} h_{ii} = p$ so that the average value is p/n. Let b_{-i} denote the least-squares estimates obtained without using the ith observation. Then simple algebra (see 32) shows that

$$\hat{y}_i = x_i'\beta = (1 - h_{ii})x_i'b_{-i} + h_{ii}y_i. \tag{25}$$

Thus $\hat{y}_i$ is a convex combination of the prediction $x_i'b_{-i}$ and the observation y_i. The ratio $h_{ii}/(1-h_{ii})$ determines the relative contribution of each part. When h_{ii} is one, the ith observation completely determines $\hat{y}_i$ and $\hat{y}_i = y_i$.

There is no general agreement on when h_{ii} is "large". Hoaglin and Welsch (1978) argue that an individual h_{ii} should not be too far from a balanced design (all $h_i = p/n$) and call the ith observation a <u>leverage point</u> when $h_{ii} > 2p/n$ (provided $n > 2p$). Belsley, Kuh, Welsch (1980) show that when the x_i (rows of X) are i.i.d. multivariate Gaussian, the distribution of $h_{ii}/(1-h_{ii})$ can be related to an F-statistic. This leads to a criterion that calls attention to the ith observation if $h_{ii} > 3p/n$. Note that these leverage criteria depend on p and n.

Huber (1981) uses (25) and suggests that when $h_{ii} > .5$ special attention is called for and observations with $h_{ii} > .2$ should be noted. These leverage criteria are independent of p and n.

A useful compromise between these two general approaches is to consider h_{ii} i=1,...,n, as a batch of data to be analyzed by exploratory data analysis (Velleman and Hoaglin, 1981). Observations with outlying values of h_{ii} would then be considered leverage points. My own simple rule of thumb is to pay attention when $h_{ii} > \min(.2, 3p/n)$.

It is also possible to compute the contribution of the individual regressors to the leverage of each observation. Let

$$\underset{\sim}{h} = \underset{\sim}{h}_{[j]} + \underset{\sim}{n}_j \qquad (26)$$

where $\underset{\sim}{h}_{[j]}$ is the vector of leverage values when x_j is omitted from the regression model. The <u>partial leverage</u>, n_j can be found from

$$n_{ij} = x_{ij}^2 \cdot [j] / \sum_{k=1}^{n} x_{kj}^2 \cdot [j] \qquad (27)$$

which is the leverage of the ith point in the partial regression plot for b_j. Data points with large partial leverage for a regressor can exert an undue influence on the selection of that regressor in most automatic regression model building methods. For some examples, see Henderson and Velleman (1981).

6. RESIDUALS

While looking for leverage points is a relatively new tool, examining various plots of the residuals, e_i, is not. Surely residuals should be plotted against index (or time), against fitted values, against proposed new regressors (it is best to adjust the new carrier for those already in the model by using the residuals from x_{new} regressed on the current model), etc. Probability plots should also be made. An excellent discussion is contained in Chambers et al (1983).

We feel that the residuals should be properly scaled. Since $\text{var}(e_i) = \sigma^2(1-h_{ii})$ two useful choices are the <u>internally studentized residual</u>

$$r_i = e_i/s\sqrt{1-h_{ii}} \qquad (28)$$

and the <u>externally studentized residual</u>

$$e_i^* = \frac{e_i}{s_{-i}\sqrt{1-h_{ii}}}, \qquad (29)$$

where s is the standard error of the regression ($\frac{1}{n-p}\sum_{i=1}^{n}(y_i - x_i'\hat{\beta})^2$) and s_{-i} is the same but with the ith observation omitted. A simple formula relates s and s_{-i}:

$$(n-p)s^2 = (n-p-1)s_{-i}^2 + \frac{e_i^2}{(1-h_{ii})}. \qquad (30)$$

Under the usual Gaussian error assumptions, e_i^* has a t-distribution with n-p-1 degrees of freedom. If a dummy variable with zero in all positions except for a one in the ith position is added to the current model (X), then e_i^* is a useful diagnostic for seeing if there should be a shift in the intercept for the ith observation. Further details are contained in Belsley, Kuh, and Welsch (1980):

Another form of residual is often useful. The <u>PRESS residual</u> is found by computing

$$y_i - x_i'b_{-i} = \frac{e_i}{(1-h_{ii})}. \qquad (31)$$

Hoaglin and Welsch (1978) have noted that when (31) is scaled by its standard error, the result is just e_i^*.

7. MEASURES OF INFLUENCE

Looking for leverage points and examining various types of residuals form an important step in regression analysis. However, we would like to know if an observation is having a disproportionately large impact on our analysis. An observation is called <u>influential</u> if its deletion would cause major changes in estimates, confidence regions, test and diagnostic statistics, etc. Usually influential observatons are outside the patterns set by the majority of the data in the context of a regression model (including the structural model, stochastic model and fitting procedure). Influential data usually arise from errors in observing or recording data, structural model failure (for example, nonlinear instead of linear) and legitimate extreme observations. Deletion is a way to find procedures to measure influential data. Data should not be deleted because they are influential, but should be flagged and carefully examined. Alternative fits or forecasts may be needed, one with and one without these data. Judgment or information external to the data will often be necessary.

There are many ways to measure influence. Perhaps the most common is to think of all of the data but the ith observation as "good" and the ith as potentially "strange." We want to find an influence function or measure to see if the ith observation really is a cause for concern. A very useful influence function is

$$\hat{\beta} - b_{-i} = \frac{(X'X)^{-1} x_i e_i}{(1-h_{ii})} \tag{32}$$

$$= (X'X)^{-1} x_i (y - x_i' b_{-i}) \tag{33}$$

or for each estimated coefficient

$$b_j - b_{j,-i} = \frac{x_{ij \cdot [j]}(y_i - x_i' b_{-i})}{\sum_{k=1}^{n} x_{kj \cdot [j]}^2} . \tag{34}$$

This can also be stated in terms of (11) rather than adjusted variables.

It is often convenient to scale this measure in some way. Since we are usually interested in changes in the estimated coefficients that are a substantial fraction of the stochastic variability of b, we divide by the standard error of b_j. To estimate the standard error we use $\hat{\sigma}\sqrt{(X'X)^{-1}_{jj}}$ with $\hat{\sigma} = s_{-i}$ since we would like an estimate of σ that is not subject to the "possibly erroneous" i^{th} observation. Other reasons for using s_{-i} are given in Welsch (1982). All this gives us

$$DBETAS_{ij} = \frac{(b_j - b_{j,-i})}{s_{-i} \sqrt{(X'X)^{-1}_{jj}}} \tag{35}$$

$$= \frac{x_{ij \cdot [j]}}{\sqrt{\sum_{k=1}^{n} x_{kj \cdot [j]}^2}} \cdot \frac{y_i - x_i' b_{-i}}{s_{-i}} . \tag{36}$$

The first term is the square root of the partial leverage (27) and the second part is related to the predicted residual (31).

There are three basic ways to decide when $|DBETAS|$ are large. The first is to note which ones are larger than say .5 or 1. That is, setting aside one observation causes a .5 or 1 standard error change in the estimation of β_j. A second method (Belsley, Kuh, and Welsch, 1980) uses the fact that for a given j, when c_{ij} is constant for all i and the h_i are balanced,

$$\sum_{i=1}^{n} (DBETAS_{ij})^2 \approx 1. \tag{37}$$

When $|DBETAS_{ij}|$ is greater than say twice the average value $1/\sqrt{n}$ we take note. A practical rule of thumb is to use min $(.5, 2/\sqrt{n})$.

The third approach is to look at the DBETAS via exploratory data analysis or contour plots. For a fixed j, $DBETAS_{ij}$ consists of the product indicated in (36). We plot the partial leverage portion on the x-axis and the predicted residual part on the y-axis. Contours of constant influence x·y = c are also plotted. Figure (2) shows such a plot for c = .5, 1.0, 1.5 etc. The symbol + denotes a positive DBETAS and Δ a negative DBETAS. Some potentially influential points have been tagged.

Often we are more interested in predictions than coefficients. A prediction is just a linear combination of the estimated parameters, say $\ell'\hat{\beta}$. It is natural to compare $\ell'\hat{\beta}$ to $\ell'b_{-i}$ and scale with a measure of the standard errors of the fit, $s_{-i} \sqrt{h_{ii}}$. However, we often do not know ℓ so we look for the worst case

$$\sup_{\ell} \frac{(\ell'\hat{\beta} - \ell'b_{-i})^2}{s_{-i}^2 \ell'(X'X)^{-1}\ell} = \frac{(\hat{\beta}-b_{-i})'X'X(\hat{\beta}-b_{-i})}{s_{-i}^2} \qquad (38)$$

$$= \frac{h_{ii}}{1-h_{ii}} \cdot (e_i^*)^2. \qquad (39)$$

From (25) we also note that the difference between the fit, $x_i'\hat{\beta}$, and the predicted fit, $x_i'b_{-i}$, is just $h_{ii}e_i/(1-h_{ii})$. When scaled by a measure of standard error of the fit, $s_{-i} \sqrt{h_{ii}}$, we get

$$DFITS_i = \frac{x_i'\hat{\beta}-x_i'b_{-i}}{s_{-i}\sqrt{h_{ii}}} = (\frac{h_{ii}}{1-h_{ii}})^{\frac{1}{2}} \cdot e_i^* \qquad (40)$$

which is the square root of (39). Notice that DFITS is the product of a leverage factor (24) and the externally studentized residual (28).

Again there are a number of approaches to deciding when $|DFITS|$ is large. We can use fraction of standard error, like .5, 1 etc. or note that

$$\sum_{i=1}^{n} DFITS_i^2 \approx p \qquad (41)$$

when $h_{ii} \sim p/n$. We do not want any observation to stray too far from the average influence so we would single out observations with $|DFITS_i| > 2\sqrt{p}/\sqrt{n}$. A reasonable rule is to use min $(1, 2\sqrt{p}/\sqrt{n})$ as a cut-off. Cook and Weisberg (1982) develop a statistic similar to $DFITS_i^2$, namely

$$D_i^2 = \frac{1}{p}(\frac{h_{ii}}{1-h_{ii}}) \cdot \frac{e_i^2}{s^2(1-h_i)} . \qquad (42)$$

They suggest that D_i^2 may be considered large when it exceeds $F_{p,n-p}^{(.5)}$ or approximately, one. This seems to be an unduly conservative cut-off in practice. D_i^2 is also troublesome because it uses s^2 instead of $s^2(i)$ and hence is not robust to errors in the ith observation. See Welsch (1982) for further discussion of D_i^2.

We prefer to look at contour plots with $[h_{ii}/(1-h_{ii})]^{1/2}$ on the x-axis and $|e_i^*|$ on the y-axis. Constant influence contours may be plotted as before. Figure 3 provides an example.

Often in inference we are interested in confidence intervals or regions. A confidence region consists of a center, a shape, and a scale. For example, in regression a confidence region might be all β satisfying

$$\frac{(\hat{\beta}-\beta)'X'X(\hat{\beta}-\beta)}{s^2} \leq a \cdot s^2$$

where a is based on the F statistic and p. Here $\hat{\beta}$ is the center, $X'X$ is the shape and s^2 is the scale.

So far we have looked at diagnostics for the center, $\hat{\beta}$. To see what happens to $X'X$ when the ith observation is considered suspect, we can look at

$$\text{trace}[X'_{-i}X_{-i}(X'X)^{-1}] = p - h_{ii} \tag{43}$$

or a ratio of volumes

$$\frac{\det[X'_{-i}X_{-i}]}{\det(X'X)} = 1 - h_{ii}. \tag{44}$$

These equations just provide more reasons to look at h_{ii} in its own right.

As for scale, we note again that

$$(n-p)s^2 = (n-p-1)s_{-i}^2 + e_i^2/(1-h_{ii})$$

so that

$$s^2/s_{-i}^2 \approx 1 + (e_i^*)^2/n-p$$

Again we have already looked at e_i^* extensively.

These measures can be combined by looking at the ratio of covariance matrix determinants

$$\text{COVRATIO} = \frac{\det s_{-i}^2(X'_{-i}X_{-i})^{-1}}{\det s^2(X'X)^{-1}} \tag{45}$$

$$= \frac{s_{-i}}{s} 2p \ (1/1-h_{ii}) \tag{46}$$

Contour plots are possible here as well. Velleman and Welsch (1981) note that h_{ii} and e_i cannot vary completely independently since

$$h_{ii} + e_i^2/[(n-p)s^2] \leq 1. \qquad (47)$$

and, of course, when $h_{ii} = 1$, $e_i = 0$.

The important point to note is that observations with large h_i decrease the size of a confidence region while observations with large $|e_i^*|$ increase it. Our goal should be to insure that we are alerted to potentially influential observations. As we can see from the above, influential observations can be both useful and harmful. How we treat them will depend on the purposes of our analysis and their relation to the rest of the data and our models. The best rule of thumb is that there may be more than one good and valid analysis of a data set. Sometimes an analysis with an influential observation and one without is the only way to adequately summarize the data.

8. GENERALIZED LINEAR MODELS

When the response variable y is Bernoulli, binomial, Poisson, etc., generalized linear models (GLM) are appropriate. A detailed discussion is contained in McCullagh and Nelder (1983). Many of the ideas discussed above can be carried over to these models as well. Basic references are Pregibon (1979, 1981).

The essential idea is to find an influence function, b-b(i), for the parameters of the model. In the GLM case, this cannot be done exactly since the computation of b requires the solution of a system of nonlinear equations via iterative procedures. However, b-b(i) can be approximated, usually by taking one interation away from b (the fully iterated solution) with the ith observation removed in an appropriate way. Various kinds of residuals can be defined as well as useful plots.

This is an extremely active area of research at the present time, especially generalizations to survival analysis (Hall et al, 1982), proportional hazards and censured data (Cain and Lange, 1983), Cox models (Storer and Crowley, 1983) and matched case control studies (Pregibon, 1984). The bibliographies in these papers should provide a good overview of work in this area.

9. INFLUENTIAL SUBSETS OF DATA

If there are two or more outliers in a clump, then influence functions based on setting aside one of the observations will not work well because we will see little change until the entire clump is set aside. The methods discussed above do generalize to subsets of data (Belsley, Kuh, Welsch, 1980; Cook and Weisberg, 1982; and Welsch, 1982), but very large amounts of computation are required.

To overcome the computational problems, we have developed a technique called bounded-influence regression (Krasker and Welsch, 1982). A bounded-influence estimator can be viewed as a procedure to find data-dependent weights (for use in weighted least-squares) so that no small subset of the data is overly influential. The weights and related statistics then become useful diagnostic tools. Examples and associated contour plots are given in Krasker and Welsch (1983). Computational details are discussed in Peters, Samarov, and Welsch (1982).

These ideas can be extended to generalized linear models in a number of ways. Some basic references are Samuels (1978), Krasker (1979), Reid (1981), Pregibon (1982), and Accomando and Pagano (1983). Much more work needs to be done in this area.

10. COMPUTATION

Computational details of many of the above methods are treated in Velleman and Welsch (1981). They also discuss how to use package programs such as SAS and MINITAB to obtain various diagnostics. The plots used here were made on the TROLL system, a large data analysis and modeling system available for license from M.I.T.

There is no reason why good diagnostics should be omitted from a packaged program. They are essential in my view. We can all demand that they be a part of the new generation of software for personal computers. Diagnostics are particularly effective on such computers because graphical tools are readily available.

ACKNOWLEDGEMENTS

The support for this research was provided, in part, by National Science Foundation Grants MCS8116778 and IST8218759 and IBM. The author would like to thank Joanne Sorrentino for her excellent typing.

REFERENCES

ACCOMANDO, W.P., and PAGANO, M. (1983), "Analyzing a Large Observational Data Set with Categorical Outcomes," Technical Report No. 335Z, Department of Biostatistics, Dana-Farber Cancer Institute, Boston, MA 02115.

BELSLEY, D.A., KUH, E., and WELSCH, R.E. (1980), Regression Diagnostics New York: Wiley.

CAIN, K.C., and LANGE, N.T. (1983), "Estimating Case Influence for the Proportional Hazards Regression Model with Censored Data," Technical Report 320Z, Department of Biostatistics, Dana-Farber Cancer Institute, Boston, MA 02115. To appear in Biometrics.

CHAMBERS, J.M., CLEVELAND, W.S., KLEINER, B., and TUKEY, P.A. (1983), Graphical Methods for Data Analysis, Massachusetts: Duxbury Press.

COOK, R.D., and WEISBERG, S. (1982), Influence and Residuals in Regression, London: Chapman & Hall.

FRIEDMAN, J.H., and STUETZLE, W.Z. (1981), "Projection Pursuit Regression," Journal of the American Statistical Association, 76, 817-823.

HALL, G.J., ROGERS, W.H., and PREGIBON, D. (1982), "Outliers Matter in Survival Analysis," Technical Report P-6761, Rand Corporation, Santa Monica, CA.

HENDERSON, H. and VELLEMAN, P.F. (1981), "Building Multiple Regression Models Interactively," Biometrics, 37, 391-411.

HUBER, P.J. (1981), Robust Statistics, New York: Wiley.

HOAGLIN, D.C., and WELSCH, R.E. (1978), "The Hat Matrix in Regression and ANOVA," The American Statistician, 32, 17-22.

KRASKER, W.S. (1979), "Efficient Bounded-Influence Estimation in Logit Models," unpublished paper, December 1979, Department of Economics, University of Michigan, Ann Arbor. (William Krasker is now at the Harvard Business School, Boston, MA.)

KRASKER, W., and WELSCH, R.E. (1982), "Efficient Bounded-Influence Regression Estimation," Journal of the American Statistical Association, 77, 595-604.

KRASKER, W.S., and WELSCH, R.E. (1983), "The Use of Bounded-Influence Regression in Data Analysis: Theory, Computation, and Graphics," Proceedings: Computer Science and Statistics: Fourteenth Symposium on the Interface, New York: Springer-Verlag, 45-51.
McCULLAGH, P., and NELDER, J.A. (1983), Generalized Linear Models, London: Chapman and Hall.
MOSTELLER, F., and TUKEY, J. (1977), Data Analysis and Regression, Massachusetts: Addison-Wesley.
PETERS, S., SAMAROV, A., and WELSCH, R.E. (1982), "Computational Procedures for Bounded-Influence and Robust Regression," Technical Report No. 30, Center for Computational Research in Economics and Management Science, MIT, Cambridge, MA 02139.
PREGIBON, D. (1979), "Data Analytic Methods for Generalized Linear Models," unpublished Ph.D. thesis, University of Toronto.
_____ (1981), "Logistic Regression Diagnostics," Annals of Statistics, 9, 705-724.
_____ (1982), "Resistant Fits for Some Commonly Used Logistic Models with Medical Applications," Biometrics, 38, 485-498.
_____ (1984), "Data Analysis Methods for Matched Case Control Studies. To appear Biometrics June 1984.
REID, N. (1981), "Influence Functions for Censored Data," Annals of Statistics, 9, 78-92.
SAMUELS, S.J. (1978), "Survival Analysis from the Viewpoint of Hampel's Theory for Robust Estimation," Institute of Statistics Mimeo Series No. 1163, Department of Biostatistics, University of North Carolina at Chapel Hill.
STORER, B.E., and CROWLEY, J. (1983), "A Diagnostic for Cox Regression and General Conditional Likelihoods," Technical Report No. 59, Department of Biostatistics, School of Public Health, Seattle, WA.
VELLEMAN, P.F., and D.C. HOAGLIN (1981), Applications, Basics, and Computing of Exploratory Data Analysis, Massachusetts: Duxbury Press.
VELLEMAN, P.F., and WELSCH R.E. (1981), "Efficient Computing of Regression Diagnostics," The American Statistician, 35, no. 4, 234-242.
WELSCH, R.E. (1982), "Influence Functions and Regression Diagnostics," Modern Data Analysis, edited by R. Launer and A. Siegel, New York: Academic Press, 148-169.

FURTHER READING

BECKMAN, R.J., and COOK, R.D. (1983), "Outliers," Technometrics, 25, 119-163.
BREIMAN, L., and FRIEDMAN, J.H. (1982), "Estimating Optimal Transformations for Multiple Regression and Correlation." To appear in Journal of the American Statistical Association.
HOCKING, R.R. (1983), "Developments in Linear Regression Methodology: 1959-1982, Technometrics, 25, 219-250.
KRASKER, W.S., KUH, E., and WELSCH, R.E. (1983), "Estimation for Dirty Data and Flawed Models," Handbook of Econometrics, 1, edited by Z. Griliches and M.D. Intrilligator, North-Holland, Amsterdam, 651-698.

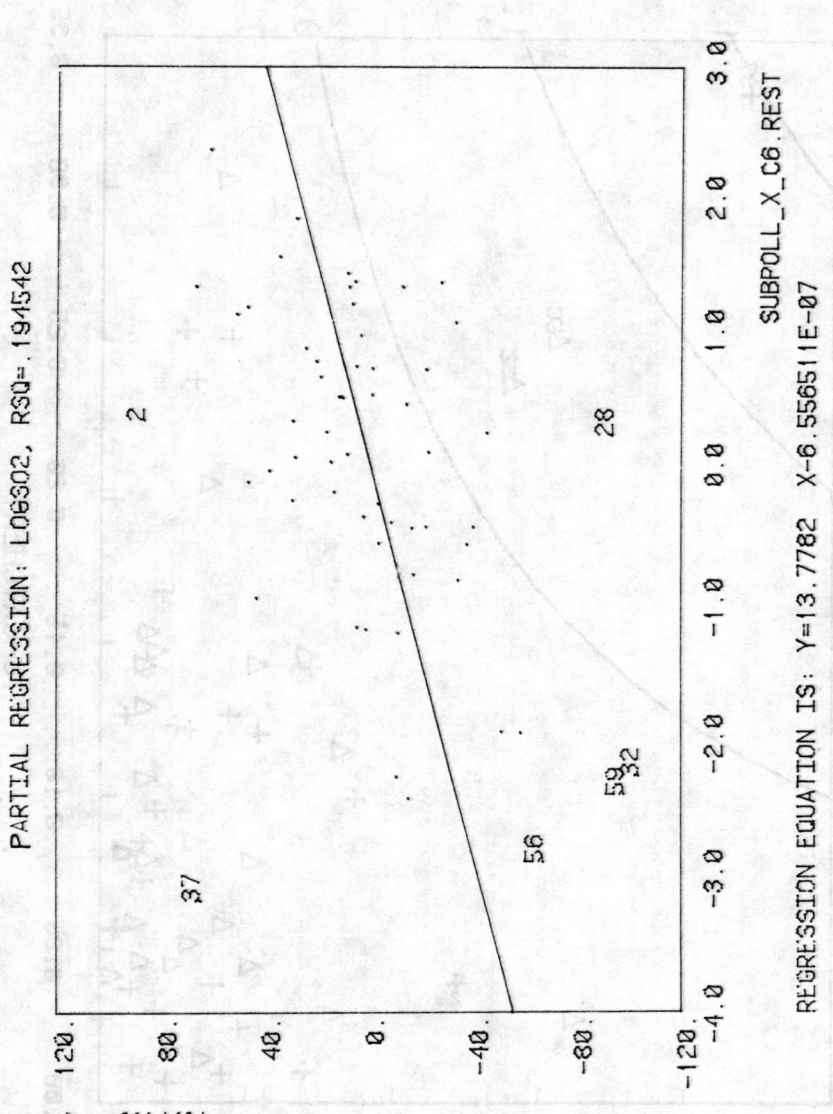

FIGURE 1.

32

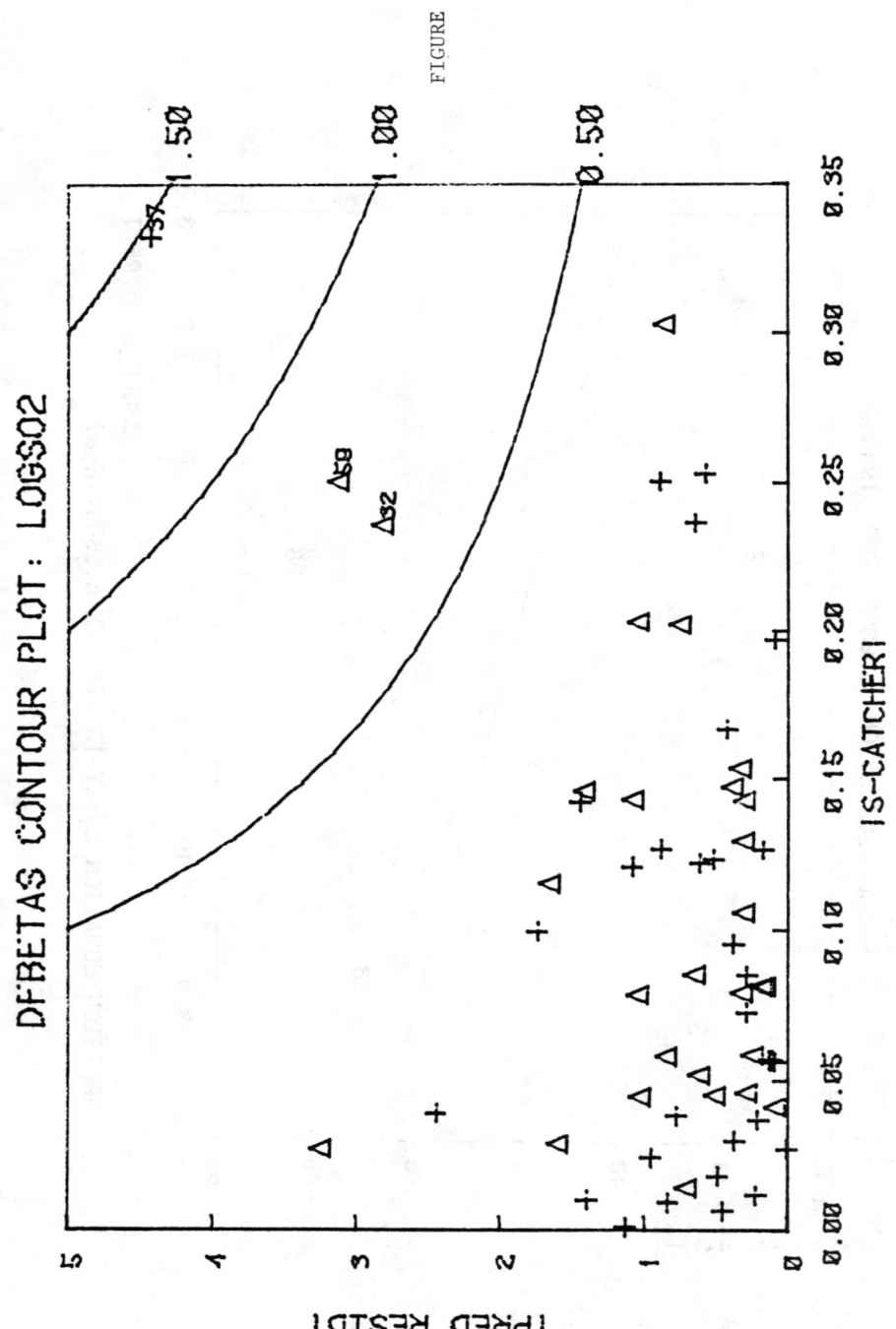

FIGURE 2.

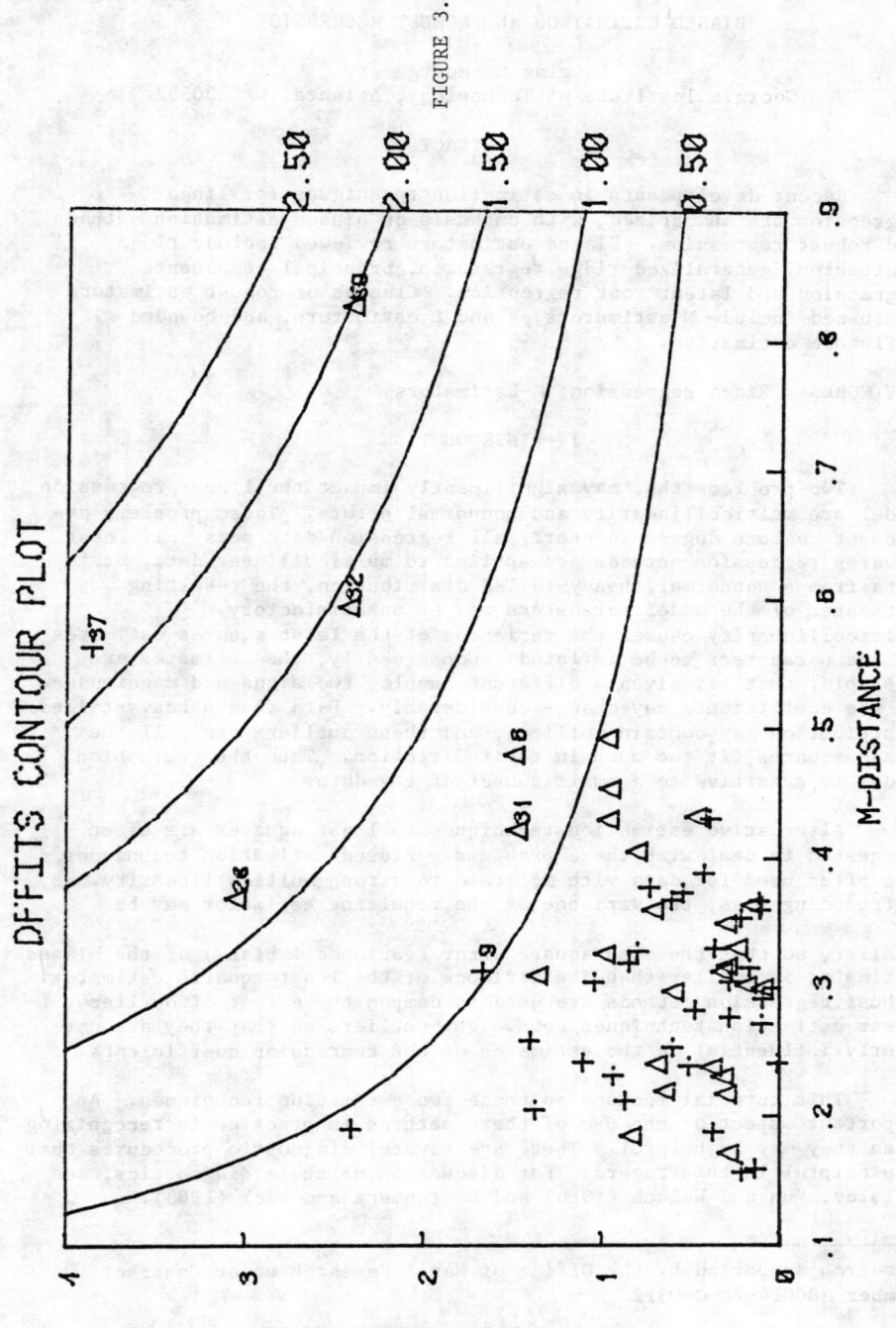

FIGURE 3.

BIASED ESTIMATION AND ROBUST REGRESSION

Douglas C. Montgomery
Georgia Institute of Technology, Atlanta, GA 30332

ABSTRACT

Recent developments in estimation techniques for linear regression are summarized, with emphasis on biased estimation methods and robust regression. Biased estimators reviewed include ridge regression, generalized ridge regression, principal components regression and latent root regression. Classes of robust estimators discussed include M estimators, R and L estimators, and bounded influence estimation.

KEY WORDS: Ridge Regression; M-Estimators.

1. INTRODUCTION

Two problems that may significantly impact the linear regression model are multicollinearity and nonnormal errors. These problems are present to some degree in nearly all regression data sets. If least squares regression methods are applied to multicollinear data, or to data from a nonnormal, heavy-tailed distribution, the resulting estimates of the model parameters may be unsatisfactory. Multicollinearity causes the variances of the least squares estimates of the parameters to be inflated. Consequently, the estimates are unstable; that is, given a different sample, the signs and magnitudes of the coefficients may change considerably. Data from a heavy-tailed distribution may contain outliers, and these outliers can pull the least squares fit too much in their direction. Thus the regression model is sensitive to a small subset of the data.

Alternative estimation techniques to least squares are often suggested to deal with these problems. Biased estimation techniques are often used for data with moderate to strong multicollinearity. By introducing bias, the variance of the resulting estimator may be smaller, so that the mean square error (variance + bias^2) of the biased estimator is smaller than the variance of the least squares estimator. Robust regression methods are used to dampen the effect of outliers. These estimation techniques downweight outliers so that they are not overly influential on the estimates of the regression coefficients.

This tutorial focuses on these two estimation techniques. An important aspect of the use of these methods in practice is recognizing when they may be helpful. There are several diagnostic procedures that are helpful in this regard. For discussion of these diagnostics, see Belsley, Kuh and Welsch (1980) and Montgomery and Peck (1983).

Research supported by the Office of Naval Research under Contract Number N00014-78-C-0312

2. BIASED ESTIMATION

2.1 Ridge Regression

When the method of least squares is applied to nonorthogonal data, very poor estimates of the regression coefficients are usually obtained. The variance of the least squares estimates of the regression coefficients may be considerably inflated, and the length of the vector of least squares parameter estimates is too long on the average. This implies that the absolute value of the least squares estimates are too large, and that they are very unstable; that is, their magnitudes and signs may change considerably given a different sample.

The problem with the method of least squares is the requirement that $\hat{\beta}$ be an unbiased estimator of β. The Gauss-Markoff property assures us that the least squares estimator has minimum variance in the class of unbiased linear estimators, but there is no guarantee that this variance will be small.

One way to alleviate this problem is to drop the requirement that the estimator of β be unbiased. Suppose that we can find a biased estimator of β, say $\hat{\beta}^*$, that has a smaller variance than the unbiased estimator $\hat{\beta}$. The mean square error of the estimator $\hat{\beta}^*$ is defined as

$$\text{MSE}(\hat{\beta}^*) = E(\hat{\beta}^*-\beta)^2 = V(\hat{\beta}^*) + [E(\hat{\beta}^*)-\beta]^2 \qquad (1)$$

or

$$\text{MSE}(\hat{\beta}^*) = \text{Variance }(\hat{\beta}^*) + (\text{Bias in }\hat{\beta}^*)^2$$

Note that the MSE is just the expected squared distance from $\hat{\beta}^*$ to β. By allowing a small amount of bias in β^* the variance of $\hat{\beta}^*$ can be made small such that the MSE of $\hat{\beta}^*$ is less than the variance of the unbiased estimator $\hat{\beta}$.

A number of procedures have been developed for obtaining biased estimators of regression coefficients. One of these procedures is ridge regression, originally proposed by Hoerl and Kennard (1970a,b). The ridge estimator is the solution to

$$(X'X+kI)\hat{\beta}_R = X'y \qquad (2)$$

or

$$\hat{\beta}_R = (X'X+kI)^{-1}X'y \qquad (3)$$

where k>0 is a constant selected by the analyst. The procedure is called ridge regression because the underlying mathematics are similar to the method of ridge analysis used earlier by Hoerl (1959) for describing the behavior of second-order response surfaces. Note that when k=0 the ridge estimator is the least squares estimator.

The ridge estimator is a linear transformation of the least squares estimator since

$$\hat{\beta}_R = (X'X+kI)^{-1}X'y$$
$$= (X'X+kI)^{-1}(X'X)\hat{\beta}$$
$$= Z_k\hat{\beta}$$

Therefore since $E(\hat{\beta}_R) = E(Z_k\hat{\beta}) = Z_k\beta$, $\hat{\beta}_R$ is a biased estimator of β. We usually refer to the constant k as the biasing parameter. The covariance matrix of $\hat{\beta}_R$ is

$$V(\hat{\beta}_R) = \sigma^2(X'X+kI)^{-1}X'X(X'X+kI)^{-1} \qquad (4)$$

The mean square error of the ridge estimator is

$$MSE(\hat{\beta}_R) = \text{Variance } (\hat{\beta}_R) + (\text{Bias in } \hat{\beta}_R)^2$$
$$= \sigma^2 \sum_{j=1}^{p} \lambda_j/(\lambda_j+k)^2 + k^2\beta'(X'X+kI)^{-2}\beta \qquad (5)$$

where $\lambda_1, \lambda_2, \ldots, \lambda_p$ are the eigenvalues of $X'X$. The first term on the right-hand side of (5) is the sum of variances of the parameters in $\hat{\beta}_R$ and the second term is the square of the bias. If k>0, note that the bias in $\hat{\beta}_R$ increases with k. However, the variance decreases as k increases.

In using ridge regression we wish to choose a value of k such that the reduction in the variance term is greater than the increase in the squared bias. If this can be done, the mean square error of the ridge estimator $\hat{\beta}_R$ will be less than the variance of the least squares estimator β. Hoerl and Kennard proved that there exists a nonzero value of k for which the MSE of $\hat{\beta}_R$ is less than the variance of the least squares estimator $\hat{\beta}$, provided that $\beta'\beta$ is bounded. The residual sum of squares is

$$SS_E = (y-X\hat{\beta}_R)'(y-X\hat{\beta}_R)$$
$$= (y-X\hat{\beta})'(y-X\hat{\beta}) + (\hat{\beta}_R-\hat{\beta})'X'X(\hat{\beta}_R-\hat{\beta}) \qquad (6)$$

Since the first term on the right-hand side of (6) is the residual sum of squares for the least squares estimates $\hat{\beta}$, we see that as k increases the residual sum of squares increases. Consequently, because the total sum of squares is fixed, R^2 decreases as k increases. Therefore the ridge estimate will not necessarily provide the best "fit" to the data, but this should not overly concern us, since we are more interested in obtaining a stable set of parameter estimates. The ridge estimates may result in an equation that does a better job of predicting future observations than would least squares (although there is no conclusive proof that this will happen).

Hoerl and Kennard have suggested that an appropriate value of k may be determined by inspection of the ridge trace. The ridge trace is a plot of the elements of $\hat{\beta}_R$ versus k, for values of k usually in the interval zero to one. Marquardt and Snee (1975) suggest using up to about 25 values of k, spaced approximately logarithmically over the interval [0,1]. If multicollinearity is severe, the instability in the regression coefficients will be obvious from the ridge trace. As k is increased, some of the ridge estimates will vary dramatically. At some value of k, the ridge estimates $\hat{\beta}_R$ will stabilize. The objective is to select a reasonably small value of k at which the ridge estimates $\hat{\beta}_R$ are stable. Hopefully this will produce a set of estimates with smaller MSE than the least squares estimates.

The ridge regression estimates may be computed by using an ordinary least squares computer program and augmenting the standardized data as follows:

$$X_A = \begin{bmatrix} X \\ kI_p \end{bmatrix} \quad Y_A = \begin{bmatrix} Y \\ 0_p \end{bmatrix}$$

where kI_p is a p x p diagonal matrix with diagonal elements equal to the square root of the biasing parameter and 0_p is a p x 1 vector of zeros. The ridge estimates are then computed from

$$\hat{\beta}_R = (X'_A X_A)^{-1} X'_A Y_A = (X'X + kI_p)^{-1} X'Y$$

Figure 1 illustrates the geometry of ridge regression for a two-regressor problem. The point $\hat{\beta}$ at the center of the ellipses corresponds to the least squares solution, where the residual sum of squares takes on its minimum value. The small ellipse represents the locus of points in the β_1, β_2 plane where the residual sum of squares is constant at some value greater than the minimum. The ridge estimate $\hat{\beta}_R$ is the shortest vector from the origin that produces a residual sum of squares equal to the value represented by the small ellipse. That is, the ridge estimate $\hat{\beta}_R$ produces the vector of regression coefficients with the smallest norm consistent with a specified increase in the residual sum of squares. We note that the ridge estimator shrinks the least squares estimator towards the origin. Consequently ridge estimators (and other biased estimators, generally) are sometimes called shrinkage estimators. Hocking (1976) has observed that the ridge estimator shrinks the least squares estimator with respect to the contours of X'X. That is, $\hat{\beta}_R$ is the solution to

$$\text{minimize}(\beta-\hat{\beta})'X'X(\beta-\hat{\beta})$$
$$\beta$$

$$\text{subject to } \beta'\beta \leq d^2$$

where the radius d depends on k.

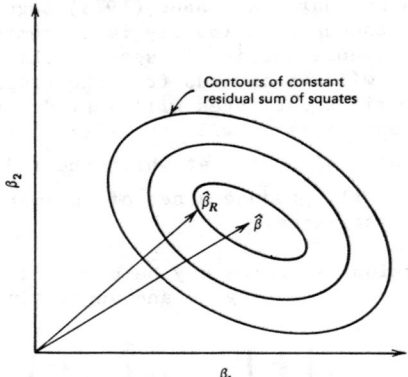

Figure 1. A Geometric Interpretation of Ridge Regression

Much of the controversy concerning ridge regresion centers around the choice of the biasing parameter k. Choosing k by inspection of the ridge trace is a subjective procedure requiring judgment on the part of the analyst. Several authors have proposed procedures for choosing k that are more analytical. Hoerl, Kennard and Baldwin (1975) have suggested that an appropriate choice for k is

$$k = p\sigma^2/\hat{\beta}'\hat{\beta} \qquad (7)$$

where $\hat{\beta}$ and σ^2 are found from the least squares solution. They showed via simulation that the resulting ridge estimator had significant improvement in MSE over least squares. In a subsequent paper, Hoerl and Kennard (1976) proposed an iterative estimation procedure based on (7).

McDonald and Galarneau (1975) suggest choosing k so that

$$\hat{\beta}_R'\hat{\beta}_R = \hat{\beta}'\hat{\beta} - \sigma^2 \sum_{j=1}^{p} \left(\frac{1}{\hat{\lambda}_j}\right) \tag{8}$$

For cases where the right-hand side of (8) is negative, they investigated letting either k=0 (least squares) or $k=\infty \left(\hat{\beta}_R=0\right)$. Neither method was better than least squares in all cases. Marquardt (1970) has proposed using a value of k such that the maximum VIF is between one and ten, preferably closer to one. Other methods of choosing k have been suggested by Dempster et al. (1977), Goldstein and Smith (1974), Lawless and Wang (1976), Lindley and Smith (1972), Mallows (1973a), and Obenchain (1975).

There is no assurance that any of these procedures will produce similar choices for k. Furthermore there is no guarantee that these methods are superior to straightforward inspection of the ridge trace.

2.2 Generalized Ridge Regression

Hoerl and Kennard (1970a) proposed an extension of the ordinary ridge regression procedure that allows separate biasing parameters for each regressor. This procedure is known as generalized ridge regression.

The discussion of generalized ridge regression is somewhat simplified if we transform the data to the space of orthogonal regressors. To do this, recall that if $\hat{\Lambda}$ is the p x p diagonal matrix whose main diagonal elements are the eigenvalues $(\hat{\lambda}_1, \hat{\lambda}_2, \ldots, \hat{\lambda}_p)$ of X'X and if T is the corresponding orthogonal matrix of eigenvectors, then

$$T'X'XT = \hat{\Lambda} \tag{9}$$

Letting $Z = XT$ and $\alpha = T'\beta$ the linear model becomes

$$y = X\beta + \varepsilon$$

$$= (ZT')(T\alpha) + \varepsilon$$

$$= Z\alpha + \varepsilon \tag{10}$$

The least squares estimator of α is the solution to

$$(Z'Z)\hat{\alpha} = Z'y \tag{11}$$

which is equivalent to

$$\hat{\Lambda}\hat{\alpha} = Z'y \tag{12}$$

or

$$\hat{\underline{\alpha}} = \underline{\Lambda}^{-1}\underline{Z}'\underline{y} \tag{13}$$

The vector of original parameter estimates is given by

$$\hat{\underline{\beta}} = \underline{T}\hat{\underline{\alpha}} \tag{14}$$

We often refer to (10) as the canonical form of the model. In terms of the canonical form, the generalized ridge estimator is the solution to

$$(\underline{\Lambda}+\underline{K})\hat{\underline{\alpha}}_{GR} = \underline{Z}'\underline{y} \tag{15}$$

where $\underline{K}$ is a diagonal matrix with elements $(k_1, k_2, \ldots, k_p)$. In terms of the original model, the generalized ridge coefficients are

$$\hat{\underline{\beta}}_{GR} = \underline{T}\hat{\underline{\alpha}}_{GR} \tag{16}$$

Now consider the choice of the biasing parameters in $\underline{K}$. The mean square error for generalized ridge regression is

$$\text{MSE}(\hat{\underline{\beta}}_{GR}) = E[(\hat{\underline{\beta}}_{GR}-\underline{\beta})'(\hat{\underline{\beta}}_{GR}-\underline{\beta})]$$

$$= \sigma^2 \sum_{j=1}^{p} \lambda_j/(\lambda_j+k_j)^2 + \sum_{j=1}^{p} \alpha_j^2 k_j^2/(\lambda_j+k_j)^2 \tag{17}$$

The first term on the right-hand side of (17) is the sum of the variances of the parameter estimates and the second term is the squared bias. The mean square error (17) is minimized by choosing

$$k_j = \sigma^2/\alpha_j^2, \quad j = 1, 2, \ldots, p \tag{18}$$

Unfortunately the optimal k depend on unknown parameters σ^2 and α_j. Hoerl and Kennard (1970a) suggest an iterative approach to determining the k_j. Beginning with the least squares solution, we obtain an initial estimate of the k_j, for example

$$k_j^0 = \sigma^2/\alpha_j^2$$

Use these initial estimates of the k_j to compute initial generalized ridge estimates from

$$\underline{\alpha}_{GR}^0 = (\underline{\Lambda}+\underline{K}^0)^{-1}\underline{Z}'\underline{y}$$

where $\underline{K}^0 = \text{diag}(k_1^0, k_2^0, \ldots, k_p^0)$. The use the initial estimates $\underline{\alpha}_{GR}^0$ to revise the estimates of the k_j:

$$k_j^1 = \sigma^2/(\alpha_{GR\,j}^0)^2 \quad j=1,2,\ldots,p$$

These new values of k_j^1 may be used to revise the estimates of the $\tilde{\alpha}$. The iterative process should continue until stable parameter estimates result. One measure of stability often used is the squared length of the vector $\tilde{\alpha}_{GR}'\tilde{\alpha}_{GR}$. Specifically, if the squared length of the vector of parameter estimates does not change significantly from iteration i-1 to iteration i, then terminate. Otherwise the iterative estimation procedure should continue. Note that there is no helpful graphical display of the coefficients such as the ridge trace in generalized ridge regression.

We may use (18) to justify our choice of the biasing parameter k in ordinary ridge regression. The value k in (7) is a weighted average of the k, from (18). Clearly if the k_j are combined to produce a single biasing parameter, we should not use an ordinary average because a small α_j would produce a large value of k inducing too much bias in the parameter estimates. However, the harmonic mean of the k_j is

$$k_h = p\sigma^2 / \sum_{j=1}^{p} \alpha_j^2 = p\sigma^2 / \sum_{j=1}^{p} \alpha_j^2 = p\sigma^2/\hat{\beta}'\hat{\beta} = k$$

as given in (7).

Hemmerle (1975) showed that Hoerl and Kennard's iterative procedure for estimating the k_j has an explicit closed form solution so that in general, iteration is unncecessary. Specifically, let

$$\hat{\tilde{\alpha}}_{GR} = B\hat{\tilde{\alpha}} \qquad (19)$$

where $\hat{\tilde{\alpha}}$ is the least squares estimator and B is a diagonal matrix of nonnegative elements $b_1, b_2, \ldots, b_p$. Hocking et al. (1976) show that Hemmerle's results are to choose

$$b_j = 0 \quad \text{if } \tau_j^2 < 4$$
$$b_j = 0.5 + [0.25 - (1/\tau_j^2)]^{1/2} \quad \text{if } \tau_j^2 > 4 \qquad (20)$$

where $\tau_j^2 = \alpha_j^2 \lambda_j / \sigma^2$. Noting that τ_j is the t-statistic associated with the jth regressor, we observe that if the t-statistic is "small" the corresponding generalized ridge coefficient is set equal to zero, while if the t-statistic is "large", the corresponding generalized ridge coefficient is a fraction b_j of the least squares coeffocient. In other words, nonsignificant coefficients are shrunk to zero, while significant coefficients are shrunk less severly. We refer to this solution as the fully iterated generalized ridge solution.

Hemmerle noted that the fully iterated generalized ridge solution often results in the introduction of too much bias (or too much shrinkage) in the final parameter estimates. Hemmerle proposed a technique to avoid this based on constraining the residual sum of squares to prevent an undesired significant increase. He recommended that a limit be placed on the total loss in R^2, and that this loss be allocated proportionally to the individual regressors. His procedure results in modified values of b_j, say b_j^*, given by

$$b_j^* = 1 - \sqrt{m}(1-b_j) \qquad (21)$$

where m is the ratio of the allowable loss in R^2 to the loss in R^2 in b_j from (20) is used. Hocking et al. (1976) object to the use of (21) because it forces all the b_j^* to be nonzero. By setting some of the b_j to zero, the strong influence of a small eigenvalue on variance inflation was eliminated. Using (21) allows the influence of that eigenvalue to return.

Unfortunately there is no clear-cut "best" choice of the k_j from generalized ridge regression. We agree with Hemmerle (1975) that fully iterated generalized ridge often results in too much shrinkage, and some type of constrained procedure is appropriate, particularly for data that is severly ill-conditioned. Constraining the maximum increase in the residual sum of squares to between one and twenty percent often works well in practice. However, more work needs to be done to develop better guidelines for choosing the parameters k_j and controlling the amount of shrinkage.

2.3 Principal Components Regresion

Biased estimators of regression coefficients can also be obtained by using principal components regression. Consider the canonical form of the model,

$$\underset{\sim}{y} = \underset{\sim}{Z}\underset{\sim}{\alpha} + \underset{\sim}{\varepsilon} \qquad (22)$$

The columns of $\underset{\sim}{Z}$, which define a new set of orthogonal regressors, say

$$\underset{\sim}{Z} = \left[\underset{\sim}{Z}_1, \underset{\sim}{Z}_2, \ldots, \underset{\sim}{Z}_p\right]$$

are referred to as principal components.

The least squares estimator of $\underset{\sim}{\alpha}$ is

$$\hat{\underset{\sim}{\alpha}} = (\underset{\sim}{Z}'\underset{\sim}{Z})^{-1}\underset{\sim}{Z}'\underset{\sim}{y} = \underset{\sim}{\Lambda}^{-1}\underset{\sim}{Z}'\underset{\sim}{y} \qquad (23)$$

and the covariance matrix of $\hat{\underset{\sim}{\alpha}}$ is

$$V(\hat{\underset{\sim}{\alpha}}) = \sigma^2(\underset{\sim}{X}'\underset{\sim}{X})^{-1} = \sigma^2\underset{\sim}{\Lambda}^{-1} \qquad (24)$$

Thus a small eigenvalue of X'X means that the variance of the corresponding orthogonal regression coefficient will be large. Since

$$Z'Z = \sum_{i=1}^{p} \sum_{j=1}^{p} Z_i Z_j' = \Lambda$$

We often refer to the eigenvalue λ_j as the variance of the jth principal component. If all the λ_j are equal to unity, the original regressors are orthogonal, while if an λ_j is exactly equal to zero this implies a perfect linear relationship between the original regressors. One or more of the λ_j near zero implies that multicollinearity is present. Note also that the covariance matrix of the standardized regression coefficients $\hat{\beta}$ is

$$V(\hat{\beta}) = V(T\hat{\alpha}) = T\Lambda^{-1}T'\sigma^2$$

This implies that the variance of $\hat{\beta}_j$ is $\sigma^2(\sum_{i=1}^{p} t_{ji}^2/\lambda_i)$. Therefore the variance of β_j is a linear combination of the reciprocals of the eigenvalues. This demonstrates how one or more small eigenvalues can destroy the precison of the least squares estimate $\hat{\beta}_j$.

The principal components regression approach combats multicollinearity by using less than the full set of principal components in the model. To obtain the principal components estimator, assume that the regressors are arranged in order of decreasing eigenvalues, $\lambda_1 \geq \lambda_2 \geq \ldots \geq \lambda_p > 0$. Suppose that the last s of these eigenvalues are approximately equal to zero. In principal components regression the principal components corresponding to near zero eigenvalues are removed from the analysis and least squares applied to the remaining components. That is,

$$\hat{\alpha}_{PC} = B\hat{\alpha} \qquad (25)$$

where $b_1 = b_2 = \ldots = b_{p-s} = 1$ and $b_{p-s+1} = b_{p-s+2} = \ldots = b_p = 0$. Thus the principal components estimator is

$$\hat{\alpha}_{PC} = \begin{bmatrix} \hat{\alpha}_1 \\ \hat{\alpha}_2 \\ \vdots \\ \hat{\alpha}_{p-s} \\ \hline 0 \\ 0 \\ \vdots \\ 0 \end{bmatrix} \begin{matrix} p-s \text{ components} \\ \\ \\ \\ \\ s \text{ components} \end{matrix}$$

or in terms of the standardized regressors

$$\hat{\underline{\beta}}_{PC} = \underline{T}\hat{\underline{\alpha}}_{PC}$$
$$= \sum_{j=1}^{p-s} \lambda_j^{-1} \underline{t}_j' \underline{X}' \underline{y} \underline{t}_j \qquad (26)$$

A simulation study by Gunst and Mason (1977) showed that principal components regression offers considerable improvement over least squares when the data are ill-conditioned. They also point out that another advantage of principal components is that exact distribution theory and variable selection procedures are available (see Mansfield, et al. (1977).

Marquardt (1970) suggested a generalization of principal components regression. He felt that the assumption of an integaral rank for the X matrix is too restrictive, and proposed a "fractional rank" estimator that allows the rank to be a piecewise continuous function. Specifically, if the rank of the X matrix is in the interval [r,r+1], then Marquardt's fractional rank estimator is

$$\hat{\underline{\alpha}}_{FR} = (1-c)\hat{\underline{\alpha}}_r + c\hat{\underline{\alpha}}_{r+1} \qquad (27)$$

for $0 \leq c \leq 1$, where $\hat{\underline{\alpha}}_r$ and $\hat{\underline{\alpha}}_{r+1}$ are the principal components estimators for α for assumed ranks r and r+1. That is, the last r - 1 elements of $\hat{\underline{\alpha}}_{FR}$ are zero, the (p-r+1)st element is $c\hat{\alpha}_{p-r+1}$ and the first p-r elements are the least squares estimators $\hat{\alpha}_1, \hat{\alpha}_2, \ldots, \hat{\alpha}_{p-r}$. Criteria for choosing r and c are discussed by Hocking et al. (1976).

2.4 Latent Root Regression Analysis

The latent root regression procedure was developed by Hawkins (1973) and Webster et al. (1974), following the same philosophy as principal components. The procedure forms estimators from the eigenvalues (or latent roots) of the correlation matrix of regressor and response variables

$$\underline{A}'\underline{A} = \begin{bmatrix} 1 & \underline{y}'\underline{X} \\ \underline{X}'\underline{y} & \underline{X}'\underline{X} \end{bmatrix}$$

Let $0 \leq \ell_0 \leq \ell_1 \leq \ldots \leq \ell_p$ and $\underline{v}_0, \underline{v}_1, \ldots, \underline{v}_p$ be the eigenvalues and eigenvectors of $\underline{A}'\underline{A}$, and denote the last p elements of $\underline{v}_j$ by $\underline{\delta}_j$ so that $\underline{v}_j' = [v_{0_j}, \underline{\delta}_j']$. The latent root estimator is

$$\hat{\underset{\sim}{\beta}}_{LR} = \sum_{j=s}^{p} \ell_j^{-1} \phi_j \underset{\sim}{\delta}_j \tag{28}$$

$$\phi_j = -S_{yy}^{1/2} v_{0_j} \sum_{q=s}^{p} v_{0_q}^2 \ell_q^{-1} \tag{29}$$

The s terms corresponding to j=0,1,...,s-1 deleted from (28) correspond to those eigenvectors for which both $|v_{0_j}|$ and ℓ_j are nearly zero. Thus like principal components regression, latent root regression attempts to identify and eliminate multicollinearities that do not aid in prediction. Latent root regression reduces to least squares when no terms are deleted (s=0).

Gunst et al. (1976) and Gunst and Mason (1977) indicate that latent root regression may provide considerable improvement in mean square error over least squares. Latent root regression can produce regression coefficients that are very similar to those found by principal components, particularly when there are only one or two strong multicollinearities in $\underset{\sim}{X}$. A number of large-sample properties of latent root regression are in White and Gunst (1979).

3. ROBUST REGRESSION

Robust regression procedures are designed to dampen the effect of observations that would be highly influential if least squares were used. That is a robust procedure tends to leave the residuals associated with outliers large, thereby making the identification of influential points much easier. In addition to insensitivity to outliers, a robust estimation procedure should be 90-95 percent as efficient as least squares when the underlying distribution is normal. Basic references in robust estimation include Andrews et al. (1972), Andrews (1974), Hill and Holland (1977), Hogg (1974, 1979a,b) and Huber (1972, 1973, 1981).

To motivate the discussion, and to demonstrate why it may be desirable to use an alternative to least squares when the observations are nonnormal, consider the simple linear regression model

$$y_i = \beta_0 = \beta_1 x_i + \varepsilon_i, \quad i = 1,2,\ldots,n \tag{30}$$

where the errors are independent random variables that follow the double exponential distribution

$$f(\varepsilon_i) = \frac{1}{2\sigma} e^{-|\varepsilon_i|/\sigma}, \quad -\infty < \varepsilon_i < 0 \tag{31}$$

The double exponential distribution, shown in Figure 2 is more pointed in the middle than the normal and tails off to zero as $|\varepsilon_i|$ goes to infinity. However, since the density function goes to zero as $e^{-|\varepsilon_i|}$ goes to zero, and the normal density function goes to zero as $e^{-\varepsilon_i^2}$ goes to zero, the double exponential distribution has heavier tails than the normal.

We will use the method of maximum likelihood to estimate β_0 and β_1. The likelihood function is

$$L(\beta_0, \beta) = \prod_{i=1}^{n} \frac{1}{2\sigma} e^{-|\varepsilon_i|/\sigma} = \frac{1}{(2\sigma)^n} e^{-\sum_{i=1}^{n}|\varepsilon_i|/\sigma} \qquad (32)$$

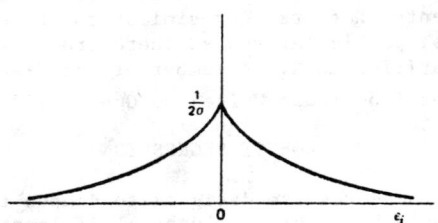

Figure 2   The Double Exponential Distribution

Therefore maximizing the likelihood function would involve minimizing $\sum_{i=1}^{n} |\varepsilon_i|$, the sum of the absolute errors. The method of maximum likelihood applied to the regression model with normal errors leads to the least squares criterion. Thus the assumption of an error distribution with heavier tails that the normal implies that the method of least squares is no longer an optimal estimation technique. Note that the absolute error criterion would weight outliers far less severely than would least squares. Minimizing the sum of the absolute erors is often called the L_1-norm regression problem. Least squares is the L_2-norm regression problem.

The L_1-norm regression problem can be formulated as a linear programming (LP) problem. Let c_i and d_i ($i = 1, 2, \ldots, n$) be the positive and negative deviations about the fitted line. Then the regression coefficients $\beta_{L_1,0}$ and $\beta_{L_1,1}$ that minimize the sum of the absolute errors are the solution to the LP problem

$$\text{minimize} \quad Z = \sum_{i=1}^{n} (c_i = d_i)$$

$$\text{subject to:} \quad y_i - \hat{\beta}_{L_1,0} - \hat{\beta}_{L_1,1} x_i + c_i - d_i = 0, \quad i=1,2,\ldots,n \tag{33}$$

$$\hat{\beta}_{L_1,0}, \hat{\beta}_{L_1,1} \text{ unrestricted in sign.}$$

The extension to multiple regression is straightforward. In general, the LP problem has n constraints (one for each observation) and p+2n variables (one variable for each model parameter and 2n variables representing the positive and negative deviations). Unfortunately, standard LP algorithms do not ensure that unbiased estimates of β are obtained. Sielken and Hartley (1973) have developed an efficient $\widetilde{LP}$ algorithm that produces an unbiased solution. For other references on L_1-norm regression, see Barrodale (1968), Barrodale and Roberts (1973), Book, Booker, Hartley and Sielken (1980), Gentle, Kennedy and Sposito (1977), and Wagner (1959).

The L_1-norm regression problem is a special case of L_p-norm regression, in which the model parameters are chosen to minimize $\sum_{i=1}^{n} |\varepsilon|^p (1 \leq p \leq 2)$. For $1 < p < 2$ this reduces to a nonlinear programming problem. Forsythe (1972) has studied this procedure extensively for the straight-line regression model via Monte Carlo simulation, using several nonnormal error distributions. He notes that p=1.5 is a good compromise choice leading to substantially better estimates than least squares when the errors are nonnormal. When the error distribution is normal, using p=1.5 results in estimates that are at worst 90 percent as efficient as least squares.

3.1 M Estimators

We have noted that the L_1-norm regression problem arises naturally from the maximum likelihood approach with double exponential errors. In general we may define a class of robust estimators that minimize a function ρ of the residuals, for example

$$\min_{\underline{\beta}} \sum_{i=1}^{n} \rho(e_i) = \min_{\underline{\beta}} \sum_{i=1}^{n} \rho[(y_i - \underline{x}_i' \underline{\beta}) \tag{34}$$

where x_i' denotes the ith row of X. An estimator of this type is called an M-estimator, where M stands for maximum likelihood. That is, the function ρ is related to the likelihood function for an appropriate choice of the error distribution. For example, if the method of least squares is used (implying that the error distribution is normal), then $\rho(z) = 1/2\ z^2$, $-\infty < z < \infty$. The M-estimator is not necessarily scale-invariant (that is, if the residuals $y_i - x_i'\beta$ were multiplied by a constant, the new solution to (34) might not be the same as the old one). To obtain a scale-invariant version of this estimator, we usually solve

$$\min_{\beta} \sum_{i=1}^{n} \rho(e_i/s) = \min_{\beta} \sum_{i=1}^{n} \rho[y_i - x_i'\beta)/s] \tag{35}$$

where s is a robust estimate of scale. A popular choice for s is

$$s = \text{median}|e_i - \text{median}(e_i)|/0.6745$$

The constant 0.6745 makes s an approximately unbiased estimator of σ if n is large and the error distribution is normal.

The minimize (35), equate the first partial derivatives of ρ with respect to $\beta_j(j=0,1,\ldots,k)$ equal to zero, yielding a necessary condition for a minimum. This gives the system of $p=k+1$ equations

$$\sum_{i=1}^{n} x_{ij}\psi[(y_i - x_i'\beta)/s] = 0, \quad j=0,1,\ldots,k \tag{36}$$

where $\psi=\rho'$ and x_{ij} is the ith observation on the jth regressor and $x_{i0} = 1$. In general the ψ function is nonlinear and (36) must be solved by iterative methods. While several nonlinear optimization techniques could be employed, iteratively reweighted least squares is most widely used. This approach is usually attributed to Beaton and Tukey (1974). To use iteratively reweighted least squares, suppose that an initial estimate $\hat{\beta}_0$ is available and that s is an estimate of scale. Then write the $p=k+1$ equations in (36)

$$\sum_{i=1}^{n} x_{ij}\psi[(y_i - x_i'\beta)/s] = \sum_{i=1}^{n} x_{ij} \frac{\psi[(y_i - x_i'\beta)/s]}{(y_i - x_i'\beta)/s} (y_i - x_i'\beta)/s = 0$$
$$j=0,1,\ldots,k \tag{37}$$

as

$$\sum_{i=1}^{n} x_{ij} w_{i0}(y_i - x_i'\beta) = 0, \quad j=0,1,\ldots,k \tag{38}$$

where

$$w_{i0} = \begin{cases} \dfrac{\psi[(y_i - x_i'\hat{\beta}_0)/s]}{(y_i - x_i'\hat{\beta}_0)/s} & \text{if } y_i \neq x_i'\hat{\beta}_0 \\ 1 & \text{if } y_i = x_i'\hat{\beta}_0 \end{cases} \quad (39)$$

In matrix notation, (38) becomes

$$X'W_0 X\hat{\beta} = X'W_0 y \quad (40)$$

where W_0 is an n × n diagonal matrix of "weights" with diagonal elements $w_{10}, w_{20}, \ldots, w_{n0}$ given by (39). We recognize (40) as the usual weighted least squares normal equations. Consequently the one-step estimator is

$$\hat{\beta}_1 = (X'W_0 X)^{-1} X'W_0 y \quad (41)$$

At the next step, we recompute the weights from (39) but using $\hat{\beta}_1$ instead of $\hat{\beta}_0$. Usually only a few iterations are required to achieve convergence, and the procedure requires only a standard weighted least squares computer program.

A number of popular robust criterion functions are shown in Table 1. Robust regression procedures can be classified by the behavior of their ψ-function. The ψ-function controls the weight given to each residual, and (apart from a constant of proportionality) is sometimes called the influence function. The behavior of several ψ functions is shown in Figure 3. For example, the ψ-function for least squares is unbounded, and thus least squares tend to be nonrobust when used with data arising from a heavy-tailed distribution. The Huber t function [Huber (1964)] has a monotone ψ-function, and does not weight large residuals as heavily as least squares. The last three influence functions actually redescend as the residual becomes larger. Ramsay's E_a function [see Ramsay (1977)] is a soft redescender; that is, the ψ-function is asymptotic to zero for large $|z|$. Andrew's wave function and Hampel's 17A function [see Andrews et al. (1972) and Andrews (1974)] are hard redescenders; that is, the ψ-function equals zero for sufficiently large $|z|$. We should note that the ρ-functions associated with the redescending ψ functions are nonconvex, and this in theory can cause convergence problems in the iterative estimation procedure. However, this is not a common occurrence. Furthermore each of the robust criterion functions requires the analyst to specify certain "tuning constants" for the ψ-functions. We have shown typical values of these tuning constants in Table 1.

Several authors [Andrews (1974), Hogg (1979a), Hocking (1978)] have noted that the starting value $\hat{\beta}_0$ used in robust estimation must be chosen carefully. Using the least squares solution can disguise the high-leverage points. The L_1-norm estimates would be a good choice of starting values. Andrews (1974) and Dutter (1977) also suggest procedures for choosing the starting values.

Table 1 Robust Criterion Functions

Criterion	$\rho(z)$	$\psi(z)$	$w(z)$	Range
Least squares	$\tfrac{1}{2}z^2$	z	1.0	$|z|<\infty$
Huber's t function $t=2$	$\tfrac{1}{2}z^2$ $|z|t-\tfrac{1}{2}t^2$	z $t\,\text{sign}(z)$	1.0 $\dfrac{t}{|z|}$	$|z|\leqslant t$ $|z|>t$
Ramsay's E_a function $a=0.3$	$a^{-2}[1-\exp(-a|z|)\\ \cdot(1+a|z|)]$	$z\exp(-a|z|)$	$\exp(-a|z|)$	$|z|<\infty$
Andrew's wave function $a=1.339$	$a[1-\cos(z/a)]$ $2a$	$\sin(z/a)$ 0	$\dfrac{\sin(z/a)}{z/a}$ 0	$|z|\leqslant a\pi$ $|z|>a\pi$
Hampel's 17A function $a=1.7$ $b=3.4$ $c=8.5$	$\tfrac{1}{2}z^2$ $a|z|-\tfrac{1}{2}a^2$ $\dfrac{a(c|z|-\tfrac{1}{2}z^2)}{c-b}-(7/6)a^2$ $a(b+c-a)$	z $a\,\text{sign}(z)$ $\dfrac{a\,\text{sign}(z)(c-|z|)}{c-b}$ 0	1.0 $a/|z|$ $\dfrac{a(c-|z|)}{|z|(c-b)}$ 0	$|z|\leqslant a$ $a<|z|\leqslant b$ $b<|z|\leqslant c$ $|z|>c$

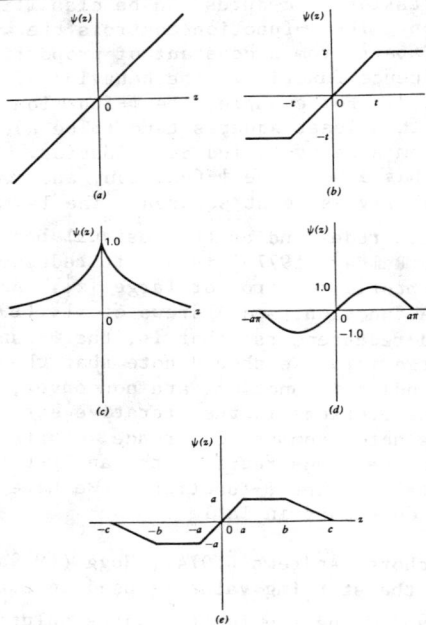

Figure 3 Robust Influence Functions. (a) Least Squares. (b) Huber's t Function. (c) Ramsay's E_a Function. (d) Andrews' Wave Function. (e) Hampel's 17 A Function.

Robust regression methods have much to offer the data analyst. They can be extremely helpful in locating outliers and highly influential observations. Whenever a least squares analysis is performed it would be useful to perform a robust fit also. If the results of the two procedures are in substantial aggreement, then use the least squares results, because inferences based on least squares are at present better understood. However, if the results of the two analyses differ, then reasons for these differences should be identified. Observations that are downweighted in the robust fit should be carefully examined.

3.2 R and L Estimation

In addition to M estimators, there are other approaches to robust regression. R estimation is a procedure based on ranks. To illustrate the general procedure, consider replacing one factor in the least squares objective function $S(\beta) = \sum_{i=1}^{n}(y_i - x_i'\beta)^2$ by its rank. Thus if R_i is the rank of $y_i - x_i'\beta$, then we wish to minimize $\sum_{i=1}^{n}(y_i - x_i'\beta)R_i$. More generally, we could replace the ranks (which are the integers $1, 2, \ldots, n$) by the score function $a(i)$, $i=1, 2, \ldots, n$, so that the objective function becomes

$$\min_{\beta} \sum_{i=1}^{n} (y_i - x_i'\beta) a(R_i)$$

If we set the score function equal to the ranks, that is, $a(i)=i$, the results are called Wilcoxon scores. Another possiability is to use median scores, that is $a(i) = -1$ if $i < (n+1)/2$ and $a_i = 1$ if $i > (n+1)/2$. Important references on R estimation in regression include Adichie (1967), Hogg and Randles (1975), Jaeckel (1972), and Jureckova (1977).

L estimators are based on order statistics. For example, suppose that we wish to estimate the location parameter of a distribution from a random sample $x_1, x_2, \ldots, x_n$. The order statistics of this sample are $x_{[1]} \leq x_{[2]} \leq \ldots \leq x_{[n-1]} \leq x_{[n]}$. The sample median would be an L-estimator, since it is a measure of location based on these order statistics. A number of other L-estimators for the location problem are described in Andrews et al. (1972). The use of L estimation in the regression context is not as simple as M and R estimation. Denby and Larsen (1977) describe slice regression, which for one regressor divides the data into groups and fits the stright line using the centroids of the groups. A stepwise-type extension of this technique could be used for multiple regresion. Moussa-Hamouda and Leone (1974, 1977a,b) propose procedures for simple linear regression with repeat observations on y at each x that involve trimming or discarding remote values of y.

While research continues on R and L estimation, we believe that currently the M estimators are more reasonable for the regression problem. They can be easily obtained from a weighted least squares computer program, and the weights in the final iteration identify influential points. Both L- and R-estimators are more difficult to obtain computationally than M-estimators. L-estimators do not always generalize clearly to multiple regression. Furthermore under certain conditions [see Jureckova (1977)] R-estimators are asyptotically equivalent to M-estimators.

3.3 Bounded Influence Estimation

Another approach to the use of iteratively reweighted least squares, is to directly implement this procedure by minimizing with respect to $\hat{\beta}$,

$$\sum_{i=1}^{n} w_i (y_i - x_i'\beta)^2$$

where the weights w_i are iteratively computed as a function some appropriate measure of influence or leverage. Mallows (1973b)(1975) and Welsch (1980) have suggested this approach, known as bounded influence estimation. Reasonable choices would include

$$w_i = w(DFFITS_i) = 1, \quad |DFFITS_i| \leq 2\sqrt{p/n}$$
$$= \frac{2\sqrt{p/n}}{|DFFITS_i|}, \quad |DFFITS_i| > 2\sqrt{p/n} \tag{42}$$

or

$$w_i = w(D_i) = 1, \quad D_i \leq 1$$
$$= \frac{1}{D_i}, \quad D_i > 1 \tag{43}$$

where

$$DFITS_i = \frac{\hat{y}_i - \hat{y}_{(i)}}{\sqrt{s^2_{(i)} h_{ii}}}$$

is the statistic proposed by Belsely, Kuh and Welsch (1980), and

$$D_i = \frac{(\hat{\beta}_{(i)} - \hat{\beta})' X'X (\hat{\beta}_{(i)} - \hat{\beta})}{pMS_E}$$

is Cook's (1977) distance measure.

Bounded influence estimators downweight observations that are influendtial in the sense of having large values of $DFFITS_i$ on D_i. Robust estimators of the type discussed previously are usually effective in downweighting outliers but are less likely to downweight high leverage cases. Bounded influence estimators should be more responsive to leverage points.

3.4 Robust Ridge Regression

Two of the more frequent problems that the regression analyst will encounter are nonnormality of the observations and multicollinearity. Although we usually think of these two problems separately, in a significant number of practical situations nonnormality and multicollinearity occur simultaneously. Several authors have suggested that either robust or biased estimation methods alone may be sufficient for dealing with the combined problem. However, since robust regression estimates are frequently unstable when the X matrix is ill-conditioned, it would be desirable to have a technique for dealing directly with both problems. Hogg (1979b) has suggested a robust form of ridge regression. Recall that the ridge estimator $\hat{\beta}_R = (X'X+kI)^{-1}X'y$ can be computed by augmenting the X,y data with p pseudo-observations

$$X_A = \begin{bmatrix} X \\ kI \end{bmatrix}, \quad y_A = \begin{bmatrix} y \\ 0 \end{bmatrix}$$

and then applying ordinary least squares to X_A, y_A, yielding

$$\hat{\beta}_R = (X_A'X_A)^{-1}X_A'y_A = (X'X+kI)^{-1}X'y$$

A robust version of ridge regression would replace least squares on X_A, y_A by an appropriate robust objective function.

Askin and Montgomery (1980) explore this approach and note that generalized ridge, principal components, and fractional rank estimators can be fit using least squares on appropriately augmented data, so robust versions of these biased estimators can be easily computed. They report that the combined estimation procedure gives stable coefficient estimates while simultaneously locating and identifying outliers. The computational properties of the procedure are also good, as the iteratively reweighted least squares algorithm usually converges to the final estimates in fewer iterations that would be required if only the robust criterion were used.

4. CONCLUSIONS

This paper has given an overview of biased and robust estimation techniques for the linear regression model. These techniques are important and useful to the regression practitioner, since most, if not all, regression data sets suffer from these problems to some extent. Good computational procedures for ridge regression and principal components regression are available (SAS and/or BMD-P), and the implementation of robust fitting methods is not difficult. While it is not recommended that these techniques be used indiscriminately, they are valuable additions to the data analyst's tool kit.

REFERENCES

ADICHIE, J.N. (1967), "Estimates of Regression Parameters Based on Rank Tests," Ann. Math Statist., 38, 894-904.

ANDREWS, D.F. (1974), "A Robust Method for Multiple Linear Regression," Technometrics, 16, 523-531.

ANDREWS, D.F., BICKEL, P.J., HAMPEL, F.R., HUBER, P.J., ROGERS, W.H., and TUKEY, J.W. (1972), Robust Estimates of Location," Princeton University Press, Princeton, N.J.

ASKIN, R.G. and MONTGOMERY, D.C. (1980), "Augmented Robust Estimators,"Technometrics, 22, 333-341.

BARRODALE, I. (1968), "L_1 Approximations and the Analysis of Data," Appl. Statist., 17, 51-57.

BARRODALE, I. and ROBERTS, F.D.K. (1973), "An Improved Algorithm for Discrete L_1 Linear Approximation," SIAM J. Numer. Anal., 10, 839-848.

BEATON, A.E. and TUKEY, J.W. (1974), "The Fitting of Power Series, Meaning Polynomials, Illustrated on Band Spectroscopic Data," Technometrics, 16, 147-185.

BELSLEY, D.A., KUH, E. and WELSCH, R.E. (1980), Regression Diagnostics: Identifying Influential Data and Sources of Collinearity, Wiley, New York.

BOOK, D., BOOKER, J., HARTLEY, H.O. and SIELKEN, R.L., JR. (1980), Unbiased L_1 Estimators and Their Covariances, ONR THEMIS Technical Report No. 64, Institute of Statistics, Texas A&M University.

COOK, R.D. (1977), "Detection of Influential Observation in Linear Regression," Technometrics, 19, 15-18.

DEMPSTER, A.P., SCHATZOFF, M. and WERMUTH, N. (1977), "A Simulation Study of Alternatives to Ordinary Least Squares," J. Amer. Statist. Assoc., 72, 77-90.

DENBY, L. and LARSON, W.A. (1977), "Robust Regression Estimators Compared Via Monte Carlo," Commun. Statist., Ay, 335-362.

DRAPER, N.R. and VAN NOSTRAND, R.C. (1979), "Ridge Regression and James-Stein Estimators: Review and Comments," Technometrics, 21, 451-466.

DUTTER, R. (1977), "Numerical Solution of Robust Regression Problems: Computational Aspects, A Comparison," J. Statist. Comput. Simul., 5, 207-238.

FORSYTHE, A.B. (1972), "Robust Estimation of Straight-Line Regression Coefficients by Minimizing pth Power Deviations," Technometrics, 14, 159-166.

GENTLE, J.M., KENNEDY, W.J. and SPOSITO, V.A. (1977), "On Least Absolute Deviations Estimators," Commun. Statist., A6, 839-845.

GOLDSTEIN, M. and SMITH, A.F.M. (1974), "Ridge-Type Estimators for Regression Analysis," J.R. Statist. Soc. Ser. B, 36, 284-291.

GUNST, R.F. and MASON, R.L. (1977), "Biased Estimation in Regression: An Evaluation Using Mean Sqaured Error," J. Amer. Statist. Assoc., 72, 616-628.

GUNST, R.F. and MASON, R.L. (1979), "Some Considerations in the Evaluation of Alternative Prediction Equations," Technometics, 21, 55-63.

GUNST, R.F., WEBSTER, J.T. and MASON, R.L. (1976), "A Comparison of Least Squares and Latent Root Regression Estimators," *Technometrics*, 18, 75-83.
HAWKINS, D.M. (1973), "On the Investigation of Alternative Regressions by Principal Components Analysis," *Appl. Statist.*, 22, 275-286.
HEMMERLE, W.J. (1975), "An Explicit Solution for Generalized Ridge Regression," *Technometrics*, 17, 309-314.
HEMMERLE, W.J. and BRANTLE, T.F. (1978), "Explicit and Constrained Generalized Ridge Regression," *Technometrics*, 20, 109-120.
HILL, R.W. (1979), "On Estimating the Covariance Matrix of Robust Regression M-Estimates," *Commun. Statist.*, A8, 1183-1196.
HILL, R.W. and HOLLAND, P.W. (1977), "Two Robust Alternatives to Least Squares Regression," *J. Amer. Statist. Assoc.*, 72, 828-833.
HOCKING, R.R. (1978), "The Regression Dilemma: Variable Elimination, Coefficient Shrinkage, or Robust Estimation," presented at the ASQC 1978 Fall Technical Conference, Rochester, N.Y.
HOCKING, R.R., SPEED, F.M. and LYNN, M.J. (1976), "A Class of Biased Estimators in Linear Regression," *Technometrics*, 18, 425-437.
HOERL, A.E. and KENNARD, R.W. (1970a), "Ridge Regression: Biased Estimation for Nonorthogonal Problems," *Technometrics*, 12, 55-67.
HOERL, A.E. and KENNARD, R.W. (1970b), "Ridge Regression: Applications to Nonorthogonal Problems," *Technometrics*, 12, 69-82.
HOERL, A.E. and KENNARD, R.W. (1976), "Ridge Regression: Iterative Estimation of the Biasing Parameter," *Commun. Statist*, A5, 77-88.
HOERL, A.E., KENNARD, R.W. and BALDWIN, K.F. (1975), "Ridge Regression: Some Simulations," *Commun. Statist.*, 4, 105-123.
HOGG, R.V. (1974), "Adaptive Robust Procedures: A Partial Review and Some Suggestions for Future Applications and Theory," *J. Amer. Statist. Assoc.*, 69, 909-925.
HOGG, R.V. (1979a), "Statistical Robustness: One View of its Use in Applications Today," *Amer. Statist.*, 33, 3, 108-115.
HOGG, R.V. (1979b), "An Introductaion to Robust Estimation," in R.L. Launer and G.N. Wilkinson (Eds.), *Robustness in Statistics*, Academic, New York, 1-18.
HOGG, R.V. and RANDLES, R.H. (1975), "Adaptive Distribution-free Regression Methods and their Applications," *Technometrics*, 17, 399-407.
HOLLAND, P.W. and WELSCH, R.E. (1977), "Robust Regression Using Iteratively Reweighted Least Squares," *Commun. Statist.*, A6, 813-828.
HUBER, P.J. (1964), "Robust Estimation of a Location Parameter," *Ann. Math. Statist.*, 35, 73-101.
HUBER, P.J. (1972), "Robust Statistics: A Review," *Ann. Math. Statist.*, 43, 1041-1067.
HUBER, P.J. (1973), "Robust Regression: Asymptotics, Conjectures, and Monte Carlo," *Ann. Statist.*, 1, 799-821.
HUBER, P.J. (1981), *Robust Statistics*, Wiley, New York.
JAECKEL, L.A. (1972), "Estimating Regresssion Coefficients by Minimizing the Dispersion of the Residuals," *Ann. Math. Statist.*, 43, 1449-1458.
JURECKOVA, J. (1977), "Asymptotic Relations of M-Estimates and R-Estimates in Linear Regression Models," *Ann. Statist.*, 5, 464-472.
LAWLESS, J.F. (1978), "Ridge and Related Estimation Procedures: Theory and Practice," *Commun. Statist.*, A7, 139-164.

LAWLESS, J.F. and WANG, P. (1976), "A Simulation of Ridge and Other Regression Estimators," Commun. Statist., A5, 307-323.

LINDLEY, D.V. and SMITH, A.F.M (1972), "Bayes Estimates for the Linear Model (With Discussion)," J.R. Statist. Soc. Ser. B, 34, 1-41.

MALLOWS, C.L. (1973a), "Some Comments on C_p," Technometrics, 15, 661-675.

MALLOWS, C.L. (1973b), "Influence Functions," presented at the NBER Conference on Robust Estimation Cambridge, MA.

MALLOWS, C.L. (1975), "On Topics in Robustness," technical memorandum, Bell Telephone Laboratories, Murray Hill, N.J.

MANSFIELD, E.R., WEBSTER, J.T. and GUNST, R.G. (1977), "An Analytic Variable Selection Procedure for Principal Component Regression," Appl Statist., 26, 34-40.

MARQUARDT, D.W. (1970), "Generalized Inverses, Ridge Regression, Biased Linear Estimation, and Nonlinear Estimation," Technometrics, 12, 591-612.

MARQUARDT, D.W. and SNEE, R.D. (1975), "Ridge Regresssion in Practice," Amer. Statist., 29, 1, 3-20.

MASON, R.L., GUNST, R.F. and WEBSTER, J.T. (1975), "Regression Analysis and Problems of Multicollinearity," Commun. Statist., 4, 3, 277-292.

MCDONALD, G.C. and GALARNEAU, D.I. (1975), "A Monte Carlo Evaluation of Some Ridge-Type Estimators," J. Amer. Statist. Assoc., 70, 407-416.

MONTGOMERY, D.C. and PECK, E.A. (1983), Introduction to Linear Regression Analysis, Wiley, New York.

MOUSSA-HAMOUDA, E. and LEONE, F.C. (1974), "The 0-Blue Estimators for Complete and Censored Samples in Linear Regression," Technometrics, 16, 441-446.

MOUSSA-HAMOUDA, E. and LEONE, F.C. (1977a), "The Robustness of Efficiency of Adjusted Trimmed Estimators in Linear Regression," Technometrics, 19, 19-34.

MOUSSA-HAMOUDA, E. and LEONE, F.C. (1977b), "Efficiency of Ordinary Least Squares from Trimmed and Winsorized Samples in Linear Regression," Technometrics, 19, 265-273.

OBENCHAIN, R.L. (1975), "Ridge Analysis Following a Preliminary Test of the Shrunken Hypothesis," Technometrics, 17, 431-441.

OBENCHAIN, R.L. (1977), "Classical F-Tests and Confidence Intervals for Ridge Regression," Technometrics, 19, 429-439.

RAMSAY, J.O. (1977), "A Comparative Study of Several Robust Estimates of Slope, Intercept, and Scale in Linear Regression," J. Amer. Statist. Assoc., 72, 608-615.

ROSEPACK (1980), "A System of Subroutines for Iteratively Reweighted Least Squares Computations," to appear in ACM Trans. Math. Softw.

SIELKEN, R.L., JR. and HARTLEY, H.O. (1973), "Two Linear Programming Algorithms for Unbiased Estimation of Linear Models," J. Amer. Statist. Assoc., 68, 639-641.

SMITH, G. and CAMPBELL, F. (1980), "A Critique of Some Ridge Regression Methods (With Discussion)," J. Amer. Statist., Assoc. 75, 74-103.

STEIN, C. (1960), "Multiple Regression," Ingrim Olkin, (Ed.), Contributions to Probability and Statistics: Essays in Honor of Harold Hotelling, Stanford University Presss, Stanford, California.

WAGNER, H.M. (1959), "Linear Programming Techniques for Regression Analysis," J. Amer. Statist. Assoc., 54, 206-212.

WEBSTER, J.T., GUNST, R.F. and MASON, R.L. (1974), "Latent Root Regression Analysis," Technometrics, 16, 513-522.

WELSCH, R.E. (1980), "Regression Sensitivity Analysis and Bounded Influence Estimation," in <u>Evaluation of Econometric Models</u>, eds. J. Kmenta and J. Ramsey, Academic Press, New York.

WHITE, J.W. and GUNST, R.F. (1979), "Latent Root Regression: Large Sample Analysis," <u>Technometrics</u>, 21, 481-488.

BAYESIAN REGRESSION AND SENSITIVITY ANALYSES

Edward E. Leamer
Department of Economics
University of California, Los Angeles, CA 90024

ABSTRACT

This paper interprets Bayesian regression analysis as a method of pooling two data sets, one real and one fictitious. The fictitious data set is selected to approximate other relevant information. Because it is impossible to select a precise fictitious data set it is necessary to study the sensitivity of the pooled estimates to changes in the fictitious data set. Several sensitivity results are discussed in this paper and two examples are provided. Weighted regression is also discussed, and t-statistics are shown to indicate the insensitivity of an estimate to reweighting of observations.

1. INTRODUCTION

Bayesian analysis is a method by which information in a given data set can be combined with other relevant evidence. The other relevant evidence is treated as if it were equivalent to another data set, and the problem of pooling the information from these two different sources is transformed into a straightforward problem of pooling two numerically defined data sets, one real and one fictitious. The serious practical defect with this approach is that the other relevant evidence cannot often be said to be equivalent to a precise fictitious data set, and Bayesian methods are consequently dismissed as "subjective" or, more accurately, "whimsical." However, it will usually be the case that the other relevant evidence can at least be said to be approximated by a fictitious data set. If a sensitivity analysis reveals that the pooled estimates are adequately insensitive to the choice of the fictitious data set, then issues concerning the quality of the approximation evaporate. If, on the other hand, the inferences are found to be very sensitive to the choice of the fictitious data set then it must be concluded that these data are useless since inferences from them are too sensitive to be believed.

All real data analyses solve in one way or another this problem of pooling information from a given data set with other information. The issue is not whether this should be done but how it should be done. The more traditional approaches of ad hoc variable omissions and computer selected variable omissions (step-wide regression) leave the exact nature of the other information unstated, and also normally involve no sensitivity analyses. In contrast, the Bayesian approach makes explicit the nature of the other information, and coupled with a sensitivity analysis can yield convincing and formally correct inferences.

The formula for pooling two data sets is

$$\underline{b}^{**} = (\underline{X}'\underline{X} + \underline{N}^*)^{-1} (\underline{X}'\underline{X}\underline{b} + \underline{N}^*\underline{b}^*)$$

where $\underline{X}'\underline{X}$ is the matrix of moments of the explanatory variables, $\underline{b}$ is the vector of least-squares estimates, and $\underline{N}^*$ and $\underline{b}^*$ are

the corresponding features of the fictitious data set. This paper
illustrates how this formula can be used in practice and focuses on
one kind of sensitivity analysis. Namely, it is assumed that the
fictitious least squares vector b* can be selected exactly, but
the fictitious moment matrix N* can only be said to lie in an
interval $L \leq N^* \leq U$, where $L \leq U$ means $U - L$ is positive semi-
definite. Corresponding to such an "interval" of prior covariance
matrices is an "interval" of pooled estimates b**. It is hoped
that the interval of covariance matrices is wide enough to be
credible and the corresponding interval of pooled estimates is
narrow enough to be useful. If, on the other hand, an incredibly
narrow interval of prior covariance matrices is required to get a
usefully narrow interval of estimates, then estimation is suspended
with this data set.

This paper includes two applications of this framework. The
first example is quadratic regression with a data set that has only
two different values of the explanatory variable. Because of this
arrangement of the data, a quadratic model cannot be estimated. A
linear model can be estimated but it implies peculiar inferences.
The alternative is to supplement the given data set with a fictitious data set that contains information about the coefficient on
the squared term. This is shown to lead to sensible inferences when
neither the linear model nor the quadratic model do. This example
is constructed to illustrate clearly the importance of information
not embodied in the data set being analyzed.

The second example is an analysis of cross state murder
rates. The conclusion is reached that any inference from these data
about the deterrent effect of capital punishment is too sensitive to
be believed.

Comments are also made about the treatment of outliers. By
including a sequence of dummy variables, one for each observation,
the observation selection problem can be interpreted as a variable
selection problem. Such a model is overparameterized but can be
analyzed if the given data are supplemented with information about
the size of the coefficients on these dummy variables. The
surprising conclusion of this analysis is that an ordinary t-
statistic gives a rather complete indication of the effects of
reweighting observations to adjust for outliers.

2. GENERAL REMARKS

Most introductions to Bayesian methods begin with a discussion
of the philosophical meaning of probability. A distinction is drawn
between frequency probabilities and subjective probabilities.
Frequency probabilities are defined as proportions of favorable
outcomes in sequences of events. Subjective probabilities are
either taken as the natural measurement of uncertainty or are
defined in the context of some decision problem under uncertainty,
often a betting problem.

Classical (sampling theory) methods are based on frequency
probabilities. For example, a 95% confidence interval computed from
a given data set takes on whatever meaning it may have because in
repeated experiments intervals "like" this one capture the true
value with relative frequency .95. Bayesian methods, on the other
hand, are based on subjective probabilities. A given 95% confidence

interval is said to have probability .95 of capturing the true value if you would bet that it does at 20 to 1 odds.

The route to Bayesian regression that begins with probabilistic foundations is quite interesting and I encourage you to explore it in, for example, Leamer (1978). Unfortunately it is a route that takes rather long to get to practical applications. For that reason I have chosen to progress quickly to the heart of the matter by describing Bayesian methods in terms of their use of fictitious data sets. I expect that many of you are having negative emotional reactions to this idea. Do we really want "scientists" to create fictitious data sets? Isn't that going in exactly the opposite direction from science which ought to be seeking truth not fiction? Are Bayesian methods a fancy way to justify superstition?

These are enormous questions that ultimately have to do with the nature of knowledge, about which many volumes have been written. In my limited space here I would like to make only two comments. First of all, your emotional reaction is partly a result of your failure to distinguish the word "fiction" from the word "fake", the latter referring to counterfeit data passed off fraudulantly as if it were the genuine article. A fictitious data set, on the other hand, is clearly presented as such. There is no intent to deceive; nor is there much possibility of deception in fact.

The second comment that I have is that all nonexperimental data sets are simply too weak to allow sensible inferences in the absence of supplementary information. There are many too many variables that might be included in an equation, especially if you allow nonlinearities in the functional form. It is therefore an unfortunate fact of life that we are going to have to use the equivalent of a fictitious data set. We can fake it, with stepwise regression or ridge regression, for example. Or we can be honest about it, as in the Bayesian analysis presented here.

3. POOLING TWO DATA SETS

Formulae for the pooling of two data sets are not difficult to derive. Typically the regression model is written $y = X\beta + u$, where y is an $n \times 1$ vector of observations, X is an $n \times k$ matrix of observations, β is a $k \times 1$ vector of unobservable "parameters" and u is an $n \times 1$ vector of unobservable "errors", which are assumed to be normally distributed with mean vector zero and covariance matrix I. The maximum likelihood estimate of the regression vector β is then the least-squares estimate

$$b = (X'X)^{-1}X'y ,$$

where the invertibility of the moment matrix $X'X$ has been assumed. The corresponding covariance matrix is

$$\text{Var}(b|\beta) = (X'X)^{-1} .$$

Exactly the same analysis applies to the case of two data sets since we may write the regression model in the analogous stacked form:

$$y^{**} = \begin{bmatrix} y \\ y^* \end{bmatrix} = \begin{bmatrix} X \\ X^* \end{bmatrix} \beta + \begin{bmatrix} u \\ u^* \end{bmatrix} ,$$

$$= X^{**}\beta + u^{**}$$

where the single asterisks refer to the second data set, and the double asterisks refer to the pooled data set. The pooled estimate of the regression vector and covariance matrix are then

$$b^{**} = (X^{**\prime}X^{**})^{-1} X^{**\prime}y^{**}$$
$$= (X'X + X^{*\prime}X^{*})^{-1} (X'y + X^{*\prime}y^{*})$$
$$= (V^{-1} + V^{*-1})^{-1}(V^{-1}b + V^{*-1}b^{*}) , \quad (1)$$

$$V^{**} = \operatorname{Var}(b^{**}|\beta) \quad (2)$$
$$= (V^{-1} + V^{*-1})^{-1} ,$$

where
$$V^{-1} = X'X$$
$$(V^{*})^{-1} = X^{*\prime}X^{*}$$
$$b = (X'X)^{-1}X'y$$
$$b^{*} = (X^{*\prime}X^{*})^{-1}X^{*\prime}y^{*}$$

4. GLOBAL SENSITIVITY RESULTS

The pooled estimate (1) is a matrix weighted average of the separate estimates b and b^*, with weights proportional to the inverse covariance matrices, V^{-1} and V^{*-1} respectively. In a Bayesian analysis the vector b^* and the matrix V^* are selected by the researcher to approximate relevant evidence not contained in the given data set, y and X. It will rarely be the case that these choices can be made with great confidence, and it is accordingly necessary to analyze the sensitivity of the pooled inference to choice of b^* and V^*.

In this section four theorems are reported that indicate the global sensitivity of the pooled estimate b^{**} to choice of the fictitious covariance matrix V^*. Proofs of these results may be found in Leamer (1982). In these global sensitivity analyses, a set of values for V^* is selected and the corresponding set of values for b^{**} is identified. If the set of values of V^* is credibly inclusive, and if the corresponding set of estimates b^{**} is usefully narrow, then it is concluded that inferences from the given data set are sufficiently insensitive, and consequently credible. If, on the other hand, in order to obtain a usefully narrow set of estimates it is necessary to select an incredibly narrow set of V^* matrices, then it is concluded that any inference from the given data set is too sensitive to be credible.

The results now to be discussed take the fictitious estimate b^* as given because there are many settings in which the ambiguity in V^* greatly swamps the ambiguity in b^*. Sensitivity analysis with respect to b^* could also be performed.

The first result which takes the real sample covariance matrix V as well as the fictitious sample covariance matrix V^* as unknown is reported for its shock value, not for its practical significance, since for most problems V will be regarded as a given, at least up to a factor of proportionality. The shock associated with Theorem 1 concerns how different matrix weighted averages are from simple averages. A simple average of two numbers must lie between the two. One might suppose that this restriction would extend in some way to the multivariate case. In fact, it essentially extends not at all, since there are virtually no restrictions on the location of a matrix weighted average.

Theorem 1. Given any pair of separate estimates, $\underline{b}$ and $\underline{b}^*$, and almost any pooled estimate $\underline{b}^{**}$, there exist values for the weight matrices $\underline{V}$ and $\underline{V}^*$ such that $\underline{b}^{**}$ is a matrix weighted average (1) of $\underline{b}$ and $\underline{b}^*$. The exceptions are values of $\underline{b}^{**}$ on the line through $\underline{b}$ and $\underline{b}^*$, not on the segment connecting $\underline{b}$ to $\underline{b}^*$.

For the problems considered in this paper, the sample covariance matrix $\underline{V}$ is given up to a factor of proportionality. The minimal restriction on the location of $\underline{b}^{**}$ is therefore an ellipsoid described in Theorem 2.

Theorem 2. Given the locations, $\underline{b}$ and $\underline{b}^*$, and the weight matrix $\underline{V}$, then, regardless of the choice of weight matrix $\underline{V}^*$, $\underline{b}^{**}$ satisfying (1) must lie in the ellipsoid

$$(\underline{b}^{**} - \underline{f})' \underline{V}^{-1}(\underline{b}^{**} - \underline{f}) \leq (\underline{b} - \underline{b}^*)' \underline{V}^{-1}(\underline{b} - \underline{b}^*)/4 \qquad (3)$$

where $\underline{f} = (\underline{b} + \underline{b}^*)/2$. Conversely, any point in the ellipsoid is a pooled estimate $\underline{b}^{**}$ for some choice of symmetric positive semi-definite $\underline{V}^*$.

Knowledge of the sample covariance matrix $\underline{V}$ therefore shrinks the set of possible pooled estimates from the whole space (Theorem 1) to the ellipsoid (3) (Theorem 2). It will usually be the case that this ellipsoid of estimates is too large to be useful and some restriction on the fictitious covariance matrix $\underline{V}^*$ is needed to narrow further the set of estimates. One possibility is to restrict $\underline{V}^*$ to be a diagonal matrix. This is appropriate when there are separate pieces of information about each of the coefficients. More accurately, the covariance matrix $\underline{V}^*$ is diagonal if the estimate of one coefficient does not depend on information regarding the value of another. If this is the case the pooled estimate can be written as a weighted average of the 2^k constrained least-squares estimates formed by imposing subsets of the k constraints $\beta_i = b_i^*$:

Theorem 3. Given the locations, $\underline{b}$ and $\underline{b}^*$, the weight matrix $\underline{V}$, and the knowledge that the weight matrix $\underline{V}^*$ is diagonal $\underline{V}^{*-1} = \text{diag}\{d_1, d_2, \ldots, d_k\}$, then the pooled estimate may be written as

$$\underline{b}^{**} = (\underline{V}^{-1} + \underline{V}^{*-1})^{-1} (\underline{V}^{-1}\underline{b} + \underline{V}^{*-1}\underline{b}^*)$$

$$= \sum_I w_I b_I ,$$

where I is a subset of the first k integers, b_I minimizes the residual sum-of-squares $(y-\underline{X}\beta)'(y-\underline{X}\beta)$ subject to the constraints $\beta_i = b_i^*$ for $i \in I$, and

$$w_I = (\prod_{i \in I} d_i)|H_I|/|V^{-1} + V^{*-1}|$$

$$\Sigma_I w_I = 1,$$

where H_I is the square matrix formed by deleting rows i and columns i of V^{-1} for all $i \in I$.

This is a very important result because it allows us to interpret stepwise regression and other variable-selection techniques in terms of implicit fictitious data sets. If $b^* = 0$ and V^* is diagonal, Theorem 3 implies that the pooled estimate can be found by first computing all possible regressions based on all possible subsets of included variables, and then taking their average. This is exactly what stepwise regression does, though the weights are in that case peculiarly selected. It is impossible to prove the converse result, but I will state it anyway, since it is my belief: The proper foundation for a procedure that involves the omission of variables is a fictitious data set with estimate $b^* = 0$. Procedures that do not make this fictitious data set explicit are fundamentally fraudulent. (More on this in the concluding remarks).

A sensitivity analysis based on Theorem 3 is numerically cumbersome since it requires the computation of the 2^k regressions. The following result is numerically tractable and widely applicable. It is based on the assumption that the covariance matrix V^* is restricted to the "interval" $L \leq V^* \leq U$ where $V^* \leq U$ means $U - L^*$ in positive definite. The variance of the linear combination of parameters $c'\beta$, where c is a $k \times 1$ vector of constants, is $c' V^* c$, and the restriction $L \leq V^* \leq U$ implies that the variance of $c'\beta$ satisfies $c' L c \leq \text{Var}(c'\beta) \leq c' U c$. Thus by choice if L and U, the prior variance of any linear combination is bounded from above and below.

Theorem 4. Given b, b^*, V and the restriction $L \leq V^* \leq U$, with L and U symmetric positive definite matrices, then b^{**} satisfying (1) must lie in the ellipsoid

$$(b^{**} - \tilde{f})' \tilde{H}(b^{**} - \tilde{f}) \leq \tilde{g} \qquad (4)$$

where

$$\tilde{H} = (V^{-1} + U^{-1})(L^{-1} - U^{-1})^{-1}(V^{-1} + U^{-1}) + (V^{-1} + U^{-1})$$

$$\tilde{f} = (V^{-1} + L^{-1})^{-1}(V^{-1}b + L^{-1}b^* + (L^{-1} - U^{-1})$$
$$(V^{-1} - U^{-1})^{-1} V^{-1}(b - b^*)/2)$$

$$\tilde{g} = (b - b^*)' V^{-1}(V^{-1}(V^{-1} + U^{-1})^{-1}(V^{-1} + L^{-1})^{-1}$$
$$V^{-1}(b - b^*)/4$$

Theorem 4 serves as a foundation for an interesting and useful sensitivity analysis as illustrated in the next section. Theorem 1 is actually a special limiting case of Theorem 4 with L equal to the zero matrix and U equal to an infinite matrix.

In order to describe the ellipsoid (4) it is necessary to select linear combinations of paramaters that are of special interest. The extreme estimates of a linear combination are given by the following result:

<u>Theorem 5</u>. The extreme values of $c'b^{**}$ over the ellipsoid (4) are

$$c' \tilde{f} \pm (c' \tilde{H}^{-1} c \tilde{g})^{1/2}.$$

These extreme values identify an interval of pooled estimates of the parameter $c'\beta$. Any pooled estimate must be within this interval provided $L \le V^* \le U$, and, conversely, any point in the interval is a pooled estimate for some value of V^* satisfying $L \le V^* \le U$.

5.0 Examples

5.1 Polynomial Regression

Polynomial regression serves as a useful introduction to a Bayesian analysis since the traditional approach which assumes a polynomial of known degree can be shown to be both inflexible and inappropriate. Consider the data illustrated in Figure 1 consisting of the following eight pairs of observations $(y, x) = (0, -1)$, $(1, -1)$, $(2, -1)$, $(3, -1)$, $(2, 1)$, $(3, 1)$, $(4, 1)$, $(5, 1)$. If it is assumed that the relationship between y and x is linear, then the estimated function is the straight line going through the means at $x = -1$ and $x = 1$. This line and the two curves representing a one standard error confidence interval are both illustrated in Figure 1. The regression line is the relatively heavy straight line, and the confidence curves are the two curves labelled $v^* = 0$ (for reasons discussed below).

These confidence intervals have a strange feature: they are shortest at the value $x = 0$, halfway between the two sampled values, $x = 1$ and $x = -1$. Thus a data analyses based on the linear model implies that the information about the function is greatest at a location where there are no observations at all! On the contrary, I have the distinct feeling when I look at this scatter of points that I am most informed about the function at the locations where the observations occur, that is $x = -1$ and $x = 1$. Something thus seems wrong with this data analysis. Maybe the problem is that I am not really committed to linearity. Suppose, then, that we allow the function to be quadratic. Now we are in another kind of trouble since there are an infinite number of quadratic curves that fit the observed data equally well; namely all curves that go through the data means at $x = -1$ and $x = 1$. Interpreted in terms of confidence intervals for the function, the intervals would be finite at $x = -1$ and $x = 1$, but infinite everywhere else. Thus by assuming the function is quadratic, I am led to the conclusion that I know something about the function <u>only</u> at the sampled points, $x = -1$ and $x = 1$.

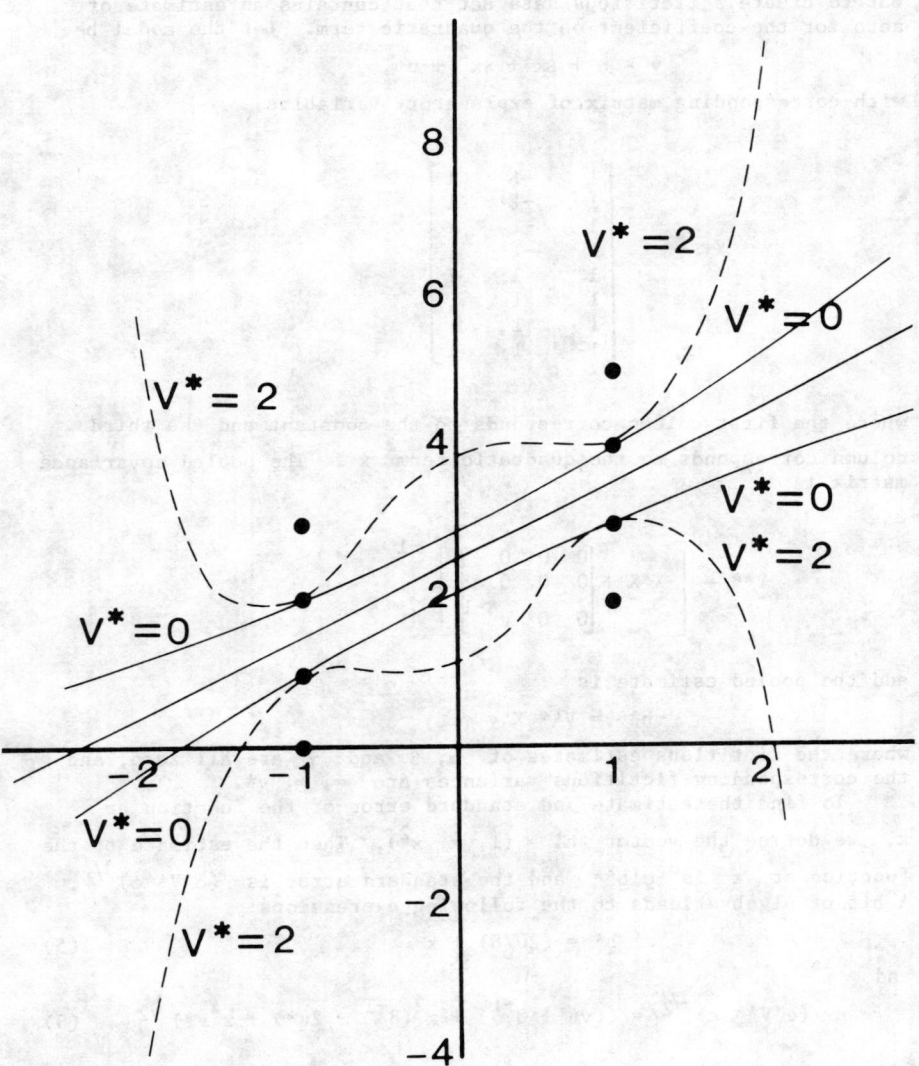

Figure 1: Confidence Bounds

Neither one of these extreme analyses seems appropriate. I don't think the data are most informative at the nonsampled point x = 0 as the linear model implies, but I also don't think that there is absolutely no information about the function there, as the quadratic model implies. Something between these two extremes seems desirable. A way to accomplish this is to use the quadratic model but to create a fictitious data set that contains an estimate of zero for the coefficient on the quadratic term. Let the model be

$$y = \alpha + \beta x + \gamma x^2 + u ,$$

with corresponding matrix of explanatory variables

$$X = \begin{bmatrix} 1 & -1 & 1 \\ 1 & -1 & 1 \\ 1 & -1 & 1 \\ 1 & -1 & 1 \\ 1 & 1 & 1 \\ 1 & 1 & 1 \\ 1 & 1 & 1 \\ 1 & 1 & 1 \end{bmatrix}$$

where the first column corresponds to the constant and the third column corresponds to the quadratic term x^2. The pooled covariance matrix is

$$V^{**} = \left[X'X + \begin{bmatrix} 0 & 0 & 0 \\ 0 & 0 & 0 \\ 0 & 0 & v^{*-1} \end{bmatrix} \right]^{-1}$$

and the pooled estimate is

$$b^{**} = V^{**} X'y ,$$

where the fictitious estimates of α, β and γ are all zero, and the corresponding fictitious variances are ∞, ∞, v^*.

To find the estimate and standard error of the function at x, we define the vector $c' = (1, x, x^2)$. Then the estimate of the function at x is $c'b^{**}$ and the standard error is $(c'V^{**}c)^{1/2}$. A bit of algebra leads to the following expressions:

$$c' \, b^{**} = (20/8) + x \tag{5}$$

and

$$(c'V^{**}c)^{1/2} = ((v^* + 8^{-1}) + x^2(8^{-1} - 2v^*) + x^4 v^*)^{1/2} \tag{6}$$

Equation (5), which includes no x^2 term, indicates that the estimated function is linear, even though the model includes a quadratic term. The reason this occurs is that the fictitious data set contains the information that the coefficient on the squared term is most likely to be zero, and there is nothing in the real data set that suggests otherwise.

Although the estimated function is the same as if a linear model were used, the standard errors given by formula (6) are quite different. The linear model corresponds to the case when $v^* = 0$, that is when the fictitious data set yields a perfect estimate of the quadratic coefficient. Then the standard errors are generated by the function $(8^{-1}(1 + x^2))^{1/2}$, which has a minimum at $x = 0$, and is graphed in Figure 1. If v^* is not equal to zero, the standard error (6) is the square root of a quartic. If $v^* < 1/16$, this function has a minimum at $x = 0$, but if $v^* > 1/16$ the function has a local maximum at $x = 0$ and minima at

$$x = \pm (1 - (1 - (16v^*)^{-1})^{1/2}.$$

An example is drawn in Figure 1 with $v^* = 2$, with local minima at $x = \pm .98$.

Another interesting result is that the standard errors at $x = \pm 1$, the sampled points, are independent of v^*. Thus the fictitious data set does not influence the inferences about the function at the sampled points, but is used only to interpolate and extrapolate to non-sampled points.

The question that I am raising in this example is a very simple one: Given the data depicted in Figure 1, what do you know about the relationship between y and x? Is it better characterized by the curves labelled $v^* = 0$, or by the curves $v^* = 2$? If you think the answer to this question is either yes or no, then you are probably a theoretical statistician. If you think the question, out of context, is entirely meaningless, then you are probably an experienced data analyst. I am of course poking fun at my colleagues who spend their time trying to find things like the optimal rule for omitting variables. Better that they should spend their time on more productive quests, like the fountain of youth. Of course it depends on the context! If this is a sample of firms and x represents the number of hours worked and y the number of shirts produced, then $v^* = 0$ seems pretty good to me. If, however, this were a study of fatigue and enthusiasm with individual data, I suspect the function may be downward sloping as enthusiasm wears off, increasing as skill level increases, and ultimately downward sloping as fatigue sets in. Lacking any sense when these three factors become operative, I might opt for $v^* = 2$.

Whatever does $v^* = 2$ mean? The formal answer is that the fictitious estimate of the quadratic coefficient is zero with a standard error of $\sqrt{v^*}$. This statement is not altogether helpful since what is desired is some way to summarize the vague notion that the curve is smooth. Yet smoothness of the curve is surely associated with the ratio of the quadratic to the linear coefficient, not the quadratic term alone. To illustrate this point consider the quadratic function $y = \alpha + \beta x + \gamma x^2$, with derivative $\beta + 2\gamma x$. Suppose that you thought that this derivative is unlikely to vary by more than ten percent over the interval from $x = o$ to $x = a$. The percentage change in this derivative over this interval is $p = 2\gamma a/\beta$ which is a function of the ratio γ/β, not the quadratic coefficient γ alone. In fact, most notions of smoothness will depend on the ratio, and the fictitious data set with information about γ alone is not altogether appropriate.

It is not my desire now to get sidetracked into a discussion of the choice of fictitious sample for polynomial regression. It is enough to recognize that it will be impossible to select a value for v^* that can be said to represent exactly the information about the smoothness of the function. For this reason, an interval of values, say $0 < v^* < 2$, may be required for a credible data analysis. Corresponding to this interval of fictitious data sets is a set of pooled estimates and standard errors. For the example under discussion, if the upper bound for v^* is set to 2, a conservative approach would report the wider confidence bounds associated with $v^* = 2$, and indicate that although these bounds could be narrowed, the information required to do so is not altogether credible.

5.2 Do Executions Deter Murder?

This section contains an analysis of the effects of executions on murder rates using data for the thirty-five states which had at least one execution between 1946 and 1950. The problem of inferring from state data the effect of executions on murder rates is similar to the problem of inferring the health effects of chemical exposures. There are a large number of other variables that might be controlled for, and the estimated effect changes greatly depending on the selection of control variables. The Bayesian approach to this problem is to select a fictitious data set that characterizes the notion that many of the possible control variables are unlikely to have much of an effect. Then a sensitivity estimate is performed to determine if the pooled estimate is adequately insensitive to the choice of the fictitious data set. Those of you who have attempted cross-sectioned epidemiological studies of low-level chemical exposures will not be surprised to learn that I conclude that any inference from these state data about the deterrent effect of capital punishment is too fragile to be believed.

The definitions of variables and the data come from McManus [4], which reports a more complete study. The least-squares estimates and standard errors are

$$M = \underset{(6.1)}{2.0} - \underset{(.20)}{.53 PC} - \underset{(.12)}{.14 PX} - \underset{(.23)}{.59 T} + \underset{(.52)}{.06 W}$$

$$+ \underset{(1.07)}{1.35 POOR} + \underset{(.10)}{.44 NW} - \underset{(.52)}{.30 URB} - \underset{(1.47)}{.66 YOUTH}, \qquad (7)$$

where the variables in natural logarithms are M = murder rate in 1950, F.B.I. estimates, PC = probability of conviction (PC = $C/\hat{Q}$ where C = convictions for murder, $\hat{Q} = M \cdot NS$, NS = state population), PX = probability of execution given conviction (average annual number of executions in 1946-1950 divided by C), T = median time served, in months, for murder by prisoners released in 1951, W = median income of families in 1949, POOR = per cent of families in 1949 with income less than $W/2$, NW = per cent nonwhite, URB = per cent urban, YOUTH = per cent 15-24 years old.

Because the variables are in logarithmic form, the least-squares estimate suggests that a doubling of the execution rate would reduce the murder rate by fourteen per cent with the fairly large standard error of twelve per cent. Or with other variables evaluated at their means, each extra execution deters 10.7 murders with a standard error of 9.2.

Different individuals will have different attitudes towards the probable importance of each of the variables. McManus [4] identifies five different fictitious data sets that differ concerning the variables that are regarded to have doubtful effects. Here, I will treat all variables as doubtful in the sense that the fictitious estimates for each coefficient is zero. Variances of the fictitious data set which reflect my feelings about the relative importance of each of the variables are reported in Table 1. The column labelled $2(v^*)^{1/2}$ is an extreme value for the size of the coefficient in absolute value in the sense that the chance of getting a larger number is approximately .05. This fictitious data set reflects the belief that convictions are more important than executions or length of sentence, that relative poverty is more important than the other social-economic effects, and that median income, race, and urbanization are least important.

TABLE 1

Variances of the Fictitious Data Set
(All fictitious estimates set to zero)

Coefficient	Variances (v^*)	$2(v^*)^{1/2}$
Prob. Conv.	.25	1.0
Prob. Exec.	.0625	.5
Sentence	.0625	.5
Median Income	.0025	.1
Pct. Poor	.25	1.0
Pct. Nonwhite	.0025	.1
Pct. Urban	.0025	.1
Pct. Youth	.0625	.5
Constant	∞	∞

Covariances of the fictitious data set are all set to zero. Many of the covariances could be argued to be either positive or negative. You might think that a large coefficient for the conviction variable would lead either to a smaller or a larger coefficient for the execution variable. Potential criminals might be thought to focus attention on either conviction risk or execution risk, but not both. Alternatively, they might be expected to be either generally aware or generally unaware of the criminal justice system. In the former case, you would select a negative covariance; in the latter case, a positive one. Even if the sign seems clear, the choice of an absolute level for a covariance is likely to tax your abilities. It is precisely this kind of problem that motivates a sensitivity analysis. Accordingly, the covariances are all set at zero, with the acknowledgement that more thought could lead to other choices, but with the hope that the bounds reported below would make that effort unnecessary, and also with the suspicion that inferences that are so sensitive to the choice of fictitious covariance matrix are not likely to be acceptable anyway.

The pooled estimates are:

$$M = -.45 - .58PC - .10PX - .72T - .03W + .57POOR + .14NW - .02URB + .18YOUTH. \qquad (8)$$

Although it might be expected that these coefficients are closer to zero than ordinary least squares since the fictitious estimates are all zero, in fact many are larger in absolute value and two have different signs. This is an illustration that pooled estimates can be surprisingly different from the separate estimates. However, the estimated effect of executions is smaller, -.10 with a standard error or .10; or, each execution deters 7.4 murders with a standard error of 7.7.

The ordinary least squares estimates are unacceptable because they ignore other relevant evidence. Normally, a researcher would omit several variables in an effort to "improve" the estimates. This is presumably done because the coefficients on the omitted variable are thought to be small. The advantage of the pooled estimates just presented is that the notion that coefficients are small is explicitly captured in a fictitious data set. But you may properly regard the pooled estimates to be incredible because they make use of a fictitious data set that can at best be said only to approximate other relevant evidence. However, a sensitivity analysis may reveal that the degree of approximation is unimportant.

A sensitivity analysis is reported in Table 2 which reports intervals of pooled estimates corresponding to intervals of covariance matrices of the fictitious sample. The intervals of covariance matrices are formed by multiplying the covariance matrix defined in Table 1 by the scalars, σ_L^2 and σ_U^2, $\sigma_L^2 \leq \sigma_U^2$. Along the diagonal of the matrix of intervals in Table 2, the upper and lower scale factors are equal,

$$\sigma_U^2 = \sigma_L^2,$$

and the covariance matrices of the fictitious sample is unique. In the center, with

$$\sigma_U^2 = \sigma_L^2 = 1,$$

the pooled estimate, which was just discussed is obtained, namely -7.4. In the lower right with

$$\sigma_U^2 = \sigma_L^2 = \infty,$$

the fictitious sample becomes irrelevant because the variances all become infinite, and we consequently obtain the least-squares estimate -10.7. At the upper left with $\sigma_U^2 = \sigma_L^2 = 0$, the fictitious sample is perfectly informative, and as a result the fictitious estimate of zero is obtained. The longest interval of estimates is in the upper right with $\sigma_L^2 = 0$ and $\sigma_U^2 = \infty$. In this case the covariance matrix is completely free, and Theorem 1 applies. Thus if you are so unsure of what fictitious covariance matrix adequately approximates the other relevant evidence

TABLE 2

INTERVALS OF ESTIMATES OF THE NUMBER OF MURDERS STIMULATED BY EACH EXTRA EXECUTION

VARIANCE MATRIX BOUNDED BETWEEN $V^* \sigma_L^2$ AND $V^* \sigma_U^2$

	σ_U						
σ_L	0	1/4	1/2	1	2	4	∞
0	0	6.9	18.0	30.9	38.4	41.5	44.2
	0	-9.8	-24.0	-38.3	-46.3	-50.2	-54.9
1/4		-2.8	8.2	21.0	28.4	31.5	34.3
		-2.8	-17.0	-31.2	-39.2	-43.1	-47.8
1/2			-5.9	6.8	14.1	17.1	20.0
			-5.9	-20.1	-28.0	-31.8	-36.6
1				-7.4	-.1	2.9	5.8
				-7.4	-15.2	-18.9	-23.9
2					-7.9	-4.96	-2.1
					-7.9	-11.7	-16.5
4						-8.7	-6.3
						-8.7	-13.2
∞							-10.7
							-10.7

that any matrix will do, then the interval of estimate goes from -54.9 to 44.2. Though this interval is long, it is not unbounded. However, the pooled estimate can be either positive or negative, and it would be impossible to say whether executions deter or encourage murders.

The figures which are in a box are based on covariance matrices between $(1/2)^2$ and 2^2 time the original matrix. Because this interval of estimates from -28.0 to 14.1 contains the origin, and because I am unable to define more precisely my fictitious sample, I am forced to conclude that these data are not useful for estimating the effects of executions. Of course, other analysts with more sharply defined fictitious samples possibly based on other data sets could find these data useful. In particular, as can be seen in Table 2, if you are sure that your fictitious sample is more diffuse than mine, in the sense that your covariance matrix is certainly greater than 2^2 times the matrix defined in Table 1, then you can only obtain a negative estimate, and can conclude that executions deter murders.

5.3 Sensitivity Analysis For Weighted Least-Squares Estimates

To this point we have considered the Bayesian alternative to procedures that omit variables from regression. Another major problem confronting empirical workers is which observations to omit. Most methods for dealing with extreme data points in the context of regression make use of the weighted regression formula

$$\hat{\beta} = (X'WX)^{-1} X'Wy$$

where W is a positive $n \times n$ diagonal matrix with weights applying to individual observations on the diagonal. Because there are many sensible ways that these weights can be selected, we ought to study the sensitivity of the estimates to the choice of weights. The sensitivity result, Theorem 4, can be readily applied to this problem.

Write the regression model as $y = X\beta + I\gamma + u$ where I is an $n \times n$ identity matrix, interpreted as a set of n "dummy" variables each selecting one of the observations. The vector u can be thought to represent the usual or "normal" errors, and the vector γ can be thought to represent the abnormal errors. Assume also that there exists information that the abnormal errors are small, summarized by a fictitious data set with estimate zero for γ and covariance matrix V^*. Then the pooled estimate analogous to formula (1) is

$$\begin{bmatrix} \hat{\beta} \\ \hat{\gamma} \end{bmatrix} = \left[\begin{bmatrix} X' \\ I' \end{bmatrix} [X \; I] + \begin{bmatrix} 0 & 0 \\ 0 & V^{*-1} \end{bmatrix} \right]^{-1} \begin{bmatrix} X' \\ I' \end{bmatrix} y$$

Using the partitioned inverse rule, the estimate $\hat{\beta}$ can be shown to be

$$\begin{aligned}\hat{\beta} &= (X'(I + V^*)^{-1}X)^{-1} X'(I + V^*)^{-1} y \\ &= (X'X)^{-1} X'(y - \hat{\gamma})\end{aligned} \quad (9)$$

where
$$\hat{\gamma} = (M + V^{*-1})^{-1} My , \quad (10)$$
$$M = I - X(X'X)^{-1} X' .$$

In words, the weighted regression estimate (9) can be written as ordinary least-squares with data y corrected for outliers. The outlier correction (10) is a matrix weighted average of the zero vector and the vector of least-squares residuals My.

Because the correction for outliers $\hat{\gamma}$ is a matrix weighted average, we can apply Theorem 4. In particular, it implies the following result taken from Leamer (1983):

Theorem 6. The extreme values of $c'\hat{\beta}$, where c is a vector of constants, $\hat{\beta}$ satisfies (9), and $V^* \leq qI$, are

$$c'\hat{\beta} = c'b(1 \pm t^{-1} q(n-k)^{1/2}/2 \, (1+q)^{1/2})$$

where t is the t-value for testing $c'b = 0$:

$$t = c'b/(c'(X'X)^{-1}c \; y'My/(n-k))^{1/2} .$$

What is surprising about this result is that the usual t value tells you all you need to know about the sensitivity of an estimate to the choice of weights and the pattern of residuals is quite irrelevant. A coefficient with a large t-value is relatively insensitive to reweighting of observations to deal with outliers. One choice for q would be one, in which case some observations are allowed to have twice the weight of others. Then the maximum percentage change in the estimate that could be induced by reweighting is $((n-k)/8)^{1/2}/|t|$. Note that constant resistance to reweighting as sample size increases requires an increasing t-statistic at a rate equal to the square root of the degrees of freedom. For example, the sign of the estimate is insensitive to reweighting if the t-statistic exceeds the critical value in the following table:

TABLE 3 Critical t-values

Degrees of Freedom (n-k)	$t = ((n-k)/8)^{1/2}$
50	2.5
100	3.5
1000	11.2

6. CONCLUSION

In conclusion, it may be useful for me to indicate how my views differ from the views of the speakers who immediately precede me in this symposium. I do so with some discomfort in the written version of this paper since I have not yet seen their papers, nor heard their lectures. I nonetheless feel confident that their main themes have already occurred in the literature, and are consequently known to me.

First of all, the Bayesian position is intellectually imperialistic. The matrix weighted average form (1) is general enough to encompass all of the estimators that have been proposed under the heading of "biased estimation." Indeed, Hoerl and Kennard (1970) acknowledge that ridge regression already existed as a Bayes estimator with a special fictitious sample, $b^* = 0$ and $V^* \propto I$. I have announced in this paper that this imperialism extends also to weighted regression, and consequently to all the estimates that have been proposed under the heading of "robust regression." This is so because a weighted regression estimator can be written as the least-squares estimate with data corrected for outliers, and because the outlier correction is a matrix weighted average analogous to the pooled estimate (1).

Although the mathematical form of the estimators proposed by Bayesians and non-Bayesians is identical, in practice there are sharp differences stemming from the fact that a Bayesian selects the fictitious sample values b^* and V^* to represent prior information whereas non-Bayesians select them to assure that the pooled estimator b^{**} has "good" sampling properties. Bayesians regard this attempt to find "optimal" values for b^* and V^* with a certain amount of incredulity, a bit of amusement and even a little irritation. A basic result in decision theory is that any estimator that takes the matrix weighted average form (1) with $\underline{V}^*$ positive definite is admissible, given quadratic loss. This means that there is no other estimator that has better sampling properties for all values of β. So you can't find optimal values of $\underline{b}^*$ and $\underline{V}^*$. Any will do, if sampling properties are the criterion. What is irritating is that so many people choose to ignore this simple result. What is amazing is that these "biased estimators" are proposed in settings in which the proposed "optimal" fictitious data set obviously and grossly distorts the other relevant information. I am now referring to the use of ridge regression with $\underline{b}^* = 0$ when zero is obviously not a sensible estimate for the coefficients.

The decisions that a data analyst must make are very great indeed, and Bayesian regression adds a multiplicity of additional choices. Conventions are ordinarily adopted in such complex settings. Two common conventions are normals for distribution and .05 for significance levels. Statisticians are also attempting to find conventions to deal with the choice of variables and the choice of observations. Stepwise regression is one convention. Ridge regression is another. These are very poor conventions, since they often lead to very inappropriate inferences. My recommendation would be to do the best you can to select the fictitious sample values, $\underline{b}^*$ and $\underline{V}^*$, to represent the available prior information, and then to study the sensitivity of inferences to changes in these choices.

REFERENCES

CHAMBERLAIN, G. and LEAMER, E. E. (1976), "Matrix Weighted Averages and Posterior Bounds," J. of the Royal Statis. Soc. B, 38, 73-84.

HOERL, A. E. and KENNARD, R. W. (1970) "Ridge Regression: Biased Estimation for Nonorthogonal Problems," Technometrics 12: 55-67.

LEAMER, E. E. (1978), Specification Searches, John Wiley, N. Y.

_____ (1982), "Sets of Posterior Means With Bounded Variance Priors," Econometrica 50: 725-736.

_____ (1983a), "Let's Take the Con Out of Econometrics," Amer. Economic Review, 73: (March), 31-43.

_____ (1984), "Global Sensitivity Results for Generalized Least Squares Estimates," J. of the Amer. Statis. Assn. 79: (December) 867-870.

LEAMER, E. E. and LEONARD, H. B. (1983), "Reporting the Fragility of Regression Estimates," Review of Economic and Statis. LXV (May) 306-317.

MCMANUS, W. S. (1980), "Murder and Capital Punishment: A Bayesian Analysis," Dept. of Economics, UCLA, mimeo.

STATISTICAL GRAPHICS ON SMALLER COMPUTERS:
THE DATA ANALYST'S NEW TOOLS*

Thomas J. Boardman
Colorado State University, Statistics Department
Fort Collins, CO. 80523

ABSTRACT

One of the objectives of a good data analysis is to learn as much as possible about the useful information contained in a data set. The exploratory analysis of data using statistical graphics can be of real help to accomplish this task. All small computers have some capability to produce graphics and many have very extensive capabilities. The purpose then of the first half of the paper will be to acquaint you with statistical graphics. To accomplish this goal we will study the Pollution and Mortality Data discussed in McDonald and Schwing (1973). The data consisted initially of sixteen measurements taken on sixty cities in the U.S. Eight more variables were added to simplify plotting, analysis, and viewing of the data. The original sixteen variables consisted of <u>four</u> climatological variables, eight demographic variables, three pollution potential variables, and the age adjusted mortality. rate. The additional variables include the longitude, latitude, and altitude for each of the cities, as well as the logarithms of the three pollutant potentials, and two grouping variables. The sixty cities have been clustered into four groups by a subjective evaluation of a panal of viewers of Chernoff Faces as described in McDonald and Ayers (1978). The presentation will begin with simple graphics for one variable (Box plots and Quartile-Quantile plots) and add additional variables to two and three dimensional figures.

The second portion of the presentation will describe some of the available packages for performing regression analyses on computers. Because of the interest in smaller computers the author will discuss some of the problems associated with performing regression analyses on microcomputers. Finally we will consider the analysis of the Pollution and Mortality Data in which we will attempt to develop a regression model for Y_i the age dependent mortality rate, and a selected collection of the other variables. The output from several commercial program packages will be reviewed. Statistical graphical evaluation of the selected model will be discussed.

Through the presentation the goal will be to see how we can use insightful and useful graphics to assist the analyst in his or her model-building exercise.

1. INTRODUCTION

In a recent article, the author (Boardman, 1982) discussed some implications for statisticians of the rapid growth rate of smaller computers. In a more recent article the author (Boardman, 1984) discussed the impact that smaller computers may have on data analysis while in a related article (Boardman, 1984) he discussed the impact

*Originally appeared in Bulletin of the International Statistical Institute, Proceedings of the 44th Session, Madrid, Spain, September 12 to 22, 1983.

that smaller computers may have on statistical education. Stuckey and Foster (1983) predicted that by 1985 "more than 10 million microcomputers will be installed". Obviously no matter how you define the class of smaller computers, we can expect that in the near future many people will have access to computing through these devices.

One of the features available on most small computers is graphics. Indeed, some of the best graphics available today are produced on small stand-alone systems. Graphics then is an integral part of the small computer environment.

A renewed interest in graphics has developed over the last several years in the United States of America. For example, the American Statistical Association (ASA) created the Committee on Statistical Graphics (CSG) which has, among other activities,:

1) created an exposition at the national ASA meetings to provide statistical graphics software venders an opportunity to display their capabilities,
2) established a set of standards for publication of statistical graphics,
3) organized tutorial and regular sessions on statistical graphics at the national ASA meetings,
4) appointed a subcommittee to determine software availability and methods for evaluation of the software, and
5) established strong contacts with the National Computer Graphics Association.

The CSG is actively considering other topics for its consideration. The main objective of the committee is to be a strong advocate for quality statistical graphics.

Another indicator of renewed interest in statistical graphics is reflected by recent publications. Probably the most authoritative book on statistical graphics has only just been released (Chambers, et al, 1983). However several other fine volumes are available as well (Tufte, 1982; Wang, 1978; and Schmid, 1983). Many articles dealing with using or creating statistical graphics have appeared recently in the literature. A number of these will be referenced subsequently in this paper. Then, too, articles have appeared in the non-statistical literature describing how researchers are using statistical graphics. One (Kolata, 1982) describes some of the work in computer motion displays under investigation at Stanford and Harvard universities. Others in the statistics profession are concerned about how people perceive graphics (Cleveland and McGill, 1983; and Bly, 1983). Carr and Littlefield (1983) are investigating using stereo scatterplots which must be viewed using special "glasses". Several of these topics will be discussed later in more detail.

If the purpose of good data analysis is to discover useful information in a data set, then well-chosen graphics should help in the process. One of the topics which is often discussed at the CSG is the need for "good and insightful" graphics in data analysis. Graphics are an indispensable tool for the analyst during the data snooping, investigation, and prospecting stages of data analysis. Of course, graphical procedures have often been used during the validation and confirmation stages as there is hardly a segment of data analysis which can not be improved by the use of "good and insightful" graphics.

The purpose of this paper is to discuss existing, enhanced and new graphics for data snooping and to describe some of the interesting changes in statistical graphics which should occur in the near future.

Most of the graphics will be available on smaller computers thus providing the user with a powerful set of tools for data analysis. While none of the graphics will be included in this paper due to space limitation and more importantly, due to AIP's requirements for publication which preclude using graphics from a smaller computer, the actual oral presentation of the paper in Mobile will show some of the enhanced and new graphics.

2. REPRESENTATIVE SET OF STATISTICAL GRAPHICS

During the last several years a number of new or enhanced versions of existing statistical graphics have been introduced. To attempt to show each of these would take too much space. Indeed most of the graphics can take various forms as well so that actual presentation of a graphic is in some sense a "sample of size one".

In order to provide the reader with sources of information on various graphics, we have prepared Table 1 below. The table lists twenty-three or so statistical plots with common names, the range of the number of variables which the plot can accommodate, references to descriptions of the graphics, and a brief comments column. The numeric codes under references refer to a selected portion of the References.

These graphics and others are under constant revision and improvement. Concerns about how one uses these graphics and new graphics of the future is the subject of the next section.

3. OPPORTUNITIES AND CONCERNS FOR STATISTICAL GRAPHICS IN THE FUTURE

Both opportunities for new and innovative statistical graphics and concerns over how these graphics are developed and then perceived by the user should result from the widespread availability of inexpensive graphics devices. The discussion below briefly addressed the joint issues of opportunities and potential concerns as more programmers implement statistical graphics on computers of the future.

3.1. Resolution and Accuracy

Preparing even simple graphics on a cathode ray tube (CRT) displays with limited resolution may result in plots which are of limited value particularly if one wishes to compare several group treatments of data on one figure. On the other hand, high quality resolution is often not needed. Useful statistical graphics have been prepared on line printers for years. However some of the refinements in statistical graphics which will be discussed below will require high resolution graphic devices.

While not all applications require high resolution, we certainly expect accuracy in all graphics output. There are two facets to accuracy: the first is that we expect that the plotting positions on the graphic must be located accurately. Any graphic which can not support this type of accuracy should be discarded. A second facet of accuracy deals with the user's perception in judgement. Cleveland and McGill (1983) discuss six levels of initial perceptual accuracy for judgement:
1) position along a common scale
2) position along a nonaligned scale
3) length, direction and angle
4) area
5) volume and curvature
6) shading and color saturation

Table 1. Sample of New, Enhanced, or Underutilized Graphics

Common Name for Statistical Graphic	Number of Variables	References[a]	Comments[b]
One Variable Graphics			
Histograms	1	1, 2, 4	NCO = normal curve overlayed
Stem and Leaf Displays	1	1, 2, 3	Tags/subs
Dot Charts	1	2, 4	Tags/subs
Probability Plots	1	2, 5	Weibull, Normal, Lognormal, etc.
Rootograms	1	1, 2	VR, NCO
Box Plots	1	1, 2	Tags/subs, VR
Data Plots	1	6	Tags/subs, VR
Quantile-Quantile Plots	1	2	Tags/subs, VR, EMP
One Variable Over Time			
Response Curve Plots	2	6, 11	Repeated measurements on subjects over time with fixed spacing on time or variable spacing
Enhanced Scattergrams			
Scattergram	2	1, 2, 3	Log scale of axes possible
Circle Scattergram	3	11	Tags/subs, area of circle∿variable 3
Q-Squared Scattergram	4	7	Tags/subs, Quartiles of Var 3 and Var 4 control Var 1 vs. Var 2 plots
Weathervane Scattergram	5	11	Tags/subs, area of circle∿Var 3, vane direction ∿ Var 4, length ∿ Var 5
Star Figures on Scattergram	14	10, 11	X=Var 1, Y=Var 2, other twelve as star plot--see Star Plot
Map Coordinate Scattergram	14	12, 11	Same except that X and Y axes are coordinates
Multi-variable Plots			
3-Dimensional Scattergram	3	6	Rotation and tilting necessary
Clock Plots	4-13	11	24 hour clock face with weathervane or stars
Andrews Plot	5-15	13, 14	Restrict range of θ if necessary
Polygon Plots	4-12	14, 8	Standard is mean for all variables
Radial Plot	4-12	14, 8	Standard is mean for all variables
Star/Kiviat Plot	4-12	10, 11	Standard is extreme for all variables
Chernoff Faces	4-20	15, 8, 14, 11	
Asymmetrical Faces	4-20	16, 6	

[a] In the References the code numbers are located to the left of the entries. The reference for code 6 is unpublished results which Boardman has developed under a research contract with Hewlett Packard. Actually all of the plots above could be references to code 6.

[b] Comments: NCO = a normal curve may be overlayed on the graph. Tags/subs = a tag variable and/or subfile may be used to designate treatment groups. VR = various refinements of the basic graphic are available. EMP = one may plot two groups using empirical quantities or one group vs. specified quantiles such as normal.

According to their research, judgements of perceptual accuracy are graduated from easiest (1) to most difficult (6). In their paper they explored results from several studies which suggest, for example, that the common practice of using divided bar charts and certain forms of pie charts can lead to serious difficulty for users to accurately judge differences between groups. In all that follows, then, we must be concerned about these two facets of accuracy.

3.2. Standards for Graphic Presentations

Under the chair of William Cleveland, the American Statistical Association's Committee on Statistical Graphics prepared suggested standards for graphics presentation (Cleveland, 1982), Cleveland (1983) also addressed this topic by reviewing the statistical graphics appearing in a recent volume of SCIENCE. As we stated in the beginning, not only must statistical graphics help the user gain insight in his or her data but they must be "good". The statistical profession must continually strive to improve the quality of our graphics presentation.

3.3. Alternagraphics

In the past most program developers have used the CRT display to show only one graphic at one time. The hardware and software technology exists now to use multiple screens/planes/windows/sections on a CRT at one time (Bly, 1983; Cleveland and McGill, 1983; Becker and Chambers, 1983; Kolata, 1982; Friedman, McDonald and Stuetzle, 1983; and Carr and Littlefield, 1983). Not only can several graphs be displayed on the screen at one time, but many layers and/or positions on the screen can be used to interact in a friendly manner with the user (Boardman, 1982). Many new forms of graphic prompts (icons) will be developed to allow the user to communicate with the computer.

3.4. Color

Most humans view life in a full spectrum of colors. Many are also accustomed to viewing color television screens. It is only natural that statistical graphics will be eventually viewed in color on most CRT displays. While some of the smaller computers currently referred to as micros may have only limited capability from color displays at the present time, this will no doubt change very soon. Several of the major scientific research journals have permitted color graphics to appear. Of course, the expense associated with high quality reproduction of color graphics can be high. One can expect that new technology and competition will help to reduce these costs.

Color is presently being used in at least three modes in statistical graphics, namely 1) for line drawing, 2) for point identification of treatments or groups, and 3) for filling or shading portions of a graphic. Statisticians are concerned (Chambers, et al, 1983; Cleveland and McGill, 1983; Bly, 1983) with how people perceive, differentiate, and interpret color displays. These researchers and others are performing human factors experiments to quantify the

results. Those who are involved in developing statistical software need to be familiar with the results for their work.
3.5. Motion and Color

One of the most exciting new developments for real time data analysis is fast changing displays. For example, if one is viewing several three dimensional figures on a screen, one can study the relationship between one set of three variables (X, Y, Z) by viewing the graphics after rapid rotation and tilting of the axes. Of course, a high resolution graphic device coupled with powerful computers are necessary to carry out this task. The capability of this type of operation now exists in smaller computers toward the top of the price range. What can we expect in the future? Tomorrow's smaller computer will have the capability to do this type of real time motion graphic and other real time operations which have not been thought of today. Color can now add insight into data analysis with real time motion. Friedman, et al (1983) decribe a work-station called ORION which makes very effective use of color. One of the displays which is part of their work-station has four three-dimensional (X, Y, Z) plots on the screen. Each of the four plots may use three different variables. Using a "mouse" one is able to designate a cluster of points on one plot using color and then identify those same points on the other three graphs using the same color. New and innovative ways of performing data analysis will be developed using motion and color together.

Years ago we went to movie theaters to view 3-D stereo films which required the viewer to wear special glasses. The effect of wearing the glasses was to view characters almost jumping out of the screen toward the audience. Carr and Littlefield (1983) have developed some very effective stereo scatterplots with full use of color and wide choice of appropriate glyphs. While, for the most part, their work is currently with stationary scatterplots, one can certainly imagine that stereo graphics could be highly effective with real time motion as well.

Other (Bly, 1983) are investigating the use of audio signals with graphical data analysis. At Lawrence Livermore Laboratories, Bly experimented with various audio dimensions such as frequency, amplitude, duration, and timbre among others. Her use of these audio tools was not merely limited to audio queueing. Rather, she attempted to add several new dimensions to data analysis using one or more of these "tools". Several comments arose after her presentation including the recommendation that a standard reference tone be sounded before the analyst determines potential difference between data points. To date we have not used audio stimulation with graphical displays in an effective manner. Perhaps the work by Bly and others will suggest new opportunities for innovative tools for data analysis.
3.6 Large Data Sets

In the future researchers may confront small computers with very large data sets. It is possible to connect microcomputers directly to devices which can record many mesurements in a relatively short period of time. Are the graphical procedures which we have discussed above appropriate for data sets with several hundred thousand observations on fifty variables? Of course, the concerns about analysis of large data sets is not limited to graphical concerns (Boardman, 1984). How does one prepare and view a scattergram, for example, which has one hundred thousand points? New and clever ways of displaying the dependence between several variables will have to be developed for large data sets.

3.7. Quality Control - An Example of One Area Needing Better Graphics

X BAR and R control charts and other similar graphical devices have been used for years by quality control personnel to monitor ongoing processes. Most of the graphics used by quality control engineers consists of plots in which only one variable is recorded. These were developed during the early 1940's. During a recent visit to our campus, J. Stuart Hunter reminded us that quality engineers usually record several variables on a given process. As an example he suggested that even when only two characteristics are recorded, a bivariate scatterplot with an elliptical rejection region would provide more information about the dependency between the two variates than separate control charts.

Opportunities exist for developing good and insightful graphics which will be of help to those in the quality profession. Today's small computer can perform the computation so that even fairly complicated multivariate calculations can be done in a timely fashion. Many of the graphics named in TABLE 1 could be used effectively by quality engineers. Very few are being used now.

4. CONCLUSION

The use of statistical graphics in data analysis is increasingly viewed as one of the fastest growing ares in statistics. Most of the graphics could be done now on fairly inexpensive computers. The graphics hardware in the future should support many new and innovative graphical tools for data analysis. Those of use who do data analysis should find these new tools to be particularly helpful.

5. ACKNOWLEDGEMENTS

The author gratefully acknowledges the contributions made by his association with members of the National Academy of Science Panel on the Future of Statistics and Computers, the speakers and discussants at the 1983 Interface Conference, J. Stuart Hunter, and colleagues at Colorado State University. All of the graphics named in TABLE 1 were implemented under a research contract which the author has with the Engineering Productivity Division of Hewlett Packard. With great pleasure the author acknowledges the contributions made by his student programmer, Chris Havelick. His ability to take the author's pencil and paper sketches and transform them into high-quality graphics on the Hewlett Packard Computer is truly remarkable.

6. SUMMARY

If one of the purposes of good data analysis is to learn as much as possible about the useful information contained in a data set, then investigative data analyses using graphics has probably the best chance of accomplishing this task. Most smaller computer systems have at least some capability to produce graphics while many have very extensive capabilities. A renewed interest in statistical graphics has already begun. The widespread availability of smaller computing devices in the future will make it possible for everyone to use graphical procedures in data analysis. Opportunities and concerns for statistical graphics of the future are discussed. A list of new and enhanced statistical graphics and appropriate references are included.

REFERENCES

[13]ANDREWS, D. F. (1972), "Plots of High-Dimensional Data," *Biometrika*, 28: 125-36.

BECKER, R. A. and CHAMBERS, J. M. (1983), "Workstations and the Human Interface," *Proceedings of the Computer Science and Statistics: 15th Symposium on the Interface*, Houston, TX, North Holland Pub. Co.

BLY, S. (1983), "Interactive Tools for Data Exploration," *Proceedings of the Computer Science and Statistics: 15th Symposium of the Interface*, Houston, TX. North Holland Pub. Co.

BOARDMAN, T. J. (1982), "The future of Statistical Computing on Desktop Computers," *The American Statistician*, 36: 49-58.

BOARDMAN, T. J. (1984), Smaller Computers: Impact on Statistical Data Analysis," Statistics: An Appraisal-International Conference to Mark the 50th Anniversary of the ISU Statistical Laboratory.

BOARDMAN, T. J. (1984), "Smaller Computers: Issues in Statistical Education," *Proceedings of the Statistical Education Section of ASA*. In preparation.

[11]BOARDMAN, T. J. (1984), "The Use of Simple Graphics to Study Hourly Data with Several Variables," ASQC Technical Supplement in Honor of Horace P. Andrews.

[9]BRUNTZ, S. M., CLEVELAND, W. S., KLEINER, B. and WARNER, J. L. (1974) "The Dependence of Ambient Ozone on Solr Radiation, Wind, Temperature, and Mixing Height," *Proceedings of the Symposium on Atmosphere Diffusion Air Poll*. Amer. Metero. Soc. 125-128.

CARR, D. B. and LITTLEFIELD, R. J. (1983), "Color Anaglyph Stereo Scatterplots-Construction Details," *Proceedings of the Computer Science and Statistics: 15th Symposium on the Interface*, Houston, TX. North Holland Pub. Co.

[2]CHAMBERS, J. M., CLEVELAND, W. S., KLEINER, B. and TUKEY (1983), *Graphical Methods for Data Analysis*, Belmont, CA. Wadsworth Pub. Co.

[15]CHERNOFF, H. (1973), "The Use of Faces to Represent Points in K-Dimensional Space Graphically," *Journal of the American Statistical Association*, 68: 361-368.

CHERNOFF, H. and RIZVI, M. (1975), "Effects on Classification of Random Permutations of Features in Representing Multivariate Data by Faces," *Journal of the American Statistical Association*, 70: 538-554.

CLEVELAND, W. (1982), "Author Guidelines for Figures," Prepared by Committee on Statistical Graphics of American Statistical Assn.

CLEVELAND, W. (1983), "Graphics in Scientific Publications," BELL Laboratories.

[12]CLEVELAND, W. S., KLEINER, B., MCRAE, J. B. and WARNER, J. L. (1976) "Photochemical Air Polution: Transport from the New York City Area into Connecticut and Massachusetts," *Science*, 191: 179-180.

[4]CLEVELAND, W. S. and MCGILL, R. (1983), "Computer Graphics and Human Factors Experimentation," *Proceedings of the Computer Science and Statistics: 15th Symposium on the Interface*, Houston, TX., North Holland Pub. Co.

[7]FILLIBEN, J. J. (1983), "Dataplot Analysis of Biomedical Data," Under review by *The American Statistician*.

[16]FLURY, B. and RIEDWYL, H. (1981), "Graphical Representation of Multivariate Data by Means of Asymmetrical Faces," *Journal of the American Statistical Association*, 76: 757-765.

FRIEDMAN, J. H., MCDONALD, J. A., and STUETZLE, W. (1983), "Interactive Real Time Graphics for Multivariate Data Analysis," *Proceedings of the Computer Science and Statistics: 15th Symposium on the Interface*, Houston, TX., North Holland Pub. Co.

[10]KOLANCE, K. W. and KIVIAT, P. J. (1973), "Software Unit Profiles and Kiviat Figures," *ACM Performance Evaluation Review*, 2: 132-20.

KOLATA, G. (1982), "Computer Graphics Comes to Statistics," *Science*, 217: 3 September, 919-920.

[8]NEWTON, C. M. (1976), "Graphical Data Analysis: Optimizing Tradeoffs Between Richness and Simplicity," *Proceedings of Statistical Computing Sec., American Statistical Association*, 238-242.

SCHMIDT, C. F. (1983), *Statistical Graphics: Design Principles and Practices*, New York, John Wiley & Sons.

STUCKEY, R. and FOSTER, J. (1983), "Meet the Micros," *SIAM News*, 16: 8-9.

TUFTE, B. R. (1982), *The Visual Display of Quantitative Information*. Cheshire, CT., Graphics Press.

[3]TUKEY, J. W. (1977), *Exploratory Data Analysis*. Addison-Wesley Pub. Co.

[1]VELLEMAN, P. F. and HOAGLIN, D. C. (1981), *Applications, Basics and Computing of Exploratory Data Analysis*, Boston, MA., Duxbury Press.

[14]WANG, P. C. C. (editor) (1978), *Graphical Representation of Multi-Variate Data*, New York, Academic Press.

[5]WILKE, M. B. and GNANADESIKAN, R. (1968), "Probability Plotting Methods for the Analysis of Data," *Biometrika*, 55: 1-17.

*Numbers are associated with the references in Table 1.

REGRESSION METHODS FOR BINOMIAL AND POISSON DISTRIBUTED DATA

E. L. Frome*

Mathematics and Statistics Research Section
Engineering Physics and Mathematics Division
Oak Ridge National Laboratory
Oak Ridge, TN 37830

ABSTRACT

Models are considered in which a rate or probability can be represented by a regression function that describes the relation between the predictor variables and the unknown parameters. Estimates of the parameters can be obtained by means of iteratively reweighted least square (IRLS). When the dependent variable is a count that follows either the Poisson or binomial distribution, the IRLS algorithm is equivalent to using the method of scoring to obtain maximum likelihood (ML) estimates. This general least squares regression approach includes linear, generalized linear, and intrinsically nonlinear regression functions. Standard statistical packages that support IRLS can be used to obtain ML estimates, their asymptotic covariance matrix, and diagnostic measures that can be used to aid the analyst in detecting outlying responses and extreme points in the model space. The results of fitting several different models to the same data set can be summarized in an ANOVA-like table using the deviance as a measure of residual variation. Five examples using data from both designed experiments and observational studies are presented to illustrate the utility of Poisson and binomial regression analysis.

1. INTRODUCTION

1.1 *Notation and Terminology*

In this paper we assume that data have been obtained on each of N (*observational or experimental*) *units*. The data for the *ith* unit consist of the following:

y_i -- the observed value of the dependent variable,
c_i -- the "sample size", and
$X_i = (x_{i1}, x_{i2}, ..., x_{im})$ -- a row vector of covariates,

where x_{i1} is value of the first covariate, etc. If the values of X_i and c_i are determined in advance by the investigator (and randomization is used) then the results are from a *designed experiment*. In many situations, primarily in human populations, the investigator is restricted to observing the

---

*Edward L. Frome is on the research staff, Mathematics and Statistics Research Section, Engineering Physics and Mathematics Division of Oak Ridge National Laboratory, P. O. Box Y, Oak Ridge, TN 37830. This research was supported by Contract W-7405-eng-26 between U.S. Department of Energy and Union Carbide Corporation, and by Contract No. DE-AC05-76OR00033 between Oak Ridge Associated Universities and the U. S. Department of Energy, Office of Health and Environmental Research. The author is grateful to the staff of the Center for Epidemiologic Research, ORAU for offering helpful suggestions, and to Donna Poole for help with manuscript preparation. By acceptance of this article, the publisher or recipient acknowledges the U. S. Government's right to retain a nonexclusive, royalty-free license in and to any copyright covering the article.

values of the covariates, and we refer to such situation as *observational studies* (see Cochran, 1983). In both situations one or more of the covariates may correspond to "causal factors" (i.e. treatments, procedures, or programs) that are of primary interest, while other x variables may affect the variation in the dependent variable and are included so that their influence can be "adjusted for" in the statistical analysis. We assume that the y_i are counts, and use either the Poisson or binomial distribution as a model for the variation in the dependent variable. The term *covariate* (or x-variable) will be used as a generic term for what are sometimes called independent variables, predictor variables, explanatory variables, or stimulus variables. Covariates may be either quantitative or qualitative in nature, and we use the term *factor* to refer to qualitative covariates. A factor can take a limited number of values called *levels*. For example, if A is a factor with k levels, then these can be coded using the integers $1,2,...,k$. The *actual* levels of the factor may be either qualitative or quantitative. The levels of a factor are used to generate "dummy variables" (indicator variables) in the covariate vector, i.e.

$$x_{i1} = \begin{cases} 1 \text{ if } level\ 1\ of\ factor\ A\ is\ present\ on\ unit\ i, \\ 0\ otherwise \end{cases}$$

This implies that there will be a parameter associated with each level of the factor.

In situations that involve a linear structure (see below) a useful notation has been developed (Wilkinson and Rogers, 1973) and adapted for use in computer programs (see e.g. GLIM-3, Baker and Nelder, 1978). In agriculturial experiments, for examples, blocks and varieties are examples of factors. In observational studies factors might include sex, socioeconomic status, geographic region, etc. Table 1 contains further examples of covariates, as well as a description of the dependent variable y and sample size for each example that we will consider in Section 2 and 3.

Table 1. List of Examples of Data Used To Illustrate
Poisson and Binomial Regression

Example	Dependent Variable y	Sample Size c	Distribution	Covariates x
1 and 2	number of chromosome aberrations	total lymphocytes scored (unit=100)	Poisson	radiation dose, exposure rate
3	number of lung cancer deaths	man-years (units=10^5)	Poisson	smoking, age
4	number of damaged lymphoblasts	total lymphoblasts examined	binomial	streptonigrin dose
5	number of mice with liver neoplasms	total mice examined	binomial	2-AAF exposure, time

Assuming that we have data $y_i, c_i, X_i, i=1,...,N$ on each of N units, the problem of interest is to determine if there is any systematic relation between the dependent variable and the covariates. We assume that the expected value of y_i is given by

$$\mu_i = c_i \, \lambda(X_i,\beta),$$

where $\lambda(X,\beta)$ is a known function, and $\beta = (\beta_1,..., \beta_p)'$ is a p-dimensional vector of unknown parameters. The regression function will in general be nonlinear in the unknown parameters, and hopefully p (the number of parameters) is much less than N. A special case of interst occurs when the regression function can be expressed as a linear combination of the covariates and the parameters, i.e.

$$\lambda(X_i,\beta) = X_i\beta = \sum_{j=1}^{P} \beta_j x_{ij}.$$

This is referred to as a (multiple) linear regression (function), and is most often encountered in the context of the "classical linear model". There is an extensive literature on the classical linear model (see Draper and Smith, 1981), and these methods are most appropriate when the y_i are continuous and follow the Normal distribution with constant variance. When the dependent variable is a count that may follow the Poisson or binomial distribution the assumptions of constant variance is clearly inappropriate, and the linear regression function is of limital value. A more general class of regression functions that involves a linear structure has developed for count data. A *generalized linear function* (GLF) is defined by

$$\lambda(X_i,\beta) = G(X_i\beta),$$

where $G(\ )$ is a monotonic differentiable function. A widely used GLF for Poisson data is the *product model* $\lambda(X_i,\beta) = \exp(X_i\beta)$, which is also called a log-linear model. A useful GLF for binomial data is $\lambda(X_i,\beta) = \exp(X_i\beta)/\left[1 + \exp(X_i\beta)\right]$. The use of GLFs for Poisson and binomial data has been discussed by Nelder and Wedderburn (1972), and the statistical program GLIM (1977) has been developed to facilitate data analysis using GLFs. Nelder and Wedderburn use the term "link function" for the inverse of the function G, i.e. for the product model the link function is log (λ) which explains why this is also called a log-linear model.

In summary, we use the term *Poisson (binomial) regression model* to refer to a situation where
   i) the y_i are independent and follow the Poisson (binomial) distribution, and
   ii) the regression function $\lambda(X_i,\beta)$ is known.

Then given the data we want to obtain estimates of the β_j, their standard deviations, and evaluate the "goodness of fit" of the regression model. In what follows we show that estimates of the β_j can be obtained using a general regression model (1.1), from which certain generalized least squares equations (1.3) are derived. A root of the system of equations can be obtained using an itertively reweighted least squares procedure which is equivalent to maximum likelihood estimation under the Poisson or binomial assumption.

1.2 General Regression Model

Consider the general regression model

$$E(y_i) = \mu(X_i,\beta) = c_i\lambda(X_i,\beta)$$
$$Var(y_i) = \sigma^2 \, V(\mu_i) \quad , \quad i = 1, ..., N \, , \qquad (1.1)$$

where y_i is the response for the *ith* unit, c_i denotes the "sample size", and V( ) is a known function that may depend on μ_i. We assume that the y_i are uncorrelated, that $\lambda(X,\beta)$ is a known function of the m dimensional row vector of covariates $X_i = (x_{i1}, ..., x_{im})$, and the p unknown parameters $\beta = (\beta_1, ..., \beta_p)'$. The *regression function* $\lambda(X,\beta)$ will in general be nonlinear (with respect to the parameters) and we assume that it is a differentiable function of β. The regression function relates the expected value of the dependent variable y to the covariates and the parameters, and given the data $\{y_i,c_i,X_i \, , \, i = 1, ..., n \}$ we want to estimate the unknown parameters.

The most widely used methods of estimation have been developed using either the maximum likelihood (ML) or the least squares (LS) principle. The assumptions underlying these principles of estimation and the properties of the estimators are well known (see e. g. Kendall and Stuart, 1946 or Rao, 1965). Following the least squares principle we define the weighted sum of squares

$$S(\beta) = \sum_{i=1}^{N} w_i [\bar{y}_i - \lambda(X_i,\beta)]^2, \qquad (1.2)$$

where $\bar{y}_i = y_i/c_i$, and w_i is a positive weight that is inversely proportional to the variance of $\bar{y}_i$. The least squares estimates are obtained by solving the p dimensional system of equations

$$P' W [\bar{y} - \lambda(\beta)] = 0, \qquad (1.3)$$

where $W = diag[w_i]$, P is an N by p matrix of partial derivatives with elements $p_{ij} = \partial \lambda(X_i,\beta)/\partial \beta_j$, $\bar{y} = (\bar{y}_1, ..., \bar{y}_N)'$, and $\lambda(\beta) = (\lambda(X_1,\beta), ..., \lambda(X_N,\beta))'$. A solution of the generalized least squares equations (1.3) can be obtained using an iterative reweighted least squares (IRLS) procedure (see the Appendix).

It is well known that when the y_i follow the Normal distribution, the IRLS procedure will yield ML estimates for the general regression mode (1.1). Moore and Seigler (1967) showed the equivalence of ML and LS for the binomial distribution, and the equivalence of the ML and LS estimation procedure for Poisson distributed data was demonstrated by Frome and Beauchamp (1968), Frome (1972), and Frome, Kutner, and Beauchamp (1973). The equivalence of IRLS and ML for certain generalized linear functions for y in the exponential family (this includes Normal, gamma, Poisson, and binomial) was established by Nelder and Wedderburn (1972). Charnes, Frome, and Yu (1976) extended this result to show that ML and IRLS are equivalent for general nonlinear regression functions when the dependent variable is in the exponential family.

Regression methods based on the classical linear model assume that the y_i are continuous variables, that the regression function is linear in the β_j -- i.e. $\lambda(X_i,\beta) = X_i\beta = \sum_j \beta_j x_{ij}$, and that the y_i have equal variances and are uncorrelated. The assumption of Normality of the distribution of y leads to an exact small-sample theory. The theory of least squares in large samples can, however, be developed using only the first- and second-moment assumptions. The classical linear model is most appropriate when the dependent variable is a continuous quantity that can take on values on the entire real line. In practice it is often used to model data that are continuous and nonnegative (e.g. weights, concentrations, etc.), provided the mean values are far from zero. In some situations the use of log y instead of y may justify the assumption of constant variance or Normality.

1.3 *Poisson and Binomial Regression Models*

There are many situations in which the dependent variable is a count and the investigator is interesting in evaluating the effect of one or more covariates on the response. The two most widely used probability models for discrete data are the Poisson and the binomial distribution, and we shall limit our discussion to these two distributions.

The Poission distribution has been widely used as a model for certain types of discrete data. Haight (1967) states that the Poisson is second in importance to the Normal distribution, both in terms of abstract theory and breadth of application. The Poisson distribution has only recently been considered in the context of regression analysis (see Frome, 1972, Frome, Kutner and Beauchamp, 1973, Kock, Athinson, and Stokes, 1984). The dependent variable y is typically a count that is made with respect to some reference quantity c that is a measure of the size of the sample. Examples are bacteria per unit volume of suspension, number of accidents per unit time, and number of abnormal chromosome per lymphocyte. Cochran (1940) was the first to propose the use of Poisson regression in the context of the analysis of variance for designed experiments. Cochran used a linear regression function $\lambda(X_i,\beta) = \sum_j \beta_j x_{ij}$ to study the effect of several factors on the yield of wireworms (per acre). He found that the exact solution to the likelihood equations was too complicated for frequent use, and proposed $\lambda(X,\beta) = (X\beta)^2$ as a more practical alternative. This "square-root-linear" model results in some simplification (an iterative procedure is still required) of the likelihood equations. Cochran [1940] also proposed the product (log-linear) model i.e. $\lambda(X,\beta) = \exp(X_i\beta)$, for Poisson distributed data.

Another important area where Poisson regression models are used is in the analysis of data on rates and survivorship in medical and epidemiologic studies. For the ith subgroup the dependent variable y_i is a count (e.g. lung cancer deaths), c_i is the total follow-up time, and X_i is a vector of covariates that describes the ith subgroup -- see Holford (1980), Frome (1983), and Breslow et al (1983).

For a general Poisson regression model y_i denotes the observed count for the ith set of covariate vector X_i, c_i is the "size" of the ith unit and $\bar{y}_i = y_i/c_i$ denotes the observed rate (i.e. bacteria per liter, failures per hour, etc.). The expected number of event is $\mu_i = c_i \lambda(X_i,\beta)$, and the regression function $\lambda(X,\beta)$ can be interpreted in a general sense as the expected rate. Under the Poisson assumption the $Var(y_i) = c_i\lambda_i$, and consequently the "Poisson weights" in (1.1) are defined by $w_i = c_i/\lambda_i$. Examples that illustrate the use of linear, log-linear, and intrinsically nonlinear models are given in Section 2.

The binomial distribution--which is one of the oldest to have been studied (see Johnson and Kotz, (1970) Chap. 3, -- has two parameters, say λ (the probability of success on a given trial) and c (the number of trials). The dependent variable y is the number of positive responses that are observed in the c trials. In a regression context we assume that c_i individuals are observed for the ith set of experimental conditions X_i and y_i show a positive response with expected value $c_i\lambda(X_i,\beta)$, where $\lambda(X_i,\beta)$ is the probability of response for each individual. For example, in a bioassay c_i would be the number of animals and y_i would be the number that respond to the ith set of experimental conditions that are defined by the covariate vector X_i. In this situation the regression function $\lambda(X_i,\beta)$ represents the probability of observing the response for each animal in the ith group, and $\bar{y}_i = y_i/c_i$ denotes the proportion that respond. The response must lie in the range $0 \leq y_i \leq c_i$, and the regression function must satisfy $0 \leq \lambda(X_i,\beta) \leq 1$. Under the binomial assumption the variance (see eq. 1.1) is given by $Var(y_i) = c_i \lambda_i (1-\lambda_i)$, and consequently the "binomial weights" in (1.2) are defined to be $w_i = c_i/[\lambda_i(1-\lambda_i)]$. The best known binomial regression function is encounted in probit analysis (Bliss, 1935). The probit regression function is often defined by $\lambda(X_i,\beta) = \Phi(\alpha+\beta d_i)$ where Φ is the Normal cumulative distribution function, $X_i = (1,d_i)$, and $\beta = (\alpha,\beta)'$. Typically, d_i is the dose (or logarithm of the dose) of a stimulant or toxin, y_i is the number of positive responses in the ith exposure group, and c_i is the number of trials. The role of the logistic function in the analysis of "binary" or "quantal" response data has been considered by Cox (1970). Note that if $c_i = 1$ for all values of i then the possible values of y_i are 0 and 1. In this case

$$E(Y_i) = prob\ (Y_i=1) = \lambda(X_i,\beta) = \exp(X_i\beta)/(1+\exp(X_i\beta))$$

The more general case with all c_i greater than one is sometimes referred to as "grouped" binary data.

Further results and an extensive bibliography concerning the use of GLFs in the analysis of discrete data are given by McCullagh and Nelder (1983). This includes topics such as log-linear models for contingency tables (see also Bishop et al, and Huberman, 1974) and the connection between log-linear and multinomial response models (see also Palmgren, 1981).

2. POISSON REGRESSION

In this section three examples will be presented to illustrate various aspects of Poisson regression analysis. Example 1 will be used to illustrate simple linear regression analysis, multiple linear regression, and the analysis of variance for Poisson distributed data. In Example 1 there are parallel counts (i.e. replication) for each set of experimental conditions, and consequently we can test for "lack of fit" of the regression function and for heterogenity of variance. Example 2 will be used to further illustrate linear regression, and to introduce nonlinear regression and regression diagnostics for Poisson data. In Example 3 we will consider a log-linear regression function and a nonlinear regression function for data obtained from an observational study.

2.1 Simple Linear Regression with Parallel Counts

Example 1: Ir-192 Dose-Response Curve. In cytogenetic dosimetry *in vitro* dose response curves are used to describe the relation between the yield of dicentric chromosome aberrations and radiation dose. Let y_{jk} denote the observed dicentric yield for the kth parallel count of human lymphocytes exposed to d_j grays -- see DuFrain *et al* (1980), Frome and DuFrain (1983). The data in Table 2a provide a numeric example of an *in vitro* dose response curve for Ir 192 and are shown graphically in Figure 1a. As a matter of convenience we define c_{jk} as the total cells scored in units of 100 cells so that the regression function $\lambda(d,\beta)$ represents the dicentric yield per 100 cells.

The dicentric yields follow the Poisson distribution and we first assume that $\lambda(d,\beta) = \beta d$, i.e. the dicentric yield can be represented by a straight line that passes through the origin. In this case the ML estimate is

$$\beta^* = \frac{\sum_j \sum_k y_{jk}}{\sum_j \sum_k c_{jk} d_j} = \frac{\sum_j c_{j.} \bar{y}_j}{\sum_j c_{j.} d_j}$$

where $c_{j.} = \sum_k c_{jk}$. This is identical to the least squares estimate with $w_j = c_{j.}/d_j$. The ML estimate is $\beta^* = 21.4$, and the deviance is 79.26 with 19 df (see Table 2c). Standardized residuals for this model

$$u_{jk} = (y_{jk} - c_{jk} \beta^* d_j)/(c_{jk} \beta^* d_j)^{1/2}$$

are listed in Table 2b and are shown graphically in Figure 1b. The residuals suggest that a dose-squared term should be included in the model, i.e. $\lambda(d) = \beta_1 d + \beta_2 d^2$ (see Section 2.2.1)

2.2 Generalized Linear Regression

There are two general forms of the regression function that are of practical interest for Poisson distributed data. They are the "multiple linear"

$$\lambda(X_i\beta) = \sum_j \beta_j x_{ij},$$

and the "log-linear"

$$\lambda(X_i\beta) = \exp(\sum_j \beta_j x_{ij}),$$

regression functions. A third regression function of historical interest is

$$\lambda(X_i\beta) = (\sum_j \beta_j x_{ij})^2,$$

and this might be called "square-root-linear". All of these GLFs can be viewed as a special case of

$$G(X_i\beta) = (\sum_j \beta_j x_{ij})^q, \quad q \neq 0$$

$$G(X_i\beta) = \exp(\sum_j \beta_j x_{ij}), \quad q = 0 \qquad (2.1)$$

Nelder and Wedderburn (1972) use the term "link function" to describe the inverse of the regression function for GLFs, i.e. they call the exponential regression function $G(X\beta) = \exp(X\beta)$ a log-linear model because $\log G(X\beta)$ is linear in the parameters. In Section 2.2.1 we will consider several multiple linear regression functions for the Ir-192 data from Example 1. Then, in Example 2, we will consider an experiment in which two independent variables, radiation dose and exposure rate, for Cs-137 gamma rays are of interest. Linear regression models will be used to investigate the effect of dose and dose rate on the frequency of chromosome abnormalities. In Section 2.3 we will consider a theoretical model derived from the

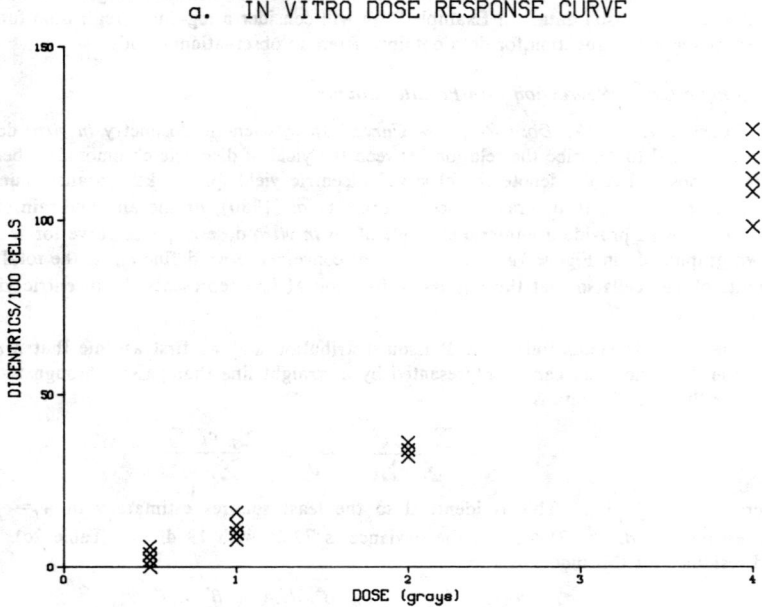
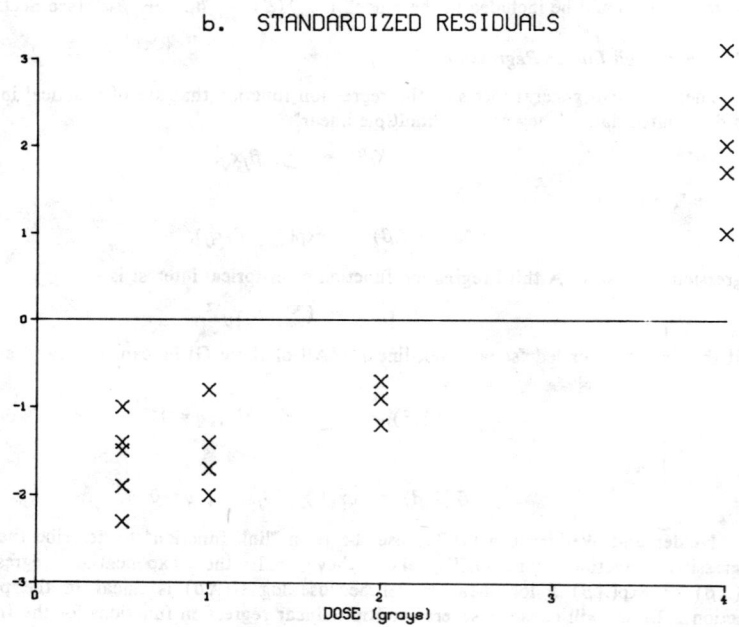

Figure 1. IRIDIUM GAMMA RAY DOSE-RESPONSE DATA.
(See Section 2.1 and Table 2a)

theory of dual radiation action (DRA) that is nonlinear in the parameters. The Cs-137 data from Example 2 is used to illustrate Poisson regression analysis for the intrinsically nonlinear DRA model.

In Section 2.2.2 an example of Poisson regression analysis for data from an observational study will be considered. The dependent variable is the number of lung cancer deaths and the covariates are age and exposure rate for individuals who were regular cigarette smokers. A log-linear model is used to obtain age adjusted estimates of "smoking effects". In Section 2.3 we will consider a nonlinear model (derived from the multi-stage theory of carcinogenesis) to further illustrate Poisson regression analysis for an intrinsically nonlinear regression function.

2.2.1 Multiple-Linear Regression

Example 1 - continued. In Section 2.1 we found that a simple one parameter model was inadequate, and suggested that a dose-squared term should be added to the regression function, i.e.

$$\lambda(X_i, \beta) = X_i \beta = (x_{i1}, x_{i2}) \begin{bmatrix} \beta_1 \\ \beta_2 \end{bmatrix}$$

where $x_{i1} = d$ and $x_{i2} = d^2$. The deviance for this model is 12.76 with 18 degrees of freedom (df) -- see line 6 in Table 2c. Table 2c contains the value of the deviance for all possible multiple linear regression models of the form

$$\lambda(d) = \beta_1 + \beta_2 d + \beta_3 d^2.$$

The next to the last line in Table 2c labelled "Dose Groups" corresponds to the "pure error sum of squares" in an ANOVA table for the standard linear models. The rows in the model matrix are based on indicator variables, i.e. for $j = 1, ..., 4$

$$x_{ij} = \begin{cases} 1 & \text{if } y_i \text{ is in dose group } j, \\ 0 & \text{otherwise.} \end{cases}$$

The ML estimates are of course

$$\beta_j^* = \frac{\sum y_i}{\sum c_j} = \bar{y}_j,$$

where the summation is over units in the jth dose group (i.e. $\sum_{k=1}^{n_j} y_{jk}$ if the double subscript notation of Section 2.1 is used). The deviance for this model is important since it provides a test of the assumption of Poisson variation. In this example the value of the deviance is 11.29 and is compared with the chi-square distribution with 16 df, indicating that the assumption of Poisson variation should not be rejected. Another measure of residual variation that can be used for this purpose is Fisher's index of dispersion

$$\sum_{j=1}^{N} \sum_{k=1}^{n_j} (y_{jk} - \bar{y}_j)^2 \bar{y}_j .$$

The value of this pooled within dose group index of dispersion is 9.51 with $\sum_j (n_j - 1) = 16 \, df$. Both of these lack of fit statistics can be viewed as a sum of squares of standardized residuals within experimental units.

If the Poisson assumption is not rejected then test statistics obtained from the Poisson ANOVA table can be compared with the appropriate chi-square distribution. For example, in Table 2c this would lead us to reject the first two models. The Poisson ANOVA in Table 2c has been given for illustrative purposes, and might be appropriate in situations of an exploratory nature where little was known about the dose-response relation. A more appropriate Poisson ANOVA table from a biologic point of view is given in Table 2d. This table indicates that the linear-quadratic model $\lambda(d) = \beta_1 d + \beta_2 d^2$ provides a better representation of these data than the linear model since the likelihood ratio statistic for $H_0 : \beta_2 = 0$ is 66.6 with 1 df. This of course agrees with the "lack-of-fit" test on line 1 of Table 2d. The "lack-of-fit" test on line 4 (which is not significant) implies that the linear-quadric regression function not only provides a significant improvement over the linear function, but that it cannot be rejected as a reasonable model for these data (see Table A.2 in the Appendix). Note that this is a true test for lack of fit of the regression function in the sense discussed by Draper and Smith (1966, Chap. 2).

The ML estimates of the parameters and their estimated standard deviations (in parenthesis) are $\beta_1^* = 3.59$ (1.68) and $\beta_2^* = 6.22$ (0.67). In summary, the IR-192 dose response data are best represented by the L-Q model, and the Poisson distribution provides an acceptable model for the variation in the dicentric counts.

Table 2   Poisson Regression Example

a) In-Vitro Dose-Response Data for Human
   Lymphocytes Exposed to 192 Ir Gamma Radiation

Dose (Grays)	Dicentric Chromosomes Per 50 Cells Scores						n_j	Total Dicentrics	Cells Scored
.5	0	1	0	2	1	3	2	7	9	350
1.0	5	6	5	4	8			5	28	250
2.0	16	17	18					3	51	150
4.0	49	59	54	56	63|		5	281	250

Notes: 1) See DeFrain et al (1980) for further discussions.

2) $E(y_{jk}) = c_i \lambda(X_i, \beta) = c_{jk} \beta d_{jk}$

3) $ML\ Estimate \beta^* = \sum_j \sum_k y_{jk} / \sum_j \sum_k c_{jk} d_{jk}$

b) Residuals for Simple Linear Regression for Dicentric Data in Table 2.1a

Dose d	Standardized Residual-- $u_{jk} = (y_{jk} - y_{jk}^*)/(y_{jk}^*)^{1/2}$						
0.5	-2.3	-1.9	-2.3	-1.4	-1.9	-1.0	-1.5
1.0	-1.7	-1.4	-1.7	-2.0	-0.8		
2.0	-1.2	-0.9	-0.7				
4.0	1.0	2.5	1.7	2.0	3.1		

c) Poisson ANOVA For Data in Table 2.1(a) All Possible Models

Model	Number of Parameters	Unexplained Variation+	df
const	1	514.2	19
d	1	79.26	19
d^2	1	18.05	19
const+d	2	23.13	18
const+d^2	2	14.58	18
d + d^2	2	12.67	18
const+d+d^2	3	11.34	17
Dose Groups	4	11.29	16
Complete	20	0.0	0

d) Poisson ANOVA For Models of Cytogenetic Interest

Model	Number of Parameters	Unexplained Variation+	df	Likelihood Ratio Statistic For Lack-of-Fit	df
d	1	79.26	19		
				66.59	1
*d+d^2	2	12.67	18		
				1.38	2
Within Dose Groups	4	11.29	16		
Complete	20				

*$\lambda(d) = 3.6d + 6.2d^2$

+ The deviance $D(\beta^*)$ is used as a measure of unexplained (i.e. residual) variation (see Appendix)

Example 2: Caesium Dose-Response Curves. The data in Table 3 (Purrott and Reeder, 1976) were obtained from an experiment (using gamma radiation from a caesium-137 source) that was designed to investigate the effect of dose rate on dicentric yield. The observed rate ($\bar{y} = y/c$) of dicentric induction (per 100 cells scored) is shown graphically in Figure 2 for each of the nine exposure rate groups. According to theoretical predictions from microdosimetry, a quadratic dose-response relation is predicted for low LET radiation, i.e. dicentric frequency is equal to $\alpha d + \beta d^2$ where d is radiation dose. From a biological point of view the two coefficients are thought of as corresponding to two different physical events. The linear term describes the induction of dicentrics by a single ionization or track, and the dose squared term which describes the induction of dicentrics by two different ionizations or tracks. Thus, the two break asymmetic exchange (dicentric) frequency is believed to be the result of these two phenomena, and is described by a second degree polynomial in dose. The validity of the quadratic model is predicated on the assumption that the absorbed dose is delivered to a "critical site" in a short period of time, i.e. at a high dose rate.

The purpose of the study by Purrott and Reeder was to test the hypothesis that the effect of decreasing the dose-rate would be to decrease the contribution of the dose-squared term, without

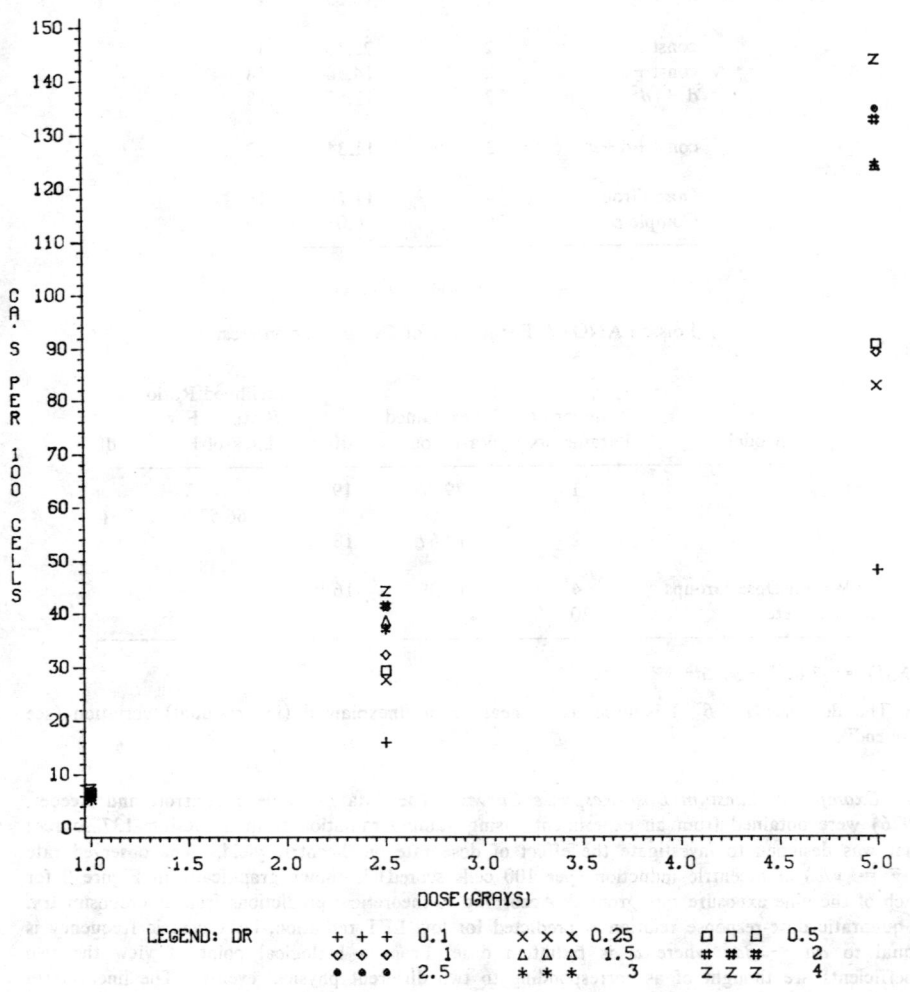

Figure 2. CYTOGENETIC DOSE RESPONSE DATA.
(See Section 2.2 and Table 3)

changing the linear term. Model 4 (see Table 4) corresponds to the most general case in which both the linear and quadratic coefficients are allowed to vary with dose rate, i.e. $\lambda_{jk} = \alpha_j d_k + \beta_j d_k^2$, where j identifies the dose rate group. For each of the models in Table 4 the regression function $\lambda(X,\beta)$ is linear in the parameters, and the procedure described in the Appendix was used to obtain the Poisson ANOVA. A test statistic for the hypothesis $\beta_1 = \beta_2 = \cdots = \beta_g$ is obtained using the difference of the deviance $D[y,\mu^*(2)] - D[y,\mu^*(3)] = 206.48$. This test statistic has an asymptotic chi-squared distribution with 8 df, if the more restrictive hypothesis is true. Consequently, we reject the hypothesis that the coefficient of the quadratic term is independent of dose rate. An alternative approach is to test for "lack of fit" of model 3. The deviance for this model is 21.52 with 17 df indicating that model 3 cannot be rejected.

It the ML estimates of the quadratic coefficients obtained from model 3 are plotted against the log of the dose rate it appears that the β_j^* increase linearly with log dose rate, and this can be described by the following regression model

$$\lambda_{jk} = \alpha d_k + \left[(\theta_1 + \theta_2 \log_{10}(r_j)\right]d_k^2. \tag{2.2}$$

The ith row of the model matrix for this $ad\ hoc$ model is $X_i = (d_i, d_i^2, d_i^2 \log_{10} r_j)$. The ML estimates and estimated standard errors for this model are given in Table 5. The value of the deviance for the model is 29.95 with 24 df, indicating that this $ad\ hoc$ model cannot be rejected for these data. This model provides a good description of effect of dose rate on dicentric yield, i.e., the quadratic component increase with the log of dose rate, and the linear component is independent of dose rate.

The results of this initial analysis (using linear regression functions to describe the effect of dose and dose-rate on dicentric yield) represents a straightforward extension of results from the standard linear model to Poisson distributed data. Although this analysis is technically correct we decided to reject this approach as being both inappropriate and misleading on biologic grounds (see Frome and DuFrain, 1983). A more appropriate analysis that utilizes a nonlinear model that was derived from the theory of dual radiation action (Kellerer and Rossi, 1972) is given in Section 2.3 of this paper.

Table 3. Cytogenetic Dose-Response Curve
Data for Continuous Exposure Experiment

Dose Rate G/hr	Dose (Grays)					
	1.0		2.5		5.0	
	c	y	c	y	c	y
.1	4.78	25	3.28	52	2.10	100
.25	19.07	102	1.85	51	1.38	113
.5	22.58	149	3.42	100	1.60	144
1.0	23.29	160	3.10	100	1.20	106
1.5	12.38	75	2.78	107	.90	111
2.0	14.91	100	2.59	107	1.00	132
2.5	15.18	99	2.49	102	3.13	419
3.0	7.64	50	2.98	110	1.82	225
4.0	13.67	100	2.43	107	1.44	206

NOTE: y = number of dicentrics, c = cells scored (100s)

Source: Purrott and Reeder (1976)

Table 4. Poisson ANOVA for Cytogenetic Data in Table 3

Regression Model		Number of Parameters	Deviance*	df
1	αd_i	1	1075.30	26
2	$\alpha d_i + \beta d_i^2$	2	228.00	25
3	$\alpha d_i + \beta_j d_i^2$	10	21.52	17
4	$\alpha_j d_i + \beta_j d_i^2$	18	11.10	9
5	Complete	27	0.0	0

* See Footnote to Table 1d

Table 5. Maximum Likelihood Estimates for Ad Hoc Model for Dose-Response Curve Data in Table 3

Parameter	Estimate	Standard Deviation
α	2.86	.305
θ_1	3.80	.141
θ_2	2.26	.144

2.2.2 Log-Linear Regression

Incidence or mortality data obtained from epidemiologic follow-up studies are often expressed as covariate stratum-specific rates, where the covariate may be age or some other presumed confounding factor. Poisson regression provides a general approach to the study of the effect of one or more covariates on disease rates -- see Frome (1983), Frome and Checkoway (1984). the attractive features of the Poisson regression approach are that summary estimates of relative risk can be obtained, an evaluation of the presence and nature of interaction is part of the analysis, and the modeling of disease rates is facilitated. Poisson regression methods are especially appropriate in follow-up studies where time-based denominators (person-years) are used to obtain disease rates in a life table type of format (Frome, 1983), or when the outcome of interest is rare so that the Poisson approximation to the binomial distribution can be used (Gart, 1978).

Example 3: Lung Cancer Mortality The data in Table 6 were obtained by Kahn (1966) in a study of lung cancer mortality in relation to cigarette smoking. The dependent variable y_{jk} is the number of lung cancer deaths for the jth level of the potential confounding variable (age) and the kth level of the "risk factor", cigarette consumption. The c_{jk} are the person-years (pys) at risk (in units of 10^5 pys) and consequently the $\bar{y}_{jk}$ are lung cancer death rates per 10^5 pys. The y_{jk} are assumed to follow the Poisson distribution with expectation $\mu_{jk} = c_{jk} \lambda_{jk}$, where λ_{jk} denotes the underlying regression function. If the covariate stratum-specific RRs are constant within each risk group, then $\lambda_{jk} = \lambda_j \phi_k$ where:

λ_j denotes the rate for the *jth* stratum level, and
ϕ_k is the summary risk ratio for risk group k ($k>1$) and $\phi_1=1$.
This is referred to as the *product model*, and for estimation purposes it can be expresed as a GLF (where $j=1,...,J$ and $k=1,...,K$)

$$\lambda_{jk} = \exp(\alpha_j + \delta_k) = \exp(X_i\beta). \tag{2.3}$$

where $\alpha_j = \log\lambda_j$ ($j=1,...,J$) and $\delta_k = \log(\phi_k)$ ($k=2,...,K$).

We have assumed that risk group 1 is the reference, or non-exposed group. Consequently, the α_j correspond to the natural logarithms of the stratum specific rates in the reference group, while the δ_k are the logarithms of the summary RR for group k (with group 1 as the reference group). In (2.3) X_i is a $p = J+K-1$ dimensional row vector of indicator variables for the *ith* cell in the table, and $\beta = (\alpha_1,...,\alpha_J,\delta_2,...,\delta_K)'$ is the p-dimensional column vector of unknown parameters. If the *ith* cell of the table corresponds to row j and column k, then the components of X_i, ($i=1,...,JK$) can be defined as follows:

$x_{im} = 1$ if $m=j$; $x_{im} = 1$ if $k>1$ amd $m = J+k-1$;

for $m=1,...,J+K-1$, and $x_{im} = 0$ *otherwise*.

When $c_{jk} > 0$ for all j and k, this situation is equivalent to a full rank parameterization of the design matrix for a two factor fixed effects ANOVA model. In practice it is not necessary to generate this matrix because its structure is implied by the levels of the factors. The ML estimates of the parameters for the product model (2.3) for the data in Table 6 are given in Table 7. The deviance is 12.5 with 10 df indicating that the product model provides a good description of these data. The ML estimates of the λ_js and ϕ_ks are given in the last row and column of Table 6b, respectively.

If the risk factor (smoking) is not important, i.e. $\delta_k = 0$(for $k=2,...,K$), then $\lambda_{jk} = \lambda_j$ and the ML estimates are $\lambda_j^* = \sum_k y_{jk}/\sum_k c_{jk}$. The deviance for this model (see line 3 Table 8) is 1037 with 25 df, and the likelihood ratio statistic (obtained by subtracting line 4 from line 3 in Table 8) for the hypothesis $\delta_2 = ... = \delta_K = 0$ is 1024.5 with 5 df indicating that the risk ratios are highly significant. Frome and Checkoway (1984) have shown that when the product model provides a good fit (as it does in this example) the ϕ_k^* can be interpreted as estimates of standardized risk ratios (SRR). The SRR for risk group k (with $k=1$ for the reference group) is defined by Miettien (1974) as follows:

$$SRR_k = (\sum_j w_j \bar{y}_{jk})/(\sum_j w_j \lambda_j),$$

where the w_j are standard population weights. If the product model provides a good fit (as indicating by the deviance), the $\bar{y}_{jk}$ in the above definition can be replaced by their ML estimates $\lambda_{jk}^* = \lambda_j^* \phi_k^*$ to obtain

$$SRR_k^* = \sum_j w_j \lambda_j^* \phi_k^*/\sum_j w_j \lambda_j^* = \phi_k^*.$$

The ϕ_k^* are estimates of the SRRs with the non-exposed group as the referent group, and the choice of the standard population weights is unimportant.

Table 6. Lung Cancer Mortality According to Cigarette Consumption and Age

a) Number of Deaths and Person-Years (pys)

Current Cigarette Smokers (cigarettes/day)

Group		Smokers	Age Occasional	1-9	10-20	21-39	40+
35-44	deaths	0	0	0	2	4	0
	pys	35164	3657	8063	59965	40643	3992
45-54	deaths	0	0	0	2	10	2
	pys	15134	1283	3129	16392	12839	1928
55-64	deaths	25	6	31	183	245	63
	pys	213858	14624	45217	151664	103020	19649
65-74	deaths	49	10	44	239	194	50
	pys	171211	10053	37130	101731	50045	8937
85-	deaths	4	1	5	15	7	3
	pys	8489	512	1923	3867	1273	232

Source: Kahn (1966)

b) Lung Cancer Deaths Rates (per 10^5 pys)

Age Group (midpoint)	0	.5	Cigarettes/day 5	15	30	45	Age Fit
40	0	0	0	3	10	0	0.4
50	0	0	0	12	78	104	3.2
60	12	41	69	121	238	321	14.0
70	29	99	118	235	388	559	25.5
80	47	195	260	389	550	1293	43.9
Smoking Effect	1.0	3.5	4.8	8.9	16.2	22.6	

*Age fit = $\exp(\alpha_j^*)$ and smoking effect = $\exp(\delta_k^*)$, where the α_j^* and δ_k^* are the ML estimates given in Table 7. The estimated lung cancer death rates per 100,000 man-years in Row j and Column k are

$$\lambda_{jk}^* = Age\ Fit\ *\ Smoking\ Effect = \exp(\alpha_j^* + \delta_k^*)$$

Table 7. ML Estimates of The Parameters for The Product
Model (2.3) for the Lung Cancer Data in Table 6

j	α_j^*	St. Dev.	k	δ_k^*	St. Dev.
1	-0.82	.42	2	1.24	.27
2	1.18	.29	3	1.56	.16
3	2.64	.12	4	2.18	.12
4	3.24	.12	5	2.79	.12
5	3.78	.20	6	3.12	.15

Table 8. Poisson ANOVA Table for Lung Cancer Mortality Data in Table 6

Model	Log (λ)	No. of Parameters	Deviance*	d.f.
1. Minimal	α	1	1438.0	29
2. Smoking effect	$\alpha + \delta_k$	6	589.7	24
3. Age effect	α_j	5	1037.0	25
4. Age and smoking	$\alpha_j + \delta_k$	10	12.5	20
5. Complete		30	0	0

* See footnote to Table 1d

2.3 Nonlinear Regression

Example2: Caesium Dose-Response Curve (continued). The *ad hoc* mode (2.2) described in the previous section can be used as an empirical description of the cytogentic dose response relation for the experimental data in Table 3. The parameters in the ad hoc model do not have a clear interpretation in terms of the quantitative effects of ionizing radiation (see Frome and DuFrain, 1983). The dual radiation action (DRA) theory described by Kellerer and Rossi (1972) utilizes concepts from microdosimetry to provide a quantitative characterization of the effect of various temporal distributions of absorbed dose on the production of chromosome aberrations (CAs). It is postulated that elementary lesions are produced at a rate that is proportional to the square of the local energy concentration produced by charged particles in certain "critical sites". The form of the dose-effect model that is appropriate here (see Kellerer and Rossi, 1972, Section 5.4) is

$$\lambda(d,t) = \kappa[\gamma d + g(t,\tau)d^2], \quad (2.4)$$

where d denotes dose, t is time, and $\lambda(d,t)$ is the yield of elementary lesions. The parameter κ is a biophysical proportionality constant that reflects the target sensitivity for the biologic system (lymphocyte). The parameter γ depends on the radiation quality and can be related to the specific energy produced in a critical site by a single ionization. The linear term in (2.4) represents the effect due to intratrack interactions and the quadratic term represents the effect of intertrack interaction. The coefficient of the d^2 term is referred to as the "reduction factor", and assuming an exponential recovery process for continuous irradiation of duration t one obtains (see Lea, 1955)

$$g(t,\tau) = \frac{2\tau}{t} - \frac{2\tau^2}{t^2}(1-e^{-t/\tau}). \tag{2.5}$$

Using (2.5) in (2.4) leads to

$$\lambda(X_i,\beta) = \kappa\left\{\gamma d_i + \frac{2\tau}{t_i}\left[1-\tau\{1-\exp(-t_i/\tau)\}/t_i\right]d_i^2\right\}, \tag{2.6}$$

where d is the absorbed dose and t is the duration of exposure at a constant dose rate.

The ML estimates of the parameters in (2.6) for the data in Table 3 were obtained using the IRLS procedure described in the Appendix. Since the DRA model is nonlinear in the parameters, the partial derivatives of (2.6) with respect to the parameters must be supplied. The ML estimates and their standard deviations are given in Table 9. The deviance for this model is 28.58 with 24 df (p = .236) indicating that the DRA model cannot be rejected. The standardized residuals in Table 10a are used to identify outlying observations, and in this example there is one large negative residual. The diagonal terms from the H matrix (see the Appendix) are given in Table 10b. There are several large h values (greater than $2p/n=0.22$) in column 3, and two of these are in the first two rows, i.e. the highest dose and the lowest exposure rates. The diagnostic quantities in Table 10 are shown graphically in Figure 3. Note that we have used scaled h values in this ($h^*=ph/n$) diagnostic plot so that two can be used as a cutpoint for large h^* values.

Table 9. ML Estimates for the DRA Model for the Cytogenetic Data in Table 3

Parameter	Estimate	Standard Deviation
κ	5.44	.208
γ	.269	.0677
τ	7.40	.857

Table 10. Regression Diagnostics for Data in Table 3 Using the Nonlinear DRA Model (2.6)

(a) Standardized Residuals $u_i = (y_i-\mu_i^*)/\mu_i^{*½}$

0.127	-0.929	1.35
-1.23	0.315	1.19
0.291	-0.627	-1.05
0.383	-0.563	-2.92
-0.927	0.914	-0.140
-0.111	1.48	0.247
-0.423	1.26	0.315
-0.293	0.144	-1.17
0.670	1.88	0.732

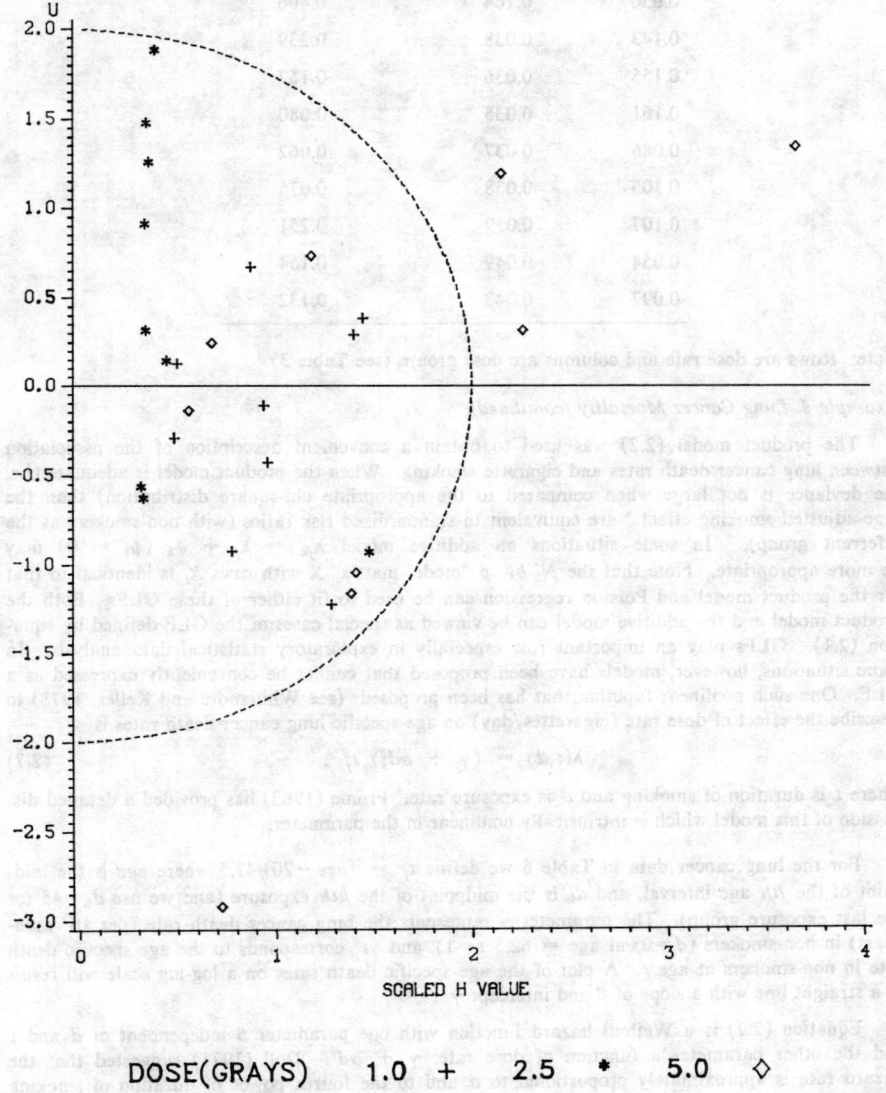

Figure 3. DIAGNOSTIC PLOT FOR DRA MODEL FOR CYTOGENETIC.
(See Section 2.3 and Table 10)

(b) Diagonal terms from the **H** matrix (p/n=0.111)

0.056	0.164	0.406
0.143	0.038	0.239
0.155	0.036	0.157
0.161	0.035	0.080
0.086	0.037	0.062
0.105	0.038	0.075
0.107	0.039	0.251
0.054	0.049	0.154
0.097	0.043	0.132

Note: Rows are dose rate and columns are dose groups (see Table 3)

Example 3: Lung Cancer Mortality (continued).

The product model (2.2) was used to obtain a convenient description of the association between lung cancer death rates and cigarette smoking. When the product model is adequate (i.e. the deviance is not large when compared to the appropriate chi-square distribution) than the "age-adjusted smoking effects" are equivalent to standardized risk ratios (with non-smokers as the referrent group). In some situations an additive model $\lambda_{jk} = \lambda_j + \phi_k$ ($\phi_1 = 0$) may be more appropriate. Note that the N by p "model matrix" **X** with rows X_i is identical to that for the product model and Poisson regression can be used to fit either of these GLFs. Both the product model and the additive model can be viewed as special cases of the GLF defined by equation (2.1). GLFs play an important role especially in exploratory statistical data analysis. In some situations, however, models have been proposed that cannot be conveniently expressed as a GLF. One such nonlinear function that has been proposed (see Whitemore and Keller, 1978) to describe the effect of dose rate (cigarettes/day) on age-specific lung cancer death rates is

$$\lambda(t,d) = (\gamma + \alpha d_k^\theta) \, t_j^\beta , \qquad (2.7)$$

where t is duration of smoking and d is exposure rate. Frome (1983) has provided a detailed discussion of this model which is intrinsically nonlinear in the parameters.

For the lung cancer data in Table 6 we define $t_j = (age-20)/42.5$ where age is the midpoint of the *j*th age interval, and d_k is the midpoint of the *k*th exposure (and we use $d_6=45$ for the last exposure group). The parameter γ represents the lung cancer death rate (per 10^5 man-years) in non-smokers ($d=0$) at age = 62.5 $t=1$), and γt^β corresponds to the age specific death rate in non-smokers at age t. A plot of the age specific death rates on a log-log scale will result in a straight line with a slope of β and intercept γ.

Equation (2.7) is a Weibull hazard function with one parameter β independent of d and t and the other parameter a function of dose rate: $\gamma + \alpha d^\theta$. Doll (1971) suggested that the hazard rate is approximately proportional to d and to the fourth power of duration of smoking (i.e. $\theta=1$ and $\beta = 4$). Note that if $\theta \neq 1$, then the exposure-effect relation will be concave ($\theta<1$) or convex $\theta>1$) toward the exposure axis. For estimation purposes we use

$$\lambda(X_i,\beta) = \left[\exp(\beta_2 + \beta_3 \, x_{i1}) + \exp(\beta_4)\right] \exp(\beta_1 x_{i1}), \qquad (2.8)$$

where $X_i = (logt_i, logd_i)$ and $\beta = (\beta, log\alpha, \theta, log\gamma)'$. The IRLS procedure (see Appendix) is used to obtain ML estimates for Kahn's data (in Table 6) and the results are summarized in Table 11. In another study (of cigarette smoking in British physicians) Doll and Hill obtained data similar to Kahn's data in Table 6 -- see Frome (1983, Table 1). The ML estimates for both of these data sets are given Table 11. The deviance for Doll and Hill's data in 59.6 with 59 df indicating a good fit, while the deviance for Kahn's data, 43.5 with 26 df suggests a considerably poor fit. Of particular concern with the data from Kahn is the estimate of θ which is less than one, indicating a concave dose-effect relation.

Table 11. Maximum Likelihood Estimates for Parameters Specified by a Nonlinear Model* for Lung Cancer and Cigarette Smoking Data

	Data source	
Parameter	Kahn	Doll and Hill
β	3.38(0.18)‡	4.46(0.33)
$\log \alpha$	2.62(0.21)	1.82(0.66)
θ	0.83(0.06)	1.29(0.20)
$\log \gamma$	2.61(0.13)	2.94(0.58)
Deviance	43.5	59.6
d.f.	26	59

* Death rate = $\lambda_{jk} = (\gamma + \alpha d_k^\theta) t_j^\beta$,
  where t = (age -20)/42.5, d = cigarettes per day

† Data from Kahn (14) and Doll and Hill (15,16)

‡ Standard deviation in parentheses

3. BINOMIAL REGRESSION

In this section two examples will be presented to illustrate various aspects of binomial regression analysis. Example 4 provides an example of a linear regression function with replication at each set of experimental conditions. In this example the ANOVA-like table for binomial data will be used to test for "lack-of-fit" of the regression function and the assumption binomial variation. Example 5 will be used to illustrate the use of nonlinear functions (probit, logistic, and Weibull), and regression diagnostics for binomial data.

3.1 Linear Regression with Parallel Counts

Example 4: Streptonigrin Dose-Response Curve The data in Table 12 were obtained by DuFrain et al (1982) as part of a study that was undertaken to investigate the potential toxicity of a chemical clastogen (streptonigrin) on somatic cells and germ cells from female rabbits. The dependent variable y_{jk} is the number of damaged cells (lymphoblasts) for the jth animal exposed to streptonigrin dose d_k. One hundred cells were examined for each animal and we let $c_{jk} = 1$ (i.e. unit = 100 cells) for each observational unit, so that $\bar{y}$ is in per cent. The data are shown graphically in Figure 4. We assume that the y_{jk}s are independent and follow the binomial distribution with expectation $\mu_{jk} = c_{jk} \lambda(X_i,\beta)$ where

$$\lambda(X_i,\beta) = \beta_1 + \beta_2 d_j \tag{3.1}$$

This is a special case of the GLF

$$\lambda(X_i,\beta) = X_i\beta = \sum_j \beta_j x_{ij}, \tag{3.2}$$

with $x_{i1} = 1$ and $x_{i2} = d_j$ if the animal is in dose group j. Equation (3.1) is a linear regression function with intercept β_1 and slope β_2, and is used to describe these data for

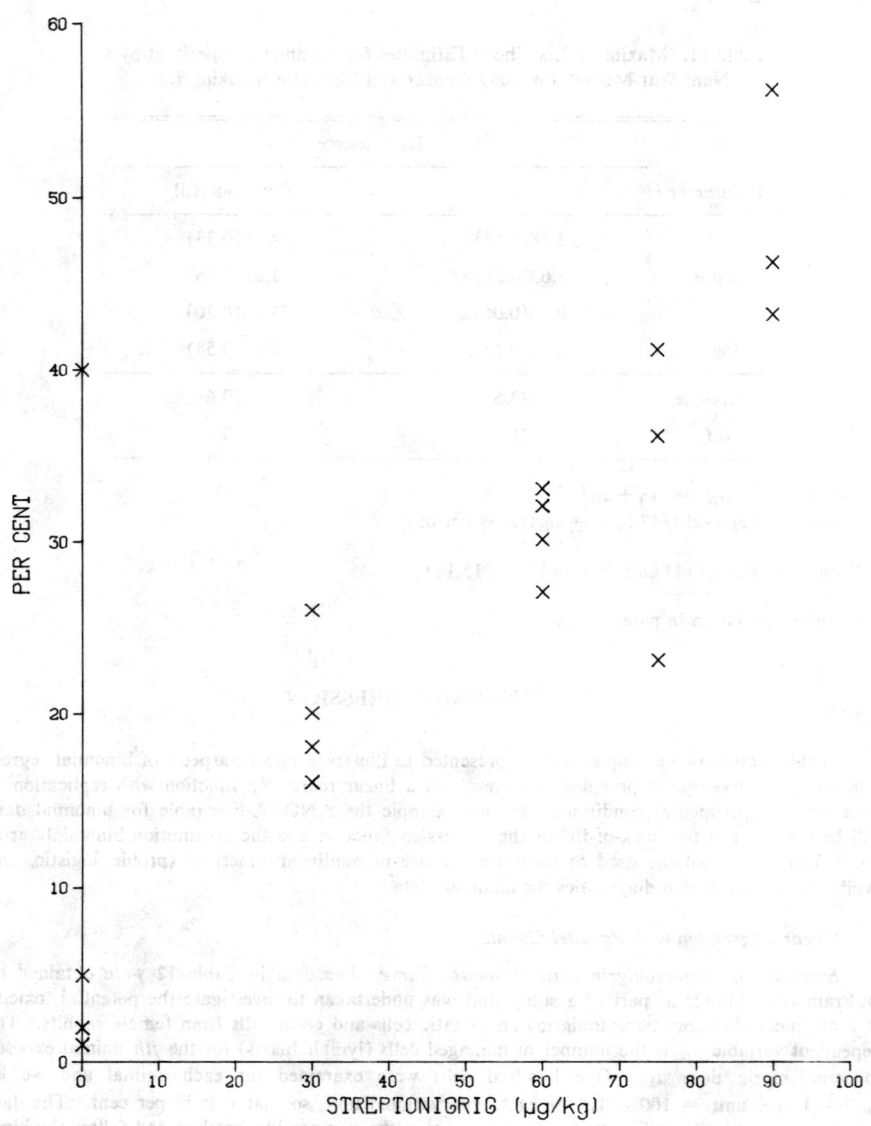

Figure 4. STREPTONIGRIN DOSE RESPONSE DATA.
(See Section 3.1 and Table 12)

$d < 100$ $\mu g/kg$. The ML estimates are obtained using IRLS (see the Appendix) and are $\beta_1^* = 2.67$ percent and $\beta_2^* = 0.478$ per cent per $\mu g/kg$, and their estimated standard deviations are 0.58 and 0.0189, respectively. The goodness-of-fit of this binomial regression model can be evaluated using the binomial ANOVA for these data (see Table 13a). Following the procedure described in the Appendix (see Table A.2) we first observe that deviance for the linear regression function (see Table 13a line 2) is 28.67 with 21 df ($p = .12$). This implies that neither the linear regression nor the assumption of binomial variation should be rejected. The assumption of binomial variation is further confirmed by the "within dose groups" deviance value of 22.64 with 18 df. The likelihood ratio statistic for "lack-of-fit" of the regression function is 6.03 with 3 df ($p = .11$) which again indicates that the linear regression function provides a reasonable model for these data. A test of the hypothesis $\beta_2 = 0$ can be based on the likelihood ratio statistic obtained as $D(\beta_0) - D(\beta_0 + \beta_1 d) = 336.6$ with 1 df ($p < 10^{-4}$).

Table 12. Streptonigrin Dose-Response Data*

Dose of Streptonigrin ($\mu g/kg$) d_j	Number of Rabbits n_j	Number of Lymphoblasts With Aberrations For Each Rabbit † y_{jk}					
0	6	5	2	1	1	2	4
30	5	20	16	26	18	16	
60	6	27	32	32	33	30	33
75	3	41	23	36			
90	3	43	46	56			

* Data obtained from R. J. DuFrain, Medical and Health Science Division, Oak Ridge Associated Universities (see DuFrain *et al*, 1982, Table 1)

† 100 Lymphoblasts were scored for each animal

For illustrative purpose consider the "linear logistic" regression function, i.e.
$$\lambda(d_i) = \exp(\beta_1 + \beta_2 d_i)/[1 + \exp(\beta_1 + \beta_2 d_i)]. \tag{3.3}$$
This can be viewed as a special case of the GLF
$$\lambda(X_i,\beta) = \exp(X_i\beta)/[1 + \exp(X_i\beta)], \tag{3.4}$$
where $X_i = (1,d_i)$. The binomial ANOVA using logistic GLF for the streptonigrin data is given in Table 13b. The values of the deviance on lines 1 and 3 are, of course, the same in both tables. The deviance for the linear logistic function (3.3) is 56.46 with 21 df ($p<10^{-4}$) indicating "lack-of-fit" of the *binomial regression model*. This could be due to heterogeneity of variance (i.e. overdispersion relative to the assumed binomial distribution within dose groups) or "lack-of-fit" of the regression function (3.3). As we noted earlier the assumption of binomial variation cannot be rejected. Consequently we can further assess the lack of fit of (3.3) by comparing the likelihood ratio statistic $D_3 - D_2 = 33.82$ with the chi square distribution with 3 df. Clearly this provides strong evidence against the linear logistic dose-response function for these data.

In summary, the streptonigrin dose-response data provides a simple example of binomial regression. A binomial regression model is based on two assumptions:

i) the y_{jk} are independent and follow the binomial distribution with expected value $\mu_{jk} = c_{jk}\lambda(X_j,\beta)$

ii) the regression function is specified.

If repeated observations are obtained at each set of experimental conditions then the appropriateness of the assumption of binomial variation can be tested. In this example we could have based our analysis on the totals at each dose -- see Table 14. The values of deviance in Table 14b differ from those in Table 13a by a constant amount, and in studies that involve a large number of experimental conditions this result can be used to simplify the analysis. In situations where heterogenity of variance is detected a detailed analysis (using standardized residuals within dose groups) should be considered. This could lead to the identification of spurious data values or the identification of additional factors that influence the dependent variable. When unexplained overdispersion occurs the analysis can be modified by introducing a heterogeneity factor (see the Appendix).

Table 13. Binomial ANOVA for Streptonigrin Data in Table 12

a) Linear Regression

	Regression Function †	df	Deviance	Likelihood Ratio Statistic	df
1.	β_1	22	365.3		
2.	$\beta_1 + \beta_2 d$	21	28.67		
				6.03	3
3.	Dose Groups	18	22.64		
				22.64	18
4.	Complete	0	0.0		

b) Linear Logistic Regression

	Regression Function †	df	Deviance	Likelihood Ratio Statistic	df
1.	β_1	22	365.3		
2.	$\beta_1 + \beta_2 d$	21	56.46		
				33.82	3
3.	Dose Groups	18	22.64		
				22.64	18
4.	Complete	0	0.0		

* The deviance $D(\beta^*)$ is used as a measure of residual (unexplained) variation (see Appendix eq. A.15).

† See explanation in text

Table 14. Streptonigrin Dose-Response Data

a) Totals from Table 12

Dose d_i	Number of Cells c_i	Number with Aberrations y_i	$\bar{y}_i$ (per cent)	$\lambda^*(d)$*
0	600	15	2.5	2.7
30	500	96	19.2	17.0
60	600	187	31.2	31.3
75	300	100	33.3	38.5
90	300	145	48.3	45.7

*$\lambda^*(d) = 2.67 + 0.478d$

b) Binomial ANOVA For Totals in Table 14a

Regression Function	df	Deviance	Likelihood Ratio Statistic	df
β_1	5	342.6		
			336.6	1
$\beta_1 + \beta_2 d$	4	6.03		
			6.03	3
complete	0	0.0		

3.2 Nonlinear Regression Functions

Example 5: Chronic Bioassay of 2-AAF The data in Table 15 are from a study at the National Center for Toxicological Research that was undertaken to investigate the effect of a chemical compound 2-Acetylaminofluorene (2-AAF) on carcinogenesis (see Farmer, et al, 1979). Mice were continuously fed 2-AAF *ad libitum* in the diet at various concentrations from weanlings until they were either sacrificed, became moribund or died. Let y_{jk} denote the number of mice with liver neoplasms, and c_{jk} the number of animals at risk in the kth dose (exposure rate) group and the jth time interval group. Then $\bar{y}_{jk}$ is the proportion of mice with liver neoplasm and the regression function $\lambda(X_i,\beta)$ will represent the probability of observing a liver neoplasm in the ith group (where $i = 8(j-1)+k$). For illustrative purposes we will begin our analysis using a linear-logistic regression function (3.4), and the factors DG ("dose" group with 8 levels) and TI (time interval with 9 levels) will be used to define the covariate vector X_i (see Example 3). This example will also illustrate how the computer program GLIM (Baker and Nelder, 1978) can be used to carry out the necessary data manipulations and computations. Figure 5 shows a listing of a GLIM-3 program that was used to (i) define the levels of the factors TI and DG; (ii) define the covariate vectors D,T, XD, and XT (that will be discussed later); and (iii) read the y_{jk} and c_{jk} into the vectors Y and N, respectively. In GLIM terminology the assumptions of binomial variation and a logistic regression function (3.4) are specified as follows:

$ERR B N $LINK G.

The values of deviance in the binomial-ANOVA table (see Table 16) were obtained using the FIT directive

$$\$FIT : DG : TI : TI + DG.$$

The diagonal terms from the H matrix (A.16) are obtained for the linear logistic function as follows:

$$\$EXTR \ \%VL \ \$CA \ H = \%VL * \%FV * (N - \%FV)/N.$$

The signed deviance residuals (see Appendix) are computed using the macro DVR (see Figure 5), and Figure 6 is a diagnostic plot of these standardized residuals and the scaled h_i for the linear logistic model TI + DG. The binomial ANOVA in Table 16 shows that most of the variation (96%) in these data can be explained by the linear logistic function (with TI and DG as factors), but the biologic interpretation of this model is not apparent.

Another approach that can be used to describe the data in Table 15 is to use (in GLIM terminology) a "probit link function". That is, we assume that $\lambda(X_i,\beta) = \Phi(X_i\beta)$, where Φ is the standard normal integral. The ANOVA table for the probit model is shown in table 17. Comparing the deviance values for the two factor models (TI + DG) suggests that the logit link provides a better representation that the probit link. We have not, however, used the values of the dose and time variables associated with the levels of the factors. Consider for example the regression function

$$\lambda_{jk} = \Phi[\delta_k + \beta\log(t_j)], \qquad (3.5)$$

where t_j = time on study (in years). This corresponds to a log-normal response time distribution for each dose group, with constant standard deviation (β^{-1}) and mean $\mu_k = -\delta_k/\beta$. The deviance for (3.5) is 133.1 with 63 df (see DG + XT in Table 17). Farmer et al (1979) used the probit model in their analysis of these data and went further by assuming that

$$\lambda_{jk} = \Phi[\alpha + \theta\log d_k + \beta\log t_j] . \qquad (3.6)$$

They limited their analysis using (3.6) to the last three time points (18, 24, and 33 months) and eliminated the control $(d_1=0)$ group. Farmer et al (1979) concluded from their limited analysis that (3.6) provided a "good fit" but indicated that the meaning of the slopes in (3.6) was not clear. Clearly, the model (3.6) cannot be considered as a reasonable representation of these data since it implies a zero response probability for unexposed animals and 45 of the mice in the zero dose group developed liver neoplasms. It would, of course, be possible to define other covariates for the probit (or the logit) function.

Another approach is to note that in both situations we have selected a cumulative distribution function (CDF) for the function G ( ), which insures that $0 \leq \lambda(X_i,\beta) \leq 1$. Another distribution function that has some appeal in this situation is the Weibull distribution, i.e.

$$\lambda_{jk} = 1 - \exp[-\delta_k t_j^\beta \ ]. \qquad (3.7)$$

This model implies that "time to tumor" follows a Weibull distribution with scale parameter δ_k and shape parameter β that is independent of dose. The tumor incidence rate for each dose group is $\beta\delta_k t^{\beta-1}$, i.e. the exposure to 2-AAP (at constant rate) has a multiplicative effect on the tumor incidence rate. This relation is predicted, for example, by the multistage theory of carcinogenesis (see Whittemore and Keller, 1978). For estimation purposes the Weibull regression function (3.7) can be written as

$$\lambda_{jk} = 1 - \exp\left[-e^{\alpha_k + \beta\log(t_j)}\right] \qquad (3.8)$$

or more generally as

$$\lambda(X_i,\beta) = 1 - \exp[-e^{X_i\beta} \ ]. \qquad (3.9)$$

Table 15. Liver Neoplasms in Dead, Moribund and Sacrificed Mice Fed 2-AFF Continuously

Months on Study	DOSE (ppm)							
	0	30	35	45	60	75	100	150
9	0[1]/199[2]	1/147	1/76	0/52	0/345	0/186	1/168	1/169
12	0/164	1/51	2/27	1/14	2/283	0/153	3/149	2/152
14	1/133	1/42	0/25	2/14	1/243	0/124	1/127	1/127
15	0/115	1/75	1/35	0/20	3/203	1/109	5/99	1/100
16	1/205	2/66	2/61	3/304	6/287	7/193	2/100	7/110
17	0/153	4/69	5/443	6/302	8/230	9/166	3/85	1/82
18	6/555	34/2014	20/1102	15/550	13/411	17/382	19/213	24/211
24	20/762	164/2109	128/1361	98/888	118/758	118/587	76/297	126/314
33	17/100	135/445	72/200	42/103	30/67	37/75	22/31	9/11

[1] Number of Bladder Neoplasms
[2] Number of mice examined in this group
Source: Farmer *et al* (1979)

Table 16. Binomial ANOVA for 2-AAF Data in Table 15 Using a Linear-Logistic Regression Function

Linear Terms	Unexplained Variation*	df
minimal	2203.6	71
TI	595.1	63
DG	1965.6	64
TI+DG	85.7	56

Table 17. Binomial ANOVA FOR 2-AAF Data in Table 15 Using a Linear-Probit Regression function

Linear Terms	Unexplained Variation	df
minimal	2203.6	71
TI	595.1	63
DG	1965.6	64
DG+XT	133.1	63
TI+DG	109.1	56

```
$SUBFILE DATA !  10 FEB 83
$M TITLE   PBEX5: J.ENV.P&T(1979,P.57) ED-01 STUDY$E
$M VLIST !   LIST OF VARIABLES
     Y= NUMBER OF LIVER NEOPLASMS  !
     N= NUMBER OF MICE!
     D= DOSE OF 2-AAF (PPM) CONTINUOUS FEEDING !
     T= MONTHS ON STUDY !
     XD= LOG( D ) !
     XT= LOG( T/12 ) -- LOG TIME IN YEARS!
     TI= TIME INTERVAL FACTOR ( ROWS OF TABLE )!
     DG= DOSE GROUP FACTOR ( COLUMNS OF TABLE )$E
$UNITS 72 $FAC TI 9 DG 8 !
$CA TI=%GL(9,8) : DG=%GL(8,1) : %R=9 !
$DATA 8 DOSE $READ 0.0 30 35 45 60 75 100 150
$CA D=DOSE(DG) : XD=%LOG(D) ! !
$DATA 9 TIME $READ 9 12 14 15 16 17 18 24 33
$CA T=TIME(TI) : XT=%LOG(T/12) !
$DATA Y $READ
 0  1 1 0 0 0 1 1
 0 1 2 1 2 0 3 2
 1  1 0 2 1 0 1 1
 0 1 1 0 3 1 5 1
 1 2 2 3 6 7 2 7
 0 4 5 6 8 9 3 1
 6  34 20 15 13 17 19 24
 20   164 128 98 118 118  76 126
 17 135 72 42 30 37 22 9
$DATA N $READ
 199   147 76 52 345 186 168 169
 164   51 27 14 283 153 149 152
 133   42 25 14 243 124 127 127
 115   75 35 20 203 109 99 100
 205   66 61 304 287 193 100 110
 153   69 443 302 230 166 85 82
 555    2014 1102 550 411 382 213 211
 762   2109 1361 888 758 587 297 314
 100    445 200 103 67 75 31 11
!     NOTE    N(33,35) CHANGED TO 200
!             N(12,30) CHANGED TO 51
$PR TITLE :: VLIST   $ERR B N $YVAR Y $DISP M !
$M DVR   $CA DV= -2*( Y*%LOG(%FV/Y) +!
   (N-Y)*%LOG( (N-%FV)/(N-Y) ) ) !
    $CA DV=%SQRT(DV) :DV=%IF( %LT(Y,%FV),-DV,DV)$E
$RETURN
```

Figure 5. GLIM-3 PROGRAM FOR 2-AAF DOSE-TIME-RESPONSE DATA.
         (See Section 3.2 and Table 15)

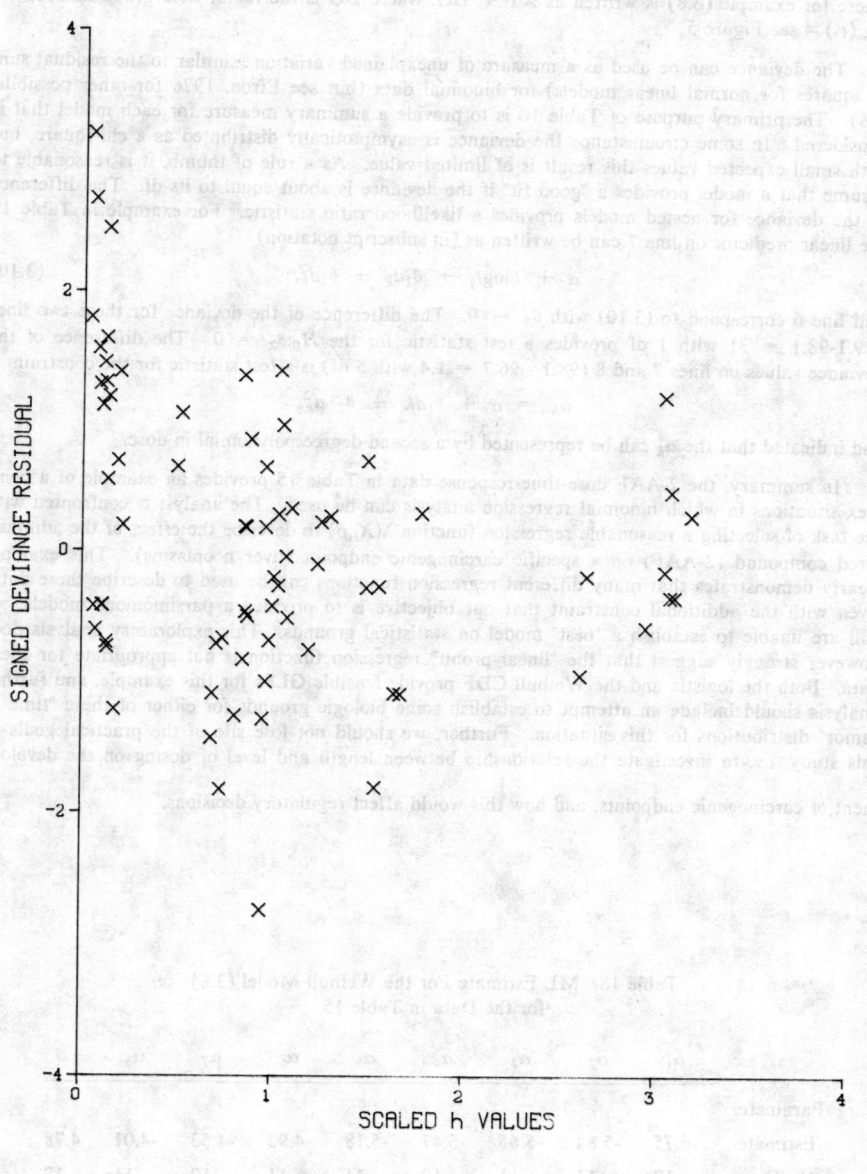

Figure 6. DIAGNOSTIC PLOT FOR 2-AAR DATA FOR LINEAR-LOGISTIC FUNCTION.
(See Section 2.3 and Figure 5)

In GLIM terminology (3.9) can be used by specifying a complementary log-log link function, i.e. $LINK C. The ML estimates of the parameters for the Weibull model (3.8) are given in Table 18. The results of fitting various Weibull regression functions using GLIM are given in Table 19, where for example (3.8) is written as XT + DG, where DG is the factor dose group and XT = log (t_j) -- see Figure 5.

The deviance can be used as a measure of unexplained variation (similar to the residual sum of squares for normal linear models) for binomial data (but see Efron, 1976 for other possibilities). The primary purpose of Table 18 is to provide a summary measure for each model that is considered. In some circumstance the deviance is asymptotically distributed as a chi-square, but with small expected values this result is of limited value. As a rule of thumb, it is reasonable to assume that a model provides a "good fit" if the deviance is about equal to its df. The difference of the deviance for nested models provides a likelihood ratio statistic. For example in Table 18 the linear predictor on line 7 can be written as (in subscript notation)

$$\alpha + \beta \log t_j + \theta_1 d_k + \theta_2 d_k^2, \quad (3.10)$$

and line 6 correspond to (3.10) with $\theta_2 = 0$. The difference of the deviance for these two lines 129.1-98.1 = 31 with 1 df provides a test statistic for the $H_0: \theta_2 = 0$. The difference of the deviance values on lines 7 and 8 (98.1 - 96.7 = 1.4 with 5 df) is a test statistic for the constraint

$$\alpha_k = \alpha + \theta_1 dk + \theta_2 d_k^2,$$

and indicated that the α_k can be represented by a second degree polynomial in dose.

In summary, the 2-AAF dose-time-response data in Table 15 provides an example of a complex situations in which binomial regression analysis can be used. The analyst is confronted with the task of selecting a reasonable regression function $\lambda(X_i, \beta)$ to describe the effect of the administered compound (2-AAF) on a specific carcinogenic endpoint (liver neoplasms). This example clearly demonstrates that many different regression functions can be used to describe these data. Even with the additional constraint that our objective is to produce a parsimonious models, we still are unable to establish a "best" model on statistical grounds. This exploratory analysis does however strongly suggest that the "linear-probit" regression function is not appropriate for these data. Both the logistic and the Weibull CDF provide feasible GLFs for this example, and further analysis should include an attempt to establish some biologic grounds for either of these "time to tumor" distributions for this situation. Further, we should not lose site of the practical goals of this study, i.e. to investigate the relationship between length and level of dosing on the development of carcinogenic endpoints, and how this would affect regulatory decisions.

Table 18. ML Estimate For the Weibull Model (3.8) for the Data in Table 15

Parameter	α_1	α_2	α_3	α_4	α_5	α_6	α_7	α_8	β
Estimate	-6.75	-5.84	-5.65	-5.47	-5.18	-4.93	-4.53	-4.01	4.78
St. Dev.	.17	.11	.11	.12	.11	.11	.12	.11	.12

Table 19. Binomial ANOVA for 2-AAF Data in Table 15 Using Linear-Weibull Regression Function

	Linear Term	p	Unexplained Variation*	df
1	minimal	1	2203.6	71
2	TI	9	595.1	63
3	DG	8	1965.6	64
4	TI+DG	16	85.6	56
5	XT	2	615.2	70
6	XT+D	3	129.1	69
7	XT+D+D2	4	98.1	68
8	XT+DG	9	96.7	63
9	XT+D+D2.TI	12	77.31	60
10	TI+D+D2.TI	19	68.67	53

* Deviance is used as a measure of unexplained variation (see Appendix).

4. SUMMARY

In this paper we have shown how the LS principle can be used as a conceptual basis for fitting regression functions to discrete data that follow the Poisson or binomial distribution. The generalized LS estimates are obtained by solving a p dimensional system of equations (1.3) using an IRLS procedures. When the weights in the IRLS procedure are based on the Poisson or binomial assumption, the IRLS algorithm will yield a root of the likelihood equations, i.e. the generalized LS estimates are also ML estimates (see the Appendix). Consequently, Poisson regression and binomial regression (i.e. ML estimation under the Poisson or binomial assumption) can be viewed as part of the regression analysis paradigm. This considerably broadens the scope of regression analysis as a "methodologic paradigm", i.e. a scientific achievement which attracts adherents from other disciplines (see Dolby, 1982). The analyst is therefore challenged to appeal to his general knowledge to develop "conjecture-based" models with data available for possible refutation. For example, the test for "lack of fit" of a regression function provides a probabilistic basis for evaluating the falsifiability of a proposed model. Dolby (1982) (in discussing the views of Karl Popper and Thomas Kuhn on the methodology of science) emphasizes the importance of "global conjectures", that are the province of the researcher in a particular field, as a basis for establishing specific (local) statistical hypothesis for attempted falsification. The alternative is the "exploratory investigation" of a specific data set that leads to an analysis that is descriptive rather than inferential. Finch (1979) points out that in some circumstances the extrapolation of a good description of the data that we have is the best we can do. The role of exploratory analysis is one of hypothesis generation, i.e. the preliminary investigation of data to uncover "good" descriptions that are relevant to the context. Attempts to test the goodness of fit of a model that is obtained in this way is misleading, and should only be used to provide guidance in an analysis. The five examples presented in this paper were selected to illustrate how regression analysis can be used for both types of data analysis, i.e. those based on global conjectures and exploratory techniques. In both situations the application of statistics to the life sciences is bound to be most fruitful when the analysis is based on a collaborative effort.

The general Poisson regression models include linear, log-linear, and intrinsically nonlinear regression functions. A numerical example from cytogenetic dosimetry was used to illustrate multiple linear regression for Poisson data. A more general dose-response model derived from the theory of dual radiation action was (DRA) also considered. The DRA regression function is intrinsically nonlinear in the parameters. Another important area where Poisson regression models are used is in the analysis of rates from observational studies. An example from an epidemiologic follow up study with the data organized into a life-table type of format was presented, and preliminary analysis was based on log-linear models. A nonlinear model, derived from the multistage theory of carcinogenesis, was then used to analyze lung cancer death rates among individuals who were regular cigarette smokers.

Binomial regression models are used for the analysis of binary (or quantal) response data, i.e. for situations where the outcome is one of two possible values (e.g. success or failure). Two numerical examples were presented that illustrate various aspects of binomial regression analysis. In the first example a linear dose response curve was used to describe the effect of streptonigrin on rabbit lymphoblasts. A linear regression function was used to describe the dose-response curve, and procedures for testing the assumption of binomial variation and lack-of-fit of the regression function were illustrated. In the second example mice were continuously fed a carcinogen (2-Acetylaminofluorene) for an extended period of time. Groups of mice were examined at various time points for each of several exposure levels and the number of mice with liver neoplasms was determined. These data were used to illustrate the application of several well-known regression functions (logistic and probit) to binomial data.

These examples of Poisson and binomial regression analysis are presented to illustrate situations in the biomedical sciences where discrete data that may follow either the Poisson or binomial distribution are encountered. The important problem in any specific situation is to determine an appropriate regression function that describes the effect of one or more covariates on the response. Historically, regression functions have been of the generalized linear type, a choice that appears to be based primarily on computational convenience. The computational requirements for the more general regression models are sufficiently complex that, in most situations, a computer based analysis is required. High quality, inexpensive portable programs (such as GLIM-3) are now widely available and can be used for all of the analyses described in this paper. The IRLS procedure (described in the Appendix) can be easily coded in any of the higher level languages (e.g. FORTRAN, Pascal) that are widely available on micro (personal) computers. Consequently, computational complexities should no longer limit the usefulness of Poisson and binomial regression models in routine data analysis.

APPENDIX

Equivalence of ML and IRLS for Poisson and Binomial Regression

The purpose of this Appendix is to show the equivalence of ML and IRLS for Poisson and binomial data. Charnes, Frome, and Yu (1976) have demonstrated the equivalence of ML and IRLS for situations where the dependent variable is from a member of the regular exponential family, and the regression function is in general nonlinear in the parameters. In the discussion that follows we will limit our discussion to the Poisson and binomial distribution.

Poisson and Binomial Regression Models

Let $y_1, y_2, \ldots, y_N$ denote the observed values of a random sample of size N from a population with density $h(y_i; \mu_i)$, where $\mu_i = \mu(X_i, \beta)$ denotes the expected value of Y_i. For the Poisson distribution

$$h(y; \mu) = e^{-\mu} \mu^y / y!, \qquad y = 0, 1, \ldots$$

and for the binomial distribution

$$h(y; \mu) = \binom{c}{y} (\mu/c)^y (1 - \mu/c)^{c-y}, \qquad y = 0, 1, \ldots, c.$$

The expected value of Y_i is expressed as

$$\mu_i = \mu(X_i, \beta) = c_i \lambda(X_i, \beta),$$

where c_i is a known constant ("sample size") and $\lambda(X, \beta)$ is a known function, that we refer to as the regression function. The regression function describes the relation between the covariate vector $X_i = (x_{i1}, \ldots, x_{im})$, and the unknown parameters $\beta = (\beta_1, \ldots, \beta_p)'$. Given the data $\{y_i, X_i, i = 1, \ldots, N\}$ the problem is to obtain estimates of the parameters $\beta_1, \ldots, \beta_p$.

Maximum Likelihood Estimation

The logarithm of the likelihood function of β is

$$L(\beta) = \sum_i \log h(y_i; \mu_i).$$

Since Y is a random variable with a densiy function of the regular exponential family

$$h(y; \mu) = \exp\{yb(\mu) - q(\mu) + g(y)\}, \tag{A.1}$$

where $E(Y) = \mu$, and $b(\ )$ and $q(\ )$ are given in Table A-1 for the Poisson and binomial distribution. Following the approach of Charnes, Frome, and Yu (1976) differentiation (with respect to μ) on both sides of $\int h(y; \mu) dy = 1$ yields

$$E(Y) = q'(\mu)/b'(\mu) = \mu \tag{A.2}$$

where $b'(\mu)$ and $q'(\mu)$ denote derivatives with respect to μ. A second differentiation of the integral, along with evaluation of the derivative of (A.2) results in

$$V(Y) = b'(\mu)^{-1}. \tag{A.3}$$

These results are summarized in Table A-1.

The kernel of the log-likelihood function can be written

$$L(\beta) = \sum_i \Big[y_i b[\mu(X_i, \beta)] - q[\mu(X_i, \beta)] \Big]. \tag{A.4}$$

The ML equations are

$$\partial L / \partial \beta_j = \sum_i b'[\mu(X_i, \beta)] (\partial \mu(X_i, \beta) / \partial \beta_j)$$
$$- \sum_i q'[\mu(X_i, \beta)] (\partial \mu(X_i, \beta) / \partial \beta_j) = 0, \quad j = 1, \ldots, p.$$

By using (A.2) and (A.3) we obtain

$$\partial L/\partial \beta_j = \sum_i [V(Y_i)^{-1}[y_i - \mu(X_i,\beta)](\partial \mu(X_i,\beta)/\partial \beta_j)] \quad (A.5)$$

$$j = 1,...,p$$

The likelihood equations are nonlinear with respect to the unknown parameters and an iterative procedure can be used to obtain a root of (A.5). A convenient computational approach to this problem is obtained by using IRLS (see below).

Iterative Reweighted Least Squares

Let $\bar{y}_i = y_i/c_i$ and w_i denote a positive weight that is proportional to the reciprocal of the variance of $\bar{y}_i$. For the binomial distribution c_i is the sample size and $\bar{y}_i$ is a proportion, i.e. the proportion of successes in c_i trials. For the Poisson distribution $\bar{y}_i$ is a "rate", (e.g. number of failures per unit time, number of events per unit area, etc.) and c_i is this "size" of the sample (e.g. number of time units, number of unit areas, etc.). Consider the following weighted sum of squares

$$S(\beta) = \sum_i w_i [\bar{y}_i - \lambda(X_i,\beta)]^2. \quad (A.6)$$

The least squares principle can be used to obtain an estimate of β by solving the system of equations

$$\sum_i w_i [\bar{y}_i - \lambda(X_i,\beta)](\partial \lambda(X_i,\beta)/\partial \beta_j), \quad (A.7)$$

$$j = 1,...,p$$

Since $\lambda(X_i,\beta)$ is in general nonlinear in the parameters an iterative procedure is required to obtain an estimate of β. On iteration $k+1$ we replace $\lambda(X_i,\beta)$ with the linear terms in a Taylor series expansion about the current estimate β^k

$$\lambda(X_i,\beta) = \lambda(X_i,\beta^k) + P_i^k \delta^k, \quad (A.8)$$

where P_i^k denotes the ith row of the $N \times p$ matrix of partial derivatives $p_{ij} = \partial \lambda(X_i,\beta)/\partial \beta_j$ evaluated at β^k, and $\delta^k = (\delta_1^k,...,\delta_p^k)'$ is the "correction vector".

Using (A.8) in (A.7) and the appropriate weights

$$w_i = \frac{c_i}{\lambda(X_i,\beta)}, \text{Poisson weights}, \quad (A.9)$$

or

$$w_i = \frac{c_i}{\lambda(X_i,\beta)[1-\lambda(X_i,\beta)]}, \text{binomial weights},$$

evaluate at β^k we obtain

$$\sum_i w_i [\bar{y}_i - \lambda(X_i,\beta^k) - P_i^k \delta^k](\partial \lambda(X_i,\beta)/\partial \beta_j)\Big|_{\beta=\beta^k} \quad (A.10)$$

Equation (A.10) can be written as

$$A(\beta^k)\delta^k = G(\beta^k), \quad (A.11)$$

where $A = P'WP$, $G = P'WZ$, $W = \text{diag}(w_i)$, and $Z = [\bar{y} - \lambda(X_i,\beta)]$, where all expressions that involve β are evaluated at the current estimate β^k.

The linear system of equations is solved for δ^k and the revised estimate $\beta^{k+1} = \beta^k + \delta^k$ is obtained. This IRLS procedure continues until some convergence criteria are satisfied.

The matrix A is the "information matrix" with elements

$$a_{js} = \sum_i (p_{ij}p_{is}w_i), \quad j,s = 1,...,p.$$

The system of equations (A.10) obtained using the least squares approach is identical to that obtained using the ML principle (A.5) -- to see this note that $w_i = c_i/v(Y_i)$ and $\partial\mu(X_i,\beta)/\partial\beta_s = c_i\partial\lambda(X_i,\beta)/\partial\beta_j$. Consequently, if the IRLS procedure converges to a stable solution (convergence is not guaranteed) it will yield a critical point of the likelihood equations. The IRLS procedure just described is equivalent to using the method of scoring to find a root of the likelihood equations (A.5). Further conditions (see Charnes, Frome, and Yu, 1976) to assure that β^* is a global maximum of $L(\beta)$ are (i) $L(\beta)$ be pseudoconcave over the parameter space, and (ii) that β^* satisfies (A.5). This will occur if $L(\beta)$ is defined over a convex parameter space and both $b[\mu(X,\beta)]$ and $q[\mu(X,\beta)]$ are concave in β over the parameter space. It will be the unique global solution of at least one of the $y_i b[\mu(X_i,\beta)], - q[\mu(X_i,\beta)]$, $i=1,...,N$ is strictly concave over the parameter space.

Table A-1
Characteristics of Poisson and Binomial Regression Models

	Poisson	Binomial
$b(\mu)$	$\log\mu$	$\log[\mu/(c-\mu)]$
$q(\mu)$	μ	$-\log(1-\mu/c)$
$E(Y)$	μ	μ
$V(Y)$	μ	$\mu(1-\mu/c)$
Regression Function	$\lambda(X,\beta)$	$\lambda(X,\beta)$
(Interpretation)	(rate)	(probability)
Regression Weight (w)	$\dfrac{c}{\lambda(X,\beta)}$	$\dfrac{c}{\lambda(X,\beta)[1-\lambda(X,\beta)]}$
$\bar{y} = y/c$	observed rate	observed proportion

Note: c is the sample size for a given observational unit with covariate vector X. The regression function is the expected value of $\bar{y}$ and is used as the dependent variable in the IRLS procedure (see text).

Covariance Matrix for the ML Estimates

The large sample covariance matrix of the ML estimators is the inverse of the information matrix $A(\beta)$ -- equation (A.11). If β^* is a stable solution of the likelihood equations (A.5) then estimates of the elements of this matrix are obtained by replacing β by the ML estimate β^*. For GLFs $\lambda(X_i,\beta) = G(\eta_i)$ where $\eta_i = X_i\beta = \sum_j x_{ij}\beta_j$ and A can be written as

$$A(\beta) = X' V X,$$

where V is diagonal with $v_i = w_i(\partial G_i/\partial\eta_i)^2 G_i^{-1}$. It can be shown (see McCullagh, 1983) that if $N^{-1} A(\beta)$ has a positive definite limit as $N\to\infty$, then

$$E(\beta^* - \beta) = O(N^{-1}) \tag{A.12}$$

and

$$N^{\frac{1}{2}} (\beta^* - \beta) \sim N_p(0, N\sigma^2 A(\beta)^{-1}) + O_p(N^{-\frac{1}{2}}) \tag{A.13}$$

The notation N_p denotes the p-variate Normal distribution and the remainder terms in (A.12) and (A.13) refers to the difference between the cumulative distributions of the statistic and the Normal. The dispersion parameter σ^2 is equal to 1 if the dependent variable follows the Poisson or binomial distribution.

Evaluating Goodness of Fit

The results of fitting a model to data can be viewed as replacing the y_i with a set of "explanations" the μ_i^* that are derived from the regression function $\lambda(X_i,\beta^*)$. A measure of the discrepancy between the y_i and the μ_i^* that is convenient for both Poisson and binomial data is the *deviance* (see Nelder and Wedderburn, 1972). The deviance component for the *ith* observation for the Poisson distribution is

$$d_i^2 = 2\left[y_i \log(y_i/\mu_i^*) - (y_i - \mu_i^*) \right], \quad (A.14)$$

and for the binomial distribution

$$d_i^2 = 2\left[y_i \log(y_i/\mu_i^*) + (c_i - y_i)\log[(c_i - y_i)/(c_i - \mu_i^*)] \right], \quad (A.15)$$

where $\mu_i^* = c_i \lambda(X_i,\beta^*)$. The deviance is then obtained by summing the individual components, i.e.

$$D(y,\mu^*) = \sum_i d_i^2 .$$

When the y_i are assumed to follow the Normal distribution the deviance is

$$\sum_i (y_i - \mu_i^*)^2,$$

i.e. the "residual sum of squares". Consequently the deviance can be used to construct a table similar to that used in standard linear model theory and referred to as an ANOVA (analysis of variance) table. The simplest model of interest (minimal model) has one parameter μ. At the other extreme is the full model which has one parameter for each observation--i.e. $\mu_i^* = y_i$, and the deviance is zero. The minimal model is usually too simple and one is interested in this model for reference purposes, since it provides the maximum value of the deviance for a given set of data.

If $\lambda(X_i,\beta)$ is a specific regression function of interest with $\beta = (\beta_1,...,\beta_p)'$, then the deviance for this model is obtained from the ML estimate β^*. The "goodness of fit" of the regression function is evaluated by comparing the observed value of the deviance with the χ^2 distribution. For Poisson or binomial data the deviance is distributed approximately as a χ^2 with $N-p$ df when the assumed regression function is appropriate.

The analysis of variance has been most widely used for Normally distributed data when two or more factors and their interactions are of interest. Extension of these methods to GLFs for dependent variables in the regular exponential family have been developed by Nelder and Wedderburn (1972). An ANOVA like table is constructed by fitting a sequence of models and recording the df and deviance for each model that is considered. For each model that is fitted to the data the difference of the deviance for that model and the previous model represents the variation accounted for by the new factor having eliminated those terms of above it in the table (see the examples in Section 2 and 3). Note that the relative importance of a specific factor depends on when it is entered into the model--this is the same problem that occurs for the classical linear model when non-orthogonality occurs. For GLFs we may fit an increasing sequence of models, say $\mu \in H_j$, $H_0 \subset H_1 \subset ... \subset H_j ...$, and the difference of the deviance has an asymptotic χ^2 distribution if the more restrictive hypothesis is true.

A special form of the ANOVA table is of interest in experimental studies that are used to investigate a specific dose response curve is shown in Table A2. We assume that the y_{jk} are independent and follow the Poisson or binomial distribution with

$$E(y_{jk}) = c_{jk} \lambda(X_j,\beta) , \quad j=1,...,N, \quad k=1,...,n_j,$$

i.e. y_{jk} is the response for the kth "parallel count" (replication) for dose group j. The dose response curve is given by $\lambda(X_j,\beta)$, and the deviance on line 2 of Table A.2 will be distributed approximately as a chi-square with $\sum_j n_j - p$ df if the *regression model* (i.e. *both* the regression function and the Poisson or binomial assumption) is appropriate. If this test statistic is large it may be due to "lack of fit" of the regression function or heterogenity of variance (under the Poisson or binomial assumption). In this situation D_3 may be compared with the chi-square distribution with $\sum_i n_i - N$ df. If this statistic is significantly large (indicating heterogenity of variance) then the ratio

$$\frac{(D_2-D_3)/(N-p)}{D_3/(\sum_j n_j-N)}$$

may be compared with the F distribution (approximate test). A significant value of this F statistic indicates lack of fit of the *regression function* $\lambda(X_i,\beta)$. We have partitioned the unexplained variation D_2 into two components since $D_2 = (D_2-D_3) + D_3$, where D_3 is equivalent to the "pure error" sum of squares in Normal regression analaysis (see Draper and Smith, 1966, chap. 2).

The values of the deviance can also be used to construct an R^2-type measure of variance explained, i.e.

$$R^2 = 100 \; (D_1-D_2)/D_1$$

is the percent of the total variation (as measured by the deviance) that is explained by the regression function $\lambda(X,\beta)$.

Table A.2

ANOVA Table for Lack of Fit Test for Poisson or Binomial Data

	Regression Function	Number of Parameters	Deviance	df
1	minimal	1	$D_1 = D(y, \mu^*1)$	$\sum_j n_j - 1$
2	$\lambda(X_i,\beta)$	p	$D_2 = D[y, \mu(X_i,\beta^*)]$	$\sum_j n_j - p$
3	Dose Groups	N	$D_3 = D[y, \mu(\lambda^*_j)^+]$	$\sum_j n_j - N$
4	Complete	$\sum_{i=1}^{N} n_i$	0.0	

$+\mu(\lambda_j^*)$ denotes a vector of fitted values based on the model $E(y_{jk}) = c_{jk} \lambda_j$. The ML estimates $\lambda_j^* = \sum_k y_{jk}/\sum_k c_{jk}$ are used to compute the deviance, i.e. $\mu_{jk}^* = c_{jk} \lambda_j^*$ on line 3.

Regression Diagnostics

An important area of regression analysis that has received considerable attention (for the standard linear model) in recent years is regression diagnostic -- see e.g. Belsey *et al* (1980). Diagnostic procedures are used to check for outlying y-values and extreme points in the "model space". Extension of these techniques to binomial (logistic) regression models and Poisson regression models have also been proposed -- see Pregibon (19810, Frome (1983), and (   ). The basic "building blocks" that are required for various diagnostic measures are standardized residuals of some type and the diagonal terms, h_i, from the matrix

$$\mathbf{H} = \mathbf{W}^{\frac{1}{2}} \; \mathbf{P} \; (\mathbf{P'} \; \mathbf{W} \; \mathbf{P})^{-1} \; \mathbf{P'} \; \mathbf{W}^{\frac{1}{2}}, \qquad (A.16)$$

where all quantities that depend on β are evaluated at the ML estimate β^* (see equation A.11). The diagonal terms from this matrix are useful in detecting extreme points in the model space that may have a substantial influence on the fitted model. Recall that for the standard linear model $H = X (X' X)^{-1} X'$, $I - H$ is the projection matrix, and large values of h_i identify extreme points in the model (design) space. For GLFs $\lambda(X_i,\beta) = G(\eta_i)$, where $\eta_i = \sum_j \beta_j x_{ij}$, and H can be written as

$$H = V^{1/2} X (X' V X)^{-1} X' V^{1/2},$$

where V is diagonal with $v_i = \{w_i(\partial G_i/\partial \eta_i)^2\}$. Note that $\sum_i h_i = p$ and that large values of h_i (say, greater than 2p/n) indicate extreme points in the model space. If u_i denotes a standardized residual the variance of u_i is approximately $1-h_i$ and "adjusted residuals" are given by $u_i/(1-h_i)^{1/2}$. Two possible definition of standardized residual are

$$u_i = (y_i - \mu_i^*)/var(y_i)^{1/2}, \text{ or}$$

$$u_i = sign(y_i - \mu_i)d_i,$$

where d_i^2 is deviance component (see A.14 and A.15).

Heterogeneity of Variance

In practical data analysis the assumption of binomial or Poisson variation may be unrealistic. Usually the variance will be greater than that predicted, a phenomenon referred to as overdispersion or heterogeneity of variance. In some situations it is reasonable to assume that $\mu_i = c_i \lambda(X_i,\beta)$ and that the variance of y is proportional to that predicted under the Poisson or binomial assumptions. In this situation the estimated parameter covariance matrix is multiplied by the dispersion parameter σ^2. An estimate of σ^2 is obtained as

$$\tilde{\sigma}^2 = \frac{1}{N-p} \sum_i \frac{(y_i - \mu_i^*)^2}{var(y_i)},$$

where $var(y) = \mu^*$ for the Poisson distribution, and $var(y) = c\lambda^*(1-\lambda^*)$ for the binomial distribution. Estimates of the regression parameters can be obtained using maximum quasi-likelihood (MQL) estimation--see Wederburn (1974), McCullagh (1983). Apart from the multiplier σ^2 the quasi-likelihoods can be treated for the most part just like ordinary likelihoods. In particular the quasi-likelihood equations are given by (A.5) so that IRLS procedure can be used to obtain the MQL estimates for overdispersed data.

REFERENCES

BAKER, R. J. and NELDER, J. A. (1978), *Generalized Linear Interactive Modelling (GLIM), Release 3,* Numerical Algorithms Group, Oxford.

BELSLEY, D. A., KUH, E., and WELSCH, R.E. (1980), *Regression Diagnostics,* New York, Wiley.

BISHOP, Y. M. M., FINEBERG, S. E. and HOLLAND, P. W. (1975), *Discrete Multivariate Analysis: Theory and Practice,* Cambridge, Massachusetts: MIT Press.

BLISS, C. I. (1935), "The Calculation of The Dosage-Mortality Curve," *Annals of Applied Biology* 22, 134-67.

BRESLOW, N. E., LUBIN, J. H., MAREK, P. and LANGHOLZ, B. (1983) "Multiplicative Models for Cohort Analysis," *Journal of The American Statistical Association* 78, 1-12.

CHARNES, A., FROME, E. L., and YU, P. L. (1976), "The Equivalence of Generalized Least Squares and Maximum Likelihood Estimation in the Exponential Family," *Journal of the American Statistical Association,* 71, 169-172.

COCHRAN, W. G. (1940), "The Analysis of Variance When Experimental Errors Follow The Poisson or Binomial Laws," *Annals of Mathematical Statistics* 11, 335-347.

COCHRAN, W. G. (1983), *Planning and Analysis of Observational Studies,* New York: J. Wiley and Sons.

COX, D. R. (1970), *The Analysis of Binary Data,* London, Methuen.

DOLBY, G. R. (1982), "The Role of Statistics in the Methodology of Science," *Biometrics* 38, 1069-1083.

DRAPER, N.R. and SMITH, H., *Applied Regression Analysis,* New York: John Wiley & Sons, Inc., 1966.

DUFRAIN, R. J., LITTLEFIELD, L. G., JOINER, E. E. and FROME, E. L. (1980), "In Vitro Human Cytogenetic Dose Response Systems," in: *The Medical Basis for Radiation Accident Preparedness* (Hubner, K. F. and Fry, S. S. eds.), Elsevier North-Holland, New York.

DUFRAIN, R. J., LITTLEFIELD, L. G., WILMER, J. L., and FROME, E. L. (1982), "Evaluation of Chemically Induced Cytogenetic Lesions in Rabbit Oocytes, II," *Mutation Research,* 94, 103-114.

EFRON, B. (1978), "Regression and ANOVA with Zero-One Data: Measures of Residual Variation," *Journal of the American Statistical Association,* 73, 113-121.

FARMER, J. H., KODELL, R. L., GREENMAN, D. L. and SHAW, G. W. (1979), "Dose and Time Response Models For the Incidence of Bladder and Liver Neoplasms in Mice Fed 2-Acetylaminofluorene Continuously," *Journal of Environmental Pathology and Toxicology* 3, 55-68.

FINCH, P.D. (1979) "Description and Analogy in the Practice of Statistics," *Biometrika* 66, 195-208.

FROME, E. L. (1972), "Nonlinear Regression and Spectral Estimation in Biomedical Data Analysis," Ph.D. Dissertation, Emory University: Atlanta, Georgia.

FROME, E. L. (1981), "Poisson Regression Analysis," *American Statistician,* 35, 262-263.

FROME, E. L. (1982), "Fisher's Exact Variance Test for the Poisson Distribution," *Appl. Statistics,* 31, 67-71.

FROME, E. L. (1983), "The Analysis of Rates Using Poisson Regression Models," *Biometrics,* 39, 665-674.

FROME, E. L. and BEAUCHAMP, J. J. (1968), "Maximum Likelihood Estimation of Survival Curve Parameters," *Biometrics,* 24, 595-605.

FROME, E. L. and CHECKOWAY, H. (1984), "The Use of Poisson Regression Models in Estimating Incidence Rates and Ratios" *American Journal of Epidemiology* (to appear).

FROME, E. L. and DUFRAIN, R. J. (1983), "Maximum Likelihood Estimation for Cytogenetic Dose-Response Curves," Research Report 123, Oak Ridge National Laboratory, Oak Ridge, Tennessee.

FROME, E. L., KUTNER, M. H. and BEAUCHAMP, J. J. (1973). "Regression Analysis of Poisson-distributed Data," *Journal of the American Statistical Association,* 68, 935-940.

GART, J. J. (1978), "The Analysis of Ratios and Cross-Product Ratios of Poisson Variates With Application to Incidence Rates," *Communication in Statistics Theory and Methods,* A7 917-37.

GOODNIGHT, J. H. and SALL, J. P. (1982), "The NLIN Procedure," *SAS Users Guide: Statistics,* SAS Institute, Cary, North Carolina.

HAIGHT, F.A., (1967), *Handbook of the Poisson Distrubition,* New York: John Wiley & Sons, Inc.

HOLFORD, T. R. (1980), "The Analysis of Rates and Survivorship Using Log-Linear Models," *Biometrics* 36, 299-305.

JOHNSON, N. L. and KOTZ, S. (1970), *Distributions inStatistics:Discrete Distributions,* New York: J. Wiley and Sons.

KAHN, H. A. (1966) "The Dorn Study of Smoking and Mortality Among U.S. Veterans: Report on Eight and One-Hlf Years of Observation," in: *Epidemiologic Approaches to The Study of Cancer and Other Chronic Diseases,* ed. Haenszel, W., *National Cancer Institute Monograph* 19, 1-125.

KELLERER, A. M. and ROSSI, H. I. (1972), "The Theory of Dual Radiation Action," *Curr. Topics Radiat. Res.,* 8, 85-158.

KENDALL, M. G. and STUART, A., (1946) *The Advanced Theory of Statistics, Vol. 2,* New York: Hafner Publishing Company.

KOCH, G. G., ATKINSON, S. S. and STOKES, M.E. (1984), "Poisson Regression," in: *Encyclopedia of Statistical Sciences,* eds Kotz, S., Johnson, L. L., and Read, A, New York: J. Wiley & Sons (in press).

LEA, D. E. (1955), *Action of Radiations on Living Cells,* 2nd Edition, Cambridge Press.

McCULLAGH, P. (1983), "Quasi-Likelihood Functions," *The Annals f Statistics,* 11, 59-67.

McCULLAGH, P. and NELDER, J. A. (1983), *Generalized Linear Models,* London: Chapman and Hall.

MIETTIEN, O. S. (1972), "Standardization of Risk Ratios", *American Journal of Epidemiology* 96, 383-88.

MOORE, R. H. and ZEIGLER, R. K. (1967), "The Use of Nonlinear Regression Methods for Analyzing Sensitivity and Quantal Response Data," *Biometrics,* 23, 563-67.

NELDER, J. A. and WEDDERBURN, R. W. M. (1972), "Generalized Linear Models," *Journal of the Roy. Statistical Society, Series A 135,* 370-384.

PALMGREN, J. (1981), "The Fisher Information Matrix For Log-Linear Models Arguing Conditionally on the Observed Explanatory Variables," *Biometrika* 68, 563-66.

PREGIBON, D. (1981), "Logistic Regression Diagnostics," *The Annals of Statistics,* 9, 705-724.

PURROTT, R. J. and REEDER, E. (1976), "The Effect of Changes in Dose Rate on the Yield of Chromosome Aberrations in Human Lymphocytes Exposed to Gamma Radiation," *Mutation Research,* 35, 437-444.

RAO, C. R. (1965), *Linear Statistical Inference and Its Applications,* New York: John Wiley & Sons, Inc.

WHITLEMORE, A. and KELLER, J. B. (1978), "Quantitative Theories of Carcinogenesis", *SIAM Review 20,* 1-30.

REGRESSION METHODS IN SURVIVAL ANALYSIS

Edmund A. Gehan
University of Texas System Cancer Center, Houston, TX. 77030

ABSTRACT

The survival time of a patient with a disease is often a major criterion for evaluating the treatment the patient received. There are two special features of survival time studies. First, survival times must be greater than or equal to zero and are often highly skewed. Secondly, "censored times" often occur when some individuals are alive or in a well state at the conclusion of study. Censored times may also occur when individuals are lost to follow-up after some period of study.

The objective of this paper is to give an introduction to the methodology for analysis of survival time studies, especially regression methods for assessing the relationships between survival time and concomitant variates, say $z_1, z_2, \ldots, z_p$. In a special case, there is only one z variable taking the value 0 or 1 in each of two samples; here a test of difference in the survival distributions in the two samples is obtained.

Any distribution of survival time can be characterized by three functions, which may be defined as follows:

<u>Survivorship function</u>, $S(t)$: probability that an individual survives longer than t, often called a survival curve.

<u>Hazard function</u>, $\lambda(t)$: limit of the probability that an individual dies in a short interval of time (Δt) given survival to time t.

<u>Probability density function</u>, $f(t)$: limit of the probability an individual dies in the short interval t to ($t+\Delta t$) per unit width (Δt).

These functions are equivalent in the sense that given any one of the functions, the other two can be derived.

In the past 10-15 years, multiple regression methodology has been adapted for use with censored survival data. Perhaps the most important paper was that of Cox (1972) who proposed a regression model which is appropriate for the analysis of survival data with and without censoring. Suppose that for each of n individuals, there is an associated survival time t_i (possibly not yet observed), and one or more explanatory or concomitant variables, say $z_1, z_2, \ldots, z_p$. For the ith individual the values of z' are $z_i' = (z_{1i}, z_{2i}, \ldots, z_{pi})$. Cox postulated the following model for the hazard function at time t:

$$\lambda(t; z) = \exp(z\beta) \lambda_0(t)$$

where $\beta' = (\beta_1, \beta_2, \ldots, \beta_p)$ is an unknown parameter vector and $\lambda_0(t)$ is an unknown hazard function for the standard conditions when all z's are zero. Cox's regression model can be re-written

$$\log_e \lambda(t; z)/\lambda_0(t) = \beta_1 z_1 + \beta_2 z_2 + \ldots + \beta_p z_p$$

It will be shown how the model can be applied to the following problems: comparison of survival distributions for two or more samples; comparison of two samples, adjusting for the presence of concomitant variables; and finding a set of z variables related to survival time. Some examples in cancer research will be given.

In Cox's regression model, $\lambda_0(t)$ is an arbitrary and unknown hazard function. Hence, the estimation of the β's does not depend upon the underlying hazard function and thus this approach has been characterized as a semi-parametric model. A parametric approach involves specifying a form of $\lambda_0(t)$ that corresponds to that for a particular survival distribution. For example, if $\lambda_0(t)$ is constant, this corresponds to an underlying exponential distribution. In this

presentation, however, $\lambda_0(t)$ will be taken as
$$\lambda_0(t) = \exp(\beta_0) kt^{k-1}$$
corresponding to a Weibull survival model. When k=1, this reduces to the exponential model. The Weibull model has been chosen for its flexibility in characterizing survival experience: if $k > 1$, the hazard function increases with time and there is positive aging; if $k < 1$, the hazard function decreases with time and there is negative aging. The term $\exp(\beta_0)$ represents an intercept parameter and k represents a shape parameter because it determines the shape of the survival curve through its influence on the hazard function.

1. INTRODUCTION

The objective of this paper is to give a general introduction to the methodology for analysis of survival time studies, especially regression methods for assessing the relationship between survival time and concomitant variates. Emphasis will be given to the development of ideas, since four textbooks published recently give a full development of the theory and many examples of applications. The texts emphasizing mostly theoretical developments are by Kalbfleisch and Prentice (1980), Lawless (1982), and Cox and Oakes (1984), while that of Lee (1982) gives more emphasis to applications.

An example will be given of a clinical trial in breast cancer in which a regression model was used to evaluate the effectiveness of adjuvant treatment (Buzdar, et al, 1978). The control group in this study involved historical control patients, so a brief discussion will be given of advantages of such a control group.

Survival time of a patient with a disese is often a major criterion for evaluating the treatment that the patient has received. Also, it is frequently desired to study the relationship of survival time to patient characteristics that are possibly related to survival. "Survival time is meant in the broad sense, so that the time may also be: length of response, disease-free survival time, or time from start of treatment to first response.

2. SPECIAL FEATURE OF SURVIVAL STUDIES

The occurrence of right-censored survival times is a special aspect of survival time studies, especially in cancer clinical trials. A common situation arising in clinical trials is shown in Figure 1.

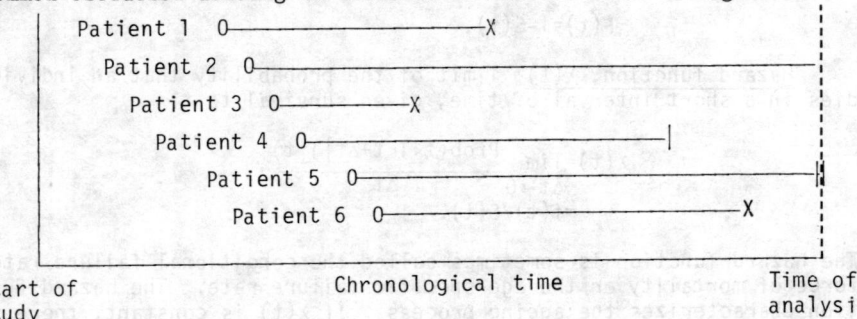

Figure 1. Schematic diagram indicating occurrence of failure and censored observations in a clinical trial (O=time of entry; X=time of failure; ⊣=time of censoring or withdrawal).

Patients (designated 1,...,6) accrue into the study at various points in chronological time, indicated by the open circles, and the data may be analyzed at some later point. Survival time is measured from time of entry into the study until failure, unless the patient is "right-censored" at the time of analysis. Patients 1, 3, and 6 represent complete survival times, which arise when failure occurs prior to the time of analysis. If the patients have not failed by that time, as for patients 2, 4, 5, their survival times are censored. Patients 2 and 5 are continuing under observation and could be observed as having failures at some later date; patient 4 represents a withdrawal prior to the time of analysis, which may mean lost to follow-up observation, refusal to continue therapy, or loss for some other reason.

An important assumption for nearly all statistical methods dealing with survival data is that time to right censoring is independent of time to failure. This means that an observation censored at t provides only the information that survival time is greater than t. If individuals who are censored have shorter or longer expected survival times than individuals remaining under follow-up beyond that point, then the usual statistical methods are inappropriate and could give misleading results. If possible, the assumption of independence should be checked in applications.

3. FUNCTIONS OF SURVIVAL TIME

Any distribution of survival time can be characterized by three functions which lead to equivalent specifications of survival distribution. Suppose there is a population of individuals, each characterized by a non-negative random variable (T) called its survival time. An outcome for a particular individual is designated as t. The functions are defined as follows:

Survivorship function, $S(t)$: probability that an individual survives t or longer,

$$S(t) = \text{Prob}(T \geq t) = \int_t^\infty f(u)du.$$

It is evident that $S(0)=1$, $S(\infty)=0$ and $S(t)$ must be a nonincreasing function of t. The cumulative distribution function is $F(t)=\text{Prob}(T<t)$ and

$$F(t) = 1 - S(t).$$

Hazard function, $\lambda(t)$: limit of the probability that an individual dies in a short interval of time, given survival to t.

$$\lambda(t) = \lim_{\Delta t \to 0} \frac{\text{Prob}(t \leq T < t+\Delta t | T \geq t)}{\Delta t}$$
$$= f(t)/S(t).$$

The hazard function is sometimes called the conditional failure rate, force of mortality or the age-specific failure rate. The hazard function characterizes the ageing process. If $\lambda(t)$ is constant, there is no ageing and failure is a random event. If $\lambda(t)$ increases (decreases) with time, then the death rate per unit time is increasing (decreasing).

Probability density function, $f(t)$: limit of the probability that an individual dies in a short interval $[t, t+\Delta t]$ per unit width (Δt).

$$f(t) = \lim_{\Delta t \to 0} \frac{\text{Prob}(t \leq T < t+\Delta t)}{\Delta t}$$
$$f(t) = -S'(t)$$

All three functions are mathematically equivalent in the sense that given any one the other two can be derived.

As an example, consider the exponential distribution which plays a central role in survival analysis similar to the Normal distribution in ordinary statistical analysis. The survival functions for the exponential distribution are:

$$\lambda(t) = \lambda$$
$$S(t) = \exp\{-\lambda t\}$$
$$f(t) = \lambda \exp\{-\lambda t\}.$$

The single parameter of the exponential distribution is λ ($\lambda > 0$). The hazard function is constant so that there is no ageing and failure is a random event. The expected survival time for an individual at the start of study is λ^{-1}; the expected remaining life-time for an individual having survived to t remains the same, independent of t. Other survival distributions represent departures from the exponential in that the hazard function changes with time. Note that the logarithm of the survivorship function yields a straight line with negative slope λ.

A survival distribution of particular importance is the two parameter Weibull distribution since the hazard function can increase, decrease, or remain constant with time (the exponential distribution is a special case). The survival functions are:

$$\lambda(t) = \lambda k (\lambda t)^{k-1}$$
$$S(t) = \exp\{-(\lambda t)^k\}$$
$$f(t) = \lambda(t) S(t).$$

The Weibull distribution becomes an exponential distribution when $k=1$; when $k<(>)1$, then hazard increases (decreases) with time. Plotting $\log(-\log S(t))$ against $\log t$ yields a straight line with slope k. The Weibull distribution is useful because it can characterize a variety of types of survival function.

Problems related to survival distributions considered in the aforementioned texts are: non-parametric estimates of survival functions given some censored and uncensored data; estimation of parameters of survival distributions; and statistical tests of difference among survival distributions (non-parametric and parametric). These problems will not be considered further here.

4. REGRESSION MODELS IN SURVIVAL ANALYSIS

In this section, an introduction is given to the development of regression models for assessing the relationship between survival time and concomitant variates. In the previous section, the characterization of survival experience in some homogeneous populations was discussed; usually, however, the survival time of a given individual may also depend upon explanatory variables. Consider survival time $T>0$

and suppose a row vector $z = (z_1,\ldots,z_p)$ of covariates has been observed. Note that z may include either or both quantitative and qualitative variables. Also, treatment group can be incroporated using indicator variables, e.g. $z_1 = 0$ for individuals receiving treatment A, $z_1 = 1$ for individuals receiving treatment B.

The exponential distribution can be generalized to obtain a regression model by allowing the failure rate to be a function of the covariates z. The hazard function at time t for an individual with covariates z can be written

$$\lambda(t;z) = \lambda(t) \, g(z\,\beta)$$

where $\lambda(t)$ is a hazard function to be specified, $z = (z_1,\ldots,z_p)$ is a row vector of covariates, and β is a column vector of regression parameters $\beta' = (\beta_1,\ldots,\beta_p)$. Feigl and Zelen (1965) considered a special case of this model with $\lambda(t) = 1$, $g(z\,\beta) = (z\,\beta)^{-1}$. In this case, the survival distribution is exponential with mean value $z\,\beta$. Feigl and Zelen derived maximum likelihood estimates of the β's and gave an example in acute leukemia.

The Feigl-Zelen model has been followed by many others, one being an exponential regression model with $\lambda(t)=\lambda$, $g(z\,\beta) = \exp\{z\,\beta\}$. The latter form for g ( ) assures a positive hazard function whatever the values of the β's. A Weibull regression model is obtained when $\lambda(t) = k(\lambda t)^{k-1}$ and $g(z\,\beta) = \exp\{z\,\beta\}$. These models are discussed by Kalbfleisch and Prentice (1980).

A semi-parametric approach was taken by Cox (1972) who used a conditional partial likelihood function to fit the model $\lambda(t) = \lambda_0(t)$, an arbitrary unspecified hazard function, and $g(z\,\beta) = \exp\{z\,\beta\}$. The model can be written

$$\log \lambda(t;z)/\lambda_0(t) = \beta_1 z_1 + \ldots + \beta_p z_p.$$

In the special case $z_1 = 0$ for individuals receiving treatment A and $z_1 = 1$ for individuals receiving treatment β, and $z_2=\ldots=z_p=0$, $\lambda(t;z)=\lambda_0(t)$ for individuals receiving treatment A and $\lambda(t;z)=\lambda_0(t)\exp\{\beta_1\}$ for those receiving treatment β. Cox's model is often referred to as a proportional hazards model, since hazards are in constant ratio, independent of time. Methods for fitting this model are given in Cox (1972) and all the texts mentioned.

This model has been utilized in a wide variety of applied problems. Problems that can be handled flexibly by use of this model are: tests of difference in survival between two groups, adjusting for other covariates; determination of important covariates related to survival in a homogeneous population; inclusion of covariates dependent on time and other problems.

5. REGRESSION MODEL FOR BREAST CANCER

This section gives an example which involves the use of regression models derived from historical control data. In a study of Buzdar et al (1978), the objective was to compare the disease-free and overall survival experience of women with stage II or III breast cancer between two groups of patients, 131 patients who received FAC-BCG as adjuvant treatment (5-Fluorouracil, Adriamycin, Cyclophosphamide and BCG immunotherapy) between January 1974 and October 1976 at the University of Texas System Cancer Center versus a historical control group of 151 patients who received surgery plus radiotherapy without adjuvant treatment between January 1972 and December 1973. All patients had stage II or III breast cancer and had at least one involved node at the time of surgery.

Before considering the regression models used for evaluating the results of this study, the main arguments for conducting historical control studies will be summarized. Gehan and Freireich (1981) indicate that the major reasons are: non-randomized (historical control) studies require fewer patients and proceed more quickly so that new knowledge is gained faster; and there is no ethical dilemma since the clinical investigator is always administering the treatment he believes best for the disease under investigation. Confirmation of results by the same or other investigators provides a way of convincing clinical scientists of the validity of conclusions reached.

Recruitment of patients to a study involving a single new treatment is easier than registering patients into a study involving randomization among two or more treatments. Randomization tends to discourage participation in the clinical trial by both referring physicians and patients.

The outstanding criticism of non-randomized clinical trials is that consciously or unconsciously patients may be selected to receive the new treatment who have a more favorable prognosis than patients receiving the standard treatment. If the patient characteristics related to prognosis are assumed known, then the technique of stratifying patients and making adjustments for prognostic characteristics by regression models may be utilized to compare results between a new and standard treatment. Designers of randomized comparative studies can rely on randomization, stratification and techniques of adjustment in conducting and analyzing their studies. The extra control obtained by virtue of randomization would be expected to provide only a slight advantage over planners of non-randomized studies. Randomization of patients guarantees comparability only on the average and patients could be non-comparable in any particular study.

Clinical studies have been conducted in cancer research since the mid-1950's and major advances have been made in acute leukemia, choriocarcinoma, lymphoma, lung cancer, osteosarcoma, breast cancer, and soft tissue sarcoma. Nearly all the major therapeutic advances were discovered in non-randomized historical control studies.

Returning to the adjuvant study of breast cancer, a summary of the results of Buzdar et al (1978) will be given.

Characteristics (and codes) available for both groups of patients were: age (years), menopausal status (1 = premenopausal, 0 = other), size of primary tumor (2 = < 3 cm, 4 = 3-5 cm, 7 = >5 cm), stage of disease (2 = II, 3 = III), location of surgery (1 = M.D. Anderson Hospital, 0 = other), number of involved nodes (2 = < 4, 7 = 4-10, 12 = > 10), race (1 = white, 2 = other), and type of surgery (1 = radical or modified-radical surgery, 0 = other). There was some evidence that prognostic characteristics were not comparable between groups. Characteristics tending to favor the control group were age and menopausal status, since there was a higher percentage of patients who were > 50 years old (68% vs 52%) and peri- or post-menopausal (75% vs 58%). Patients in the FAC-BCG group were favored with respect to size of tumor, stage of disease, and type of surgery. There was a lower percentage of patients in the FAC-BCG group with primary tumors > 5 cm (15% vs 26%) and stage III disease (25% vs 40%) and a higher percentage receiving radical or modified-radical mastectomy (85% vs 55%). The groups were comparable in the distribution of the number of nodes involved.

Cox's regression model was fitted to the disease-free survival data from the 151 control patients in forward stepwise fashion and the following regression model was obtained:

$$\log \frac{\lambda(t;\underline{z})}{\lambda_o(t)} = 0.1110 \text{ (number of nodes-6.2)} + 0.8122 \text{ (stage - 2 4)} + 0.8720 \text{ (menopausal status-0.3)}.$$

The significance levels of the patient characteristics entering the model were: number of nodes ($P < 0.01$), stage of disease ($P = 0.02$), and menopausal status ($P < 0.01$). Favorable characteristics were: small number of nodes, stage II disease, and postmenopausal status.

Cox's regression model was then fitted to the combined data from FAC-BCG and control patients to test whether adjuvant treatment affected disease-free survival. Type of treatment (0 = control, 1 = FAC-BCG) was included as a variable in the model, in addition to number of involved nodes, stage of disease, and menopausal status. The equation obtained by fitting the model to the combined group of control and FAC-BCG patients was:

$$\log \frac{\lambda(t;\underline{z})}{\lambda_o(t)} = -1.6982 \text{ (treatment-0.5)} + 0.0965 \text{ (number of nodes-6.3)} + 0.8616 \text{ (menopausal status-0.3)} + 0.6563 \text{ (stage-2.3)}$$

The first variable entering the model in the forward stepwise fitting process was type of treatment ($P < 0.01$), menopausal status ($P = 0.01$), menopausal status ($P = 0.01$), and stage of disease ($P = 0.01$). Type of treatment was the variable most significantly related to disease-free survival, with patients in the control group having 5.5 times the risk of relapse per unit time as patients receiving FAC-BCG.

The regression model fitted to data from the control group only (i.e., with the variables number of nodes, stage of disease, and menopausal status) was applied to the data from both the control and FAC-BCG groups to obtain the hazard ratio, $\lambda(t)/\lambda_o(t)$, in each group of patients. The median hazard ratio, $\lambda(t)/\lambda_o(t)$, was .875 in the control group and it was also .875 in the FAC-BCG group, suggesting that differences in the distribution of prognostic factors between groups tended to balance out so that the overall risk of relapse per unit time was comparable. Hence, an overall comparison in disease-free survival was made between groups. As of March 1984 the estimated percentages of patients disease-free at 7 years were 55% in the FAC-BCG group and 35% in the control group; there was a highly statistically significant difference between the disease-free survival curves ($P < 0.01$). Comparisons of the FAC-BCG and the control groups were also made within subgroups of patients comparable in hazard ratio and individual patient characteristics, and these analyses confirmed the highly statistically significant overall difference between FAC-BCG and control patients. Overall survival at 7 years is 66% for FAC-

BCG and 48% for the control group (P <.01). A Cox regression analysis has confirmed the significance of the difference in survival, adjusting for prognostic factors.

In the course of analyzing this study, several further questions were investigated to assure the validity of the results obtained. The questions were: Does the significance of the treatment effect depend upon the particular set of three patient characteristics related to disease-free survival? A second question considered was whether the control group in this study was different from control groups in other studies, since a beneficial effect of adjuvant treatment might have been observed because of the adverse experience in the control group. Thirdly, patients in the control group were analyzed to determine whether there was evidence that prognosis was improving with time, possibly suggesting that the benefit for adjuvant treatment was simply the next step in an evolving process. Further analyses given in Gehan, Smith and Buzdar (1980) confirmed that the answers to all of these questions was "no", further evidence of the benefit of adjuvant treatment. These results indicate the precautions that should be taken in analyzing and interpreting historical control studies.

REFERENCES

BUZDAR, A., GUTTERMAN, J., BLUMENSCHEIN, G. R., HORTOVAGYI, G. N., TASHIMA, C. K., SMITH, T. L., HERSH, E. M., FREIREICH, E. J.,GEHAN, E. A. (1978), Intensive post-operative chemo-immunotherapy for patients with stage II and III breast cancer. Cancer 41: 1064-1075.

COX, D. R. (1972), Regression Models and life tables (with discussion) Journal of the Royal Statistical Society B, 34: 187-202.

COX, D. R. and OAKES, D. (1984), Survival Data Analysis, Methuen & Co., London.

GEHAN, E. A. and FREIREICH, E. J. (1981) Cancer clinical trials: a rational basis for use of historical controls. Seminars in Oncology 8, 4, 430-436.

GEHAN, E. A., SMITH, T. L., BUZDAR, A. U. (1980), Use of prognostic factors in analysis of historical control studies. Cancer Treatment Reports 64Ñ 2-3, 373-379.

KALBFLEISCH, J. and PRENTICE, R., The Statistical Analysis of Failure Time Data, John Wiley and Sons, N. Y.

LAWLESS, J. F. (1982) Statistical Models and Methods for Lifetime Data John Wiley and Sons, N. Y.

LEE, E. T. (1980), Statistical Methods for Survival Data Analysis, Lifetime Learning Publications, Belmont, CA.

ANIMAL EXPERIMENTS AND CLINICAL TRIALS

David A. Schoenfeld, Ph.D.
Dana-Farber Cancer Institute and Harvard School of Public Health
Boston, MA. 02115

ABSTRACT

Special statistical methods have been developed for toxicity experiments on animals and for clinical trials. The lecture will discuss those methods which involve determining the effect of dependent variables (covariates) on an independent variable (outcome). Some of these methods involve only one covariate and thus are not part of multiple regression but we thought that their importance justified their inclusion in this series of lectures.

Animal experiments are often used for evaluating the toxicity of a substance. In the standard experiment, the animals are divided into equal groups and one group is not exposed to the substance and the other groups are given increasing doses of the substance. The animals are then watched for toxic outcomes such as death, weight loss, and tumor development. Four common statistical problems in these experiments will be discussed.

1) How does one test for and estimate a toxic effect? Tests of trend are far more powerful than one way analysis of variance or chi-squared tests. Tests based on a parametric model that does not fit the data will be more powerful than these omnibus tests. In the absence of a parametric model, isotonic regression can be used to estimate dose effects and will often provide the best test of trend.

2) How does one extrapolate the results of the experiment to doses below the lowest dose? In carcinogenicity testing, the experiment is conducted at doses far above the usual human exposure. One method that is often used is based on the multi-hit model. An alternative approach is to extrapolate linearly from an upper confidence bound on the lowest dose. This bound can be found parametrically or by isotonic regression.

3) How does one extrapolate the results to humans? One experiments with small mice, with fast metabolisms that have a short lifetime and extrapolates to large humans, with slow metabolisms who live long lives. It is well established that using the surface areas of animals to equate dosages is better than using mass. It is usually thought that the long life of the human is balanced by his slow metabolism so exposure duration can be measured in life times for both species.

4) How are sample sizes decided upon? Since power and significance level are usually chosen to be a standard value, sample size is controlled by the difference that must be detected by the experiment. One must determine the level of toxicity that when extrapolated to human exposures, would pose a threat.

Clinical trials are used to test therapies for safety and efficacy. There is a tendency to assume that the statistical considerations of all clinical trials are the same. However, there are several types of trials that require different considerations than the "Phase III" trial to compare a new therapy to a standard therapy. In pilot studies, the main purpose may be to determine if there are any frequent severe side effects. In Phase II studies, the main purpose is to determine whether a drug may be effective. A false positive claim for the treatment may be better than a false negative claim. In dose finding studies, the purpose of the study is to determine which dose is best. Simple formulae can be used to find sample sizes for each of these studies and in each case, the sample size will be considerably smaller than that in a Phase III study.

An important question is whether a randomized control group is necessary in a clinical trial. The answer to this question depends on the role of the clinical trial in the national research effort. For instance, a single institution clinical trial of a new technique will usually be replicated before the technique reaches usual medical practice. Thus, a single institution study that makes a false positive claim may not create problems. On the other hand, if the new treatment is rejected as ineffective, it may never be tried again. Since historical controls allow a four-fold reduction in sample size, they could be used in such a study. On the other hand, covariate adjustment ordinarily cannot substitute for randomized controls in a confirmatory clinical trial. One problem with covariate adjustment when used with historical controls is that misspecification of the model and changes in the definition of the covariates over time can change the estimate of the treatment effect and cause misleading results.

1. ANIMAL EXPERIMENTS

The most common design of an animal toxicology experiment is to divide the animals into groups and give each group a different dose of the test substance. Suppose that each group has n_i animals for $i=1,\ldots,m$ and group i is given dose d_i. Assume that $d_1 \leq d_2,\ldots,\leq d_m$. In most experiments one dose, say d_1, is zero and the corresponding group is known as the control group. For each animal, a response y_{ij} is noted. In some experiments, the response is binary, say survival and in others it is continuous. Define $y_{i.} = \sum_{j=1}^{n_i} y_{ij}/n_i$. Furthermore, let μ_i be the expected value of y_{ij}. Then the plot of μ_i versus d_i is the theoretical dose response curve. The data from the experiment can be used to estimate this curve and provide a measure of the uncertainty of the estimate. However, the statistical analysis of the data usually focuses on some aspect of the dose response curve.

1.1 Testing for a Toxic Effect

One aspect of the dose response curve of primary interest is the first value of d_i where $\mu_1 = \mu_i < \mu_{i+1}$. This value of d_i, say d*, is estimated by testing the hypotheses $H_i: \mu_1 = ,\ldots,=\mu_i$ and letting d* be the last dose for which this hypothesis is accepted. Then d* is called the "maximum safe dose". The problem with the concept of "maximum safe dose" is that if the μ_i increase gradually, then the maximum safe dose will depend on the number of animals tested in each group and on the statistical test used to test H_i as well as the value of $\mu_1,\ldots,\mu_m$. Thus, the determination of a maximum safe dose, is only appropriate if there is a very strong belief that there is a dose d* which is nontoxic. That is, that the dose response curve has a threshold. There are varied methods used to test the hypotheses $H_1,\ldots,H_m$ to determine the maximum safe dose, unfortunately most of the methods that are in common practice are adaptations of methods that were developed for other purposes. The most common method is first to test H_m, the hypothesis that all the μ_i are equal using a one way analysis of variance (or equivalently a chi-squared test if y_{ij} is binary) and then simultaneously test $H_1,\ldots,H_{m-1}$ using a multiple comparison procedure like Dunnett's or Duncan's procedures. These procedures will produce a set of $\{d_i\}$ where μ_i is significantly greater than μ_1. Then d* is chosen to be the dose below the minimum of this set.

The preliminary test of H_m is completely unnecessary since the multiple comparison procedures also test H_m. Furthermore, chi-squared test and one way analysis of variance are very inefficient for testing for a non-zero dose response curve. To understand this phenomena, think of $\mu_1,\ldots,\mu_m$ as a vector μ in m dimensional space. The power of a test for H_m is the probability of rejecting H_m. If μ is a constant vector, then the power is .05; otherwise, it is some function of μ. Different tests of H_m will have different power functions and for the common tests, a relatively high power function at one value of μ will be compensated by a relatively low power at another value of μ. In toxicology experiments, one can be reasonably sure that $\mu_1 \leq,\ldots,\leq \mu_m$ so we require a test which will have high power in a corner of m space that contains 1/m! of the volume of that space. A one way analysis of variance (when the μ_i are equal) is designed to have equal power for all points that are equidistant from the subspace where $\underline{\mu}$ is the constant vector. Thus, its power is uniform throughout m space and to achieve power in this large volume, its power is low in the corner of interest to the experimenter. Dunnetts test is designed to be powerful when all but one of the components of μ are equal and Duncans test is designed to be powerful when some subject of $\{\mu_1,\ldots,\mu_m\}$ has a different value than the remainder. Both tests have high power in a different region than $\mu_1 \leq,\ldots,\mu_m$.

An excellent procedure for finding d* if y_{ij} is continuous, and the n_i are equal to n was developed by Williams (1971). First one finds the isotonic regression of $y_1,\ldots,y_k$. The isotonic regression, denoted by $\mu_1^*,\ldots,\mu_k^*$, of $y_1,\ldots,y_k$ with weights $w_j =$

n_j is the maximum likelihood estimate of the vector $(\mu_1,\ldots,\mu_k)$, subject to the condition that $\mu_1 \leq,\ldots,\leq \mu_k$. To calculate the isotonic regression, inspect $y_1,\ldots,y_k$ and if $y_j > y_{j+1}$, replace both y_j and y_{j+1} by $(w_j y_j + w_{j+1} y_{j+1})/(w_j + w_{j+1})$. The process is repeated until the sequence is in ascending order. Whenever adjacent values are equal, they are combined as a group. This, if at any step $y_j = y_{j+1} > y_{j+2}$, then at the next stop y_j, y_{j+1} and y_{j+2} are replaced by $(w_j y_j + w_{j+1} y_{j+1} + w_{j+2} y_{j+2})/(w_j + w_{j+1} + w_{j+2})$. The resulting sequence is $\mu_1^*,\ldots,\mu_k^*$. This and other algorithms for isotonic regression are presented in Barlow et al. (1972). Williams statistic rejects the null hypothesis when $(\mu_m^* - y_{1.})/$ $[2 \sum_i \sum_j (y_{ij} - y_{i.})^2 / \{n\, m(n-1)\}]^{1/2}$ is larger than a critical value which Williams has tabulated. When the y_{ij} are binary but n_i are large, the same procedures can be used by replacing $y_{i.}$ by $\sin^{-1}\sqrt{y_{i.}}$ and the denominator by $(1/2n)^{1/2}$.

Those methods could be extended to continuous variables where the n_i are unequal or to situations when the n_i are small and y_{ij} are binary. However, since they cannot presently be used in these situations, an alternative method of estimating d* if y_{ij} are continuous is to form the t statistics, say t_k, $k=2,\ldots,m$, to test for $\beta=0$ when $\mu_i = (i-\bar{i})\beta + \mu_0$ for $i=1,\ldots,k$, and $\bar{i} = \sum i\, n_i / \sum n_i$. Assuming that the variances of $y_{i.}$ have been made homogeneous by some suitable transformation, the same estimate of σ^2 can be used in each t statistic. Thus, each statistic will have $\Sigma(n_i - 1)$ degrees of freedom. Then each t statistic is tested for significance starting with t_k until a null hypothesis is detected. For binary y_{ij}, the procedure can be applied although in this case, the normal distribution is used and

$$t_k = \sum_i n_i (i-\bar{i}) \sin^{-1}\sqrt{y_{i.}} / \{.25 \sum_i n_i (i-\bar{i})^2\}^{1/2}$$

If the n_i are small, a permutation test based on $\sum_i n_i (i - \bar{i})\, y_{i.}$ can be used (Schoenfeld, 1983).

1.2 Estimating the Dose Response Curve

If one wishes to estimate a dose response curve without assuming any particular parametric model, then the method of isotonic regression can be used. The method gives the maximum likelihood estimates of $\mu_1,\ldots,\mu_m$ under the sole assumption that $\mu_1 \leq \mu_2,\ldots,\leq \mu_m$. One novel use of this method is when there is one observation per dose and the data is a censored exponential survival time. That is, suppose $t_1,\ldots,t_m$ are either death or follow-up times and let $\delta_i = 1$ if t_i is a death time and $\delta = 0$ otherwise. Finally, assume that the probability of death before time t given dose d_i is $1 - e^{-\theta_i t}$ where $\theta_1 \leq,\ldots,\leq \theta_m$. Thus, survival decreases with increasing dose. Then use the previous algorithm with $y_i = \delta_i/t_i$ and $w_i = t_i$. Call the result $\hat{\theta}_1,\ldots,\hat{\theta}_m$. Because the likelihood of negative exponential survival times is the same as that of poison counts a result of Barlow et al, pg. 43 can be used to show that $\hat{\theta}_1,\ldots,\hat{\theta}_m$ are the maximum likelihood estimates. For display purposes, it may be better to plot log $(.5)/\theta_i$ $i=1,\ldots,m$, which are the maximum likelihood estimates of the median survival. Figure 1 shows the results of this calculation for a clinical example. In this case, d_i is the patients age. Notice that the graph for each treatment is a step function with plateaus where survival times are pooled. The advantage of this display is that it shows the nonlinear association of survival with age and also shows that for old patients chemotherapy was not of value while for younger patients, there was a large difference in the survival of the treatment groups.

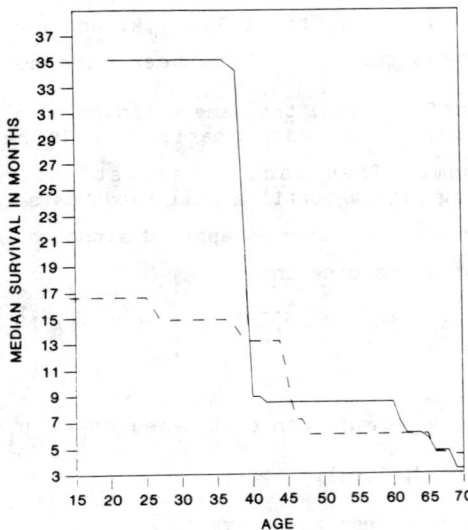

Figure 1. Median Survival by Age for Control (dashed line) and BCNU (solid line).

Confidence intervals can be found around a dose response curve can be found by isotonic regression (Schoenfeld, 1982). When the dose response curve does not rise too steeply, these intervals are smaller than the intervals about each mean based on the standard error. The intervals compare favorably with those of a linear model.

1.3 Low Dose Extrapolation

An animal experiment to determine whether a chemical is carcinogenic is usually conducted with 50 rodents in each of three groups. One group receives normal feed, one group receives close to their maximum safe dose and one group receives half that dose. In this context, "safe" generally means a dose which will cause no appreciable reduction in lifetime and will cause no more than a 10% decrease in body weight. The animals are followed for their natural lives and the number of animals with tumors are counted. In some situations, the times until tumors arise are also analyzed. This experiment is conducted on male and female mice, and rats, thus, there are four independent experiments. When a test of trend in any experiment rejects the null hypothesis, the compound is considered a carcinogen. If it is a food additive, the law specifies that it should be banned although exceptions, such as saccharin, are made. If the compound is not a food additive, the maximum safe dose for humans is computed. In this context, "safe" is defined as that dose that would produce only one cancer in one million people.

Doses that are "safe" to the animals by the definition of no weight loss or survival are usually tens of thousands of times higher than the doses that are "safe" to a human population in that they would produce a cancer rate of less than 10^{-6}. In fact, if this were not the case, the experiment would be completely unable to detect carcinogenicity. Thus, one must extrapolate from the number of tumors among 50 animals at the doses given in the experiment down to the very low doses that might produce one cancer in one million animals. Converting this dose to humans is discussed in the next section.

The extrapolation from animal doses to human exposure levels depend on the model that is chosen for the dose response curve. To simplify the discussion, suppose that there is no tumors in the control group and that $P(d)$ is the probability of a tumor at dose d. The "one hit" model assumes that $P(d) = 1-e^{-\theta d}$. This is also called a linear model because for small d $P(d)$ is approximately equal to θd. One can show that if the mechanism of carcinogenesis is the same as that which produces cancer in nature and the two sources add, then this model is appropriate. A generalization of this model is the "multi-hit" model (Guess and Crump, 1977) which is defined by $P(d) = 1-\exp\left(-\sum_{i=1}^{n} \theta_i d^i\right)$, $\theta_i \geq 0$. The "one-hit" and "multi-hit" models assume that a cell becomes cancerous when one molecule of the carcinogen hits a receptor by chance ("one hit") or when the receptor requires n molecules "multi-hit". Actually, these idealized biological models need not hold for the "one-hit" and "multi-hit" models to be useful. The "multi-hit" model is fairly general and might fit several different situations.

Furthermore, if $\theta_1 \neq 0$, then for small doses, the one-hit model gives the same results as the multi-hit model.

Since there is sampling variation in the estimation of the parameters of these models, one often calculates lower confidence bound on the maximum safe dose. A conservative approach to this problem is to assume that although a multi-hit model may be appropriate in the experimental range, the more conservative linear model will hold for low doses. Thus, Farmer, Kodell and Gaylor (1982), find the upper confidence bound on $P(d_1)$ using the multi-hit model and then extrapolate linearly from this point downward. This will yield a lower confidence bound on the maximum safe dose. The upper confidence bound on $P(d_1)$ could also be found using isotonic regression. The following is an example of these methods used to determine the safe dose of the solvent Dioxane which was present in the defoliant agent orange.

To simplify the discussion, consider only the data for the frequency of hepatocellular carcinoma in male mice. In the NCI experiment, the mice received .5% and 1% Dioxane in their drinking water. The control group had 2/49 cancers and the two treatment groups had 18/50 and 24/47 cancers, respectively (NCI, 1978). We assume that the probability of a spontaneous tumor is independent of that of an induced tumor. Table 1 shows the results using three different models. The logit and Weibull models were fit using Risk 81, a program of Kover and Krewski (1981) and the one-hit model was fit by least squares using MLAB (Knott, 1979). A multi-hit model with a quadratic term was also fit but this term had a negative coefficient indicating that the quadratic term should be zero. The logit and Weibull models with independent spontaneous tumor rates have three parameters so they fit the data perfectly. The one-hit model had expected rates of (.05, .33, .53) compared to the observed rates of (.04, .36, .51). Notice that in this case all the models except the probit give the same safe dose. The lower confidence bounds are lower in the three parameter models than in the two parameter models. Based on this analysis, Dioxane would pose a threat to humans at any concentration at which it would be detectable.

TABLE 1 Results of a Risk Assessment of Dioxane

Name	Formula	Safe Dose	Lower Confidence Bound
Probit	$1-\Phi(2.67+.59 \log d)$	$.3 \times 10^{-5}$	$.6 \times 10^{-9}$
Logit	$\{1+.07d^{.59}\}^{-1}$	$.5 \times 10^{-8}$	$.7 \times 10^{-15}$
Weibull	$1-\exp\{-19.8d^{.73}\}$	1.2×10^{-8}	$.3 \times 10^{-18}$
One Hit	$1-\exp\{-73d\}$	1.4×10^{-8}	1.2×10^{-8}
Linear	$66d$	1.5×10^{-8}	1.3×10^{-8}

1.4 The man to mouse problem

If d is a dose with a given effect on an animal species, what dose d* will have an equivalent effect on a human? The ratio d*/d is called the application factor. This question is essentially insoluble because the answer may depend on the compound in question. Given the impossibility of this question, the only reason that animal experiments are useful is that one only needs a very rough approximation of d^*/d. In experiments of acute toxicity, safety factors like dividing the dose by 100 are used after an animal dose has been converted to a safe human dose. In carcinogenicity studies, the variation in maximum safe dose determined by the model for low dose extrapolation is much greater than the variation of d^*/d so again only a rough approximation of the application factor is necessary.

In converting doses from animal to human terms, it is necessary to account for differences in size and duration of exposure. When a dose is given in feed, it is usually expressed in terms of percentage of feed. If the human exposure would also be in food, then this method of expressing dose rates automatically accounts for the different size of men and mice. On the other hand, if gavage (injection into the mouse's stomach) is used, the dose rate for mice might be expressed as mg/kg. In going from species to species in animal experiments, it has been shown that using surface area to measure doses predicts toxic response better than using mass (Harwood, P.D., 1963).

Both humans and mice are exposed to toxins for the duration of their lifetimes. In most extrapolations, this is considered an equivalent exposure even though the lifetimes of men and mice are of very different lengths. The concept is that the faster metabolism of the mouse compensates for its shorter lifetime.

1.5 How are Sample Sizes Decided

To determine sample size for a dose response experiment, we determine the power desired, usually .8. The significance level of a test, is usually .05. Then we must specify the difference that we wish to detect. The specification of this difference say Δ, is the most difficult judgement in a sample size determination. In an efficacy study, Δ can be taken to be the smallest benefit of the compound that would be important clinically if the compound were given to humans. In a toxicity study, Δ would be the smallest effect that would pose a significant threat when extrapolated to human exposures.

The latter consideration is often overlooked in carcinogenicity studies. Suppose that a carcinogenicity bioassay is conducted on a compound which has a human exposure of d_0 and the maximum safe dose in animals (safe in terms of no decrease in survival or weight) is d_1. Using the one hit model and assuming no spontaneous tumors, we wish to show that the drug is a carcinogen if

$$P(d_0) = 10^{-6} = 1 - e^{-\theta d_0}.$$

Thus, $\theta = -\log(1 - 10^{-6})/d_0$, and

$$P(d_1) = 1 - (1 - 10^{-6})^{d_1/d_0} \approx 10^{-6} d_1/d_0.$$

Suppose that to test for a carcinogen effect, we use a test of trend based on $S = y_3 - y_1$, where y_1, y_2, y_3 are the proportion of tumors among the n animals in the control, half dose and full dose groups respectively. (This test is equivalent to the test based on $\sum_{i=1}^{3} i\, y_i$). The best power is obtained when $P(0) = 0$. In this case, the null hypothesis is rejected when y_3 is large. The exact test would reject when $\binom{n - ny_3}{ny_3} / \binom{2n}{n} < .05$. This is approximately equal to the criteria $(.5)^{n\, y_3} < .05$ or $n\, y_3 \geq 5$. The power of the procedure is then $\beta = 1 - \sum_{i=0}^{4} \binom{n}{i} (10^{-6} d_1/d_0)^i (1 - 10^{-6} d_1/d_0)^{n-i}$. For instance, for $n = 50$ to be an adequate sample size, $(d_1/d_0) \geq 1.3 \times 10^5$ ($\beta = .80$). The situation is worse when there are spontaneous tumors. Using a normal approximation, we would reject when $S > 1.645 \{y_1(1-y_1)/n + Y_3(1-Y_3)/n\}^{1/2}$, supposing a 50% spontaneous tumor rate. $\beta = 1 - \Phi(1.645 - \sqrt{2n}\, \{P(d_1)\})$ where in this case $P(d_1)$ corresponds to the increase in the tumor rate due to dose d_1. Thus, to achieve 80% power with $n = 50$, $d_1/d_0 > 2.5 \times 10^5$. Therefore, tests for carcinogenesis with 50 animals will only be adequate when the maximum safe dose for animals is of the order of 10^5 times the human exposure.

2. CLINICAL TRIALS

Sample Size Considerations for Clinical Trials

There have been several papers on finding sample sizes for the Phase III study designed to test a new therapy against an established standard. However, most clinical trials conducted at a single institution should have different statistical considerations. The most common type of single institution clinical trial is the pilot study. A small number of patients are given a new treatment. The purpose of the trial is to determine the complications of the new treatment and to get some idea of its efficacy. In order to calculate a sample size, it is necessary to formalize the purposes of such a study and calculate the sample size necessary to accomplish these purposes. The following is a brief discussion of how this is done. For a more complete description, see Schoenfeld (1980).

In terms of complications, the goal of a pilot study might be to uncover all frequent complications of treatment. Thus, if a complication would occur p proportion of the time in many patients, it is necessary to treat enough patients on the pilot study so that the probability of seeing the complication at least once is high.

Thus, if a complication is not seen at all, one can be reasonably sure that it will not occur as much as p percent of the time in a large study. Let β be the probability of seeing at least one occurrence of the complication in n patients. Then $\beta = 1-(1-p)^n$ and thus $n = \log(1-\beta)/\log(1-p)$. For $\beta = .8$ $n \approx 1.6/p$. Thus, if $p = .05$, $n = 32$, and if $p = .10$, $n = 16$.

Determining the sample size needed to show efficacy in a pilot study is more difficult. Assume that treatment effect is measured in terms of success or failure. Then there is a success rate p_1 where the new therapy would be an important improvement over the standard therapy and there is a success rate p_0 where the new therapy would not be any better than the standard therapy. If n patients are treated on the new therapy than one decides to test the treatment further if the observed success rate r is better than a predetermined cut-off value r_0.

The operating characteristics of the pilot study are defined by two numbers. The first is the probability of continuing to test the treatment if the rate is p_1. This probability β is given by

$$\beta = \Sigma \binom{n}{i} p_0^i (1 - p_0)^{n-i}$$

where the summation is over i greater than $n \cdot r_0$. The second is the probability of continuing to test the treatment if the rate is p_0. This probability α is given by

$$\alpha = \Sigma \binom{n}{i} p_0^i (1 - p_0)^{n-i}.$$

To determine whether a pilot study with sample size n is feasible, calculate α and β for various values of r_0 and see whether β is sufficiently high and α is sufficiently low. Since the purpose of a pilot study is to determine whether a treatment should be tested further, an α value of .05 is too stringent. It is important to realize that r_0 must be high enough so that if r is greater than r_0, there will be enthusiasm about the new treatment. The decision to test the new treatment further may be made by different investigators than the ones who designed the study or it may be made by funding agencies.

As an example of these considerations, consider the statistical considerations for a pilot study of a combination of alkylating agents combined with autologous marrow support, for malignant melanoma. The therapy involved a month of hospitalization, so it was much more expensive than standard therapy. The statistical considerations that appear in the protocol are as follows:

> We are attempting to develop a therapy for malignant melanoma which has a 70-80% response rate (CR + PR) on patients with advanced disease in the hope that such a therapy would be curative in patients with a smaller tumor burden. Since the response rate we observe on this trial may vary from the true response rate. We will consider STAMP I sufficiently active to warrant further study if the observed response rate is at least (10/15) 67%. With 15 patients treated, we will have a 72% to 94%

chance of seeing this many responses if the true response rate is between 70% and 80%. With conventional therapy with a response rate of less than 40% we would have less than 4% chance of seeing this many responses.

Another type of study that has different sample size considerations is the dose finding study. One may conduct a clinical trial in which doses d_1, d_2... are given. The purpose of the study is to pick the best dose for a future clinical trial against a standard treatment. One wants a small probability of making an incorrect choice of dose if one dose is better than the others. For instance, suppose the endpoint is treatment success and one will pick the treatment with the best success rate. Suppose there are m treatments which we order by increasing theoretical success rate, i.e., $p_1 \leq p_2 \leq ,..., \leq p_m$. The $\sin^{-1}\sqrt{}$ transformation transforms each observed success rate s_i into an approximate normal variable with variance $1/4n$ and mean $\sin^{-1}\sqrt{p_i}$. The probability that the mth mean will be correctly chosen is greater than

$$\beta = \int_{-\infty}^{\infty} \Phi(x)^{m-1} \phi(x-\delta) \, dx$$

where $\delta = \sqrt{4n}\{\sin^{-1}\sqrt{p_{(m)}} - \sin^{-1}\sqrt{p_{(m-1)}}\}$, Φ is normal probability distribution function and ϕ is the normal density function. (This is shown by noting that the probability is minimized for fixed δ by letting $p_1 = p_2 =,..., = p_{m-1}$). To achieve $\beta = .8$, δ must equal 1.2, 1.7, 1.9, 2.1 for m equal to 2, 3, 4, 5, respectively. Thus, if m = 3, p_{m-1} = .5 and p_m = .65, then n = 31. Notice that to detect a difference between p_m = .65 and p_{m-1} = .5 at β = .8 and α = .05, would require n = 133 patients per arm. Thus, it requires much fewer patients to choose the best treatment than to show that one treatment is better than another. When survival is an endpoint, a similar calculation can be performed. In this case, $\delta = (np)^{1/2} \log R$ where p is the probability that a patient in the trial will be observed to die and R is the hazard ratio of the best treatment to the next best. If R=1.25, m=3, p=.75, then n=77 per arm. An example of a dose finding study is the RTOG-ECOG clinical trial of three hyperfractionation regimens whose statistical consideration section is as follows:

Assuming an exponential distribution of survival times, we will need to observe 126 total treatment failures to choose the dose level which yields the longest median survival. We can be 80% confident with this sample size that we have selected either the best dose level, or one within a 30% range of the best median survival. For example, if the true median survival times for dose levels 1, 2, and 3 were 8 months, 12 months, and 10 months, respectively, then we can be 80% confident that dose level 1 would not be chosen (since it is more than 30% worse than the best dose). We could not be so confident, however, in a choice between levels 2 and 3. We have specified that an error in choosing a dose level with a median survival within 30% of the best dose is not a major concern.

One of the most important questions in designing a clinical trial is whether to have a randomized control group. To achieve a specified power in a study with a randomized control group requires four times as many patients as would be required by an uncontrolled study. In an uncontrolled study, the test statistic is based on $Y_1 - \mu_0$ where μ_0 is the known historical mean, while in a controlled study, the statistic is based on $y_1 - y_0$, the difference in two means. Since two means are being estimated rather than one, the latter quantity has twice the variance of the former. Thus, to get the same power y_1 and y_0 must be estimated with twice the sample size y_1 would require in a comparison with μ_0. Many studies conducted by a single institution will be replicated before the results will be generally accepted. In this situation, the problems with an historically controlled study may be offset by the large increase in power that the study allows.

In order to decide when to use historical controls and how to use them properly, it is necessary to understand the pitfalls of their use. Historical controls will give misleading results when the patient population has changed over time and the characteristics that have changed have not been properly controlled for in the analysis or when there is a patient characteristic that affects outcome whose measurement has changed over time. It is well know that changing patient characteristics will bias an historical comparison. What is not well understood is that if prognostic characteristics are not modeled properly or the measurement of a characteristic has changed, biases can occur even if the variable has been used in the analysis.

Suppose, for example, that a clinical trial is conducted for a disease in which age is an important factor. Let y be the outcome measure and let x be age and let

$$p(y = 1|x) = \exp(3.33 - x|15)/\{1 + \exp(3.33 - x|15)\}.$$

Thus, $p(y=1|x=50) = .5$ and $p(y=1|x=30) = .79$, $p(y=1|x=70) = .21$. Suppose that in the historical control group, the ages are uniformly distributed between 30 and 70. If patients in the treatment group are 10 years younger than those in the control group then the success rate will go from 50% to 64% even if the treatments have the same effect. Now suppose the investigator decides to control for the effect of age and forms two age groups, those under 50 and those over 50. In the historical control group, the success rates in these two age groups will be 65% and 34%, respectively. In the treatment group, those two age groups will have success rates of 72% and 42% and it will again appear as if the new treatment is better than the historical control although the difference in this case will be smaller. Thus, to properly control for a covariate in a historically controlled trial, it is necessary to correctly specify how the endpoint depends on the covariate. A similar problem does not occur in randomized trials where one can show that misspecification has a negligible effect on the significance level of a test for a treatment effect (Lagakos and Schoenfeld, 1984).

Another problem that can occur in historically controlled trials is that changes in the definition of covariates can bias a treatment comparison. Suppose x in the previous example were not age but some laboratory measurement whose meaning had changed over time. Suppose a patient in the historical control group that had a measurement of x would now have a measurement of 10 + x. In that case, patients under 50 in the treatment group would have a success rate of 73% and those over 50 in the treatment group would have a success rate of 42% which are again both superior to the 65% and 34% in the control group. This problem is not solved by using the logistic model to correct for x since the main problem is not caused by misspecification but by a change in the measurement in x.

Therefore, when historical controls are used, care must be taken to determine that the model fits the data and that the covariates that are measured in the treatment group had the same meaning when they were measured in the control group. Tests of goodness of fit for survival data have been developed by Schoenfeld (1980, 1982) and test of fit for response data have been developed by Tsiatis (1980).

ACKNOWLEDGMENT

This research was partially funded by Public Health Service Grants CA-23415, ESO-2709 and by a grant from the Mellon Foundation.

REFERENCES

BARLOW, R. E., BARTHOLOMEW, D. J., BREMNER, J. M. and BRUNK, H. D. (1972) *Statistical Inference Under Order Restrictions*, Wiley & Sons, New York.

FARMER, J. H., KODELL, R. L. and GAYLOR, D. W. (1982) "Estimation and Extrapolation of Tumor Probabilities From a Mouse Bioassay with Survival Sacrifice Components," *Risk Analysis*, 2: 27-34.

GUESS, H. A. and CRUMP, K. S. (1977) "Can We Use Animal Data to Estimate 'Safe' Doses for Chemical Carcinogens," *in Environmental Health: Quantitative Methods*, SIAM, Philadelphia, 13-30.

HARWOOD, P. D. (1963) "Therapeutic Dosage in Small and Large Mammals," *Science*, 139: 684-685.

KNOTT, G. D. (1979) "MLAB - A Mathematical Modeling Tool", *Computer Programs in Biomedicine*, 10(3): 271-280.

KOVER, J. and KEWSKI, D. (1981) "Users Instructions for Risk 81: A Computer Program for Low Dose Extrapolation of Quantile Response Toxicity Data," Department of Health and Welfare, Vanier, Ontario Canada.

LAGAKOS, S. W. and SCHOENFELD, D. A. "Consequences of Misfit for Covariate Models of Censored Survival Data," *Biometrics* in press.

NATIONAL CANCER INSTITUTE, (1978) *Bioassay of 1, 4-Dioxane for Possible Carcinogenicity CAS No. 123-91-1 NCI-CG-TR-80*, National Cancer Institute Carcinogenesis, Technical Resport Series No. 80, DHEW Publication No. (NIH) 78-1330.

SCHOENFELD, D. (1980) "Chi-Squared Goodness-of-Fit Tests for the Proportional Hazards Regression Model," *Biometrika*, 67: 145-153.

SCHOENFELD, D. (1980) "Statistical Considerations for Pilot Studies," *International Journal of Radiation Oncology, Biology and Physics*, 6: 371-374.

SCHOENFELD, D., (1982) "Residuals for the Proportional Hazards Regression Model," Biometrika, 69(1): 239-241.

SCHOENFELD, D. (1983) "Analysis of Categorical Data: Logistic Models," Statistics in Medical Research: Methods and Issues with Applications to Clinical Oncology, ed. by Mike, V. and Stanley, K., John Wiley and Sons. New York.

SCHOENFELD, D., (1983) "Confidence Intervals for Normal Means Under Order Restrictions, with Applications to Dose Response Curves," Technical Report 291Z, Dana-Farber Cancer Institute, Boston, MA.

TSIATIS, A. A. (1980) "A Note on a Goodness-of-Fit Test for the Logistic Regression Model," Biometrika, 67: 250-251.

OVERVIEW OF DISCRIMINANT ANALYSIS

Peter A. Lachenbruch
University of Iowa, Iowa City, IA. 52242

ABSTRACT

The basic object of Discriminant Analysis is to allocate an object (patient) to one of two or more groups (diagnostic categories) on the basis of a multivariate observation, $\underline{x}$. This observation may consist of basic patient information such as age, sex, height, weight, and additional information more specifically related to the disease in question. Thus, in a cancer study items such as number of positive nodes, size of tumor, etc. might be used to determine if the patient would be in a group which survives 3 or more years.

The earliest work was due to R. A. Fisher who proposed to find the **linear** combination of observations which best separated two groups of observations. This separation was best in the sense of maximizing the between groups difference of the linear combination relative to its standard deviation. This became known as the linear discriminant function (LDF). Extending this to multiple groups leads to an eigenvector solution. Three years later, Welch showed that the allocation rule which minimized the total error rate is based on the ratio of the probability density functions. The proof of this is identical to that of the Neyman-Pearson lemma. However, this assumes knowledge of the density functions. For the multivariate normal case where the parameters are estimated, this rule is identical to the Fisher LDF. This also leads to an easy generalization to multiple groups. The allocation rule for multivariate normal distributions with the same covariance matrix is to assign to the first group if

$$(\underline{x} - .5(\underline{x}_1 + \underline{x}_2))' \underline{S}^{-1} (\underline{x}_1 - \underline{x}_1) > C$$

where $\underline{x}_i$ is the mean in the ith group, $\underline{S}$ is the sample covariance matrix, and C is a cutoff point chosen to minimize the loss due to misclassification. If the costs of misclassification are equal, then $C = \ln(1-p)/p$ where p is the prior probability of group 1.

The usual assumptions of an LDF include:
1. Multivariate Normal Distributions
2. The same covariance matrix within groups
3. Correct classification of the training samples
4. Simple random samples

If these assumptions are not met, the LDF is not optimal and may be seriously in error. A series of robustness studies have been performed examining the first three of these assumptions. The effects of non-normality depend upon the sort of non-normal distribution is present. If the distribution is highly skewed, the performance of the LDF deteriorates markedly. Discrete non-normality does not seem to be too detrimental.

The effect of unequal covariance matrices depends on the separation of the means. If they are close the information provided by the covariance matrices is important and should be used. This leads to a quadratic discriminant function. This implies a need for a larger training sample size to allow good estimation of both covariance matrices.

If the training samples are incorrectly allocated, the means will be "closer" than they would be otherwise, but the estimated covariance matrix is also "smaller" and so the performance of the LDF does not deteriorate greatly. However, if one uses a Quadratic Discriminant Function, very bad things can happen; if there is initial misclassification the error rates increase dramatically, and the QDF is quite unreliable. (The QDF is the theoretically best rule when the covariances are unequal.)

A number of methods are available to evaluate a discriminant function. The most common criterion used is the error rate in each group, or the total error rate. This may be estimated by:

1. Use the Mahalanobis distance to estimate the true distance between groups (D^2) and compute $P(z < -D/2)$ where z is a standard normal variable. This is dangerous if the data are not normal.
2. Resubstitute the observations from the training sample into the LDF and count the misclassifications. This is biased and may be a cause for concern if the samples are small relative to the number of variables. For large data sets this is quite adequate.
3. The Leave-one-out (or Jackknife) method sequentially omits one observation from the data set, recomputes the LDF and classifies that observation. The number of errors is used as an estimate of the error rate. It attempts to protect against the bias of the resubstitution method in small data sets, but has a larger variance due to correlation among the pseudo-values.
4. The Bootstrap method computes many LDFs by taking repeated samples with replacement from the training sample and estimating the error rate from each. This method combines the best features of the resubstitution and leave-one-out methods, but it can be expensive.

Generally, methods which make use of distributional assumptions in the data are suspect.

Most of the major statistical packages have complete Discriminant Analysis programs. These programs include stepwise selection of variables, testing for the equality of covariance matrices, error rate estimation, and graphical display of the discriminant functions. The programs are:

1. BMDP7M

2. SAS PROCs - CANDISC, DISCRIM, NEIGHBOR, STEPDISC

3. SPSS[x] PROCEDURE DISCRIMINANT

4. P-STAT DISCRIM

1. INTRODUCTION

The fundamental goal of discriminant analysis is to allocate an object or patient to one of two or more groups on the basis of a multivariate observation $\underset{\sim}{x}$. The observation may consist of basic patient information such as height, weight, age, sex, socioeconomic status, and more specific information relative to the diseases under consideration. This additional data might consist of white blood count, number of positive lymph nodes, tumor size, and so forth.

I shall use a study of mass screening for breast cancer by absolute temperature thermography as an example (Haberman, 1980). I served as a consultant on this project, and the good features present here should be attributed to her and her colleagues. The primary goal of this study was to demonstrate the feasibility of using an absolute temperature thermographic sensing device as a mass screening method for breast cancer. This method registered the temperature in a given part of an image of a breast (called a pixel) and combined the numbers to decide whether the screened woman should be recalled for further tests. The question of what pixels to combine and how leads to the consideration of the choice of variables. Selection of variables is a key problem in any application. Once the choice of variables has been made, the choice of an allocation function may be addressed. Many such functions are possible: linear, quadratic, products of variables may be included.

In this study, the first phase was devoted to developing a prototype instrument, calibration devices, and to getting the bugs out of the procedure so that the performance of the system could be properly evaluated. For each breast, an array of numbers giving the absolute temperature in each region of the breast was obtained by the device. In the (apparently) usual thermographic method an infrared photograph is taken and then interpreted by a radiologist. All of the thermographic information from this system was obtained automatically, with no intervention needed from an operator other than to position the woman and make certain the skin temperature had stabilized to room temperature prior to sensing.

Once a set of thermographic images was obtained, and screening by other methods was accomplished, the problem was how to combine the thermographic data in such a way as to best separate those women who had breast cancer from those who did not. In this study, a combination of mammography, physical examination, and conventional thermography was used to recall women for further examination and biopsy if necessary. Histologic examination of biopsy was necessary to confirm a tumor. Thus, it was possible for a cancer to be missed. However, since women were screened repeatedly, a missed cancer (say in an early stage) was likely to be picked up at a later visit. About 10,000 women were screened as part of the Breast Cancer Detection Demonstration Project (BCDDP), to develop the variables that could be used to discriminate cancers from non-cancers, and to develop the allocation rules by which women were assigned to a "cancer" or "non-cancer" group. It should be emphasized that this assignment was not a "final" diagnostic category, but rather one in which the woman was recalled for further examination if results on the screening test suggested it. This is the appropriate use of screening - the error rates involved usually make it far too hazardous to venture anything like a diagnosis at this stage of examination. Two allocation rules were developed using a sample of 561 normals, 208 benign and 97 cancer cases. The first rule was to be used for subjects of any age, and the second one was for subjects under 50 years of age.

Where are we?

Thus far, we have discussed the basic object of an allocation problem, the choice of variables, and the selection of a training sample. In addition, such an allocation procedure may not be the final word on diagnosis, but more a suggestion for further study. This is particularly true in a screening program in which the vast majority of screened women do not have the disease. Finally, there may be subgroups in which different allocation rules might be applied.

Variables for the breast cancer study

In developing variables to detect suspicious patterns on the thermograms, the variables were grouped into general categories:
Total Breast - "one or both breasts may be generally warmer than normal, or one breast may be warmer than the other."
Quadrant - each quadrant of a breast may be compared with other quadrants, or with the corresponding quadrant on the contralateral breast.
Areola - warm regions, or asymmetric areas were important.
Vascularity - "pronounced difference in vascularity is highly suspicious for cancer."
Vein structure - may depend on quadrant, tortuousity, length.
Hot regions - are evidence of unusual heat concentration below the surface, or a network of capillaries.
Cold regions - usually assessed relative to hot veins or region evidence.
Blood perfusion
Note that in all of these classes of variable, asymmetries are especially noted. Overall, a total of 52 variables were constructed from the thermographic images. The report notes that the
"quality of these variates is dependent upon the following items: Environmental controls, Calibration of the instrumentation, Quality of the digital thermal imagery, Proper delineation of the breast extent."
These factors are, I suspect, well-known to this audience.
The subjects were divided into three classes defined as:
NORMAL - no previous history of breast cancer, no previous mastectomy, no breast symptoms, and "negative breast physical and mammography examinations at the visit during which the thermographic data were collected."
BENIGN DISEASE - no personal history of breast cancer, no previous mastectomy, confirmation of benign disease by pathological examination.

CANCER - no personal history of breast cancer or previous mastectomy, confirmation of breast cancer by pathological examination. Thus, these were all first cancer in women with both breasts.

A further complication is that breast cancer is not a single disease, but a group of several diseases which may have different growth rates, and different presenting signs and symptoms. This suggests that an approach based on mixtures of distributions might be helpful. This was considered in this problem, but it was found that using normal theory with unequal covariance matrices did as well as the considerably more complex procedure.

2. SOME THEORETICAL BACKGROUND

The earliest work in discriminant analysis was by R. A. Fisher (1936) who suggested that a linear combination of observation might be used to separate two groups. The choice of linear combination was to be based on that function which gave the greatest difference between groups relative to the within-groups standard deviation. This function can be shown to be proportion to

$$\underline{S}^{-1}(\bar{\underline{x}}_1 - \bar{\underline{x}}_2)$$

where S is the pooled estimate of the (assumed common) covariance matrix, and $\bar{\underline{x}}_i$ is the estimate of the mean in the $i^{\underline{th}}$ group. This formulation led to connections with other statistical problems including the Mahalanobis distance, Hotelling's T^2, and multivariate analysis of variance. This procedure was later shown to minimize the total error rate for the multivariate normal distribution by Welch (1939). In fact, Welch's methods showed how to generate optimal rules for <u>any</u> distribution. The proof of this was indentical to that for the Neyman-Pearson lemma. This requires complete knowledge of the underlying density functions including both the functional form and the values of any parameters in the function. This does not often come to the statistical analyst with the poser of the problem. Welch's procedure allows one to estimate the parameters of the function, and this leads, for the multivariate normal distribution to the function

$$D(\underline{x}) = (\underline{x} - .5*(\bar{\underline{x}}_1 + \bar{\underline{x}}_2)) \underline{S}^{-1}(\bar{\underline{x}}_1 - \bar{\underline{s}}_2) .$$

The allocation rule then becomes, assessing the unknown observation to group 1 if

$$D(\underline{x}) > \ln((1-p)/p)$$

and to group 2 otherwise. In this formula, p is the a priori probability of group 1. For a rare disease such as breast cancer this may be .0001 or less. This rule minimizes the total probability of misclassificiation which is the weighted sum of the probabilities of error, the weights being the a priori probabilities of each group. Thus, a premium is placed on correctly allocating the larger population. The total error rate is defined as

$$T = pP_1 + (1-p)P_2$$

where P_i is the error rate in group i. In the breast cancer example, a quite large error rate in the "rescreen group: can be tolerated to obtain a minimum total error rate. However, this defeats the purpose of the allocation in this context. The relative costs of misclassification can be considered to modify this problem. Alternatively, the cutoff point can be set so that one error rate is fixed at an acceptable level, and the other one determined. For example, one might decide that no more that 5% of the breast cancer cases should be misclassified. The cutoff point is chosen so that error rate among the "recall" cases is obtained and the error rate in the normals determined. Sometimes the groups will be sufficiently well separated that both error rates are below the target rate.

The assumptions which are required for the linear discriminant function are:
1. Both populations have an underlying multivariate normal distribution of the variables. This assumption is almost
always violated, and certainly is in the breast cancer screening problem. However, studies have indicated that the effect of violations of this assumption are not severe unless the data are badly skewed. One speculation is that the first order expansion of the distributions agree fairly well with that of the multivariate normal distribution.
2. Both groups have the same covariance matrix. The effects of violation of this assumption depend upon the amount of difference and the separation of the means of the groups.
3. The training samples (i.e. those samples which are used to estimate the parameters of the multivariate normal distributions) are correctly classified. For the linear discriminant function, the effects seem to be minimal, but for some other procedures, a substantial loss of discriminating ability may occur.
4. The training samples are simple random samples of the populations to be allocated. Thus, a sample which overrepresented a rare cancer might lead to an allocation rule which was not very useful for other, more common, breast cancers.

Figure 1 shows the pattern one might expect if the covariances are equal. There are two ellipsoids of concentration with the same shape, and a (hyper) plane separates them. The farther apart the means are, the better the separation.

Figure 2 shows the pattern one might expect if the covariance matrices are not the same. In this case the variance of the first variable in group 1 is about equal to the variance of the second variable in group 2, and the variance of the second variable in group 1 is about equal to the variance of the first variable in group 2. The correlation is positive in the first group and negative in the second.

After estimating the parameters of the linear discriminant function by maximum likelihood methods (this means using sample means and sample covariances to estimate the population parameters), an important question still remains: How good is the allocation rule we have just developed? There are two broad classes of answers to this question. The first class is based on theoretical considerations, usually involving the underlying distributional assumption of normality.

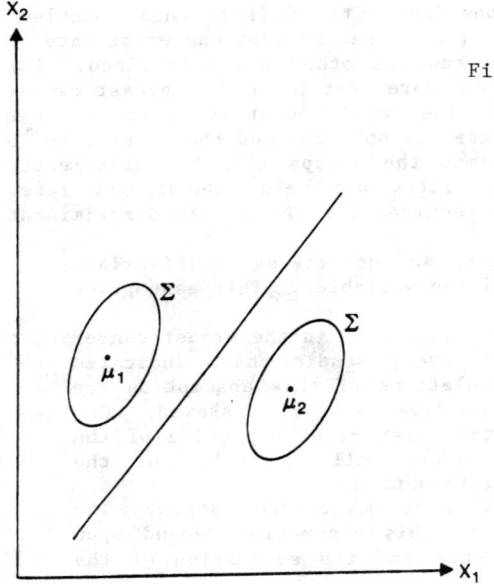

Figure 1. Illustration of the LDF Decision Boundary

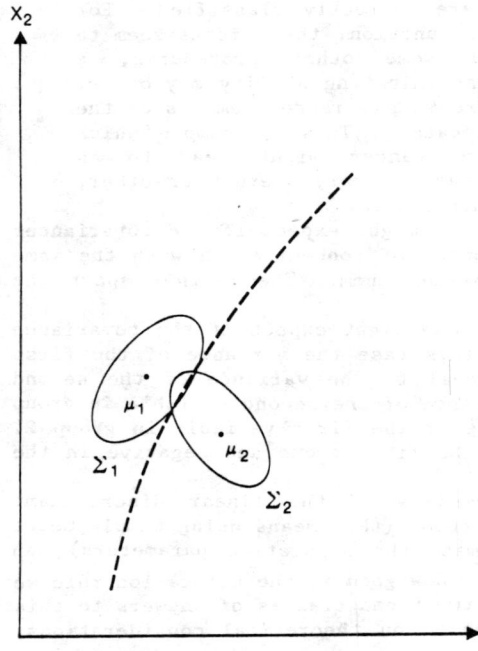

Figure 2. Illustration of the QDF Decision Boundary

It is easy to show that the error rate in each group is a simple function of the population Mahalanobis distance between the two groups. If the a priori probabilities are equal to .5, then each error rate is equal to

$$P_i = \Phi(-\delta/s)$$

i=1,2, where $\Phi(\cdot)$ is the cumulative normal distribution function and $\delta^2 = (\mu_1-\mu_2) \underset{\sim}{\Sigma}^{-1} (\mu_1-\mu_2)$ is the Mahalanobis distance. Essentially all theoretically based methods are elaborations of this. These methods all suffer from a potential lack of robustness when the distributions are not normal. While the discriminant function is quite robust, the error rate estimates seem not to be. The second class of methods is based on empirical estimates of the error rates. A seemingly obvious method of estimating error rates is to use the allocation rule on some handy sample of the two groups you wish to allocate. What could be handier than the samples which were used to develop the allocation rule itself? This method is called the resubstitution method, or the apparent error rate method. It is the method that is common to all of the major statistical packages. It is, however, biased and one should worry about this bias if the sample size is "small" relative to the number of variables. I get concerned if the number of observations isn't at least 2 or 3 times as large as the number of variables. For small numbers of variables, the ratio should be larger still. Two other empirical methods have been proposed: the leave-one-out method (called the jackknife method by BMDP7M) sequentially omits one observation from consideration, recalculates the rule, and allocates the omitted observation. This reduces the bias, but has a larger variance than the resubstitution method. For large samples, the resubstitution method is better, for small samples, the leave-one-out method is better. The bootstrap method computes many functions by repeatedly sampling the training samples (with replacement) as if they were a population. This seems to combine the best properties of the two previous methods. The cost of the bootstrap method is much higher than the leave-one-out method which is slightly more expensive that the resubstitution method.

3.SELECTION OF VARIABLES

In any study, there are more variables than can be effectively used in an allocation rule. These variables must be entered into the classifier in such a way as to extract the available information in a parsimonious manner. There are many methods of selection of variables. A simple one is to perform t-tests on each variable and order the variables on the basis of these tests. This is an effective method if the number of variables to be selected is small and the variables are not highly correlated. Correlation among variables has the effect of reducing the effect of some, and increasing the effect of other variables. For this reason, stepwise selection procedures have been developed.

Some variables should be included because biological theory indicates they are important. The statistical packages described later offer the possibility of forcing a variable to be entered, allowing variables to enter in a stepwise manner, or a combination of these methods. The t selection method is not supported.

Transformation of variables is sometimes needed. I have found that a transformation to remove skewness is the most important one that can be used. These transformations are usually a square root or logarithm.

For the breast cancer data, the investigators decided to use 5 to 10 variables, so that there were about 10 observations per variable in the smallest group. The authors tested normality, and equality of covariance matrices. After this they considered allocation rules based on likelihood ratio rules and on concentration ellipsoid models. The likelihood ratio rules are the familiar discriminant functions (linear or quadratic). The concentration ellipsoid models were used because the breast cancers were really a group of diseases, and a mixture model would be too complex. The concentration ellipsoid is based on the normal population and is chosen to include a large proportion of the normal group. This approach has problems because it does not lend itself to an effective variable selection procedure, and because a quadratic rule seems to work substantially better. A non-parametric procedure, the K-nearest neighbor rule, also was used. This procedure examines the K observations "closest" to the unknown one, and allocates it to the group which has the majority of the K observations. This rule allocates observations to the observation which has the largest non-parametric density estimate.

4. RESULTS IN THE BREAST CANCER STUDY

The two variables used to allocate women were the maximum of the warmest vein temperature in either breast, and the difference in the warmest area sizes between the two breasts. These were transformed by logarithms to obtain closer agreement to normality. Only the Cancer and the Normal groups are considered here, although the Benign group must be considered in practice. Recall that by adjusting the cutoff point we can have any level of false positive or false negative we desire. In this case, the investigators found a 10 per cent false positive rate led to a 70 per cent false negative rate. To detect 80 per cent of the cancers, a 45 per cent false positive rate was needed. In a large screening program, this leads to many recalls, and most of the women recalled will not have breast cancer. Figure 3 shows the distribution of the 181 normals and 27 cancers in women under 50 years of age. It is rather difficult to separate them visually, although the Cancer group seems to be aimed toward the upper right hand portion of the graph. Figure 4 gives the operating characteristic curve of the allocation rule based on the linear discriminant function using this data. The points labeled with A, are the observed values at various cutoff points, while the solid curve is the theoretical curve based on normal theory. The dashed line is the line on which error rates are equal; thus about 38% of each group would be incorrectly allocated if such a rule were used. Figures 5 and 6 give a histogram of the discriminant scores. We see the cancer patients scores tend to be smaller than the normals.

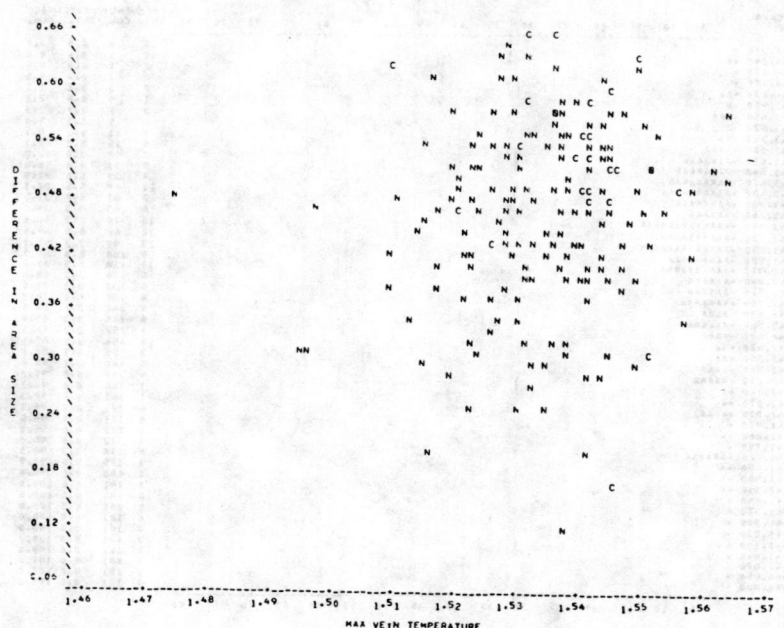

Figure 3. Scatter Plot for Subjects < 50 Years of Age

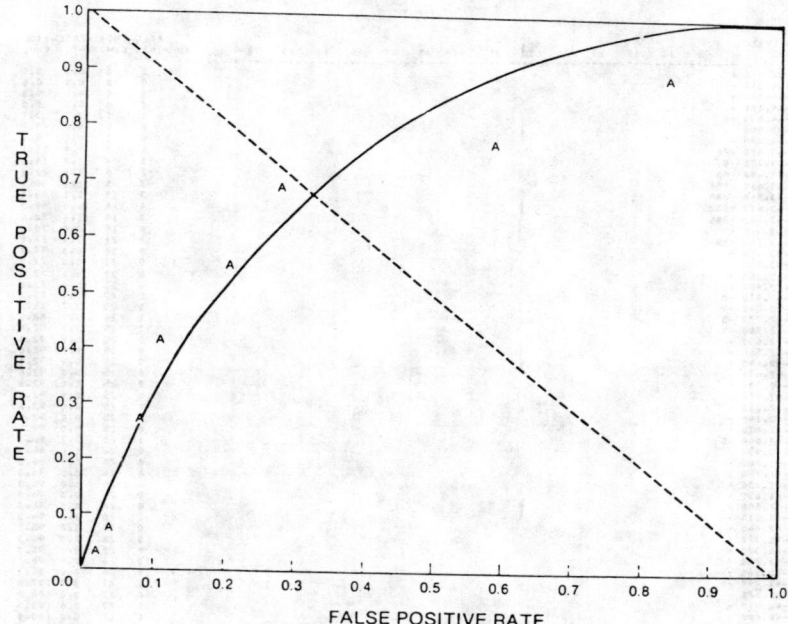

Figure 4. Operating Characteristics Curve for Subjects < 50 Years of Age

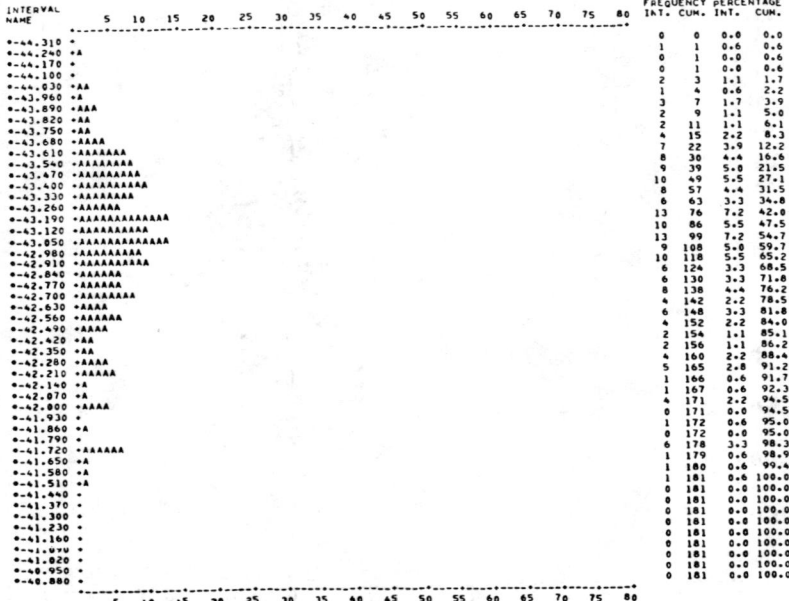

Figure 5. Histogram of LDF for Normals < 50 Years of Age

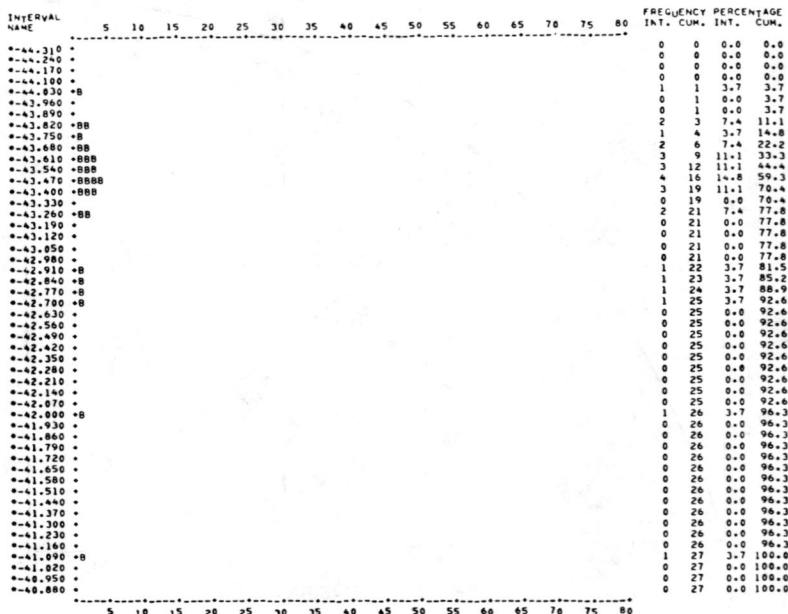

Figure 6. Histogram of LDF for Cancers < 50 Years of Age

Figures 7 to 10 give the corresponding data for the full set of 561 Normal and 97 Cancer subjects. Similar results hold, although the magnitudes of these scores are different. The data are far from clean, and the authors are quite open in discussing the limitations of their study. Among the problems they mention are:
1. Hardware failures: Scanner response not as good early in the study as later; Flat or clipped images; Split images; Grainy images; Tape read errors
2. Calibration errors: Bad hohlraum fit; Cold clipping (i.e., cold areas not recorded properly)
3. Procedural errors: Incorrect positioning, poor definition of breasts.

Many of these problems were corrected as the study progressed, so the frequency and magnitude of errors was reduced in later years. What is the effect of the "better data" in the last part of the study on the allocation rule? Of course, none can say for certain, but I am comfortable with the assumption that these errors affected all women equally.

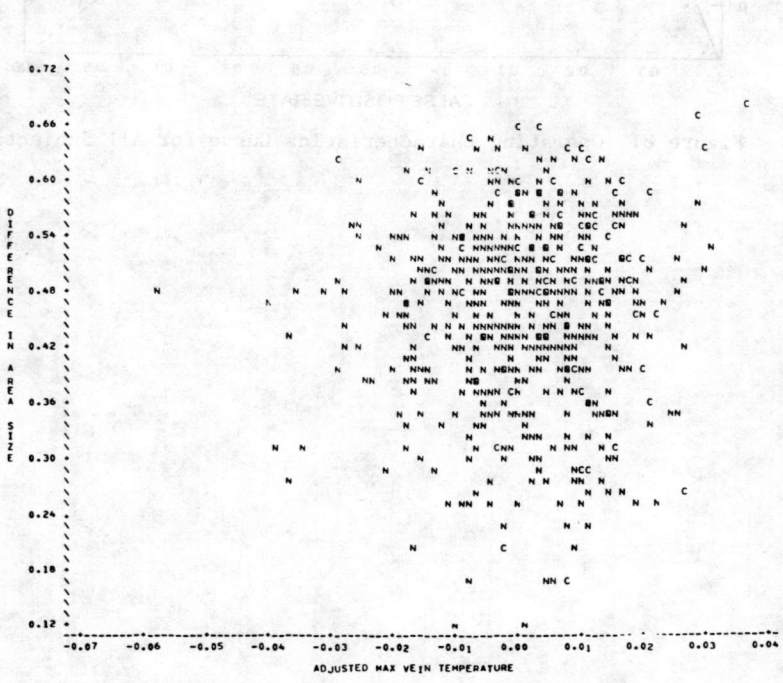

Figure 7. Scatter Plot for All Subjects

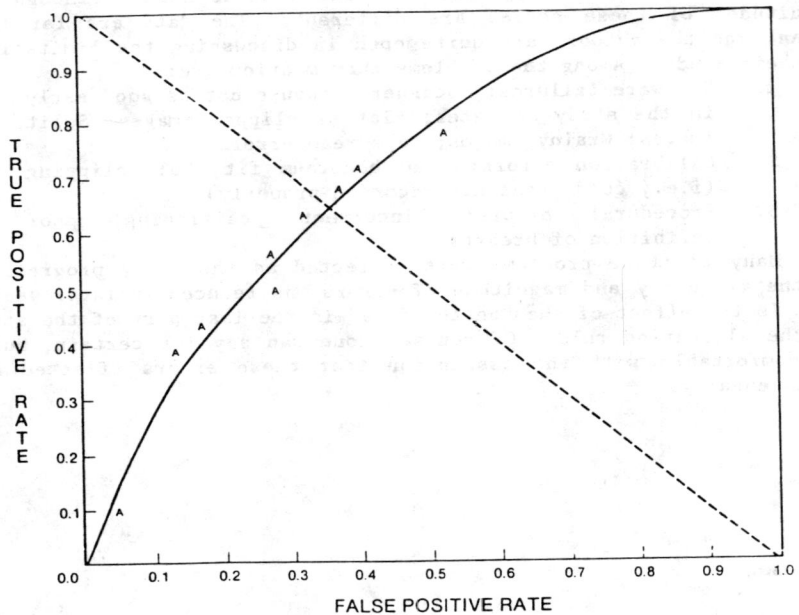

Figure 8. Operating Characteristics Curve for All Subjects

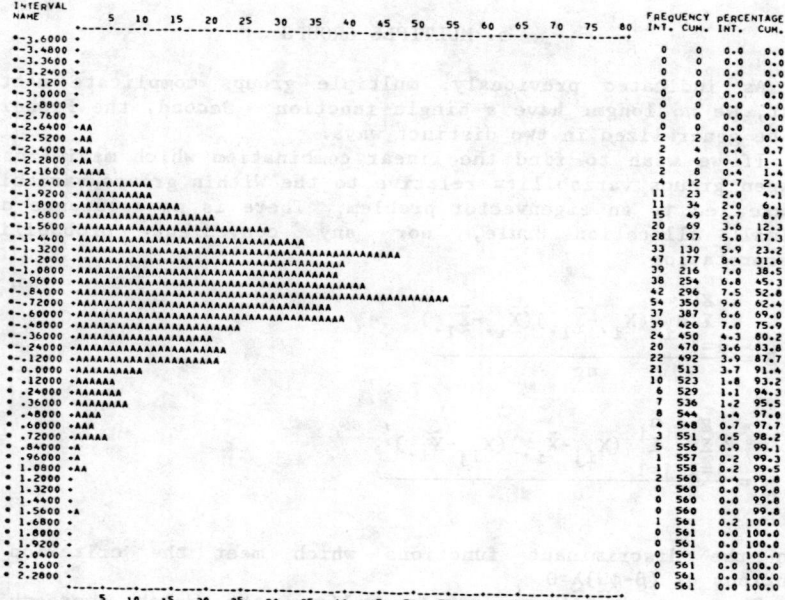

Figure 9. Histogram of LDF for Normals, All Ages

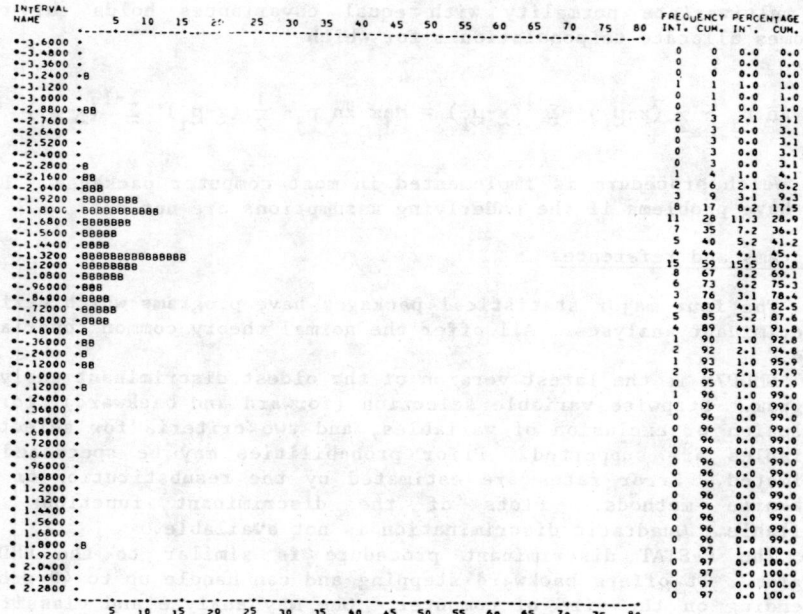

Figure 10. Histogram of LDF for Cancers, All Ages

5. MULTIPLE GROUPS

As indicated previously, multiple groups complicate matters. First, we no longer have a single function. Second, the Fisher LDF may be generalized in two distinct ways.

If we wish to find the linear combination which maximizes the Between groups variability relative to the Within groups variability we are led to an eigenvector problem. There is no assurance of an optimal allocation rule, nor any convenient probabilistic interpretation.

Let $\underset{\sim}{B} = \dfrac{\sum_{i=1}^{g} n_i (\bar{X}_i - \bar{X}_1)(\bar{X}_i - \bar{X}_1)'}{ng}$

and $\underset{\sim}{W} = \dfrac{\sum_{i=1}^{g} \sum_{j=1}^{n_i} (X_{ij} - \bar{X}_i)(X_{ij} - \bar{X}_i)'}{n}$.

Then the discriminant functions which meet the criterion are solutions of $(\underset{\sim}{B} - \phi \underset{\sim}{W})\lambda = 0$.

The second alternative generalizes the Welch approach and allocates an unknown observation to distribution i such that

$$p_i f_i(\underset{\sim}{x}) = \underset{j}{\text{Max}}\, p_j f_j(\underset{\sim}{x}).$$

If multivariate normality with equal covariances holds the rule becomes allocate to population i for which

$$\ell n\, p_i - \tfrac{1}{2}(\underset{\sim}{x} - \underset{\sim}{\mu}_i)' \underset{\sim}{\Sigma}^{-1} (\underset{\sim}{x} - \underset{\sim}{\mu}_i) = \underset{j}{\text{Max}}\, \ell n\, p_j - \tfrac{1}{2}(\underset{\sim}{x} - \underset{\sim}{\mu}_j)' \underset{\sim}{\Sigma}^{-1} (\underset{\sim}{x} - \underset{\sim}{\mu}_j).$$

The Welch procedure is implemented in most computer packages. Both may have problems if the underlying assumptions are not met.

Programs and references

The four major statistical packages have programs which perform discriminant analyses. All offer the normal theory common covariance case.

BMDP7M is the latest version of the oldest discriminant analyses program. Stepwise variable selection (forward and backward), forced inclusion or exclusion of variables, and two criteria for selecting variables are supported. Prior probabilities may be specified or estimated. Error rates are estimated by the resubstitution or the jackknife methods. Plots of the discriminant functions are available. Quadratic discrimination is not available.

The P-STAT discriminant procedure is similar to the BMDP7M program. It offers backward stepping and can handle up to 40 groups depending on the size of computer. One may analyze and classify a data set (this seems to be the resubstitution method), classify a known data set with a previously generated function (this is a cross-validation method), or one may classify an unknown data set.

SAS offers four procedures: DISCRIM, NEIGHBOR, CANDISC, STEPDISC. The DISCRIM procedure will calculate a linear or quadratic discriminant function on a fixed set of variables. Prior probabilities may be specified. Classification can be done on the training sample or a test sample. NEIGHBOR does a nearest neighbor discriminant analysis. The order of the neighbors may be specified and either Mahalanobis or Euclidian distance may be used. STEPDISC performs a stepwise analysis similar to BMDP7M.

SPSS-X permits forward selection and stepping. The Eigenvector (canonical vector) method is used for multiple groups. Plots are possible and the coefficients may be saved for further use.

There are many books which discuss discriminant analysis. Most multivariate analysis texts have chapters on the subject and there are at least two monographs available. Hand (1981) offers a variety of methods including distribution-free, parametric, linear discriminant practices, and discrete methods. Lachenbruch (1975) considers primarily normal theory methods and discusses evaluating error rates, robustness, non-parametric methods among others. Mardia, et al. (1979) is a general text, well-written with a good chapter on discriminant analysis.

REFERENCES

FISHER, R. A. (1936), "The Use of Multiple Measurements in Toxonomic Problems," Ann. Eugen., 7: 179-188.

HABERMAN, J. (1980), "Mass Screening for Breast Cancer by Electronic Infrared Pattern Recognition," Final Report Phase II. University of Oklahoma Health Sciences Center.

HAND, D. J. (1981), Discrimination and Classification, John Wiley and Sons, New York.

LACHENBRUCH, P. A. (1975), Discriminant Analysis, Hafner Pub. Co., New York.

MARDIA, K. V., KENT, I. T. and BIBBY, J. M. (1979), Multivariate Analysis, Academic Press, New York.

WELCH, B. L. (1939), "Note on Discriminant Functions," Biometrika, 31: 218-220.

ESTIMATE OF RADIONUCLIDE INTAKES FROM
REPETITIVE BIOASSAY MEASUREMENTS

K.W. Skrable, G.E. Chabot, and C.S. French
University of Lowell, Lowell, MA 01854

ABSTRACT

This paper discusses, compares, and gives examples of the different fitting procedures that may be used to evaluate bioassay data and includes the example output of a calculator and computer program, "BIO", that gives: (1) an unweighted least squares fit that assumes the variances in the bioassay measurements to be equal, which tends to give an overestimate of the intake, (2) a chi square fit that uses point estimated variances assumed to be proportional to the measurements themselves, which tends to give an underestimate of the intake, and (3) an iterative chi square fit in which the variance of a measurement is assumed to be proportional to the intake and its expectation value estimated from the previous fit. Iteration in procedure (3) continues until the chi square statistic is minimized, and the final estimated intake is usually between the extremes of the other two procedures. The expectation function that is used to fit the data is an intake retention function that is based upon reference man metabolic models and parameter values appropriate to the known physical and biochemical characteristics of the radionuclide. The paper includes a detailed discussion of the derivation of various intake retention functions that may be of interest in the estimation of intakes from bioassay data.

1. INTRODUCTION

Bioassay is the final quality control procedure that is used to assure adequate protection of workers from internal radiation. Because of the usually poor and limited bioassay data base that is available in an exposure incident and the complex pattern that can be associated with the uptake from compartments in the respiratory and GI tracts, the recycling of a radionuclide in systemic compartments, and its final excretion from the body, in most situations it is not possible to resolve an individual's own metabolic model and associated parameter values from the bioassay data. If the physical and biochemical characteristics of radionuclides are known in terms of ICRP classifications, then such information can be used in the derivation of an intake retention function, which gives the fraction of an intake of a radionuclide expected to be present in a particular in vivo or in vitro bioassay compartment of interest.

There are various quantities of interest in bioassay and internal radiation protection, and various retention functions have been associated with these quantities. The exact meaning of these quantities is important to understanding the mathematics used in the derivation of the various intake retention functions described here. Quantities of interest include intake, uptake, deposition, and content defined:

intake = quantity of a radioelement taken into the body, e.g., by inhalation or ingestion,

uptake = quantity of a radioelement absorbed into the systemic circulation, e.g., by injection into the blood or by absorption from compartments in the respiratory or GI

tracts,
deposition = quantity of a radioelement deposited, e.g., in the respiratory tract following an acute inhalation intake, and
content = quantity of a radioelement present in some compartment of interest.

In general, upper case letters will be used to represent quantities for stable elements while lower case letters will be used to represent quantities for radioelements. Subscripts will be used to identify compartments.

Associated with the first three quantities are corresponding retention functions. Each type of retention function applies to the content of a particular compartment relative to the given referenced quantity, i.e., the fraction of an <u>intake</u>, an <u>uptake</u>, or a <u>deposition</u> expected to be present in the particular compartment at some time post that <u>intake</u>, <u>uptake</u>, or <u>deposition</u>. Consider an acute inhalation intake of radioactive aerosols having a particle size distribution characterized by an activity median aerodynamic diameter (AMAD) of 1 micron in which 33% of the inhaled activity is expected to deposit in the lungs. Thus, the fraction of the intake expected to be present in the lungs at t=0, i.e., the value of the "intake retention function" for the lungs at t=0 is 0.33, while the fraction of the total lung deposition that is expected to be present in the lungs at t=0, i.e., the "deposition retention function" for the lungs is unity at t=0.

Thus, the acute intake retention function gives for a bioassay compartment of interest the fraction of a single intake that would be expected to be present at some time post that intake.

The acute uptake retention function for a compartment gives the fraction of an acute uptake that would be expected to be present in that compartment at some time post the uptake. The values for the uptake retention functions for the systemic whole body and the central extracellular fluid compartment, by definition, must equal unity at t=0. The values for the uptake retention functions for the peripheral tissue and organ compartments must be zero at t=0.

If the fundamental metabolic process that describes the removal of a radioelement from a compartment is described by simple linear first order kinetics, then the deposition retention function for this compartment would be given by a single exponential term having a coefficient of unity. This simple deposition retention function would then give the fraction of a deposition that would be expected to be present some time later. Such deposition retention functions are not difficult to envision for the stomach and various compartments of deposition within the respiratory tract, but it is difficult to envision an instantaneous deposition within cells of systemic organs and tissues. If the fundamental retention of a systemic organ or tissue is simple, then its deposition retention function would be given by a single exponential term having a coefficient of unity, and it would give the fraction of a single deposition that would be expected to be present at some time post the deposition neglecting any recycling of the radioelement. Because of recycling, such deposition retention functions would not describe the actual content of organs or tissues, but they would describe the fundamental metabolic processes and the fundamental retentions. Provided there is no direct excretion from an organ, the fundamental biological rate constant in the exponential of a simple organ deposition retention function would describe the transfer of an element to the central extracellular fluid compartment. While the deposition retention function of a single organ could be given by a single exponential term, because of recycling, the uptake retention function for that organ will often be given by a sum of exponential terms with constant coefficients. The algebraic sum of the coefficients in the uptake retention function for a peripheral organ, by definition, must equal zero. Each of a number of peripheral organ and tissue compartments could have a simple stable element deposition retention function described by a single exponential term that contains a characteristic rate constant describing the total biological removal. If significant recycling of the stable element also occurs, then the uptake retention function for each peripheral compartment and the central extracellular fluid compartment would contain the same number of exponential terms (one plus the number of peripheral organ/tissue compartments). The eigenvalue or effective rate constants in these exponential terms would not equal the fundamental rate constants that describe the actual retention of a deposition within an organ. The uptake retention function of each compartment and the uptake retention function of the systemic whole body, which is the sum of the uptake retention functions for the central and all peripheral compartments, would have the same exponential terms but different coefficients. Only if there is negligible recycling do the eigenvalue or effective rate constants equal the fundamental rate constants. In such case the stable element uptake retention function for the central compartment would be given by a single exponential term containing the fundamental rate constant that describes removal by all pathways. The stable element uptake retention function for each peripheral organ compartment then would contain an exponential term with the fundamental total removal rate constant for the central compartment and one other exponential term with the fundamental rate constant describing total biological removal from that organ, i.e., the rate constant in its deposition retention function. In such case, individual exponential terms in the uptake retention function for the systemic whole body could be associated with specific tissues or organs. If significant recycling does occur, the eigenvalue or effective rate constants in the uptake retention function for the systemic whole body would not equal any of the fundamental rate constants that describe the actual retention of a deposition within a compartment. In such case, no single exponential term of the systemic whole body uptake retention function can be associated with a particular structured physiological compartment.

The description of the distribution and retention of elements in the ICRP 30 report has many conceptual errors that relate to this last point. Because of this situation, care must be exercised in using the information in the ICRP 30 report for the development of bioassay models that are needed to make an initial assessment of the significance of bioassay data.

Intake retention functions that are based upon reference man metabolic models can be used to make a rapid <u>initial</u> assessment of the significance of bioassay results in terms of an estimated intake. This assessment is needed to identify those exposures that require further investigation and action from those that do not require such action and expenditure of valuable time and resources. The intake that is estimated from the quotient of the measured content of a bioassay compartment and the value of the compartment's intake retention function can be compared to either the current Nuclear Regulatory Commission (1984) 10 CFR 20 quarterly intake limit or the ICRP Publication 30 Annual Limit of Intake (ALI) recommended by the International Commission on Radiological Protection (1979a) to determine the significance of the bioassay result. The quotient of the estimated intake of a radioelement by its respective stochastic ALI value when multiplied by 5 rem gives the committed effective whole body internal radiation dose equivalent of an exposed worker, which then can be added to the worker's external radiation exposure to determine his total radiation exposure status.

When a number of bioassay measurements are used to evaluate an internal radiation exposure incident, the intake retention function derived from ICRP 30 or other current metabolic models can be used as the expectation fitting function for an estimation of the intake. The amount of the intake then is that which gives the "best fit" of the individual measurements to their respective expectation values. Because of the unknown and usually large contribution of biological variance, the actual variance of each datum will often greatly exceed the counting or analytical variance associated with the measurement itself. Thus, the proper weighting factor for each datum, normally taken as the inverse variance, will not be known. This paper discusses and compares three fitting procedures that can be used in the estimation of radionuclide intakes from repetitive bioassay measurements.

2. INTAKE RETENTION FUNCTIONS

2.1 Catenary Pathways from Intake to Excretion

The metabolism of elements in the respiratory and GI tracts and the systemic whole body is described here by linear first order kinetics, which can be simply depicted as one way transfers between compartments in various catenary systems beginning with some intake compartment and ending with some excretion compartment as depicted in Figure 1. The term catenary refers here simply to a chain of compartments in which material is transfered from one compartment to another by simple linear first order kinetics as characterized by appropriate translocation rate constants. The metabolism of all elements can be described in this way, which can be shown to be mathematically consistent with the recycling of elements within the systemic whole body. Because the metabolism can be described in terms of these simple one way transfers between compartments composing various catenary systems, a recursive catenary kinetics equation can be used to obtain the expected content of all in vivo and in vitro bioassay compartments. An excretion compartment is simply treated as the last compartment in each contributing catenary system. Removal of a radioelement from an excretion compartment is described entirely by its radioactive decay constant while total removal from an in vivo compartment is described by the sum of the decay constant and all applicable biological removal rate constants. It is recognized that this model and description of the metabolism is a gross oversimplification of the actual metabolism. Provided that appropriate values are chosen and lacking more specific details, the model can provide reasonable values for the intake retention functions applicable to the estimation of intakes from bioassay data and the estimation of internal radiation doses. Further discussion of the applications and limitations of models, in particular the ICRP 30 models, is beyond the scope of this paper. Suffice it to say that care must be exercised in the interpretation of the ICRP 30 metabolic models when these models are applied to a description of the short term kinetics that is needed for the estimation of intakes from bioassay data.

In addition to providing an assessment of the significance of bioassay results, intake retention functions can be used in the design and conduct of a bioassay program including: (1) the identification of those bioassay procedures that have sufficient sensitivity for the detection of an appropriate investigation level of interest, (2) the determination of appropriate derived investigation levels, (3) the determination of the minimum frequency of monitoring required for the detection of an investigation level, (4) the determination of the appropriate ALI value for the intake of a radioelement involving mixtures of different chemical compound classes and particle size distributions different from the standard 1 micron AMAD particle size distribution used in the calculation of the ICRP 30 ALI value, and (5) in cases involving significant exposures, the determination of special bioassay procedures that can be used to confirm and make better estimates of the intake and committed dose equivalent.

From Figure 1, pathways can be determined for inhalation intakes, ingestion intakes, instantaneous uptake, or exponential uptake, e.g., from a wound site. If several chains lead to a particular compartment of interest, then the total intake retention function for that compartment is simply obtained by summing the contributions from all chains. If an organ or a system of organs and tissues, such as the systemic whole body, is modeled as if it were comprised of a number of independent pseudo catenary compartments, then the total intake retention function for that organ or system is obtained from the sum of the intake retention functions for the individual pseudo catenary compartments that compose the organ or system.

There are a number of catenary systems or chains for each type of intake; each chain begins with an intake compartment and ends with an excretion compartment. Pathways involving inhalation intakes and depositions into various regions of the respiratory tract are shown on the upper left, and pathways involving ingestion intakes are shown on the right in Figure 1. Inhalation involves, in general, eight compartments of deposition in the respiratory tract, each of which initiates a separate catenary system. Ingestion begins with one compartment of deposition, the stomach.

Arrows leaving a compartment show the specific compartmental biological or pseudo biological removal pathways, which are characterized by specific biological translocation rate constants applicable to the particular pathways shown by the arrows. In addition to biological removal, radioelements are removed from each catenary compartment by radioactive decay as characterized by the decay constant for the radioelement. For a radioelement, this removal pathway is implicit for each compartment that is shown in Figure 1.

Compartments associated with the respiratory tract are shown on the upper left hand side of Figure 1. The arrows designated by D_{NP}, D_{TB}, and D_P that lead respectively into the nasal passage region (NP), tracheobronchi region (TB), and pulmonary region (P) of the respiratory tract represent depositions, and the respective fractional depositions, in the three regions of the respiratory tract. The activity deposition fractions for an inhalation intake of 1 microm AMAD aerosols are respectively 0.3, 0.08, and 0.25 for the NP, TB, and P regions of the respiratory tract.

Compartments a, c, and e shown on the left hand side of the respiratory tract schematic are cleared via pathways that lead directly to the systemic circulation. Compartment h represents a clearance pathway to the lymph nodes, which are deemed to be comprised

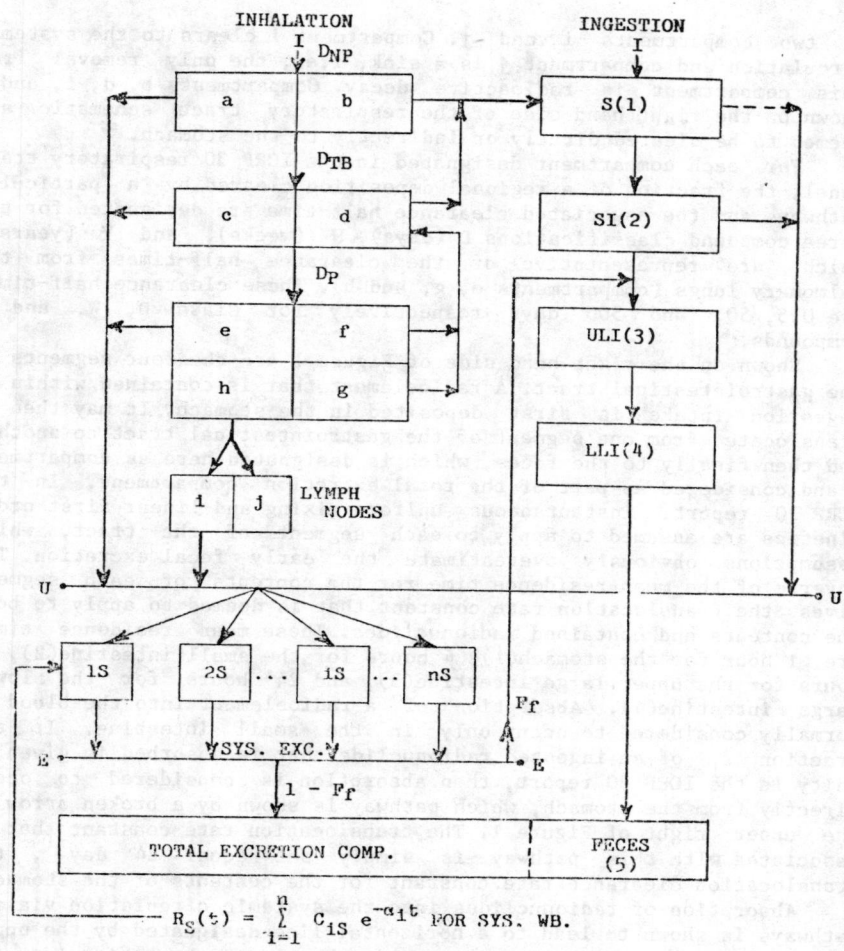

Figure 1. CATENARY PATHWAYS FROM INTAKE TO EXCRETION

Respiratory tract compartments are a through j. Gastrointestinal tract compartments are 1-4. $R_S(t)$ defines the stable element uptake retention function for the systemic whole body which is expressed by a sum of exponential terms (i=1 to n) with constant coefficients. Each ith exponential in $R_S(t)$ is treated as a deposition retention function of a pseudo catenary compartment, which is modeled to be cleared directly to systemic excretion E at an instantaneous fractional rate given by the eigenvalue rate constant α_i of the exponential term. The effective fraction of an uptake U that passes into the ith pseudo catenary compartment of the systemic whole body is given by the coefficient C_{iS} of the exponential term in $R_S(t)$.

of two compartments i and j. Compartment i clears to the systemic circulation and compartment j is a sink, i.e., the only removal from this compartment is radioactive decay. Compartments b, d, f, and g shown on the right hand side of the respiratory tract schematic are deemed to be cleared dirctly or indirectly to the stomach.

For each compartment designated in the ICRP 30 respiratory tract model, the fraction of a regional deposition cleared by a particular pathway and the associated clearance half-time are designated for the three compound classifications D (days), W (weeks), and Y (years), which are representative of the clearance half-times from the pulmonary lungs (compartments e, g, and h). These clearance half-times are 0.5, 50, and 500 days respectively for Class D, W, and Y compounds.

Shown on the right hand side of Figure 1 are the four segments of the gastrointestinal tract. A radioelement that is contained within an ingestion intake is first deposited in the stomach. It may then be translocated from one segment of the gastrointestinal tract to another and then finally to the feces, which is designated here as compartment 5 and considered as part of the total excretion compartment. In the ICRP 30 report, instantaneous uniform mixing and linear first order kinetics are assumed to apply to each segment of the tract, which assumptions obviously overestimate the early fecal excretion. The inverse of the mean residence time for the contents of each segment gives the translocation rate constant that is deemed to apply to both the contents and contained radionuclides. These mean residence times are 1 hour for the stomach(1), 4 hours for the small intestine(2), 13 hours for the upper large intestine(3), and 24 hours for the lower large intestine(4). Absorption of a radioelement into the blood is normally considered to occur only in the small intestine. If the fraction f_1 of an ingested radionuclide that is absorbed is given as unity in the ICRP 30 report, then absorption is considered to occur directly from the stomach, which pathway is shown by a broken arrow to the upper right of Figure 1. The translocation rate constant that is associated with this pathway is simply set equal 24 day^{-1}, the translocation clearance rate constant for the contents of the stomach.

Absorption of radionuclides into the systemic circulation via all pathways is shown to lead to a horizontal line designated by the upper case letter U, which represents uptake, i.e., absorption into the systemic circulation, which itself is identified by the lower case letter s. All catenary pathways combine at this point and then divide into n pseudo catenary compartments represented by the n exponential terms in the systemic whole body stable element uptake retention function $R_s(t)$, which is shown at the bottom of Figure 1. Although this may be difficult to envision physiologically, it is mathematically correct. This can be shown by deriving the stable element intake retention function $I_s(t)$ for the systemic whole body from the convolution of the uptake rate with the uptake retention function and then comparing this result with that obtained from this catenary model. Both approaches give the same expressions for $I_s(t)$.

The translocation rate constant that describes transfer to one of the pseudo catenary compartments from a compartment that feeds the systemic circulation is simply obtained by multiplying the total translocation rate constant, which describes transfer from the feed compartment to the systemic circulation, by the coefficient of the exponential term in $R_s(t)$ that pertains to the particular pseudo catenary compartment of interest. For example, consider an ingestion intake where $f_1 = 1$ so that the translocation rate constant $k_{1,s}$ of 24 day^{-1} describes transfer from the stomach to the systemic circulation. To obtain the translocation rate constant $k_{1,iS}$ that describes transfer from the stomach(1) to the iS pseudo catenary compartment of the systemic whole body, the translocation rate constant k_1 of 24 day^{-1} is simply multiplied by the coefficient C_{iS} that pertains to the particular iS pseudo catenary compartment. This pseudo catenary compartment's stable element deposition retention function, thus, is simply the exponential $\exp(-\alpha_i t)$.

Because the uptake retention function $R_S(t)$ embodies, in principle, all of the dynamic processes that describe the metabolism of stable elements in the systemic whole body, including the recycling of elements, the intake retention function $I_S(t)$ also embodies all of these metabolic processes plus the metabolic processes that occur in all of the compartments that feed the systemic circulation. The fact that each exponential term of the uptake retention function for the systemic whole body can be treated as a deposition retention function for a pseudo catenary compartment is very fortunate. It greatly simplifies the derivation of intake retention functions from the general catenary kinetics equation shown as equation (1) below.

Atoms of a radioelement that leave one of the pseudo catenary compartments of the systemic whole body are shown to go directly to systemic excretion designated here by a horizontal line identified by the upper case letter E. The translocation rate constant that describes the fractional rate of excretion from each iS pseudo catenary compartment is simply the eigenvalue rate constant α_i of the ith exponential term of $R_S(t)$, i.e., the rate constant in the exponential $\exp(-\alpha_i t)$ that represents the pseudo deposition retention function for the iS pseudo catenary compartment. The horizontal excretion bar, i.e., the line identified by E, is necessary for designating what fraction of systemic excretion leaves the systemic whole body via the fecal excretion pathway and what fraction leaves by all other pathways, e.g., via urine, sweat, exhalation, hair, fingernails, etc. The fraction F_f of systemic excretion via the fecal pathway is shown to enter the top of the upper large intestine. The primary systemic fecal excretion pathway is probably that involving biliary excretion, which actually passes into the duodenum or first part of the small intestine. Because the small intestine is the GI tract segment from which absorption into the blood is generally assumed to take place, the systemic fecal excretion pathway is shown here to effectively bypass the small intestine. Thus, the fraction F_f of systemic fecal excretion should be considered as an effective value. Although this model does simplify the mathematics, it is not realistic on physiological grounds. Because of the general lack of data for the systemic fecal excretion pathway, this simplifying assumption seems reasonable for developing bioassay models. The biliary excretion pathway, in fact, is quite complicated and may include deposition of a radioelement within liver parenchyma cells, excretion into intrahepatic bile passages finally leading into the hepatic duct with the bile, temporary holdup due to storage of bile in the gall bladder, and final excretion with the bile into the duodenum. Part or all of the radioelement may then, in fact, be reabsorbed into the systemic circulation by absorption from the small intestine into blood, which first passes into the liver via the portal vein before circulating through the rest of the body.

Because excretion of a radioelement from liver parenchyma cells will in general depend on the content of these cells, which will depend on the uptake retention function for the liver, the fraction F_f

of total systemic excretion via the biliary pathway is probably a complicated function of time post uptake. Thus, the fraction $(1 - F_f)$ of total systemic excretion via all other pathways, including the urinary pathway, will in general depend on the time post uptake. Because of a general lack of data, constant excretion fractions are often assumed. This may be a serious weakness of the excretion intake retention function shown here for a specific pathway.

All systemic excretion as well as direct fecal excretion is shown to eventually end up in the compartment that is designated, "total excretion compartment". This compartment is treated as any other catenary compartment; it is the last compartment of all catenary systems. The only removal pathway from this total accumulated excretion compartment is, for a radioelement, radioactive decay. Thus, the total removal rate constant k_E that describes removal from this compartment is set equal to the decay constant λ for a radioelement or zero for a stable element.

By applying the general catenary kinetics equation to the appropriate chains, single acute inhalation intake retention functions can be obtained for specific organs, organ systems, and excretion by summing the functions for the appropriate compartments. This includes: (1) the nasal passage (compartments a and b), (2) the lungs (compartments c - j), (3) the GI tract (compartments 1 - 4), (4) the systemic whole body (pseudo catenary compartments 1S - nS), (5) a specific systemic organ x, e.g., the thyroid, whose stable element uptake retention function $R_x(t)$ is designated by a sum of n exponential terms with constant coefficients (pseudo catenary compartments 1x - nx), (6) accumulated total systemic excretion (compartment E), (7) accumulated total fecal excretion (compartment 5), (8) accumulated urinary excretion (F_u x the intake retention function for compartment E if F_u is constant), and (9) the total body (the sum of (1) - (4) above). Similar intake retention functions can be obtained for ingestion intakes and instantaneous uptakes. To apply the general catenary kinetics equation, the only requirement is the identification of the appropriate chains, the specification of the biological translocation rate constants and total biological removal rate constants that apply to compartments within each chain, and the specification of the decay constant λ for the radioelement. Because the same recursive kinetics equation is used for each chain, numerical values are easily obtained from programmed solutions on a calculator or computer. Once the single acute intake retention functions have been determined for the radioelements, other functions such as continuous intake functions, incremental excretion intake functions for single or continuous intakes, and excretion rate intake functions for single or continuous intakes can be obtained by simply replacing the exponentials in the terms for the single intake functions by another time function. Because the coefficients of the exponentials in each single acute intake retention function are constants that are independent of the decay constant or half-life of the radioelement, values for these other intake retention functions could be calculated by a subroutine in a calculator or computer program solution of the kinetics equation.

2.2 Radioelement Intake Retention Function $i_n(t)$ for n^{th} Catenary Compartment

Figure 1 shows the catenary pathways from intake to excretion. The concise catenary kinetics equation shown here as equation 1 can be applied to all of those catenary pathways that lead to an nth compartment of interest in order to obtain the radioelement intake retention function $i_n(t)$ for that compartment:

$$i_n(t) = \sum_C F_C \prod_{p=1}^{n-1} k_{p,p+1} \sum_{j=1}^n \frac{e^{-k_j t}}{\prod_{\substack{p=1 \\ p \neq j}}^n (k_p - k_j)}, \tag{1}$$

where
$i_n(t) \equiv \langle q_n(t)/I \rangle$ = fraction of a single acute intake I expected to be present at time t in n^{th} compartment,
$\langle q_n(t) \rangle \equiv$ expectation content of nth compartment at time $t = I i_n(t)$,
$C \quad \equiv$ one of the chains that leads to the nth catenary compartment of interest,
$F_C \quad \equiv$ fraction of intake deposited into the first compartment of chain "C",
$k_{p,p+1} \equiv$ rate constant that describes transfer of element from p^{th} to $(p+1)^{th}$ compartment, and
$k_j \quad \equiv$ total rate constant that describes total removal of radioelement from the jth compartment = $K_j + \lambda$, where K_j is the total biological removal rate constant and λ is the decay constant of the radionuclide.

The subscript n on $i_n(t)$ is a general numerical index for the compartment of interest. For a particular catenary pathway that leads to the nth compartment of interest, the numerical index n could have one value and for another pathway some other value or perhaps the same value. For example, consider the intake retention function $i_d(t)$ for compartment d in the TB region of the respiratory tract. Three catenary systems, as shown in Figure 1, contribute to the fraction of an inhalation intake that is present in compartment d: (1) direct deposition for which n=1, (2) translocation to compartment d of a deposition in compartment f of the pulmonary region for which n=2, and (3) translocation to compartment d of a deposition in compartment g of the pulmonary region for which n=2. The ICRP 30 respiratory tract model provides the parameter values describing these processes and needed for the application of the catenary kinetics equation. The parameter values generally depend on the particle size and the compound classification of the inhaled aerosol. Particles containing a radioelement that are translocated from compartments f and g to compartment d are assumed to instantaneously mix with the contents of d, which is cleared by translocation to the stomach. The translocation to the stomach(1) is described by a translocation rate constant $k_{d,1}$ that can be obtained from the clearance half-time of 0.2 days given in the ICRP 30 report for all compound classes. The catenary kinetics equation shown here would be summed over three separate chains C to obtain the intake retention function $i_d(t)$ for compartment d in the TB region of the lungs. The radioelement intake retention function $i_L(t)$ for the lungs could be obtained by repeating this procedure for the other lung compartments c, e, f, g, h, i, and j, all of which may or may not be applicable for each chemical compound classification. For example, for the highly transportable compound Class D, the clearance pathways designated by f and g are not applicable. To keep the same form of a calculator or computer solution of the kinetics equation, these compartments would simply be assigned depositions of zero for

Class D compounds.

Symbols for parameter values that are needed for the calculation of the value for the intake retention function $i_n(t)$ are defined after equation (1). Their meaning and their application to this equation will be reinforced by referring to the example intake retention function $i_d(t)$ for compartment d of the TB region of the lungs. Numerical values that will be given are for 1 micron AMAD Class Y aerosols. The radioelement acute intake retention function $i_d(t)$ gives the fraction of a radioelement contained in a single acute inhalation intake that is expected to be present in compartment d at time t post that acute intake.

The index C refers to one of the three catenary systems that contribute to the retention of compartment d. The symbol F_C is the fraction of the activity or of the number of radioactive atoms in the inhalation intake that is expected to be deposited and cleared from the first catenary compartment of each chain C that feeds the compartment of interest. These fractions are obtained from the product of the regional deposition fraction and the fraction within the region that is cleared via a particular pathway to the compartment of interest. They are in the order of the pathways identified above: (1) $D_{NP} F_d = (0.08 (0.99)$ for direct deposition in compartment d, (2) $D_P F_f = (0.25) (0.4)$ for translocation of the deposition from f to d, and (3) $D_P F_g = (0.25) (0.4)$ for translocation of the deposition from g to d.

The symbol $k_{p,p+1}$ applies to either the stable or radioactive element; it represents a fundamental biological rate constant. It is the translocation rate constant or instantaneous fraction of an element transferred per unit time from the pth catenary compartment to the next or $(p+1)^{th}$ catenary compartment down the chain. The symbol $\prod_{p=1}^{n-1} k_{p,p+1}$ represents the continuous product of translocation rate constants, $k_{1,2} k_{2,3} \ldots k_{n-1,n}$, which describes the translocation from the 1st compartment of deposition to the nth compartment of interest in a particular chain C. If the upper index n-1 on this continuous $\prod$ product happens to be less than unity, then this simply means that there is no feed compartment and the $\prod$ product is set equal to the value of unity. For the case of direct deposition in compartment d where n=1 and n-1=0, the $\prod$ product is "empty" and is assigned the value of unity. In the case of translocation from either compartments f and g to compartment d, n=2 and n-1=1; so, the continuous $\prod$ product is then given by the single translocation rate constant $k_{1,2}$. This translocation rate constant $k_{1,2}$ for transfer from compartment f to compartment d or $k_{f,d}$ has the value of 0.693/1 day. For the other chain involving transfer from compartment g to d, $k_{1,2}$ takes on the value of $k_{g,d}$, which has the value of 0.693/500 days for Class Y aerosols.

The symbol k_j is the total rate constant that describes removal of the radioelement from the jth compartment of a catenary system. It is given by the sum of the total biological removal rate constant K_j and the decay constant λ for the particular radioelement. (Note: upper case letters generally refer to quantities specified for the stable element and lower case letters refer to quantities specified for the radioelement.) If the only biological removal process from the jth compartment is translocation to the $(j+1)^{th}$ compartment, then the total biological removal rate constant K_j is simply the translocation rate constant $k_{j,j+1}$; then k_j simply equals $k_{j,j+1} + \lambda$. This is the situation for all compartments shown in Figure 1 except for the small intestine contents designated as the catenary compartment 2. The two arrows that leave the box designated SI(2) represent translocation from the small intestine(2) to the upper large intestine(3) and translocation from the small intestine(2) to the systemic circulation "s" by absorption into the blood. Thus, the total biological removal rate constant K_2 for the small intestine(2) is given by the sum of $k_{2,3}$ and $k_{2,s}$. This sum depends on the value of $k_{2,3}$ and the value of f_1 given in the ICRP 30 report. This total biological removal rate constant K_2 for the small intestine can be shown to be given by $k_{2,3}/(1-f_1)$. If the fraction f_1 of the stable element absorbed into the blood is zero for a particular chemical compound class, then K_2 simply equals the value of $k_{2,3}$ of 6 day^{-1}. If the fraction f_1 is unity, then K_2 would also equal $k_{2,3}$ which is moot because then absorption into the blood is modeled as if it takes place entirely from the stomach. In this latter case, there would be no translocation from the stomach(1) to the small intestine(2) and $k_{1,2}$ would be set equal to zero. The total biological removal rate constant K_1 for the stomach would then be set equal to the value of $k_{1,s}$ of 24 day^{-1} representing translocation into the systemic circulation or blood shown by the dotted arrow leaving the stomach S(I).

The contribution of a particular catenary pathway or chain to the intake retention function $i_n(t)$ is shown to depend on a sum of exponential terms containing the total removal rate constants beginning with the 1st catenary compartment and ending with the nth catenary compartment of interest. The exponentials $\exp(-k_j t)$ within the summation from j = 1 to n are divided by a continuous product $\Pi(k_p-k_j)$ of all the total removal rate constant differences from the 1st compartment up to the nth compartment of interest. The subtracted rate constant k_j always corresponds to the one in the exponential $\exp(-k_j T)$ and the index p of the other rate constant k_p runs from the index 1 for the intake compartment up to the index n for the compartment whose retention function is being determined. the condition that the index p not be allowed to equal the index j on the continuous product $\Pi(k_p-k_j)$ means that there are (n-1) rate constant differences in this continuous product, which corresponds to the number of translocation rate constant in the continuous product $\Pi k_{p,p+1}$ of translocation rate constants shown outside the sum of exponential terms. Thus, the intake retention function $i_n(t)$ is a dimensionless function or pure fraction as required. Because each total removal rate constant k in the continuous product $\Pi(k_p-k_j)$ of total removal rate constant differences contains the decay constant, the decay constant λ cancels so that this continuous Π product depends only on the total biological removal rate constant differences. To maintain computational accuracy, the biological total removal rate constants, upper case letter K, rather than the radioelement total removal rate constants, lower case letter k, should be used in the calculation of the continuous product of rate constant differences. Except for the small intestine, these total biological removal rate constants K would be given by the translocation rate constants $k_{p,p+1}$ shown in the product outside the summation of exponential terms. Each Π product of rate constnt differences, in addition, would include the total biological removal rate constant K_n for the nth catenary compartment of interest.

Because the value of F_C and the continuous product of translocation rate constants can be multiplied into each exponential of the summation to yield a numerical coefficient when divided by the continuous Π product of rate constant differences for the particular exponential term and because all of these factors are independent of the half-life of the radioelement, the resulting numerical coefficients of the exponential terms also apply to the stable element. Thus, the stable element intake retention function $I_n(t)$ can be obtained from this same equation and coefficients by simply replacing each radioelement total removal rate constant k_j in the exponentials by the biological total removal rate constant K_j. This shows, also, that the radioelement intake retention function $i_n(t)$ can be obtained from the stable element function $I_n(t)$ by simply multiplying it by $\exp(-\lambda t)$. This, tabular or plotted values of the stable element intake retention function $I_n(t)$ can be obtained for all radioelements of interest at a particular facility and later converted to the particular radioelement function $i_n(t)$ when needed for the evaluation of a positive bioassay result that is found in a worker who is known to have been involved in an acute exposure incident.

If exact parameter values that are given in the ICRP 30 report or elsewhere are used to calculate the value if $I_n(t)$, sometimes it happens that two or more of the rate constants K_p and K_j in the Π product of rate constant differences are equal to one another, e.g., the rate constants K_h and K_i for comprtments h and i in the respiratory tract model. This results in rate constant differences of zero and singularities that can't be handled directly by this catenary kinetics equation. We have derived exact kinetics expressions to handle the situation of any number of equal rate constants but not involving more than one type of singularity. A simpler way of handling this situation is to simply increment one of the rate constants by an arbitrarily small amount, e.g., by multiplying it by a factor 1.00001 close to unity. This approximate method has been shown to give results essentially equal to results obtained from the mathematically exact expressions. To avoid problems of computational accuracy with the particular calculator or computer that is used to obtain numerical results, the incrementing factor should be neither too near nor too far from unity. Our experience has shown that the fractional part by which a rate constant is incremented should be about half the significant digits available in the computer or calculator that is used.

Even though the radioelement single acute intake retention function $i_n(t)$ can be obtained by simply multiplying the stable element function $I_n(t)$ by $\exp(-\lambda t)$, the radioelement function $i_n(t)$ has been presented in equation (1) instead of the stable element function $I_n(t)$ because we are primarily interested in the retention of radioelements and because other types of intake retention functions of interest can be obtained from the radioelement function $i_n(t)$ by simply replacing the exponential $\exp(-k_j t)$ in $i_n(t)$ by other time functions. Because the coefficient C_j of each jth exponential $\exp(-k_j t)$ in $i_n(t)$, which is given by

$$C_j \equiv (F_C \Pi k_{p,p+1})/\Pi(k_p - k_j), \qquad (2)$$

applies to both stable and radioactive elements, the same coefficients are used in the expressions for these other intake retention functions. Only the exponential function $\exp(-k_j t)$ is changed so that calculator or computer solutions should be constructed to first calculate the value of each coefficient and then go to a subroutine that calculates the vale of the time function applicable to the particular intake retention function of interest, e.g., single acute intake, continuous intake, incremental urine, instantaeous urinary excretion rate, etc.

This catenary kinetics equation can be applied to each catenary compartment in a particular pathway to obtain the intake retention function of each compartment in the chain. The exponentials or deposition retention functions of all feed compartments are propagated down the chain in the intake retention function of each succeeding catenary compartment, and a recursion can be noted in the coefficients for the propagated exponentials. This recursion can be used to simplyfy the calculations. To obtain the coefficient of a particular exponential in a receiving compartment's function from the coefficient of the same exponential in a feed compartment's function, multiple the coefficient of the exponential term in the feed compartment by the translocation rate constant for the feed compartment and divide by the difference of the total biological removal rate constant for the receiving compartment and the total biological removal rate constant in the exponential term of interest. The coefficient of the exponential term of the receiving compartment itself is simply obtained from the negative sum of all the newly generated coefficients of the other exponential terms in the expression for the intake retention function for the receving compartment.

To obtain the retention function of a specific bioassay compartment of interest, the first step in the application of this recursive kinetics equation is to identify all of the catenary pathways that lead to the bioassay compartment. The second step is to specify te numerical values of all parameters applicable to each catenary pathway. The third step is the repetitive application of the kinetics equation to obtain the total retention from the sum of the contributions of all catenary pathways. Example applications of equation (1) are given below for specific bioassay compartment of interest.

2.3 Radioelement Ingestion Intake Retention Function $i_S(t)$ for the Systemic Whole Body

The explicit ingestion intake retention function $i_S(t)$ for the systemic whole body may be obtained by reference to equation (1) and Figure 1, which shows the catenary pathways from intake to excretion. For an ingestion intake, the only pathway to the systemic circulation is absorption into the blood from the small intestine, whole absorption rate constant $k_{2,s}$ is given by $f_1 k_{2,3}/(1-f_1)$. This single catenary pathway divides into n pathways corresponding to the n pseudo catenary compartment iS (i=1 to n) that are identified by the individual exponential terms of the systemic whole body stable element uptake retention function $R_S(t)$, which is shown at the bottom of Figure 1. The effective fraction of the uptake of a radioelement that passes into a particular pseudo catenary compartment iS is simply the coefficient C_{iS} of the ith exponential $\exp(-\alpha_i t)$ in $R_S(t)$. Thus, the product $k_{2,s} C_{iS}$ gives the effective transloction rate constant $k_{2,iS}$ that describes transfer from the SI(2) to the iS pseudo catenary compartment of the systemic whole body. In addition to radioactive decay, removal of a radioelement from each pseudo catenary compartment is shown to go directly to excretion. The eigenvalue rate constant α_i is the effective translocation rate constant that describes transfer from each ith pseudo catenary compartment to the systemic excretion compartment designated by E.

The pathways from ingestion to systemic excretion and the

expressions for the radioelement intake retention functions are shown as follows:

For $f_1 / 1$:

$$1 \longrightarrow 2 \rightleftarrows \begin{matrix} 1S \\ iS \\ \vdots \\ nS \end{matrix} \longrightarrow E.$$

$$i_S(t) = \sum_{i=1}^{n} k_{1,2} (k_{2,s} C_{iS}) \left(\frac{e^{-k_1 t}}{(k_2-k_1)(k_{iS}-k_1)} \right.$$

$$\left. \frac{e^{-k_2 t}}{(k_1-k_2)(k_{iS}-k_2)} \quad \frac{e^{-k_{iS} t}}{(k_1-k_{iS})(k_2-k_{iS})} \right) \qquad (3)$$

For $f_1 = 1$, compartment 2 is eliminated from the pathway; so $k_{1,s} = 24$ day^{-1} and $i_S(t)$ is expressed:

$$i_S(t) = \sum_{i=1}^{n} (k_{1,s} C_{iS}) \left(\frac{e^{-k_\alpha t}}{(k_{iS}-k_1)} + \frac{e^{-k_{iS} t}}{(k_1-k_{iS})} \right), \qquad (4)$$

where $k_{iS} \equiv \alpha_i + \lambda$.

The expression for the single acute ingestion intake retention function $i_S(t)$ for the systemic whole body is shown to be comprised of a sum (i=1 to n) over the individual intake retention functions $i_{iS}(t)$ of the n pseudo catenary compartments of the systemic whole body. For the case where the fraction f_1 of absorption into the blood is not unity, all absorption is assumed to take place in the small intestine(2). Thus, for an ingestion intake, there are three exponential terms in the intake retention function $i_{iS}(t)$ of each iS pseudo catenary compartment of the systemic whole body. The three exponentials are the deposition retention functions for the catenary compartments: stomach(1), small intestine(2), and the systemic pseudo catenary compartment (iS). The translocation rate constants $k_{1,2}$ and $k_{2,iS} = k_{2,s} C_{iS}$ are contained in the continuous product of translocation rate constants that describe transfer to the iS pseudo catenary compartment of interest. Each exponential is divided by the continuous product of all possible total removal rate constant differences that are applicable to each exponential beginning with the stomach(1), small intestine(2), and ending with the iS pseudo catenary compartment of the systemic whole body. The radioelement total removal rate constant k_{iS} for the iS pseudo catenary compartment is given by the sum of the stable element eigenvalue rate constant α_i and the decay constant λ.

For the case where $f_1=1$, absorption ito the blood is modeled as if it takes place entirely in the stomach; so, the term involving the small intestine(2) is eliminated from the explicit expression for $i_S(t)$.

2.4 Radioelement Ingestion Intake Retention Function $i_x(t)$ for Systemic Organ on Tissue x

In addition to the intake retention function $i_S(t)$ for the systemic whole body, we can obtain the intake retention function $i_x(t)$ for a systemic organ or tissue x whose uptake retention function $R_x(t)$ for the stable element is expressed by a sum of exponential terms with constant coefficients. It is to be noted that a true uptake retention function $R_x(t)$ and not the typical ICRP 30 pseudo deposition retention function must be available and expressed as a sum of exponential terms with constant coefficients. Such uptake retention functions would, in principle, incorporate the dynamic process of recycling and would be applicable to both the long term and short term kinetics. The procedure for obtaining the intake retention function $i_x(t)$ for a systemic organ is exactly the same as that used to obtain the intake retention function $i_S(t)$ for the systemic whole body. The stable element uptake retention function $R_x(t)$ for the organ or tissue x, which is understood to be expressed by a sum (i=1 to n) of exponential terms, simply replaces the stable element uptake retention function $R_S(t)$ for the systemic whole body depicted at the bottom of Figure 1. Because the retention of a peripheral organ or tissue compartment of an uptake is by definition zero at t=0, the coefficients C_{ix} in $R_x(t)$ must sum to zero. This means that some of the coefficients C_{ix} are negative. Yet for an ingestion intake, the effective translocation rate constant $k_{2,ix}$ that describes translocation from the small intestine(2) to the ix pseudo catenary compartment of organ x is given by the product $k_{2,s} C_{ix}$. This can be proven to be mathematically correct even though negative translocation rate constants would not have any physical meaning. The radioelement total removal rate constant k_{ix} that describes removal from one of the ix pseudo catenary compartments is again given by the sum of the stable element eigenvalue rate constant α_i and the decay constant λ. However, it is not proper now to show excretion as the biological removal pathway from the organ pseudo catenary compartments even though this may happen, e.g., from liver parenchyma cells to bile. Thus, no specific receiving compartment can be specified for excretion from the pseudo catenary compartments of the organ x because, in most situations, excretion from cells would be to the central extracellular fluid compartment. When the uptake retention function $R_x(t)$ of organ or tissue x is expressed by:

$$R_x(t) = \sum_{i=1}^{n} C_{ix} \exp(-\alpha_i t), \qquad (5)$$

the pathways from ingestion to all of the ix(i=1 to n) pseudo catenary compartments of organ x are depicted for the case $f_1 \neq 1$:

$$1 \longrightarrow 2 \begin{array}{c} \nearrow 1x \xrightarrow{\alpha_1} \\ \rightarrow ix \xrightarrow{\alpha_i} \\ \vdots \\ \searrow nx \xrightarrow{\alpha_n} \end{array}$$

The explicit expression for $i_x(t)$ is thus given by

$$i_x(t) = \sum_{i=1}^{n} k_{1,2} (k_{2,s} C_{ix}) \left(\frac{e^{-k_1 t}}{(k_2-k_1)(k_{ix}-k_1)} - \frac{e^{-k_2 t}}{(k_1-k_2)(k_{ix}-k_2)} - \frac{e^{k_{ix} t}}{(k_1-k_{ix})(k_2-k_{ix})} \right). \qquad (6)$$

where $k_{ix} \equiv \alpha_i + \lambda$.

The intake retention function $i_x(t)$ is shown to be comprised to a sum (i=1 to n) over the individual intake retention functions $i_{ix}(t)$ of the n pseudo catenary compartments of the systemic organ or tissue x. All interpretations that applied to $i_S(t)$ for the systemic whole body, except for the fact that excretion can't be modeled from these pseudo catenary compartments, also apply here for $i_x(t)$.

2.5 Radioelement Ingestion Intake Retention Function $i_E(t)$ for the Total Systemic Excretion Compartment E

The explicit ingestion intake retention function $i_E(t)$ for the accumulated systemic excretion compartment(E) may be obtained by reference to equation (1) and Figure 1, which shows the catenary pathways from intake to excretion. Except for the systemic excretion compartment(E) itself, the catenary compartment applicable here are the same as those that were used to obtain the ingestion intake retention function $i_S(t)$ for the systemic whole body. The translocation rate constants applicable to the pathway through the iS pseudo catenary compartment of the systemic whole body are: $k_{1,2}$ that describes translocation from the stomach(1) to small intestine(2), $k_{2,s}\, c_{iS} = k_{2,iS}$ that describes translocation from the small intestine(2) to the iS pseudo catenary compartment of the systemic whole body, and α_i that describes translocation to the total accumulated systemic excretion compartment(E) shown here by the horizontal line E. Provided that the fraction F_p of systemic excretion via a particular pathway is constant in time, the accumulated systemic excretion via a particular pathway, e.g., the urinary pathway, could be obtained by simply multiplying the accumulated excretion function by this fraction F_p. The product $\alpha_i F_p$, thus, would represent the translocation rate constant that describes transfer from the iS pseudo catenary compartment to the systemic excretion compartment p. The total removal rate constants k in the exponentials of the deposition retention functions of the catenary compartments involving the pathway through the catenary compartment iS are k_1, k_2, $k_{iS} = \alpha_i + \lambda$, and $k_E = \lambda$. Thus there are four exponential terms in $i_E(t)$. Pathways through all pseudo catenary compartments of the systemic whole body would involve each of the n pseudo catenary compartments; so, the total retention of the accumulated excretion compartment(E) would be obtained by summing the contributions from each pathway depicted for the case where $f_1 \neq 1$:

$$1 \longrightarrow 2 \longrightarrow \begin{array}{c} 1S \\ \vdots \\ iS \\ \vdots \\ nS \end{array} \longrightarrow E.$$

The explicit expression for $i_E(t)$ is thus given by

$$i_E(t) = \sum_{i=1}^{n} k_{1,2}\,(k_{2,s}\,c_{iS})\,\alpha_i \left[\frac{e^{-k_1 t}}{(k_2-k_1)(k_{iS}-k_1)(k_E-k_1)} \right.$$

$$\frac{e^{-k_2 t}}{(k_1-k_2)(k_{iS}-k_2)(k_E-k_2)} \frac{e^{-k_{iS} t}}{(k_1-k_{iS})(k_2-k_{iS})(k_E-k_{iS})}$$

$$\left. \frac{e^{-k_E t}}{(k_1-k_E)(k_2-k_E)(k_{iS}-k_E)} \right], \qquad (7)$$

where $k_E \equiv \lambda$, the decay constant for the radioelement.

The intake retention function $i_E(t)$ is shown to be comprised of a sum (i=1 to n) over the contributing pseudo catenary compartments iS of the systemic whole body. The recursive nature of this explicit catenary kinetics expression when compared to the recursions previously noted for the general catenary kinetics equation shows that the accumulated systemic excretion compartment(E) is treated just like any other pseudo catenary compartment. For the radioelement function $i_E(t)$, the total removal rate constant k_E applicable to the excretion compartment is simply the decay constant λ. The other total removal rate constants shown in the product of rate constant differences also include the decay constant; so, the product of the rate constant differences in the denominator of the exponential $\exp(-k_E t)$ that is applicable to the excretion compartment(E) reduce to the continuous product of the stable element total removal rate constants, K_1, $K_2 K_{iS}$. This product of total removal rate constants can be expressed also by $k_{1,2}$ $(k_{2,s}/f_1)\alpha_i$. When the continuous product of translocation rate constants $k_{1,2}(k_{2,s} C_{i,S})\alpha_i$ are multiplied into each exponential term, the coefficient of the exponential $\exp(-k_E t)$ for the excretion compartment can be shown to be given simply by the product $f_1 C_{iS}$. Because the sum of the coefficients C_{iS} of the systemic whole body uptake retention function $R_S(t)$ must be unity by definition, the summation of the coefficients of the exponential $\exp(-k_E t)$ over all pseudo catenary compartments gives f_1, the fraction of the stable element that is absorbed into the blood. Because the value of $i_E(t) = 0$ at $t=0$, the sum of the coefficients of all other exponential terms over all pseudo catenary compartments must yield the negative of f_1.

Because the coefficients of the exponential terms of $i_E(t)$ also apply to the stable element, when the radioelement total removal rate constants k in each exponential are replaced by their corresponding total biological removal rate constants K, the stable element intake retention function $I_E(t)$ for the accumulated systemic excretion compartment is obtained. The replacement rate constants would be given specifically: k_1 by $K_1 = k_{1,2}$, k_2 by $K_2 = k_{2,3}/(1-f_1)$, k_{iS} by $K_{iS} = \alpha_i$, and k_E by $K_E = 0$. Thus, the stable element intake retention function $I_E(t)$ would have a constant term given by the fraction f_1 of an ingestion intake of a stable element that is absorbed into the blood and a sum of exponential terms. The exponential terms contain the biological total removal rate constants K_1, K_2, and K_{iS} for the catenary compartments in each chain that feeds the systemic excretion compartment. Because the exponential terms ultimately go to zero for a time t long compared to the biological half clearance times for the feed compartments, the stable element intake retention function $I_E(t)$ approaches the fraction f_1 of the ingestion absorbed into the blood. This evaluation of the intake retention function $i_E(t)$ has given further insight into the application of the general catenary kinetics equation and further reinforcement of the fact that the excretion compartment can be treated like any other catenary compartment.

2.6 Radioelement Accumulated Urine Intake Retention Function $i_u(t)$

The accumulated urine single acute intake retention function $i_u(t)$ for a radioelement is defined as the fraction of a single acute intake I expected to be present in the accumulated urine at time t post intake:

$$i_u(t) \equiv \langle q_u(t)/I \rangle, \qquad (8)$$

where
$\langle q_u > t) \rangle \equiv$ expectation content of the accumulated urine compartment at time t post a single acute intake.

If the fraction F_u of total systemic excretion by the urinary pathway is constant, then the radioelement accumulated urine function $i_u(t)$ can be obtained from the product of F_u and the radioelement intake retention function $i_E(t)$ for the total systemic excretion compartment E:

$$i_u(t) = F_u\, i_E(t). \tag{9}$$

Although accumulated urine would not normally be collected following an acute exposure incident, if the exposure involves a transportable radioelement and is deemed to be serious, such bioassay procedure could give the best estimate of the intake. The intake I can be estimated simply by dividing the measured content $q_u(t)$ of accumulated urine by the value of the radioelement accumulated urine intake retention function $i_u(t)$ at the time t of the bioassay measurement of $q_u(t)$. Alternatively, the measured content $q_u(t)$ can be corrected for radioactive decay back to the time of intake and then divided by the stable element intake retention function $I_u(t)$ evaluated at the time t:

$$I = q_u(t)/i_u(t) = q_u(t)/I_u(t)\, \exp(-\lambda t), \tag{10}$$

or

$$I = q_u(t)\, \exp(\lambda t)/I_u(t). \tag{11}$$

3. TRANSFORMATIONS OF SINGLE ACUTE INTAKE RETENTION FUNCTIONS

The radioelement single acute intake retention function $i(t)$, which gives the fraction of a single acute intake I of a radioelement expected to be present at time t in a particular bioassay compartment of interest can be expressed by a sum of exponential terms with constant coefficients:

$$i(t) \equiv \langle q(t)/I \rangle = \Sigma_i C_i\, \exp(-k_i t), \tag{12}$$

where
  $\langle q(t) \rangle \equiv$ the expectation content $= I\, i(t)$, and
  $C_i \equiv$ constant numerical coefficient of ith exponential term.

The subscript on $i(t)$ and $q(t)$ identifying the particular bioassay compartment of interest has been dropped for simplicity.

This single acute intake retention function $i(t)$ can be transformed into other intake retention functions of interest by simply replacing the exponential $\exp(-k_i t)$ in the expression for $i(t)$ by other appropriate functions of time. The derivations are discussed by Skrable (1983). The replacement functions for each exponential $\exp(-kt)$ are shown below, where the subscript i on the rate constant k has been dropped for simplicity.

3.1 Continuous Intake Retention Function $p(t)$

To obtain the fraction $p(t)$ of the total quantity $\dot{I}T$ of a constant continuous intake expected to be present at time t in the bioassay compartment of interest, replace the exponential $\exp(-kt)$ in each term of the single acute intake retention function $i(t)$ by

$$(1-\exp(-kT))\, \exp(-k(t-T))/kT, \tag{13}$$

where
  $t \equiv$ time post onset of continuous intake to time of the bioassay measurement,
  $p(t) \equiv \langle q(t)/\dot{I}T \rangle$, the continuous intake retention function,
  $\dot{I} \equiv$ constant continuous intake rate,
  $T \equiv$ continuous intake interval ($T \leq t$), and

$\langle q(t) \rangle \equiv$ expectation content of compartment = $\dot{I}T\, p(t)$.

An experimental estimate of the content $q(t)$ of the bioassay compartment at time t post the onset of the continuous intake can be used to estimate the total intake $\dot{I}t$ over the continuous intake interval T:

$$\dot{I}T = q(t)/p(t), \qquad (14)$$

or the intake rate $\dot{I}$:

$$\dot{I} = q(t)/T\, p(t). \qquad (15)$$

Equation (15) can be used to obtain an estimate of the intake rate $\dot{I}$ from a single measurement of the content $q(t)$ of the bioassay compartment at time (t-T) post the end of a single continuous intake interval T. If several measurements are made of the content $q(t_i)$ of the bioassay compartment in situations when a worker is chronically exposed at a more or less constant rate $\dot{I}$ and if these measurements apply to different intake intervals T_i and different measurement times t_i, then an expectation function F_i given by the product of each T_i and $p(t_i)$ can be used to obtain an estimate of the intake rate $\dot{I}$ from a fit of all available bioassay data.

3.2 Number of Transformations $\langle D(T) \rangle$ or Decays Expected Over a Time Interval T Post a Single Acute Intake I

To obtain the number of transformations $\langle D(T) \rangle$ expected to occur over a time interval T in a particular source organ or tissue following a single acute intake I, replace the exponential $\exp(-kt)$ in each term of the single acute intake retention function $i(t)$ for the source organ or tissue by

$$(1-\exp(kT))/k, \qquad (16)$$

and multiply the resulting function by the intake I. When the intake I has the activity units of Bq, T has the units of years, and each k has the units of reciprocal years, then the expected number of transformations $\langle D(T) \rangle$ has the units of Bq-years, which unit equals 3.15E7 transformations. It can be shown that the expected number of transformations $\langle D(T) \rangle$ also can be obtained from the product of the intake I, the period T, and the continuous intake retention function $p(T)$, which is evaluated at t=T:

$$\langle D(T) \rangle = IT\, p(T). \qquad (17)$$

The number of transformations in the various source organs and tissues in the body and other appropriate factors available from the ICRP 30 report can be used to calculate the committed dose equivalent expected to be delivered to various target organs and tissues over the period T following an acute intake I. For the purpose of establishing the Annual Limit of Intake (ALI) and the Derived Air Concentration (DAC) of a radionuclide, the period T is taken as 50 years by the ICRP.

The first and probably most tedious step in the calculation of the ALI and DAC secondary standards that are recommended by the ICRP is the calculation of the expected transformations $\langle D(T) \rangle$ in the various source organs of reference man. Equation (17) when incorporated as an option in a calculator or computer program provides a simply way of deriving ALI and DAC values based upon site specific information on the physical and biochemical characteristics of radionuclides in the working environment.

3.3 Incremental Urinary Excretion Function $i_u^*(t)$ For a Single Acute Intake I

To obtain the fraction $i_u^*(t)$ of a single acute intake I that is expected to be present at time t in an incremental sample of urine that is collected over the time interval t from $(t-\Delta t)$ to t, replace the exponential $\exp(-kt)$ in each term of the accumulated urinary excretion function $i_u(t)$ by

$$-\exp(-k(t-\Delta t))\ \exp(-\lambda \Delta t)\ (1-\exp(-(k-\lambda)\Delta t), \qquad (18)$$

where
$i_u^*(t) \equiv \langle q_u^*(t)/I \rangle$ = incremental urinary excretion function for a single acute intake I,
$\langle q_u^*(t) \rangle \equiv I\ i_u^*(t)$ = expected content of incremental sample at time t, and
$\Delta t \equiv$ the collection interval for the urine sample.

It is to be noted that the collection interval Δt can extend over any period of time following a single acute intake, including $\Delta t = t$. When $\Delta t = t$, $i_u^*(t)$ then equals $i_u(t)$; although, this may not be immediately obvious from the equations.

A measurement of the content $q_u^*(t)$ of a single incremental sample of urine provides an estimate of the intake I:

$$I = q_u^*(t)/i_u^*(t). \qquad (19)$$

If a number of measurements of the content $q_u^*(t_i)$ of incremental samples of urine at various times t_i post a single acute intake are available then the incremental urinary excretion function $i_u^*(t_i)$ can be used as the fitting function F_i to obtain an estimate of the intake using all of the available incremental urine data. The time interval Δt of collection need not be the same value for each incremental sample.

3.4 Instantaneous Urinary Excretion Rate Function $\langle \dot{e}_u(t)/I \rangle$ for a Single Acute Intake I

To obtain the fraction $\langle \dot{e}_u(t)/I \rangle$ of a single acute intake I expected to be excreted per unit time at time t post the intake, replace the exponential $\exp(-kt)$ in each term of the accumulated urinary excretion function $i_u(t)$ by

$$-(k-\lambda)\ \exp(-kt), \qquad (20)$$

where
$(k-\lambda) \equiv K$, the total biological removal rate constant, and
$\langle \dot{e}_u(t) \rangle \equiv I \langle \dot{e}_u(t)/T \rangle$ = expected urinary excretion rate.

When only spot urine concentration data are available, the excretion rates can be estimated from the product of the measured concentrations and the volume of urine expected to be excreted per unit time. Any single estimate of the excretion rate $\dot{e}_u(t)$ can be used to estimate the intake I:

$$I \equiv \dot{e}_u(t)/\langle \dot{e}_u(t)/I \rangle \qquad (21)$$

If several excretion rate estimates $\dot{e}_u(t_i)$ are made at various times t_i post a single acute intake, then values of $\langle \dot{e}_u(t_i)/I \rangle$ can be used as the expectation function F_i to obtain an estimate of the intake I from a fit of all available urinary excretion rate estimates. Because of large variations in the volume of urine excreted, because of the fact that the measured concentration is actually that of an incremental sample, and because of the large variations in the radionuclide urinary excretion rate at early times post an intake, this urinary excretion rate bioassay method for estimating intakes may not be very accurate. For this reason, the analysis of incremental samples of urine is expected to give more accurate and reliable estimates of intakes following an acute exposure incident.

3.5 Incremental Urinary Excretion Function $P_u^*(t)$ For a Continuous Intake

To obtain the fraction $P_u^*(t)$ of the total quantity $\dot{I}T$ of a continuous intake that is expected to be present at time t post the onset of the continuous intake in an incremental sample of urine collected over the time interval Δt from $(t-\Delta t)$ to t, replace the exponential $\exp(-kt)$ in each term of the accumulated urinary excretion function $i_u(t)$ by

$$-\exp(-\lambda\Delta t)(1-\exp(-kT))\exp(-k(t-\Delta t-T))(1-\exp(-(k-\lambda)\Delta t))/kT, \quad (22)$$

for $t \geq T+\Delta t$, and where
$P_u^*(t) \equiv \langle q_u^*(t)/\dot{I}T \rangle$ = incremental urinary excretion function for a continuous intake $\dot{I}T$, and
$\langle q_u^*(t) \rangle \equiv \dot{I}T\, P_u^*(t)$ = expected content of incremental sample at time t.

An experimental estimate of the content $q_u^*(t)$ of an incremental sample of urine can be used to estimate the total intake $\dot{I}T$ over the continuous intake interval T:

$$\dot{I}T = q_u^*(t)/P_u^*(t), \quad (23)$$

or the intake rate I:
$$\dot{I} = q_u^*(t)/T\, P_u^*(t) \quad (24)$$

The function $TP_u^*(t_i)$ can be used as the fitting function F_i to obtain an estimate of the intake rate $\dot{I}$ from a number of measurements of the content $q_u^*(t_i)$ of incremental urine samples.

3.6 Instantaneous Urinary Excretion Rate Function $\langle \dot{e}_u(t)/\dot{I}T \rangle$ For a Continuous Intake $\dot{I}T$

To obtain the fraction $\langle \dot{e}_u(t)/\dot{I}T \rangle$ of the total quantity $\dot{I}T$ of a continuous intake that is expected to be excreted per unit time at time t post the onset of the continuous intake, replace the exponential $\exp(-kt)$ in each term of the accumulated urinary excretion function $i_u(t)$ by

$$-(1-\exp(-kT))\exp(-k(t-T))(1-\frac{\lambda}{k})/T, \quad (25)$$
for $t \geq T$.

The expectation function $\langle \dot{e}_u(t)/\dot{I} \rangle$, which gives the ratio of the expected urinary excretion rate relative to the intake rate $\dot{I}$, is obtained:

$$\langle \dot{e}_u(t)/\dot{I} \rangle = T \langle \dot{e}_u(t)/\dot{I}T \rangle. \qquad (26)$$

Depending upon the type ob bioassay data that is available, intake retention functions or their transformed functions can be used as the expectation fitting function F_i to obtain estimates of intakes from all of the available bioassay data.

4. ALTERNATIVE METHODS OF FITTING BIOASSAY DATA

When significant exposures and intakes occur, the variance of a bioassay measurement is most often weighted by biological variance rather than the variance associated with the analytical procedure. This is particularly true for excretion rates that are estimated from spot urine concentration data. In the case of repetitive bioassay measurements following a single acute intake, the magnitude of the measurements and their associated variances about their expectation values can vary considerably. Because each datum has a different expectation value and associated variance, it is not appropriate to treat each datum with equal weight, which is the assumption in the Least Squares Fitting (LSF) procedure. To a first approximation, the variance σ_i^2 may be assumed to be proportional to $\langle Y_i \rangle$:

$$\sigma_i^2 = k \langle Y_i \rangle, \qquad (27)$$

where k is a constant and $\langle Y_i \rangle$ is the expectation measurement given by the product of the intake parameter A and the expectation function F_i that is applicable to the particular bioassay measurement:

$$\langle Y_i \rangle = AF_i = A \langle Y_i/A \rangle. \qquad (28)$$

When the measurement is the urinary excretion rate $\dot{E}_u(t)$ following a single acute intake I, then the expectation variance applies to this excretion rate:

$$Y_i \equiv \dot{E}_u(t_i), \qquad (29)$$
$$F_i \equiv \langle \dot{E}_u(t_i) \rangle / I; \text{ so}, \qquad (30)$$
$$\sigma_i^2 \equiv kI \langle \dot{E}_u(t_i)/I \rangle. \qquad (31)$$

The assumption that the variance associated with a bioassay measurement is proportional to the expectation value for that measurement itself can be justified if it is assumed that bioassay represents a binomial experiment for which the probability of detecting an atom of the intake is small compared to unity and that the total variance, including biological variance, is proportional to the product of the atoms of the intake and the expectation function F_i.

When it is assumed that the variance associated with a bioassay measurement is proportional to its expectation value, then an iterative chi square fit of the data can be performed to obtain the best estimate of the intake parameter value A, which is that value of A that minimized the chi square statistic defined by

$$\chi^2 = \sum_{i=1}^{N} (Y_i - \langle Y_i \rangle)^2 / \sigma_i^2, \text{ or} \qquad (32)$$

$$\chi^2 = \sum_{i=1}^{N} (Y_i - AF_i)^2/\sigma_i^2, \text{ where} \tag{33}$$

$$\sigma_i^2 = k \langle Y_i \rangle = k\, AF_i. \tag{34}$$

To simplify the notation, the indices on the summation symbol and all other symbols within the summation will be dropped; so, the summations in all expressions are understood to be over the data points from i=1 to N.

When the chi square statistic is minimized with respect to the intake parameter A, the value for the parameter A is obtained from the resulting equation:

$$A = (\Sigma YF/\sigma^2)/(\Sigma F^2/\sigma^2), \tag{35}$$

which can be seen to be independent of the constant k in the expression for the variance σ^2:

$$\sigma^2 = k \langle Y \rangle = k\, AF. \tag{36}$$

Thus, the constant k can have any arbitrary value, e.g., unity. Although the value of k does not influence the value of the intake parameter A, it does influence the value of the chi square statistic itself. The value of the chi square statistic, if we knew the value of k, would be useful in the evaluation of the fit. Because we do not really know the value of k, we set it equal to unity. The numerical value of the chi square statistic then is simply k times larger than the value that would obtain if we had known k and had included it as part of σ^2.

The theoretical standard error in the intake parameter value A is obtained by propagation of the error on the equation for A neglecting systematic errors in the fitting function F and considering only the theoretical variances in the measurements. When this procedure is followed, the theoretical variance σ_A^2 th is obtained by

$$\sigma_A^2 \text{ th} = 1/(\Sigma F^2/\sigma^2), \tag{37}$$

which does not depend on the actual deviations $(Y - \langle Y \rangle)$ of the measurements from their expectation values but only on the a priori theoretical variances and values for the expectation function F.

For N bioassay measurements, the theoretical variance σ_Y^2 th of all the measurements about their expectation values, i.e., about the line of best fit, can be obtained by

$$\sigma_y^2 \text{ th} = N/(\Sigma 1/\sigma^2). \tag{38}$$

The value of the chi square statistic can be obtained by expanding the equation that defines the chi square statistic:

$$\chi^2 = \Sigma(Y-\langle Y \rangle)^2/\sigma^2 = \Sigma(Y-AF)^2/\sigma^2, \text{ or} \tag{39}$$

$$\chi^2 = \Sigma(Y^2/\sigma^2) - 2A\,\Sigma(FY/\sigma^2 + A^2\,\Sigma F^2/\sigma^2. \tag{40}$$

The numerical value of the chi square statistic, if k were known, would give a measure of the goodness of fit of the bioassay measurements to their assumed expectation values, which would really be a test of the validity of the expectation function F. The square of the deviation of a measurement Y from its expectation value $\langle Y \rangle$ relative to its expectation variance σ^2 is expected on the average to be unity. For a large number N of measurements, the numerical value of the chi square statistics is expected to approach (N-1), the number of degrees of freedom ν. The reduced chi square statistic, χ_ν^2, thus is expected to approach unity:

$$\chi_\nu^2 = \chi^2/\nu = \chi^2/(N-1). \tag{41}$$

If a numerically large value of the χ_ν^2 statistic is obtained in a fit of bioassay data (i.e., $\chi_\nu^2 \gg 1$ when the fit incorporates the value of k along with its appropriate units), then it may indicate one of two possibilities or both:

(1) the wrong expectation function F has been used to fit the data, or
(2) the actual experimental variances are greater than the assumed values.

If a large χ_ν^2 value is obtained, or if k is not known and is therefore set equal to unity, then the theoretical standard errors obtained for the measurements Y and the intake parameter value A should be adjusted to experimental standard errors. It can be shown that when the theoretical variances are multiplied by the value obtained for the χ_ν^2 statistic, the experimental variances are obtained that reflect the actual deviations of the measurements Y from their expectation values:

$$\sigma_A^2 \text{ exp} = \sigma_A^2 \text{ th } \chi_\nu^2, \text{ and} \tag{42}$$

$$\sigma_Y^2 \text{ exp} = \sigma_y^2 \text{ th } \chi_\nu^2. \tag{43}$$

If the variances σ_i^2 of the measurements Y_i are assumed to be some constant variance σ^2, then the intake parameter value A can be estimated from a simple least squares fit (LSF) of the data:

$$\chi^2 \equiv \sum_i (Y_i - F_i A)^2 / \sigma_i^2 = (\Sigma (Y_i - F_i A)^2) / \sigma^2. \tag{44}$$

When the chi square statistic is minimized with respect to the intake parameter A, the constant variance σ^2 cancels from the equation and A is obtained by

$$A = (\sum_i Y_i F_i) / (\sum_i F_i^2). \tag{45}$$

Again, the index i will be dropped from other equations that follow. The experimental variance σ_Y^2 exp of individual measurements Y about their expectation values gives an estimate of the constant variance σ^2 that is assumed for the least squares fit. It is obtained by

$$\sigma_Y^2 \text{ exp} = (\Sigma (Y - FA)^2)/(N-1), \text{ or} \tag{46}$$

$$\sigma_Y^2 \text{ exp} = (\Sigma Y^2 - 2A\Sigma Y F + A^2 \Sigma F^2)/(N-1). \tag{47}$$

When it is assumed that only the individual measurements Y have experimental errors that propagate on the intake parameter value A (i.e., the systematic errors in the expectation function F are neglected), then the variance σ_A^2 in the calculated intake parameter value A can be estimated by propagation of the error on the equation for A, which gives:

$$\sigma_A^2 = \sigma_Y^2 \text{ exp}/(\Sigma F^2). \tag{48}$$

4.1 Three Fits to the Uranium Excretion Data of a Worker Accidentally Exposed to Aerosols of Natural Uranium

To demonstrate the use of the various fitting procedures and the equations above, three fits to the uranium excretion data of a worker who was exposed to aerosols of natural uranium are presented Table 1 and Figure 2. In the process of removing contaminated dust collector bags from inside a uranium ore concentrates baghouse, the worker, who wore protective clothing and an air line respirator, accidentally inhaled aerosols of the uranium concentrates. The concentrates had been identified previously through simulated lung solubility tests as consisting of approximately 60% ICRP 30 Class D and 40% ICRP 30 Class W transportability. Spot urine samples were obtained at various times post the exposure. The uranium concentration $C_u(t)$ was determined in each sample and multiplied by the standard daily urine excretion of 1.4 liters/day in order to estimate the excretion rate $\dot{E}_u(t)$ in units of micrograms/day.

The ICRP 30 metabolic models for the respiratory and GI tracts and the systemic whole body uptake retention function $R_S(t)$ obtained from information in the ICRP 30 report (1979b) were used to generate values for the urinary excretion rate expectation function applicable to the 1 micron particle size and to each compound class. All systemic excretion, per the ICRP 10 recommendation (1968), was assumed to be in the urine. To obtain the total instantaneous fraction $\langle \dot{E}_u(t)/I \rangle$ of the inhaled uranium expected to be excreted per day at the various times t post the exposure when the spot urine samples were taken, the values for the urinary excretion rate expectation function for each compound class were weighted by their respective fraction of the intake and then summed.

Because the proportionality constant k is set equal to unity, the values that are obtained for the χ^2 statistic and the χ_ν^2 statistic are k times the values that would otherwise be obtained if we had known the appropriate value for k for doing the fits. Because the χ_ν^2 statistic rapidly approaches unity if the right expectation function F and variances σ^2 are used in the fit, the value of k is simply the value for the χ_ν^2 statistic that is obtained from the fit when k is set equal to unity. The theoretical variance σ_Y^2 th in the measurements and the theoretical variance σ_I^2 th in the estimated intake are, therefore, converted to their corresponding experimental variances by multiplying them by the value obtained for the χ_ν^2 statistic when k is set equal to unity:

$$\sigma^2_Y \text{ exp} = \sigma_A^2 \text{ th } \chi_\nu^2, \text{ and} \tag{49}$$

$$I \text{ exp} = I \text{ th } \chi_\nu^2. \tag{50}$$

Because the units that are used to express bioassay measurements are arbitrary, the numerical value obtained for the chi square statistic will normally not be meaningful.

To begin the iterative chi square fitting procedure, an initial guess to the correct intake I is needed to calculate the expectation variances,

$$\sigma^2 = \langle \dot{E}_u(t) \rangle = I \langle \dot{E}_u(t)/I \rangle, \qquad (51)$$

in the measurements of the excretion rates. This guess could be any value; however, a more efficient method involves making a guess close to the final intake I that will be estimated from the iterative procedure. One method is to use an initial intake I that is estimated from the unweighted least squares fit. Another method is to use an initial intake I that is estimated from the chi square fit that relies on point estimated variances, i.e., procedure 2. Either method works quite well.

When these fitting procedures are used for the uranium excretion data, the results shown in Table 1 are obtained. When the expectation values $\langle E_u(t) \rangle$ obtained for the three fits are compared to the experimental values $E_u(t)$ of the excretion rates, it can be seen that the LSF procedure gives the highest expectation value and highest estimated intake. For the least squares fit procedure, the highest observed value $E_u(t)$ of 1078 g/day is very close, on a relative basis, to its expectation value $\langle E_u(t) \rangle$ of 1035 µg/day. The only other data point shown in the table for this fit that is closer on a relative basis is the value shown at 7.6 days. This is a common feature of the fit obtained in a LSF procedure. To minimize the sum of the squares of the deviations, the highest observed values would have to be closer on a relative basis to their expectation values. This shows that the least squares fit tends to be partically weighted by the highest observed values, which gives the highest estimated intake I of (14.4 $\pm$ 0.8) mg ($\pm 1\sigma$). In the chi square fitting procedure, each datum is effectively weighted by its inverse variance. Because the variance of a measurement is assumed here to be proportional to the measurement itself, the smaller measurements tend to weight the fit, which results in a lower estimate of the intake than the value obtained in the LSF procedure. In the chi square fit that used point estimated variances made equal to the measurement $E_u(t)$ itself, the observed measurements that are higher than their expectation values are weighted less and the observed measurements that are lower than their expectation values are weighted more than they should be weighted. This fitting procedure, therefore gives the lowest estimate of the intake I of (11.9 $\pm$ 1.5) mg ($\pm$ 1σ). In the iterative chi square fit that used expectation variances made equal to the expectation measurement $\langle E_u(t) \rangle$, a more appropriate weight is given to each datum. This fitting procedure gives an estimated intake I of (13.4 $\pm$ 1.4) mg ($\pm$ 1σ), which is between the values obtained for the other two procedures. The experimental standard errors in the estimated intakes for the chi square fits are larger than the standard error estimated for the LSF procedure. However, the estimated experimental standard errors in the excretion rates, which give a measure of the goodness of fit of the observed excretion rates to their expectation values, are less for the chi square fits than the value obtained from the LSF procedure. The values of intake obtained in the LSF and the iterative chi square procedures are within 1 standard error of their mean I of (13.9 $\pm$ 1.6) mg. We believe that the iterative chi square fitting procedure is fundamentally the better procedure. The LSF procedure tends to give overestimates of the intake while the chi square fit based upon point estimated variances made equal to the measurement itself tend to give underestimates.

4.2 Total Body Committed Effective Dose Equivalent for Uranium Inhalation Exposure

The activity distribution of the uranium isotopes with respect to the chemical compound classes, the fraction of the respective ALI's, and the total body committed effective dose equivalent are summarized in Table 2.

TABLE 2
TOTAL BODY COMMITTED EFFECTIVE DOSE EQUIVALENT
FOR URANIUM INHALATION EXPOSURE

ISOTOPE	CLASS D ACTIVITY (Bq)	I/ALI	CLASS W ACTIVITY (Bq)	I/ALI	TOTAL I/ALI
U-234	106	1.5 E-3	71	2.4 E-3	3.9 E-3
U-235	5.02	7.2 E-5	3.45	1.2 E-4	1.9 E-4
U-238	106	1.3 E-3	71	2.4 E-3	3.7 E-3

Grand Total I/ALI = 7.8 E-3
(Total I/ALI) (5000 mrem) = 39 mrem.

The ICRP 30 report provides estimates of the 50 year committed dose equivalent and values weighted by cancer and hereditary disease risks post the intake of a unit activity of 1 Bq of a radionuclide. Values for inhalation intakes are calculated only for the 1 micron AMAD particle size but do consider different f_1 values and chemical classifications D, W, and Y as well as the contribution of ingrowth daughter radionuclides. Values of weighted dose equivalent are used to derive the Annual Limit of Intake (ALI) and Derived Air Concentration (DAC) when stochastic effects are limiting. This case corresponds to an effective whole body dose equivalent of 0.05 Sv or 5 rem and allows direct comparison of internal doses with whole body doses from external radiation.

The total intake of natural uranium has been estimated here as 14.4 mg, which corresponds to 177 Bq of U-234, 8.37 Bq of U-235, and 177 Bq of U-238. These activities are comprised of 60% Class D and 40% Class W uranium, which give the distribution and fractions of the respective ALI's for stochastic effects shown in the table. The total fraction of the ALI for all uranium isotopes and chemical classifications is calculated as 7.8 E-3. This corresponds to a whole body committed effective dose equivalent of 39 mrem. Although this exposure would not warrant much further investigation or action, considerable time was spent by the licensee, the Nuclear Regulatory Commission, and others in the evaluation of this exposure. Although the exposure evaluated in this report corresponds to 80 MPC-hours, the whole body committed effective dose equivalent is estimated to be only 39 mrem, and the basic chemical toxity limit of 900 µg for the burden of uranium in the kidneys was not exceeded. Although the spot urine sampling data did provide adequate information for estimating the intake, twenty four hour samples would have been preferable in this known exposure incident. Such data could be evaluated according to the accumulated urinary excretion model, which will provide more accurate results and the best estimate of the intake. As a minimum, the exposed worker should be asked to record the time of urination. This information then could be used to obtain the incremental time interval

Δt of urine collection in the bladder applicable to each sample. This information then would allow evaluation of the data by the incremental model, which also is more accurate than the excretion rate model that is applied to spot urine concentration measurements.

SUMMARY

Bioassay is the final quality control procedure that is used to assure adequate protection of workers from internal radiation exposure. Intake retention functions that are based upon standard man metabolic models can be used to make a rapid initial assessment of the significance of bioassay results in terms of an estimated intake. This assessment is needed to identify those exposures that require further investigation and action from those that do not require such action and expenditure of valuable time and resources. The intake that is estimated from the quotient of the measured content of a bioassay compartment and the value of the compartment's intake retention function can be compared to either the current NRC quarterly intake limit or the ICRP 30 Annual Limit of Intake to determine the significance of the bioassay result. The quotient of the estimated intake of a radioelement by its respective stochastic ALI value when multiplied by 5 rem gives the effective whole body internal radiation dose equivalent of an exposed worker, which then can be added to the worker's external radiation exposure to determine his total radiation exposure status. This total effective whole body dose equivalent gives a measure of the total committed risk from combined external and internal exposures in a given year of practice. In cases of significant internal exposures, special followup bioassay procedures should be used to confirm and make better estimates of the intake and committed dose equivalent.

When a number of bioassay measurements are used to evaluate an internal radiation exposure incident, an intake retention function derived from ICRP 30 or other current metabolic models can be used as the expectation fitting function for an estimation of the intake. The amount of the intake then is that which gives the "best fit" of the individual measurements to their respective expectation values as determined from the product of the estimated intake and the value for the appropriate expectation intake retention function. This simple fitting procedure which relies on an a priori expectation function derived from reference man metabolic models and parameter values provides only an estimate of the intake and not the exposed individual's own metabolic model. Because of the usually poor and limited bioassay data that is available in an exposure incident and the complex pattern that can be associated with the uptake from compartments in the respiratory and GI tracts, the recycling of a radionuclide in systemic compartments, and its final excretion from the body, in most situations it will not be possible to resolve an individual's metabolic parameter values from the bioassay data. If the physical and biochemical characteristics of inhaled aerosols are known in terms of ICRP classifications, then such information can be used in the derivation of the applicable intake retention function.

Because of the unknown and usually large contribution of biological variance, the actual variance of each datum will often greatly exceed the counting or analytical variance associated with the measurement itself. Thus, the proper weighting factor for each datum, normally taken as the inverse variance, will not be known. Different fitting procedures that may be used to evaluate bioassay data include

(1) an unweighted least squares fit that assumes the variances to be equal, which tends to give an overestimate of the intake, (2) a chi square fit that uses point estimated variances, which tends to give an underestimate of the intake, and (3) an iterative chi square fit in which the variance of a measurement is assumed to be proportional to its expectation value estimated from the previous fit. Iteration in procedure (3) continues until the chi square statistic is minimized, and the final estimated intake is usually between the extremes of the other two procedures.

REFERENCES

International Commission on Radiological Protection (1968), <u>Evaluation of Radiation Doses to Body Tissues from Internal Contamination due to Occupational Exposures</u>, ICRP Publication 10, 87, Oxford: Pergamon Press.

International Commission on Radiological Protection (1979a), <u>Limits for Intakes of Radionuclides by Workers</u>, ICRP Publication 30, New York: Pergamon Press.

International Commission on Radiological Protection (1979b), <u>Limits for Intakes of Radionuclides by Workers</u>, ICRP Publication 30, Part 1, 102-103, New York: Pergamon Press.

Nuclear Regulatory Commission (1984), Title 10, Code of Federal Regulations, Part 20, <u>Standards for Protection Against Radiation</u>, U.S. Nuclear Regulatory Commission, Washington, D.C. 20555.

SKRABLE, K. W. (1983), "Retention Functions and Their Applications to Bioassay and Internal Dose Estimation," mannual and presentation at Health Physicis Society 1983 Summer School on Internal Radiation Dosimetry, June 12-17, 1983, University of Maryland at Balitmore County, Catonsville, Maryland.

REGRESSION METHODS IN CLINICAL RADIOBIOLOGY

Howard D. Thames, Jr., Susan L. Tucker
University of Texas System Cancer Center, Houston, TX. 77030

Shelley L. Rasmussen
University of Lowell, Lowell, MA. 01854

Jerry W. McLarty
University of Texas Health Science Center at Tyler, Tyler, TX. 75701

ABSTRACT

Regression approaches are described for two problems in radiobiology. In the first instance we are concerned with the responses of cell populations to multiple, equal doses of radiation, and the tissue responses that result from cell killing. In particular, if equal proportions of cells are killed by each of the repeated, equal doses ("equal effect per fraction"), it is possible to extrapolate from known models of single-dose response to models of response to fractionated doses. When the data are plotted in certain transformed coordinates, and fit linearly, the significance of a regression coefficient against zero is a test of the equal-effect-per-fraction hypothesis. If the hypothesis cannot be rejected, a regression estimate may be obtained of the repair capacity of the tissue. A likelihood approach to this problem yields more reliable confidence limits.

The second application involves the problem of estimating equal-effect dose ratios, called RBEs, after experimental determination of equi-effective doses under two differeing conditions. The regression approach to this problem involves the estimation of equal-effect dose pairs, by numerical inversion of dose-response functions obtained from least-squares fits to the data. These pairs are then fit to a quadratic model using least squares. The significance of the quadratic coefficient against zero is a test of the hypothesis of dose-invariant RBE.

INTRODUCTION

Experimental radiotherapy is devoted to the characterization of the effects of ionizing radiation on tumors and normal tissues, with the aim of improving the results of radiotherapy of human cancer. Two aspects of radiotherapy that lead to problems susceptible to a regression approach are: (1) Doses are fractionated, i.e., given as many small doses separated by several hours to several days, rather than as one or two large doses. Therefore an understanding is required of the responses of tumors and normal tissues to fractionated radiation, and of whether there is a difference that might be exploited therapeutically; (2) Doses are of different radiation types (such as gamma rays, X rays, neutrons, etc.), and may be given in conjunction with drugs designed to modify radiation effect in some way, such as radioprotectors and radiosensitizers. When contemplating a change in radiation beam type or the use of drugs that modify radiation action, radiotherapists naturally want an understanding of how the radiation dose should be changed so that no increased toxicity to normal tissues results. This leads to the problem of estimating the ratio of equally effective doses, given the responses under the two conditions of exposure.

For each of these problems, the typical regression approach is to manipulate mathematical models of radiation response so that regression coefficients (or their ratio) may be interpreted biologically. A further twist is to attach a special interpretation to a value (such as zero) of a coefficient and to use the significance of the coefficient as a test of that interpretation. Since the usual satistical assumptions for linear least squares regression are often violated, simulation is used throughout as a check on the validity of the methods.

TISSUE RESPONSE TO FRACTIONATED RADIATION

Normal tissue responses to radiation injury are measured in terms of altered appearance (skin color assayed by qualitative scores), functional deficit (increased breathing rate or urination frequency), or dichotomous effects (paralysis, death, or other binary response). The connection between cell killing and organ response is made by assuming that a specified level of tissue damage corresponds to a unique level of cell survival. In this way, the effect of different patterns of dose delivery (fractionation) on cell survival may be translated into predictions of isoeffect doses for organ injury. In practice, the dependence of tissue isoeffect dose on fractionation is used to make inferences about the dose-survival parameters of the underlying target-cell population.

To illustrate, suppose that n equally spaced (in time) doses of size x_n reduce log cell survival according to

$$\ln[f(x_n)^n] = -n(\alpha x_n + \beta x_n^2) \tag{1}$$

in which $f(x_n)$ = surviving fraction of cells after each dose x_n. If we denote by $D_n = nx_n$ the total dose given in n "fractions" of size x_n, then Eq. (1) becomes

$$E = (\alpha + \beta x_n)D_n \tag{2}$$

where $E = -\ln[f(x_n)^n]$. If a tissue isoeffect corrsponds precisely to effect E in terms of cell survival, and if doses $D_n = nx_n$ can be determined experimentally for various values of n, then linear regression of reciprocal isoeffect dose against dose per fraction:

$$1/D_n = \alpha/E + (\beta/E)x_n \quad \text{(LQ model)} \tag{3}$$

yields estimates of α/E and β/E, and of their ratio, β/α. In the sequel this model will be called the "linear-quadratic", or LQ, model.

This ratio has been shown to be of particular importance for radiotherapy, in that it is significantly larger for dose-limiting normal tissues than for tumors. Therefore, a means of increasing the therapeutic differential between these is suggested, in terms of dose fractionation (Thames, et al 1983). This is illustrated in Fig. 1, where the dependence of tissue isoeffect dose on size of dose per fraction is shown. Slowly responding tissues (usually dose-limiting in radiotherapy) are characterized by steeper slopes (solid curves), or equivalently larger β/α ratios, than are tissues that respond acutely, and many tumors as well (dashed curves). Therefore, small doses per fraction will preferentially spare slowly responding normal tissues, allowing the total dose to be increased with increased tumor control as a result.

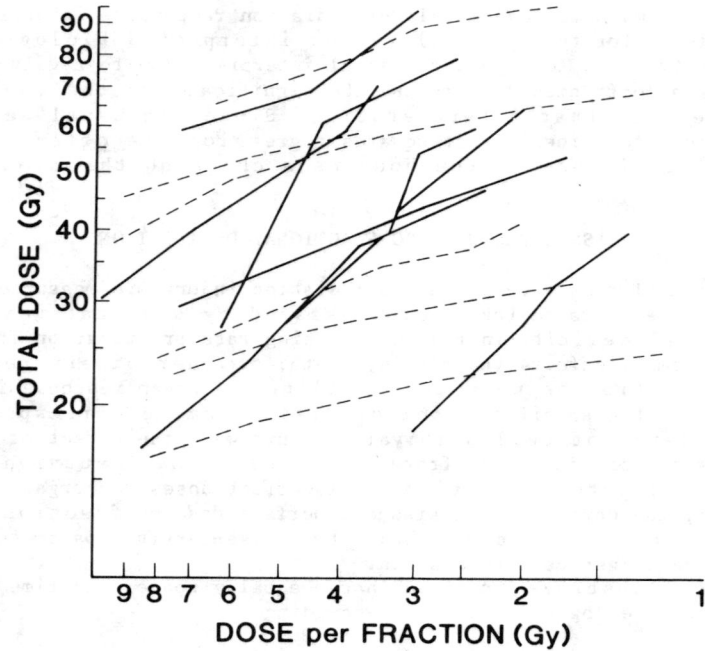

Fig. 1. Total dose for isoeffect as a function of dose per fraction, determined experimentally for various normal tissues. The curves describing late responding normal tissues (solid) have steeper slopes than those for rapidly responding normal tissues and many tumors (dashed) with the implication of a systematic difference in the ratio of survival parameters, β/α (adapted from Thames et al. 1982).

The method of analyzing tissue responses given by Eq. (3) may be called reciprocal-dose (RD) regression. Its shortcomings may be divided into those arising from biological assumptions, as opposed to those stemming from the estimation technique.

(1) Biological

The principal biological assumption on which Eqs. (1)-(3) rest is that of equal effect per dose (given the adequacy of the survival model, $f(x)$), since effects such as repopulation of survivors during treatment would complicate matters. Regression approaches have been developed to test this proposition in both the cell-survival (Thames and Withers 1980) and organ-response (Tucker, 1984) settings. Suppose the function $s_i(x)$ represents perturbations occurring between the ith and $(i+1)$st dose fractions, in the sense that Eq. (1) is modified to

$$\ln [\text{surviving fraction}] = n \ln f(x) + \sum_{i=1}^{n} \ln s_i(x). \quad (4)$$

Now suppose that the experimental design is such that the same dose per fraction is also used in m fractions, so that the fractional dose $x = x_n = x_m$ is given in regimens of n and m fractions. Eq.(4) may then be rewritten in terms of m; since $\ln f(x)$ is the same in both equations, it may be eliminated, resulting in

$$n \ln \lambda(m,x) - m \ln \lambda(n,x)$$
$$= (n-m) \ln k + n m [\mu(m,x) - \mu(n,x)]. \quad (5)$$

In Eq. (5) k is the number of cells at risk prior to the first dose, the λ's are measures of cell survival in the n- and m- fraction regimens, and the μ's are means of the first m or n $\ln s_i$'s:

$$\mu(n,x) = \sum_{i=1}^{n} \ln s_i(x). \quad (6)$$

Neglecting the interdependence of the quantities on the left- and right-hand hand sides of Eq. (5), the hypothesis of equal effect per fraction may be tested as follows. The left-hand side may be determined experimentally; when these observations are regressed on the variables n-m and nm, estimates of ln k and $\mu(m,x)-\mu(n,x)$ result. If the second of these does not differ significantly from zero, we may infer that the mean of the $\ln s_i$'s (Eq. (6)) is independent of the number in the sample. This can only occur if $\ln s_i$ = constant, which is the hypothesis of equal effect per fraction.

This regression approach may be questioned on several counts, namely the appearance of n and m in both the dependent and independent variables, normality assumptions, etc. The failure of normality can be corrected to some extent by the use of weighting terms. It has been shown (Tucker and Thames 1983) that the variance of the observable quantity $n \ln\lambda(m,x)-m \ln\lambda(n,x)$ on the left-hand side of Eq. (5) is approximately proportional to

$$n^2(2 + \lambda(m,x))/[2pN\lambda(m,x)] + m^2(2 + \lambda(n,x))/[2pN\lambda(n,x)]. \quad (7)$$

Therefore, the reciprocal square root of the quantity given by Eq. (7) is an appropriate weighting factor.

The validity of the weighted regression approach was tested using a simulation study in which the hypothesis of equal effect per fraction was made to fail in various ways. The first step involved an enumeration of all possible pairs of fractionation regimens (n,x) and (m,x). For technical reasons concerning the nature of the biological assay, there are restrictions on the common-dose fraction number pairs (n,m) that can be selected. It was observed that the acceptable (n,m) pairs comprise the points with integral coordinates that lie inside a certain triangle in the (n,m)-plane designated the "design triangle" (Fig. 2). Therefore, an experimental design for the (weighted) test of equal effect per fraction consists of a selection of points (i.e. pairs n,m) from the design triangle, together with a choice for the common dose per fraction for each pair of fraction numbers. The results of the simulation study showed that the success of the regression test of equal effect per fraction was dependent upon the experimental design.

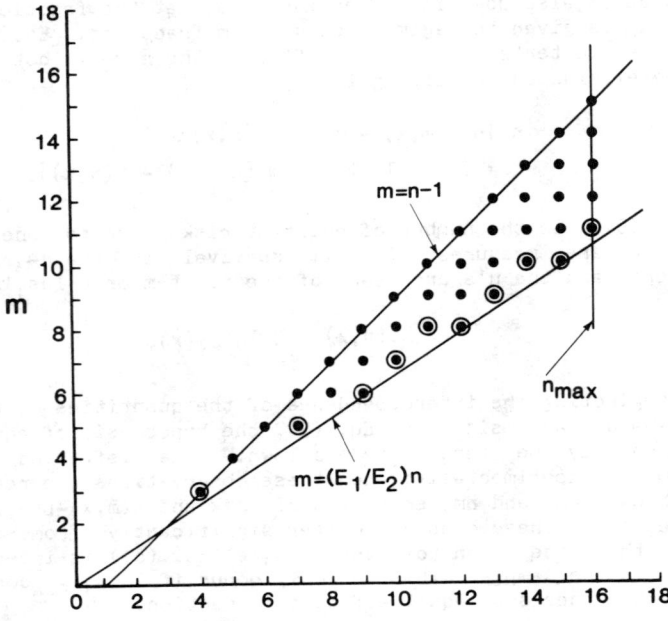

Fig. 2. Design triangle: integer pairs (n,m) for which there is some common dose per fraction x with total doses nx and mx resulting in observable survival levels. The circled points near the lower boundary of the triangle comprise the optimal design.

However, a certain experimental design was found to be optimal in two senses: first, that the hypothesis of equal effect per fraction was most likely to be rejected when it did not hold for the data, and second, that the variance of the estimate of ln k was approximately minimized when equal effect per fraction did hold. The optimal design corresponds to the (n,m) points lying along the lower edge of the design triangle, as shown in Fig. 2.

The foregoing illustrates the regression-cum-simulation approach outlined in the Introduction. Manipulation of survival models leads to formulations that allow estimation of parameters using a least squares approach; simulation is used to establish the domain in which the approach seems to work.

The organ response setting differs from the cell survival setting in that there is no measure of injury that is directly proportional to cell survival $\lambda(n,x)$. Despite this limitation the technique described above can be modified to provide a test for the fit of the multifraction linear-quadratic model. A measure of the quantity β/α for multifraction isoeffect data is meaningful provided that for every fractionation regimen, the number n of fractions and the isoeffective fractional dose x_n are related according to the expression

$$n(\alpha x_n + \beta x_n^2) = \text{constant} = E. \qquad (8)$$

If Eq. (8) does not hold, then a non-zero correction term γ_n must be added to obtain equality for one or more values of n:

$$n(\alpha x_n + \beta x_n^2) + \gamma_n = E. \tag{9}$$

The γ_n might represent the combined effects of proliferation and incomplete repair of sublethal injury occurring during a course of fractionated irradiation.

If Eq. (9) is also used to describe regimens for which no perturbing factors are present (i.e. $\gamma_n = 0$) then its left-had side can be equated for any two regimens consisting of distinct numbers n and m of dose fractions. This leads to the expression

$$D_n - D_m = (\beta/\alpha)(D_m x_m - D_n x_n) + (1/\alpha)(\gamma_m - \gamma_n) \tag{10}$$

where $D_n = nx_n$, and similarly for D_m.

A graphical test for the fit of the multifraction LQ model follows immediately from Eq. (10). Suppose that D_{n_1} through D_{n_k} are the experimentally-observed isoeffect doses corresponding to regimens of n_1 through n_k doses, respectively. If the LQ model is valid for these data, then $\gamma_{n_i} = \gamma_{n_j}$ for each choice of n_i, n_j (or equivalently, the constant "E" in Eq. (9) can be adjusted so that $\gamma_{n_i} = 0$ for all i), and Eq. (10) implies that a plot of $D_n - D_m$ vs $D_m x_m - D_n x_n$ results in a straight line through the origin with non-negative slope β/α. However, if non-zero perturbation terms γ_{n_i} are present, the data plotted in this manner will not lie on a line, or will lie on a line with non-zero intercepts.

The method suggested by Eq. (10) for testing the fit of the LQ model has been applied to experimental data and appears to be successful in identifying data for which the model is appropriate. That is, for regimens in which proliferation was known to be a factor, the data were not well-described by a straight line through the origin (Fig. 3b and 3c, next page), but for regimens having interfraction intervals that are known to be sufficient for essentially complete repair of sublethal injury, but with sufficiently few fractions to preclude the possibility of proliferation, the data lay very close to a straight line with zero intercept (Fig. 3a).

The limitations of the method are similar to those of regression methods described previously. First, the coordinates of a data point are correlated; in particular, the "independent" variable is subject to measurement error. Second, the assumption of equal variance is not met, so that regression estimates of β/α as the slope of the best-fitting line are not strictly appropriate. Third, the test of the fit of the model is purely a graphical one, since it is not meaningful to examine the significance level of the intercept.

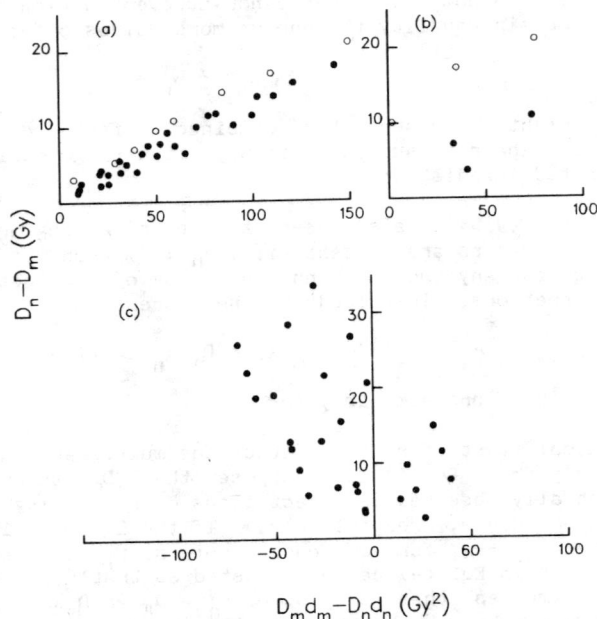

Fig. 3. Graphs of D_n-D_m vs. $D_m d_m - D_n d_n$ for each pair (D_n, D_m) of isoeffect doses from an assay of response of the colonic epithelium (Withers and Mason 1974; Tucker et al. 1983) to multiple doses separated by intervals of (a) 3 hours, (b) 12 hours, or (c) 24 hours. Near-linearity of data in (a) indicates a good fit to the LQ model, with the possible exception of protracted (20-fraction) regimens where proliferation might have occurred (open circles). The non-linearity of the data in (b) suggests failure of the LQ model because of proliferation occurring in the more protracted regimens, in particular the 10-fraction data (open circles). The model clearly fails for the data in (c).

(2) <u>Numerical</u>

To appreciate the numerical problems associated with the RD method (Eq. (3)), it is necessary to develop in more detail the way Eq. (3) is used in practice. We require pairs (x_n, D_n), where D_n is the dose sufficient to elicit a given tissue response when given in n fractions of size x_n. A typical situation is one in which we estimate ED50(n) =dose effective in eliciting a response in 50% of subjects, given in n fractions. Experimentally, a range of total doses is given, by varying x_n, and the proportion of responders N(resp)/N(subj) is recorded for each dose (Fig. 4). The ED50 may be estimated using a logistic model:

$$\text{prob\{response\}} = e^{(A + B \ln \text{dose})}/[1 + e^{(A + B \ln \text{dose})}] \equiv e^Q/[1 + e^Q]$$

for which log likelihood is given by

$$\ln L = \sum_k \{N(resp)_k \, Q_k - N(subj)_k \ln(1 + e^{Q_k})\}. \quad (11)$$
(dose groups)

We have $\ln ED50 = -\hat{A}/\hat{B}$, with confidence limits computed from the asymptotic covariance matrix using Fieller's theorem.

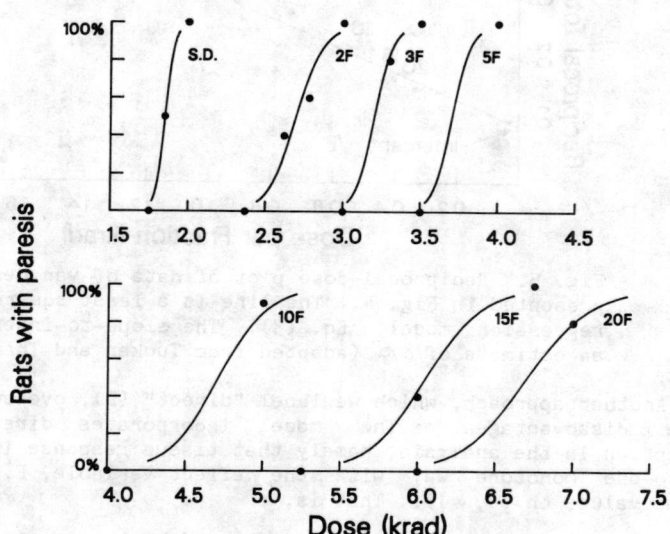

Fig. 4. Proportion of rats with hind-leg paralysis resulting from exposure of the spinal cord to single (S.D.) or multiple doses of radiation (van der Kogel 1979). For each curve, the likelihood estimate of ED50 (effective dose for 50% of the animals) is obtained (adapted from Tucker and Thames 1983).

This procedure is carried out for at least 3 values of n as pictured in Fig. 4, resulting in pairs (x_n, D_n), with $D_n = ED50(n)$ and $x_n = ED50(n)/n$, which are used with Eq. (4) to estimate α/E and β/E (Fig. 5-see below). The deficiencies of the RD methods are as follows:

(1) "Throw away" data: if ND dose groups are used for each of the NFX values of n, there is a total of ND*NFX observations. But the least squares estimation of α/E and β/E is reduced to NFX-2 degrees of freedom.

(2) Bias: there is measurement error in $x_n = ED50(n)/n$, which will lead to underestimation of the slope, and consequently of β/α. This is aggravated by negative correlation between the slope and the intercept.

(3) Poor estimates of confidence intervals: all information about the uncertainty in the individual (total of NFX) estimates of ED50 is lost in the reciprocal-dose regression.

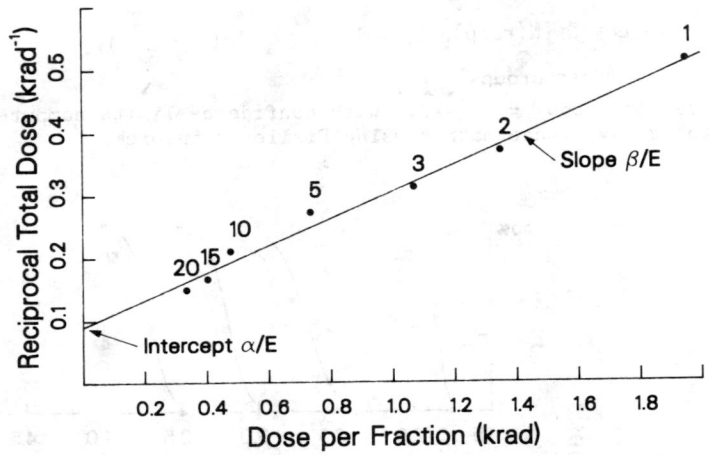

Fig. 5. Reciprocal-dose plot of data of van der Kogel (1979) presented in Fig. 4. The line is a least squares fit of the regression model, Eq. (3). The slope-to-intercept ratio is an estimate of β/α (adapted from Tucker and Thames 1983).

Another approach, which we label "direct" (D), overcomes many of these disadvantages. The model incorporates directly the key assumption in the analysis, namely that tissue response increases in a one-to-one monotone way with the effect variable, E (which may now assume values on $(0, \infty)$). That is,

$$\text{prob\{response\}} = e^{(b_0 + b_1 E)}/[1 + e^{(b_0 + b_1 E)}] \equiv e^Q/[1 + e^Q],$$

where $E = (\alpha + \beta x)D$. Hence Q has the form

$$Q = b_0 + b_1 D + b_2 x D,$$

and all of the ND*NFX observations of proportion of responders for each pair (x,D) are used in the likelihood fitting (Eq. (11)). Further, the confidence interval for β/α will reflect the uncertainty in the individual ED50s.

A simulation study was conducted to compare the RD and D methods for "small" and "large" sample sizes. In this context "small" means NFX = 3 (minimum for least squares regression) and ND = 5, i.e. 15 observations. For the large sample size, NFX = 7 and ND = 7. No obvious bias resulted from the study (means and mean squared errors were roughly the same). The main difference occurred in the number of "wayward" estimates of confidence limits. In the small sample run, 95% confidence limits did not include the true value 18% of the time with the RD method, vs. 2% with the D method.

ESTIMATION OF EQUAL-EFFECT DOSE RATIOS

Attention has been devoted in the preceding to the response of cells in the tissues to fractionated radiation doses, where a single cell-survival response $f(x)$ is applicable. This curve describes the survival response to a particular type (or quality) of ionizing radiation. In practice there are many types of radiation in use therapeutically, and drugs as well. The question naturally arises whether there are differences in biological effect per unit dose between two different radiation qualities, and if so, how should they be quantitated?

The relative biological effectiveness (RBE) of two radiation conditions is defined as the ratio of equally effective doses. For concreteness, suppose that irradiation with doses x (under condition 1) and w(under condition 2) elicits the same biological response. Then the RBE is given by

$$RBE = x/w.$$

Labels are usually chosen such that RBE $\geq$ 1, although this is not always possible.

An important subset of such comparison problems is that in which RBE is found to be approximately independent of effect level, or equivalently, dose-independent. In this case it is said to be a dose-modifying factor (DMF)

$$RBE = DMF \ (x/w \ constant).$$

This practice arose from the identification of oxygen as a radiation sensitizer, in which role it is usually a DMF, which allows the dose under hypoxic conditions to be computed as a constant multiple of the equi-effective dose given in the presence of oxygen.

The estimation problem is as follows. Let doses x_i ($i = 1, ..., n$) under irradiation condition 1 result in biological effects (e.g., log survival) y_{ij} ($j = 1, ..., n_i$), and doses w_i ($i = 1, ..., k$) under condition 2 result in effects z_{ij} ($j = 1, ..., k_i$) (cf. Fig. 6). Find the set of equi-effect doses (x,w), and determine whether x/w = constant, independent of dose.

While non-parametric approaches to this problem have been described (Kellerer and Brenot 1973), in the following we focus on parametric models. The earliest of the regression approaches to this problem is that of Pike and Alper (1964). These authors restricted attention to the high-dose range, where the survival curves may be approximated linearly, i.e.

$$\ln f(x) = \ln N - a \ x. \qquad (12)$$

Equal-effect dose pairs (x,w) under radiation conditions 1 and 2 must satisfy

$$\ln N_1 - a_1 \ x = \ln N_2 - a_2 \ w$$

or

$$x/w = RBE = a_2/a_1 + (\ln N_1 - \ln N_2)/(a_1 w).$$

The null hypothesis (i.e., DMF hypothesis) x/w = constant is equivalent to

$$H_0 : \ln N_1 = \ln N_2$$

with the two-sided alternative

$$H_A: \ln N_1 \neq \ln N_2.$$

A test of H_0 is provided by fitting the model

(1)
$$y = \ln N_1 - a_1 x$$
$$z = \ln N_2 - a_2 w$$

as opposed to the model

(2)
$$y = \ln N - a_1 x$$
$$z = \ln N - a_2 w$$

and asking whether the improvement in fit gained with separate intercepts is greater that what could be expected if model (2), with common N, were correct. The test statistic is

$$F(1, \Sigma(n_i + k_i) - 4) = (R_2 - R_1)/[R_1/(\Sigma(n_i + k_i) - 4)]$$

where R_1 and R_2 are the residual sums of squares for models 1 and 2. If we cannot reject, the estimate of the DMF is

$$DMF = a_2/a_1$$

with confidence limits obtained from Fieller's theorem.

As a practical matter, the low-dose response is of most clinical interest, so that exclusion of the low-dose range might render the RBE estimate of questionable significance in therapeutic applications. This difficulty is set aside by a recently developed test of the DMF hypothesis for the entire dose range (Thames and Rasmussen 1978). The dose-response pairs (x,y) and (w,z) are fit using any monotonic function, $\ln f$. Two commonly used models are the linear-quadratic (LQ) and two-component (TC), and we might choose (based on quality of fit)

$$f(x) = -\alpha_1 x - \beta_1 x \qquad \text{(LQ model)}$$

$$f(w) = -\alpha_2 w + \ln[1 - (1 - e^{-\beta_2 w})^N]. \qquad \text{(TC model)}$$

The fit of the LQ model is shown in Fig. 6. Note that at high doses the TC model is asymptotically linear as in Eq. (12), with a $= \alpha_2 + \beta_2$.

Next, the pairs (x_i, y_{ij}) and (w_i, z_{ij}) are transformed into estimated equal-effect dose pairs $(\hat{w}_{ij}, x_i)$ and $(w_i, \hat{x}_{ij})$ as follows. For each survival level y measured after dose x under radiation

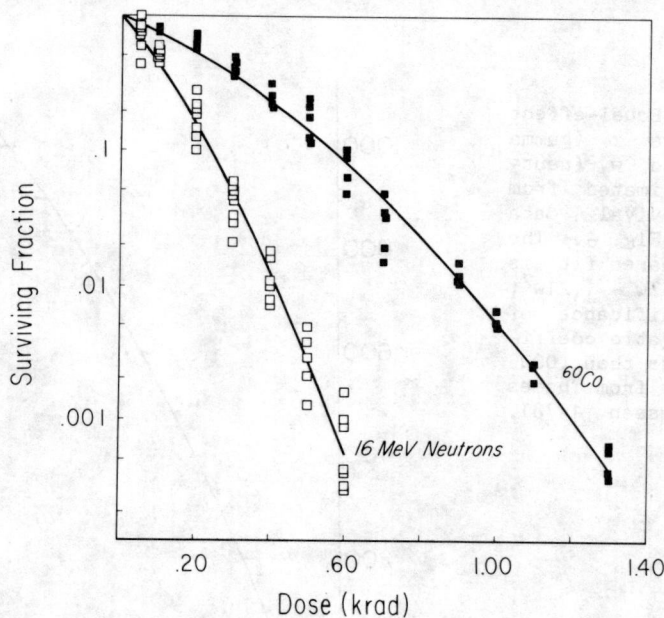

Fig. 6. Survival of chinese hamster ovary cells as a function of cobalt-60 and 16-Mev neutron dose. Curves are least-squares fits of the data to the linear-quadratic model (adapted from Thames and Rasmussen 1978).

condition 1, we use the inverted response curve 2 to obtain an estimate $\hat{w}$ of the equal-effect dose under condition 2 (cf. Fig. 6):

$$\hat{w} = f_2^{-1}(y)$$

Similarly, each of the observed log surviving fractions z measured under radiation condition 2 is converted into an estimate $\hat{x}$ of the corresponding equal-effect dose under condition 1, by inverting response curve 1:

$$\hat{x} = f_1^{-1}(z)$$

If the RBE is a DMF, then

$$x/w = \text{constant} = r \quad \text{i.e.} \quad x = rw.$$

If not, then a better fit of the equal-effect pairs will be provided by

$$x/ = r + sw \quad \text{i.e.} \quad x = rw + sw^2. \tag{13}$$

A regression approach is used with the quadratic model Eq. (13) fit to all dose pairs $(x,\hat{w})$ and $(\hat{x},w)$ in the set of equal-effect doses. Appreciable curvature (cf. Fig. 7) implies $s \neq 0$. Thus, the significance level against

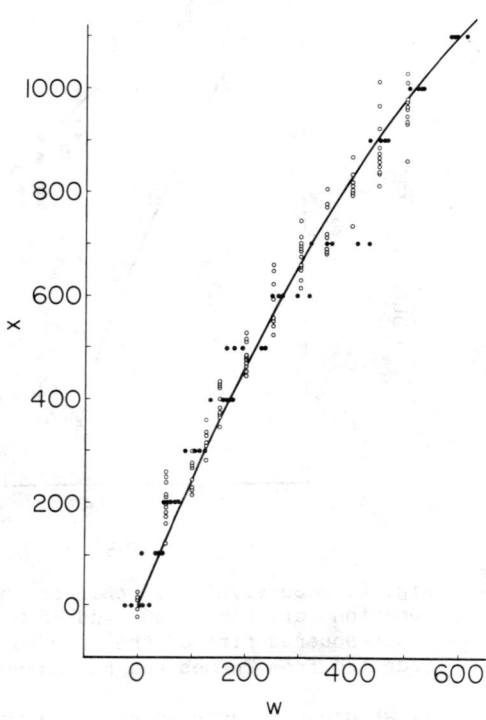

Fig. 7. Equal-effect dose pairs x (gamma rays) and w (neutrons) estimated from the survival data shown in Fig. 6. The least squares fit is $x = 2.57w - 1.21w^2$, with significance of the quadratic coefficient less than .0005 (adapted from Thames and Rasmussen 1978).

zero of the regression coefficient s is a test of the DMF hypothesis:

$$H_o : s = 0 \text{ (DMF)}$$

This method may not be construed as an exact test of the constant-RBE hypothesis. First, it is not known to what extent the true survival curves are well approximated by the models f_1 and f_2. The standard test of bias in the fit of such models does not permit any conclusions concerning this point.

A second source of inexactness is the possible inappropriateness of the regression (least squares) approach to the estimation of r and s. Monte Carlo simulations were used to investigate the appropriateness of regression. "Exact" values of the survival parameters were chosen, along with doses used under the two irradiation conditions. Then 100 simulated sets of data were generated from Gaussian deviates for various values of the standard deviation. Least squares estimates $\hat{r}$ and $\hat{s}$ of r and s were calculated for each data set. The distribution of $\hat{s}$ is shown in Fig. 7. The $\hat{s}$-values

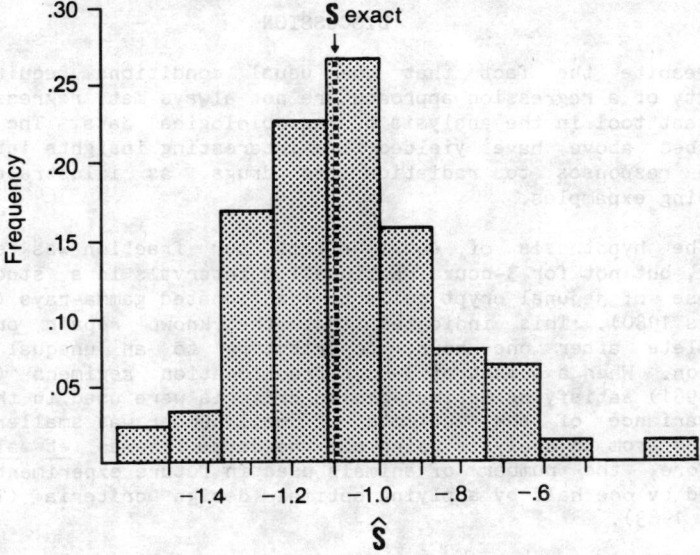

Fig. 8. Frequency of $\hat{s}$-values when Gaussian deviates were used to generate simulated data. The "exact" value of s was -1.101, while the sample mean was -1.095. Normality could not be rejected (adapted from Thames and Rasmussen 1978).

are scattered about the "exact" s-value -1.101. The sample mean of the $\hat{s}$-values was -1.095. The p-value resulting from the Kolmogorov-Smirnov test for normality of the distribution of $\hat{s}$ was 0.79. Although experimental in nature, so that no firm conclusions can be drawn from them, these results indicate that (a) the distribution of the $\hat{s}$-values is roughly symmetric with peaks near the "exact" s, and (b), Kolmogorov - Smirnov two-tailed p-values were large enough that the hypothesis of normally distributed s's cannot be rejected.

If we can reject H_o, then we will want to estimate the RBE with associated confidence limits as a function of dose. A progagation-of-error approach can be taken (McLarty and Thames, 1977). For a chosen set of doses w_i, i =1, ..., n, at which the value of the RBE is desired, we compute from the fitted survival models f_1 and f_2:

$$x_i = f_1^{-1}(f_2(w_i)), \; i = 1,\ldots, n,$$

and therefore

$$RBE(w) = f_1^{-1}(f_2(w))/w.$$

A similar result holds in terms of x:

$$RBE(x) = x/f_2^{-1}(f_1(x)).$$

Confidence limits are calculated from Fieller's theorem, with variances and covariances computed by propagation of error.

DISCUSSION

Despite the fact that the usual conditions required for the validity of a regression approach are not always met, regression is an important tool in the analysis of radiobiological data. The techniques described above have yielded some interesting insights into cell and tissue responses to radiation and drugs, as illustrated by the following expamples.

The hypothesis of equal effect per fraction was rejected for 1-hour, but not for 3-hour fractionation intervals in a study of the response of jejunal crypt cells to fractionated gamma-rays (Thames and Withers 1980). This indicates that the known repair process was incomplete after one hour, contributing to an unequal effect per fraction. When a subset of 3-hour fractionation regimens (Thames et al. 1981) satisfying optimal-design criteria were used in the analysis the variance of the estimate of cell number was smaller than that obtained from an analysis using all the data (Thames et al. 1981). Therefore, the number of animals used in future experiments could be reduced by one half by applying optimal design criteria (Tucker and Thames 1983).

The goodness of fit of the linear quadratic model might not always be adequately reflected in reciprocal-dose plots (Fig. 5). There are fractionation data for which the model appears to give a good fit in the reciprocal-dose plot, but which, when plotted in the transformed coordinates shown in Fig. 3, are clearly incompatible with the model (Tucker 1984). Since the β/α ratio obtained from reciprocal-dose analysis is used clinically to calculate treatment doses when fractionation is altered (Withers et al. 1983), this graphical test provides a useful safeguard against improper use of the model.

Equal-effect dose ratios were studied in the response jejunal crypt cells to fractionated radiation in presence and absence of the radioprotector, WR-2721 (Travis et al. 1984). It was found that the sensitizer enhancement ratio (ratio of equi-effective doses in the presence and absence of WR-2721) was an increasing function of dose, i.e., the DMF hypothesis was rejected. The clinical implication is that WR-2721 may lose effectiveness in the low range of doses used in conventional fractionated radiotherapy.

ACKNOWLEDGMENTS

We gratefully acknowledge the expert assistance of Ms. C. Seifert. The research reported herein was supported in part by grants CA-29026, CA-11430, and CA-11138 from the NCI, DHHS.

REFERENCES

DOUGLAS, B. G., and FOWLER, J. R. (1976), "The effect of multiple small doses of X-rays on skin reactions in the mouse and a basic interpretation," Radiat. Research, 66: 401-426.

KELLERER, A. M. and BRENOT, J. (1973), "Nonparametric determination of modifying factors in radiation action," Radiat. Research 56: 28-39.
MCLARTY, J. W. and THAMES, H. D., JR. (1977), "A parametric method for estimation of dose-modifying factors: low-dose extrapolation limit, Radiat. Research, 69: 1-15.
PIKE, M. D. and ALPER, T. (1964), "A method for determining dose-modifying factors," Br. Journal. Radiol. 37: 458-462.
THAMES, H. D., JR., PETERS, L. J., WITHERS, H. R. and FLETCHER, G. H. (1983), "Accelerated fractionation vs. hyperfractionation: rationales for several treatments per day," Int'l. Journal Radiat. Oncol. Biol. Phys. 9: 127-138.
THAMES, H. D., JR. and RASMUSSEN, S. L. (1978) "A test for dose-modifying factors," Radiat. Research, 76: 308-324.
THAMES, H. D., JR. and WITHERS, H. R. (1980), "Test for equal effect per fraction and estimation of initial clonogen number in microcolony assays of survival after fractionated irradiation," Br. Journal of Radiol. 53: 1071-1077.
THAMES, H. D., JR., WITHERS, H. R., MASON, K. A. and REID, B. O. (1981) "Dose-survival characteristics of mouse jejunal crypt cells," Int'l. Journal Radiat. Oncol. Biol. Phys. 7: 1591-1597.
TRAVIS, E. L, THAMES, H. D., JR., KISS, I., WATKINS, T. L., and TUCKER, S. L. (1984), "Radioprotection by WR-2721 of mouse jejunum in the clinical dose range," submitted to Int'l. Journal Radiat. Oncol. Biol. Phys.
TUCKER, S. L., WITHERS, H. R., MASON, K. A., and THAMES, H. D., JR. (1983), "A dose-surviving fraction curve for mouse colonic mucosa," Eur. Journal Cancer Clin. Oncol. 19: 443-437.
TUCKER, S. L. and THAMES, H. D., JR. (1983), "Optimal design of multi-fraction assays of colony survival in vivo," Radiat. Research, 94: 280-294.
TUCKER, S. L. (1984), "Tests for the fit of the linear-quadratic model to radiation isoeffect data," submitted to Int'l. Journal Radiat. Oncol. Biol. Phys.
VAN DER KOGEL, A. J. (1979), "Late effects of radiation on the spinal cord: dose-effect relationships and pathogenesis." Thesis, University of Amsterdam, Radiobiological Institute TNO, Rijswijk, The Netherlands.
WITHERS, H. R. and MASON, K. A. (1974), "The kinetics of recovery in irradiated colonic mucosa of the mouse," Cancer, 34: 896-903.
WITHERS, H. R., THAMES, H. D., JR. and PETERS, L. J. (1983), "A new isoeffect curve for change in dose per fraction," Radiotherapy Oncology, 1: 187-191.

CLINICAL DOSE RESPONSE MODELS. I.
REGRESSION DIAGNOSTICS AND BIASED ESTIMATION

Donald E. Herbert, Ph.D.
University of South Alabama, College of Medicine
Dept. of Radiology, Mobile, Alabama 36688

ABSTRACT

The construction of multiple regression models of dose-response from data obtained in non-experimental studies of clinical data is correctly regarded as a (usually) frustrating and even hazardous enterprise. However, it is quite often the case that such studies are an important source of information on the radiation response(s) of greatest clinical interest in the target system, the cancer patient. The nature and degree of relevance of the more precisely estimated regression models of various dose-response relations in the surrogate system most immediately accessible to designed experiments (mice) are often questionable - where not indeterminant.

For the clinical investigator the problem of experimental studies on surrogate systems is that of interpretation (of the model). The problem of non-experimental studies on target systems is that of estimation (of the model). Stratification of cases to produce homogeneous prognostic groups assures that the sizes of samples obtained in most such non-experimental studies will be small. The presence of the ethical, socio-economic, medical-legal, logistic, etc. constraints that frustrate the application of proper experimental designs to the target system assures the presence in the joint distributions of the observations, $[\underline{y}, X]$, of the samples obtained in non-experiments of such hazards, or artefacts, as 1) multicollinearity in the treatment variables, 2) outlying responses, 3) extreme levels of the treatment variables and 4) random errors of measurement in the treatment variables. The first five lectures described the Normal theory linear regression model, $\underline{y} = X\underline{\beta} + \underline{\varepsilon}$, $-\infty < y_i < \infty$, and several post-hoc salvage operations by which the presence of hazards 1) - 3) could be recognized and their characteristic effects on the estimates $\hat{\underline{\beta}}$, $\hat{y}_i$, $Var(\hat{\underline{\beta}})$, $Var(\hat{y}_i)$, RSS, etc., could be identified, measured and, in some instances - in some sense and to some degree - ameliorated.

The present lecture provides an account of the application of these - and two additional - salvage operations to the rather common case of a small sample of non-clinical, dose-response data emcumbered by the joint effects of the simultaneous presence of "all of the above" artefacts and from which it is required to construct an empirical dose-response model that will predict the response - which is binary, y = 0, 1 - in a region of interest and will also provide some information on the nature and size of the respective roles of the treatment variables in evoking and/or modulating that response.

The salvage operations by which the clinically useful information is to be retrieved from the sample include regression diagnostics, data augmentation, ridge regression, mixed estimation, validation and errors in variables estimation for a multivariable probit model of the radiation response of head and neck cancer. The present lecture makes use of the methods presented in Lectures 1 - 10. The value of the clinical data discussed in Lecture 13 in motivating a careful study of Lectures 1 - 10 exceeds its evidential value in issues of patient management.

1. INTRODUCTION

"All models are wrong, but some are useful" (Box 1979).

"A regression is constructed using prior knowledge, data, models and a fitting (estimation) process of some form. It is important to know when the resultant regression depends heavily on a small part of the prior knowledge, on a small part of the data, or on the exact choice of model or fitting process" (Welsch 1984).

The purpose of this Lecture is to present an exposition of one application of the insights and methods presented in the first ten Lectures of this Symposium to the construction of a multiple regression model of the ablation of tumor in a set of patients afflicted with cancer of the head and neck. The available data are non-experimental and thus include <u>all</u> of the several model-sample interactions for which diagnosis and treatment were discussed separately in the first five lectures. In addition, the data also are encumbered by the presence of a random error of measurement of non-negligible size in one of the independent, or predictor, variables, a common artefact of data obtained in non-experimental studies which was not addressed in the first ten lectures.

Because the effective use of statistical methods in the acquisition and analysis of medical - especially clinical - data seems to be not infrequently (See Altman 1982) an acquired taste we have felt constrained to provide some motivation and justification as well as an exposition of the applications of these insights, methods, etc. We hope to show that the methods of the first ten lectures should have a strong appeal to the "reasonable and prudent" clinical investigator.

We must also enter some disclaimers. First as to the data. They have been previously published in papers concerned with issues of clinical management of cancer by radiation. It is of interest to note that these data were selected because they are typical in most of their important features of much of the data upon which papers on clinical radiation responses are regularly published.

Next, the analyses were done on an HP 9845T system, a PC (See Boardman, Lecture <u>6</u>). Most of the software required for the various analyses was written by the author, e.g., regression diagnostics, multiple regression, probit regression, mixed estimation, ridge regression, discriminant analysis, etc. in HP Basic language. It has been validated on problems taken from the published papers in the appropriate statistics literature. However, the nature and degree of its' "portability" is uncertain. Moreover, it must, for the moment at least, be considered propriety. Of course, there is equivalent software available in the "open market". (This same disclaimer holds for Lectures <u>13</u>, <u>14</u>, <u>17</u> and <u>19</u>.)

By way of motivation and justification of the present study we assume an instance of a rather common enterprise: A clinical investigator is interested in obtaining a regression model that 1) predicts well the level of specified radiation <u>response</u> to specified adjustable <u>inputs</u> within the region, say R_3, of a system, say S_1, (See Figure <u>1</u>) in a unique state specified by a set of <u>covariates</u> and 2) provides information about the roles of each of these <u>inputs</u> and covariates in <u>evoking</u> and/or <u>modulating</u> the response of interest. We assume further that there is an immediate requirement for such a model and that some large (but unspecified) loss(es) will be incurred if the model is incorrect.

In this Lecture the following notation will be used: <u>Column</u> vectors may be upper or lower case and are only underlined, e.g., $\underline{y}$, $\underline{X}_j$. <u>Row</u> vectors are always lower case and are underlined <u>and</u> also include an apostrophe, e.g., $\underline{x}_i'$, $\underline{u}_h'$. <u>Matrices</u> with more than one row and column are always upper case and carry no subscripts other than 0, e.g., X, U, $X_0'X_0$. <u>Scalars</u> are either upper or lower case, e.g., z, z_i, x_1, x_2, R_j^2, λ_k, ν_{jk}, etc. Least Squares and Maximum Likelihood estimates are indicated by a carat, eg., $\hat{\beta}$, $\hat{\alpha}$, $\hat{V}^{-1}$, $\hat{z}_i$, $\hat{y}_i$, etc.

The response(s) of interest are the probability of occurrence of the binary event, $S = E_1$ <u>and</u> $\overline{E}_2$ – a treatment success, where E_1 is ablation of cancer, E_2 is the occurrence of a specified "complication" (e.g., necrosis) of normal tissues within and enclosing the tumor and $\overline{E}_i$, i = 1, 2 and $\overline{S}$ are the respective complementary events. As a first <u>approximation</u> we may assume $P(S) = P(E_1 \& \overline{E}_2) = P(E_1) - P(E_2)$. The system, S_1, is the (patient + tumor) in a unique state specified by a level, $\underline{u}_g'$ of a set of covariates, $U = (\underline{U}_1, U_2, \ldots U_m)$, that defines the prognostic stratum. In general, the model describes the observed response, or yield, y, as a function of <u>treatment</u> variables, X, for a system in a unique state specified by <u>covariates</u>, U. The model is <u>explicit</u> in X and <u>implicit</u> in U. We take as a basis for discussion a fairly general representation, the <u>mechanistic</u> model, $y_i = \emptyset(\underline{\theta}; \underline{x}_i') + \varepsilon_i,;\ \theta = \Theta(\underline{u}_h')$ where y_i is the level of a (scalar) response at the ith level of the treatment variables, $\underline{x}_i'$, for a system in the state defined by the hth level of covariates. $\underline{x}_i'$ is a (1 x p) vector of p treatment (continuous) variables, $\underline{u}_h'$, is a (1 x m) vector of q (discrete) covariates, ε_i is an unspecified scalar for which $E(\varepsilon_i) = 0$. Thus, $E(y_i) = \emptyset(\underline{\theta}; \underline{x}_i')$. Note that the response is <u>nonlinear</u> in $\underline{\theta}$. With no great loss of generality we can assume that all the covariates are dichotomies (0, 1). Thus, there are 2^m unique states of the system each specified by a level, u_h, $1 \leq h \leq 2^m$. y_i is a stochastic variable, $\underline{x}_i'$, $\underline{u}_h'$ are non-stochastic variables; all three are observable. $\underline{\theta}$ is an unknown (t x 1) parameter vector. However, the form, $\emptyset(\ )$, is <u>not</u> unknown and we may assume that it is non-linear in $\underline{\theta}$ which is an <u>implicit</u> function of $\underline{u}_h'$: $\underline{\theta}_h = \Theta(\underline{u}_h')$. However, it will be useful to describe the dependence of $\underline{\theta}$ on the level, $\underline{u}_h'$, of the covariates rather more explicitly by an example. If there are 3 covariates, each at 2 levels, then there are $2^3 = 8$ unique states of the system each with a corresponding $\underline{\theta}_h$ vector:

$$U = \begin{array}{ccc} \underline{U}_1 & \underline{U}_2 & \underline{U}_3 \\ 0 & 0 & 0 \\ 0 & 0 & 1 \\ 0 & 1 & 0 \\ 0 & 1 & 1 \\ 1 & 0 & 0 \\ 1 & 0 & 1 \\ 1 & 1 & 0 \\ 1 & 1 & 1 \end{array} \quad (1)$$

It is, of course, possible that although each state is unique not all of the $\underline{\theta}$ vectors will be distinct. In a classic paper, Myers, Axtell and Zelen (1966) describe an exponential survival model, $S = e^{-\lambda t}$, for a system in which each unique state is specified by 4 dichotomous covariates and for which there are $2^4 = 16$ different states and thus 16 different levels of the (scalar) parameter, λ. The dependence of the hazard function, λ, on the 4 covariates can be described by the regression model, $\ln \lambda_i = \underline{u_i}'\underline{\beta} + \varepsilon_i$, i = 1, ..., 16. Note for future reference that in this case, the response, λ_i, is a continuous variate and the "treatment" variables are discrete - the form of the data is that of Analysis of Variance. (See Gehan, Lecture 8, for a more detailed discussion and more recent examples.)

An important consequence of the multiplicity of states of the system is the resultant "fragmentation" of the data it produces. As an example, assume that the cumulative experience of an institution includes as many as 200 cases of cancer at a given anatomical site (a large series) and there are 3 dichotomous covariates (say sex, stage, grade). Then in each homogeneous state, or prognostic stratum, will include on average only about 25 cases and only small samples will be available (locally) for the estimation of θ.

The adjustable inputs are the treatment variables: total radiation dose, D (centigray or rad), total duration of treatment, T (days), total number of treatment fractions, N. There are other treatment variables of interest, e.g., dose/fraction D/N, but for the purposes of the present exposition these three are sufficient.

Note that the initial selection of the treatment variables is based on various a priori considerations. Fisher has perhaps given the best description of the most fundamental of these:

"... in the state of knowledge or ignorance in which genuine research intended to advance knowledge, has to be carried on ... we are usually ignorant which, out of innumerable possible factors, may prove ultimately to be the most important, though we may have strong presuppositions that some few of them are particulrly worthy of study. We have usually no knowledge that any one factor will exert its effects independently of all others that can be varied, or that its effects are particularly simply related to variations in those other factors. On the contrary, when factors are chosen for investigation, it is not because we anticipate that the laws of nature can be expressed with any particular simplicity in terms of these variables, but because they are variables which can be controlled or measured with comparative ease" (Fisher 1958). Both treatment variables and covariates "... can be controlled or measured with comparative ease."

However, the a priori selection of the variables to be included in the model is not restricted only by what can be easily measured and controlled. For, as Gunst has observed, "One use of a priori information is in the initial selection of predictor variables and specification of the model. Injudicious use of a priori information at this stage can lead to (a) bias if important predictor variables are not included or are misspecified in the model or (b) multicollinearities if redundant predictor variables are included or (c) increased variance of estimate if irrelevant predictor variables are included" (Gunst 1980).

Refer now to Figure 1. The region, R_1 circumscribes the set of levels of treatment over which the response can be anticipated with greater or lesser precision on the basis of the received wisdom on S_1 as represented for instance in the empirical models of Ellis (1969) for the responses E_1 and E_2 and of Supe, Nagalaxmi and Meenaksi (1983) for the response E_1. ("Modern science is an institutionalized social phenomenon. There is an institutional accumulation and propagation of knowledge by way of schools, universities, text-books, encyclopedias, etc." ... "[forming] the consensus that is the social basis of knowledge." (See Wald 1969).)The limits on the variables shown in the Figures are the ranges of the respective marginal distributions of the observations in the data upon which these models are based. Because of the presence of various <u>constraints</u> on the levels of treatment (vide infra) the volume of the <u>joint</u> distribution of (D, T, N) in the data base is much less than that delimited by the respective marginal distributions. <u>Mechanistic</u> models such as Cohen's (1983) for E_1 and E_2 in S_1 is based upon samples in which the marginal distributions of D, T, N and D/N are quite similar to those which bound R_1 in Figures <u>1a</u>, <u>1b</u> and <u>1c</u>.

The region of overlap of the region of interest R_2 with the region, R_1, which (broadly) delimits current received wisdom is small; most of the former is <u>terra incognito</u>. One is reluctant to extrapolate from R_1 to R_2 (for, as noted by old cartographers, "Here there be dragons"): - "... as we move in the space of the [treatment] variables, the mechanism may change or estimation errors may become serious so <u>unchecked</u> [unbridled?] extrapolation is <u>never</u> safe. ... even a mechanistic model should preferably be used only to suggest regions where <u>further experimentation</u> might be fruitful" (Box, Hunter, and Hunter 1978).

Thus, the most suitable answers to our (rather modest) questions would seem to require that an experiment be performed. Given the nature of our interest and our level of confidence in the currently available prior information on y_i and $\emptyset(\underline{\theta}; x_i')$ it seems appropriate to initially estimate a first-order response surface over the region R_3. See Myers 1971. (It may be fruitful to consider the system, S_1, as simply a <u>Black Box</u> consisting in the underlying phenomenon, or <u>measurement - generating</u> process, and the communication link, or the <u>measurement process</u> itself (Bury 1975).) We thus take the model to be $y_i = g(x_i'\underline{\beta}) + \varepsilon_i$ where y_i, $\underline{x_i'}$ and ε_i are as described before, $\underline{\beta}$ is a (px1) vector and g( ) is a (presently to be specified) <u>link function</u> such that $E(y_i) = g(x_i'\underline{\beta})$. Note that the response is <u>linear</u> in $\underline{\beta}$ in this model. Note that the first-order response model is answering to our initial requirements. A simple 2^p factorial design is appropriate (See Myers 1971 and Davies 1967). Note that this will not provide for second-order, effects, i.e., quadratic and interaction terms in the treatment variables.

Since it is known from toxicological experiments as well as the general Weber-Fechner theory, that the level of a biological response, $\underline{y}$ is (often) proportional to the logarithm of the level of the stimulus, or of the toxic agent(s), $\underline{X_j}$, we take the three treatment variables to be $\underline{X_1} = \log_{10}(\underline{D})$, $\underline{X_2} = \log_{10}(\underline{T})$ and $\underline{X_3} = \log_{10}(\underline{N})$. We then have $2^3 = 8$ levels of treatment, $\underline{x_i'}$, $i = 1, \ldots, 8$, wherein each variable takes one of two levels.

By an obvious transformation $\underline{x_j} = 2(\underline{X_j} - \underline{\bar{m}_j})/d_j$, the upper level of each variable is coded as +1 and the lower level as -1 (See Myers 1971). There are k = p+1 terms in the equation of a response surface of first degree in p variables. Thus, we have the design, or model, matrix X:

$$X = \begin{array}{cccc} \underline{X_0} & \underline{X_1} & \underline{X_2} & \underline{X_3} \\ 1 & -1 & -1 & -1 \\ 1 & 1 & -1 & -1 \\ 1 & -1 & 1 & -1 \\ 1 & 1 & 1 & -1 \\ 1 & -1 & -1 & 1 \\ 1 & 1 & -1 & 1 \\ 1 & -1 & 1 & 1 \\ 1 & 1 & 1 & 1 \end{array} \quad (2)$$

(See also p 446).

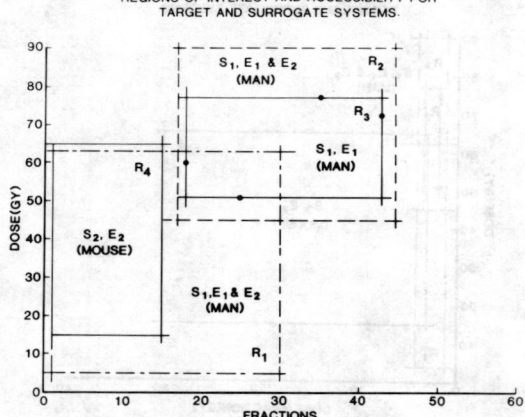

Figure 1a. Regions of a priori information, R_1, (clinical) interest, R_2, feasibility, R_3, and accessibility, R_4, for target, S_1, and surrogate, S_2, systems. D-N plane.

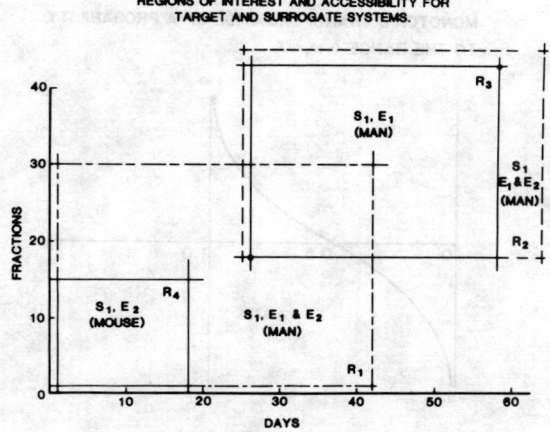

Figure 1b. Regions of a priori information, R_1, (clinical) interest, R_2, feasibility, R_3, and accessibility, R_4, for target, S_1, and surrogate, S_2, systems. N-T plane.

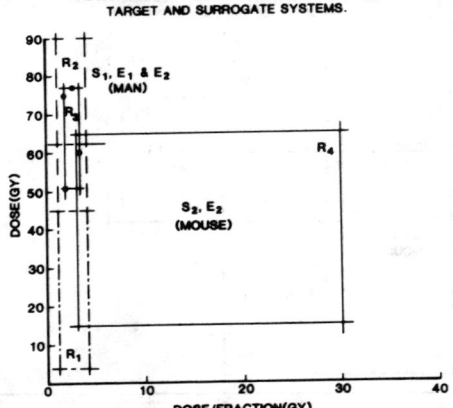

Figure 1c. Regions of <u>apriori information</u>, R_1, (clinical) <u>interest</u>, R_2, <u>feasibility</u>, R_3, and <u>accessibility</u>, R_4, for <u>target</u>, S_1, and <u>surrogate</u>, S_2, systems. D-D/N plane.

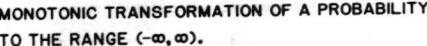
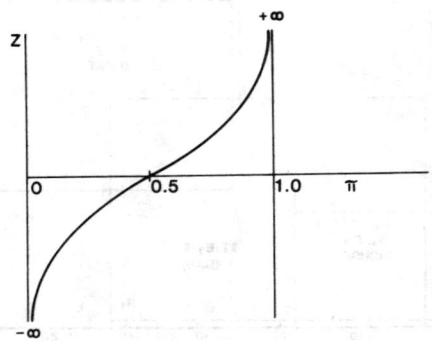

Figure 2. Mapping of $0 \leq \pi \leq 1$ into $-\infty < z < \infty$. Probit transformation.

The matrix is <u>orthogonal</u> and thus the main effect of each treatment variables, $\underline{X}_j$, on the response is described unambiguously by the sample estimate, $\hat{\beta}_j$, $j = 0, \ldots, 3$ of the respective coefficient. There are several criteria for choosing the levels of $\underline{X}_j$. Since the size of the confidence ellipsoid on $\underline{\hat{\beta}}$ is proportional to $\text{Det}[X'X)^{-1}]$ (independent of $\underline{\beta}$) one criterion for an optimal design to estimate $\underline{\hat{\beta}}$ is to select treatment levels, $\underline{x}_i'$ so as to minimize this determinant. This choice will also minimize the maximum value of $\text{Var}(\hat{y}_i) = \text{Var}(\underline{x}_i'\underline{\hat{\beta}})$ over the region of the design.

Moreover, there would be an optimal number, $n_i \gg 1$, say 30 to 50, patients randomly allocated to each of the $2^3 = 8$ treatment levels. Note that the <u>form</u> of the equation, the <u>number</u>, <u>nature</u> and <u>levels</u> of the treatment variables, the <u>number</u>, n_i, allocated to each level, etc., are determined from non-sample, a priori, information on the conditional response and the a priori desiderata of good experimental design.

An example of a 3x4 factorial design (which permits estimation of second-order terms in $\underline{x}_i'\underline{\beta}$) with $n_i \simeq 25$, $i = 1, \ldots, 12$ is discussed in detail in Lecture 17. (The example is taken from Finney 1971.) The vertices of the region R_3 in Figures <u>la</u>, <u>lb</u> and <u>lc</u> can be taken to represent possible candidate levels for a 2^3 factorial experiment. However, within the region R_3 - as is also the case in R_1 - there are levels of the treatment variables which must be <u>excluded</u> from direct observation because of the occurrence (either presumed or demonstrated) of uninteresting or untoward levels of the responses, inconvenience, etc. That is, there are ethical, socio-economic, medical-legal, etc., <u>constraints</u> on the treatment variables. These can in general be described by linear relations between the treatment variables. For example, in R_1, R_3 and R_4, $7N - 5T = 0$ describes the approximate relation that subsists between N and T for treatments given Monday through Friday of each week. Because of such constraints, observations can be (and have been) made only in the vicinity of the diagonal running from lower left to upper right of these regions in Figures <u>3a</u> and <u>3b</u>. The (approximate) limits of this <u>feasible</u> region are indicated by the black circles on the boundary of R_3. The presence of these constraints will lead to the problems of <u>estimation</u> for regression models that were discussed in Lectures <u>1</u> - <u>4</u>.

One is thus led to consider experiments on a <u>surrogate</u> system, S_2, for which the ethical, medical-legal, socio-economic, etc., constraints on the treatment variables that obtain for S_1 may be replaced by others, characteristic of S_2, which, at least in principle, will not compromise the requirements of good experimental design. We call this the <u>accessible</u> region, R_4. (Because of the difference in constraints between R_3 and R_4 it is commonly observed that the "light is better" over the latter.)

As remarked, it is frequently the case that other concerns such as convenience, etc., give rise to other constraints which restrict the observations on the surrogate system to the region, R_4, which while in principle <u>accessible</u> to designed experiments may be quite remote from either R_3 or R_2. "Discrepancies usually exist between the fractionation schedules used in radiobiological experiments and in clinical radiotherapy" ... "... most biological investigations have been carried out with fraction sizes which are considerably larger than those usually employed in clinical practice." ... "It is ... very difficult to use small fraction sizes in radiobiological studies." (See Ang, et al 1983). An example is the region of the mouse experiment which is discussed in Lecture <u>17</u>. The region for this experiment is shown by R_4 in Figure <u>1</u>.

In attempting to exploit the results of experiments on a <u>surrogate</u> system S_2 in a <u>accessible</u> region, R_4 to inform the anticipations of conditional response of the <u>target</u> system S_1 in the region of <u>interest</u>, R_2, the usual problems of <u>extrapolation</u> evident in Figure <u>1</u> are exacerbated by those of <u>interpretation</u> - the so-called "mouse-to-man" problem of toxicology. (See Schoenfeld, Lecture <u>9</u>). The two frustrations of <u>interpretation</u> and <u>extrapolation</u> of the experimental results obtained on a surrogate system in a remote region taken together comprise the "scaling problem" of engineering design. The "mouse-to-man" problem is non-trivial. It was accurately characterized by Schneiderman (1967) as "... a leap made largely in the dark." More recently Pochin (1970) has (accurately) described experimental results as "mouse-bound". Schoenfeld (Lecture <u>9</u>) has remarked on the intrinsic differences in size and basal metabolic rates (BMR) between man and the mouse surrogates used in drug screening. Another striking intrinsic difference that is especially relevant to tumor control is the distinct difference in the respective <u>functional forms</u> - Gompertz and Exponential - of the survival function for tumor-bearing mice and cancer patients.

For the normal, non-tumor bearing, mouse as well as man, the survival function is Gompertz; the age-specific death rate, or hazard function, is time-dependent: $\lambda(t) = e^{\lambda_0 + \lambda_1 t}$. Thus, both species "age" at a rate that is in part a function of the BMR: the doubling times of $\lambda(t)$ are 8.5 years for man (after age 30) and 2.8 months for the mouse.

For mice bearing transplanted tumors the hazard function is still time-dependent, the increase in the hazard function corresponding to the increase in tumor burden so that, "Death frequently occurs when tumors have grown so that the preponderance of the body mass is measurably tumor" (Jones 1956). However, for the tumor-bearing man the survival function is Exponential; the hazard function is independent of time: $\lambda(t) = \lambda_0$; (λ_0 is, of course, a function of specified covariates as has been described by Myers, Axtell and Zelen (1966) and others.) "The death rate early in the disease is strikingly similar to death rates in the middle course or at the end of the disease ... the argument is advanced for the tentative theory that the causes of death in human cancer have less to do with the extensive growth of cancer than they do with some other explanation of the metabolic state in cancer" (Jones 1956).

Because the scaling problem is absent in data obtained in non-experimental studies, one is tempted to construct <u>empirical</u> models of response from retrospective data (and some are even tempted to construct <u>mechanistic</u> models of such data.) But, as long ago as the 1920's it was recognized by Fisher that the inferences which could be made from the analysis of routinely recorded observations are often of limited validity. Indeed, it was precisely to overcome the ambiguities and uncertainties which encumber such inferences that Fisher developed the concepts and methods of designed experiments in which <u>randomization</u>, <u>replication</u>, <u>blocking</u> and <u>orthogonality of design</u> were essential elements. (See also p 446)

For, as well as comprising a non-random allocation from the target population, the joint distributions of the observations (cases) on the response and treatment variables obtained in non-experimental studies (frequently stigmatized as "happen-stance data") are characterized by the presence of time-trends, of errors of observation and measurement (random as well as systematic) of <u>extreme</u> levels of treatment, <u>outlying</u> responses and <u>multicollinearity</u>. (See Lectures <u>1-4</u>). These are, in part, the consequence of the <u>constraints</u> described above. It is most important to note that the latter - extreme cases, outliers and multicollinearity - both in kind and degree are, for a given sample of data, characteristic of a given model. They should be regarded as characteristic model-sample interactions. A different form of model, including transformation of variables, may change either - or both - the number and kind of idiosyncracy (See Cook and Wang 1983).

By way of definition we recall that (Hocking and Pendleton, 1983): "<u>Extreme</u> cases are those for which the input vector is, in some sense, far from the rest of the data." <u>Outliers</u> are "... those observations for which the inputs are reasonable but the response is abnormally large or small as compared to other cases with similar inputs," and <u>multicollinearity</u> "... refers to a situation in which there is an exact (or nearly exact) linear relation among two or more of the input variables." Data in which a significant degree of multicollinearity is present are said to be <u>ill-conditioned</u> (See also Welsch, Lecture <u>3</u> and Montgomery, Lecture <u>4</u>).

According to Huber (1977) as many as 5% - 10% of the observations in non-experimental studies are either outliers or extreme cases. The pernicious effects of such features on the conclusions from regression analysis of non-experimental data evoked Brownlee's well-known astringent iconoclasm: "The justification sometimes advanced that a multiple regression analysis on observational data can be relied upon if there is an adequate theoretical background is utterly specious and disregards the unlimited capability of the human intellect for producing plausible explanations by the carload lot." (See Brownlee 1965).

Thus, one might think that such questions as the roles played in evoking responses E_1 or E_2 in S_1 by such variables as D, T and N belong to that class of questions labelled "transscientific", since both experimental and non-experimental studies appear to be unsatisfactory. "... questions which can be asked of science and yet <u>which cannot be answered by science</u>. I propose the term <u>transscientific</u> for these questions since, though they are, epistemologically speaking, questions of fact and can be stated in the language of science, they are unanswerable by science; they transcend science." (Weinberg 1972).

However, it has also been observed, from an equal eminence, that, "The proposition that some inherent logical incompetence attaches to an inference based on observations, as distinguished from experimental evidence seems to have little to commend it beyond the great positiveness with which it is sometimes asserted." (See Cornfield 1959). Indeed it is much to the point to recall that the first draft of one of the more resounding, "success stories", of radiation oncology - the radiation treatment of Hodgkins disease - was based largely on Kaplan's careful analysis of non-experimental data. (See Kaplan 1966).

Leamer has more recently (1982) remarked that, "One should not jump to the conclusions that there is a substantive difference between drawing inferences from experimental as opposed to non-experimental data." And in still more recent (1984) comment which holds out some hope to those investigators who may study happenstance data (or in

Fischhoff's locution, "Those condemned to study the past." (See Tversky and Kahneman 1982)): "... all non-experimental data sets are simply too weak to allow sensible inferences in the absence of supplementary information." (italics added) But, as we have seen, experimental data sets also require supplementary information - that which informs their design and thus ante-dates the acquisition of the data. (Indeed for the efficient design of experiments in non-linear situations - the response is a non-linear function of the parameter vector Θ - it is necessary that there be some prior information on the values of the parameters, Θ_j, themselves, since Var(Θ) is proportional to Det$[(F'F)^{-1}]$ where F is a matrix of derivatives $\partial \emptyset/\partial \Theta_j$ and hence a function of Θ and one criterion of good design is to minimize Det$[(F'F)^{-1}]$. (See Box and Lucas 1959). This is, of course, cognate to the criterion for the design of experiments in "linear situations" where the response is a linear function of two parameter vectors: minimize Det$[(X'X)^{-1}]$.) It will be seen that the "supplementary information" required for non-experimental data sets post-dates the acquisition of the data. It is thus, "after the fact" and therefore may be regarded as a post hoc, "salvage operation". The required supplemental information is of two sorts: 1) information on the distribution of the parameter, β, of the model and 2) information on the nature and size of the effects of the model-sample interactions, the idiosyncracies, of the distribution of [y, X] - extreme cases, outlying responses, multicollinearity, etc. - on the sample estimates of β and of methods for ameliorating them.

Given a sample of size n with observation matrix [y, X] where y is (nx1) and X is (nxk) it is the objective of multiple regression analysis to obtain optimal estimates of the unobservable matrices, Θ or β, from the observation matrix [y, X]. Optimal usually signifies that the estimates, $\hat{\Theta}$ or $\hat{\beta}$, minimize RSS = $(y - \hat{y})'W(y - \hat{y})$ where $\hat{y}_i$ is the appropriate estimate of response at x_i' and W is a weight matrix. The estimates, $\hat{y}$, $\hat{\Theta}$ or $\hat{\beta}$ must also be plausible, that is, consistent with non-sample or a priori information, as well as with the sample information. (As we shall see, "optimal" estimates are not invariably "plausible", as well.) In the following development we follow McCullagh and Nelder (1983) and Baker and Nelder (1978).

We consider first the so-called Normal theory model. The response variable, y, is a continuous random variable. It has an independent Normal distribution with expected value E(y) = μ and constant variance σ^2: $y \sim N(\mu, \sigma^2)$. The sample values of y are y_i(i = 1, ..., n) where $y_i \sim N(\mu_i, \sigma^2)$. $y_i = \mu_i + \varepsilon_i$, E($y_i$) = μ_i and E(ε_i) = 0. Note that the response variable can take any value on the real line: $-\infty, \infty$. y_i has a systematic component and a random component, the latter having a Normal distribution, $\varepsilon_i \sim N(0, \sigma^2)$. The systematic component of y is considered to be a sum, or a linear combination, of k systematic components, x_j, the treatment variables (and perhaps other co-variates). This sum can be represented as the linear predictor, η_i = $x_i'\beta$, where x_i' is a (1xk) matrix of treatment variables and β is a (kx1) matrix of (unknown) weights. The link between the response y_i and the ith level of treatment variables is E(y_i) = μ_i = η_i. This can be rewritten as $y = X\beta + \varepsilon$ where X is an (nxk) model matrix. Note that y and X are known. y is a random variable. X is fixed. β and ε are unknown. ε is a random variable. β is fixed. However, the distribution function of ε is known: $\varepsilon \sim N(0, \sigma^2 I)$. y and X are known because they can be observed. β and ε are unknown because they cannot be observed.

Least squares estimates, $\hat{\beta}$ of β are obtained by minimizing the

sum of squares, $(\underline{y} - X\underline{\beta})'(\underline{y} - X\underline{\beta})$ with respect to $\underline{\beta}$:
$$\partial/\partial \underline{\beta} \, [(\underline{y}-X\underline{\beta})'(\underline{y}-X\underline{\beta})] = 0 \qquad (3)$$
This leads to the so-called Normal equations
$$X'(\underline{y}-\hat{\underline{y}}) = 0 \qquad (4)$$
where $\hat{\underline{y}} = X\hat{\underline{\beta}}$. The solution gives the estimates $\hat{\underline{\beta}} = (X'X)^{-1}X'\underline{y}$ and $Var(\hat{\underline{\beta}}) = \hat{\sigma}^2(X'X)^{-1}$, $\hat{y}_i = \underline{x}_i'\hat{\underline{\beta}}$ and $Var(\hat{y}_i) = \underline{x}_i'(\hat{\sigma}^2 X'X)\underline{x}_i$ where $\hat{\sigma}^2 = (\underline{y}-X\hat{\underline{\beta}})'(\underline{y}-X\hat{\underline{\beta}})/(n-k) = RSS/(n-k)$. If the estimates, $\hat{\underline{\beta}}$, are approximately Normally distributed, the determinant $Det[Var(\hat{\underline{\beta}})]$ is proportional to the volume contained within any specific ellipsoidal probability contour for $\underline{\beta}^*$ about $\underline{\beta}$ (the true value of the parameter vector) in the space of the parameters. Note that for any other estimates, say $\hat{\underline{\beta}}^*$ of $\underline{\beta}$, we have $RSS^* = (\underline{y}-X\hat{\underline{\beta}})'(\underline{y}-X\hat{\underline{\beta}}) + (\hat{\underline{\beta}}^*-\hat{\underline{\beta}})'X'X(\hat{\underline{\beta}}^*-\hat{\underline{\beta}}) > RSS$. However, although the estimates, $\hat{\underline{\beta}}^*$, may not describe the <u>sample</u> [y, X] with as great a <u>fidelity</u> as do $\hat{\underline{\beta}}$, they <u>may</u> be closer to $\underline{\beta}$ than are $\hat{\underline{\beta}}$. The reason that $\hat{\underline{\beta}}^*$ may be a "better" estimate of $\underline{\beta}$ than $\hat{\underline{\beta}}$ is that the former incorporates "good" <u>non</u>-sample information on the sign, size, etc., of β_j that is not included in the latter since $\hat{\underline{\beta}}$ includes only sample information (and <u>dis</u>information) in the unconstrained minimization of RSS. We shall have more to say on this matter presently. Note that $\hat{\underline{\beta}}$ is an <u>unbiased</u> estimate of $\underline{\beta}$: $E(\hat{\underline{\beta}}) = \underline{\beta}$. However, there are several circumstances in which the Least Squares estimates of $\underline{\beta}$ may be biased owing to the failure of one more of the assumptions concerning [$\underline{y}$, X], etc.: $E(\hat{\underline{\beta}}) = \underline{\beta} + \underline{b}$. The bias, b_j, may either <u>inflate</u> or <u>deflate</u> the estimates. For example, if there is random error present in the measurement of one or more of the treatment variables, $\underline{X}_j$, then $E(\hat{\underline{\beta}}) = \underline{\beta} - \underline{b}_1$ where b_1 can be estimated as $b_1 \simeq n(X'X)^{-1}\hat{D}\underline{\beta}$ where $\hat{D}$ is an a priori estimate of the dispersion matrix of the measurement error. Obviously, if $X'X$ is nearly singular the presence of even small random errors of measurement in $\underline{X}_j$ can result in the presence of a large bias, b_1, in the estimates, $\hat{\underline{\beta}}$.

Moreover, although $\hat{\underline{\beta}}$ is an unbiased estimate of $\underline{\beta}$, $E(\hat{\underline{\beta}}) = \underline{\beta}$, the Euclidean norm, $\hat{\underline{\beta}}'\hat{\underline{\beta}}$ is <u>not</u> an unbiased estimate of $\underline{\beta}'\underline{\beta}$: We note that $E[(\hat{\underline{\beta}}-\underline{\beta})'(\hat{\underline{\beta}}-\underline{\beta})] = \sigma^2 Trace[X'X)^{-1}]$ or, equivalently $E(\hat{\underline{\beta}}'\hat{\underline{\beta}}) = \underline{\beta}'\underline{\beta} + \sigma^2 Trace[(X'X)^{-1}] = \underline{\beta}'\underline{\beta} + \sigma^2 \sum_{p} \lambda_j^{-1}$. Thus, if one or more of the eigenvalues, λ_j, of $X'X$ is very small - $X'X$ <u>non</u>-orthogonal - then $\hat{\underline{\beta}}'\hat{\underline{\beta}}$ is greatly <u>inflated</u> with respect to $\underline{\beta}'\underline{\beta}$ and the distance from $\underline{\beta}$ to $\hat{\underline{\beta}}$ will tend to be large.

"Estimated coefficients that are large in absolute value have been observed by all who have tackled live non-orthogonal data problems" (Hoerl and Kennard 1970). The comment has been appropriately rephrased by Leamer (1978): "... estimated coefficients that are far from a priori likely coefficients have been observed by all who have tackled live non-orthogonal data problems." Leamer's locution seems more generally appropriate since it is easier to recognize when the estimated parameter vector has the wrong <u>direction</u> than the incorrect <u>length</u>. That is, <u>qualitative</u> a priori information on the <u>sign</u> of β_j is more commonly available than <u>quantitative</u> a priori information on the <u>size</u>, $|\beta_j|$, of the components of $\underline{\beta}$. For instance, although it is possible (See Hoerl and Kennard 1970) to obtain estimates, say $\hat{\underline{\beta}}_R$ of $\underline{\beta}$ that <u>minimize</u> the sum of squares, $RSS = (\underline{y}-X\hat{\underline{\beta}}_R)'(\underline{y}-X\hat{\underline{\beta}}_R)$ subject to the constraint $\hat{\underline{\beta}}_R'\hat{\underline{\beta}}_R \leq c^2$, it is usually difficult to obtain a satisfactory a priori specification of c^2. It is less difficult to obtain non-sample information on the sign of β_j. However, see Oman (1983). Note that β_R is the Ridge regression estimate of $\underline{\beta}$ (See Montgomery, Lecture <u>4</u>). We shall return to this in a later section of the present paper.

The determinant, $\sigma^2 \text{Det}[(X'X)^{-1}]$, is proportional to the volume included within any specified probability ellipsoid for values of the parameter about $\underline{\beta}$. Since $\sigma^2 \text{Det}[(X'X)^{-1}] = \sigma^2 \Pi_j^p \lambda_j^{-1}$ if one or more of the eigenvalues λ_j of X'X is very small $\text{Var}(\hat{\underline{\beta}})$ is greatly <u>inflated</u>. There are several circumstances which may lead to small eigenvalues of X'X. Prominent among these is <u>multicollinearity</u>. Thus, the presence of strong linear relations among the treatment variables will lead to inflated estimates, $\hat{\underline{\beta}}'\hat{\underline{\beta}}$ and $\text{Var}(\hat{\underline{\beta}})$. Obviously, this implies that $\hat{\underline{\beta}}$ may be very far from $\underline{\beta}$ and quite imprecise as well.

It is evident that $\hat{\underline{\beta}}$, $\text{Var}(\hat{\underline{\beta}})$ and σ^2 will be sensitive to the presence of collinearity in X, to outlying responses, y_i, and to extreme levels of treatment, $\underline{x_i}'$ (vide infra). In the present context it is useful to assume that the Normal theory linear model, $y = X\beta + \varepsilon$, presents a response surface (See Davies 1967 and Myers 1971). It will also be useful to comment upon a few general features of the linear model and its philosophy and methodology which are also applicable to the non-linear models we shall presently discuss. First, the systematic part of a response surface model is a Taylor series, to terms of degree not greater than second, in the treatment variables; it is a second-order, <u>empirical</u>, approximation to an unknown (perhaps unknowable) "true" model, $\emptyset(\underline{\theta}; X)$ in the region defined by the observation matrix. $\underline{\beta}$ is (kxl) where k $\geq$ m, the dimension of $\underline{\theta}$. This indicates one of the advantages - <u>parsimony</u> - which a <u>mechanistic</u> model <u>may</u> enjoy over an <u>empirical</u> model. For the latter, the model matrix X is an (nxk) matrix of treatment variables. In general, the latter are continuous varibles such as dose. However, X may also include piece-wise continuous variables (splines, See Wald, 1974) and discrete (dichotomies, See Neter and Wasserman 1974)) covariates. For p treatment variables there are k = (p+1) columns in X for a first-order model and k = (p+1)(p+2)/2 columns in X for a second-order model. Thus, for p = 3, k = 4 columns in X for a first-order model and k = 10 columns for a second-order model. As a general rule in order to obtain adequate estimates of $\underline{\beta}$, the <u>minimum</u> sample size, n must exceed 20k or 100 whichever is the <u>greater</u> (See Lindeman, Merenda and Gold 1980). It is evident that <u>parsimony</u> in a model can in principle confer considerable financial advantage to a study. There are other rewards for parsimony as well, as we shall subsequently discuss.

Perhaps the most parsimonious description of the canon of <u>parsimony</u>, sometimes described as (William of) Occam's Razor, is that of Sir William Hamilton: "Neither more, nor more onerous, causes are to be assumed than are necessary to account for the phenomena." As Pearson (1892) has observed, "This 'simplicity of nature' is, of course, pure dogma but the regula philosophandi which forbids us to revel in superfluous causes is fundamental to our view of science as an economy of thought." (See Pearson 1957). One of the penalties incurred by such "revels" is a decrease in the precision of the estimated response, $\hat{y}_i$ (as noted above in the remarks by Gunst, 1980). A general (a very approximate) estimate of the average variance of the estimated response over all n observations is, $\text{Var}(\hat{y}_i) = w\sigma^2/n$, where w is the number of <u>components</u> in the parameter vector - $\underline{\beta}$, for the <u>empirical</u> model, $\underline{\theta}$ for the <u>mechanistic</u> models - estimated from the sample. (See Box, Hunter and Hunter 1978).

Another penalty has been described by Crocker (1972): "The expectation of R^2 [the multiple correlation coefficient] in the null case was shown ... to be $E(R^2) = p/(n-1)$ where p is the number of predictors and n is the sample size. Values of R near 1.0 are thus easily obtained by chance as the number of predictors approaches (oneless than) the sample size - a common problem where data are limited."

β, as well as θ, refers to a single, unique, state of the system, characterized by a given set of levels of the features which define the states. For θ, the dependence is _implicit_. For β, however, this dependence is _explicit_. For example, for a qualitative variable with r levels, $(r-1)$ dichotomous variables $u_j = 0$ or 1 $(j = 1, ..., r-1)$ are included in the model matrix, X, and β is augmented by $(r-1)$ components, β_j. If there are _interactions_ between the covariates and treatment variables, say $u_j x_k$, then still other components must be included in β, of course.

ϵ is an _un_observable (nx1) so-called error vector. ϵ represents everything that affects the response, y, excluding $X\beta$. Leamer (1978) has made this more explicit: $\epsilon = Q\gamma$ where Q is an unspecified nxq matrix of covariates, qualitative as well as quantitative, and γ is (qx1) vector with fixed components. Box, Hunter and Hunter (1978) have aptly characterized iterative model building in multiple regression as moving varibles from the Q matrix to the X matrix - and vice versa.

The concept of a "white-noise" sequence has been used by Box, Hunter and Hunter (1978) to provide an operational definition of an adequate statistical model, say $y = \emptyset(\theta, X)$. An adequate model of a set of observations [y, X] exists if there is a set of _plausible_ parameters, θ, such that $y_i - \emptyset(\theta, x_i) = \epsilon_i$ is a white-noise sequence. By white-noise sequence is meant that $E(\epsilon) = 0$, $Var(\epsilon) = \sigma^2$ (a constant) and the sequence $[\epsilon_i]$, $i = 1, ..., n$ is uncorrelated with itself or any other known phenomenon. By plausible is meant that θ is consistent with prior information. In other words, an adequate model is a _plausible mapping_ of [y, X] into ϵ: [y, X] $\xrightarrow{\theta}$ ϵ.

Let us now consider a binary response: $\overline{Ca}$ (cancer absent) or Ca (cancer present). These may be coded as $y_i = 0$ or 1, respectively, for Normal theory model of this response, $y_i = \mu_i + \epsilon_i$, and $E(y_i) = \mu_i = \Pi_i = P(y=1) = P(Ca| x_i')$ and it is evident that for some levels of x_i', $\Pi = x_i'\beta$, will give values of μ_i which do _not_ satisfy the condition, $0 \le \mu_i \le 1$. Moreover, $Var(y_i) = \mu_i(1-\mu_i)$, so the condition $Var(y_i) = \sigma^2$ a constant, does _not_ hold. An obvious maneuver is to impose restrictions on either x_i' or β, or both, such that $0 < x_i'\beta < 1.0$, but this is usually less desirable than it is obvious. An alternative solution lies in the relation between conditional, joint and marginal probabilities: $P(Ca| x_i') = P(Ca \& x_i')/[P(\overline{Ca} \& x_i') + P(Ca \& x_i')]$. If the conditional frequency distribution of x_i' on the response, Ca and $\overline{Ca}$, are multivariate Normal with means $\mu_1 > \mu_0$ and common covariance matrix, $\Sigma_0 = \Sigma_1 = \Sigma$, it can be shown that $P(Ca| x') = [1+e^{-x'\delta}]^{-1}$ where $\delta_0 = -\ln(q_0/q_1) - (\mu_1 - \mu_0)'\Sigma^{-1}(\mu_1 + \mu_0)$ and $(\delta_1, \delta_2, ...)' = (\mu_1 - \mu_0)'\Sigma^{-1}$ where q_0 is the _a priori_ probability of $y_i = 1$. This is simply the discriminant function model of the logistic function. (See Cornfield, Gordon and Smith 1961; Lachenbruch 1975). We note that the parameter vectors δ and β are simply proportional: $\delta = c\beta$, where c is a function of the data.

We now take up still another alternative, namely, a monotonic transformation that _maps_ the interval (0, 1) onto the real line ($-\infty$, ∞). See Figure 2.

We consider a regression model for the quantal, dose-response study at hand. In this study the reponse is a binary variable: $y_i = 0$

or 1 accordingly as the event $\bar{E}_1$ or E_1 occurs at the ith dose level, $\underline{x}_i'$, where E_1 denotes cancer absent, $\overline{Ca}$, a "success" and $\bar{E}_1$ denotes cancer present, Ca, a "failure". More generally, the response variable y_i represents the number of successes out of n_i independent trials where Π_i, the probability success in a single trial and $\underline{x}_i'$, the dose-level at each trial vary with $i (i=1, \ldots, n)$. n_i is the number at risk at $\underline{x}_i'$. Thus, $P(y_i=1) = \Pi_i$ and $P(y_i=0) = 1-\Pi_i$ and $y_i \sim B(n_i, \Pi_i)$ where $B(n_i, \Pi_i)$ is the Bernoulli distribution with <u>index</u> n_i and <u>parameter</u> Π_i. Note that y_i can take only integer values over the finite range, $0, 1, \ldots, n_i$. Thus, $y_i = n_i \Pi_i + \varepsilon_i$. As in the Normal theory model the random variable y_i has a <u>systematic</u> component, $n_i \Pi_i$, and a <u>random</u> component, ε_i: $E(y_i) = n_i \Pi_i$, $Var(y_i) = n_i \Pi_i (1-\Pi_i)$, $E(\varepsilon_i) = 0$, $Var(\varepsilon_i) = n_i \Pi_i (1-\Pi_i)$.

In the more common quantal dose-response studies, the dose, $\underline{x}_i'$, is an "agent" whose administration is completely under experimental control. However, in the non-experimental study that we now consider, the "dose" is not closely controlled (See Cornfield, et al 1961). Moreover, the number at risk, n_i, which is a pre-determined constant in the controlled case is a random variable when the dose is uncontrolled and one which takes only the values zero and one at each possible level, $\underline{x}_i'$. For a given set of n levels of treatment $n_i = 1$, $1 \leq i \leq n$. Then $y_i = \Pi_i + \varepsilon_i$ and $E(y_i) = \Pi_i$, $Var(y_i) = Var(\varepsilon_e) = \Pi_i (1-\Pi_i)$, $E(\varepsilon_i) = 0$. The linear predictor, $\eta_i = \underline{x}_i'\underline{\beta}$, is related to $E(y_i)$ by the <u>link function</u> $g(\Pi_i) = \eta_i$. Two commonly used link functions are the <u>logit</u>, $g_1(\Pi) = \ln[\Pi/(1-\Pi)]$ and the <u>probit</u>, $g_2(\Pi) = \Phi^{-1}(\Pi)$ where $\Phi^{-1}(\Pi)$ is the inverse of the standard Normal integral. In the discussions to follow we denote these link functions as z_1 and z_2, respectively, or where no ambiguity is likely as simply z. In this context the link function for the Normal theory model, where $-\infty < y_i < \infty$, is the <u>identity</u>: $\mu_i = \eta_i$.

We have previously noted that for a <u>binary</u> response where y_i takes only the values 0 and 1 the identity link function may not be useful since it implies that equal increments to the levels of the treatment variables will produce equal increments in μ_i which is obviously not realistic over the range of μ_i and may also lead to values outside the range (0, 1). For both the logit and probit link functions equal changes in the treatment variable produce equal changes in z_i, $i = 1, 2$. Moreover, z_i, unlike y_i, may take any value on the real line: $-\infty < z_i < \infty$.

Estimates of $\underline{\beta}$ for either the logit or probit link can be obtained by solving the Maximum Likelihood (ML) equations for which the matrix formulation is $X'(\underline{y}-\hat{\underline{\Pi}}) = 0$. These are formally similar to the so-called normal equations for the Normal theory model. However, the models for a binary response variable ($y_i = 0, 1$) are <u>non-linear</u> in $\underline{\beta}$ and iterative methods are required to solve them (Pregibon 1981). As is the case for any iterative estimation method, good initial estimate, $\hat{\underline{\beta}}(0)$, of $\underline{\beta}$, are crucial (Draper and Smith 1981). We have found that in general the <u>unit weight</u> regression model gives satisfactory initial estimates; e.g., for, $\underline{z} = X\underline{\beta}$, $X = (\underline{1}, \underline{X}_1, \underline{X}_2, \underline{X}_3)$, $\hat{\underline{\beta}}(0) = (-1, 1, -1, -1)$ (See Lecture 19).

If we introduce the concept of a pseudo-response vector, z^*, then (at the final iteration) the ML estimates, $\hat{\underline{\beta}}$, $Var(\hat{\underline{\beta}})$ and RSS can be represented as Generalized Least Squares estimates: $\hat{\underline{\beta}} = (X'\hat{V}^{-1}X)^{-1} X'\hat{V}^{-1}z^*$, $Var(\hat{\underline{\beta}}) = (X'\hat{V}^{-1}X)^{-1}$ and RSS $= (\underline{y}-\hat{\underline{\Pi}})'W(\underline{y}-\hat{\underline{\Pi}})$ where $\hat{V}^{-1}$ and W are diagonal weight matrices. On the null hypothesis that the model is correct, RSS $\sim \chi^2(n-k)$. Also, $\hat{V} = Diag[\hat{\Pi}_i(1-\hat{\Pi}_i)/\phi^2(\hat{\eta}_i)]$ and $W = Diag[1/$

$\{\hat{\Pi}_i(1-\hat{\Pi}_i)\}]$ and $\emptyset( ) = \phi'( )$. More precisely, the model, $z^* = X\underline{\beta} + \varepsilon^*$, can be transformed to $T\underline{z}^* = TX\underline{\beta} + T\underline{\varepsilon}^*$ where $T'T = \hat{V}^{-1}$ and then $\underline{\hat{\beta}}$, Var$(\underline{\hat{\beta}})$, RSS obtained by Ordinary Least Squares methods. T is an (nxn) diagonal matrix with non-zero elements consisting of the square root of the weights of the respective responses at the final iteration. See Theil 1971; Johnston 1972.

As is the case for the Normal theory linear models, the estimates $\underline{\hat{\beta}}$, Var$(\underline{\hat{\beta}})$ and RSS are sensitive to the presence of <u>collinearity</u> in X, to <u>outlying</u> responses, y_i, and to <u>extreme</u> levels of the treatment variables, $\underline{x}_i'$ (vide infra). Pregibon (1981) has remarked that the effects of these idiosyncracies of the observation matrix [$\underline{y}$, X] may be worse for non-linear models than for linear models. (We shall later present some evidence that tends to corroborate his anticipations.)

We consider only the <u>probit</u> model in the discussions to immediately follow in which it will be helpful to refer to Figure <u>2</u> which describes the mapping of the finite range $0 \leq \Pi \leq 1.0$ into the infinite range $-\infty < z < +\infty$ by the probit link function, $z = \phi^{-1}(\Pi)$.

It can be shown that for the probit model, the pseudo-response vector can be represented by a first-degree Taylor series, $z^* = \eta + (y-\Pi)/\emptyset(\eta)$ where $\emptyset(\eta) = d\Phi/d\eta$.

For the probit model of the binary response, $\hat{\beta}$ is an unbiased estimate of β: $E(\hat{\beta}) = \beta$. However, this is an asymptotic ($n \longrightarrow \infty$) result. But for n small, $E(\hat{\underline{\beta}}) = \underline{\beta} + \underline{b}_2$. The effect of the <u>small sample size</u> is to <u>inflate</u> the ML estimates of $\underline{\beta}$. Schaefer (1983) has noted that, "For small samples the ML estimates may have substantial bias and, if no account is made of the bias, could lead to incorrect conclusions concerning the effects of risk factors" - as much as 10% to 30% for n < 50. He has accordingly developed a one-step approximation to the bias, b_2:

$$\underline{\hat{b}}_2 = (-0.5)*[(X'\hat{V}^{-1}X)^{-1}X'\hat{V}^{-1}\{(1-2\hat{\Pi}_k)\underline{x}_k'(X'\hat{V}^{-1}X)^{-1}\underline{x}_k\} \qquad (5)$$

where $\{a_k\}$ describes a <u>vector</u> whose k^{th} element is a_k. Thus, the corrected ML estimate of $\underline{\beta}$ is $\underline{\hat{\beta}} - \underline{\hat{b}}_2$. (It should be noted that although the corrected estimates are routinely computed in our current software, they appear explicitly only in Table <u>17</u> of the present paper.)

The presence of random errors of measurement in one or more of the treatment variables will also lead to biased ML estimates of $\underline{\beta}$ for the probit model. As with the LS estimates, $\hat{\underline{\beta}}$, for the Normal theory model, $E(\hat{\underline{\beta}}) = \underline{\beta} - \underline{b}_1$, that is, the ML estimates are <u>deflated</u> by the presence of random errors of measurement. Carroll, Spiegelman, Lan, Bailey and Abbott (1984) have recently published a paper on a errors-in-variables model of binary response. We present a heuristic development of such a model later in this paper.

Although the (asymptotic) estimate of the covariance matrix of the ML estimate of $\underline{\beta}$ is given by Var$(\underline{\hat{\beta}}) = (X'\hat{V}^{-1}X)^{-1}$ in studies where $n_i > 1$, i.e., grouped data conditions may arise in which $(X'\hat{V}^{-1}X)$ <u>underestimates</u> Var$(\underline{\hat{\beta}})$. In such cases, Var$(\underline{\hat{\beta}}) = \sigma^2(X'\hat{V}^{-1}X)^{-1}$ where the <u>over</u> dispersion factor is σ^2 = RSS/(n-k). Most commonly such cases occur when observations <u>between</u> groups are <u>uncorrelated</u> but observations <u>within</u> groups are positively correlated. (See Cox 1970), Finney 1971 and McCullagh and Nelder 1983.) We shall encounter an example of this effect in Lecure <u>14</u>.

It was remarked in Lectures <u>1-4</u> that the maximum likelihood and least squares estimates of $\underline{\beta}$ are sensitive to the presence of <u>outlying</u>

responses, y_i, extreme levels of the predictor or treatment variables, x_i', and multicollinearity in X, the degree of sensitivity depending on the model. The possibility of post hoc salvage operations to retrieve a given non-experimental study was mentioned. We now take up in brief discussions, several of these maneuvers. It will be sufficient to use only the Normal theory model for our principal example. There is a common format to these operations, namely, that of forming matrix-weighted averages (See Leamer 1978 and Lecture 5).

2. DATA AUGMENTATION

"The most direct and obvious method of improving data conditioning is through the collection and use of additional data points that provide the needed independent variation relative to the original data. This answer is rarely useful to ... many users of least squares who typically have short data series. New data are obtainable in adequate numbers only at substantial cost, either in terms of the time one must wait for new observations to occur or in terms of the collection costs needed to obtain a less collinear sample. Further, even if new data are obtained there is typically no guarantee that they will be consistent with the original data or they they will indeed provide independent information. Applied statisticians are often unable to control their experiments, whereas nature often closely replicates hers. [Not only natural constraints are stationary over time. For instance, Figure 3a shows the presence of the same clinical (ethical?) constraints in the additional (T_3) data as were in the initial (T_2) data.] The introduction of new data therefore, is not likely to   be a fix of much practical importance in many applications of least squares. When it is possible however, to provide new data, the corrective action is simple and straight forward"
(Belsley, Kuh and Welsch 1978).

The maneuver of pooling the original [y_1, X_1) collinear data with the additional data chosen to reduce the effects of the collinearity etc. on the estimates of $\underline{\beta}$ may be represented as follows: (See Theil 1971 and Johnston 1972).

$$\begin{bmatrix} \underline{y}_1 \\ \underline{y}_2 \end{bmatrix} = \begin{bmatrix} X_1 \\ X_2 \end{bmatrix} \underline{\beta} + \begin{bmatrix} \underline{\varepsilon}_1 \\ \underline{\varepsilon}_2 \end{bmatrix} \qquad (6)$$

Then the unbiased estimate of $\underline{\beta}$ is

$$\underline{\hat{\beta}}^* = [1/\sigma_1^2 \; X_1'X_1 + 1/\sigma_2^2 \; X_2'X_2]^{-1} \; [X_1'\underline{y}_1 + X_2'\underline{y}_2] \qquad (7)$$

and
$$\text{Var}(\underline{\hat{\beta}}^*) = [1/\sigma_1^2 \; X_1'X_1 + 1/\sigma_2^2 \; X_2'X_2]^{-1} \qquad (8)$$

That is, $\underline{\hat{\beta}}^*$ is a matrix-weighted average of the original and additional data. Dykstra (1966, 1971) and Gaylor and Merrill (1968) describe methods by which [$\underline{y}_2$, X_2] can be systematically selected in an optimal manner. The criterion is to maximize $|X'X|$. This minimizes both the volume of the confidence region for the regression coefficients and the maximum variance of a predicted value, $\text{Var}(\hat{y}_i)$. The extremes are achieved by obtaining responses at the extremes of the region of observations. For a first-order model in p treatment variables these are the observations at the 2^p corners of the polytope. These methods may salvage "... undesigned, nonorthogonal experimental data in those cases in which it is desired to separately estimate and test the effects of the independent variables without discarding the existing data. Moreover, these methods provide ... better estimates of the regression coefficients than discarding the original data and conducting an orthogonal experiment ..." (Gaylor and Merrill 1968) with the same number of data points.

3. BAYES ESTIMATION (See Montgomery and Peck 1982)

<u>Classical theory</u> assumes that the <u>form</u> of the model, $\underline{y} = X\underline{\beta} + \underline{\varepsilon}$, is known but the elements of the (kx1) parameter vector, $\underline{\beta}$, although <u>fixed</u>, are quite unknown. The only information on $\underline{\beta}$ is that within the observation matrix, $[\underline{y}, X]$ and represented by the Least Squares elements of the estimates, $\hat{\underline{\beta}}$ and $Var(\hat{\underline{\beta}})$. It also assumes that the elements, ε_i, of the (nx1) error vector, $\underline{\varepsilon}$, are <u>random variables</u> but otherwise unknown. However, the <u>distribution</u> of $\underline{\varepsilon}$ is known: $\underline{\varepsilon} \sim N(0, \sigma^2 I)$.

<u>Bayesian theory</u> assumes that the elements of $\underline{\beta}$ are <u>random variables</u> and that the joint distribution of these variables is known, a priori. "Bayesian inference is different from classicial inference in that it makes use of information that is not contained in the sample. Bayesian theory is concerned with the optimal pooling of sample information with non-sample information." (See Leamer 1978) A discussion of Bayes Estimation in terms of Normal Theory models will suffice.

"... if prior information about $\underline{\beta}$ can be described by a p-variate normal distribution with mean vector $\underline{\beta}_0$ and covariance matrix Σ_0, then the Bayes estimator of $\underline{\beta}$ is
$$\hat{\underline{\beta}}_B = [1/\sigma^2 \, X'X + \Sigma_0^{-1}]^{-1}(1/\sigma^2 \, X'\underline{y} + \Sigma_0^{-1}\underline{\beta}_0)"$$ (Montgomery and Peck (1982).

Since $X'X\hat{\underline{\beta}} = X'\underline{y}$ we have
$$\hat{\underline{\beta}}_B = [1/\sigma^2 \, X'X + \Sigma_0^{-1}]^{-1}(1/\sigma^2 \, X'X\hat{\underline{\beta}} + \Sigma_0^{-1}\underline{\beta}_0) \tag{9}$$
or, the Bayes' estimator is a matrix-weighted average of <u>a priori</u> information $(\underline{\beta}_0, \Sigma_0)$ and <u>sample</u> information $(\hat{\underline{\beta}}, \sigma^2(X'X)^{-1})$. Or, "The posterior mean is the matrix-weighted average of a prior location vector and a sample location vector" (See Leamer 1978). It is obvious that
$$Var(\hat{\underline{\beta}}_B) = (1/\sigma^2 \, X'X + \Sigma_0^{-1})^{-1} \tag{10}$$

The Bayes' estimator of $\underline{\beta}$ for the Generalized Least Squares model is an obvious generalization of the Bayes' estimator for the Least Squares model. The <u>sample</u> information on $\underline{\beta}$ is represented by $\hat{\underline{\beta}}$ and $Var(\hat{\underline{\beta}}) = (X'\hat{V}^{-1}X)^{-1}$. The <u>a priori</u> information is represented by the first two moments, $\underline{\beta}_0$ and Σ_0, of the a priori distribution which is p-variate Normal. The Bayes estimator is, again, the matrix-weighted average of the sample and non-sample information:
$$\hat{\underline{\beta}}_B = [X'\hat{V}^{-1}X + \Sigma_0^{-1}]^{-1}(X'\hat{V}^{-1}X\hat{\underline{\beta}} + \Sigma_0^{-1}\underline{\beta}_0) \tag{11}$$
and
$$Var(\hat{\underline{\beta}}_B) = [X'\hat{V}X + \Sigma_0^{-1}]^{-1} \tag{12}$$

However, "Two major drawbacks of this [Bayesian] approach are that the data analyst must make an explicit statement about the form of the prior distribution and the statistical theory is not widely understood" (See Montgomery and Peck 1982). Moreover, "The critical defect of a Bayesian analysis of data is that prior distributions are both personally difficult to specify and also subject to variation among interested people" (See Leamer 1978).

Perhaps a greater limitation is that it is not often the case that the investigator can acquire either from his theories and experiments or from his introspections, plausible estimates of all, or even some, of the elements of $\underline{\beta}$. However, it is not at all unlikely that such sources will provide estimates of the form and parameters of certain relations between $q < k$ of the elements of $\underline{\beta}$, e.g., ratios, $\beta_j/\beta_m = \rho$, or differences, $\beta_j - \beta_m = \delta$, that can be expressed by the non-stochastic constraints, $\underline{r} = R\underline{\beta}$, or, if less precisely known, by the stochastic constraints, $\underline{r} = R\underline{\beta} + \underline{v}$, $E(\underline{v}) = 0$, $Var(\underline{v}) = \Psi$. If still less precisely known, perhaps as inequalities, $\underline{r} < R\underline{\beta}$.

4. CONSTRAINED ESTIMATION

"... theory often suggests that the coefficients of a relation should obey a linear restriction" (Johnston 1972). "A completely specified model usually contains a large number of collinear explanatory variables, and least-squares estimates that result are rarely 'acceptable'. Various constraints on the parameters may be imposed to improve the estimate, and one among many constrained least squares estimates is usually selected to convey the data evidence"
(Leamer 1978).

If the constraints can be represented by the linear form $\underline{r} = R\underline{\beta}$ where $\underline{r}$ is (qx1) and R is (qxk) then the constrained estimates, $\underline{\hat{\beta}}_c$, of $\underline{\beta}$ can be written as

$$\underline{\hat{\beta}}_c = \underline{\hat{\beta}} + (X'X)^{-1}R'[R(X'X)^{-1}R']^{-1}(\underline{r} - R\underline{\hat{\beta}}) \tag{13}$$

and

$$Var(\underline{\hat{\beta}}_c) = \sigma^2 A(X'X)^{-1}A' \tag{14}$$

where

$$A = I - (X'X)^{-1}R']R(X'X)^{-1}R']R.$$

Note that $RSS_c = (\underline{y} - X\underline{\hat{\beta}}_c)'(\underline{y} - X\underline{\hat{\beta}}_c) = (\underline{y} - X\underline{\hat{\beta}})'(\underline{y} - X\underline{\hat{\beta}}) + (\underline{r} - R\underline{\hat{\beta}})'[R(X'X)^{-1}R']^{-1}(\underline{r} - R\underline{\hat{\beta}})$ and $Det[Var(\underline{\hat{\beta}}_c)] < Det[Var(\underline{\hat{\beta}})]$. That is, the constraint inflates RSS and deflates $Var(\underline{\hat{\beta}})$. The constrained estimates, $\underline{\hat{\beta}}_c$, are obtained by minimizing $(\underline{y} - X\underline{\beta})'(\underline{y} - X\underline{\beta})$ subject to the constraint $\underline{r} = R\underline{\beta}$. The method of Legrange multipliers is required (See Johnston 1972).

It is necessary to determine whether the constraint $\underline{r} = R\underline{\hat{\beta}}$ is consistent with the sample estimate, $\underline{\hat{\beta}}$. This can be tested by the statistic

$$\tau = (\underline{r} - R\underline{\hat{\beta}})'[R(X'X)^{-1}R']^{-1}(\underline{r} - R\underline{\hat{\beta}}) \tag{15}$$

which is distributed, on the null hypothesis as chi-squared with q degrees of freedom.

5. MIXED ESTIMATION

"... it is not uncommon to find statisticians discarding their maintained hypothesis after the estimation has been undertaken" ... "The difficulty seems to be that the investigator has a priori knowledge which he cannot conveniently incorporate in the maintained hypothesis and which he therefore omits ..." "In particular, the a priori knowledge that a regression slope is positive ... cannot readily be built into a maintained hypothesis by conventional methods. This kind of a priori knowledge is precisely the major source of rejections of maintained hypotheses; ..." "... it seems clear that it is logically more consistent to incorporate such knowledge in the maintained hypothesis right at the beginning than to exclude it from the hypothesis and reject it afterwards when the results contradict the omitted knowledge" (Theil and Goldberger 1961).

Although the methods of mixed estimation were developed by Theil and Goldberger (1961) a more succinct description is provided in the textbook by Montgomery and Peck (1982): "... mixed estimation ... uses prior or additional information to augment the data directly instead of through a prior distribution. Mixed estimation starts with the usual regression model $\underline{y} = X\underline{\beta} + \underline{\varepsilon}$ and assumes that the analyst can write a set of q < k prior, stochastic, restrictions on $\underline{\beta}$ such that

$$\underline{r} = R\underline{\beta} + \underline{v} \tag{16}$$

where $E(\underline{v}) = 0$, $E(\underline{r}) = R\underline{\beta}$, $\text{Var}(\underline{v}) = \Psi$. R is a (qxk) matrix of known constraints of rank q and $\underline{r}$ is a (qx1) vector of random variables. If we augment $\underline{y}$ and X to give

$$\begin{bmatrix} \underline{y} \\ \underline{r} \end{bmatrix} = \begin{bmatrix} X \\ R \end{bmatrix} \underline{\beta} + \begin{bmatrix} \underline{\varepsilon} \\ \underline{v} \end{bmatrix} \tag{17}$$

and apply Least Squares, we obtain the unbiased mixed estimator:

$$\hat{\underline{\beta}}_{ME} = [(1/\sigma^2)X'X + R'\Psi^{-1}R]^{-1}((1/\sigma^2)X'\underline{y} + R'\Psi^{-1}\underline{r}) \tag{18}$$

and, of course,

$$\hat{\underline{\beta}}_{ME} = [(1/\sigma^2)X'X + R'\Psi^{-1}R]^{-1}((1/\sigma^2)X'X\underline{\beta} + R'\Psi^{-1}\underline{r}) \tag{19}$$

and

$$\text{Var}(\hat{\underline{\beta}}_{ME}) = [(1/\sigma^2)X'X + R'\Psi^{-1}R]^{-1} \tag{20}$$

$\hat{\underline{\beta}}_{ME}$ is usually referred to as the *posterior* estimate, $\hat{\beta}^{**}$, of $\underline{\beta}$.

As in the case of constrained estimation, the consistency or compatibility of the sample and non-sample information on $\underline{\beta}$ must be assessed before the pooled estimate, $\hat{\beta}^{**}$, can be interpreted or used in prediction. The compatibility is assessed by the statistic γ which is distributed as chi-squared on q degrees of freedom on the null hypothesis of consistency:

$$\gamma = (\underline{r}-R\underline{\beta})'[\sigma^2 R(X'X)^{-1}R' + \Psi]^{-1}(\underline{r}-R\underline{\beta}) \tag{21}$$

Moreover, the Mixed estimation methods provide an estimate of the proportion, Θ_p, of a priori information in the posterior estimates, $\hat{\beta}^{**}$ of $\underline{\beta}$:

$$\Theta_p = m^{-1}\text{Trace}\{R'\Psi^{-1}R[(1/\sigma^2)X'X + R'\Psi^{-1}R]^{-1}\} \tag{22}$$

where m = # columns of X.

Note that the Mixed estimates devolve into Constrained estimates as $\Psi \longrightarrow [0]$, the null matrix (Actually, it is usually sufficient to set $\Psi = 10^{-10}I$, where I is the (qxq) identity matrix).

Note also that $\text{RSS}_{ME} = (\underline{y}-X\hat{\underline{\beta}})'(\underline{y}-X\hat{\underline{\beta}}) + (\hat{\underline{\beta}}_{ME}-\hat{\underline{\beta}})'(X'X)(\hat{\underline{\beta}}_{ME}-\hat{\underline{\beta}}) = \text{RSS} + (\hat{\underline{\beta}}_{ME}-\hat{\underline{\beta}})(X'X)(\hat{\underline{\beta}}_{ME}-\hat{\underline{\beta}})$.

Note the profound similarity of forms of the mixed estimation to those of data augmentation (vide supra). To this Leamer has contributed the following remark: "A schizophrenic attitude toward probabilities allows Theil and Goldberger (1961) to introduce prior information into classical inference in this way" - an illuminating observation. (In this usage "schizophrenic" is not an evaluative word.)

6. DATA

"A common way to deal with data which are inconsistent with one's own position is to deny their statistical validity" (Mazur 1973).

"We must, however, recognize that the fitting of equations to observational data (as opposed to data from carefully designed experiments) is, at best, a risky business" (Hocking 1983).

Table 1 presents the a priori information on β for an empirical, or response surface, model of the binary response of a tumor $(E_1/\overline{E}_1)$ = (Ablation/Recurrence) in a patient of a specified prognostic stratum. We do not claim that it is necessarily "correct", or even factual, but it seems to represent the received wisdom in the matters at hand. However, as Leamer (1983) has observed, "What is a fact? A fact is merely an opinion held by all, or at least by a set of people you regard to be a close approximation to all." (One also recalls Napoleon's wistful locution: "What is History but a fable agreed upon?")

Note that the a priori information on β specifies (very approximately) the <u>direction</u> but <u>not</u> the <u>magnitude</u>, say $\beta'\beta \leq c^2$, of the parameter vector, β. That is, β is constrained, a priori, to a specified <u>octant</u> of parameter space: $\beta_1 > 0$, $\beta_2 < 0$, $\beta_3 < 0$.

The observations consist in 45 patients irradiated for head and neck cancer at a single institution. Of these, 17 were at stage T_2 and 28 at stage T_3. For our expository purposes we consider that only one covariate, the T-stage, with only two levels, is sufficient to uniquely specify the state of the system. Thus, we have only $2^1 = 2$ prognostic strata. As observed previously the number within each stratum is about 20, typical for clinical data from a single institution—and it required over 18 years to accumulate this total of these patients at this institution. The response is binary, $\overline{Ca}/Ca$ or $E_1/\overline{E}_1$. The number of positive responses, E_1 (ablation), is 31. The number of negative responses, $\overline{E}_1$ (recurrence), is 14. The time at risk for recurrence was long enough to assure that all of the events $\overline{E}_1$ had occurred. Scattergrams of the data in the X_1, X_2 and N, T planes are presented in Figures <u>3a</u>, <u>3b</u> and <u>3c</u>. The <u>minimum</u> dose was 5083 centigray given in 25 fractions in 32 days (N/T = 0.78, D/N = 203). The <u>maximum</u> dose was 7700 centigray given in 32 fractions in 43 days (N/T = 0.74, D/N = 241).

These data have formed the basis for several previously published (peer-reviewed) papers on the management of head and neck cancers by radiation. They are quite typical, in all respects, of the genre. However, from the previous discussions in this and other Lectures in the Symposium, it is apparent that their evidential value for issues of cancer patient management will be exceeded by their motivational value - in stimulating a thoughtful examination of the statistical issues presented in this Symposium. In particular, it will be evident that, in greater or lesser degree the insights and methods presented in each of the first ten Lectures are necessary to the understanding and assessment of the information on clinical radiation response that is contained in these 45 patients.

Scattergrams of the observations on the conditional radiation response of head and neck cancer are presented in Figures <u>3a</u>, <u>3b</u> and <u>3c</u>. (The extreme levels in Figures <u>3a</u> and <u>3c</u> are the levels marked with black circles in Figures <u>1a</u> and <u>1b</u>, respectively.) Figure <u>3a</u> presents the superposition of the scattergrams of the conditional (stage) joint distributions of the treatment variables D and T for each of the two stages, T_2 and T_3. The curves for the respective <u>management</u> equations, $D = 2100T^{0.31}$ (T_2) and $D = 2050^{0.31}$ (T_3) are also presented. These curves suggest that the patients in each stage

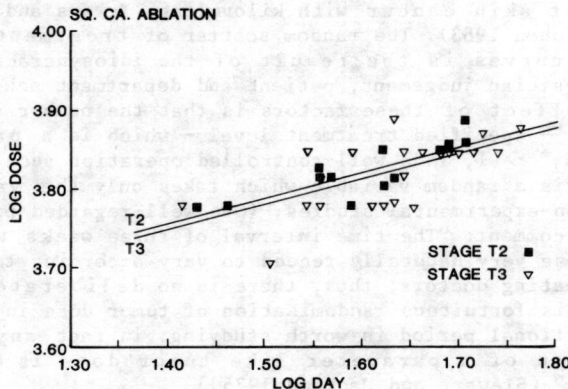

Figure 3a. Scattergrams of conditional distributions of treatment levels of head and neck cancer patients on <u>stage</u>. (n = 45) D-T plane.

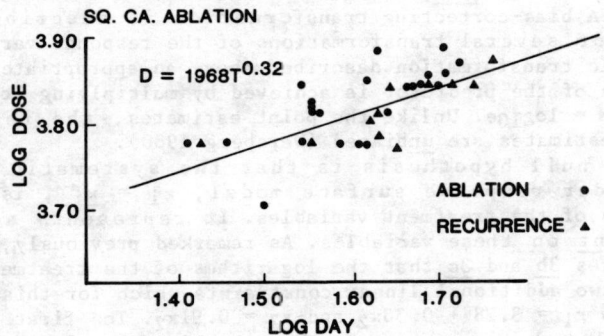

Figure 3b. Scattergrams of conditional distributions of treatment levels of head and neck cancer patients on <u>response</u>, $\overline{E}_1/E$. (n = 45) D-T plane. Graphs in Figures 3a and 3b are plots of <u>management equations</u>: $D = k_0 T^{k_1}$.

were managed according to a slight variation on a common theme in treatment philosophy, that usually attributed to Strandqvist for the treatment of skin cancer with kilovoltage X-rays and for which $D = kT^{0.33}$ (See Cohen 1983). The random scatter of treatments about the respective curves is the result of the idiosyncracies in patient response, physician judgement, patient and department schedules, etc. The joint effect of these factors is that the number of patients at risk at the ith specified treatment level - which is a predetermined constant, $n_i \gg 1$, in a well-controlled operation such as a designed experiment - is a random variable which takes only the values 0 or 1 in these non-experimental studies. (One well-regarded published study includes the comment: "The time interval of three weeks was retained but tumor dose very naturally tended to vary according to the whims of different treating doctors; thus, there is no deliberate randomization, but this fortuitous randomization of tumor dose in cases treated during transitional period is worth studying; in fact any deliberate randomination of a parameter like tumor dose is ethically very difficult ..." (Stewart and Jackson 1975))

It should be remarked that hitherto the coefficients in the __management__ equations have invariably been estimated from the transformed variables, $\log_{10}D$, $\log_{10}T$, of a sample, but the predicted values, the __point__ estimate, D_i, have been calculated from the power law equation: $D = 10^{\gamma_0} T^{\gamma_y}$. Unfortunately, this estimate is biased and will be, on the average, too small. The re-transformed predictor, $10^{\gamma_0} T^{\gamma_1}$, gives an estimate of the __median__ of the predicted values rather than the __mean__, as would be the case if no transformation were involved. A bias-correcting transformation is described in Miller (1984) for several transformations of the response variable. For the logarithmic transformation described above an appropriate approximate inflation of the predictor is achieved by multiplying it by the factor $10^{\sigma^2/2m}$ $M = \log_{10}e$. Unlike the point estimates, the re-transformed interval estimates sre __un__biased (Weisberg 1980).

The null hypothesis is that the systematic part of the first-order response surface model, $z_i = \underline{x}_i'\underline{\beta}$, is linear in the logarithms of the treatment variables. It represents a __biological constraint__ on these variables. As remarked previously, it is evident from Figures __3b__ and __3c__ that the logarithms of the treatment variables satisfy two additional linear constraints which for this sample may be written as $x_1 = 3.28 + 0.33x_2$ and $x_3 = 0.91x_2$. The first represents a __medical-legal__ constraint - the so-called Strandqvist relation. The second represents a __socio-economic__ constraint - treatments are usually given Monday-Friday. For such samples, the parameter β in the equation of interest, "... ceases to be estimable" (See Theil 1963), since the $X'\hat{V}^{-1}X$ matrix is ill-conditioned owing to the presence of multicollinearity in the data - multicollinearity "... refers to a situation in which there is an exact (or nearly exact) linear relation among two or more of the input variable" (See Hocking and Pendleton 1983). Now, "Collecting additional data has been suggested as the best method of combatting multicollinearity." But, "... note that collecting additional data is not a viable solution to the multicollinearity problem when the multicollinearity is due to constraints on the model or in the population" (Mongomery and Peck 1982).

One of the better comments on the maneuver - "collecting additional data" - can be found in Belsley, Kuh and Welsch (1980) and was quoted earlier in this paper.

The present sample provides a good example of both the achievements and the disadvantages and difficulties that are characteristic of this most obvious maneuver - data augmentation - for "combatting multicollinerity" and in the present case the effects of the presence of extreme - and hence perhaps influential- treatment levels, x_i', and small sample size as well. It is evident in Figures 3a and c that there is a high degree of multicollinerity present in the X matrix for stage T_2, that there are also several extreme levels, x_i', and that the sample size is small, $n_1 = 18$. Thus, these data were pooled with a larger set of similar observations on stage T_3, $n_2 = 27$, to give the larger, n=45, but heterogeneous sample, $T_2 + T_3$, in which not all of the systems included in the pooled sample are in the same

A visual appreciation of Figure 3a discloses the degree to which the pooling maneuver was successful in reducing the effects of extreme levels of treatment; it was obviously rather successful. A more quantitative assessment is available for the reduction of the effects of multicollinearity. It will be recalled that the object of data augmentation is to maximize Det[X'X] since this will minimize both the volume of the confidence region on $\hat{\beta}$ and minimize the maximum value of Var($\hat{y}_i$) over the region of observation. For T_2, Det(X'X) = $1.9223*10^{-10}$. For $T_2 + T_3$, Det(X'X) = $9.7674*10^{-9}$. The ratio is 50.8111. Although the pooling maneuver has achieved over a fifty-fold inflation of Det[X'X] it is clear from Figure 3a that there remains a significant degree of multicollinearity. The maneuver was obviously more effective in reducing the collinearity in D and T than that in N and T for reasons noted above. And, of course, the size of the pooled sample, n < 100, is such that these data still have only heuristic value even if the treatment variables were orthogonal because of the large bias and variance of the estimates of β obtained from small samples.

Multicollinearity of about the same kind and degree was present in the T_3 data as in the T_2 data so it was perhaps not the most appropriate sample, but this is a fairly typical experience. We note that the constraints - socio-economic and medical-legal - that are responsible for the multicollinearity are present in the population as well as in the samples so that simply acquiring more appropriate additional data from extant records may not be "that easy". If it were possible to generate additional records according to the procedures and criteria - maximize Det| X'X| - recommended by Dykstra (1966, 1971) and Gaylor and Merrill (1968) this will impose additional costs in time, money, hazard, etc. - and, of course, it may simply not be possible if it contravenes these (or other) constraints.

Thus, we are led to consider Bayes-like methods employing supplementary information, that in Table 1b, for, as remarked by Leamer (1984) "... all non-experimental data sets are simply too weak to allow sensible inferences in the absence of supplementary information."

However, in Lecture 14 we will discuss an important instance where data-augmentation was used successfully to reduce the degree of multicollinearity in the city-specific (Hiroshima and Nagasaki) joint distributions of gamma and neutron radiation doses. The Hiroshima and Nagasaki data must be pooled in order to obtain plausible estimates of the parameter vectors in the LQ-L, L-L and Q-L models of the radiation induced cancer incidence and mortality rates from which are extracted the estimates of excess risks that are presented in the so-called BEIR III Report. (Data augmentation is necessary in this case because at present there are no biased estimation methods

Table 1. Non-Sample (A Priori) Information on $\underline{\beta}$ for an Empirical Model of the Binary Response of Ca (Ablation of Primary Tumor) in a Given Prognostic Stratum to Clinical Irradiation Schedules, X.

a) Number and Dimension of $\underline{\beta_j}$ (Semi-Classical Probit Model)
$\beta_j \neq 0$. $j = 0, 1, 2, 3$. $y_i \sim B(n_i, \Pi(\underline{x_i}'\underline{\beta}))$
$\underline{X_0} = \underline{1}$, $\underline{X_1} = \log_{10}(\underline{D})$, $\underline{X_2} = \log_{10}(\underline{T})$, $\underline{X_3} = \log_{10}(\underline{N})$

b) Size and Sign of $\underline{\beta}$ (Semi-Bayesian)
$\beta_0 < 0$, $\beta_1 > 0$, $-1 < \beta_j/\beta_1 < 0$. $j = 2, 3$.

Table 2. Sample Information on Probit Model of Binary Response of Cancer to Irradiation.

a) Probit Model: $z = -30.31 + 11.66x_1 - 13.48x_2 + 5.53x_3$
$\qquad\qquad\quad$ (-1.138) $\;(1.412)$ $\;\;(-1.072)$ $\;(0.464)$
$n = 45$. RSS $= 52.879$. $P(\chi^2 > \text{RSS}| 41) = 0.101$
$\sigma^2 = \text{RSS}/41 = 1.290$
$\text{RSS} = \sum_{i=1}^{n} t_i^2$. $t_i = (y_i - n_i \hat{\Pi}_i)/\sqrt{n_i \hat{\Pi}_i (1-\hat{\Pi}_i)}$
where $\hat{\Pi}_i = F(\hat{z}_i)$ and $n_i = 1$, $i = 1, \ldots, n$

b) Sign, size and significance of elements of $\hat{\underline{\beta}}$:
i) $\beta_1 < 0(?)$; $\beta_2 = 0(?)$; $\beta_3 = 0$
ii) $\beta_2/\beta_1 = -1.16$. $\beta_3/\beta_1 = 0.47$

The respective statistics for the precision of the estimates of the coefficients are shown in parentheses: $\hat{\beta}_j / \sqrt{\text{Var}(\hat{\beta}_j)}$. Note that the sample information, $\hat{\underline{\beta}}$, is inconsistent with the a priori information on $\underline{\beta}$ in Table 1a and Table 1b.

Table 3. Empircal (Probit) Models of Tumor Ablation for Reduced Sample $(n = 44)$. $X = (\underline{1}, \underline{X_1}, \ldots)$.

a) M_1. $X = (\underline{1}, \underline{X_1}, \underline{X_2}, \underline{X_3})$:
$z = -60.318 + 25.027x_1 - 30.057x_2 + 9.747x_3$
$\;\;\;\;(-1.661)$ $\;\;(2.017)$ $\;\;\;(-1.879)$ $\;\;(0.737)$[a]
RSS $= 36.114$. PRESS $= 41.348$. $P(\chi^2 > \text{RSS}| 41) = 0.646$.
VIF: $(2.00, 21.90, 20.10)$.
$r = 0.399$. $\delta = 2.21$

b) M_2. $X = (\underline{1}, \underline{X_1}, \underline{X_2})$:
$z = -56.695 + 23.465x_1 - 19.747x_2$
$\;\;\;\;(-1.679)$ $\;\;(2.047)$ $\;\;(-2.580)$
RSS $= 38.432$.[b] PRESS $= 41.223$. $P(\chi^2 > \text{RSS}| 40) = 0.585$.
VIF: $(1.98, 1.98)$.

c) M_3. $X = (\underline{1}, \underline{X_1}, \underline{X_3})$:
$z = -33.171 + 13.847x_1 - 12.841x_3$
$\;\;\;\;(-1.229)$ $\;\;(1.632)$ $\;\;(-2.332)$
RSS $= 49.116$. PRESS: 59.932. $P(\chi^2 > \text{RSS}| 40) = 0.178$.
VIF: $(1.82, 1.83)$.

---

[a] $(\hat{\beta}_i / \sqrt{\text{Var}(\hat{\beta}_i)})$
[b] $\Delta df = 1$, $\Delta \text{RSS} = 38.432 - 36.114 = 2.2328$ (<3.84. No significant difference in "fit" between M_1 and M_2.)

Table 3d. Comparison of Binary, Discriminant Function and Probit
          Models of Tumor Ablation for Reduced Sample (n=44).
          $X = (\underline{1}, \underline{X}_1, \underline{X}_2)$.

i) <u>Probit Model (M_2)</u>. $-\infty < z_j < \infty$.
   $z = -56.695 + 23.465 x_1 - 19.747 x_2$
       (-1.679) (2.047) (-2.580)
   RSS = 38.432. $\hat{r}_1 = -0.842$. $r^2 = 0.150$.
ii) <u>Discriminant Function Model</u>. $0 \leq P(y_j = 1| x_j') \leq 1.0$
    $L = -58.777 + 24.486 x_1 - 20.698 x_2$
                  (1.898) (3.046)[a]
    $D^2 = 1.036$. $R^2 \cong c^* D^2 = 0.184$.[b] $\hat{r}_1 = -0.845$.
    Average Misclassification Rate: $\Phi(-D/2) = 0.305$
iii) <u>Binary Model</u>. $y_j = 0$ or 1
     $y = -9.944 + 4.356 x_1 - 3.682 x_2$
         (-1.212) (1.756) (-3.002)
     $R^2 = 0.184$. $\bar{R}^2 = 0.144$. $\hat{r}_1 = -0.845$.

---

[a] $Var(\delta_i) \cong s^{11}(n_1^{-1} + n_2^{-1})$. $i=1,2$. (See Kleinbaum and Kupper, 1978)
[b] $c^* = (n_1 n_2/n_1+n_2)/[n_1+n_2-2 + (n_1 n_2/n_1+n_2) D^2] = 0.1779$

Table 4. Collinearity Diagnostics. Sq. Ca. Ablation. Full Model.
         (Probit 2)

Model Matrix, $X = (\underline{1}, \underline{X}_1, \underline{X}_2, \underline{X}_3)$. n=44.
a) Correlation Matrix, $X_0'X_0 (p=3)$:
                           1.0, 0.703, 0.670
                                1.0,   0.974
                                       1.0
   i) Eigenvalue, Eigenvector (λ_j, $\underline{V}_j$):
      $\lambda_1 = 2.574$, $\underline{V}_1 = (-0.525, 0.605, 0.598)$
      $\lambda_2 = 0.402$, $\underline{V}_2 = (-0.850, 0.334, 0.408)$
      $\lambda_3 = 2.432 \times 10^{-2}$, $\underline{V}_3 = (0.047, -0.723, 0.690)$
   ii) Condition Number, $\kappa_0 = \lambda_1/\lambda_3 = 105.83$
   iii) Proportion of Variance, $\lambda_1/\Sigma\lambda_j = 0.858$
   iv) $\Sigma\lambda_j^{-1} = 43.00 > 5p = 15$
   v) Variance Inflation Factors, VIF: (1.995, 21.901, 20.102)
b) Normal Equations Matrix, $X'X (k=4)$.
   $\lambda_1 = 1557.376$, $\lambda_4 = 2.464 \times 10^{-3}$.
   Condition Number, $\kappa_1 = \sqrt{\lambda_1/\lambda_4} = 795.03$

Table 5. Concordance of Model and Reduced Sample.

	Frequencies				
	Expected		Total	Observed	
	0	1		0	1
< 0.1	0	0	0	0	0
0.1 - 0.6	7.77	6.23	14	7	7
0.6 - 0.9	5.05	12.95	18	6	12
> 0.9	0.26	11.74	12	0	12
Total	13.07	30.92	44	13	31

available for Poisson regression models. vide infra.)

It is often useful to summarize a bivariate distribution by the <u>contour ellipse</u> (See Hald 1952). If the sample estimates of the respective means, standard deviations, and correlation coefficients of the variables x_1 and x_2 are $\bar{x}_1$, $\bar{x}_2$, s_1, s_2 and r_{12} then

$$c^2 = 1/(1-r_{12}^2) \ [(x_1-\bar{x}_1/s_1)^2 - 2r_{12}(x_1-\bar{x}_1/s_1)(x_2-\bar{x}_2/s_2) + (x_2-\bar{x}_2/s_2)^2] \quad (23)$$

is distributed as chi-squared with 2 degrees of freedom, $c^2 \sim \chi^2(2)$, <u>provided</u> x_1 and x_2 have a bivariate Normal distribution. The equation represents an ellipse in the (x_1, x_2) plane with center at $(\bar{x}_1, \bar{x}_2)$. The probability that any observation (x_1^*, x_2^*) lies within the ellipse is simply $P(\chi^2 < c^2 | 2)$. Alternatively, such an ellipse will include 100P% of the sample - if the sample has a bivariate Normal distribution. Such summaries are especially useful if the distribution of (x_1, x_2) is <u>conditional</u>, as in the present case, since the relative positions of the respective ellipses describes how uniquely the complementary conditions are distinguished by values of x_1 and x_2.

Figure <u>3d</u> presents a super-position of the respective 0.80 contour ellipses for the joint distribution of <u>X₁</u> and <u>X₂</u> on the conditions $\overline{Ca}$ (=E_1) and Ca (=$\bar{E}_1$).

The center of each ellipse is marked as "+". Inspection discloses that $\underline{X_1}$ and $\underline{X_2}$ are distributed approximately bivariate Normal on each of the respective conditions, since about 80% of the observations are included within the respective ellipses. This is consistent with the respective chi-squared probability plots for the two sub-groups E_1 and $\bar{E}_1$ (See Hald 1952 and Lecture <u>19</u>).

The observation labelled #1 in the Figure <u>2d</u>, a <u>recurrence</u>, $\bar{E}_1$, at a treatment level, $\underline{x}'$, quite similar to two others for which the response is E_1 provides a good instance of an <u>outlying observation</u>. (We shall discuss observations #19 and #34 later.) Outliers are "... those observations for which the imputs are reasonable but the response is abnormally large or small as compared to other cases with similar inputs." The presence of outliers and extreme cases in the sample distribution just as the presence of multicollinearity, will greatly influence the estimates of $\underline{\beta}$ in the dose-response model. Thus, their identification is of considerable interest to the data analyst.

It is useful to note that this representation of the quantal response of these patients is cognate to that of a classification or discrimination problem (See Lachenbruch, Lecture <u>10</u>) and the conditional response E_1 at a given treatment level becomes the complement of the conditional risk of cancer, $\bar{E}_1$, given the levels of "etiological factors", X = (<u>D</u>, <u>T</u>, <u>N</u>), i.e., the treatment level, $\underline{x_i}$, is now a set of <u>classification</u> variables. This gives rise to the <u>discriminant function model</u> of dose-response for non-experimental studies.

Cornfield (1959, 1961) has demonstrated that the conditional risk of a binary event, $P(E_1 | X)$ can be represented by a multivariate logistic function for which the argument is the ratio of the two conditional multivariate density functions $f_0(X)$ and $f_1(X)$: $P(E_1 | X) = [1 + \{(1-P)/P\} \ f_0(X)/f_1(X)]^{-1}$. P and (1-P) are the respective <u>unconditional</u> probabilities of the events E_1 and $\bar{E}_1$. In the present study, $P(E_1 | X)$ is thus the conditional probability of ablation, E_1. If the conditional distributions are multivariate Normal with different mean vectors and equal covariance matrices

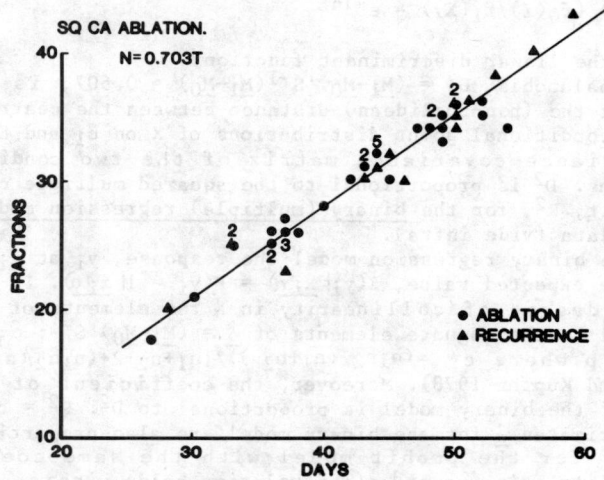

Figure 3c. Scattergrams of conditional distributions of treatment levels of head and neck cancer patients on response, $\bar{E}_1/E_1$. (n = 45) N-T plane.

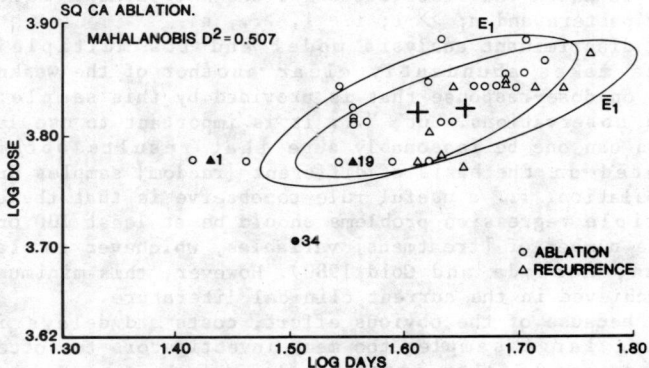

Figure 3d. Super position of scattergrams of conditional distributions of treatment levels of head and neck cancer patients on response, $\bar{E}_1/E_1$, together with respective 0.80 contour ellipses. (n = 45) D-T plane.

$$(1-\Pi/\overline{\Pi})(f_0(X)/f_1(X)) = e^{-\ln L} \tag{24}$$

where L is the linear discriminant function.

The Mahalanobis, $D^2 = (\underline{M}_1-\underline{M}_0)'S^{-1}(\underline{M}_1-\underline{M}_0) = 0.507$, is the sample estimate of the (non-Euclidean) distance between the centroids, $\underline{M}_1$ and $\underline{M}_0$, of the conditional joint distributions of X on $\overline{E}_1$ and E_1. S is the common variance-covariance matrix of the two conditional joint distributions. D^2 is proportional to the squared multiple correlation coefficient, R^2, for the binary (multiple) regression model, $\underline{y} = X\underline{\alpha} + \underline{\varepsilon}$, of such data (vide infra).

In the binary regression model the response, y_i at $\underline{x}_i'$ is either 0 or 1 and the expected value, $E(y_i|\underline{x}_i'\underline{\alpha}) = P(y_i = 1|\underline{x}_i'\underline{\alpha})$. In the absence of a high degree of collinearity in X the elements of $\underline{\alpha}$ are simply proportional to the cognate elements of $\underline{\hat{\delta}} = (\underline{M}_1-\underline{M}_0)'S^{-1}$: $\hat{\alpha}_j = c*\hat{\delta}_j$, $j = 1, \ldots, p$ where $c* = (n_1n_2/(n_1+n_2))/[n_1+n_2-2+(n_1n_2/(n_1+n_2))D^2]$ (See Kleinbaum and Kupfer 1978). Moreover, the coefficient of determination, R^2, of the binary model is proportional to D^2: $R^2 = c*D^2$.

The estimates $\underline{\hat{\alpha}}$ for the binary model are also proportional to the estimate $\underline{\hat{\beta}}$ for the probit model with the same coefficient of proportionality, $c*$: $\underline{\hat{\alpha}} \simeq c*\underline{\hat{\beta}}$. (The relation holds within ± 5%.) Note then that the ratios of cognate elements are equal. For example, $\hat{r}_1 = \hat{\alpha}_2/\hat{\alpha}_1 = \hat{\beta}_2/\hat{\beta}_1$. See Table 3e. However, this degree of equality is unusual. (It is also unusual for δ_j to be equal to β_j, of course.)

The structure of the sample is that appropriate for a discrimination between complementary groups, E_1 and $\overline{E}_1$, since the n treatment levels have a conditional, continuous multivariate Normal distribution on E_1 and $\overline{E}_1$ and $n_i = 1$; $i = 1, \ldots, n$ rather than for a bioassay in which the distribution of the n treatment levels forms a discrete pattern and $n_i \gg 1$; $i = 1, \ldots, n$.

The discriminant analysis model and its multiple regression analogue makes abundantly clear another of the weaknesses of the evidence on dose-response that is provided by this sample: There are only 45 observations. But, "... it is important to use large samples. Only then can one be reasonably sure that results obtained can be replicated on the basis of different [random] samples drawn from the same population. ... a useful rule to observe is that the sample size in multiple regression problems should be at least 100 or at least 20 times the number of [treatment] variables, whichever is larger" (See Lindeman, Morenda and Gold 1980). However, this minimum size is not always achieved in the current clinical literature.

But because of the obvious effort, costs and delays incurred in acquiring large samples too many investigators too often base their analysis on samples that are too small. Tversky and Kahneman (1982) provide some interesting insights into this sub-group and its practices:

"The law of large numbers guarantees that very large samples will indeed be highly representative of the population from which they are drawn. ... People's intuitions about random sampling appear to satisfy the law of small numbers, which asserts that the law of large numbers applies to small numbers as well."

"... the believer in the law of small numbers practices science as follows:
1) He gambles his hypothesis on small samples without realizing that the odds against him are unreasonably high.
2) He has undue confidence in early trends (e.g., the data of the first few subjects) and in the stability of observed patterns (e.g.,

the number and identity of significant results. He over-estimates significance.
3) In evaluating replications, his or others', he has unreasonably high expectations about the replicability of significant results. He underestimates the breadth of confidence intervals."
4) He rarely attributes a deviation of results from expectations to sampling variability, because he finds a causal 'explanation' for any discrepancy. Thus, he has little opportunity to recognize sampling variation in action. His belief in the law of small numbers, therefore, will forever remain intact."

Since it was necessary to pool the observations on two different stages (T_2 and T_3) of disease, it is now required to determine whether either the average level of response, β_0, or its rate of change, β_1, with respect to radiation dose, D, is a function of stage of disease. In other words, we must determine whether, $\underline{\beta} = \underline{\beta}(u_h')$, since the size of the tumor differs between T-stages and it is known, a priori, that the dose required to elicit a given level of ablation increases with the size of the tumor. One way of examining for such dependence is to introduce an indicator or dummy variable, S, where S = 0 (State T_2), S = 1 (Stage T_3). The model is then

$$z = \beta_0 + \beta_1 x_1 + \beta_2 x_2 + \beta_3 x_3 + \beta_4 S + \beta_5 S x_1. \qquad (25)$$

Therefore, for Stage T_2 we have the model

$$z = \beta_0 + \beta_1 x_1 + \beta_2 x_2 + \beta_3 x_3 \qquad (26)$$

and for State T_3 we have the model

$$z = (\beta_0 + \beta_4) + (\beta_1 + \beta_5) x_1 + \beta_2 x_2 + \beta_3 x_3. \qquad (27)$$

In this way one may include in the first order response surface the dependence of the coefficient vector, $\underline{\beta}$, on the prognostic features of the patient: $\underline{\beta} = \underline{\beta}(u_h')$.

If the intercept and slope of the dose-response equation are significantly dependent on the stage of disease then $\hat{\beta}_4$ and $\hat{\beta}_5$ should exceed the respective standard error, $\sqrt{Var\hat{\beta}_j}$, $j = 4, 5$. This was not the case for these data and these qualitative variables are remanded to the Q matrix. Thus, for these data we can be concerned only with describing the dependence of the response of the heterogeneous system ($T_2 + T_3$) on the levels of treatment: dose, days and fractions.

The (pooled) sample information on the response surface model of ablation of head and neck tumor described in Figure $\underline{3}$ is summarized in Table $\underline{2}$. The residual sum of squares, $RSS = \sum_{i=1}^{n} t_i^2$, where $t_i = (y_i - n_i \hat{\pi}_i)/\sqrt{n_i \hat{\pi}_i(1-\hat{\pi}_i)}$ is a summary measure of model-sample interaction. Division of the residual, $(y - n_i \hat{p}_i)$, by an estimate of its standard deviation, $\sqrt{n_i \hat{p}_i(1-\hat{p}_i)}$ gives t_i the standardized residuals. (In Normal theory, division of each residual by an estimate of its standard deviation is called a Studentized residual, $\overset{\approx}{e}_i/\hat{\sigma}\sqrt{1-h_i}$, where $\hat{\sigma}$ is an estimate of the square root of the mean RSS and h_i is the hat matrix diagonal. See Lectures $\underline{1} - \underline{3}$.) The data are apparently consistent with the model, $P(RSS > \chi^2 | 41) = 0.10$. However, for ungrouped binary data, $y = 0$ or 1, $n = 1$, the goodness of fit statistic, RSS, is distributed only approximately as a chi-squared statistic for which $E(\chi^2) = \nu$, $Var(\chi^2) = 2\nu$ where $\nu = n-m = n-k-1$ is the number of degrees of freedom. The sampling distribution of RSS still has expected value, ν, but the dispersion is increased so that the variance exceeds 2ν. If RSS $\sim \nu$, or RSS $< \nu$, the sample presents no evidence against the hypothesis – unless, of course, that RSS is "... so small as to suggest that a neglected feature of the experiment has caused unusual regularity or even that the data have been faked!" (See Finney 1971). Thus, the use of the summary measure, RSS, to decide on the concordance of model and data is not without hazard in the present case. We shall discuss other criteria which are designed to answer other questions below. However, as a criterion of <u>concordance</u> of model and data we shall refer to the sampling <u>distribution</u> of RSS. Moreover, for binary data the <u>difference</u> ($RSS_1 - RSS_2$) of the respective sums of squares for rival models with degrees of freedom ($n-k_1$) and ($n-k_2$), $k_1 > k_2$ is distributed as chi-squared with k_1-k_2 degrees of freedom.

The over-dispersion factor, σ^2 in Table <u>2</u> cannot be ignored and <u>none</u> of the coefficients differs significantly from zero for this model of these data. Given the evidence of the scattergrams in Figure <u>3</u> this latter effect appears to be the (familiar) consequence of the presence of collinearity in the joint distribution of X, since the existence of a linear relationship between the response and treatment variables can be established, but <u>not</u> the individual influence of each factor. One remedy is to <u>omit</u> one or more treatment variables – or to construct a new variable as a linear <u>combination</u> of two or more. For either of these the parameter is denoted as $\underline{\beta}*$. Either maneuver will increase the ratios, $\hat{\beta}*_j / \sqrt{Var(\hat{\beta}*_j)}$ for the remaining variables but also <u>inflate</u> the value of RSS by $(\underline{\beta}* - \underline{\hat{\beta}})'(X'X)(\underline{\beta}* - \underline{\hat{\beta}})$, as we have shown.

7. REGRESSION DIAGNOSTICS.

a) <u>ROWS OF X. OUTLIERS AND EXTREME VALUES.</u>

"The fact that a small subset of the data can have a disproportionate influence on the estimated parameters or predictions is of concern to users of regression analysis, for if this is the case, it is quite possible that the model-estimates are based primarily on the data subset rather than on the majority of the data." ... "If there is evidence of influential data, several corrective actions are possible ... At the very least, excessively influential data should be mentioned in any discussion of the model fitting and estimation process" (Belsley, Kuh and Welsch 1980).

"Outliers and influential observations ... are always judged relative to some model, either implicit or explicit" (Cook and Wang 1983).

"As a rough definition, we refer to <u>outliers</u> as those observations (cases) for which the inputs are reasonable but the response is abnormally large or small as compared to other cases with similar inputs. <u>Extreme cases</u> are for which the input vector is, in some sense, far from the rest of the data. Such cases are also referred to as high leverage points. In either case, these observations greatly influence the least squares estimates. Detecting these observations and examining the degree to which they influence the results is surely of interest in any regression analysis" (Hocking and Pendleton 1983).

"For the normal theory linear model, much is known about the effect on a least-squares fit of outlying responses and extreme design points ... An important extension of these dignostic approaches is to nonlinear regression models, where presumably the effects of outliers and leverage points could be worse" (Pregibon 1981).

"An influential observation is one which either indirectly or together with several other observations, has a demonstrably larger impact on the calculated values of various estimates (coefficients, standard errors, t-values, etc.) than is the case for most other observations. One obvious means for examining such an impact is to delete each row, one at a time, and note the resultant effect on the various calculated values. Rows whose deletion produces relatively large changes in the calculated values are deemed influential" (See Belsley, Kuh and Welsch 1980).

We first examine the joint distribution of the two key diagnostics, the standardized residuals, t_i, and the diagonals, h_i, of the hat matrix, $H = X\hat{V}^{-1/2}\overline{(X'\hat{V}^{-1}X'\hat{V}^{-1/2}}$, in Figure 4a. Observations for which $|t_i| > 2.0$ are outliers. Observations for which $h_i > 2m/n = 0.178$ are extreme. "These measures should readily identify observations that are not well explained by the data as well as those dominating some important aspect of the fit" (See Pregibon 1981). In Figure 4a we note that the t_i and h_i are uncorrelated for these data (the correlation coefficient is $r = -0.05$). There is one observation for which $|t_i| > 4.0$ and four observations for which $h_i > 0.18$.

Five out of forty-five or 11% of the observations in the sample are anomolous - in some sense. However, it has been remarked previously that it is commonly found that 5% - 10% of the observations in a data set are "wrong" - outlying responses and/or extreme values of the treatment variables. See Huber 1977.

We turn from the joint distribution to consideration of the joint effects of outliers and extreme cases on the sample estimates, $\hat{\beta}$, $Var(\hat{\beta})$, etc. The combined influence of the ith row on the estimate $\hat{\beta}$ of the coefficient vector is described by the beta-shift statistic, the diagnostic, $D_i^1(X'\hat{V}^{-1}X, k) = (1/k)[y_i - n_i\hat{p}_i)^2/n_i\hat{p}_i(1-\hat{p}_i)] h_i/(1-h_i)^2 = h_i/k(t_i/1-h_i)^2 D_i^1(X'\hat{V}^{-1}X, k)$ joins the two key diagnostics t_i and h_i in an economical way that will identify rows of $[y, X]$ for, which either t_i or h_i are large or both are moderate but act jointly on $\hat{\beta}$. Thus, D_i^1 highlights those rows of $[y, X]$ that have the largest influence on the estimate, $\hat{\beta}$. D_i^1 is a measure of the amount by which the deletion of case #i, shifts the sample estimate of β. It is a second-order Taylor series approximation to the squared distance from $\hat{\beta}$ to $\hat{\beta}_{(i)}$ where the latter denotes the estimate of $\hat{\beta}$ obtained from the same sample as the former but with the ith case deleted. This distance is relative to the fixed geometry of $(X'\hat{V}^{-1}X)$ and thus, the "beta shift" can be described in the space of the family of joint confidence ellipsoids of $\hat{\beta}$. That is, D_i^1 is proportional to the generalized distance, $(\hat{\beta}_{(i)} - \hat{\beta})'(X'\hat{V}^{-1}X)(\hat{\beta}_{(i)} - \hat{\beta})$. Comparison of kD_i^1 to the percentage points of the $\chi^2_{(k)}$ distribution gives a very rough estimate of the confidence ellipsoid for $\hat{\beta}$ to which the estimate $\hat{\beta}_{(i)}$ is displaced. Or better, compare D_i^1 to the $F(k, n-k)$ distribution.

In Normal theory, $D_i \geq 1.0$, which corresponds to displacements of $\hat{\beta}_{(i)}$ to beyond a 0.50 confidence ellipsoid, is used to identify influential observations on the basis of beta shift, i.e, $D_i = 1.0$ is the "cut-off". However, for probit or logit models, D_i^1, the Taylor series approximation, underestimates the squared generalized distance, $(\hat{\beta}_{(i)} - \hat{\beta})$. For example, for case #1 $D_1 = 0.672$ and $4D_1 = 2.6880$. $P(\chi^2 < 4D_1|k) = 0.389$. (Alternatively, $P(F < 0.672|4, 41) = 0.385$) This

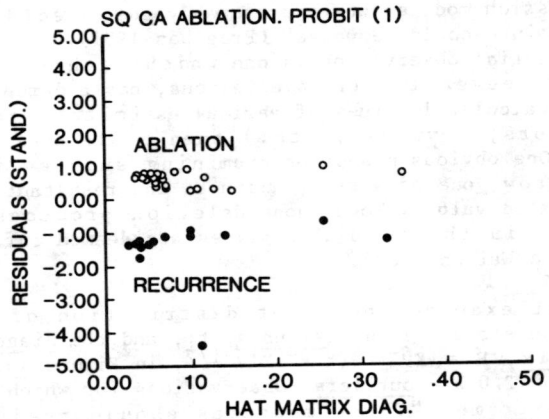

Figure 4a. Scattergrams of conditional distributions of Studentized residuals, t_i, and hat matrix diagonals, h_i, for response surface (probit) model of ablation, E_1. (n = 45) There is one <u>outlier</u> and four <u>extreme levels of treatment</u>. h_i and t_i are <u>un</u>correlated.

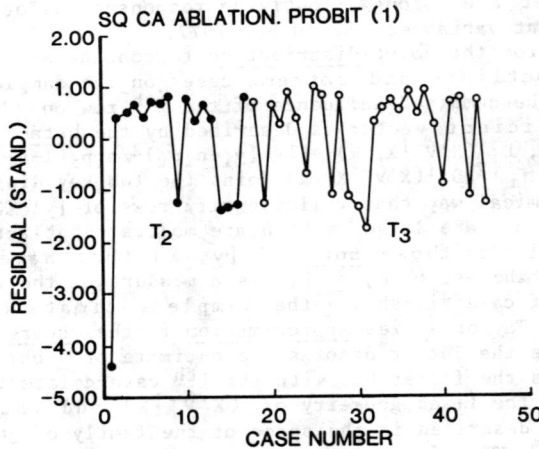

Figure 4b. Case plot of Studentized residuals, t_i, for response surface (probit) model of ablation, E_1. (n = 45) There is one <u>outlier</u>: $t_1 = -4.40$. It occurs in Stage T_2.

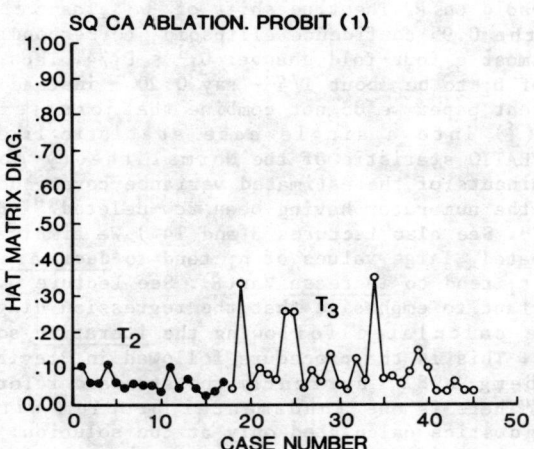

Figure 4c. Case plot of hat matrix diagonals, h_i, for response surface (probit) model of ablation, E_1. (n = 45) The two most <u>extreme levels of treatment</u> $h_{19} = 0.34$, $h_{34} = 0.35$, occur in T_3.

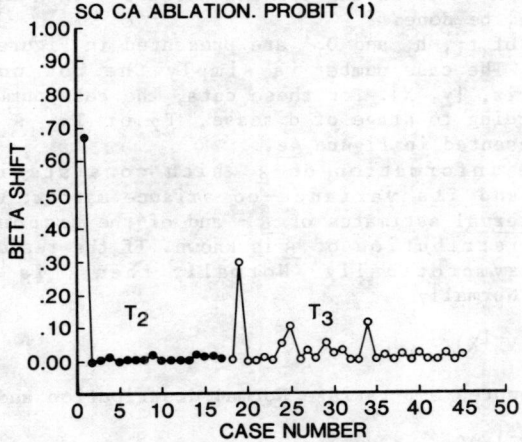

Figure 4d. Case plot of beta shift, D_i^1 (one-step approximation) for response surface (probit) model of ablation, E_1. (n = 45) There is one most <u>influential case</u>: $D_1^1 = 0.68$ (the <u>outlier</u>, case #1). However, cases #19, #25 and #34 have some influence, also.

implies that deletion of case #1 will shift $\underline{\hat{\beta}}(1)$ to the 0.39 confidence ellipsoid on $\underline{\hat{\beta}}$. The true shift of $\underline{\hat{\beta}}(1)$, as we demonstrate below, is to the 0.95 confidence ellipsoid, corresponding to a value of $D_1 = 2.56$, almost a four-fold change: $D_1{}^1 \sim D_1/4$. Thus, we take the critical value of D_i to be about $1/4$ - say 0.20 - instead of 1.0.

In the present paper we do not combine the joint effects of h_i and t_i on $Var(\underline{\hat{\beta}})$ into a single case statistic like $D_i{}^1$ (See for example, the COVRATIO statistic of the Normal theory model. "... a ratio of determinants of the estimated variance-covariance matrices of the parameters, the numerator having been row-deleted." (Belsley, Kuh and Welsch 1978. See also Lectures 3 and 14.) We simply note that, as would be anticipated, large values of h_i tend to decrease $Var(\underline{\hat{\beta}})$ and large values of t_i tend to increase $Var(\underline{\hat{\beta}})$. See Lecture 3.

It is important to emphasize that the regression diagnostics, t_i, h_i and $D_i{}^1$ are calculated following the iterative solution of the Normal equations. This is the procedure followed in Pregibon 1981 and Cook and Weisberg 1982, currently the standard references in this field. However, "There is one fundamental problem with nonlinear regression dignostics calculated only at the solution: data that are influential during specific iterations can cause the minimization algorithm to find a local minimum different from the one that would have been found had the influential data been modified or set aside. It is therefore advisable to monitor influential data as an algorithm proceeds" (See Belsley, Kuh and Welsch 1978 and Lecture 3). When one is more interested in the construction of a "useful model" of radiation response than in the exposition of the several aspects of its construction - as in the present lecture - then such monitoring should, of course, be done.

Case plots of t_i, h_i and $D_i{}^1$ are presented in Figures 4b, 4c, and 4d, respectively. The case number is simply the row number in the observation matrix, $[\underline{y}, X]$. For these data, the case numbers have been partitioned according to stage of disease, T_2 or T_3. A probability plot of t_i is presented in Figure 4e.

The sample information on $\underline{\beta}$ which consists in the point estimator, $\underline{\hat{\beta}}$, and its variance-covariance matrix, $Var(\underline{\hat{\beta}})$, can be combined into interval estimates of $\underline{\beta}$ - and of the response, $\underline{y}$, - if the sampling distribution of $\underline{\hat{\beta}}$ is known. If the residuals, t_i, are distributed (asymptotically) Normally then $\underline{\hat{\beta}}$ is distributed (asymptotically) Normally

$$\underline{\hat{\beta}} \sim N_p(\underline{\beta}, X'\hat{V}^{-1}X)$$

where $N_p(-, -)$ denotes a p-variate Normal distribution and

$$(\underline{\hat{\beta}} - \underline{\beta})'(X'\hat{V}^{-1}X)(\underline{\hat{\beta}} - \underline{\beta})$$

is distributed as chi-squared with p degrees of freedom. Since it is obvious that the distribution of residuals is far from Normal we do not construct interval estimates of $\underline{\beta}$. Figures 4b and 4e disclose the presence of that most common anomaly - a single "outlier"; in these data, case #1. The value of $D_i{}^1$ suggests the profound effect of case #1 on the estimate of $\underline{\beta}$ - the beta shift in Figure 4d. We find that for this observation, cancer recurred - the event E_1, $y = 0$ - at a treatment regimen of 6000 cgy in 20 fractions over 27 days - 300 cgy per fraction in less than four weeks - for which, on the basis of the estimate, $\underline{\hat{\beta}}$, the predicted probability of recurrence is only 0.049. It seems likely that the <u>recorded</u> failure to control disease at this regimen is due either to errors of transcription, etc., or to a "topographic miss"; perhaps the dose "seen" by the tumor differed substantially from the nominal dose recorded in the patient chart.

It is evident from Figure <u>4d</u> that the case numbers #1, #19 and #34 have the greatest influence on the estimates of β. These observations are marked in the scattergrams of Figures <u>5a</u>, <u>5b</u>, <u>5c</u> and <u>5d</u>, which present the scattergrams of the sample (Figures <u>5a</u> and <u>5b</u> recapitulate Figures <u>3b</u> and <u>3c</u>, respectively). These distributions cover the <u>feasible region</u> of R_3 in Figure <u>1</u>. As is evident from its definition, D_i^1 depends <u>jointly</u> on the key diagnostics, t_i and h_i. The large value of D_i^1 occurs because case #1 is an <u>outlier</u>: t_i is large. The large values of D_{19}^1 and D_{34}^1 arise because these cases are at <u>extreme</u> levels of X; h_{19} and h_{34} are large.

It is perhaps worthwhile examining why the observation #1 is <u>not</u> an extreme point of the distribution ($h_1 = 0.11$) although observations #19 and #34 are such ($h_{19} = 0.33$ and $h_{34} = 0.35$). It would seem from Figure <u>5</u> that perhaps all three should be: #1 seems even more "remote" from the centroid of the observations than are #19 and #34. A <u>part</u> of the explanation may be found in the definition of h_i in terms of the eigenvalues and eigenvectors of the <u>centered</u> matrix, $_0X'_0X$, of the distribution of X; $h_i = 1/n + \sum_{k=1}^{p} u_{ik}^2/\mu^k$, $i = 12, \ldots, n$ where μ_j, U_j are the jth eigenvalue and eigenvector, respectively, of the matrix, $_0X'_0X$. Evidently, h_i is large if 1) the ith observation is remote from the <u>centroid</u> of the joint distribution of X, that is, $\underline{x_i}'\underline{x_i}$ is large <u>and</u> 2) $\underline{x_i}'$ has a <u>large</u> projection on the eigenvector corresponding to a <u>small</u> eigenvalue of $_0X'_0X$. It is evident in Figures <u>5a</u>, <u>5b</u>, <u>5c</u> and <u>5d</u> that for observation #1, $\underline{x_1}'\underline{x_1}$ is large, but $\underline{x_1}'$ has a large projection in the direction of the <u>largest</u> eigenvalue, μ_1, of $_0X'_0X$. However, for observations #19 and #34, $\underline{x_i}'\underline{x_i}$ is large <u>and</u> $\underline{x_i}'$ also has a large projection in the direction of one of the smaller eigenvalues, μ_2, of $_0X'_0X$. See Cook and Weisberg 1982.

Another part of the explanation lies in the differences in the respective estimated responses - and hence <u>weights</u>, $\hat{V}^{-1}$ - at each of the three levels of $\underline{x_i}'$, $i = 1$, 19 and 34. For observations #1, #19 and #34 the estimates responses are 0.95, 0.62 and 0.64, respectively. For generalized least squares estimates, the hat matrix diagonals are a joint measure of the <u>position</u> and <u>weight</u> of an observation. (See also the discussion of h_i for the Poisson models of the cancer incidence and mortality rates in Lecture <u>14</u>.) For the matrix $H = V^{-1/2}X(X'V^{-1}X)^{-1}X'V^{-1/2}$, we have $h_1 = 0.111$, $h_{19} = 0.332$ and $h_{34} = 0.350$. For the matrix $H = X(X'X)^{-1}X'$, we have $h_1 = 0.179$, $h_{19} = 0.299$ and $h_{34} = 0.333$. For each matrix the size-adjusted cut-off is 0.178. We note from Figure <u>4d</u> that observation #34 has little influence on β: $D_{34}^1 < 0.20$. However, deletion of observation #19 might have a considerable effect: $D_{19}^1 > 0.20$. We shall have occasion to examine these observations again.

Since observation #1 may well be an error we now delete it and refit the full ($k = 4$) model to the reduced ($n = 44$) sample. Table <u>3a</u>, <u>b</u> and <u>c</u> presents the respective estimates of β for the rival model matrices, $X = (\underline{1}, \underline{X_1}, \underline{X_2}, \underline{X_3})$, $X = (\underline{1}, \underline{X_1}, \underline{X_2})$ and $X = (\underline{1}, \underline{X_1}, \underline{X_3})$. Table <u>3d</u> presents a comparison of the rival probit, discriminant function and binary regression models of the conditional response, $P(y_i = 1| \underline{x_i}')$, for the reduced matrix, $X = (\underline{1}, \underline{X_1}, \underline{X_2})$, of the reduced

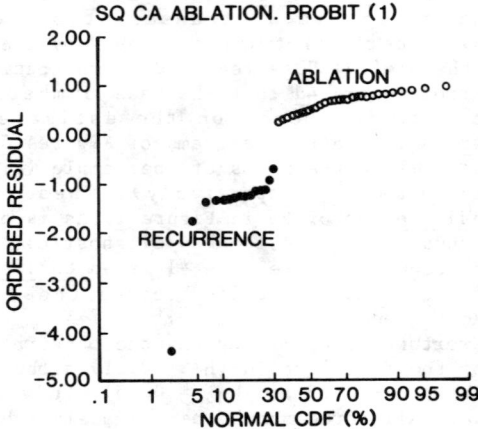

Figure 4e. Normal probability plot of Studentized residuals, t_i, for response surface (probit) model of ablation, E_1. (n = 45) The distribution is heavy-tailed and includes one "contaminant", (case #1).

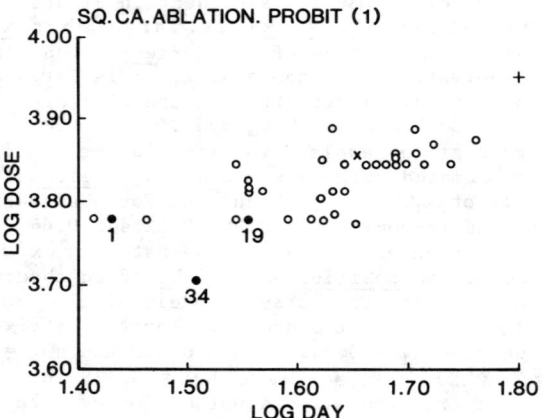

Figure 5a. Scattergram of distribution of treatment levels of head and neck cancer patients. (n = 45) Case #1 is an outlying response, (t_1 = -4.40). Cases #19 and #34 are extreme levels of treatment (h_{19} = 0.33, h_{34} = 0.35). Case statistics refer to response surface (probit) model. "x" is a point of interpolation. "+" is a point of extrapolation. D-T plant. See sections 10 and 11.

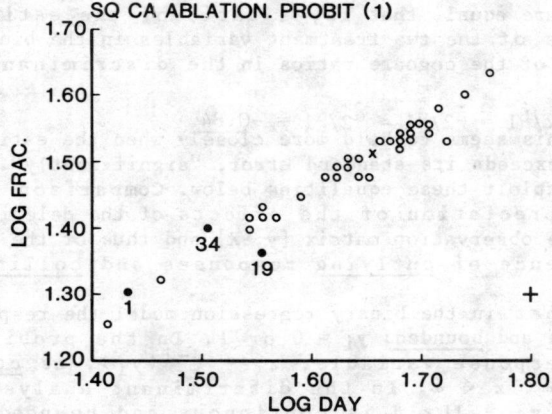

Figure 5b. Scattergram of distribution of treatment levels of head and neck cancer patients. (n = 45) Case #1, #19 and #34 as in Figure 5a. "x" and "+" as in Figure Figure 5a. N-T plane.

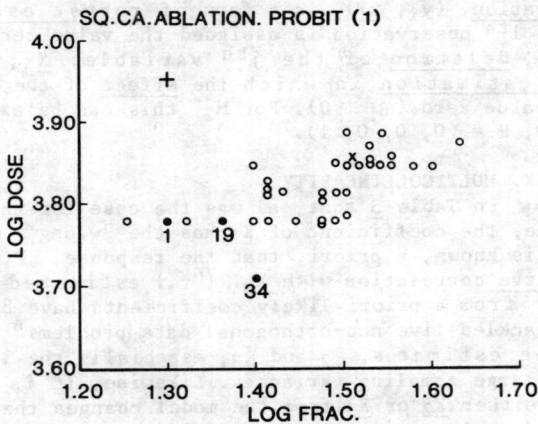

Figure 5c. Scattergram of distribution of treatment levels of head and neck cancer patients. (n = 45) Cases #1, #19 and #34 as in Figure 5a. "x" and "+" as in Figure 5a. D-N plane.

sample. Note that the ratios of the respective estimates of cognate coefficients are equal, that is, a ratio of the estimates of the coefficients of the two treatment variables in the binary model is a good estimate of the cognate ratios in the discriminant and probit models.
$$r_1 = \hat{\beta}_2/\hat{\beta}_1 = \hat{\delta}_2/\hat{\delta}_1 = \hat{\alpha}_2/\hat{\alpha}_1 = -0.84.$$
In general, this seems to hold more closely when the estimate of each coefficient exceeds its standard error, "significantly". We shall have occasion to exploit these equalities below. Comparison with Table 2 gives an appreciation of the effects of the deletion of rows and columns of the observation matrix [y, X] and thus of the effects on β of the presence of outlying responses and collinear treatment variables.

Note that in the binary regression model the response variable, y_i is discrete and bounded: $y_i = 0$ or 1. In the probit regression model the response variable, $z_i = \phi^{-1}(y_i)$, is continuous and unbounded: $-\infty < z < \infty$. In the discriminant analysis model the response, $P(y_i = 1 | x_i')$ is continuous and bounded: $0 \leq P \leq 1.0$. Comparison of the estimates of β for the full model for $n = 44$ with those for $n = 45$ in Table 2 is consistent with the value of $D_1^1 = 0.672$ in Figure 4d: $(\hat{\beta}_{(1)} - \hat{\beta})$ is quite large. Comparison of Table 3a, $\hat{\beta}_2 = -30.057$, with Table 2, $\hat{\beta}_2 = -13.48$, and with Table 3b, $\hat{\beta}_2 = -19.747$ shows that deletion of case #1 (row of X) produces a change in the sample estimate of the effect of the variable X_2 that is a similar size to the change produced by the deletion of variable X_3 (column of X): $\Delta \hat{\beta}_2 = -16.58$ and -10.310, respectively. (We note that deletion of the ith observation, (y_i, x_i'), is a form of robust estimation: The weight of the ith observation is assigned the value zero ($v^{11} \equiv 0$). In the same sense, deletion of the jth variable, X_j, is a form of constrained estimation in which the effect of the jth variable is assigned the value zero ($\beta_j \equiv 0$). For M_2, this can be expressed as $r = R\hat{\beta}$ where $r = 0$, $R = (0, 0, 0, 1)$.

b) COLUMNS OF X. MULTICOLLINEARITY

Note now in Table 3 that, as was the case for the full model of the full sample, the coefficient of X_3 has the "wrong" sign: $\hat{\beta}_3 > 0$. However, it is known, a priori, that the response, $P(y_i = 1 | x_i')$, has a strong negative correlation with X_3. ("... estimated coefficients that are far from a priori likely coefficients have been observed by all who have tackled live non-orthogonal data problems" (Leamer 1978).) Moreover, the estimates, $\hat{\beta}_2$ and $\hat{\beta}_3$, especially the latter, $(\hat{\beta}_3/\sqrt{Var(\beta_3)}) < 1)$ have large sampling variances. Likewise, it is evident that deletion of either X_2 or X_3 from the model changes the estimate of β, often by considerable amount. As is well-known, each of these features is an effect of the presence of strong multicollinearity in the observations and, "It is recommended that one should be extremely cautious about any and all substantive conclusions based on a regression analysis in the presence of multicollinearity" (Chatterjee and Price 1977).

We turn now to systematically assess the nature and degree of multicollinearity for the full model ($k = 4$) that is present in the reduced ($n = 44$) data set.

The collinearity diagnostics for the full model, M_1, of the reduced ($n = 44$) sample are presented in Table 4.

"The ill-effects that result from regression based on collinear data are two: one computational, one statistical" (See Belsley, Kuh and Welsch 1980).

The _statistical_ ill-effect is the inflation of the estimates $\hat{\beta}$ and Var($\hat{\beta}$). The _computational_ ill-effect is the amplification (inflation) of the "noise" present in the system. Both ill-effects can be well described in terms of the so-called Variance Inflation Factor, the VIF_j of the treatment variables, $\underline{X}_j$, "We believe that the VIFs and the procedures based on the eigenvalues of the correlation matrix are the best currently available _multicollinearity_ diagnostics. They are easy to compute, straightforward to interpret, and useful in investigating the specific nature of the multicollinearity." (Montgomery and Peck 1982). But see Belsley, 1984. The VIFs are the Variance Inflation Factors. (Snee 1973). There is a VIF for each component of $\hat{\beta}$. A conservative rule of thumb is that the untoward effects of the presence of multicollinerity in the distribution of X becomes a statistical problem when any $VIF_j > 5.0$ (Snee 1973).

For the jth component, $\hat{\beta}_j$, $VIF_j = 1/(1-R_j^2) = \sum_{k=1}^{p} v_{jk}^2/\lambda_k$. $j = 1$, ..., p where R_j is the multiple correlation of the jth treatment variable with all the other treatment variables and λ_k, $\underline{V}_k = (v_{1k}, ..., v_{pk})$ the kth eigenvalue and eigenvector, respectively, of the _correlation_ marix, $X_0'X_0$. X_0 is a matrix of _standardized_ (_centered_ and _scaled_) treatment variables. (It sould be noted that general inspection of the correlation matrix will not disclose any collinearity that is more complex than simple pairwise correlation.) VIF_j is the jth diagonal element of $(X_0'X_0)^{-1}$ and it is evident that the VIF_j will be _large_ for those coefficients for which the cognate variable has a large component in the direction of the eigenvector of a _small_ eigenvalue of $X_0'X_0$. (The eigenvalue spectrum of $X_0'X_0$ is _uniform_ if the variables are orthogonal: $\lambda_1 = \lambda_2 = \lambda_3 = 1$ and $VIF_1 = VIF_2 = VIF_3 = 1$.) A large h_i may also be a small eigenvalue effect.

The variance inflation factor, VIF_j, can be interpreted as the ratio of the variance of $\hat{\beta}_j$ to what that variance would have been if $\underline{X}_j$ were _uncorrelated_ with the remaining columns of X: This is one of the statistical ill-effects of collinearity in the data - an inflated Var($\hat{\beta}$) sharply circumscribes the use of Least Squares and Maximum Likelihood estimates of β in such _statistical enterprises_ as estimation, control, prediction, model identification, etc.

The correlation matrix, $X_0'X_0$, for the (reduced) head and neck sample and its' eigenvalues, λ_j, and eigenvectors, V_j, are presented in Table 4 together with the functions, λ_1/λ_3, $\lambda_1/Trace[X_0X_0]$, $\Sigma\lambda_j^{-1}$, and $VIF_j = \Sigma\lambda_k^{-1}v_{kj}^2$. The first function is the condition number, κ_0, and the second measures the proportion of the variation in the distribution of treatment variables, X, that is accounted for by the largest eigenvector, V_1. Other useful "rules of thumb" - based on the λ_j - for identifying the presence of multicollinearity in a data set as follows: if any of the eigenvalues are less than 0.01, $\lambda_j < 10^{-2}$, $j = 1, ..., k$) _or_ if the sum of the reciprocals of the eigenvalues exceeds _five_ times the number of treatment variables ($\sum_{j=1}^{p} \lambda_j^{-1} > 5*p$) then the variables are collinear. Otherwise, the variables are regarded as effectively orthogonal. (Chatterjee and Price 1977)

For this sample, these canonical measures precisely describe as well as confirm the high degree of collinearity in the distribution of X that is disclosed by examination of Figures 4a, 4b and 4c. We note that the variables, $\underline{X}_2$ ($log_{10}(\underline{T})$) and $\underline{X}_3$ ($log_{10}(\underline{N})$) have large components on the eigenvector, $\underline{V}_3$, corresponding to the smallest eigenvalue, λ_3, hence VIF_2 and VIF_3 are large. It is evident that we have a rather severe multicollinearity problem with these data.

We note that we should expect the estimates, $\hat{\underline{\beta}}$, to be exceedingly sensitive to rounding errors and to the presence of random errors of measurement in X. It can be shown (See Belsley, Kuh and Welsch 1980) that if the condition number, $\kappa_1 = \sqrt{\lambda_1/\lambda_p}$, of the normal equations matrix, X'X, is of the order of magnitude 10^r and if the data are known to be significant digits then a small change in the data in the last place "... can (but need not) affect the solution [here $\hat{\underline{\beta}} = (X'X)^{-1}X'\underline{z}*$] in the (d-r)th place." (It will be recalled that $z*$ is the pseudo-observation for the probit model.) For the reduced sample $2 < r < 3$. The data on $X'\underline{z}*$ are known to (at most) $d = 3$ significant digits. Therefore a small change in dose (say) will affect the estimates of $\underline{\beta}$ in either the zeroth or first place. We will discuss this at greater length in the section on Errors-in-Variables estimates below.

Moreover, variable selection will be a very "dicey" enterprise as well (See Marquardt and Snee 1975 and Myers in Lectures $\underline{1}$ and $\underline{2}$).

We must next evaluate the case statistics, t_i, h_i and D_i^1 for the full model of the reduced data set. Case plots of these are presented in Figures $\underline{6a}$, $\underline{6b}$ and $\underline{6c}$, respectively. A Normal probability plot of t_i is presented in Figure $\underline{6d}$. However, it will be useful to introduce the interval estimates of $\underline{\beta}$ before we proceed to these diagnostics.

8. POINT AND INTERVAL ESTIMATES OF $\underline{\beta}$ FOR M_1 (n = 44)

Interval estimates of $\underline{\beta}$ are of considerable use both for statistical inference on those models in which the components, β_j, have a substantive interpretation in terms of some theoretical conjecture or model (See Duncan 1978) and also in providing a convenient geometry in which to examine the effects on $\underline{\beta}$ of various post hoc "salvage operations" on the sample; for example, deletion of $\underline{rows}$ and $\underline{columns}$ of the observation matrix [$\underline{y}$, X]. In this latter context, Oberchain has argued that, "... estimates [of $\underline{\beta}$] are of relatively little interest when they are so 'extreme' that they lie outside of the least squares region of, say, 90 percent confidence" (Oberchain 1979). We find his argument (characteristically) persuasive and therefore only $\underline{joint}$ 0.90 interval estimates of the parameter vector, $\underline{\beta}$, are presented in the present paper.

Oberchain has also offered an alternative and useful interpretation of such joint confidence regions; to wit, that they circumscribe a region in parameter space wherein one can believe what one pleases about the value of $\underline{\beta}$ with a specified probability of $(1-\alpha)$ of being wrong.

We present in Figure $\underline{7}$ a plot of the $\underline{approximate}$ simple $(1-\alpha/2) = 0.95$ and simltaneous, or joint, $(1-\alpha) = 0.90$ interval estimates on β_2 and β_3 from the reduced data (See McCullagh and Nelder 1983). The broken dashed line from the point $(-1, -1)$ to the point b $(-30.06, 9.75)$ describes the projection of the sequence of estimates $\hat{\underline{\beta}}(q)$, $0 \leq q \leq 7$, obtained by the Iterative Reweighted Least Squares solution of the Normal equations: $X'(\underline{y} - \hat{\underline{\mu}}) = 0$. Note that the initial estimate of $\underline{\beta}$ is the so-called Unit Weight regression: $\hat{\underline{\beta}}(0) = (-1, 1, -1, -1)$. See Lecture $\underline{17}$. The Maximum Likelihood estimates of (β_2, β_3) for the full model are at point b: ($\hat{\beta}_2$, $\hat{\beta}_3$) = $(-30.06, 9.75)$. That is, the plane of the Figure is at $\hat{\beta}_1 = 25.03$. The region bounded by the ellipse is one

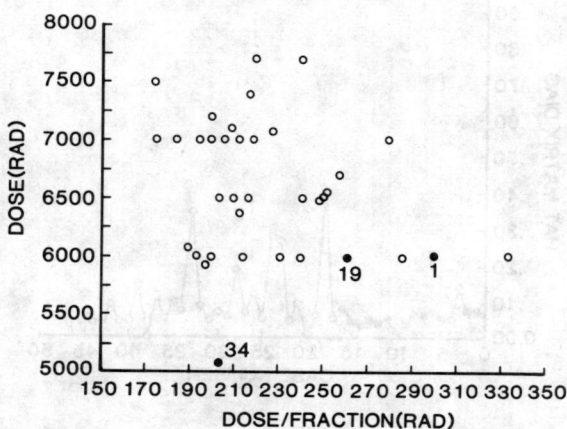

Figure 5d. Scattergram of distribution of treatment levels of head and neck cancer patients (n = 45). Cases #1, #19 and #34 as in Figure 5a. D-D/N plane.

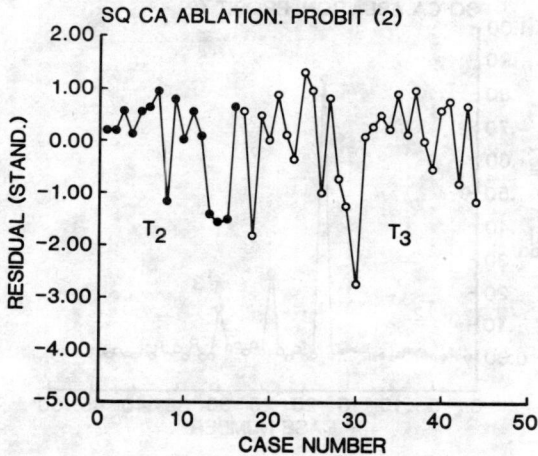

Figure 6a. Case plot of Studentized residuals, t_i, for response surface (probit) model, M_1, of ablation, E_1. (n = 44) There is one outlier: t_{30} = -2.75. Note that although for Normal theory models small residuals, r_i, imply large values of h_i, this is not necessarily the case for probit models. Compare cases #18 and #30. (See Figure 6b).

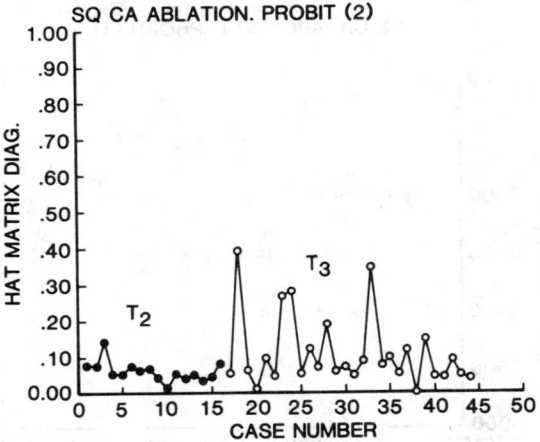

Figure 6b. Case plot of hat matrix diagonals, h_i, for response surface (probit) model, $\underline{M_1}$, of ablation, E_1. (n = 44) There are five extreme levels of treatment. The most extreme is, h_{18} = 0.40. (Case #19 in Figure $\underline{5}$).

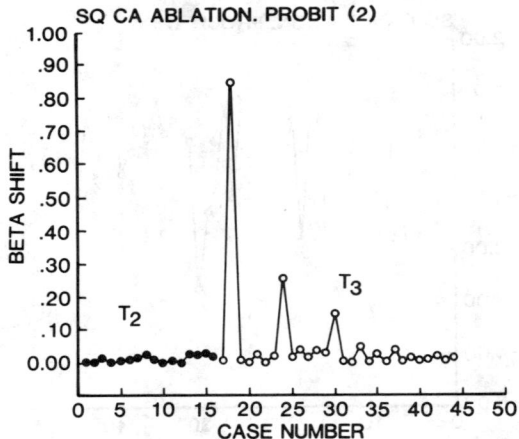

Figure 6c. Case plot of beta shift, D_i^1 (one-step approximation) for response surface (probit) model, $\underline{M_1}$, of ablation, E_1. (n = 44) There is one influential case: D_{18} = 0.83 (the extreme level of treatment, case #18).

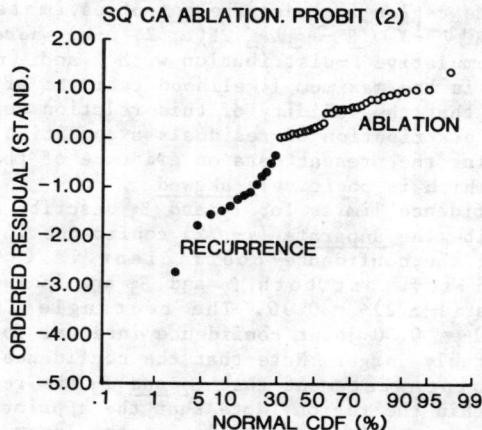

Figure 6d. Normal probability plot of Studentized residuals, t_i, for response surface (probit) model, $\underline{M}_1$, of ablation, E_1. (n = 44) The distribution has a positive skew.

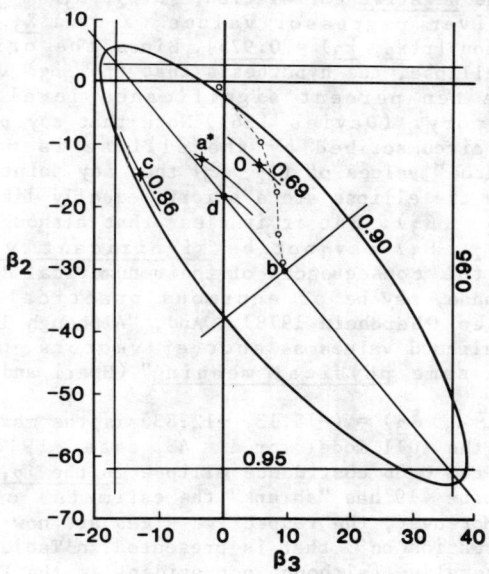

Figure 7. Interval estimates, $\hat{\beta}$. 0.90 simultaneous confidence intervals on (β_2, β_3) for model $\underline{M}_1$. The Figure describes the beta-shifts corresponding to 1) deletion of outlier, $D_1{}^1$: $0 \longrightarrow b$ (n = 45 to n = 44); 2) deletion of extreme level, D_{18}: $b \longrightarrow c$ (n = 44 to n = 43); 3) deletion of collinear variable, VIF_3: $b \longrightarrow d$ (n = 44); 4) "cascaded" RR and EV transformations of $\hat{\beta}$ to correct for effects of collinearity in N and T and random errors of measurement in D: $b \longrightarrow a^*$. The broken line describes the estimates, $\hat{\beta}_2$ and $\hat{\beta}_3$, at successive iterations.

which may be expected, with confidence P = 0.90, to include the true values, (β_2, β_3). The region includes the set of estimates of β such that $(\beta - \hat{\beta})'(X'\hat{V}^{-1}X)(\beta - \hat{\beta}) \leq 2F(\alpha; 2, n-k)$ where F(.) is α-percentile of the cumulative F-distribution with 2 and (n-k) degrees of freedom and $\hat{\beta}$ is the maximum likelihood estimate of β. For Figure 7, (n-k) = 40. Note that the validity of this relation depends on the Normality of the distribution of residuals a condition which is only approximately true for the present data on evidence of the probability plot of Figure 6d, which is positively skewed.

The simple confidence limits for β_2 and β_3 describe a rectangle. These lines describe the separate (1-α/2) confidence intervals for β_2 and β_3 taken singly; the confidence coefficient is (1-α/2) = 0.95 since the probability that both β_2 and β_3 are included within the respective limits is (1-α/2)^2 = 0.90. The rectangle is in fact an alternative (1-α)^2 = 0.90 joint confidence interval for (β_2, β_3) but the area is considerably larger. Note that the confidence coefficient applies to the joint statement that β_2 and β_3 are represented by a point, (β_2, β_3), within the region. Note that the a priori information on β (Table 1) constrains (β_2, β_3) to lie in the lower left quadrant. Note that although neither β_2 nor β_3 is significantly different from zero at P = 0.95 the 0.90 confidence ellipse cuts both axes and does not include (β_2, β_3) = (0, 0). Thus, we may not conclude that both β_2 and β_3 are simultaneously zero.

"The oblateness and orientation of the ellipse disclose that β_2 and β_3 have a large negative correlation, [r(β_2, β_3) = -0.865]. This is due to the given regressor values, X_2 and X_3, having a large positive correlation [r(x_2, x_3) = 0.975]. Since the origin does not lie within the ellipse, the hypothesis that (β_2, β_3) is (0, 0) cannot be accepted at the ten percent significance level under normal distribution theory. (Davies 1961) Note that any point (β_2*, β_3*) within the region circumscribed by the ellipse is more likely to represent the "true" values of (β_2, β_3) than any point without it and that all points on the ellipse are a priori, equally likely to be the true values ($\hat{\beta}_2$, β_3). But it is clear that although these several estimates of ($\hat{\beta}_2$, β_3) may not be significantly different, statistically, the consequences of their numerical differences, for estimates of response, may be of enormous practical - clinical - significance (See Oberchain 1978). And, "Although long vectors [β] give the same likelihood values as shorter vectors, they will not always have the same physical meaning" (Hoerl and Kennard 1970). (italics added)

Point c, ($\hat{\beta}_2$, $\hat{\beta}_3$) = (-15.13, -12.83) is the maximum likelihood estimates of β in the full model for n = 43, case #19 deleted. This corresponds to the 0.86 confidence ellipse in the $\hat{\beta}_2$, $\hat{\beta}_3$ plane. Note that deletion of case #19 has "shrunk" the estimates of ($\hat{\beta}_2$, $\hat{\beta}_3$) in the full model. Moreover, the respective signs are now consistent with the a priori information on β that is presented in Table 1: $\hat{\beta}_2$ < 0, $\hat{\beta}_3$ < 0. Or, more generally, (although not evident in the Figure) -1 < $\hat{\beta}_j$/$\hat{\beta}_1$ < 0. j = 2, 3, $\hat{\beta}_1$ > 0. However, because deletion of #19 has increased the VIFs, the ratios $\hat{\beta}_j/\sqrt{Var(\hat{\beta}_j)}$ are nearer to zero. (Of course, this also, is not evident in the Figure.) Point 0, ($\hat{\beta}_2$, $\hat{\beta}_3$) = (-13.48, 5.53) is the maximum likelihood estimate of β in the full model for n = 45, case #1 included, Case #19 included. Point d, ($\hat{\beta}_2$, $\hat{\beta}_3$) = (19.75, 0), on the 0.24 confidence ellipse, is the maximum likelihood estimate of β in the full model for n = 44, Case #19 included, case #1 deleted and variable X_3 deleted (model M_2, Table 3). The figure shows quite vividly the exquisite sensitivity (or "fragility" (Leamer 1983)) of the maximum likelihood estimates to the presence of extreme treatment levels (large h_{19}), outlying responses (large t_1), and collinear variables (Large VIF_3) in the observation matrix [$\underline{y}$, X].

Point a* on the 0.44 confidence ellipse represents the posterior estimates, $\hat{\beta}$**, of $\hat{\beta}$. $\hat{\beta}$** includes the corrections to $\hat{\beta}$ for the effects of a) the anomalous outlying response in observation #1, b) the small size (n = 44) of the reduced sample c) the collinearity of $\underline{X}_2$ and $\underline{X}_3$) and d) the presence of random errors of measurement of variance $g_1^2 = 0.01^2$ in $\underline{X}_1$ (corresponding to a standard deviation of 100 centigray in the measurement of D). The estimate, $\hat{\beta}$**, will be discussed in considerable detail in Sections 11-14 of this Lecture.

We note in Figures 6a and 6d the presence of one "outlier" - case #30 - and five observations for which $h_i > 2k/n = 0.18$. One of the latter, #18, has a very marked effect on the estimates, $\hat{\beta}$. $D_{18}^1 = 0.85$ (This is observation #19 in Figures 3d, 5a, 5b and 5c. This value of the Cook's D-one-step approximation implies that deletion of case #18 shifts the sample estimate (n = 43) of $\underline{\beta}$ to the 0.50 confidence ellipsoid on $\hat{\beta}$ for n = 44. A better estimate of Cook's D for this case is $D_{18} = 1.28$ which implies a shift of $\hat{\beta}$ to the 0.70 confidence ellipsoid. This estimate is obtained by direct evaluation of the RHS of $D_{18} = k^{-1}(\hat{\beta}^+ - \hat{\beta})'(X'\hat{V}^{-1}X)(\hat{\beta}^+ - \hat{\beta})$ where $\hat{\beta}$ and $(X'\hat{V}^{-1}X)^{-1} = Var(\hat{\beta})$ are the estimates for the sample with n = 44 and $\hat{\beta}^+$ is the sample estimate for n = 43. Note that Figure 7 presents the projections of $\hat{\beta}$ on the (β_2, β_3) plane. Thus, deletion of case #18 (#19) shifts $\hat{\beta}$ (for n=43) to the 0.70 confidence ellipsoid on the estimates obtained for n = 44. The projection of $\hat{\beta}$ onto the (β_2, β_3) plane lies on the 0.86 confidence ellipse.

However, there seems to be somewhat less reason to suspect that case #19 may be in error than for case #1. Moreover, as is evident from an examination of Figure 5b, deletion of this case will increase the correlation between $\underline{X}_2$ and $\underline{X}_3$ and thus inflate the $Var(\hat{\beta}_j)$; the respective VIF_j become (1.99, 27.83, 25.21) compared with (1.99, 21.90, 20.10). Therefore, we do not delete it.

Note that it would seem from Figure 6c that, although the results of laboratory experiments may be frequently - and accurately - stigmatized as "mouse-bound", the conclusions from non-experimental studies may often be, "case-controlled". However, unless the full set of diagnostics are examined, this aspect of the estimates of $\underline{\beta}$ will be missed. The fidelity of the model to the sample is described not only by the summary statistic, RSS = Σt_i^2, but also, and perhaps more uniquely in non-experimental studies, by the profiles of the model-sample interactions: the index plots of t_i, h_i and D_i^1. They provide the "signature" of the sample on the estimates of the model.

The Normal probability plot of the residuals, t_i, for n = 44 in Figure 6d is more nearly linear than that of Figure 4e. This suggests that the distribution of t_i is more nearly Normal for full model of the reduced (n = 44) than it is for the full (n = 45) sample. However, there is a marked positive skewness in the distribution. In order to assess the "Normality" of a univariate distribution with sample mean $\bar{x}$ and standard deviation, s, it is a common practice to plot on Normal probability paper the empirical cumulative distribution of the observations described by the ordered pairs, x(i), (i - 0.5)/n, where x(i) is the ith order statistic in a sample of size n and then determine "by eye" whether the scattergram is adequately described by a straight line with intercept $\bar{x}$ and slope (1/s) (Hald 1952).

Several probability plots of random samples from the unit Normal distribution, N(0, 1) with sample sizes $8 \leq n \leq 384$ are presented in Daniels and Wood (1971) and are useful as "training sets" by use of which investigators may sharpen their appreciation of Normal shapes (Seber 1977).

However, the usefulness of such probability plots is determining whether a given distribution is Normal, or not, has hitherto been strongly dependent on the sample size, n: "... samples of 8 tell us almost nothing about normality, whereas samples of 384 seem very stable, except for their few highest and lowest points. Sets of 16 show shocking wobbles; sets of 32 are visibly better behaved; and sets of 64 nearly always appear straight in their central regions, but fluctuate at their ends (Daniels and Wood 1971). For this reason, "It is recommended that n be at least 20 and preferably greater than 50" (Seber 1977).

However, Filliben (1975) has constructed a test statistic, the probability plot correlation coefficient, r_F, which is useful over the range of sample sizes $3 \leq n \leq 100$ for which he has prepared a table of quantiles of the sampling distribution of r_F. However, "It is to be noted that the use of the probability plot correlation coefficient should complement (rather than replace) the use of probability plots themselves" (Filliben 1975).

For the probability plot of Figure 6d, we have $r_F = 0.950$. This lies below the 0.005 quantile of the sampling distribution, and hence we reject H_0. Thus, interval estimates of β and y_i are approximate.

Figures 8a, 8b and 8c, respectively, present plots of the residuals, t_i, against the estimated response, $\hat{z}$, and against the treatment variables X_1 and X_2. Such plots are part of good practice in any regression analysis. The first plot has a shape characteristic of a binomial response, ($y_i = 0$ or 1) variable in that the greatest dispersion of t_i occurs – more-or-less – in the vicinity of $z_i = 0$, or $\Phi(z_i) = 0.50$. The other two plots are relatively featureless. One may tentatively conclude that the full model has apparently mapped these data into "binomial noise". Thus, the model M_1 includes all of the information on radiation response (conditional on D, T and N) that was present in the (reduced) sample; none has "leaked" into the residuals (Box, Hunter and Hunter 1978). A successful model is one that leaves a patternless set of residuals. "The argument is that if we can detect pattern we can find a better model; the practical problem is that a finite set of residuals can be made to yield some pattern if we look hard enough so that we have to guard against over-interpretation" (McCullagh and Nelder 1983). However, such conclusions are tempered in the presence of such a high degree of multicollinearity (See Myers, Lectures 1 and 2).

It may seem to some that the regression diagnostics provide more information on model-sample interactions than an investigator wants – or needs – to know. Cook and Weisberg (1982) have cautioned that, "For the unwary, there is the inherent danger that is caused by the recent explosion of available methods of criticism. If every recommended diagnostic is calculated for a single problem the resulting 'hodge podge' of numbers and graphs may be more of a hinderance than a help ... Life is short and we cannot spend an entire career on the analysis of a single set of data." However, as Fisher once remarked in a Presidential Address to the Royal Statistical Society, "The errors arising from too much information are not likely to exceed those arising from too little."

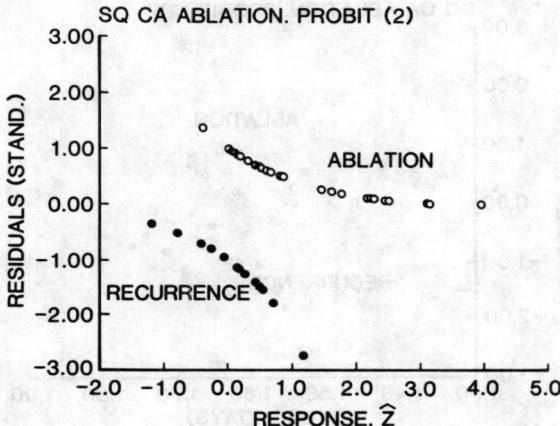

Figure 8a. Scattergrams of conditional distributions of Studentized residuals, t_i, and estimated response, $\hat{z}_i$, for model $\underline{M}_1$. (n = 44) The general shape is characteristic of a binomial response variable.

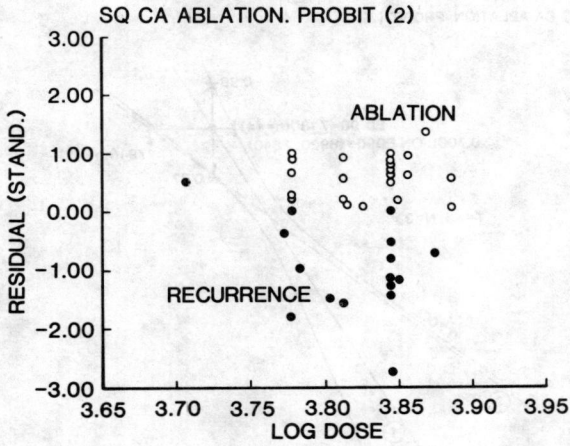

Figure 8b. Scattergrams of conditional distributions of Studentized residuals, t_i, and $\underline{X}_{1i}$, for model $\underline{M}_1$. (n = 44).

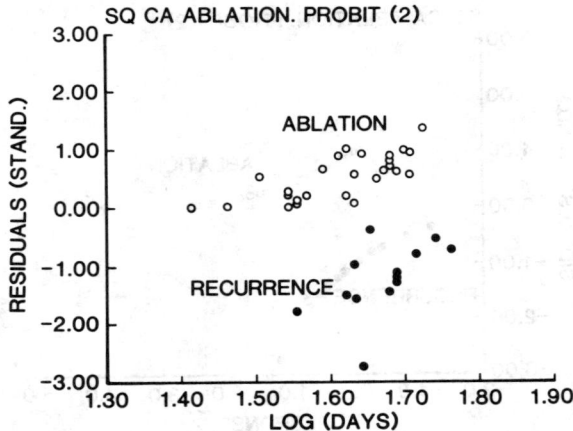

Figure 8c. Scattergrams of conditional distributions of Studentized residuals, t_i, and X_{2i}, for model M_1. (n = 44). Save for the outlier, #30, the distributions of Figures 8b and 8c are featureless, suggesting that the model, M_1, has "mapped" the data into (binomial) "noise".

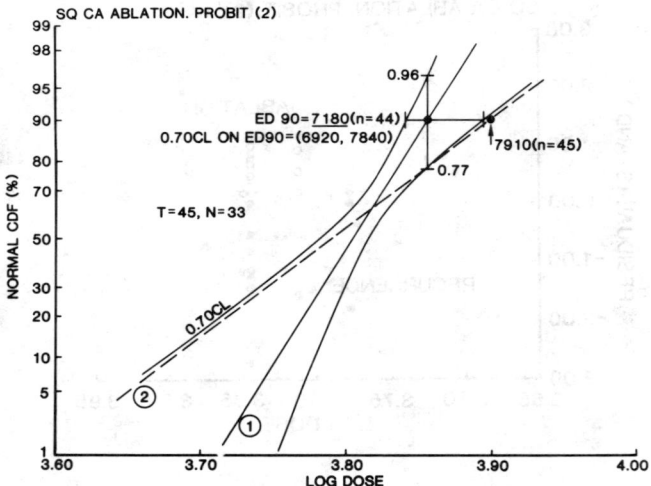

Figure 9. Interval estimates, $\hat{z}_i(\underline{x}')$, and $\hat{x}_i'(\Pi)$. 0.70 confidence intervals on $\hat{z}_i(\underline{x}')$ and inverse estimates $\hat{x}_i'(\Pi)$ for M_1 (n=44). Curves 1 and 2 are the dose-response curves for the response surface models for n = 44 and n = 45, respectively. The respective levels of ED90 are 7180 rad and 7910 rad. The 0.70 CL on $\hat{z}_i$ at the interpolation point "x" in Figures 5a to 5c are (0.77, 0.96). The 0.70 confidence limits on ED90 for model M_1 are (6920 rad, 7840 rad).

Although the RSS for this model of these data is <u>less</u> than the number of degrees of freedom and hence the data offer no compelling evidence <u>against</u> the model, it is prudent to assess the concordance of model and data by yet another measure. To this end, the 44 observations are grouped according to which of four ranges of response, $\phi(\hat{z})$, is predicted by the model. Then the expected and observed frequencies within each range are compared. This comparison is presented as Table <u>5</u>. The concordance is so obviously satisfactory that it is unnecessary to compare the frequencies by a formal test of the null hypothesis. (See Finney 1971).

Thus, we have the <u>provisional</u> empirical model, M_1, of the response over the region of observations. However, we note that the estimates, $\hat{\underline{\beta}}$, are dominated by the presence of case #18 and strongly affected by cases #24 and #30. Although these assessments have disclosed that this model is not, in the large, inconsistent with these observations, the analysis is still incomplete. We have 1) no measure of the "utility" of the model for prediction of response, z, 2) no measure of the stability of the predictive power of the estimates, $\hat{\underline{\beta}}$, in "new data" and 3) no satisfactory measures of the roles of each factor (D, T, N) separately in evoking or modulating this response and 4) no measure of its performance in extrapolation. One measure of the utility of the prediction equation in interpolation can be obtained by comparing the <u>range</u> of the predicted values to their <u>average standard error</u>. In the general case, a model is not likely to be an adequate predictor unless the range of fitted values is large relative to their average standard error (See Box and Wetz 1973 and Montgomery and Peck 1982). For the binary response data one appropriate measure is $\gamma_m = [n_{(n)}\hat{\Pi}(n)) - n_{(1)}\hat{\Pi}(1)]/\sqrt{\Sigma n_i \hat{\Pi}_i (1-\hat{\Pi}_i)/n}$ where $n_{(i)}\hat{\Pi}(i)$ is the ith order statistic of the set of fitted values, $n_i \hat{\Pi}_i$. For the equation $z = \hat{\beta}_0 + \hat{\beta}_1 x_1 + \hat{\beta}_2 x_2 + \hat{\beta}_3 x_3$ we have $\gamma_m = 2.212$. Obviously, if the range of fitted values exceeds the average standard error by only a factor of 2, the equation is not very <u>useful</u> in any iterpolation of response.

A second method of assessing the utility of the proposed model, $z = \beta_0 + \beta_1 x_1 + \beta_2 x_2 + \beta_3 x_3$, is to compare the observed <u>binary</u> response with that predicted by the model at each of the treatment regimens, $\underline{x_i}'$, included in the sample. Since the response is binary it will be convenient to present the comparison as the four-fold table in Table <u>6</u> for which it is stipulated that the predicted response is 0 or $\overline{1}$ according as $\hat{z}_i \leq z_0$ or $\hat{z}_i > z_0$, respectively, where $z_0 = 0$ is the threshold. Clearly, the model does not predict the observed response very accurately since the correlation coefficient for observed and estimated binary reponses is $r = 0.399$. See Anderberg 1973. Note that this value, $r^2 = 0.159$, is consistent with the value of $R^2 = 0.198$ of the binary regression model and $D^2 = 1.036$ of the discriminant function model. See Table <u>3d</u>. It is, of course, possible to inflate the correlation between the observed and predicted response by choosing a different threshold, z_0. (This method of assessment of performance is based on the discriminant function model of the response. See Lecture <u>19</u>.)

9. POINT AND INTERVAL ESTIMATES OF RESPONSE AT A GIVEN LEVEL OF TREATMENT, $\hat{\Pi}(\underline{x_i}')$, AND OF TREATMENT REGIMEN REQUIRED TO EVOKE A GIVEN LEVEL OF RESPONSE, $\hat{\underline{x}}'(\Pi_i)$.

"The size of the confidence limits is inversely proportional to the quality of the data used to make the estimate and directly proportional to the amount of extrapolation involved. This important information is lost if the confidence limits and best estimates are not routinely reported. The width of the confidence interval is one of the best measures ... to evaluate the quality of the estimates of potential risks. It is important to distinguish between those situations in which the risk is precisely estimated and those in which it is not" (Park and Snee 1983).

"What is often overlooked in the application of regression models is their great dependence on the validity of the model employed, generally one of a linear nature" (See Mantel 1983).

For present expository purposes we assume the validity of the model M_1. We defer a discussion of model validation to a later point in the paper.

An important objective of the construction of empirical models is the prediction of response at a specified level, (D, N, T) of the treatment variables. The <u>point</u> estimate of the expected value of response at $\underline{x_i}'$ is $\hat{\Pi}_i = \Phi(\hat{z}_i)$ where $z_i = \underline{x_i}'\underline{\beta}$. It is also frequently required to estimate the dose, x_1, required to elicit a specified level of response, $\Pi_i = \Phi(z_i)$. The <u>point</u> estimate is $\hat{x}_1(\Pi) = (z_1 - \hat{\beta}_0)/\hat{\beta}_1 - (\hat{\beta}_2/\hat{\beta}_1)x_2 - (\hat{\beta}_3/\hat{\beta}_1)x_3$ where $z_i = \Phi^{-1}(\Pi_i)$. <u>Interval</u> estimates of each of these quantities are also important. For the estimate, $\hat{z}_i$, the $(1-\alpha)$ confidence limits are easily obtained. These are

$$\hat{z}_i \pm u(1-\alpha/2)\sqrt{Var(\hat{z}_i)} \qquad (28)$$

or,

$$\underline{x_i}'\hat{\underline{\beta}} \pm u(1-\alpha/2)\sqrt{\underline{x_i}'Var(\hat{\underline{\beta}})\underline{x_i}} \qquad (29)$$

or,

$$\underline{x_i}'\hat{\underline{\beta}} \pm u(1-\alpha/2)\sqrt{(1/\Sigma n_i w_i) + V} \qquad (30)$$

where
$$V = v_{11}(x_1-\bar{x}_1)^2 + v_{22}(x_2-\bar{x}_2)^2 + v_{33}(x_3-\bar{x}_3)^2 + 2v_{12}(x_1-\bar{x}_1)(x_2-\bar{x}_2)$$
$$2v_{13}(x_1-\bar{x}_1)(x_3-\bar{x}_3) + 2v_{23}(x_2-\bar{x}_2)(x_3-\bar{x}_3). \qquad (31)$$

v_{jk} is the jk^{th} element of $Var(\hat{\underline{\beta}})$, and $\bar{x}_j$ is the mean of x_j and $n_i w_i$ is the weight of the ith observation. i, .., n = Σn_i, j, k = 1, ... p. This expression for V makes very vivid one reason for requirement for "parsimony" in any model, since the addition of a treatment variable always increases, or at least, never decreases, the variance of the estimated response, $Var(\hat{z})$ - vide supra.

For the estimate, $\hat{x}_{1i}(\Pi)$, the $(1-\alpha)$ confidence limits are obtained from the cognate confidence limits given by Fieller's theorem for the <u>ratio</u>, m, of random variables, a and b, on the right-hand side of the following equation (See Finney 1971):

$$(x_1-\bar{x}_1) = m = a/b = [z - \bar{z}-\hat{\beta}_2(x_2-\bar{x}_2) - \hat{\beta}_3(x_3-\hat{x}_3)]/\hat{\beta}_1. \qquad (32)$$

For this ratio the variance of the <u>numerator</u>, a, is

$$v_{11}^* = 1/\Sigma n_i w_i + v_{22}(x_2-\bar{x}_2)^2 + 2v_{23}(x_2-\bar{x}_2)(x_3-\bar{x}_3) + v_{33}(x_3-\bar{x}_3)^2. \qquad (33)$$

The variance of the <u>denominator</u>, b, is

$$v_{22}^* = v_{11}. \qquad (34)$$

The covariance is

$$v_{23}^* = -v_{12}(x_2-\bar{x}_2) - v_{23}(x_3-\bar{x}_3). \qquad (35)$$

The confidence limits are obtained as the solutions, $\bar{x}_1 + x_L$, $\bar{x}_1 + x_U$, where

$$x_L, x_U = [m-g\ v_{12}^*/v_{22}^* \pm u(1-\alpha/2)b^{-1} \{v_{11}^* -2m\ v_{12}^* + m^2\ v_{22}^*$$
$$- g\ (v_{11}^* - v_{12}^{*2}/v_{22}^*)\}^{1/2}]\ (1-g)^{-1} \qquad (36)$$

where $g = u^2(1-\alpha/2)v_{22}^*/b^2$.

It should be noted that the <u>interval</u> estimate of the ratios of regression coefficients such as β_j/β_1, $j=2$, 3, which occur in the so-called isoeffect equations, or the ratio, α/β, for the so-called linear-quadratic model, which is frequently discussed in the recent literature of clinical radiobiology (see Lecture <u>17</u>) can - and should - be obtained directly from Fieller's theorem. However, it will be recalled that interval estimates are rarely presented in the current literature of radiation oncology.

The interval estimates, $\hat{z}, \hat{x}_1(\Pi)$, depend on the covariance matrix, $Var(\hat{\beta})$ and the character of the <u>sampling distribution</u> of residuals. If that distribution is Normal, then $u(1-\alpha/2)$ is the $(1-\alpha/2)$ quantile of a Normal distribution, i.e., $u_{0.975} = 1.96$, provided that, for the model in question, RSS/(n-k) $\simeq 1.0$. If the distribution is Normal, <u>but</u> RSS/(n-k) $\gg 1.0$, although the model is "correct" - no obvious pattern is present in the residual plots - then $u(1-\alpha/2)$ is the $(1-\alpha/2)$ quantile of the Student's t distribution with (n-k) degrees of freedom and $Var(\hat{\beta})$ is replaced by $\sigma^2 Var(\hat{\beta})$ where σ^2 is the over-dispersion factor, $RS\overline{S}/(n-k)$ (See Finney 1971 and McCullagh and Nelder 1983). Thus, it is important to determine (approximately) the nature of the distribution function of the residuals, as for instance, by a (Normal) probability plot of the residuals, as in Figures <u>4e</u> and <u>6d</u>.

For <u>ungrouped</u> observations, $n_i = 1$ $(i=1, \ldots n)$ it is also desirable to replace the degrees of freedom, n-k, by the effective degrees of freedom, $(n-k)/[1+0.5\overline{\gamma}_2]$ where $\overline{\gamma}_2$ is average standardized kurtosis of the distribution of $\hat{y}_i$ (See McCullagh and Nelder 1983). However, for the full model of the reduced data, the required correction to (n-k) could be neglected.

In Figure <u>9</u> we present the dose-response curve at T = 45 days, N = 33 fractions. (The curve is appropriate for patients treated at 5X/week with the first treatment given on M, T or W.) This is curve #1. The confidence limits on $\hat{\Pi}$ for $(1-\alpha) = 0.70$ are superimposed. The curves are traces of the (planar) response surface and hyperbolic error surface with an (x_2, x_3) plane through the weighted mean $(\overline{x}_1, \overline{x}_2, \overline{x}_3)$ of the joint distribution of X. Note that the width of the interval estimates of $x_1(\Pi)$ is a minimum at $\hat{z} = 0.361(\Pi = 0.64)$ which occurs at $\overline{x}_1 = 3.826 (D = 6693)$. It is evident from the Figure as well as from the expressions for each <u>interval estimate</u> given above that the width of each increases <u>quadratically</u> with the distance from the <u>centroid</u> (weighted mean), $\overline{x}_1, \overline{x}_2, \overline{x}_3 = (3.826, 1.657, 1.513)$, of the joint distribution of X. The point estimate of response at $D = 10^{x_1} = 7180$ rads (217 rad/fraction) is $\Pi = 0.90$. The 0.70 CL are (0.77, 0.96). The 0.70 CL on $D(0.90) = 10^{x_1(0.90)}$ are 6920, 7840). These were estimated by Fieller's Theorem. The <u>lower</u> 0.95 CL on D(90) is 6834 rad. (The upper 0.95 CL is 56,380 rad!) The level of confidence, 0.70, is quite low since $Var(\hat{\beta})$ is <u>inflated</u> by the narrow ranges of $\underline{X}_j$ and high degree of collinearity present in the joint distribution of X - as well as the small sample size, n = 44.

Curve #2 in the Figure <u>7</u> is the estimated dose-response curve, $z = X\hat{\beta}$, for $(T, N) = (45, 33)$, when $\hat{\beta}$ is the sample estimate of $\underline{\beta}$ with the "outlier", case #1, <u>included</u> (n = 45). The estimate, $x_1(0.90) = 7910$, has been inflated by the presence of this anomalous observation.

More generally, the presence of this "outlier" has deflated the sample estimate of β_1 in the dose-response equation. The same effect can also be produced by <u>errors of identification</u>, either misclassification of response at completion of treatment (false positive, false negative) or misallocation (for instance, incorrect staging) of patients to a treatment schedule. This will be briefly discussed below.

10. INTERPOLATION

"Interpolation means prediction for new cases with predictor variables not too different from the values of the predictor variables in the construction sample" (Weissberg 1980).

In using the model to obtain point and interval estimates of response, $\hat{z}_i = \underline{x}_i'\hat{\beta}$, at a level of treatment, $\underline{x}_i'$, <u>not</u> included in the original, or <u>Construction</u>, sample from which the estimate, $\hat{\beta}$, was obtained, care must be exercised to recognize <u>extrapolation</u> beyond the region of the treatment variables, X, which, in a sense to be subsequently specified, is "covered" by the <u>joint</u> distribution of the treatment variables in the sample, since it is possible that a model which fit the observations very well in the region of the sample may perform poorly outside that region. It is, of course, obvious that the <u>variance</u> of the estimates, $Var(\hat{z}_i)$, will be large if $\underline{x}_i'$ is remote from the centroid of the sample. However, it is also possible for the estimate, $\hat{z}_i$, to be biased as well, since the model may only be valid within the region of X spanned by the sample observations. (This possibility exists for <u>mechanistic</u> as well as <u>empirical</u> models.)

In a multiple regression model of non-experimental data it is obviously easy to extrapolate, inadvertently, since a treatment level may lie well within the <u>marginal</u> distributions of each of the treatment variables but lie well beyond the region "covered" by the <u>joint</u> distribution of the sample owing to the presence of a strong linear relation between two or more of the treatment variables. See Figure 5.

The diagonal elements, h_i, of the hat matrix are useful in specifying the region "covered" by the joint distribution of the sample since h_i is a measure of the (Mahalanobis) <u>distance</u> of the ith observation from the <u>centroid</u> of the distribution of X. Note that the contours of constant $h_i = \underline{x}_i'(X'X)^{-1}\underline{x}$, are <u>ellipsoids</u>. Thus, if $h_{(n)}$ is the largest value of h_i in the construction sample, (the nth order statistic, See Hald 1952), the set of all points $\underline{x}_j'$ such that $\underline{x}_j'(X'X)^{-1}\underline{x}_j \leq h_{(n)}$ describes an ellipsoid which includes all the observations in the original sample. It defines the so-called Regressor Variable Hull (RVH) described by Montgomery and Peck (1982). New treatment levels, $\underline{x}_k'$, for which $\underline{x}_k'(X'X)^{-1}\underline{x}_k > h_{(n)}$ are <u>extrapolations</u>; levels for which $\underline{x}_k'(X'X)^{-1}\underline{x}_k \leq h_{(n)}$ are <u>interpolations</u>. The predicted responses, $\hat{z}$, of the former must be used with more caution. Obviously, if $h_{(n)}$ is much larger than the next largest value of h_i, say $h_{(n-1)}$, then that observation may be an <u>extreme point</u> of the original sample. In such a case it may be more prudent to use $h_{(n-1)}$, $h_{(n-2)}$, ..., etc. to define the ellipsoid for interpolation.

Note that for the probit regression we have used $h_{(n)}$ for the matrix $X'(X'X)^{-1}X$ of the binary regression, $\underline{y} = X\underline{\alpha}$, where y takes only the values 0, (no response) or 1 (response) rather than for the matrix $\hat{V}^{-1/2}X'(X'\hat{V}^{-1}X)^{-1/2}X\hat{V}^{-1/2}$. The centroid of the former is at $(\bar{x}_1, \bar{x}_2,$

$\bar{x}_3$) = (3.823, 1.630, 1.487) or (6653, 43, 31) whereas the latter is at (6699, 45, 32).

The treatment (D, T, N) = (7180, 45, 33) at which $\hat{\pi}$ = 0.90, described above, is at the point "x" in Figures 5a, b and c. For this case, h_j = 0.056. For this sample, $h_{(n)} = \bar{h}_{33}$ = 0.340, and $2k/n$ = 0.182. Thus, this regimen, $\underline{x}_j'$, is well within the region of X "covered" by the sample RVH and the bias in $\hat{\pi}(\hat{z})$ should be small (although the variance, Var($\hat{z}_i'$), is quite large).

11. MODEL VALIDATION

"Which of the models, $M_1, \ldots, M_m$, best explains a given set of data? This is a fundamental question confronting research workers. In fact, subject to restrictions of various sorts, this question has been the problem of empirical science for many years. While this question is of interest, it is not the crucial one. In most circumstances, a more pertinent one (certainly with a more directly useful answer) is: Which of the models, $M_1, \ldots, M_m$, yields the best predictions for future observations from the same process which generated the given set of data? While this question is much more difficult to answer than the first, it bears more directly on the reason for having models at all." (Geisser and Eddy 1979).

"Users have often been disappointed by procedures such as multiple regression equations, that 'forecast' quite well for the data on which they were built. When tried on fresh data, the predictive power of these procedures fell dismally."

...

"Even when the specific variables to be used in a multiple regression have been picked in advance, so that the form is determined, the coefficients are chosen from infinitely many combinations of possibilities to make the results of substituting in the formula fit the data as close as possible. Thus, testing the data on the data that gave it birth is almost certain to over-estimate performance for the optimizing process that chose it from among many possible procedures will have made the greatest possible use of any and all idiosyncracies of those particular data. Sometimes we say that 'Optimization capitalizes on chance!' As a result, the procedure will likely work better for these data than for almost any other data that will arise in practice. The apparent degree of fit will be closer than the true degree of fit, on the average" (Mostellar and Tukey 1977).

"Methods to determine the validity of regression models include comparison of model predictions and coefficients with theory, collection of new data to check model predictions, comparison of results with theoretical model calculations and data splitting or cross validation in which a portion of the data is used to estimate the model coefficients and the remainder of the data is used to measure the predictive accuracy of the model" (Snee 1977).

There are several aspects of a model; one is the substantive interpretation or identification of the individual coefficients estimates as $\hat{\beta}_j$; another is the prediction of response, estimates as $\hat{y}_j = \hat{\underline{x}}_j'\hat{\underline{\beta}}$. After a model has been developed and estimates, $\hat{\underline{\beta}}$ and Var($\underline{\beta}$), obtained from a given construction sample it is important that the estimates of $\underline{\beta}$ and the predictions of response, y_i, be validated. One method for doing this consists in the comparison of the sample estimates with prior information on $q \leq k = p + 1$ elements of the (unobservable) matrix $\underline{\beta}$. When the prior information can be described by a linear form, or constraint, on $\underline{\hat{\beta}}$, $\underline{r} = R\underline{\hat{\beta}}$, where $\underline{r}$ is qx1 and R is (qxk), this comparison may be made by a method due to Wold. This is a test of the null hypothesis of rank q, $H_0: \underline{r} = R\underline{\hat{\beta}}$. The test statistic is distributed as chi-squared with q degrees of freedom:

$$\gamma = (\underline{r} - R\underline{\hat{\beta}})'[R\mathrm{Var}(\underline{\hat{\beta}})R']^{-1}(\underline{r} - R\underline{\hat{\beta}}). \tag{37}$$

For example, one might wish to test the hypothesis that the sample estimates of the coefficients of a linear model of a binary tissue response is consistent with prior information on these coefficients. Suppose, for instance, that the systematic part of the model is

$$z = \underline{x}'\underline{\beta} = \beta_0 + \beta_1 x_1 + \beta_2 x_2 + \beta_3 x_3 \tag{38}$$

and that the prior information can be specified by the ratio, $\rho_1 = (\beta_3/\beta_1)$. Then the constraint matrices, $\underline{r}$ and R, of the **equality** have the forms, $\underline{r} = 0$, $R = [0, \rho_1, 0, -1]$. The test statistic is distributed as chi-squared with q=1 degree of freedom. We do not at this point elaborate further on this topic since it will be treated in more detail under the more general topic of Mixed Estimation. But it is evident from Table 1 that although the relevant a priori information on $\underline{\beta}$ **cannot** be described by a linear form, $\underline{r} = R\underline{\beta}$, the sample and a priori information on $\underline{\beta}$ are clearly inconsistent. We also here note that the method of Ridge Regression (see Lecture 4) which will be used to implement Mixed Estimation, is appropriate to some problems in which the a priori information on $\underline{\beta}$ can only be represented as an **inequality**, say, $\underline{r} < R\underline{\beta}$ and the **incompatibility** of sample and non-sample information can be attributed to (or is consistent with) the presence of multicollinearity in the sample.

Note that the interval estimates discussed above refer to the estimates of the expected values, $E(\hat{\beta})$ and $E(\hat{x}_1(\Pi))$, respectively, that will be obtained in repeated observations of response over the initial sample distribution of treatment variables, X, i.e., in subsequent, **exact** repetitions of the same experiment. However, in non-experimental studies it is rare that the **exact** distribution of treatment variables in a given sample is repeated - or repeatable. Let us therefore assume that although the same levels of treatment may not be repeated, the subsequent treatment regimens will be drawn from the same population as the initial sample, i.e., the distribution of treatment variables is **stable over time**. As we have discussed, one of the important applications of a multiple regression equation is to predict the values of the response variable at a given set of levels of treatment variables. Now, it is commonly observed in practice that the predictive performance of an equation of given form is better - often much better - in the initial sample from which the estimates are obtained than in a subsequent, independent sample from the same population (See Mostellar and Tukey 1977).

This effect is often referred to as the "shrinkage" of the multiple correlation coefficient, R^2, and this "shrinkage" occurs even though the predictions of response for second sample are made within the region of X covered by the joint distributions of the first sample, for example, within the region that is circumscribed by the $h_{(n)}$ ellipsoid.

Another method of model validation is to collect additional data and see how well the model predicts the new data. This is often not possible. [And even if acquisition is possible, the new data may be inappropriate. Moreover, by its acquisition one must incur additional delays, additional costs in money and effort as well as time, additional risks, etc. The problems are the same as for Data Augmentation.] One way to simulate the collection of new data is to split the data in hand into two subsets. One subset, called the "estimation data", is used to estimate the coefficients in the model. The remaining subset, called the "prediction data" is used to measure the prediction accuracy of the model. (Marquardt and Snee 1975).

"The DUPLEX algorithm, developed by R. W. Kennard, is recommended for dividing data into the estimation set and prediction set when there is no obvious variable such as time to use to split the data." (Snee 1975). The determinant $|X'X|$ of $X'X$ is a measure of the volume of X-space covered by the sample. If X_1 and X_2 are the model matrices for the data in the estimation and prediction sets, respectively, then $(|X_1'X_1|/|X_2'X_2|)^{1/p} \simeq 1.0$ is the criterion of commensurability of the two sets to be achieved by the split. Here, $p = k-1$ where, $\underline{\beta}$, the parameter vector of the model is (kx1). It may also be desirable to compare the respective VIF_j and λ_j, V_j for the two sets.

It is important to emphasize that in these maneuvers, the region of the space of the treatment variables, X, spanned by the sample must be the same in the "fresh data" as in the data from which the equations were "built" – i.e., the data from which the initial sample estimates of β were obtained. In other words, we are not concerned with "extrapolation" of the model in this maneuver.

In general, a minimum total sample size is required for such "data splitting" maneuvers to be effective. For Normal Theory models, this minimum size is approximately, $n = 2k + 25$. The minimum will be larger for the models of binomial response, since the standard error of the estimates of a quantal response are larger than the standard error of estimates of a quantitative response for a given sample size and range of treatment variables.

In the event that the sample is too small for the data-splitting maneuver to be effective, still another change can be rung on the cross-validation maneuver. This procedure requires the division of the sample (size n) into a "construction" sub-sample (size n-1) and a "validation" sub-sample (size 1) in all (n) possible was. Each possible subset of (n-1) observations forms the <u>construction</u> sub-sample and each observation, in turn, forms a <u>validation</u> sub-sample. This is the method of the PRESS statistic developed by Allen (1974). See Myers, Lectures <u>1</u> and <u>2</u>. For Normal theory models, PRESS is the sum of squares of <u>deleted</u> residuals.

$$\text{PRESS} = \sum_{i=1}^{n} (y_i - \hat{y}(i))^2 = \sum_{i=1}^{n} r(i)^2 \qquad (39)$$

where $\hat{y}(i) = \underline{x}_i'\underline{\beta}(i)$, is the estimate of y at $\underline{x}_i'$ when $\underline{\beta}$ is estimated on a sample (of size n-1) from which the ith observation $[y_i, \underline{x}_i']$ has been deleted. $r(i)$ is the ith deleted residual. It can be shown that PRESS can be constructed from the two key regression diagnostics, the residuals, r_i, and the hat matrix diagonals, h_i, since $r(i) = r_i/(1-h_i)$. That is,

$$\text{PRESS} = \sum_{i=1}^{n} (y_i - \hat{y}_i)/(1-h_i)^2 \qquad (40)$$

In this form PRESS is just a weighted sum of squares of the <u>ordinary</u> residuals, r_i. The residuals that correspond to influential points (large h_i) are weighted the more heavily. In using PRESS to discriminate between alternative models, that model is chosen which has the smallest value of PRESS, since PRESS is a measure of the degree to which a model will "hold its predictive power" in new data. It should be kept in mind that using PRESS as a criterion for model selection will result in a preference for models that fit relatively well at extreme values of X - the sample "edges" (See Cook and Weisberg 1982).

In Normal theory models, it has been fruitful to define a multiple correlation coefficient from PRESS: R_p^2 = 1-PRESS/SSY, where SSY is the total sum of squares of the response variable in the sample. R_p^2 gives another measure of the "shrinkage" of the multiple correlation coefficient referred to above. R_p^2 is a measure that is similar in intent (and size) to the adjusted multiple correlation coefficient, $\bar{R}^2$ (See Montgomery and Peck 1982).

For the models of Binomial response we use the Cook-Weisberg (1982) specification of the PRESS statistic:

$$\text{PRESS} = \sum_{i=1}^{n} t_i^2/(1-h_i) \qquad (41)$$

PRESS is a weighted sum of squares of the standardized residuals - the components of χ^2 - in which the respective weights are now $1/(1-h_i)$. Thus, those observations for which h_i is large - extreme levels of treatment variables - are weighted the more heavily. Models for which PRESS is small should be more stable than those for which it is large, since the respective estimates, $\hat{\beta}_j$, are less affected by the idiosyncracies of the sample - the model-sample interactions are weaker. Thus, in using PRESS to discriminate between alternative models, that model is best for which PRESS is least. We note that the elements of PRESS, $t_i^2/(1-h_i)$, are just the changes in RSS produced by deletion of the ith observation (Cook & Weisberg 1982).

Table 7 presents several model selection criteria for the present models (M_1-M_3) and samples (n=44, 45). The choice depends on how the model will be used; in particular whether it is to be used in interpolation or extrapolation. "Because further research is obviously needed on the properties of the various measures, it is recommended that several of the measures should always be calculated when comparing the different models"(Seber 1977).

Table 7 presents the RSS and PRESS statistics for each of the three models of the response for both the full (n=45) and reduced (n=44) samples, together with the respective upper tail areas of the sampling distribution of the cognate chi-squared statistics. It can be seen that on the basis of the (approximate) sampling distribution of RSS, all three models are consistent with the data. For n=44 the difference between the respective RSS for models, M_1 and M_3, is significant, but that between M_1 and M_2 is not significant. This evidence suggests that M_2 is the "correct" model of the observations in the reduced sample. Examination of the distribution of the respective PRESS statistics discloses that model M_3 will be found to be inconsistent with "new data" and that model M_2 "holds its predictive power" better than either M_1 or M_3. The Table shows that deletion of the outlier - case #1 - improves the stability of the "predictive power" of model M_1. The estimates, $\hat{\beta}$, obtained from reduced sample should give better predictions in new data.

The previous selection of the model, $z = \beta_0 + \beta_1 x_1 + \beta_2 x_2$, over the alternative model, $z = \beta_0 + \beta_1 x_1 + \beta_2 x_2 + \beta_3 x_3$ was made in part, on the basis of the Neyman-Pearson hypothesis testing theory (See

Table 6.
Measures of Association between Observed and Predicted Response. Sq. Ca. Ablation. Probit. n = 44.

$$z = -60.318 + 25.027x_1 - 30.057x_2 + 9.747x_3$$

		Predicted Status		
		($z \leq 0$)	($z > 0$)	
		0	1	Totals
True	0	5	8	13
Status	1	2	29	31
	Totals	7	37	44

a) Correct Classification Rate: 0.77
b) False Positive Rate: 0.62
c) False Negative Rate: 0.07
d) Average Misclassification Rate: 0.36
e) Correlation, Observed & Predicted: $r(y_i, \Phi(\hat{z}_i)) = r = \underline{0.399}$ ($r^2 = 0.159$).

Table 7. Model Identification. MSE of Prediction and Concordance Criteria. AIC[a] (Information Theory) and PRESS[b] (Cross-Validation). RSS[c] (Concordance).

a) $z = \beta_0 + \beta_1 x_1 + \beta_2 x_2 + \beta_3 x_3 + u$. Sq. Ca. Ablation
   i) n=45: AIC = $\underline{54.122}$. PRESS = $\underline{58.774}$. RSS = $\underline{52.879}$ (P = 0.101)
   ii) n=44: AIC = $\underline{47.360}$. PRESS = $\underline{41.348}$. RSS = $\underline{36.114}$ (P = 0.646)

b) $z = \beta_0 + \beta_1 x_1 + \beta_2 x_2 + u$. (n = 44)
   AIC = $\underline{45.972}$. PRESS = $\underline{41.223}$. RSS = $\underline{38.432}$ (P = 0.585)
   (minimum)   (minimum)

c) $z = \beta_0 + \beta_1 x_1 + \beta_3 x_3 + u$. (n = 44)
   AIC = $\underline{49.612}$. PRESS = $\underline{52.932}$. RSS = 49.116 (P = 0.178)

---

[a] AIC = $-2\ln L + 2p$

[b] PRESS = $\sum_{i=1}^{n} t_i^2/(1-h_i) = \sum_{i=1}^{n} t_{(i)}^2$

[c] RSS = $\sum_{i=1}^{n} t_i^2$ $P = P(\chi^2 > RSS| df)$

Table 8. Sample and Non-Sample Information on $\underline{\beta}$ for Model M_1: $z = \beta_0 + \beta_1 x_1 + \beta_2 x_2 + \beta_3 x_3$.

a) Sample (n = 44).
   $\hat{\beta}_2/\hat{\beta}_1 = -1.201$, $\hat{\beta}_3/\hat{\beta}_1 = 0.389$, $\hat{\beta}_1 = \underline{25.027}$, $\hat{\beta}_0 = 60.218$
   $\hat{\beta}_1/\sqrt{Var(\hat{\beta}_1)} = 2.017$ ($\beta_1 > 0$), $\hat{\beta}_2/\sqrt{Var(\hat{\beta}_2)} = 1.879$ ($\beta_2 = 0$?).
   $\hat{\beta}_3/\sqrt{Var(\hat{\beta}_3)} = 0.737$ ($\beta_3 = 0$?).
b) Non-Sample (a priori)
   $\beta_j \neq 0$; j = 0, 1, 2, 3.
   $-1 < \beta_j/\beta_1 < 0$, j=2, 3. $\beta_1 > 0$, $\beta_0 < 0$.

Finney 1971). By way of review it will be recalled that in this theory, the null hypothesis, H_0, is that the model in question is "true" and the criterion for accepting or rejecting the hypothesis on the basis of the evidence of the sample is the sampling distribution of the chi-square statistic for (n-k) degrees of freedom. It is implicit in such use of this theory that the RSS of the model of this sample has a sampling distribution described by chi-squared. This is the standard variable selection practice in the classical methods of probit analysis, for example. However, it has been pointed out by Akaike (1974, 1977), Atkinson (1980) and Schwarz (1978) that the practical utility of the hypothesis testing procedure as a method of model building, or identification, on the basis of finite samples is quite limited. Akaike has proposed an alternative procedure with a different and more pragmatic motivation: control of the error - or loss - incurred by the subsequent use of the identified model to forecast response. Moreover, Akaike has questioned whether, given the limited data that is usually available, it even makes sense to speak of the "true model" of the process that generated the observations (See also Leamer 1982). It will be recalled that in the hypothesis tests we have described above, RSS is distributed as chi-squared with (n-k) degrees of freedom on the null hypothesis that the model, say M_1, is "correct". For example, H_0: M_1 "correct". The alternative procedure is based on Akaike's Information Criterion (AIC) which is developed from Boltzman"s concept of entropy and provides an explicit expression for the principle of parsimony in data analysis. It is a method for selection of the optimal dimension of a linear model. Note that it does not address the question of optimal functional form, e.g., linear vs non-linear models.

"Akaike's criterion stemmed from a recognition that unreserved maximization of likelihood provides an unsatisfactory method of choice between models that differ appreciably in their parametric dimensionality" (Stone 1976).

For the ith model, $AIC_i = -2\ln L_i + 2p_i$. L_i is the maximum likelihood function for the ith model and p_i is the number of adjustable parameters. p_i is a correction term without which we would be maximizing L_i, "... models with parameters of high dimensionality are given a severe handicap by this correction term" (Stone 1976). $-2\ln L_i$ is distributed (asymptotically) as chi-squared with (n-k) degrees of freedom. The first term on the RHS of the above equation specifies the penalty assessed for a poor fit of the ith model, M_i, to the data. The second term exacts a penalty for the unreliability of M_i. (See previous remarks by Mostellar and Tukey.) Also, Akaike (1977) has pointed out that when the concordance of the ML estimates of β with the sample are measured by the Pearson chi-squared statistic, χ^2, that $AIC(p) = \chi^2 - 2(n-k) = RSS - 2(n-k)$. That is, the AIC is a joint measure of the concordance, RSS, of the sample [y, X] and the model M_i and a factor which describes the relation between the size of the sample n and the dimension of the model k.

The model which gives the minimum of the AIC represents the best compromise and hence the most parsimonious model. As we have noted above, "parsimony", in a model is held to be a desideratum, in part because of philosophic concerns (Occam's Razor) and in part because of such pragmatic considerations as the inflation of $Var(\hat{\beta})$ and $Var(\hat{z}_i)$ as the model dimension, p, increases (partly because of the increased opportunity for redundancy and hence multicollinearity). The AIC procedure has been shown to be asymptotically equivalent to a cross-validation procedure (Stone 1976). We have previously noted that the latter procedure can be simulated by the use of the PRESS statistic in the case of small samples (n < 25 + 2k for Normal theory models). (The AIC procedure is also equivalent to the procedure for a generalized C_p statistic (See Myers, Lectures 1 and 2) or the generalized information criterion of Atkinson (1980). It will be recalled that for Normal theory regression models $C_p = RSS_p / \hat{\sigma}^2 - n + 2p$. Akaike has pointed out (correctly) that the AIC criterion is usually less subjective than C_p, in that the estimate, $\hat{\sigma}^2$, required for C_p is usually subjective.)

The AIC_i for each of the alternative models is presented in Table 7. It is evident that this criterion selects the same model as does the classical procedure, as well as that based on the PRESS statistic, namely, M_2, or $z = \beta_0 + \beta_1 x_1 + \beta_2 x_2$, for which $\beta_3 \equiv 0$. But, this model is not consistent with the prior information on β given in Table 1. There is a strong a priori belief in M_1 for which $\beta_3 \neq 0$. However, since the variance inflation factor, VIF for the full model, M_3, discloses that a high degree of collinearity is present in the joint sample distribution of X_1, X_2 and X_3, since N and T are highly correlated (see Figure 2c). X_3 is largely redundant in this sample and its' omission, as in the model, M_2, would not be expected to inflate the residual sum of squares, RSS, by a significant amount. (See Table 3d).

As remarked above, in classical Sampling Theory inferences on the data are cast in the form of the Neyman-Pearson Test of the null hypothesis, H_0. In the present case of model selection, the problem is to discriminate on the evidence of the sample and with specified probabilities of Type I and Type II errors, between the null hypothesis, H_0: M_1 and [y, X] are consistent, with the alternative hypothesis, H_1: M_1 and [y, X] are inconsistent. Let us consider, instead of the probabilities of errors in taking a decision on a null hypothesis, H_0, the concept of the probability of the hypothesis itself, say $P(H_0)$ (See Leamer 1978). Such a concept does not arise in Sampling Theory because of its requirement for a repetitive element in the definition of probability. However, the concept arises quite naturally in Bayesian Theory. Thus, in the present case we are led to consider the probability of a model, say $P(M_i)$. Or, more precisely, the concept of the posterior probability, say $P(M_i|y)$, of a model, M_i, given the set of observations, [y, X]. This is a useful way of comparing rival models of a given sample. Moreover, it provides a fruitful perspective on one of the applications of a priori information, namely, informing the selection of the form of the relation between treatment and response. Later, we shall also consider the application of prior information in fixing the relation between two or more of the elements of the parameter, β, in a model of specified form. Assume however, that the immediate concern is to discriminate between models M_1 and M_2, given the sample, [y, X]. It is assumed that M_1 and M_2 are mutually exclusive and exhaustive: $P(M_1) + P(M_2) = 1.0$. Assume that the losses incurred for making either incorrect choice are equal. Then the Bayesian selection rule is to choose that model, M_i, which has the highest posterior probability, $P(M_i|y) = (P(y|M_i)P(M_i)/P(y)$ where $P(y) = P(y \& M_1) + P(y \& M_2)$ and $P(M_i)$ is the a priori probability of M_i: $0 < P(M_i) \leq 1$. The estimates of $P(M_i)$ are obtained from prior experience - either experimental or non-experimental observations - or from relevant theory, or from introspection. $P(y|M_i)$ is the average likelihood of y given M_i. Note that when the form and parameters of the model M_i are derived from experiments on target populations other than patients, $P(M_i)$ includes the strength of the (prior) belief about the validity of the interspecies extrapolation - the so-called "mouse to man" extrapolation of the model, M_i.

It is instructive to cast such a binary decision rule in the form of a posterior odds ratio, $P(M_2|\underline{y})/P(M_1|\underline{y})$: $P(M_2|\underline{y})/P(M_1|\underline{y}) = BP(M_2)/P(M_1)$ where $P(M_2)/P(M_1)$ is the a priori odds ratio and B is Leamer's Bayes Factor: $B = P(\underline{y}|M_2)/P(\underline{y}|M_1)$ (See Leamer 1978). The evidence of the sample, $\underline{y}$, is said to favor the model, M_2, if $B > 1$. This form of the rule forces a response to the most pertinent question: How strong is the evidence of the sample for the model, M_i, relative to the prior odds?

Just as in the case of the AIC, the sample evidence for the model can be described by a joint measure of the concordance, of the data $[\underline{y}, X]$ and model, M_i, the residual sum of squares, RSS_i, and a factor which describes the relation of the dimension of the model, p_i, and the size of the sample n. Thus, we have

$$B = (RSS_1/RSS_2)^{n/2} n^{(p_1-p_2)/2} \qquad (42)$$

As a rather trivial example consider the models, M_1 and M_2, for which we have

$$B = (47.360/45.972)^{44/2} \, 44^{1/2} = 12.76 \qquad (43)$$

Thus, the prior odds for M_1, in which the level of response does depend on the number of fractions, N, (as well as dose, D and time, T) must exceed 13:1 in order to dominate the sample evidence for M_2 in which it does not. The apriori evidence for M_1 is dominant.

It is of interest to note that the difference in the respective values of AIC_i for two models, M_i, i=1, 2 has a form rather similar to the Bayes Factor, B:

$$\Delta AIC = \ln[(L_2/L_1)^2 e^{(p_1-p_2)}] \qquad (44)$$

12. EXTRAPOLATION

"Interpolations tend to be accurate and reliable, while extrapolations are less accurate and unreliable" (Weisberg 1980).

"Never make predictions, especially about the future" (Casey Stengel).

"In achieving optimum fit to the estimation data, least squares often destroys good prediction of new data (possibly outside the region covered by the estimation data)" (Marquardt 1980).

The use of response surface models to estimate response, $z = \underline{x}'\hat{\underline{\beta}}$, at treatment regimens, $\underline{x}'$, not included within the initial set of observations from which the estimates $\hat{\underline{\beta}}$ and $Var(\hat{\underline{\beta}})$ are obtained but still within the space covered by these observations has been discussed previously. In this use, it is assumed that the level of treatment, $\underline{x}'$, is drawn from the same distribution - or that the distribution is stable over time - to interpolate response within the ellipsoid defined by $h_{(n)}$.

Another use is to forecast response at treatment regimens which lie outside the region defined by the RVH, i.e., at regimens drawn from a quite different distribution. For instance, the X matrix of the initial sample can be considered to be representative of those drawn from standard fractionation regimens. Now, let us assume that an estimate of the response, $\hat{z}$, is required at the regimen (D, T, N) = (9000, 63, 20), a <u>hypofractionation</u> regimen at the boundary of the region of interest R_2. (See Figure 1). This regimen is at the position marked by "+" in Figures 5a, 5b and 5c. At this regimen the estimated response for the full model, $M_1(\underline{X}_1, \underline{X}_2, \underline{X}_3)$, is $\hat{z_i} = -2.76(\hat{\Pi}_i = 0.003)$. For the reduced model, $M_2(\underline{X}_1, \underline{X}_2)$, $\hat{z_i} = 0.56$ ($\hat{\Pi}_i = 0.71$). For the reduced model, $M_3(\underline{X}_1, \underline{X}_3)$, $\hat{z_i} = 4.88(\hat{\Pi}_i = 1.00)$. The responses estimated by the models, M_1 and M_3, are each <u>implausible</u> – <u>inconsistent</u> with a priori information ("clinical experience") on the level of $\hat{\Pi}$, since "one intuitively feels" that the response at the hypofractionation regimen must exceed zero but be less than one – as well as mutually inconsistent. Moreover, M_3 is also inconsistent with the a priori information on $\underline{\beta}$ namely, $\beta_2 \ne 0$; M_1 is similarly inconsistent: $\hat{\beta}_2/\hat{\beta}_1 < -1$, $\hat{\beta}_3/\hat{\beta}_1 > 0$. See Table 8. And, although the level of response, P, estimated by M_2 is "plausible" – consistent with non-sample information – the model is <u>in</u>consistent with the (other) non-sample information on $\underline{\beta}$ since $\hat{\beta}_3 = 0$ for M_2. See also Table 1 and 3. Thus, although model M_1, provides the best fit (of the three rivals) to the sample observations and seems to interpolate well enough, (See Figure 5a), the estimates of β_2 and β_3 and, of course, of the response, $\Pi(\underline{x_i}'\underline{\beta})$, in regions of X-space <u>beyond</u> that "covered" by the sample RVH are inconsistent with the relevant non-sample information.

We have noted earlier that several of the inconsistencies between the sample and prior information on $\underline{\beta}$ for M_1 that are disclosed in Tables 1, 2 and 4 – the sign, size and significance of $\hat{\beta}_2$ and $\hat{\beta}_3$ – are consistent with the effects of collinearity, i.e., the presence of a large correlation coefficient for N and T – or $\underline{X}_2$ and $\underline{X}_3$. See Table 4. It will be instructive to express the sample estimates of the parameter vector of a regression model in terms of the sample estimates of the location, scale and shape parameters of the joint distribution of the response and treatment variables. In this way the characteristic effects of the presence of collinearity in the distribution of the treatment variables on the sample estimates of the parameter vector of the regression model will be disclosed. Those effects are most vividly displayed in a Normal theory model. Therefore, by way of example, we examine the estimated <u>management equation</u>, vide supra, for these data which, in terms of x_2 and x_3, may be written as $x_1 = \hat{\gamma}_0 + \hat{\gamma}_1 x_2 + \hat{\gamma}_2 x_3$, or, $x_1 = 3.271 + 0.459x_2 - 0.132x_3$. Here x_1 is the response variable and x_2 and x_3 are treatment variables. Note that if x_3 is <u>deleted</u> we have $x_1 = 3.284 + 0.330x_2$ or, $D = 1923T^{0.33}$. Note also that $0.459 - 0.132 = 0.327$. We may represent the sample estimates, $\hat{\gamma}_j$, $0 \le j \le 2$, in terms of the location, scale and shape parameters of the joint distribution of x_1, x_2 and x_3 in the following way:

$$\hat{\gamma}_0 = \bar{x}_1 - \hat{\gamma}_1\bar{x}_2 - \hat{\gamma}_2\bar{x}_2$$
$$\hat{\gamma}_1 = [(r_{12} - r_{13}r_{23})/(1-r_{23}^2)](s_1/s_2)$$
$$\text{Var}(\hat{\gamma}_1) = [(1-R^2)/(1-r_{23}^2)(n-3)] \ (s_1^2/s_2^2)$$
$$\hat{\gamma}_2 = [(r_{13} - r_{12}r_{23})/(1-r_{23}^2)] \ (s_1/s_3)$$
$$\text{Var}(\hat{\gamma}_2) = [(1-R^2)/(1-r_{23}^2)(n-3)] \ (s_1^2/s_3^2)$$
$$R^2 = (r_{12}^2 + r_{13}^2 - 2r_{12}r_{13}r_{23})/(1-r_{23}^2)$$

$\bar{x}_j$, s_j, $1 \le j \le 3$ are the mean and standard deviation, r_{jk}, $1 \le j, k \le 3$, $j \ne k$, is the Pearson product-moment correlation coefficient and R^2 is the coefficient of determination. See Table 4. Several effects of collinearity ($r_{23} \longrightarrow 1.0$) in the treatment variables, are at once apparent: As $r_{23} \rightarrow 1.0$, 1) the separate effects of x_2 and x_3 on the response, x_1, can no longer be described by the estimates $\hat{\gamma}_1$ and $\hat{\gamma}_2$, respectively; 2) the variance of each estimates, $Var(\hat{\gamma}_i)$ is inflated; 3) if $r_{12} > r_{13} > 0$, then $\hat{\gamma}_1 > 0$ and $\hat{\gamma}_2 < 0$; 4) R^2 is inflated.

We have now completed the first step in dealing with the ill effects of collinearity, namely, identification and diagnosis of the presence, nature and degree of collinearity in the X matrix.

"... the second step, correction, requires the generation of additional information. Just how this information is to be obtained depends largely on the tastes of an investigator and on the specifics of a particular problem. It may involve additional primary data collection, the use of extraneous parameter estimates from secondary data sources, or the application of subjective information through constrained regression, or through Bayesian estimation procedures" (Farrar and Glauber, 1967).

However, we defer a discussion of the fruitful use of additional primary data information to reduce collinearity to Lecture 14 in which the collinearity of D_γ and D_n in the joint distribution at each of the two cities, Hiroshima and Nagasaki, is reduced by pooling the observations in the two samples. (Recall that we have pooled the observations on stages T_2 and T_3 (primary data) in the present paper in order to reduce the collinearity in X as well as the number and "strength" of extreme levels of treatment, x_i. The maneuver was rather effective in reducing the correlation between X_1 and X_2 but not that between X_2 and X_3.)

"The most obvious and, at the same time, least helpful approach to multicollinearity is to collect more data ..." "Multicollinearity, however, arises primarily in studies in which the investigators can exercise little or no control over the structure or quality of the predictor [treatment] variables ..." The naive advice simply to gather more data, then, often is impossible to follow, and when possible it invariably is expensive. A related approach, however, is to utilize statistical information about one or more parameters in the model derived from another source of data" (Thisted 1980). It is especially the case with clinical studies, that the advice to gather more data may be impossible to follow because the observations having the desired relation to the extant data may require unethical treatments of patients ("First do no harm ...")?

One common alternative to data augmentation (adding rows to [y, X]) for reducing the effects of collinearity on the sample estimates of β is simply to delete one member of a pair of correlated variables (deleting columns of [y, X]). It will be useful in the sequel to formulate this procedure as a regression under a linear, nonstochastic, constraint, $r = R\beta$, where the deletion of, say, x_3 corresponds to the constraint, $\beta_3 = 0$. Then we have $r = 0$, $R = [0, 0, 0, 1]$. The linear form $r = R\beta$, implements the a priori information on the conditional response which is represented by the hypothesis, H_0: $z = \beta_0 + \beta_1 x_1 + \beta_2 x_2$, or alternatively, H_0: $\beta_3 = 0$. The test statistic, $(r - R\hat{\beta})'[R(X'\hat{V}^{-1}X)^{-1}R']^{-1}(r-R\hat{\beta})$ is distributed as chi-squared with 1 degree of freedom.

However, as Gunst has cautioned (vide supra) excluding a relevant variable may lead to biased estimates of β (an error of specification). This will, of course, also lead to biased estimates, $\hat{z}$, of conditional response as well as inflated values of RSS and $Var(\hat{z})$. It will be useful to demonstrate both the nature and the degree of such bias in the case of the alternative linear models, $z = \beta_0 + \beta_1 x_1 + \beta_2 x_2 + \beta_3 x_3$ and $z = \beta_0 + \beta_1 x_1 + \beta_2 x_2$ described in Table 3. These may be rewritten as $z = X_1 \underline{\beta_1} + X_2 \underline{\beta_2}$ and $X_1 \underline{\beta_1}^*$, respectively, where X_1 and X_2 are (nx3) and (nx1) matrices of treatment variables. $\underline{\beta_1}$ is a (3x1) matrix of unknown parameters, and $\underline{\beta_2}$ is a (1x1) matrix of unknown parameters. We then have, to a close approximation, the estimates

$$\begin{pmatrix} \hat{\beta}_0^* \\ \hat{\beta}_1^* \\ \hat{\beta}_2^* \end{pmatrix} = \begin{pmatrix} \hat{\beta}_0 \\ \hat{\beta}_1 \\ \hat{\beta}_2 \end{pmatrix} + \beta_3 \begin{pmatrix} A_0 \\ A_1 \\ A_2 \end{pmatrix} \tag{45}$$

where A is the <u>alias matrix</u>, $A = (X_1' \hat{V}_1^{-1} X_1)^{-1} X_1' \hat{V}^{-1} X_2$, or

$$\begin{pmatrix} -57.509 \\ 24.000 \\ -20.482 \end{pmatrix} = \begin{pmatrix} -60.318 \\ 25.027 \\ -30.057 \end{pmatrix} + 9.747 \begin{pmatrix} 0.288 \\ -0.105 \\ 0.982 \end{pmatrix} \tag{46}$$

Note that the estimated coefficient of x_2 in M_2 the <u>reduced</u> model, $z = \hat{\beta}_0 + \hat{\beta}_1 x_1 + \hat{\beta}_2 x_2$, is approximately, the <u>sum</u> of the coefficients of x_2 and x_3 in M_1, the <u>full</u> model, $z = \hat{\beta}_0 + \hat{\beta}_1 x_1 + \hat{\beta}_2 x_2 + \hat{\beta}_3 x_3$. Note as well that deletion of the <u>column</u> X_3 produces only small changes in the estimates, $\hat{\beta}_0$ and $\hat{\beta}_1$, but a large change in $\hat{\beta}_2$ since $\underline{X}_2$ and $\underline{X}_3$ are highly correlated. (See "Alias Matrix" in Draper and Smith 1966.)

It is evident that unless the coefficients corresponding to the deleted variables, X_2, are zero, <u>or</u> the variables, X_1, included in the reduced model are orthogonal to the omitted variables, X_2, the estimates of the coefficients of the subset model are <u>biased</u>. Hence, the selection of a model by deletion of variables is <u>not</u> a maneuver to be practiced in the presence of a high degree of collinearity in the sample observations (Marquardt and Snee 1975). This concern will be important in Lecture 17 in which rival models of <u>experimental data</u> are compared.

As remedies for the multicollinearity problem neither of the maneuvers available from classical least squares methodology are really satisfactory: <u>Data augmentation</u> (addition of <u>rows</u> of [y, X]) is usually infeasible and where feasible may introduce another problem, namely, the presence of subsets (of one or more observations) which are not well explained by the model or which dominate some aspect of the fit. An instance of this is evident in Figures 3a and 3d. Augmentation of stage T_2 data with that of stage T_3 introduces an <u>extreme</u> observation, #34: $h_{34} = 0.35$ (Figure 3c). <u>Variable deletion</u> (omission of <u>columns</u> of [y, X]) always discards sample information on conditional response, may frequently introduce considerable <u>bias</u> into the estimates of $\underline{\beta}$ and may conflict with strong a priori beliefs on $\underline{\beta}$. Thus, we are led to consider what may be called non-least squares methodology (see Weissberg 1980), which makes explicit use of non-sample, or a priori, as well as sample, information on $\underline{\beta}$. (The former has been referred to by Leamer (Lecture 5) as "fictious data".) All of the methods we now consider provide estimates of $\underline{\beta}$ that are matrix-weighted averages of <u>sample</u> and <u>non-sample</u> information on $\underline{\beta}$ (See Leamer 1978, 1984 and Lecture 5).

13. SOME SEMI-BAYESIAN METHODS

Therefore, let us next try to "... break the multicollinearity deadlock," through the introduction of non-sample information on $\underline{\beta}$ by a semi-Bayesian method which does not require the specification of the distribution function of the prior information on β.

Let us assume that the a priori, or non-sample, information on β obtained from the secondary source can be represented by the stochastic constraint $\underline{r} = R\beta + \underline{v}$, where $\underline{r}$ and $\underline{v}$ are random variables: $E(\underline{v}) = 0$ and $Var(\underline{v}) = \Psi$. For instance, in the present case it may be that the prior information consists in the (non-sample) estimates, β_3' > 0, $Var(\beta_3') > 0$. Then $r = \beta_3'$, $R = [0, 0, 0, 1]$ and $\Psi = Var(\beta_3')$. Then the sample and a priori information on β can be combined (in an obvious manner) in a matrix-weighted average to give the posterior estimates, $\hat{\beta}^{**}$, $Var(\hat{\beta}^{**})$, where for the generalized linear model

$$\hat{\beta}^{**} = [X'\hat{V}^{-1}X + R'\Psi R]^{-1}(X'\hat{V}^{-1}X\hat{\beta} + R'\Psi r)$$
$$Var(\hat{\beta}^{**}) = [X'\hat{V}^{-1}X + R'\Psi^{-1}R]^{-1}. \tag{47}$$

The compatability of the prior and sample information is assessed on the evidence of the statistic, γ, distributed as chi-squared with q degrees of freedom on the null hypothesis that the two sets of information are consistent: $\gamma = (\underline{r}-R\hat{\beta})$ $[R(X'\hat{V}^{-1}X)R' + \hat{\Psi}]^{-1}$ $(\underline{r}-R\hat{\beta})$. Moreover, the Mixed Estimation methods provide an estimate of the proportion, Θ_p, of a priori information in the posterior estimates, where $\Theta_p = k^{-1}Trace\{ R'\Psi^{-1}R[X'\hat{V}^{-1} X + R'\Psi^{-1}R]^{-1}\}$. (k = # columns of X).

Note that Θ_p is in the nature of a regression diagnostic, rather similar to D_i^1: Θ_p measures the degree of "influence" of the a priori information $(\underline{r} = R\beta, \Psi)$ on the estimate, $\hat{\beta}^{**}$, in the same way that D_i^1 describes the joint effects of t_i and h_i for the ith observation of the sample on the estimate, $\hat{\beta}$ (See Welsch quotation in Introduction).

It will be useful in the sequel to recall that the estimates obtained under a stochastic constraint, $\Psi \neq [0]$, approach those obtained under a non-stochastic constraint with the same matrices $\underline{r}$ and R as $\Psi \longrightarrow [0]$ where [0] is the null matrix. In general, it is sufficient to take $\Psi = 1*10^{-10}I$, where I is the (qxq) identity matrix.

By way of introduction we present some Mixed Estimates for the full model of the reduced sample in Table 9. In Table 9bi, the constraint, $\underline{r} = R\beta + v$, is non-stochastic; $\Psi \equiv [0]$, where [0] is a null matrix. In Table 9bii the constraint is stochastic: $\Psi \neq [0]$. Note that in Table 9bi the Mixed Estimates with $\Psi = 1*10^{-10}I$ are equivalent to deletion of X_3 or the Constrained Estimates with $r = \beta_3 = 0$. Table 9bii-bv presents a sensitivity analysis with $1 \leq \Psi \leq 49$. See Lecture 5. We obtain the a priori estimate $r = \beta_3 = -10$ from model M_3 of the reduced sample in Table 3. We have arbitrarily "shrunk" the sample estimate, $\hat{\beta}_3 = -12.841$, to compensate for the collinearity of X_1 and X_3 (See Figure 5c and Table 4). For Tables 9bii to 9bv, $\Psi = Var(\beta_3)$ and the precision, $\beta_3/\sqrt{Var(\beta_3)}$, of the a priori information is -1.429, -2.000, -3.333, -10.0, respectively. Note that the proportion, Θ_p varies weakly with Ψ^{-1} and corresponds to the determination of about 1 in 4 of the elements of $\hat{\beta}^{**}$ by the non-sample information. Note that the a priori information is compatible with the sample information for all Ψ. Note that RSS^{**} varies directly with Ψ^{-1} as well for a given set of constraints $\underline{r} = R\beta + v$. Note that $\hat{\beta}_3^{**}$ varies by a factor of ~ 2 as Ψ varies by a factor of ~ 50. We shall next consider more complicated constraints.

It is not infrequently the case that the non-sample information on β consists in a relation between two (or more) coefficients, such as a difference, $\beta_j - \beta_k$, or ratio, β_j/β_k, rather than a value, β_j, of

Table 9. Some Mixed Estimates of $\underline{\beta}$ for Full Model of Reduced Sample.

a) Sample Estimates. $\hat{\underline{\beta}}$ and Var($\hat{\underline{\beta}}$) (n = 44). See Table 3a.

$\hat{\underline{\beta}}$ = -60.318, 25.027, -30,057, 9.747
    (-1.661) (2.017) (-1.879)(0.737)

$$\text{Var}(\hat{\underline{\beta}}) = \begin{matrix} 1318.843, & -442.573, & 296.152, & -76.966 \\ & 153.951, & -115.894, & 30.185 \\ & & 255.815, & -182.910 \\ & & & 174.908 \end{matrix}$$

RSS = 36.115

b) A Priori Estimates of β_3: $\underline{r} = R\underline{\beta} + \underline{v}$. $E(\underline{v}) = 0$. $\text{Var}(\underline{v}) = \Psi$.

  i) r = 0. R = (0, 0, 0, 1). $\Psi = 1*10^{-10}$. (Constrained Estimation)
     $\hat{\underline{\beta}}** = -56.029, 23.345, -19.865, 5.572*10^{-12}$
         (-1.563) (1.914) (-2.473)(5.572*10^{-7})
     $\gamma = 0.543$. $P(\chi^2 > \gamma | 1) = 0.461$
     $\theta_p = 0.25$. RSS** = 36.658

Compare with Table 3b (Model M_2)

From Table 3c, Model M_3, $r \simeq \hat{\beta}_3 = -12.841$.

  ii) r = -10.0. R = (0, 0, 0, 1). $\Psi = 49.0$.
      $\hat{\underline{\beta}}** = -53.531, 22.365, -13.926, -5.679$
          (-1.489) (1.827) (-1.350)(-0.918)
      $\gamma = 1.741$. $P(\chi^2 > \gamma | 1) = 0.187$
      $\theta_p = 0.20$. RSS** = 37.856

  iii) r = -10.0. R = (0, 0, 0, 1). $\Psi = 25.0$.
       $\hat{\underline{\beta}}** = -52.716, 22.045, -11.990, -7.531$
           (-1.468) (1.804) (-1.275)(-1.610)
       $\gamma = 1.951$. $P(\chi^2 > \gamma | 1) = 0.162$
       $\theta_p = 0.22$. RSS** = 38.066

  iv) r = -10.0. R = (0, 0, 0, 1). $\Psi = 9.0$
      $\hat{\underline{\beta}}** = -52.054, 21.786, -10.418, -9.034$
          (-1.451) (1.785) (-1.212)(-3.088)
      $\gamma = 2.120$. $P(\chi^2 > \gamma | 1) = 0.145$
      $\theta_p = 0.24$. RSS** = 38.235

  v) r = -10.0. R = (0, 0, 0, 1), $\Psi = 1.0$
     $\hat{\underline{\beta}}** = -51.678, 21.638, -9.525, -9.888$.
         (-1.442) (1.774)(-1.176)(-9.916)
     $\gamma = 2.217$. $P(\chi^2 > \gamma | 1) = 0.140$
     $\theta_p = 0.249$. RSS** = 38.332

one or more of the coefficients. Moreover, such information may be too vague to be represented by an equality constraint, e.g., only the <u>sign</u> of the slope, say, $\beta_1 > 0$, is known. Table <u>1</u> describes just such a case in which the a priori information on the ratios, β_2/β_1 and β_3/β_1, can only be represented as inequalities. However, we first consider two cases in which there is a priori information on <u>two</u> of the elements of $\underline{\beta}$ which can be represented by constraints of the form $\underline{r}$ = $R\underline{\beta} + \underline{v}$. For one case we have $\Psi = [0]$. For the second, $\Psi \neq [0]$.

In the two cases we now examine there is a priori information on $\underline{\beta}$ from two different <u>secondary</u> data sources. One of these was provided by Ellis (1971) and the other by Supe, Nagalaxmi and Meenaksi (1983). For each of these sources the prior information on $\underline{\beta}$ can be summarized by the ratios, $r_1 = \hat{\beta}_2/\hat{\beta}_1$ and $r_2 = \hat{\beta}_3/\hat{\beta}_1$. The Ellis relations, which specify the exponents in the Tumor Standard Dose, TSD, are represented by a non-stochastic constraint. We have assumed that the Ellis exponents of T and N, 0 and 0.24, respectively, are simply <u>stipulated</u> values, distilled from clinical experience – and hence are not burdened with any ambiguity of estimate: $\Psi = [0]$. Supe, Nagalaxmi and Meenaksi(1983) have apparently made the same assumption. In his 1971 paper Ellis states that the regression analysis of his data gives a value of 0.22 for the exponent, but he presents no standard error: D = $2500T^{0.22}$. (However, our analyses of the same data gives, D = $2420T^{0.24}$, with 0.95 confidence limits on the exponent of (0.215, 0.266).) The Supe, et al, relations which specify the exponents in the Tumor Significant Dose, TSD, can be represented by a stochastic constraint, $\Psi \neq [0]$, since that paper presents ranges for the respective ratios as follows: $0.04 < r_1 < 0.08$ and $0.16 < r_2 < 0.20$. We have assumed that the respective ranges, 0.04, describe 2*1.96 = 3.92 standard deviations, hence $\Psi = 1.04*10^{-4}I$ where I is a (2x2) identity matrix.

More generally, if the a priori information on $\underline{\beta}$, $\underline{r} = R\underline{\beta} + \underline{v}$, is extracted from <u>sample</u> estimates, $\hat{\underline{\beta}}_0$ and $Var(\hat{\underline{\beta}}_0)$, obtained from some other set of data then one estimate of $Var(\underline{v})$ is $\Psi = RVar(\hat{\underline{\beta}}_0)R'$.

The Mixed Estimation analyses for the Ellis and Supe, et al, constraints are presented in Tables <u>10a</u> and <u>10b</u>, respectively. Both the Ellis TSD and the Supe, et al, TSD give values of γ which are significant at 0.05; thus, the information from neither of these secondary data sources is compatible with that from the primary sources – the sample – and hence, cannot be used to form posterior estimates, $\hat{\underline{\beta}}^{**}$, for which the effects of collinearity of N and T in the sample on the estimate of $\underline{\beta}$ have been reduced by the introduction of additional information on $\underline{\beta}$ (<u>augmenting</u> [$\underline{y}$, X] with [$\underline{r}$, R]).

We have introduced into our study the received information on $\underline{\beta}$ in the <u>same</u> form $\underline{r} = R\underline{\beta}$ but in two <u>different</u> contexts. One is <u>Sampling Theory</u> in which the sample and non-sample information are simply <u>compared</u> in a test of the null hypothesis, $H_0: \underline{r} = R\underline{\beta}$. The test statistic is $\tau = (\underline{r}-R\underline{\beta})'(RVar(\underline{\beta})R')(\underline{r}-R\underline{\beta})$. The second context is <u>Bayesian</u> in which the sample and non-sample information on $\underline{\beta}$ could be <u>combined</u> to give the Mixed Estimates, $\hat{\underline{\beta}}^{**}$. (Of course, these posterior estimates make little sense if the sample and non-sample information on $\underline{\beta}$ are <u>incompatible</u>.)

14. BIASED ESTIMATION. RIDGE REGRESSION

"It has been shown that when X'X is such that it has a nonuniform eigenvalue spectrum, the estimates of $\underline{\beta}$ in $\underline{y} = X\underline{\beta} + \underline{\varepsilon}$, based on the

Table 10. Non-Stochastic and Stochastic Prior Information on $\underline{\beta}$.

$\underline{r} = R\underline{\beta} + \underline{v}$. $E(\underline{v}) = 0$, $Var(\underline{v}) = \Psi$

a) Ellis Tumor Standard Dose (TSD)
   $TSD = DN^{-0.24}$
   $\underline{r} = \begin{pmatrix} 0 \\ 0 \end{pmatrix}$, $R = \begin{pmatrix} 0, & 0, & 1, & 0 \\ 0, & 0.24, & 0, & 1 \end{pmatrix}$, $\Psi = 1.0*10^{-10} I$ [a]
   Compatibility, $\gamma(2) = 6.82$. Reject H_0.[b]

b) Supe, et al. Tumor Significant Dose (TSD)
   $TSD = DN^{-0.18} T^{-0.06}$
   $\underline{r} = \begin{pmatrix} 0 \\ 0 \end{pmatrix}$, $R = \begin{pmatrix} 0, & 0.06, & 1, & 0 \\ 0, & 0.18, & 0, & 1 \end{pmatrix}$, $\Psi = 1.04*10^{-14} I$
   Compatibility, $\gamma(2) = 6.81$. Reject H_0.[b]

[a] $I = 2 \times 2$ identity matrix.
[b] H_0: Sample and Non-Sample Information on $\underline{\beta}$ are Compatible.

Table 11.

a) <u>Sample Information</u>. Sq. Ca. Ablation. (n=44)

   i) Probit Regression Model. $-\infty < z < \infty$.
      $z = -60.318 + 25.027 x_1 - 30.057 x_2 + 9.747 x_3$. RSS = 36.11
      $\hat{\beta}_2/\hat{\beta}_1 = -1.201$. $\hat{\beta}_3/\hat{\beta}_1 = 0.389$. $\hat{\beta}_0 < 0$. $\hat{\beta}_1 > 0$.
      $\hat{\beta}_1/\sqrt{Var(\hat{\beta}_1)} = 2.017$. $\hat{\beta}_2/\sqrt{Var(\hat{\beta}_2)} = 1.879$.
      $\hat{\beta}_3/\sqrt{Var(\hat{\beta}_3)} = 0.437$. $r^2 = \underline{0.159}$.

   ii) Binary Regression Model. $\underline{y} = 0, 1$. Unweighted. (Surrogate Model).
      $y = -10.216 + 4.544 x_1 - 6.776 x_2 + 3.093 x_3$. RSS = 7.36
      $\hat{\alpha}_2/\hat{\alpha}_1 = -1.391$. $\hat{\alpha}_3/\hat{\alpha}_1 = 0.681$. $\hat{\alpha}_0 < 0$. $\hat{\alpha}_1 > 0$.
      $\hat{\alpha}_1/\sqrt{Var(\hat{\alpha}_1)} = 1.815$. $\hat{\alpha}_2/\sqrt{Var(\hat{\alpha}_2)} = -1.652$
      $\hat{\alpha}_3/\sqrt{Var(\hat{\alpha}_3)} = 0.791$
      $R^2 = 0.197$. $\overline{R}^2 = 0.137$.

   iii) Correlation of <u>estimates</u> of response for <u>probit</u> and <u>surrogate</u> models:
      $r = (\hat{y}_i, \Phi(\hat{z}_i)) = 0.919$.

   iv) Binary Regression Model. $y = 0, 1$. Weighted
      $y = -8.514 + 3.904 x_1 - 7.662 x_2 + 4.567 x_3$. RSS = 33.28.
      $\hat{\alpha}_2/\hat{\alpha}_1 = 1.963$. $\hat{\alpha}_3/\hat{\alpha}_1 = 1.170$.
      $\hat{\alpha}_1/\sqrt{Var(\hat{\alpha}_1)} = 2.109$. $\hat{\alpha}_2/\sqrt{Var(\hat{\alpha}_2)} = -2.138$.
      $\hat{\alpha}_3/\sqrt{Var(\hat{\alpha}_3)} = 1.319$.
      $R^2 = 0.293$. $\overline{R}^2 = 0.240$.

b) <u>Apriori Information on β</u>.

   $\beta_0 < 0$, $\beta_1 > 0$, $\beta_j < 0$. $j = 2, 3$.
   $-1 < \beta_j/\beta_1 < 0$. $j = 2, 3$.

criterion of minimum residual sum of squares can have a <u>high probability</u> of being <u>far removed</u> from $\underline{\beta}$" (Hoerl and Kennard 1970). (italics added)

"Ridge regression is a form of Bayesian estimation. In the rare instance in which the a priori distribution of the regression parameters is explicitly known, the corresponding Bayesian formulation should be used. Estimators such as ridge and generalized inverse are the methods of choice when the a priori information is less precise"
(Marquardt 1980)

Let us now consider still another methodology for a post hoc, salvage operation which is appropriate for those studies in which the a priori information cannot be specified in terms of a single-valued expectation $\underline{r} = R\underline{\beta} + \underline{v}$, $E(\underline{v}) = 0$ and covariance matrix, $E(\underline{vv}') = \Psi \neq [0]$, but rather, only in terms of an inequality such as $\underline{r} < R\underline{\beta}$. See for instance, Table 1. For there are many important circumstances in which "... ones' beliefs are not always easily reduced to prior distributions and ... it is better to incorporate imprecise information approximately and thoughtfully than to insist rigidly or formulating a prior or on following the dictates of invariance." In such circumstances, "... an explicit semi-Bayesian analysis is in order" (Thisted 1980).

The method of Ridge regression (RR) has been widely used in recent years as an alternative to variable selection to obtain reduced variance estimates of $\underline{\beta}$ from samples in which there is a high degree of collinearity in the joint distribution of treatment variables. One method for reducing the variance inflation of collinearity described previously is simply to eliminate one member of the offending pair of collinear variables. But, "Large prediction biases can result from elimination of 'non-significant' predictors. It is better to use a little bit of all the variables than all of some variables and none of the remaining ones. This is what biased estimators do" (Marquardt and Snee 1975). Moreover, "... we felt that by keeping all the terms in the model and reducing the variable correlations with Ridge we were able to (i) Obtain a model which predicts well, and (ii) Learn more about the roles of all the variables in the model" (Ibid).

Darlington (1978) has put the case for Ridge regression in still another way. Least squares (including generalized least squares) methods give estimates, $\underline{\hat{\beta}}$, that maximize the correlation of the estimated, $\hat{y} = X\underline{\hat{\beta}}$, and observed, $\underline{y}$, responses in the initial sample, or, alternatively, $\underline{\hat{\beta}}$ minimizes the residual sum of squares, RSS = Σe_i^2. Ridge regression methods give estimates, $\underline{\beta}^{**}$, that maximize the correlation of the estimated, $\hat{y} = X\underline{\beta}^{**}$, and observed, $\underline{y}$, responses in a subsequent sample (in "new data"). The reader will recall a similar parallelism in the RSS and PRESS statistics. RSS gives a measure of concordance of model and data in the initial sample. PRESS = $\Sigma[e_i/(1-h_i)]^2$ gives a measure of concordance of model and data in subsequent samples (in "new data").

Oman's (1983), "... approach rather sharply exhibits the relation between ridge estimation and <u>extrapolation</u> from collinear data." (italics added) We shall discuss this approach briefly below.

Over 700 papers on Ridge regression have been published between 1970 and 1983. It seems to many that they are especially - uniquely - appropriate when the only available a priori information on $\underline{\beta}$ is too vague to be represented by equality restrictions in the form, $\underline{r} = R\underline{\beta} + \underline{v}$, that is required for Mixed Estimation methods - as in the present case. (See Table 4). However, there are several criticisms of both the foundation and methods of Ridge regression. Some of the more interesting and cogent can be found in the writings of Smith and Campbell (1980), Coniffe and Stone (1973) and Leamer (1978).

It will be recalled from Lecture 5 of this symposium that Leamer makes the (mildly) invidious distinction between "fictitious" and "fake" ancillary data sets. Either sort may be pooled with the set of data that comprise the sample in hand, to obtain posterior estimates of β that are matrix-weighted averages of the respective estimates obtained from the two data sets. In that Lecture it is then proposed that Bayesian estimates are constructed from (merely) "fictitious" data sets but that Ridge estimates are concocted from "fake" data sets. The stigma seems to attach to that (acknowledged) feature (Hoerl and Kennard 1970) of the Ridge maneuver, namely, that in the cognate Bayesian or Mixed Estimation models the a priori estimate of the parameter is zero: $\beta = 0$. (This is, of course, the null hypothesis of Sampling Theory for the model, $H_0: \beta = 0$.)

Despite the cogency and sincerity of such arguments, an earnest investigator is frequently confronted with both <u>weak</u> (in part because collinear) <u>data</u> and <u>weak</u> (because not much is "known" about the elements of β - beyond their number, dimensions and sign) <u>prior</u> information and a clearly evident and insistent requirement to construct a predictive regression model from <u>these</u> data. In such circumstances one can, taking counsel from G. K. Chesterton ("Things worth doing are worth doing badly.") as well as from Hope, make a fruitful use of Ridge regression methods. We proceed to provide a <u>heuristic</u> instance of this. In this enterprise we have found some comfort in the remarks of Gunst (1980): "Practitioners who frequently use biased regression techniques, in this case, ridge regression, have one additional piece of prior knowledge: Ridge Regression has been successfully applied too frequently, when little prior information is available, for its use to be restricted only to data sets for which the formal Bayesian priors are known to be valid."

For Normal theory models, $\underline{y} = X\underline{\beta} + \underline{\varepsilon}$, the Least Squares estimators are $\hat{\underline{\beta}} = (X'X)^{-1}X'\underline{y}$ and $Var(\hat{\underline{\beta}}) = s^2(X'X)^{-1}$. $\hat{\underline{\beta}}$ is an unbiased estimator of $\underline{\beta}$. The Ridge estimators are, $\hat{\underline{\beta}}** = (X'X + kI)^{-1}X'\underline{y}$ and $Var(\hat{\underline{\beta}}**) = (X'X + kI)^{-1}X'X(X'X + kI)^{-1}$ where I is the (pxp) identity matrix, and $0 < k < 1$ is a constant. $\hat{\underline{\beta}}**$ is a <u>biased</u> estimate of $\hat{\beta}$ and k is the biasing parameter. $\hat{\underline{\beta}}**$ is a linear transformation of $\hat{\underline{\beta}}$; it, "shrinks" $\hat{\underline{\beta}}$. $\hat{\underline{\beta}}** = (X'X + kI)^{-1}(X'X)\hat{\underline{\beta}}$. Recall that $E(\hat{\underline{\beta}}'\hat{\underline{\beta}}) = \underline{\beta}'\underline{\beta} + \sigma^2 \Sigma \lambda_j^{-1}$, and thus for collinear data, for which $\lambda_p \longrightarrow 0$, the estimates, $\hat{\underline{\beta}}$, are inflated. Moreover, one or more of the $\hat{\beta}_j$ may have the "wrong" sign. (The Ridge estimators are obtained in the <u>correlation</u> basis and transformed to the <u>natural</u> basis: $\hat{\underline{\beta}}_0^{**} \longrightarrow \hat{\underline{\beta}}^{**}$. (Marquardt and Snee 1975).)

Despite the rich, even effulgent, literature on Ridge estimators for Normal theory models, there has to date (apparently) only been one paper which developes Ridge estimators for the generalized least squares model, $\hat{\underline{\beta}} = (X'\hat{V}^{-1}X)^{-1}X'\hat{V}^{-1}\underline{y}$, which is of interest to us. This is the recent paper by Schaefer (1984) which presents Ridge estimators for logistic regression models.

However, we have examined the methods developed by Schaefer for our head and neck cancer data, and found that - for our data - these yielded Ridge estimators which were inconsistent with the (received) prior information on β. Therefore, we were constrained to develop an alternative, heuristic, procedure. Moreover, we shall also use the same maneuver to obtain a heuristic Errors-in-Variable (EV) estimate of β. And we have also made use of this procedure to "cascade" the two characteristic RR and EV <u>linear transformations</u> of the sample estimate of the parameter vector in order to obtain a first-order approximation to β in which the effects of multicollinearity <u>and</u> random errors of measurement in the model matrix X are much reduced.

Our (heuristic) maneuver is simple. It consists in exploiting the <u>approximate</u> proportionality ($\pm 5\%$) which exists between the Least Squares estimates of the parameter vector, α, of the binary regression model, $y = X\alpha + u$, and Maximum Likelihood estimates of β of the probit regression model, $z = X\beta + \epsilon$, of the same observation matrix [y, X] when the X'X matrix is <u>not</u> ill-conditioned: $\hat{\alpha} \sim c^*\hat{\beta}$, where c^* is a function of D^2, the Mahalnobis distance between the response groups (0, 1) within the sample and the sample size. The binary regression model is a <u>surrogate</u> for the probit regression model in our maneuver. It can be regarded as a variety of linearization, or better, a dichotomization, of the response variable, z, in the probit regression model. See Figure <u>1</u>.

We do <u>not</u>, of course, maintain that our "engineering" maneuver has achieved more than a good first approximation to the solution. (But, as a heuristic it may be that our maneuver is more <u>accessible</u> to the physicist than are some of its rivals.) The further development of the canonical methods to deal with the problems of multicollinearity and random errors of measurement in the model matrix for probit and logit models of binary response awaits the further work of Schaefer and of Carroll, Spiegelman, Lan, Bailey and Abbott (1984). vide infra. But, maybe the basic justification for our methods is, again, expressed in the apothegm of G. K. Chesterton quoted above. Or perhaps more charitably: "If you can't solve a problem in the form it presents with, then transform it to a form which you can solve." (The maxim of the practicing engineer of mature years - or the applied mathematician.)

The proximate rationale of the method lies in the observation that it is the ill-conditioning of the matrix X'X, rather than of $(X'\hat{V}^{-1}X)$, which is the fundamental cause of the inflation of the Least squares (or Maximum Likelihood) estimates, $\hat{\beta}$ and Var($\hat{\beta}$). For this reason, Cox (1970) has recommended that the degree of collinearity of the treatment variables, X, be assessed, and if necessary, reduced, before the data are used to construct a generalized linear model of binary response: "Therefore, it may be good to have a preliminary calculation of the formal 'correlation' matrix of the regressor variables, followed if necessary by a linear transformation of the regressor variables to ones more nearly orthogonal in the usual least squares sense." As we shall see, our remedy is based on the correlation matrix of the regressor variables but devolves into a linear transformation of <u>parameters</u> - Ridge Regression - rather than of <u>variables</u> as recommended by Cox. In this respect, the remedy is <u>formally</u> similar to the remedy recommended by Gordon (1974) - deletion of one or more collinear variables - since, as we have seen, this also can be represented as a transformation on the parameters: $r = R\beta$.

The effects of such features of the joint distribution of [y, X] as extreme points, outlying responses, collinear variables and random errors of measurement on the Least Squares (LS) estimates of α in the linear model, $y = X\alpha$, where the binary response, y, is represented by a (0, 1) indicator variable should be at least qualitatively the same as their effect on the Maximum Likelihood estimates of β in the probit model, $\underline{z} = X\underline{\beta} + \underline{\varepsilon}$, where $\underline{z}$ is the probit transform of the binary response.

We now proceed to describe our "engineering approach" to these problems. The Least Squares estimates of the parameter vector in the binary model are (of course):

$$\hat{\underline{\alpha}} = (X'X)^{-1}X'\underline{y} \text{ and } Var(\hat{\underline{\alpha}}) = \hat{\sigma}^2(X'X)^{-1}$$

where $\hat{\sigma}^2 = (\underline{y} - X\hat{\underline{\alpha}})'(\underline{y} - X\hat{\underline{\alpha}})/(n-k) = RSS/(n-k)$. The estimates $\hat{\alpha}_j$ and $\hat{\alpha}_j/\sqrt{Var(\hat{\alpha}_j)}$ are presented in Table 11aii for k=4, X = ($\underline{1}$, $\underline{X}_1$, $\underline{X}_2$, $\underline{X}_3$) together with the coefficient of determination, $R^2 = 0.197$, and the adjusted coefficient of determination, $\overline{R}^2 = 0.136$ (See Montgomery and Peck 1982). Comparison with the cognate estimates of Table 3, where k=3, X = ($\underline{1}$, $\underline{X}_1$, $\underline{X}_2$), shows two of the effects of the collinearity of $\underline{X}_2$ and $\underline{X}_3$: 1) $\hat{\alpha}_3/\sqrt{Var(\hat{\alpha}_3)} < 1.0$ and 2) the adjusted coefficients of determination are equal. Since $\underline{X}_3$ is redundant it contributes no information on response.

The estimates for the probit model for X = ($\underline{1}$, $\underline{X}_1$, $\underline{X}_2$, $\underline{X}_3$) first presented in Table 3 are also included in Table 11. We notice that the corresponding components of the respective parameter vectors are no longer approximately proportional: $\hat{\alpha} \neq c*\hat{\underline{\beta}}$. This was not the case for X = ($\underline{1}$, $\underline{X}_1$, $\underline{X}_2$), described in Table 3. The difference in the two cases is a consequence of the strong collinearity of $\underline{X}_2$ and $\underline{X}_3$.

We notice however, that the ratios, $\hat{\alpha}_2/\hat{\alpha}_1$ and $\hat{\alpha}_3/\hat{\alpha}_1$, for the binary regression model differ from the received a priori information in the same manner as do the ratios for the probit regression model: both lie outside the interval (-1, 0). Evidently the effects of the collinearity in $\underline{X}_2$ and $\underline{X}_3$ are roughly the same for sample estimates of either α or β. Moreover, the correlation coefficient of the respective estimates of response of the two models is fairly high: $r(\hat{y}, \Phi(\hat{z})) = 0.92$. (It will be recalled that $E(\hat{y}_i) = P(y_i = 1|\underline{x}_i'\hat{\underline{\alpha}})$.) Thus, the unweighted binary regression model seems to be an appropriate surrogate for the probit regression model for these data.

We also present for comparison the weighted Least Squares estimates for the binary regression model. These are $\hat{\underline{\alpha}}* = (X'\hat{V}^{-1}X)^{-1}X'\hat{V}^{-1}\hat{y}$ $Var(\hat{\underline{\alpha}}*) = \hat{\sigma}(X'\hat{V}^{-1}X)^{-1}$ where $\hat{V} = Diag[\hat{y}_i(1-\hat{y}_i)]$, $i=1$, ..., n. The weighted estimates are correct since the response is a binary variable and hence the variance, σ^2, is not homogeneous. However, the unweighted estimates are more appropriate as a surrogate.

The collinearity diagnostics for the binary model are the same as for the probit model and were presented in Table 4. Plots of the case statistics - Studentized residuals (See Belsley, Kuh and Welsch 1978), hat matrix diagonals and Cook's D - are presented in Figures 10a-10d for the surrogate (unweighted estimates of $\underline{\alpha}$) model. Although these plots bear a satisfying resemblance to the cognate plots of Figure 6 for the probit model, comparison of the two plots of Cook's D offers some support for Pregibon's earlier conjecture: "An important extension of these diagnostic approaches is to non-linear regression models, where presumably the effects of outliers and leverage points could be worse" (See Pregibon 1981). We note that the deletion of case #18 shifts the sample estimate of $\underline{\beta}$ to the 0.70 confidence ellipsoid on $\hat{\underline{\beta}}(D_{18}^1 = 0.85, D_{18} = 1.28)$ but shifts the sample estimate of $\underline{\alpha}$ to only the 0.13 confidence ellipsoid on $\hat{\underline{\alpha}}$ ($D_{18} = 0.31$).

Note that the probability plot in Figure 10d suggests a heavy-tailed distribution for the Studentized residuals (See Montgomery and Peck 1982).

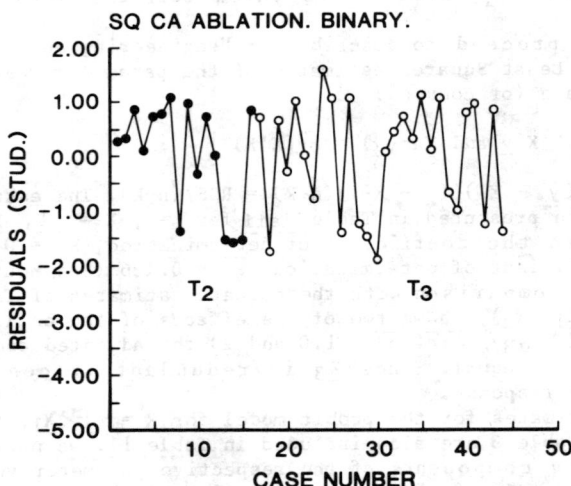

Figure 10a. Case plot of Studentized residuals, e_i^*, for binary regression model of ablation, E_1 (n = 44). There is one outlier: e_{30} = -2.04. Compare Figure 6a.

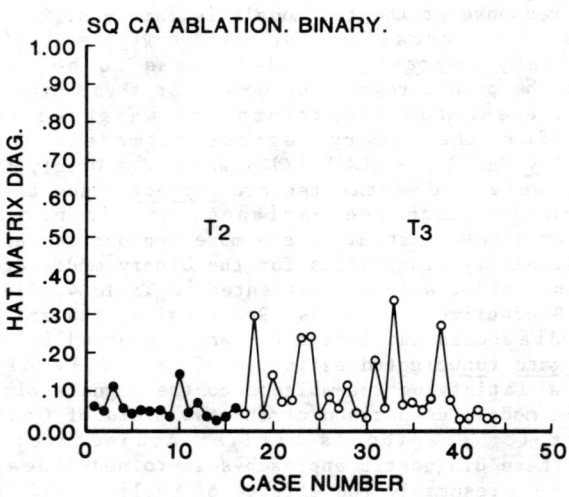

Figure 10b. Case plot of hat matrix diagonals, h_i, for binary regression model of ablation, E_1 (n = 44). There are six extreme levels of treatment. The two most extremes are h_{18} = 0.30 and h_{33} = 0.33. Compare Figure 6b.

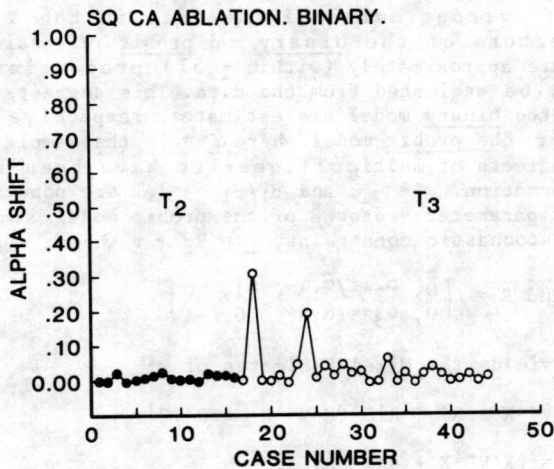

Figure 10c. Case plot of alpha shift, D_i, for binary regression model of ablation, E_1 (n = 44). There is one influential case: $D_{18} = 0.31$. Compare Figure 6c.

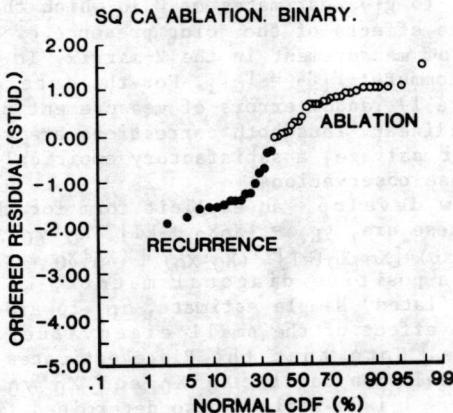

Figure 10d. Normal probability plot of Studentized residuals, e_i^*, for binary regression model of ablation, E_1 (n = 44). The distribution is heavy tailed. Compare Figure 6d.

We have remarked earlier - and instanced in Table 3d - that, in the absence of strong multicollinearity in the X matrix, the parameter vectors of the binary and probit regression models of a given sample are approximately (within ± 5%) proportional: $\hat{\underline{\alpha}} \simeq c^*\underline{\beta}$ where c^* can be estimated from the data. This suggests $\hat{\alpha}_2^{**}/\hat{\alpha}_1^{**}$ and $\hat{\alpha}_3^{**}/\hat{\alpha}_1^{**}$ for the binary model are estimates, respectively, of β_2/β_1 and β_3/β_1 for the probit model where $\hat{\underline{\alpha}}^{**}$ is the sample estimate of $\underline{\alpha}$ in which the effects of multicollinearity have been reduced by a linear transformation: $\hat{\underline{\alpha}}^{**} = \Gamma\hat{\underline{\alpha}}$, and β_j/β_1, j=1,2 are population values.

Thus, the parameter vector $\underline{\beta}$ of the probit model should satisfy the following stochastic constraint, $\underline{r} = R\underline{\beta} + \underline{v}$ where

$$\underline{r} = \begin{pmatrix} 0 \\ 0 \end{pmatrix} \text{ and } R = \begin{pmatrix} 0, & \hat{\alpha}_2^{**}/\hat{\alpha}_1^{**}, & -1, & 0 \\ 0, & \hat{\alpha}_3^{**}/\hat{\alpha}_1^{**}, & 0, & -1 \end{pmatrix} \quad (48)$$

The procedure yields the Mixed Estimates of $\underline{\beta}$:

$$\hat{\underline{\beta}}^{**} = [X'\hat{V}^{-1}X + R'\psi^{-1}R[^{-1}(X'\hat{V}^{-1}X\hat{\underline{\beta}} + R'\psi^{-1}\underline{r}) \quad (49)$$

and

$$Var(\hat{\underline{\beta}}^{**}) = [X'\hat{V}^{-1}X + R'\psi^{-1}R]^{-1} \quad (50)$$

There is additional, ancillary, information in the statistics γ and Θ_p described above. γ is the measure of the compatibility of sample and non-sample information and Θ_p is an estimate of the proportion of the latter present in $\hat{\underline{\beta}}^{**}$.

More generally, we have $\hat{\underline{\alpha}}^{**} = \Gamma_m\hat{\underline{\alpha}}$, m = 1, 2, 3 where Γ_1 is the linear transformation that represents Ridge estimation, Γ_2 is the linear transformation that represents Errors-in-Variables estimation and $\Gamma_3 = \Gamma_2\Gamma_1$ represents the "cascading" of the two transformations of $\hat{\underline{\alpha}}, \Gamma_1$ and Γ_2, to give estimates of $\underline{\alpha}$ in which there are first-order corrections to the effects of the joint presence of collinearity and random errors of measurement in the X-matrix. It is assumed that the transformations commute: $\Gamma_1\Gamma_2 = \Gamma_2\Gamma_1$. For the data on head and neck cancer there are 1) random errors of measurement present in X_1 and 2) X_2 and X_3 are collinear. Thus both corrections are required if one is to retrieve, or salvage, a satisfactory empirical model of radiation response from those observations.

Let us now develop an explicit form for the Ridge regression estimate of $\underline{\alpha}$. These are, $\hat{\underline{\alpha}}_{R_0} = [X_0'X_0 + kI]^{-1}X_0'y_0 = [X_0'X_0 + kI]^{-1}(X_0'X_0)\hat{\underline{\alpha}}_0$; $Var(\hat{\underline{\alpha}}_R) = [X_0'X_0 + kI]^{-1}(X_0'X_0)[(X_0'X_0 + kI]^{-1}$. Note that the effect of adding a positive diagonal matrix, kI, to $X_0'X_0$ is to "shrink" the (inflated) sample estimate, $\hat{\underline{\alpha}}_0$, towards the origin, thus, counteracting the effect of the small eigenvalues of $X_0'X_0$ on the sample estimate. Note that the Ridge estimates of $\underline{\alpha}$ are initially obtained in a correlation basis: $X_0'X_0$ and $X_0'y_0$ are correlation matrices, $0 < k < 1$ is a scalar, also determined from the data. These estimates are subsequently (re-)transformed to the so-called natural basis, $\hat{\underline{\alpha}}_R/\hat{\underline{\alpha}}_R$, for estimation and inference (Marquardt and Snee 1975). Note that $\hat{\underline{\alpha}}_R$ is a biased estimate of $\underline{\alpha}$ and that for $k \geq 0$, $RSS_k = RSS_0 + (\hat{\underline{\alpha}}_R - \hat{\underline{\alpha}})'X'X(\hat{\underline{\alpha}}_R - \hat{\underline{\alpha}})$ (as would be expected since $\hat{\underline{\alpha}}$ is chosen to minimize $RSS_0 = (\underline{y} - X\hat{\underline{\alpha}})'(\underline{y} - X\hat{\underline{\alpha}})$ for the sample).

We have previously stipulated that the response surface model is to be valid only over the region of interest, R_2; that is, the range of values of the treatment variables, x_i', is bounded. Moreover, the response, y_i, in the binary model is bounded above and below: $0 \leq E(y_i) \leq 1.0$. Oman (1983) has observed that for a Normal theory model, say, $y_i = \underline{x}_i'\underline{\alpha} + \varepsilon_i$, $\varepsilon_i \sim N(0, \sigma^2 I)$ these two assumptions imply that $\underline{\alpha}$ is <u>bounded</u> as well. This bound has the form, $\underline{\alpha}'\underline{\alpha} \leq c^2$, a constant. The bound is a <u>constraint</u> on the Euclidean norm of $\underline{\alpha}$. The <u>constrained</u> maximum likelihood estimator, $\hat{\underline{\alpha}}_{ML}$ of $\underline{\alpha}$ then has the form of a <u>Ridge</u> estimator:

$$\hat{\underline{\alpha}}_{ML} = (X'X + kI)^{-1}(X'X)\hat{\underline{\alpha}} = \hat{\underline{\alpha}}_R \qquad (51)$$

Oman (1983) obtains <u>explicit</u> bounds on $\underline{\alpha}$ (implicit bounds were described by Marquardt and Snee (1975) for example) from the bounds on $\underline{x}_i'$ and $E(y_i)$. The relation between the two bounded regions can be understood in terms of the <u>duality</u> between (treatment variable) X-space and (parameter) $\underline{\alpha}$-space. Compare Figures <u>5b</u> and <u>7</u>, for example. (One recalls the relation between a pulse and its Fourier Transform.)

We have remarked that some statisticians object to the logical foundations of Ridge regression. Among those for whom these foundations present no problem, some controversy exists over the method for selecting an optimal value of k. The principal concern over the foundations of the Ridge estimator we have already mentioned: It can be readily shown that the Ridge estimator is a Bayes estimator for which the first two moments of the prior distribution are $\underline{\beta}_0 = \underline{0}$ and $\Sigma_0 = \sigma_0^2 I$. Then $k = \hat{\sigma}^2/\sigma_0^2$ where the (Normal Theory) sample estimate of $\underline{\beta}$ is $\hat{\underline{\beta}} = (X'X)^{-1}X'y$ and $Var(\hat{\underline{\beta}}) = \hat{\sigma}^2(X'X)^{-1}$ (See Leamer 1978; Montgomery and Peck 1982). (An examination of the estimates of $\underline{\alpha}$ presented in Table <u>11</u> discloses that the objections to a zero prior, $\overline{\underline{\alpha}} = 0$, should perhaps have less than its usual cogency - for these data.)

Since one of the effects of collinearity is to inflate the estimates of the Euclidean norm of $\underline{\alpha}_0$ ($E(\hat{\underline{\alpha}}_0'\hat{\underline{\alpha}}_0) = \underline{\alpha}_0'\underline{\alpha}_0 + \sigma^2 \Sigma \lambda_j^{-1}$) one rather obvious criterion for an optimal value of k is that $\hat{\alpha}_R'\hat{\alpha}_R = \hat{\underline{\alpha}}_0\hat{\underline{\alpha}}_0 - \hat{\sigma}^2\Sigma\lambda_j^{-1}$ (See McDonald and Galarneau 1975). Another optimum is that proposed by Hoerl, Kennard and Baldwin (1976): $k = \hat{p}\sigma^2/\hat{\underline{\alpha}}_0'\hat{\underline{\alpha}}_0$ (See also Schaeffer 1983). However, several statisticians (Marquardt and Snee 1975; Montgomery and Peck 1982) prefer to use the graphical method - the so-called ridge trace - that was described by Hoerl and Kennard in their initial paper (1970). The ridge trace is a plot of $\hat{\underline{\alpha}}_R$ vs k for $0 \leq k \leq 1$. If there is multicollinearity present, some of the elements of $\hat{\alpha}_{R_0}$ will vary dramatically. At some value of $k > 0$, $\hat{\underline{\alpha}}_{R_0}$ will stabilize. This gives an "optimal" value for k - which is usually larger than that selected by either of the previous methods.

"One of the big advantages of ridge regression is that a graphical display, called 'Ridge Trace', can help the analyst to see which coefficients are sensitive to the data. Thus, sensitivity analysis is an aim of ridge regression" (Marquardt and Snee 1975). Marquardt (1980) has noted that the value of selected by the ridge trace generally increases as the value of the adjusted coefficient of determination, $\overline{R}^2$, decreases. We note from Table <u>11</u> that for the binary regression model, $R^2 = 0.198$, $\overline{R}^2 = 0.136$ which implies that a rather large value of k will be required to stabilize these estimates.

The ridge trace for the binary regression model is presented in Figure <u>11</u>. Note that the plot refers to $\hat{\underline{\alpha}}_R$ - the natural basis: at k=0, $\hat{\underline{\alpha}}_R = \hat{\underline{\alpha}}$. (In the Figure it has been convenient to denote $\hat{\underline{\alpha}}_R$ as the posterior estimate, $\hat{\underline{\alpha}}^{**}$.) The elements of $\hat{\underline{\alpha}}_R$ have stabilized in the vicinity of k = 0.10. Note that the Ridge Regression <u>transformation</u>, $\hat{\underline{\alpha}} \longrightarrow \hat{\underline{\alpha}}_R$, has <u>not</u> "shrunk" $\hat{\alpha}_1$ as much as it has $\hat{\alpha}_2$ and $\hat{\alpha}_3$ - as would be expected, since the largest multicollinearity is that between N and T. See Table <u>4</u>. The respective variance inflation factors - given by the diagonal elements of $Var(\hat{\underline{\alpha}}_R) = [X_0'X_0 + kI]^{-1}(X_0'X_0)[X_0'X_0' + kI]^{-1}$ - are $VIF_1 = 1.254$, $VIF_2 = 1.131$ and $VIF_3 = 1.143$. The residual sum of squares, RSS_k, (in the correlation basis) exceeds RSS_0 by only a factor of 1.03. Thus, one object of the exercise - to reduce the effects of the multicollinearity in the X matrix seems to have been achieved. Note that the MG criterion requires $0.01 < k < 0.02$ and the HKB criterion requires $k = 0.037$. However, the change in sign of $\hat{\alpha}_3^{**}$ that is required by prior information - see Table 1 - does not occur for $k < 0.05$.

However, for reasons that will soon become apparent, we prefer $k = 0.12$ over $k = 0.10$ and the summary of the ridge estimates - in the natural basis, $\hat{\alpha}^{**}$ - are presented in Table 12 for that value. Note that the ratios, $\hat{\alpha}_j^{**}/\hat{\alpha}_1^{**}$, $j=2, 3$, now lie within the interval $(-1, 0)$ and thus are consistent with prior information. See Table 1.

The Mixed estimates of $\underline{\beta}$ for the probit regression model, with the constraint, $\underline{r} = R\underline{\beta} + \underline{v}$, defined by the ridge estimates $\hat{\alpha}_R$ of the parameter vector $\underline{\alpha}$ for the binary, surrogate, model of the same sample are presented in Table 13. As would be anticipated, the prior information on $\underline{\beta}$, that was articulated, or formalized, through the use of the binary surrogate is consistent with the sample information on $\underline{\beta}$. However, for this choice of $\psi = 10^{-2}I$ and $k = 0.12$, the sample information contributes only half the information on $\underline{\beta}$ represented by the posterior estimates, $\underline{\beta}^{**}$: $\Theta_s = 1-\Theta_p = 0.50$. Note that $\beta_j^{**}/\sqrt{Var(\hat{\beta}_j^{**})} > 2.0$, $j=1, 2, 3$ as would be anticipated since the sample data have been augmented by the prior information. Note as well that the posterior estimates of the response, $\hat{z}_i^{**} = \underline{x}_i^t\underline{\beta}^{**}$ predict the response over the RVH of the reduced sample as well - $r = 0.47$ - as do the estimates, $\hat{z}_i = \underline{x}_i^t\hat{\underline{\beta}}$, for model M_1, i.e., it interpolates as well as any - and better than most. See Table 3.

Outside the RVH, at the treatment level, $\underline{x}_i$' marked "+" the response estimated from the Ridge estimates, $\underline{x}_i^t\underline{\beta}^{**}$ -is $\hat{\pi}_i = 0.938$, a "plausible" level of response which conforms with our a priori anticipations. It will be recalled that the estimated response for M_1 at this level of the treatment variables was $\hat{\pi}_i = 0.003$. Thus, as observed by Marquardt and Snee (1975) and Darlington (1976) the biased estimator, $\underline{\beta}^{**}$, performs better in extrapolation than does the unbiased estimator, $\hat{\underline{\beta}}$. (It is, as well, a "reduced variance estimator".)

15. ERRORS-IN-VARIABLES ESTIMATES

"As in linear regression, there are at least two good reasons to estimate the error-free regression. First, measurement processes may improve, making the errors-in-variables estimates more valuable. Second, it can be meaningful to investigate the true regression coefficient; ... [there are cases in which] the errors-in-variables estimates are physically sensible but the least squares estimates are not" (See Carroll, Spiegelman, Lan, Bailey and Abbott 1984).

"For the prior-dependent problem the least-squares estimate is shrunk towards the origin by adding to the X'X matrix a positive diagonal matrix. For the errors-in-variables model the least-squares estimate is blown away from the origin by subtracting from the X'X matrix a positive diagonal matrix" (See Leamer 1978).

One of the basic assumptions in constructing estimates of the parameter, $\underline{\beta}$, in the Normal theory model $\underline{y} = X\underline{\beta} + \underline{u}$ is that the elements of the observation matrix, $[y, X]$, are known without error. In such cases the least sqaures estimates, $\hat{\underline{\beta}}$, are unbiased: $E(\hat{\underline{\beta}}) = \underline{\beta}$. However, the observed treatment variables, $\underline{X}_j$, may include

Table 12. Least Squares Regression on Binary Variable, y=0, 1: $y=x'\underline{\alpha}$, a surrogate equation.

a) k=0 (Least Squares Estimate)
$y = -10.216 + 4.544x_1 - 6.776x_2 + 3.093x_3 = \underline{x}_i'\underline{\alpha}$
    (-1.239) (1.815) (-1.652) (0.791)[a]
$R^2 = 0.197.$ $\bar{R}^2 = 0.136$[b]
Sample Information on $\underline{\alpha}$:
$\hat{\alpha}_2/\hat{\alpha}_1 = -1.491$, $\hat{\alpha}_3/\hat{\alpha}_1 = 0.681$

b) k=0.12[c] (See Figure 12 point k = 0.12)
$y = -5.646 + 2.879x_1 - 2.230x_2 - 0.685x_3 = \underline{x}_i'\hat{\underline{\alpha}}_R$
$R^2 = 0.148.$ $\bar{R}^2 = 0.084.$
Ridge Information (Natural Basis, [$\underline{y},X$]) on $\underline{\alpha}$:
$\hat{\alpha}_2^{**}/\hat{\alpha}_1^{**} = -0.775$, $\hat{\alpha}_3^{**}/\hat{\alpha}_1^{**} = -0.238$

c) Estimated response in extrapolation: (D, T, N) = (9000, 63, 20).
(See Figure 5, point "+".)

	$\hat{y}_i$
k = 0	-0.416[d]
k = 0.12	0.835[e]

[a] $(\hat{\alpha}_i/\sqrt{\text{Var}(\hat{\alpha}_i)})$. [b] Adjusted R^2 (Montgomery and Peck 1982).
[c] From Ridge Trace (See Figure 11.) [d] $\hat{y}_i = \underline{x}_i'\hat{\underline{\alpha}}$. [e] $\hat{y}_i = \underline{x}_i'\hat{\underline{\alpha}}_R$

Table 13. Mixed Estimates of $\underline{\beta}$ for Probit Model.

a) A priori Information (from $\hat{\underline{\alpha}}^{**}$)
$\underline{r} = R\underline{\beta} + \underline{v}$. $E(\underline{v}) = 0$. $\text{Var}(\underline{v}) = \Psi$.
$\underline{r} = \begin{pmatrix} 0 \\ 0 \end{pmatrix}$, $R = \begin{pmatrix} 0, & 0.775, & 1, & 0 \\ 0, & 0.238, & 0, & 1 \end{pmatrix}$, $\Psi = 1.0*10^{-2}I$

b) Mixed Estimates, $\hat{\underline{\beta}}^{**}$. (See Figure 7, point "a*")
$\gamma(2) = \chi^2(2) = 1.33$. Accept H_0
$\Theta_p = 0.50$
$\underline{\beta}^{**} = [-36.938, 17.092, -13.239, -4.067]$
        (-2.321),(2.345),(-2.345),(-2.343)
$r(\underline{y}, \phi(\underline{z})) = 0.468$. $r^2 = 0.219$.

Table 14. Approximations to Effects of Random Measurement Errors in Dose on Least Squares Estimates of $\underline{\alpha}$ in Binary Model, $\underline{y} = X\underline{\alpha} + \underline{\varepsilon}$ of Sq. Ca. Ablation.

Standard Errors, g_1[a] of Measurement of Dose	$\hat{\alpha}_1$	Bias, b[b]	Ratio $b_1/\hat{\alpha}_1$	$b_1/\sqrt{\text{Var}(\hat{\alpha}_1)}$
a) 0.01	4.54	0.60	0.13	0.240
b) 0.02	4.54	2.42	0.53	0.967

[a] A priori estimate of variance-covariance matrix of measurement errors:
$\hat{E} = \begin{pmatrix} 0 & 0 & 0 & 0 \\ 0 & \hat{g}_1^2 & 0 & 0 \\ 0 & 0 & 0 & 0 \\ 0 & 0 & 0 & 0 \end{pmatrix}$
where $g_1^2 = \text{Var}(\log_{10}D) = (0.4343)^2 \text{Var}(D)/D^2$. $\sqrt{\text{Var}(\hat{\delta}_1)} = 150$, D = 6500, $g_1 = 0.01$, etc.
[b] $\underline{b} = (n-m)(X'X)^{-1}\hat{E}\underline{\alpha} = (\hat{b}_0, \hat{b}_1, \hat{b}_2, \hat{b}_3)$

random errors of measurement, $\underline{\delta}_j$, such that $\underline{X}_j = \underline{X}_j + \underline{\delta}_j$ or, $X = X^* + \Delta$, where X^* is the true, error-free, model matrix. Let us assume that Δ has a joint distribution with mean zero and diagonal dispersion matrix, nE, where we have $E = \text{Diag}[\gamma^2, \ldots \gamma^2_k]$. We do not consider a non-zero mean since our linear model includes the constant, β_0, and the presence of a constant error in $\underline{X}_j$ will simply change the estimate of $\underline{\beta}_0$ (Davies and Hutton 1975). In this case the estimates $\underline{\hat{\beta}} = (X'X)^{-1}X'\underline{y}$ are biased: $E(\underline{\hat{\beta}}) = \underline{\beta} - \underline{b} \simeq \underline{\beta} - n(X'X)^{-1}E\underline{\beta}$. Moreover, the sample estimate of the coefficient of determination, R^2, is also attenuated; it is "shrunk".

We shall consider a transformation of the sample estimate, $\underline{\hat{\beta}}$, that will provide a first-order correction for the presence of random errors of measurement in dose - $\underline{X}_1$. See Theil 1971; Seber 1977.

An estimate of E is, $\hat{E} = \text{Diag}[0, \hat{g}_1^2, \hat{g}^2, \ldots \hat{g}_p^2]$ where g_j^2 is the <u>non-sample</u> estimate of the error variance in the measurement of the jth treatment variable, $\underline{X}_j$; covariances of the measurement errors are assumed to be either zero or unknown. It is important to note that even when the <u>nominal</u> - recorded - value of a treatment variable can be determined with negligible error it is often the case that the variable is standing as a <u>proxy</u> for some other factor which is a more immediate "cause" but which cannot be measured directly and so the nominal value is burdened with a random measurement error. Thus, the nominal radiation dose recorded in the patient's chart may differ by a <u>random amount</u> from that which is the "cause" of the observed "effect" - the response - in the target tissue. See Leamer 1978.

Beaton, Rubin and Barone (1976) have introduced a useful rule of thumb for assessing the sensitivity of the estimates of $\underline{\beta}$ for a given model, $\underline{y} = X\underline{\beta} + \underline{u}$, to the presence of random errors of measurement in X. This criterion is the Perturbation Index, $PI = \text{Tr}[X'X]^{-1}E]$. If PI $\ll$ 1 then the effect of the random measurement error on the estimates, $\underline{\hat{\beta}}$, can be ignored. The PI can be written in the following form, $PI = \Sigma(g_j/s_j)^2 VIF_j$, where s_j is the <u>marginal</u> variance of $\underline{X}_j$. It is evident that the degree of bias in $\hat{\beta}_j$ is determined, in part, by the ratio of the variance, g_j^2, of the distribtion of the <u>error of measurement</u> of $\underline{X}_j$ to the variance of the cognate <u>marginal distribution</u> of $\underline{X}_j$. If this ratio is small, say < 0.10, then the bias due to measurement error in $\underline{\hat{\beta}}$ <u>may</u> be negligible. One maneuver for reducing this ratio, and hence the size of the bias in $\hat{\beta}_i$, is to reduce the variance, g_i^2, of the measurement error by grouping the individual cases into groups of (average) size m, say, $g_i\#^2 = g_i^2/m$. However, the pair-wise correlations in the aggregated data, $[\underline{y}\#, X\#]$, that result from such a maneuver are <u>inflated</u> with respect to the unaggregated data, $[\underline{y}, X]$. This gives 1) a biased (spuriously high) estimate of the concordance of model and data - an inflated R^2 - and 2) an inflated variance, $Var(\underline{\hat{\beta}})$. The latter is a consequence of the increased degree of collinearity in $X'X$ (inflation of VIF_j) and of the inefficiency in multivariate grouping (Johnston 1972). It should be noticed that the inflation of R^2 which occurs when an equation is estimated from grouped data means that the "shrinkage" of R^2 referred to above may be quite dramatic when the equation is used as a predictor of response in a second, independent, <u>un</u>grouped sample. See Brown 1963.

For a given size of the variance of the errors of measurement the degree of collinearity in the joint distribution of X will determine the degree of bias in the estimates, $\underline{\hat{\beta}}$, that is induced owing to the <u>interaction</u>, $(X'X)^{-1}\hat{E}$, between these two features of the sample - the <u>degree of precision</u> in the measurement of $\underline{X}_j$ and the <u>degree of</u>

Table 15. Indices of Sensitivity of Least Squares Estimates of $\underline{\alpha}$ to Random Measurement Errors in X. Binary Model, $\underline{y} = X\underline{\alpha} + \underline{\varepsilon}$ of Sq. Ca. Ablation.

Standard Errors, g_i, of Measurement of Dose	$(m_1 n)^{1/2}$	Davies and Hutton Sens. Indices. p_1 [a]	c_1 [b]	c_2 [c]	Perturbation Index [d]
0.01	6.63	17.13	0.70	0.17	0.150
0.02	6.63	8.56	1.41	1.09	0.600

[a] $p_1 = \hat{\sigma}/\sqrt{\Sigma g_i^2 \text{Var}(\hat{\beta}_i)}$. Should greatly exceed $(m_1 n)^{1/2}$
[b] $c_1 = \sqrt{n} \, \Sigma g_i |\hat{\beta}_i| / \hat{\sigma}$. Should <u>not</u> exceed 1.0
[c] $c_2 = n\sqrt{\Sigma g_i^2 \hat{\beta}_i^2}/p_1 \hat{\sigma}$. Should <u>not</u> exceed 1.0
[d] $PI = \text{Tr}[(X'X)^{-1}\hat{E}]$. Should be much less than 1.0

Table 16. Least Squares Estimates of α, Corrected for Bias due to Random Errors of Measurement in X. (Theil). Sq. Ca. Ablation

Standard Error, g_1, of Measurement of Dose		Uncorrected Estimates of α	Corrected[a] Estimates of α	Ratio
0.01[b]	$\hat{\alpha}_0 =$	−10.22	−12.37[c]	1.22
	$\hat{\alpha}_1 =$	4.54	5.24	1.15
	$\hat{\alpha}_2 =$	−6.78	−7.12	1.05
	$\hat{\alpha}_3 =$	3.09	3.20	1.04

[a] $\hat{\underline{\alpha}}_{EV} \cong (X'X - (n-m)\hat{E})^{-1}(X'X)\hat{\underline{\alpha}}$
[b] $\delta \sim 150$ centigray in 6500 centigray, or about 2% error in dose.
[c] Note that the standardized estimate of the bias, $b_j/\sqrt{\text{Var}(\hat{\alpha}_j)}$, where $\underline{b} \propto (n-m)(X'X)^{-1}\hat{E}\underline{\hat{\alpha}}$, in the sample estimate, $\underline{\hat{\alpha}}$, due to the presence of random errors of measurement in the model matrix, X, is a <u>regression diagnostic</u>. It is quite similar to DFBETAS which provides an estimate of the effects on $\hat{\alpha}_j$ due to the joint presence of outlying responses, y_i, and extreme levels of treatment, $\underline{x_i}'$, in the observation matrix $[\underline{y}, X]$. (See Lectures 3 and 14).

The respective formulae are quite similar.
i) <u>Absence</u> of ith observation (Belsley et al 1980):
$\underline{\hat{\alpha}}_{(i)} = \underline{\hat{\alpha}} - [e_i/(1-h_i)] \, (X'X)^{-1}\underline{x_i}$
ii) <u>Presence</u> of random errors of measurement in X (Seber 1977).
$\underline{\hat{\alpha}}_E \cong \underline{\hat{\alpha}} - n(X'X)^{-1}\hat{E}\underline{\hat{\alpha}}$
For instance, the <u>size</u> of the effects on $\hat{\underline{\alpha}}$ of measurement error $g_1 = 0.01$ in x_1 is comparable to that of deletion of the outlier, $e_{30} = 0.774$.

In the formulae above $\underline{\hat{\alpha}}$ is the LS estimate of $\underline{\alpha}$ obtained from a sample of size n; $\underline{\hat{\alpha}}_{(i)}$ is the LS estimate of $\underline{\alpha}$ obtained when the ith observation is deleted; $\underline{\hat{\alpha}}_E$ is the LS estimate of $\underline{\alpha}$ obtained from a sample of size n in which there are random errors of measurement present.

collinearity in the joint distribution of X. Thus, one method for reducing the effects of measurement error on the estimates of $\underline{\beta}$ is to choose the frequency distribution of the variables so that the range within each variable is large and the correlation between each variable is small. This is one of the benefits conferred by good experimental design. See Lecture 17.

The size of bias, $\underline{b}$, which results from the joint presence in X of multicollinearity and measurement error is presented in Table 14 for the binary model of our sample for two levels of precision of measurement. It is important to determine the degree of sensitivity of the estimates, $\hat{\underline{\alpha}}$, to random errors of measurement in each X_j. Davies and Hutton (1975) have provided a set of indices similar to the perturbation index, PI, described above. These are presented in Table 15 for the binary model of our sample. Suppose that $m_1 \leq p$ of the treatment variables have non-zero errors of measurement with E being diagonal as above and g_j^2 being the error variance for the jth variable. First compute $\rho_1 = \hat{\sigma}(\sum_{j=1}^{p} g_j^2 Var(\hat{\beta}_j))^{-1/2}$. If ρ_1 does not exceed $(m_1 n)^{1/2}$ then at least some of the estimates, $\hat{\beta}_j$, are likely to have very little meaning. ρ_1 is a measure of the "distance" of the matrix X'X from singularity. If $\rho_1 \leq (m_1 n)^{1/2}$ then compute $c_1 = n^{1/2} \sum_{j=1}^{p} g_j |\beta_j| / \hat{\sigma}$. If c_1 is markedly less than 1 then the measurement errors can be ignored. c_1 is a measure of the maximum bias likely to be introduced by measurement error. If $c_1 \geq 1$ then compute $c_2 = n(\Sigma g_j^2 \hat{\beta}_j^2)^{1/2}/(\rho_1 \hat{\sigma})$. If $c_1 > 1$ then the bias in $\hat{\underline{\beta}}$ that is due to the presence of random errors of measurement in the sample distribution of X is non-negligible. Thus, from Tables 14 and 15 it would seem that if the standard deviation of the distribution of random errors of measurement of dose is about 150 rads, the resultant bias is fairly large, but is still tractable - a first-order correction is warranted. However, if it is as large as 300 rads, then the utility of sample estimates of $\underline{\alpha}$ is questionable. That is, the first-order corrected estimates, $\hat{\underline{\alpha}}_{EV}$, may not be valid. Thus, the usefulness of this transformation to be described is probably limited to measurements of dose for which the standard error of $\log_{10} D$ is about $g_3 = 0.015$, say 225 centigray in 6500 centigray or about 3.5%.

A cue to a maneuver by which reduced bias estimates of $\underline{\beta}$ can be obtained from a sample in which there were random errors of measurement present in X is found in the Ridge regression methods. It will be recalled that the Normal theory estimates, $\hat{\underline{\beta}}_0 = (X_0'X_0)^{-1}X_0'y_0$, that were obtained from a sample in which there was multicollinearity present in X were transformed to give the Ridge estimates; $\hat{\underline{\beta}}_{R_0} = [X_0'X_0 + kI]^{-1}(X_0'X_0)\hat{\underline{\beta}}_0$. Although for $k > 0$ $\hat{\underline{\beta}}_R$ is a biased estimate of $\underline{\beta}$, $\hat{\underline{\beta}}_{R_0}' \hat{\underline{\beta}}_{R_0} = L_0^2(k)$ is a reduced bias estimate of the Euclidean norm, $\underline{\beta}_0' \underline{\beta}_0$. An optimal value of k for the transformation is that value for which $L_0^2(k) = \hat{\underline{\beta}}_0'\hat{\underline{\beta}}_0 - \hat{\sigma}_0^2 \Sigma \lambda_j^{-1}$ (See McDonald and Galarneau 1975). k is obtained by solving the above scalar eqation.

In a similar manner a reduced bias estimate, $\hat{\underline{\beta}}_{EV}$ of $\underline{\beta}$ can be obtained by solving the following matrix equation: $E(\hat{\underline{\beta}}) = \underline{\beta} - n(X'X)^{-1}E\underline{\beta}$. An approximate solution gives $\hat{\underline{\beta}}_{EV} = [X'X - nE]^{-1}(X'X)\hat{\underline{\beta}}$ (See Theil 1971). There is an obvious similarity to the Ridge regression transformation of $\hat{\underline{\beta}}_0$ - as remarked by Leamer (1978). Thus, it seems reasonable that the estimate of $Var(\hat{\underline{\beta}}_{EV})$ can be obtained by analogy with $Var(\hat{\underline{\beta}}_{R_0})$: $Var(\hat{\underline{\beta}}_{EV}) = [X'X-nE]^{-1}(X'X)[X'X-nE]^{-1}$. The Errors-in-Variables estimates, $\hat{\underline{\alpha}}_{EV}$ for the parameter of the binary

Table 17a. Schedule of Corrections to Estimated Coefficient Vector for Empirical Model. Sq. Ca. Ablation.

1) Initial Model (n=45)
   $z = -30.313 + 11.662x_1 - 13.482x_2 + 5.534x_3$ (Probit)
   $\quad(-1.138)\quad(1.412)\quad(-1.072)\quad(0.464)$[a]

2) Correction for "Outlier", $t_1 = -4.40$. Observation #1 Deleted (n=44)
   $z = -60.318 + 25.067x_1 - 30.057x_2 + 9.474x_3$ (Probit)
   $\quad(-1.661)\quad(2.017)\quad(-1.879)\quad(0.737)$
   $y = -10.216 + 4.544x_1 - 6.776x_2 + 3.093x_3$ (Binary)
   $\quad(-1.239)\;(1.815)\;(-1.652)\;(0.791)$[b]

3) Correction for "Extreme Level", $h_{18} = 0.39$. Observation #18 was not deleted since this would increase VIF_2 and VIF_3. (But See Figures 7 and 12, point c.)

4) Correction for Small Sample Bias
   $z = -53.867 + 22.196x_1 - 26.584x_2 + 8.821x_3$ (Probit)

5) Correction for "Collinearity" in N and T. Ridge Estimation.
   $y = -5.646 + 2.879x_1 - 2.230x_2 + 0.685x_3$ (Binary - RR. k = 0.12)
   (See Figure 12, k = 0.12)

6) Correction for Random Errors of Measurement in D. Errors-in-Variables Transformation applied to 2) above.
   $y = -12.471 + 5.240x_1 - 7.119x_2 + 3.196x_3$ (Binary - EV. $s_D = 100$)
   (See Figure 12, point "a".)

7) Cascading Corrections 1) - 5)
   $z = -39.667 + 17.231x_1 - 12.699x_2 - 3.221x_3$ (Binary - Probit, ME)
   $\quad(-2.0613)\;(2.0896)\;(-2.0799)\;(-2.0772)$
   (See Figure 7, point "a*".)

---

[a] $(\hat{\beta}_i / \sqrt{Var(\hat{\beta}_i)})$
[b] $(\hat{\alpha}_i / \sqrt{Var(\hat{\beta}_i)})$

Table 17b. Estimates of the Probability of Response, $P(E_1 = F(\underline{x}_0' \hat{\beta})$, for different models and biased and unbiased estimates of $\underline{\beta}$.

Regimen $\underline{x}_0'$	Model			
	M_1[a]	M_2[a]	M_3[a]	M_1[b]
(7180, 45, 33)[c]	0.90	0.87	0.76	0.81
(9000, 65, 20)[d]	0.00	0.71	1.00	0.92

[a] Unbiased estimates, $\hat{\beta}$, of $\underline{\beta}$.
[b] Biased estimates, $\hat{\beta}^{**}$, of $\underline{\beta}$. (Mixed estimates. See Table 17a7.)
[c] Interpolation. h = 0.056. (eufractionation). Figure 1a, "x".
[d] Extrapolation. h = 10.324. (hypofractionation). Figure 1a, "+".

regression model are given in Table 16.

16. CORRECTIONS FOR THE SIMULTANEOUS PRESENCE OF MULTICOLLINEARITY IN N AND T AND RANDOM ERRORS OF MEASUREMENT IN D. AN EXTRAPOLATION MODEL

In a sample in which both collinearity and random errors of measurement are simultaneously present in X it seems reasonable to cascade the respective corrections in order to obtain approximate estimates, $\hat{\alpha}_{REV}$, of α in which the effects of both idiosyncracies were much reduced. This can be represented as $\hat{\alpha}_{REV} = [X'X - nD]^{-1}(X'X)\hat{\alpha}_R$ where $\hat{\alpha}_R$ is the Ridge estimate in the so-called natural basis (See Marquardt and Snee 1975). Because the errors-in-variables transformation inflates all of the components of $\hat{\alpha}_R$ it is necessary to use a somewhat larger than "optimal" value of the bias parameter, k. For the sample in hand we chose k = 0.12 rather than k = 0.10.

It is frequently the case that estimates of response, $\hat{y}_i$, are required at treatment levels x_i' that lie outside the so-called Regressor Variable Hull (RVH) which is defined by the ellipsoid, $h_{(n)}$, where $h_{(n)}$ is the largest value of the hat matrix diagonals for a sample of size n (See Montgomery and Peck 1982).

It has previously been remarked by Marquardt and Snee (1975) and by Darlington (1978) that the Ridge estimates of β perform better in extrapolation than do the unbiased estimates, $\hat{\beta} = (X^TX)^{-1}X'y$. That is, $x_m'\hat{\beta}_R$ is a more plausible estimate of true response, y_m, at the treatment level, x_m; that is, $x_m'\hat{\beta}$, when x_m lies outside the RVH. Moreover, we have seen from the definition of the PRESS and AIC statistics that the presence in the sample of outlying responses and extreme levels of treatment may give sample estimates of β that are more characteristic of the sample than of the target population and hence, do not predict well in new data. Thus, it seems reasonable that in order to obtain a model that performs well in extrapolation the several idiosyncratic features of the initial sample must be corrected - the respective corrective maneuvers must be applied in "cascade". Let us review each of these features and their characteristic effects on the sample estimates of the parameter vector, β.

The examination of the regression analysis of these data on the full probit model, $z = \beta_0 + \beta_1 x_1 + \beta_2 x_2 + \beta_3 x_3$, has disclosed that the concordance of model and data, RSS, and the estimates, $\hat{\beta}$ and Var($\hat{\beta}$), are strongly affected - either inflated or deflated - by the presence of various features of the distribution of the observations, [y, X] - outlying responses, extreme levels of the treatment variables, collinearity of the variables, as well as the number of observations - the sample size n. We summarize the general nature of these effects below:
1) The ML estimates of β are biased by the small size (n < 50) of the sample. In general, such bias inflates $\hat{\beta}$. For the full model of the reduced sample (n = 44) the ratios of corrected to uncorrected estimates, $\hat{\beta}_j$, j=0, 1, 2, 3, of the probit model are, respectively, 0.893, 0.885, 0.884, 0.905. The bias would be still larger for a smaller sample.
2) The ML estimates, $\hat{\beta}$, Var($\hat{\beta}$), are biased by the presence of the strong correlation (columns of X) of N and T. In general, such bias inflates $\hat{\beta}$ and Var($\hat{\beta}$). For the full model of the reduced sample the ratios of the corrected (Ridge, k = 0.12) to uncorrected (k = 0) estimates, α_i, i=0, 1, ..., 3, of the binary model are, respectively, 0.553, 0.634, 0.329, 0.221.

3) The presence of <u>extreme</u> levels of treatment, x', (a <u>row</u> effect) biases the estimates, $\hat{\beta}$ and Var($\hat{\beta}$). In general, such bias <u>inflates</u> $\hat{\beta}$ and <u>deflates</u> Var($\hat{\beta}$). For the full model of the reduced sample (n=44) and the (further) reduced sample with observation #18 (h_{18} = 0.391) <u>deleted</u> (n = 43) the ratios of the estimates of $\hat{\beta}_j$, j=0, 1, 2, 3 of the <u>probit</u> models obtained from the two samples are 0.980, 1.085, 0.503, 1.316. As previously observed, these estimates are <u>controlled</u> by case #18.

4) The presence of <u>random errors</u> of measurement of dose, D, biases the estimates $\hat{\beta}$ and Var($\hat{\beta}$). In general, such bias <u>deflates</u> $\hat{\beta}$ and <u>inflates</u> Var($\hat{\beta}$). For the full model of the reduced sample the ratios of the corrected to the uncorrected estimates of $\hat{\alpha}$ for the <u>binary</u> model are, respectively, 1.22, 1.15, 1.05, 1.04, for a standard error of measurement of dose of 150 centigray.

5) The presence of <u>outliers</u> in the response, y_i, biases the estimates $\hat{\beta}$ and Var($\hat{\beta}$) and the measure of concordance, RSS. The bias may either <u>inflate</u> or <u>deflate</u> $\hat{\beta}$; Var($\hat{\beta}$) and RSS are always <u>inflated</u>. For the full model, the ratios of the estimates $\hat{\beta}_j$, j = 0, 1, 2, 3 of the <u>probit</u> model obtained with the inclusion (n=45) and omission (n=44) of observation #1 are, respectively, 0.503, 0.465, 2.229, 1.761.

We have now described the methods by which the effects of each of these on the estimate, $\hat{\beta}$, may be corrected singly, e.g., <u>additions</u> or <u>deletions</u> of rows and columns of [y, X], and transformations of estimates of $\underline{\beta}$. It is however, the case - as we have seen - that each of the features of the distribution of [y, X] that gives rise to each of the respective effects are often present simultaneously in many - if not most- samples of non-experimental data, and therefore the estimates, $\hat{\beta}$, are encumbered by the <u>joint</u> effects of all of them. Thus, it is required to correct the sample estimates, $\hat{\beta}$, for the joint effects of small sample size, multicollinearity, outlying <u>responses</u>, errors of measurement, extreme levels of <u>treatment</u>, etc.

The effect on $\hat{\beta}$ and Var($\hat{\beta}$) of the anomalous response at observation #1 in the initial sample (n=45) - an "outlier" - was <u>corrected</u> by deleting that observation (thereby slightly increasing, no doubt, the degree of bias due to the already small sample size). The effects of the observation #18 in the reduced sample (n=44) - a high leverage point, h_{18} = 0.391 - were <u>ignored</u>, as remarked earlier, since these effects would be corrected by the subsequent Ridge Regression transformation of $\hat{\underline{\alpha}}$ (vide infra) to reduce the effects of the correlation in N and T. The Ridge Regression and the Errors-in-Variables transformations were successively applied to the estimates of $\underline{\alpha}$ in the so-called <u>surrogate</u> model: $\underline{y} = X\underline{\alpha} + \underline{\varepsilon}$: In the correlation basis for that model we have

$$\hat{\underline{\alpha}}^{**} = [X_0'X_0 + kI]^{-1}X_0'X_0\hat{\underline{\alpha}}_0. \qquad (51)$$

This <u>transformation</u> corrected $\hat{\underline{\alpha}}$ for the effects of the correlation of N and T. The estimates, $\hat{\underline{\alpha}}_0^{**}$, were transformed to the <u>natural basis</u> to give the corrected estimates $\hat{\underline{\alpha}}_R$. Those corrected estimates were next transformed to the Errors-in-Variables estimates:

$$\hat{\underline{\alpha}}_{EV} = [X'X - (n-m)E]^{-1}(X'X)\hat{\underline{\alpha}}_R. \qquad (52)$$

This <u>transformation</u> corrected for the bias due to the random error present in the measurement of radiation dose, D.

Since the Errors-in-Variables transformation, $\hat{\underline{\alpha}}_{EV} = [X'X - (n-m)E]^{-1}(X'X)\hat{\underline{\alpha}}$, <u>inflates</u> the Least Square estimates of each component of $\underline{\alpha}$, it is necessary to select a somewhat larger than "optimal" value of the biasing parameter, k, in order to provide a compensatory degree of <u>deflation</u> by the Ridge Regression transformation of $\hat{\alpha}$. At k = 0.12, the respective VIFs for each component are 1.166, 0.904, and 0.918, indicating that the Ridge Regression estimates have somewhat over-compensated the effects of multicollinearity on Var($\hat{\alpha}$).

It is important to determine that the binary regression model, $\underline{y} = X\underline{\alpha}$, is a faithful, as well as useful, <u>surrogate</u> for the probit regression model, $\underline{z} = X\underline{\beta}$. 1) The multicollinearity diagnostics and the respective distributions, e.g., index plots, etc., of the cognate case statistics, r_i, h_i and $D_i{}^1$ are similar for both models and 2) The correlation coefficient of the respective estimates of response, Π_i, of the two models is $r = 0.92$. (It will be recalled in the linear model, $\underline{y} = X\underline{\alpha}$, of the binary response, $y = 0, 1$, the expected value of the response at the treatment $\underline{x}'$ is just the probability that $y = 1$, or, $E(\hat{y}) = P(y = 1) = x'\hat{\underline{\alpha}}$.

Estimates, $\underline{\beta}^+$, corrected for the bias, $\underline{b}$, present in the Maximum Likelihood estimates, $\underline{\hat{\beta}}$, as a result of the finite sample size, n, are easily obtained: $\underline{\beta}^+ = \underline{\hat{\beta}} - \underline{b}$. Then the ratios, $(\hat{\alpha}_2/\hat{\alpha}_1)$, $(\hat{\alpha}_3/\hat{\alpha}_1)$, for the corrected estimates of $\underline{\alpha}$ (corrected for joint presence of <u>collinearity</u> in the distribution of N and T and <u>random error</u> in the measurement of D) were taken to define the stochastic constraints, $\underline{r} = R\underline{\beta} + \underline{v}$, on the estimates, $\underline{\beta}^+$. For these constraints the variance-covariance matrix of v was selected to be $Var(\underline{v}) = \Psi = 10^{-2}I$, where I is a (2x2) identity matrix. The variance-covariance matrix for $\underline{\beta}^+$ taken to be that for $\underline{\hat{\beta}}$: $Var(\underline{\hat{\beta}}^+) = Var(\underline{\hat{\beta}})$. The Mixed Estimates $\hat{\beta}_j^{**}, \hat{\beta}_j^{**}/\sqrt{Var(\hat{\beta}_j^{**})}$ are summarized in the following equation (See also Table 17):

$$\hat{z} = -39.667 + 17.231x_1 - 12.699x_2 - 3.221x_3$$
$$(-2.0613)\ (2.076)\quad (-2.0799)\ (-2.0772) \tag{53}$$

The so-called iso-effect curve is $x_1(0.90) = 2.376 + 0.737x_2 + 0.187x_3$, or $D(0.90) = 238T^{0.737}N^{0.187}$. The measure of compatibility of sample and non-sample information on $\underline{\beta}$ is $\gamma = 0.892$. Since γ is distributed (asymptotically) as chi-squared with 2 degrees of freedom on the null hypothesis of compatibility, the two are obviously consistent. The proportion of non-sample information included in these posterior estimates is $\Theta_p = 0.50$. Objections may be raised to this last statement on the grounds that the Ridge estimate, $\hat{\underline{\alpha}}^{**}$, from which we obtained the constraint, $\underline{r} = R\underline{\beta} + \underline{v}$ is "sample information". While it is true that $\hat{\underline{\alpha}}^{**}$ is estimated from the sample, it represents, or better, it <u>articulates</u>, the non-sample information on $\underline{\beta}$ represented in Table 1. Ridge regression is used to <u>implement</u> or deploy, <u>not</u> define, the apriori information on β, via Mixed Estimation.

We have been able to use the RR transformation on $\hat{\underline{\alpha}}$ for the surrogate model to generate the matrices r and R required for the constraint $\underline{r} = R\underline{\beta}$. It is less clear how we should choose the moment matrix, Ψ, required for the matrix-weighted average, $\hat{\underline{\beta}}^{**}$, for the probit model. The case is quite similar to that of Bayesian regression discussed by Leamer in Lecture 5. "Because it is impossible to select a precise ficticious data set it is necessary to study the sensitivity of the pooled estimates to changes in the ficticious data set." The moment matrix, Ψ, "... can only be said to lie in an interval $L \leq [\Psi] \leq U$, where $L \leq U$ means U-L is positive definite. Corresponding to such an 'interval' of prior covariance matrices is an 'interval' of pooled estimates $[\underline{\beta}^{**}]$. It is hoped that the interval of covariance matrices is wide enough to be credible and the corresponding interval of pooled estimates is narrow enough to be useful. If, on the other hand, an incredibly narrow interval of prior covariance matrices is required to get a usefully narrow interval of estimates, then estimation is suspended with this data set."

We find that for $10^{-4}I \leq \Psi \leq I$, where I is a (2x2) identity matrix and r, R are as given in Table 13a we obtain essentially the posterior estimates $\hat{\beta}_j^{**}$ and $\hat{\beta}_j^{**}/\sqrt{Var(\hat{\beta}_j^{**})}$ presented in Table 17a - 7 for which $\Psi = 10^{-2}I$. Over this interval of Ψ we find $q_p \simeq 0.50$.

The posterior estimates, $\hat{\beta}_2^{**}$ and $\hat{\beta}_3^{**}$, are represented by point a* in Figure 7. Note that it lies on the 0.44 confidence ellipse and thus (β_2^{**}, β_3^{**}), although vastly different numerically, do not differ "statistically" from the maximum likelihood estimates, ($\hat{\beta}_2$, $\hat{\beta}_3$), that are represented by point b in the Figure. (The estimate $\hat{\beta}^{**}$ lies on the 0.12 confidence ellipsoid on $\hat{\beta}$.) As we shall presently see, however, there are significant consequences for the numerical differences.

Note also that obtaining the Ridge estimates of $\hat{\beta}$ indirectly from the binary model has not inflated the RSS appreciably over what might have been achieved directly.

Note also that the trace of the posterior estimates, $\hat{\beta}_{(k)}^{**}$ as a function of the bias parameter, k, for the surrogate binary regression model seems to be a plausible approximation to the ridge trace that might have been obtained from the direct Ridge transformation of the parameter, $\hat{\beta}$, of the probit model, e.g., the RSS(k) increases only very slowly for rather large decreases in $\hat{\beta}_j^{**}(k)$. ("As k increases without bound from zero, $\hat{\beta}_k$ traces a curved path through the parameter space from $\hat{\beta}_0$ to $\underline{0}$. The path is uniquely determined so that the distance from $\bar{\beta}_k$ to $\underline{0}$ is diminished as rapidly as possible while the residual sum of squares is increased as slowly as possible" (Swindel 1976).) Or, $\hat{\beta}^{**}(k)$, $0 \leq 0.12$, lies along the "ridge" that would be defined by the family of concentric (1-α) confidence ellipses on ($\hat{\beta}_2$, $\hat{\beta}_3$). See Figure 12.

The above equation of the response, $\hat{z}$, now includes corrections for each of the defects of the sample described previously. The schedule of maneuvers to achieve these corrected estimates of β are summarized in Table 17. It is evident that the sign, size and significance of these posterior estimates of the parameter β are now consistent with both the a priori information (opinion) on β, and the sample information on β. Let us now examine whether the predictions, of response, z, of the corrected model are more consistent with prior information on the matter than are those of the uncorrected model, since the construction of an extrapolation model was the initial, "object of the exercise".

The predicted responses at the centroid and edge of the sample distribution of X for standard fractionation regimens (x_i' within RVH of sample) are, respectively, $\hat{z}_i = 0.417$ ($\hat{\Pi}_i = 0.66$) and $\hat{z}_i = 0.893$ ($\hat{\Pi}_i = 0.81$). The predicted response for the hypofractionation regimen (x_i' not within RVH of sample) is $\hat{z}_i = 1.428$ ($\hat{\Pi}_i = 0.92$). The latter seems more "plausible" - more consistent with our a priori anticipations from past experience.

We have noted earlier that the estimated levels of response, $\hat{\Pi}_i$, at (D, T, N) = (9000, 63, 20) for models M_1, M_2 and M_3 (see Table 3) are, respectively, $\hat{\Pi}_i = 0.003, 0.712$, and 1.00, as extrapolation models M_1 and M_3 give estimates of response that are clearly "impluasible". Although model M_2 gives a "plausible" estimate of response, it is strictly valid only for levels of the treatment variables x_i' for which the relation $N = 0.714T$, obtains. But for the extrapolation in question, $N = 0.317T$. Hence, we tend to prefer the estimate $\hat{\Pi}_i = 0.920$ corresponding to $\hat{z}_i = x_i'\hat{\beta}^{**}$ - our extrapolation

model.
The predictive value of this extrapolation model over the RVH of the reduced sample (n=44) as assessed by the correlation coefficient, r = 0.407, for the observed, y = 0, 1 and predicted, z > 0, z ≤ 0, responses is similar to that for the uncorrected model, r = 0.399.

It will be instructive to examine both the RR, EV and RR + EV transformations on $\hat{\underline{\alpha}}$ from a perspective provided by the geometry of X'X and in the context of another "salvage operation" - deletion of observations. Figures 12a and 12b are the 0.90 confidence ellipses for (α_1, α_2) and (α_2, α_3), respectively, for the full surrogate (binary) model, $\underline{\alpha} = (\alpha_0, \alpha_1, \alpha_2, \alpha_3)$, of the reduced data set (n = 44).

Note that in Figures 12a and 12b the variance-covariance matrix of the estimate $\hat{\underline{\alpha}}$ of the surrogate model is $Var_1(\hat{\underline{\alpha}}) = \hat{\sigma}^2(X'X)^{-1}$ and not $Var_2(\hat{\underline{\alpha}}) = \hat{\sigma}^2(X'X)^{-1}(X'\hat{V}^{-1}X)(X'X)^{-1}$ where $V = Diag[\hat{y}_i(1-\hat{y}_i)]$, $i=1,\ldots, 44$, is the "correct" variance-covariance matrix of the binary observations. Although the transformed matrix, $Var_2(\hat{\underline{\alpha}})$, is correct (statistically) it is not appropriate for our surrogate model (as remarked previously). As defined above the 0.90 confidence ellipse includes the set of values of α such that

$$(1/\hat{\sigma}^2)(\underline{\alpha} - \hat{\underline{\alpha}})'(X'X)(\underline{\alpha} - \hat{\underline{\alpha}}) \leq 2F(\alpha; 2, n-k) \quad (54)$$

where (n-k) = 40. The (α_2, α_3) ellipse in Figure 12b is cognate to the (β_2, β_3) ellipse in Figure 7: The respective eccentricities are ε (binary)= 0.988 and ε(probit) = 0.965. In each Figure point c represents the maximum likelihood estimates, $(\hat{\beta}_2, \hat{\beta}_3)$, and least squares estimates, $(\hat{\alpha}_2, \hat{\alpha}_3)$, for the full model of the sample with case #18 (#19) deleted (n=43). Note that in both the binary, (surrogate)model and in the probit model the effect of the deletion of case #18 is to "shrink" toward the respective origins, the estimators $\hat{\underline{\alpha}}$ and $\underline{\beta}$ thereby improving their consistency with the a priori information on the directions of $\underline{\alpha}$ and $\underline{\beta}$ ($-1 < \alpha_j/\alpha_1 < 0$; $\alpha_1 > 0$ and $-1 < \beta_j/\beta_1 < 0$; $\beta_1 > 0$) as well as reducing the respective Euclidean norms, $\underline{\alpha}'\underline{\alpha}$ and $\underline{\beta}'\underline{\beta}$.

Figures 12a and 12b show the trace of the ridge estimates, $\hat{\underline{\alpha}}^{**}(k)$ in the (α_2/α_3) and (α_1, α_2) planes, respectively. It is evident in Figures 12a and 12b that the Ridge Trace indeed follows the "ridge" that would be defined by the respective families of concentric $(1-\alpha)$ confidence ellipses in each Figure. At k = 0.00 the ridge estimates coincide with the maximum likelihood estimates, $\hat{\underline{\alpha}}^{**}(0.00) = \hat{\underline{\alpha}}$ and at k = 0.10 the ridge estimate, $\hat{\underline{\alpha}}^{**}(0.10)$, has stabilized and is now consistent with the a priori information in Tables 1 and 11. We repeat Swindels' remarks: "As k increases without bound from zero, $[\hat{\underline{\alpha}}^{**}]$ traces a curved path through the parameter space from $[\hat{\underline{\alpha}}]$ to $\underline{0}$. The path is uniquely determined so that the distance from $[\hat{\underline{\alpha}}]$ to $\underline{0}$ is diminished as rapidly as possible while the residual sum of squares increases as slowly as possible" (Swindel 1976). Hence, the ridge trace lies (approximately) along the major axis of the confidence ellipses in Figures 12a and 12b. It is evident from the ridge trace of Figures 11 and 12 that the least squares estimates of α_2 and α_3 are quite sensitive to the data since only a small perturbation of the Normal equations matrix, X'X ⟶ X'X + kI, k = 0.12, has induced profound changes in the estimates of α_2 and α_3. This evidence from the Ridge Trace is consistent with the evidence from the Regression Diagnostics since deletion of case #18 changed the least squares estimates, $\hat{\underline{\alpha}}$, by about the same amount. This can be seen in Figures 12a and 12b where the estimates of $\hat{\underline{\alpha}}$ for n = 43 - point c - are quite close to the ridge regression estimates for k = 0.10. (In Figure 12a point c is at the 0.35 confidence ellipse. In Figure 12b, it is at the 0.39 confidence ellipse.)

There is a similar kind of degree of sensitivity of the maximum likelihood estimates, $\hat{\beta}$, to the joint presence of collinearity in the treatment variables (columns of X) and to extreme levels of these variables as well as to outlying responses (rows of [y, X]). These effects are seen in Figure 7 in which the respective "beta-shifts" are presented in the geometry of the $(X'\hat{V}^{-1}X)^{-1}$ matrix. Shown is the 0.90 confidence ellipse on the subset (β_2, β_3) of $\underline{\beta}$: 1) The shift from point 0 to point b is produced by deletion of the outlying response ($t_1 = -4.40$). 2) The shift from point b to point c is produced by deletion of the extreme level of treatment ($h_{18} = 0.40$). 3) The shift from point b to point d is produced by deletion of $\underline{X}_3$ (VIF$_3$ = 20.10). 4) The shift from point b to point a* is produced (via Mixed Estimation) by "cascading" the Ridge Regression and Errors-in-variables transformations. (Recall that 1) is for n = 45 and 2) - 4) are for n = 44.) We note that the point a* is on the 0.44 confidence ellipse in (β_2, β_3) - subspace but the cognate point in $\underline{\beta}$-space is on the 0.12 confidence ellipsoid.

Points a and b in Figures 12a and 12b describe the errors-in-variables estimates of $\underline{\alpha} - \hat{\underline{\alpha}}_{EV} = [X'\overline{X} - nD]^{-1}(\overline{X}'X)\hat{\underline{\alpha}}$ - for $g_1^2 = 0.01^2$ and 0.02^2, respectively. (These correspond to standard deviations of random errors of measurement of dose of 150 and 300 rads.) Note that the Errors-in-Variables transformation, $\hat{\underline{\alpha}} \longrightarrow \hat{\underline{\alpha}}_{EV}$, "inflates" $\hat{\alpha}_1$ much more than it does either $\hat{\alpha}_2$ or $\hat{\alpha}_3$ - as would be expected since the random errors of measurement are largest for D (N and T are presumably observed without error). Compare the respective effects on $\hat{\alpha}_1$, $\hat{\alpha}_2$ and $\hat{\alpha}_3$ of the RR (the points k = 0.01 and 0.12 on the Ridge trace) and the EV (the points a and b on the dashed line) transformations of $\hat{\underline{\alpha}}$ in Figures 12a and 12b. It is evident that the so-called "alpha-shifts" of the least squares estimates produced by transformations of $\hat{\underline{\alpha}}$ corrresponding to k = 0.10 ($\hat{\underline{\alpha}}_{RR}$) and $g_1^2 = 0.02^2$ ($\hat{\underline{\alpha}}_{EV}$) are each of a size similar to that produced by deletion of observation #18 (#19 in full sample), point c, or deletion of variable $\underline{X}_3$, $(\hat{\alpha}_2, \hat{\alpha}_3) = (-3.68, 0)$. Thus, the effects on the least squares sample estimate, $\hat{\underline{\alpha}}$, of the presence of extreme levels, $\underline{x_i}'$, multicollinearity, and random errors of measurement in $\underline{X}_1$ seem to be quite similar.

Points a* and b* describe the "cascaded", RR + EV, estimates of $\underline{\alpha} - \hat{\underline{\alpha}}_{EV} = [X'X - nD]^{-1}(X'X)\hat{\underline{\alpha}}_R$ - for $g_1^2 = 0.01^2$ and $g_1^2 = 0.02^2$, respectively. $\hat{\underline{\alpha}}_R$ is the Ridge regression estimate of $\underline{\alpha}$ for k = 0.12. It will be recalled that k = 0.12 was selected over k = 0.10 as the optimal value for the bias parameter in order to compensate for the inflation of estimates of α_2 and α_3 produced by the EV transformation.

17. ERRORS OF IDENTIFICATION AND TIME TRENDS IN THE SAMPLE

We now discuss briefly two other idiosyncracies of non-experimental data. Each may be considered as a "contamination" of the sample.

We have noted that the presence of the prominant "outlier" in the initial sample, observation #1, exerted a profound effect upon the estimates of Π in the first-order response surface model, that is, the beta-shift diagnostic, D_1^1 was very large. We included in the set of probable causes of the anamalous reported response (recurrence of disease, E_1, at an intense level of treatment, $\underline{x_i}$') the occurrence of an error of identification, either at allocation to treatment

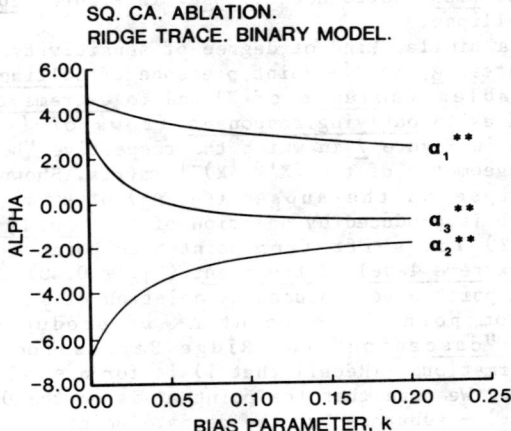

Figure 11. Ridge trace (natural basis) for binary model of ablation, E_1 (n = 44). The ridge transforms of each of the least squares estimates, $\hat{\alpha}_j$, j = 1, 2, 3 has stabilized at k = 0.10. Note that $\hat{\alpha}_2$ and $\hat{\alpha}_3$ have "shrunk" more than $\hat{\alpha}_1$.

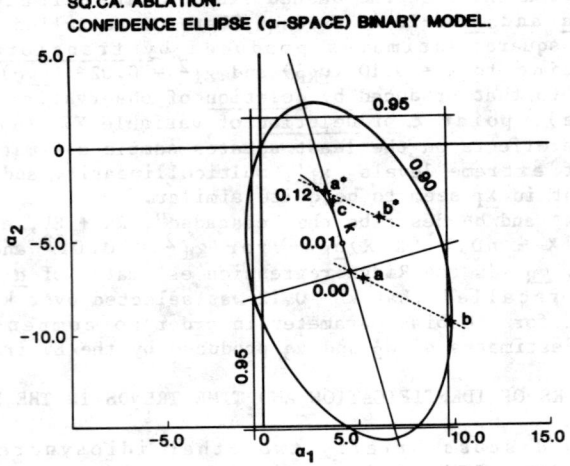

Figure 12a. Interval estimates, $\hat{\underline{\alpha}}$. 0.90 simultaneous confidence intervals for (α_1, α_2) for the binary model of ablation, E_1. The Figure describes the "alpha-shifts" corresponding to 1) deletion of extreme level, case #18 (point c), 2) Ridge Regression (RR) Transformations of $\hat{\underline{\alpha}}$, for $0 \le k \le 0.12$, 3) Errors-in-Variables (EV) transformations of $\hat{\underline{\alpha}}$ for s = 150 and s = 300 rad, (points a and b), 4) "Cascade" of RR and EV transformations (points a* and b*).

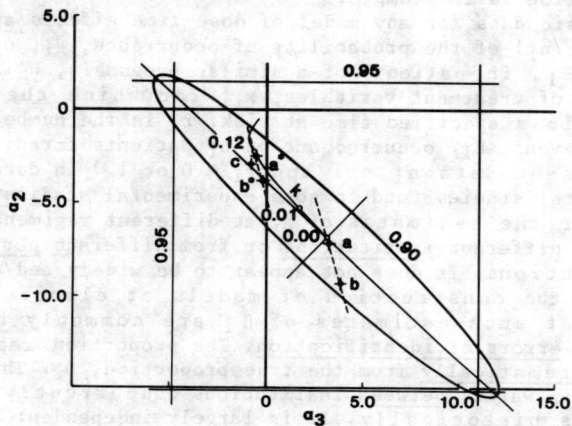

Figure <u>12b</u>. Interval estimates, $\hat{\underline{\alpha}}$. 0.90 simultaneous confidence intervals for (α_2, α_3) for the binary model of ablation, E_1. See legend under Figure <u>12a</u> for description. Note in both Figures that the RR transformation "shrinks" $\hat{\alpha}_2$ and $\hat{\alpha}_3$ more than $\hat{\alpha}_1$ and that the EV transformation "inflates" $\hat{\alpha}_1$ more than $\hat{\alpha}_2$ or $\hat{\alpha}_3$. (The ridge regression estimator, $\hat{\underline{\alpha}}_R$, is commonly described as a linear transformation of the least squares estimator, $\hat{\underline{\alpha}}$, in the so-called <u>correlation</u> basis:

$$\hat{\underline{\alpha}}_R = (X'X + kI)^{-1}(X'X)\hat{\underline{\alpha}}.$$

It is perhaps useful to note that the maneuver may also be considered as a variety of the data-augmentation measures described by Dykstra (1966, 1971). But in ridge regression, it is the X'X matrix rather than the X matrix that is augmented. And, as Leamer 1984 has remarked, the augmenting data in the latter maneuver are more substantial (less fictitious) than those in the former. See also Montgomery and Peck 1982.)

(incorrect staging of case #1) or at evaluation of response to treatment (false negative). Let us now generalize the treatment of the effects on the estimates of Π of the presence of both sorts of errors of identification in the sample.

The basic data for any model of dose-time effects are the sample estimates, (r_i/n_i) of the probability of occurrence, Π_i, of the binary event, say E_1, in patients of a similar prognosis, u_h' irradiated at each of a set of treatment variables, x_i', for which the event, E_1, occurs within a specified time at risk. r_i is the number of patients in which the event, E_1, occurred out of n_i patients irradiated at x_i'. (In the present data $n_i = 1$ and $r_i = 0$ or 1.) In data obtained in non-experimental studies (and in some experimental studies) it is not uncommon for the estimates of Π_i at different regimens, x_i', to be obtained from different <u>institutions</u> or from different <u>periods</u> at the same institutions. It does not appear to be widely and/or explicitly recognized in the construction of models of clinical radiation response that such estimates of Π_i are commonly <u>biased</u> by the occurrence of <u>errors of identification</u>. The proportion reported, Π_i^*, differs <u>systematically</u> from the true proportion, Π_i. The difference, $\Delta\Pi_i = \Pi_i^* - \Pi_i$, varies <u>between</u> institutions concurrently and <u>within</u> institutions historically. It is largely independent of the sample size, n_i. The existence of this bias is well-known to those who design randomized clinical trials and has been a strong argument <u>against</u> the use of <u>historical</u> controls, instead of <u>concurrent controls</u>, within the same institution. (The use of historical controls is supported, of course, by strong practical arguments, not the least persuasive of which is the fact that it requires only 1/4 of the number of patients to achieve the same <u>power</u> and <u>significance</u> for a test of the null hypothesis as is required with <u>concurrent</u> controls. Lecture <u>9</u>.)

The size of the bias, $\Delta\Pi_i$, is dependent jointly upon the size of the <u>errors of identification</u> at <u>allocation</u> to therapy and at <u>evaluation</u> of response to therapy, say (α, β) and (γ, δ), respectively, which obtain at a given institution during a given period and upon the distribution of patients over the several prognostic strata, u_h', of a given system (patient disease) which prevailed at the institution during that period.

It can be shown that the expression below describes the result of the occurrence of these errors in "cascade". This causes the reported estimate, Π_i^*, to differ from the true estimate, Π_i, of the yield of ablation, E_1, in patients of prognostic stratum, u_h', irradiated at treatment regimen, x_i'; $h = 1, 2$. The observed probability of occurrence of E_1 at x_i' in the stratum 1 is

$$\Pi_i^* = \delta + \{(1-\delta-\gamma)/[1+(n_2/n_1)(\beta/(1-\alpha))]\}[\Pi_1+(n_2/n_1)(\beta/(1-\alpha))\Pi_2] \quad (55)$$

δ — Probability of a <u>false positive</u>, that is, of identifying a recurrence, $\overline{E}$, of disease as ablation, E, at evaluation.

γ — Probability of a <u>false negative</u>, that is, of identifying an ablation, E, as recurrence, $\overline{E}$, at evaluation.

α — Proportion of stratum u_1' incorrectly identified and <u>not</u> allocated to regimen x_i'.

$1-\alpha$ — Proportion of stratum u_1' correctly identified and allocated to regimen x_i'.

β — Proportion of stratum u_2' incorrectly identified as stratum u_1' and allocated to x_i'.

n_1 — Number of patients in stratum u_1' at institution.

n_2 — Number of patients in stratum u_2' at institution.

Note that n_1 and n_2 are functions of the referral pattern, which in turn is a function of the staffing, funding, siting, etc. of the institution during a given period. Note also that (α, β) and (δ, γ) can be reduced by the use of discriminant analysis (See Lachenbruch, Lecture 10).

Π_1 – True yield of ablation at $\underline{x_i}'$ in stratum $\underline{u_1}'$.
Π_2 – True yield of ablation at $\underline{x_i}'$ in strutum $\underline{u_2}'$.
$n = n_1(1-\alpha) + \beta n_2$ – Number of patients irradiated at $\underline{x_i}'$.
$\underline{x_i}$ – Irradiation schedule.

It can be readily shown that the bias may become non-negligible. The following values of error rates do not appear to be inconsistent with those obtained in the few studies which have been made of current identification procedures: $\delta = 0.10$, $\gamma = 0.20$, $\beta = 0.15$, $\alpha = 0.15$. (For current laboratory and imaging techniques the total error rates are about 0.30.) The ratio n_2/n_1 is taken to be 1.30 and $\Pi_1 = 0.70$ and $\Pi_2 = 0.40$. Then it is readily shown that $\Pi_i^* = 0.55$. The bias is $\Delta\Pi = -0.15$. The variance of the biased estimate is $\Pi^*(1-\Pi^*)/n$ and variance of the unbiased estimates is $\Pi(1-\Pi)/n$. In general, the presence of the errors of identification inflates the variance of Π_h^*.

The temptation to uncritically "pool" the data on the probability of occurrence of an event, E_i at various treatment regimens obtained from disparate institutions and/or periods must be tempered by the recognition that the subsequently derived associations between Π_i and $\underline{x_i}'$, such as are described by the response surface may be, in substantial part, illusory. In particular, the errors of identification may introduce second-order terms into the model, $z_i = \underline{x_i}'\beta$. Note that the effects of errors of identification of response, y, and errors of measurement of dose, $\underline{X_1}$, upon the dose-response curve are similar: the estimated slope is less than the true slope. The size of the coefficients of these second-order terms may be (incorrectly) identified as clinical evidence of the occurrence of such phenomena as "supralethality", "reoxygenation", etc.

Each observation, $(r/n)_i$, is marked by the circumstances of its place and time of origin. Among these are the idiosyncracies in the sizes of the error rates (α, β) and (δ, γ) which characterize the identification procedures which obtain at allocation to therapy and at evaluation of response to therapy, respectively, and the distribution of patients over the prognostic strata, $\underline{u_h}'$, which is fixed by the local referral pattern during the period that the patients are being allocated to treatment. Considerably more effort should be expended in the quality control of patient data, in particular, in the acquisition of estimates of the errors of identification which are characteristic of each diagnostic procedure and in the documentation of the distribution of patients over the prognostic strata from which the patients are allocated. These estimates are regularly absent from the data of experimental as well as non-experimental studies.

Another idiosyncracy of such data which also has perhaps been too seldom recognized is the "age" with respect to the date of the study of each observation in the sample. It is, in general, the case that in order to achieve an approprate value of the sample size, say n = 100 or 20 p, it is necessary to cumulate, or "pool", data over time as well as across stages, instututions, etc. It may require several years, say ten to twenty, (even at large institutions) to accumulate a sufficient number of observations (even with pooling across prognostic strata, etc.) to reduce the bias and variance of the ML estimates of $\underline{\beta}$ to acceptable levels. But it seems not at all unlikely that the observations on patient response obtained ten to twenty years prior to the study are not as appropriate for prediction of response in current patients as are those on more recent patients because of secular changes in the patients, in their management, in their positions on the respective "learning curves" of primary treatment and ancillary support, at a given institution, etc. There are <u>time-trends</u>, as well as multicollinearity, outlying response, extreme levels of treatment, random errors of measurement, etc. present in each sample of non-experimentl data. Thus, each observation should be discounted for its "age" by assigning a <u>weight</u>, say $e^{-\zeta(t_0-t_i)}$, for the ith case in the sample much as is currently done in the forcasting from time-series (See Brown 1963) - or in robust regression. t_0 is the date at which the response surface is constructed. t_i is the date at which the ith case completed treatment. Estimates of the possible effects of the "age" of case on $\underline{y}$ could be obtained by including the "age" ($t_0 - t_i$) as a covariate in the model matrix, X.

18. DISCUSSION

We have presented an empirical probit model, $\underline{z} = X\underline{\beta} + \underline{\varepsilon}$, of a set of clinical observations on a binary response - ablation/recurrence of a head and neck tumor - that were accumulated in a non-experimental study. The data have been previously published and thus may be taken to be typical of such studies. Such data are characterized by small sample sizes, the absence of randomization in the relation of the sample to the target population, by the presence of collinearity of the treatment variables, by influential observations at extreme levels of the treatment variables, by anomalous or outlying levels of response, by errors of measurement in one or more of the treatment variables, etc. These characteristics of the sample give rise to characteristic <u>model-sample interactions</u> which frustrate the efforts to construct either explanatory or predictive models of the data. This frustration has been fruitful of several colorful expressions (See Brownlee 1965) of dissatisfaction with non-experimental data.

We have introduced methods which describe and measure the nature and degree of the <u>model-sample interactions</u> which circumscribe the validity of the inferences on the causal mechanisms which generated these observations, as well as the predictive performance of the model. The methods are the Regression Diagnostics. We have also introduced some methods for ameliorating some of the effects of some of these interactions on the estimates $\hat{\underline{\beta}}$ and Var$(\hat{\underline{\beta}})$ for the model. These are the Bayesian methods of Mixed Estimation, Ridge Regression and Errors-in-Variables Estimation. These methods were augmented by application of the cruder procedures of <u>deletion</u> to outlying observations and to <u>collinear</u> variables in order to reduce the effects of these features on the estimates $\hat{\underline{\beta}}$ and Var$(\hat{\underline{\beta}})$ of the model.

In the application of the methods of Ridge Regression and of Errors-in-Variables Estimation to the correct, probit, models of the binary response data it was necessary to introduce a surrogate model, $\underline{y} = X\underline{\alpha} + \underline{u}$, of the data in which the response $\underline{y}$ takes only the values 0 or 1, since versions of these methods that are appropriate to the probit model have not yet been developed. The effective RR and EV transformations of the Maximum Likelihood estimates, $\hat{\underline{\beta}}$, for the probit model were achieved by Mixed Estimation methods.

ACKNOWLEDGEMENT

We acknowledge, with thanks, the dedication, as well as expertise, of Marie Woodall in the preparation of the manuscript.

REFERENCES

AKAIKE, H. (1974), A New Look at the Statistical Model Identification, IEEE Trans. on Auto. Cont. AC-19(6) 716-723.
────── (1977), On Entropy Maximization Principle, in Applications of Statistics, P. R. Krishnaiah, ed. North-Holland, 27-41.
ALTMAN, D. G. (1982), Statistics in Medical Journals, Statis. in Med. 1: 59-71.
ANDERBERG, M. R. (1973), Cluster Analysis for Application. Academic Press. N.Y.
ANG, K. K., VAN DER KOGEL, A. J. and VAN DER SCHUEREN, E. (1983), Int. Journal Radiation Oncology, Biology, Physics, 9: 1487-1491.
ATKINSON, A. C. (1980), A Note on the Generalized Information Criterion for Choice of a Model, Biometrika, 67: 413-418.
BAKER, R. J. and NELDER, J. A. (1978), The GLIM System Release 3: Generalised Linear Interactive Modelling, Numerical Algorithms Group 7 Banbury Rd., Oxford, OX2 6NN.
BEATON, A. E., RUBIN, D. B. and BARONE, J. L. (1976) The Acceptability of Regression Solutions: Another Look at Computational Accuracy, Journal of the Amer. Statis. Assn. 71: 158-168.
BELSLEY, D. A. (1984), Demeaning Conditioning Diagnostics Through Centering, American Statistician, 38(2): 73-82.
BELSLEY, D. A., KUH, E. and WELSCH, R. E. (1980) Regression Diagnostics: Identifying Influential Data and Sources of Collinearity, John Wiley & Sons, N.Y.
BOARDMAN, T., Lecture 6, Statistical Graphics on Smaller Computers: The Data Analyst's New Tools, See This Volume.
BOX, G.E.P. (1979) Robustness in the Strategy of Scientific Model Building, in Robustness in Statistics, R. Launer and G. N. Wilkinson eds., Adacemic Press, N.Y.
BOX, G.E.P., HUNTER, W. G. and HUNTER, J. S. (1978) Statistics for Experimenters, John Wiley & Sons, N.Y.
BOX, G.E.P. and LUCAS, H. L. (1959) Design of Experiments in Nonlinear Situations, Biometrika, 46: 77-90.
BOX, G.E.P. and WETZ, J. (1973) Criteria for Judging Adequacy of Estimation by an Approximating Response Function, Technical Report No. 9, Dept. of Statis., University of Wisconsin, Madison, WI. 53706.
BROWN, R. G. (1963) Smoothing, Forecasting and Prediction of Discrete Time Series, Prentice-Hall, Englewood CLiffs, N.J.
BROWNLEE, K. A. (1965) Statistical Theory and Methodology in Science and Engineering, 2nd Ed., John Wiley & Sons, N.Y.
CARROLL, R. J., SPIEGELMAN, C. H., LAN, K.K.G., BAILEY, K. T. and ABBOTT, R. D. (1984) On Errors-in-Variables for Binary Regression Models, Biometrika, 71: 19-25.
CHATTERJEE, S. and PRICE, B. (1977) Regression Analysis by Example, John Wiley & Sons, N.Y.
COHEN, L. (1983) Biophysical Models in Radiation Oncology, CRC Press, Boca Raton, FL.
CONNIFFE, D. and STONE, J. (1973) A Critical View of Ridge Regression, The Statistician, 22(2): 181-187.
COOK, R. D. and WANG, P. C. (1983) Transformations and Influential

COOK, R. D. and WEISBERG, S. (1982) Residuals and Influence in Regression, Chapman and Hall, N.Y.
CORNFIELD, J. (1959) Principles of Research, Amer. Journal of Mental Deficiency, 64: 240-252.
CORNFIELD, J., GORDON, T. and SMITH. W. (1961), Quantal Response Curves for Experimentally Uncontrolled Variables, Institut. Int'l. de Sttistique, Rome, Bulletin 38: 97-115.
COX, D. R. (1970) The Analysis of Binary Data, Methuen & Co. Ltd., London.
CROCKER, D. C. (1972) Some Interpretations of the Multiple Correlation Coefficient, The Amer. Statistician, 31-33.
DANIEL, C. and WOOD, F. S. (1971) Fitting Equations to Data: Computer Analysis of Multifactor Data for Scientists and Engineers, Wiley-Interscience. N.Y.
DARLINGTON, R. B. (1978) Reduced-Variance Regression, Psychological Bulletin, 85(6): 1238-1255.
DAVIES, O. L., ed. (1961) Statistical Methods in Research and Production. 3rd ed., Hafner Pub. Co., N.Y.
_____, ed. (1967) The Design and Analysis of Industrial Experiments, 2nd ed., Hafner Pub. Co., N.Y.
DAVIES, R. B. and HUTTON, B. (1975) The Effect of Errors in the Independent Variables in Linear Regression, Biometrika, 62(2): 383-391.
DUNCAN, G. T. (1978) An Empirical Study of Jackknife-Constructed Confidence Regions in Nonlinear Regression, Technometrics, 20(2): 123-129.
DYKSTRA, O. (1966) The Orthogonalization of Undesigned Experiments, Technometrics, 8(2): 279-290.
_____ (1971) The Augmentation of Experimental Data to Maximize $|X'X|$, Technometrics, 13(3): 682-688.
ELLIS, F. (1969) Dose, Time and Fractionation: A Clinical Hypothesis, Clinical Radiology, 20: 1-7.
FARRAR, D. and GLAUBER, R. R. (1967) Multicollinearity in Regression Analysis: The Problem Revisited, The Review of Economics and Statis. 49: 92-109.
FILLIBEN, J. J. (1975) The Probability Plot Correlation Coefficient Test for Normality, Technometrics, 17(1): 111-117.
FINNEY, D. J. (1971) Probit Analysis, 3rd ed., University Press, Cambridge.
FISHER, R. A. (1958) Statistical Methods for Research Workers, 13th ed. Oliver & Boyd, London.
FROME, E., Lecture 7, Regression Methods for Binomial and Poisson Distributed Data, See This Volume.
GAYLOR, D. W. and MERRILL, J. A. (1968) Augmenting Existing Data in Multiple Regression, Technometrics, 10(1): 73-81.
GEHAN, E., Lecture 8, Regression Methods in General Survival Analysis, See This Volume.
GEISSER, S. and EDDY, W. (1979) A Predictive Approach to Model Selection, Journal of the American Statistical Assn. 74: 153-160.
GORDAN, T. (1974) Hazards in the Use of the Logistic Function with Special Reference to Data from Prospective Cardiovascular Studies, Journal of Chronic Diseases, 27: 97-102.
GUNST, R. F. (1980) Comment on G. Smith & F. Campbell paper, A Critique of Some Ridge Regression Methods, Journal of the American Statistical Assn. 75: 98-100.

HALD, A. (1952) Statistical Theory with Engineering Applications, John Wiley & Sons. N.Y.
HERBERT, D. E., Lecture 14, Clinical Radiocarcinogenesis.Applications of Regression Diagnostics and Bayesian Methods to Poisson Regression Models, See This Volume.
_____, Lecture 17, Clinical Dose-Response Models. II. Probit and Logit Models of Experimental and Non-Experimental Data, See This Volume.
_____, Lecture 19, Clinical Diagnostic Models. Discriminant Analysis of In Vitro NMR Measurements on Normal and Malignant Tissues, See This Volume.
HOCKING, R. R. (1983) Developments in Linear Regression Methodology: 1959-1982, Technometrics, 25(3): 219-249.
HOCKING, R. R. and PENDLETON, O. J. (1983) The Regression Dilemma, Communications in Statistics - Theory & Methods, 12(5): 497-527.
HOERL, A. E. and KENNARD, R. W. (1970) Ridge Regression: Biased Estimation for Nonorthogonal Problems, Technometrics, 12(1): 55-67.
HOERL, A. E., KENNARD, R. W. and BALDWIN, K. F. (1975) Ridge Regression: Some Simulations, Communications in Statistics, 4(2): 105-123.
HUBER, P. J. (1977) Robust Statistical Procedures, Society for Industrial and Applied Mathematics, Philadelphia, PA.
JOHNSTON, J. (1972) Econometric Methods, 2nd ed., McGraw-Hlll, N.Y.
JONES, H. B. (1956), Demographic Consideration of the Cancer Problem. Trans. N. Y. Academy of Sciences, 18(4): 298-333.
KAPLAN, H. S. (1966) Evidence for a Tumoricidal Dose Level in the Radiotherapy of Hoddgkin's Disease, Cancer Research, 26: 1221-1224.
KLEINBAUM, D. G. and KUPPER, L. L. (1978) Applied Regression Analysis and Other Multivariable Methods, Duxbury Press, North Scituate, MA.
LACHENBRUCH, P. A. (1975) Discriminant Analysis, Hafner Press, N.Y.
_____, Lecture 10, Overview of Discriminant Analysis, See This Volume.
LEAMER, E. E. (1978) Specfication Searches: Ad Hoc Inference with Nonexperimental Data, Wiley-Interscience, N.Y.
_____ (1982) Sets of Posterior Means with Bounded Variance Priors, Econometrica, 50(3): 725-736.
_____ (1983) Let's Take the Con out of Econometrics, The American Economic Review, 73: 31-43.
_____ (1984) Global Sensitivity Results for Generalized Least Squares Estimates, Journal of the Amer. Statistical Assn. 79: 867-870.
_____, Lecture 5, Bayesian Regression and Sensitivity Analysis, See This Volume.
LINDEMAN, R. H., MERENDA, P. F. and GOLD, R. Z.(1980) Introduction to Bivariate and Multivariate Analysis, Scott, Foreman & Co., Glenview, IL.
MANTEL, N. (1983) Cautions on the Use of Medical Databases, Statistics in Medicine, 2:(355-362.
MARQUARDT, D. W. (1980) Comment - A Critique of Some Ridge Regression Methods, Journal of the Amer. Statistical Assn., 75: 87-91.
MARQUARDT, D. W. and SNEE, R. D. (1975) Ridge Regression in Practice, The American Statistician, 29(1): 3-20.
MAZUR, A. (1973) Disputes Between Experts, Minerva: A Review of Science, Learning and Policy, H: 243-262.
MCCULLAGH, P. and NELDER, J. (1983) Generalized Linear Models, Chapman and Hall, N.Y.
MCDONALD, G. C. and GALARNEAU, D. I. (1975) A Monte Carlo Evaluation of Some Ridge-Tupe Estimators, Journal of the Amer. Statistical Assn 70: 407-416.

MILLER, D. M. (1984) Reducing Transformation Bias in Curve Fitting, The American Statistician, 38(2): 124-126.
MONTGOMERY, D., Lecture 4, Biased Estimation and Robust Regression, See This Volume.
MONTGOMERY, D. and PECK, E. (1982) Introduction to Linear Regression Analysis, John Wiley & Sons, N.Y.
MOSTELLER, F. and TUKEY, J. (1977) Data Analysis and Regression, Addison-Wesley, Reading, MA.
MYERS, M. H., AXTELL, L. M. and ZELEN, M. (1966) The Use of Prognostic Factors in Predicting Survival for Breast Cancer Patients, Journal of Chronic Diseases, 19: 923-933.
MYERS, R. H. (1976) Response Surface Methodology, Edwards Brothers, Inc. Distributors, Ann Arbor, MI.
_____, Lecture 1 and 2, Regression Analysis - Basics and Current Frontiers, See This Volume.
NETER, J. and WASSERMAN, W. (1974), Applied Linear Statistical Models, Richard H. Irvin, Publ., Homewood, IL.
OBENCHAIN, R. L. (1978) Good and Optimal Ridge Estimators, Annals of Statistics, 6(5): 1111-1121.
_____ (1977) Classical F-Tests and Confidence Regions for Ridge Regression, Technometrics, 19(4): 429-438.
_____ (1980) Comment on G. Smith and F. Campbell paper, A Critique of Some Ridge Regression Methods, Journal of the Amer. Statistical Assn 75: 95-96.
OMAN, S. D. (1983) Regression Estimation for a Bounded Response Over a Bounded Region, Technometrics, 25(3): 251-261.
PARK, C. N. and SNEE, R. D. (1983) Quantitative Risk Assessment: State-of-the-Art for Carcinogenesis, The American Statistician, 37(4): 427-441.
PEARSON, K. (1957) The Grammer of Science, Meridian Books, Inc. N.Y.
POCHIN, E. E. (1970) The Development of the Quantitative Bases for Radiation Protection, British Journal of Radiology, 43: 155-160.
PREGIBON, D. (1981) Logistic Regression Diagnostics, Annals of Statistics, 9(4): 705-724.
SCHAEFER, R. L. (1983) Bias Correction in Maximum Likelihood Logistic Regression, Statistics in Medicine, 2: 71-78.
SCHAEFER, R. L., ROI, L. D. and WOLFE, R. A. (1984) A Ridge Logistic Estimator, Communication in Statistics - Theory & Methods, 13(1): 99-113.
SCHNEIDERMAN, M. A. (1967) Mouse to Man: Statistical Problems in Bringing a Drug to Clinical Trial, 5th Berkley Symposium on Mathematical Statistics and Probability: Biology and Problems of Health, IV: pp 855-866.
SCHOENFELD, D., Lecture 9, Animal Experiments and Clinical Trials, See This Volume.
SCHWARZ, G. (1978) Estimating the Dimension of a Model, Annals of Statistics, 6(2): 461-464.
SEBER, G.A.F. (1977) Linear Regression Analysis, John Wiley & Sons, N.Y.
SMITH, G. and CAMPBELL, F. (1980) A Critique of Some Ridge Regression Methods, Journal of the Amer. Statistical Assn., 75: 74-81.
SNEE, R. D. (1973) Some Aspects of Nonorthogonal Data Analysis: Part I Developing Prediction Equations, Journal of Quality Technology, 5(2): 67-79.
_____ (1973) Some Aspects of Nonorthogonal Data Analysis: Part II. Comparison of Means, Journal of Quality Technology, 5(3): 109-122.

_____ (1977) Validation of Regression Models: Methods and Examples, Technometrics, 19(4): 415-428.
STEWART, J. G. and JACKSON, A. W. (1975) The Steepness of the Dose Response Curve Both for Tumor Cure and Normal Tissue Injury, The Laryngoscope, LXXXV: (7) 1107-1111.
STONE, M. (1976) An Asymptotic Equivalence of Choice of Model by Cross-Validation and Akaike's Criterion, Science, 8: 44-47.
SUPE, S. J., NAGALAXMI, K. V. and MEENAKSI, L. (1983) Tumor Significant Dose, Medical Phyics, 10(1): 51-56.
SWINDEL, B. F. (1976) Good Ridge Estimators Based on Prior Information Communications in Statistics- Theory & Methods, A5(11): 1065-1075.
THEIL, H. (1963) On the Use of Incomplete Prior Information in Regression Analysis, Journal of the Amer. Statistical Assn. June, 401-414.
_____ (1971) Principles of Econometrics, John Wiley & Sons, N.Y.
THEIL, H. AND GOLDBERGER, A. S. (1961) On Pure and Mixed Statistical Estimation in Economics, International Economic Review, 2(1): 65-78.
THISTED, R. A. (1980) Comment on Smith and Campbell paper, A Critique of Some Ridge Regression Methods, Journal of the Amer. Statistical Assn., 75: 81-86.
TVERSKY, A. and KAHNEMAN, D. (1982) Belief in the Law of Small Numbers in Judgement Under Uncertainty: Heuristics and Biases. D. Kahneman, P. Slovic & A. Tversky, eds., University Press, Cambridge, 23-31.
WEINBERG, A. M. (1972) Science and Trans-Science, Minerva, 10: 209-222.
WEISBERG, S. (1980) Applied Linear Regression, John Wiley & Sons, N.Y.
WELSCH, R., Lecture 3, Introduction to Regression Diagnostics, See This Volume.
WOLD, H. O. (1969) Nonexperimental Statistical Analysis From the General Point of View of Scientific Method, Institut Int'l. De Statis., Rome, 42(1): 391-424.
WOLD, S. (1974), Spline Functions in Data Analysis, Technometrics, 16 (1): 1-11.

NOTES ADDED IN PROOF

1. In the empirical models that have been considered in this Lecture the time has been included only as a "dose factor", T, (or $\underline{X}_2$) a column of the model matrix, X. However, it is important to note that time also affects the level of response, $\underline{y}$, by another route namely as the time to occurrence, $t > 0$, of the binary event, E. The typical picture is that after completion of the treatment, $\underline{x}_i'$, at say $t = t_0$, the response, (r_i/n_i), rises with time, $t > t_0$, by a sigmoid curve reaching a maximum stable level of response, the "endpoint response", characteristic of $\underline{x}_i'$ at say $t = t_1$. (Hewlett and Plackett 1979) The elapsed time, $t_1 - t_0$, may be minutes, hours, days, etc., and, "It is perhaps natural to expect that an individual subject will respond the sooner the more the dose ... exceeds its individual tolerance ... Often this is so, but not always." (Hewlett 1974)

Thus, the estimates of the intercept, β_0, and slope, β, of the dose-response curve, $z = \beta_0 + \beta_1 x_1$, where z is the probit of response and $\underline{x} = \log_{10}$dose are functions of the time of observation, $t^* > t_0$. If $\hat{\beta}_0$ and $\hat{\beta}_1$ are the estimates of β_0 and β_1 at t_1 (end-point response) and $\hat{\beta}_0^*$ and $\hat{\beta}_1^*$ are estimates at time of observation, then in general for $t^* < t_1$ we have $\hat{\beta}_0^* > \hat{\beta}_0$ and $\hat{\beta}_1^* < \hat{\beta}_1$. It is generally observed that as $t^* \longrightarrow t_1$ from below the estimated dose-response curve is shifted to lower doses and becomes steeper. (Beard 1949) Theoretically, the dose-response curves at times of observation $t^* <$

t_1 and $t^* > t_1$ differ from that at $t^* = t_1$ in a rather more complicated manner that has been described by Hewlett: "If it is assumed that dose does not influence the statistical distribution of time to [occurrence of a quantal response] then the general effect of time of observation on the relation between [response] and dose is readily elucidated."
...
If, ... probit [response] at end point is assumed linear in log-dose ... the relations would be <u>curvilinear</u> for observations made both before and after this [time to end-point]. For those before, the curves would be <u>convex</u> toward the probit axis; for those after, <u>concave</u>. (Hewlett 1974)

Note that this implies that for $t^* \neq t_1$ there is a second-order term in log dose in the dose-response equation: $z = \beta_0 + \beta_1 x_1 + \beta_2 x_2^2$, where $\beta_2 < 0$ for $t^* < t_1$ and $\beta_2 > 0$ for $t^* > t_1$. If β_2 is small in absolute value the curvilinear relation may be difficult to observe, of course.

Note that the effect on the estimates of the slope and intercept of the dose-response curve is similar to that produced by an outlying response due say to an error of transcription, observation, "set up" etc. See Figure <u>9</u>. What is required is some single process of computation for the simultaneous estimation of the parameters of the joint distribution of tolerance "doses", <u>x</u>', for the binary event, E, and times to occurrence of E. (Hewlett 1974)

The possibility that the time of observation, t^*, exceeded t_1 and thus introduced a second-order term, x_1^2, into the empirical dose-response equation is considered in Lecture <u>17</u>.

2. It may be useful to consider the regression diagnostics in the context of <u>similarity</u>, or dissimilarity, measures. The <u>observations</u> (y_i, $\underline{x_i}'$) - <u>rows</u> of [<u>y</u>, X] - are classified in terms of their proximity to each other. The outliers (extreme y_i) and extreme levels of treatment (extreme $\underline{x_i}'$) are defined in terms of characteristic (often non-Euclidean) <u>distance</u> measures. The <u>variables</u> ($\underline{X_i}$, $\underline{X_k}$) - <u>columns</u> of X - are classified in terms of their <u>association</u> with each other. Collinear variables are (broadly) defined in terms of an <u>angular "distance"</u> measure - the pair-wise product-moment correlation coefficient.

3. "The fact that a small subset of the data can have a disproportionate influence on the estimated parameters or predictions is of concern to users of regression analysis, for, if this is the case, it is possible that the model estimates are based primarily on this data subset rather than on the majority of the data." (Belsley, et al 1980). We note that the data subset may be of size one. We also note that it is useful to determine whether the pseudo-observation matrix, [$Q\underline{r}$, QR], where $Q'Q = \Psi^{-1}$, which describes the a priori information on $\underline{\beta}$ in the Mixed Estimation method can be regarded as such a data subset. We examine this possibility in Lecture <u>14</u>.

ADDITIONAL REFERENCES

BEARD, R.L. (1949) Time of Evaluation and the Dosage-Response Curve. <u>Journal of Economic Entomology</u>, 42:(4) 579-85.

HEWLETT, P.S. (1974) Time from Dosage to Death in Beetles, Tribolium Castaneum, treated with Pyrethrins or DDT, and its bearing on Dose-Mortality Relations. J. stored Prod. Res. 10: 27-41.

HEWLETT, P.S. and PLACKETT, R.L. (1979) <u>The Interpretation of Quantal Response in Biology</u>, Edward Arnold, Pub. London.

CLINICAL RADIOCARCINOGENESIS. APPLICATIONS OF REGRESSION DIAGNOSTICS AND BAYESIAN METHODS TO POISSON REGRESSION MODELS

Donald E. Herbert
University of South Alabama, College of Medicine
Dept. of Radiology, Mobile, Alabama 36688

ABSTRACT

Several rival models (LQ-L, L-L, Q-L and others) of the lifetime incidence and mortality rates of cancer in the Japanese populations present in Hiroshima and Nagasaki in early August of 1945 (the LSS sample) are examined in the context of a general discussion of appropriate methodologies for mapping any estimates of radiation dose into meaningful estimates of radiation risk. Each operation is seen as implemented by the methods of mixed estimation which combines the sample information $\hat{\beta}$, Var($\hat{\beta}$) on the (unknown) parameter, β, of the regression model, $y = X\beta + \varepsilon$, with non-sample, or a priori information, $r = R\beta + v$, $E(v) = 0$, $Var(v) = \Psi$, to give posterior estimates, $\hat{\beta}**$, Var($\hat{\beta}**$). In general, the degree of success in retrieving plausible posterior estimates, $\hat{\beta}**$, from the LSS data that is presented in the BEIR III Report can be shown to be achieved only by rather dubious choices for the second-moment matrix, Ψ. It is shown that it is fruitful to regard the principal methods (mixed estimation) by which plausible estimates of the respective coefficient vectors, β, of the several (Poisson) models - and hence, defensible estimates of risk - were constructed from the LSS data, to be in the nature of post hoc, "salvage operations", required by the rather bizarre distribution of the observations on dose-response which comprise the LSS data on radiogenic cancer risk.

The methods of Poisson regression (iterative reweighted Least Squares) are augmented by those of Regression Diagnostics in order to "... readily identify observations that are not well explained by the model as well as those dominating some important aspects of the fit" (Pregibon, 1981). There are evidently many of both sorts in the LSS data from which were constructed the risk estimates presented in the BEIR III Report.

A heuristic development of Errors-in-Variables models for dose-response for radiogenic cancer risk is also presented as an effort to motivate the development of a more appropriate methodology for reducing the bias induced in the sample estimates of β by the presence of random errors in the measurements of the radiation dose.

1. INTRODUCTION

"What is the range of probability estimates, and probability distributions, of effects vs dose level for models of dose response that currently represent available scientific data?" (Alexander 1980)

"The gap between our scientific knowledge and our societal needs appears to be continually widening. In a third of a century of inquiry, embodying among the most extensive and comprehensive scientific efforts on the health effects of an environmental agent, much of the practical information necessary for the determination of radiation protection standards for public health policy is still lacking. It is now assumed that any exposure to radiation at low levels of dose carries some risk of deleterious effects. However, how low this level may be, or the probability, or magnitude of the risk, still are not known" (Fabrikant, 1979).

"One of the most insidious and nefarious properties of scientific models is their tendency to take over, and sometimes supplant, reality." (Chargaff, 1963)

"All models are wrong, but some are useful" (Box 1979).

"A regression is constructed using prior knowledge, data, models and a fitting (estimation) process of some form. It is important to know when the resultant regression depends heavily on a small part of the prior knowledge, on a small part of the data, or on the exact choice of model or fitting process" (Welsch 1984).

We have obtained the Hiroshima and Nagasaki LSS data on the induced rates of the following radiogenic cancers: 1) leukemia (incidence), 2) cancer ($\overline{s}$ leukemia) (mortality) and 3) breast cancer (incidence). These data were not included in the Report itself. The data on leukemia incidence rates and breast cancer incidence rates can be found in the Proceedings of the 13th International Conference on Environmental Toxicology (Land, 1981). The required observations on the non-leukemia cancer mortality rates have never been published in the open literature so far as we can determine. We have obtained these data through the generousity of Charles Land (1982). These data provide one of the bases for the estimates of excess risk of radiogenic cancer presented in the so-called BEIR III Report (NRC/NAS 1980) (Tables V-16 to V-21 of that report).

These observations are based on the T65D estimates of the radiation doses delivered to the populations at Hiroshima and Nagasaki in 1945. These are inconsistent with the estimates of these doses made more recently by workers at Lawrence Livermore National Laboratory (and others) (Loewe and Mendelsohn, 1981). Because of this inconsistency the _evidential_ value of the LSS data in issues of radiation risk assessment is now regarded by many as questionable. (But it should be remarked that these data apparently form part of the basis of the soon-to-be-issued radioepidemiological tables.)

However, the _motivational_ value of the LSS data in vivifying a discussion of the normative statistical procedures by which estimates of radiation dose are mapped into estimates of radiation risk is undiminished. It is this motivational value that is exploited in the present paper, which is a commentary on the several procedures by which estimates of the low- and high-LET components of a radiation dose, X, have been mapped into estimates of excess cancer risk using the regression model, $y = X\beta + \varepsilon$, where y is a Poisson rate parameter. (The probability of occurrence of r binary events in a group of n persons at risk for T years is $P(r|nTy) = e^{-(ynT)} (ynT)^r/r!$.) The radiation-induced occurrence of tumor, as well as ablation thereof, are both binary events. Both follow a _conditional_ Bernoulli distribution, $B(n, \pi)$, with index n and parameter, $\pi = \pi$(radiation dose, ...). However, the occurrence of cancer is a _rare_ binary event; the parameter π is small: $0 \leq \pi \ll 1$. For ablation of tumor the range of the parameter is much greater: $0 \leq \pi \leq 1$. Hence, for _carcinogenesis_, the Bernoulli distribution can be _approximated_ by the _Poisson_ distribution, $P(\Theta)$, with parameter, $\Theta = n\pi$. ($P(\Theta)$ is a good _approximation_ to $B(n, \pi)$ for $n > 10$ and $\pi < 0.1$. See Molina 1942;

Johnson and Kotz 1969.) As we have seen in Lecture 13, $z=\Phi^{-1}(.)$, is an appropriate transformation of π (Finey 1971). ($\Phi(.)$ is the Normal distribution function.)

Typically, the respective ranges of radiation dose within which the two binary events are evoked are also different: 0 - 100 rad (occurrence) and 5000 - 10000 rad (ablation). There are other distinctions. The former event typically occurs with a latency of years and in the absence of gross tissue damage. The latter event occurs with a latency of weeks and in the presence of gross tissue damage. In the Poisson regression model the parameter, Θ is a Bernoulli parameter π, a linear function of radiation dose, etc.: $\Theta = X_\beta + \underline{\varepsilon}$. In the probit regression model, the transform, z, of the Bernoulli parameter, π, is a linear function of radiation dose, etc.: $\underline{z} = X_\beta + \underline{\varepsilon}$. The respective weights are $n_i T_i / \hat{\Theta}_i$ and $n_i \hat{\emptyset}_i^2 / \hat{\pi}_i (1-\hat{\pi}_i)$. $\emptyset_i$ is the unit Normal density function corresponding to $\Phi(\hat{z}_i) = \hat{\pi}_i$, the two known parameters, n_i and T_i are the number at risk and the duration of risk of the event at the ith level of radiation dose, etc., respectively. For both, $E(\varepsilon_i) = 0$ and $Var(\varepsilon_i)$ are functions of i. In the present lecture it will be convenient to denote the Poisson parameter, Θ_i, as the rate constant y_i.

Most of the insights and methods required for the construction, evaluation and interpretation of Poisson regression models of the LSS data have been presented in general form in Lectures 1-2, 3, 5, 7 and 13. Moreover, because of the central position of authority which the BEIR III report presently occupies in the received wisdom of radiation risk management, the estimates, $\hat{y}_i$, $\hat{\beta}$, etc., themselves retain an intrinsic, as well as a motivational interest, despite the several current controversies concerning radiation dose levels, form of dose-response function, etc. Therefore, the format of Lecture 14 has been altered from that of the other Lectures in order both to take advantage of the preceeding expositions and to present as concisely as possible the results of the application of Poisson regression methods to the LSS data. These are presented in summary form following a brief restatement and specialization of those methods of the preceding Lectures that are immediately applicable. This should stimulate an examination of the additional, as well as graphical, expositions of the methods and results that follow in Appendices I-III.

2. METHODS

We have applied to the LSS data the iterative, reweighted Least Squares methods of Poisson regression analysis described by Frome, et al, (1973) to construct estimates of the so-called LQ-L, L-L and Q-L models described in the BEIR III Report. In our notation systematic parts of the rival models are written as the polynomials in dose:

$$y = \beta_0 + \beta_1 D\gamma + \beta_2 D\gamma^2 + \beta_3 D_n + \beta_4 C. \quad \text{(LQ-L)} \quad (1)$$
$$y = \beta_0 + \beta_1 D\gamma + \beta_3 D_n + \beta_4 C. \quad \text{(L-L)} \quad (2)$$
$$y = \beta_0 + \beta_2 D\gamma^2 + \beta_3 D_n + \beta_4 C. \quad \text{(Q-L)} \quad (3)$$

C is a dichotomous variable: C = 1 (Hiroshima); C = 0 (Nagasaki). In addition, estimates of $\underline{\beta}$ for several other rival models of these conditional rates, y, are presented. All such estimates may be obtained as iterative reweighted Least Squares estimates: $\hat{\beta} = (X'\hat{V}^{-1}X)^{-1}X'\hat{V}^{-1}y$; $Var(\hat{\beta}) = (X'\hat{V}^{-1}X)^{-1}$; $RSS = (y-X\hat{\beta})'\hat{V}^{-1}(y-X\hat{\beta})$. $\hat{V}^{-1}$ = Diag $[n_i T_i / \hat{y}_i]$ is the weight matrix at the final iteration. ($\hat{\beta}$ is a Generalized Least Squares (GLS) estimate.) n_i is the number of persons at risk for T_i years following exposure to the ith level of radiation dose, $(D\gamma, D_n)_i$. The disease experiences of irradiated individuals have been <u>aggregated</u> into groups of size n_i and $(D\gamma, D_n)_i$ is the <u>mean</u> dose of the ith group.

If, as well as the estimates $[\hat{y}, Var(\hat{y})]$, $[\underline{\hat{\beta}}, Var(\underline{\hat{\beta}})]$, etc., the respective approximate sampling distributions are also known, the corresponding <u>set</u> estimates, as well as <u>point</u> estimates, $\hat{y}$, $\underline{\hat{\beta}}$, etc., may be constructed for a given regression model. It is appropriate to reiterate the importance of set, or <u>interval</u>, estimates since these seem to have been largely omitted from the BEIR III Report:

"The size of the confidence limits is inversely proportional to the quality of the data used to make the estimate and directly proportional to the amount of extrapolation involved. This important information is lost if the confidence limits and best estimates are not routinely reported. The width of the confidence interval is one of the best measures risk assessors, and risk managers have to evaluate the quality of the estimates of potential risks. It is important to distinguish between those situations in which the risk is precisely estimated and those in which it is not."(Park and Snee, 1983)

"Interval estimation of $\underline{\Theta}$ [the parameter vector of the model] is central to statistical inference in a non-linear regression model because of the substantive interpretation that can be placed on the parameters of a theoretically derived non-linear model" (Duncan, 1978).

Both <u>point</u> and <u>set</u>, or interval, estimates of the coefficient vector, $\underline{\beta}$, the conditional response rate $\underline{y}$, the neutron RBE, et cetera, are presented in Appendix III. These interval estimates are obtained using the methods of Frome, et al, (1973) of Fieller (1971) and of others.

Current criteria for model selection and validation include, in addition to the residual sum of squares, RSS, 1) the Akaike/Atkinson Generalized Information Statistic (Akaike 1974; Atkinson 1981), a criterion which, "rewards parsimony" in the construction of models, and is an alternative to Neyman-Pearson hypothesis test for selection of model dimensions, or the C_p statistic (See Lectures <u>1</u> and <u>2</u> and Montgomery and Peck 1982) and 2) the Allen PRESS statistic (Allen 1974), a cross-validation measure for small samples.

It is important to validate a regression model before using it - or arguing its "validity" (Montgomery & Peck, 1982). Most desirably, this is done by evaluating its predictive performance in "new" data; that is, data which were not used for the initial estimation of form, parameters, etc. And it will be recalled that, "Users have often been disappointed by ... multiple regression equations that 'forecast' quite well for the data on which they were built. When tried on fresh data, the predictive power of the [equations] fell dismally" (Mostellar and Tukey, 1977). But when new data are unavailable and the initial sample is not large enough for the various cross-validation procedures - in which the initial sample is <u>split</u> into a "construction" sample and a "validation" sample (Snee 1977) - to be feasible, the PRESS statistic can (should) be used (Allen, 1974; Montgomery & Peck, 1982). PRESS, "... corresponds to division of the sample (size n) into a 'construction' subsample (size n-1) and a 'validation' subsample (size 1) in all (n) possible ways" (Stone, 1974). Both AIC and PRESS provide measures of the "portability" of the estimated model. The Akaike Information Criterion, AIC, is defined as AIC = $-2L(p) + 2p$ where $L(p)$ is the maximum $\log_e$ likelihood of the model with p independently adjusted parameters. For first-order regression models p is the number of explanatory variables. In using the criterion to discriminate between rival models of a finite sample of observations the rule is to choose that model for which AIC is <u>least</u> - the so-called minimum AIC estimate, or, MAICE (Akaike, 1974). This rule selects the model for which the mean squared error (bias^2 + variance) of prediction is a <u>minimum</u>. It is evident that AIC <u>explicitly</u> "rewards parsimony" - an important desideratum in the construction of multiple regression models. See Lecture <u>13</u>.

It can be shown that an alternative estimate is AIC = RSS + 2ν where RSS is the sum of squares of Pearson residuals distributed as chi-squared with ν = (n - k) degrees of freedom for the model. Also, another related measure is AIC* = $n\ln[RSS/(n-k)] + 2p$ (Atkinson 1980), where k = p+1.

The PRESS statistic is defined as PRESS = $\sum_{i=1}^{n} e^2_{(i)}$ where $e_{(i)}$ is the <u>predictive</u> residual. That is, $e_{(i)} = y_i - x_i'\hat{\beta}_{(i)}$ where $\hat{\beta}_{(i)}$ is the estimate of β obtained from the sample from which the ith observation is omitted. It can be shown that $e_{(i)} = (y_i - x_i\hat{\beta})/(1-h_i) = e_i/(1-h_i)$. Thus, PRESS is a function of the key diagnostics, e_i and h_i, vide infra. For Poisson regression models, the residuals, e(i), are <u>weighted</u>, weights, $v^{ii} = \sqrt{n_i T_i / \hat{y}_i}$, in the PRESS statistic as are the Pearson residuals, $e_i^\#$, for the RSS statistic: PRESS = $\Sigma e^{\#2}_{(i)}$ and RSS= $\Sigma e^\#_i{}^2$ and $h_i = [V^{-\frac{1}{2}}X'(X'V^{-1}X)XV^{-\frac{1}{2}}]_{ii}$ (Belsley et al 1980, Pregibon 1981).

In using PRESS to discriminate between rival models of a finite sample, the rule is to choose that model for which PRESS is <u>least</u>. This rule selects the model which will best retain its predictive performance in new data (Montgomery & Peck, 1982).

Multiple regression methods were developed to obtain optimal (in various senses, i.e., minimum residual sum of squares, RSS) estimates, $\hat{\beta}$, of the parameters in the linear model, $y = X\beta + \varepsilon$, from sets of observations obtained in <u>designed</u> experiments. In such experiments the location, scale and shape of the joint distribution of an optimal number of observations, [y, X], can be set to optimal (in various senses, i.e., maximum |X'X|, a D-optimal design (Montgomery & Peck, 1982) levels. Observations obtained in non-experiments usually have distributions which are less than optimal and which give rise to characteristic <u>model-sample interactions</u> that distort the estimates, $\hat{\beta}$ and Var($\hat{\beta}$), obtained from such data. Thus, "We must ... recognize that fitting of equations to observational data (as opposed to data from carefully designed experiments) is, at best, a risky business" (Hocking, 1983). (See also Brownlee's classic warning quoted in Appendix I.) Thus, it is most important, "to assess the suitability of a given data set for estimating a specific linear regression model by least squares", (Belsley, et al, 1980) in such cases.

The characteristic <u>model-sample interactions</u> for each model of each LSS sample are assessed by the single-row regression diagnostics of Belsley, Kuh and Welsch (1980): h_i, RSTUDENT, DFBETAS, DFFITS, COVRATIO. These measures, "... readily identify observations that are not well explained by the model as well as those dominating some important aspect of the fit" (Pregibon, 1981). Although these diagnostics are defined in terms of the [y, X] matrix, satisfactory approximations can be constructed from the Least Squares (LS) estimates of transformed model, Py = PX + Pε where P'P = $\hat{V}^{-1}$; $p_{ii} = \sqrt{n_i T_i / y_i}$ (Theil 1971). Since, "At the very least, excessively influential data should be mentioned in any discussion of the model fitting and estimation process", (Belsley, et al, 1980) several of the regression diagnostics for some of these models of this data are presented in Appendix II. It should be remarked that even a cursory examination of these diagnostics will disclose that the estimates, $\underline{y}$, $\hat{D\gamma}(y_i)$, $\underline{\hat{\beta}}$, $Var(\underline{\hat{\beta}})$, RSS, PRESS, etc. for each of the rival models are dominated by characteristic subsets of the observations. The fact that a small subset of observations in the LSS sample can be a disproportionate influence on such estimates should be of concern to the "consumers" of the BEIR III Report, "... for if this is the case then it is quite possible that the model-estimates are based primarily on this data sub-set rather than on the majority of the data" (Belsley, et al, 1980).

Uncertain a priori beliefs and estimates based on theory, previous experiments (often on surrogate systems), surveys, introspection, etc., are frequently decisive for model selection for non-experiments since the sample evidence for any model of such data is often quite equivocal. This is another kind of risk to which the modelling of "weak data" (such as that obtained in non-experiments) is especially vulnerable: "One of the most insidious and nefarious properties of scientific models is their tendency to take over, and sometimes supplant, reality" (Chargaff, 1963). Thus, it is often of interest to examine the explicit role of such uncertain non-sample information in discriminating between two alternative models, M_0, M_1, of a sample, [$\underline{y}$, X]. This can be represented by the respective prior probabilities, $P(M_i)$, i=0,1. Given the sample evidence [$\underline{y}$, X], the posterior odds of the model, M_0, are $P(M_0|y)/P(M_1|y) = B \ P(M_0)/P(M_1)$ where the Bayes' factor, $B = P(y|M_0)/P(y|M_1)$ is estimated from the sample, where $B = (RSS_0/RSS_1)^{n/2} \ n^{(k_0-k_1)/2}$. The sample evidence is said to favor the alternative model M_1 if $B < 1$ (Leamer, 1978).

Note that the concept of the probability of a model (hypothesis) does not arise in classical, Sampling Theory, statistics owing to the requirement of a repetitive element in the classical theory of probability. However, such a concept arises quite naturally in Bayesian statistics. Note that in classical statistics, model discrimination devolves into acceptance, or rejection, of a null hypothesis, H_0, on the evidence of a function - the quantile of the sampling distribution - of the difference, $\Delta RSS = RSS_1 - RSS_0$, between the respective residual sums of squares. However, Bayesian criteria are a function of the ratio, $RSS_0/RSS1$.

In obtaining estimates of $\underline{\beta}$ from non-experimental data, "... the least squares estimates that result are rarely 'acceptable'. Various constraints on the parameters may be imposed to 'improve' the estimate, and one among many constrained least squares estimates is usually selected to convey the data evidence" (Leamer, 1978). Minimization of the (weighted) sum of squares, RSS, $(\underline{y}-X\underline{\beta})'\hat{V}^{-1}(\underline{y}-X\underline{\beta})$ subject to the "certain" constraint, $r = R\underline{\hat{\beta}}$, requires the method of Lagrange multipliers (Theil 1971; Leamer 1978). This yields the constrained estimates

$$\underline{\hat{\beta}}* = \underline{\hat{\beta}} + (X'\hat{V}^{-1}X)^{-1}R'[R(X'\hat{V}^{-1}X)^{-1}R']^{-1} \ (r-R\underline{\hat{\beta}}) \quad (4)$$

$$Var(\underline{\hat{\beta}}*) = A(X'\hat{V}^{-1}X)^{-1}A' \quad (5)$$

$$RSS* = RSS + (r-R\underline{\hat{\beta}})' \ [R(X'\hat{V}^{-1}X)^{-1}R']^{-1} \ (r-R\underline{\hat{\beta}}) \quad (6)$$

where

$$A = I - (X'\hat{V}^{-1}X)^{-1}R'[R(X'\hat{V}^{-1}X)^{-1}R']^{-1}R \quad (7)$$

The constraint increases the precision of estimate of $\underline{\beta}$, but inflates the residual sum of squares: $|Var(\underline{\hat{\beta}})*|/|Var(\underline{\hat{\beta}})| < 1$; $RSS* - RSS > 0$.

The role in stabilizing the estimates, $\underline{\hat{\beta}}$, for a linear model of non-experimental data, of uncertain prior information on one or more elements of the vector $\underline{\beta}$ which can be represented by the set of q stochastic constraints, $r = R\underline{\beta} + \underline{v}$, $E(\underline{v}) = 0$, $Var(\underline{v}) = \Psi$, is assessed by the proportion, θ_p, of the a priori information on $\underline{\beta}$, represented by the posterior estimates, $\hat{\underline{\beta}}**$, of $\hat{\underline{\beta}}$ (Theil, 1971).

$$\hat{\underline{\beta}}** = [X'\hat{V}^{-1}X + R'\Psi^{-1}R]^{-1} (X'V'\underline{y} + R'\Psi^{-1}\underline{r}) \qquad (8)$$
$$Var(\hat{\underline{\beta}}**) = [X'\hat{V}^{-1}X + R'\Psi^{-1}R]^{-1} \qquad (9)$$
$$RSS** = RSS + (\hat{\underline{\beta}}** - \hat{\underline{\beta}})'(X'\hat{V}^{-1}X)(\hat{\underline{\beta}}** - \hat{\underline{\beta}}) \qquad (10)$$
$$\theta_p = m^{-1} Trace\{R'\Psi^{-1}R[X'\hat{V}^{-1}X + R'\Psi^{-1}R]^{-1}\}. \quad (m=\# \text{ columns of } X)(11)$$

$\Psi = E(vv')$ is the (q x q) dispersion matrix of the a priori information on $\underline{\beta}$. The compatibility of the sample and non-sample information on $\hat{\underline{\beta}}$ is assessed by the statistic $\gamma = (\underline{r}-R\hat{\underline{\beta}}) [R(X'\hat{V}^{-1}X)R' + \Psi]^{-1} (\underline{r} - R\hat{\underline{\beta}})$ which is distributed as chi-squared on q degrees of freedom. $\hat{\underline{\beta}}**$ is a __matrix-weighted average__ of a priori and sample information on $\underline{\beta}$. Obviously, incompatible information should __not__ be, "pooled", in order to stabilize estimates, etc. It will be useful in the present work to note that the mixed estimates devolve into constrained estimates as $\Psi \longrightarrow [0]$, a null matrix. Again, $|Var(\hat{\underline{\beta}}**)| / |Var(\hat{\underline{\beta}})| < 1$, $RSS** - RSS_0 > 0$. (See also Leamer 1978 and Lecture __5__.)

It is useful to note that the mixed estimates, $\hat{\underline{\beta}}**$ and $Var(\hat{\underline{\beta}}**)$ can also be obtained by __augmenting__ the (transformed) sample matrix [Py, PX] with the matrix [Qr, QR] of prior information and applying Least Squares methods, where $P'P = \hat{V}^{-1}$ and $Q'Q = \Psi^{-1}$. [Qr, QR] may be considered to be pseudo-observations (Belsley, et al, 1980). Then, the regression diagnostics described by Belsley, Kuh and Welsch may be used to identify "observations" that are "... not readily explained by the model as well as those dominating some important aspect of the fit." (See Appendix II, Figure __4__.)

The work of Loewe and Mendelsohn, 1981 (and others) has apparently, as remarked above, disclosed the presence of fairly large systematic errors in the measurements of $D\gamma$ and D_n and both Hiroshima and Nagasaki. Presumably, the revised dosimetry will reduce or eliminate these errors. However, random, as well as systematic, errors were (and are still) present in the measurements of radiation dose in the LSS data. S. Jablon has previously determined that the standard errors of the T65D measurements of the dose at Hiroshima and Nagasaki are about 30% of the respective doses (Jablon, 1971). These errors were included (apparently) in the estimates of dose in the data from which the models described in the BEIR III Report were constructed. But, it is well-known that the presence of __random__ error, say $\underline{\delta}$, in the measurement of the predictor variables in the matrix X will induce __bias__ in the Least Squares estimates of $\underline{\beta}$ in the linear model, $\underline{y} = X\underline{\beta} + \underline{\varepsilon}$, of a sample of size n (Montgomery & Peck, 1982). That is, $E(\hat{\underline{\beta}}) \cong \hat{\underline{\beta}} - n(X'X)^{-1}D\hat{\underline{\beta}}$, where $E(\underline{\delta}) = 0$ and $Var(\underline{\delta}) = D$ are a priori __estimates__ of the first two moments, respectively, of the (joint) distribution of $\underline{\delta}$(Seber 1977). Obviously, the size of the __bias__ depends on the __size__ and __shape__ of the joint distribution of X as well as the dispersion, D. The so-called Perturbation Index, PI, is a useful criterion of the strength of this interaction of X and D(Beaton, Rubin and Barone): $PI = Trace[(X'X)^{-1}D] = \sum_{j=1}^{m} VIF_j (d_j^2/s_j^2)$ where VIF_j and s_j^2 are the Variance Inflation Factor (See Lectures __1-4__ and __13__.) and marginal variance of the jth variable, respectively, and d_j^2 is the measurement error variance. For the bias due to random errors of measurement to be negligible, PI should be much less than one, say $PI \leq 10^{-1}$. One maneuver for reducing these effects on $\hat{\underline{\beta}}$ of random errors of measurement is __aggregation__ of the data (Brown, 1983) a maneuver which unfortunately also __inflates__ the collinearity in the X matrix and __deflates__ the RSS, both of which lead to difficulties in subsequent inferences and exploitations of regression models of the data. Another maneuver is to obtain Errors-in-Variables estimates of $\underline{\beta}$ (Theil 1971). Approximate Errors-in-Variables estimates may be obtained as a linear transformation of the Least Squares estimates: $\hat{\underline{\beta}}^{**} \cong (X'X - n\hat{D})^{-1}(X'X)\hat{\underline{\beta}}$ (Theil 1971; See also Lecture 13. Note the similarity to the transformation of $\hat{\underline{\beta}}$ which yields the Ridge Regression estimates: $\hat{\underline{\beta}}(0)^{**} = (X_0X_0 + kI)^{-1} (X_0X_0) \hat{\underline{\beta}}(0)$, where $0 < k < 1$ is the bias parameter and the equation holds in the correlation basis (Montgomery & Peck 1981). See also Lecture 13. (X_0X_0 is a correlation matrix, etc.) kI and D convey the characteristic non-sample information on the unknown, $\underline{\beta}$, and the known, X, matrices, respectively. The bias which these errors in X may have introduced into the estimates of $\underline{\beta}$ for the several models, LQ-L, L-L and Q-L, does not seem to have been prominently discussed in the BEIR III Report. It would seem that perhaps it should have been - if only to demonstrate that such bias is negligible for these estimates.

3. RESULTS

The results of these extended and augmented analyses of the LSS data are now presented. Only some of these results which seem to be of interest (perhaps concern?), and may be of subsequent use, despite the new and different estimates of the radiation dose at the two cities, are briefly discussed below. A rather more lengthy discussion can be found in the paper, "Model or Metaphor? More Comments on the BEIR III Report", published in the Proceedings of the Health Physics Society 16th Midyear Topical Symposium, Jan. 1983, pp357-390. The LSS sample estimates of $\underline{\beta}$ for the models of choice of the three responses are presented in Table 1. It is evident that the sample information on $\underline{\beta}$ for the non-leukemia cancer mortality rates and the breast cancer incidence rates must be augmented by non-sample information before these models can be used in risk estimation. The non-sample information used in the BEIR III Report is represented by the respective linear constraints, $\underline{r} = R\underline{\beta}$, and Ψ. (One must note that in the Tables to follow the figure in the parentheses immediately beneath the coefficient estimates, $\hat{\beta}_j$, is the precision of estimate: $\hat{\beta}_j/\sqrt{Var(\hat{\beta}_j)}$. Also, the rates are normalized to 10^5 person-years.) Note that $\hat{\beta}_4$, the estimate of the weight of the city-variable, C_i is "significant" only for the breast cancer incidence rate. Note that β_3, the estimate of the weight of the neutron dose, D_n, is "not significant" only for the breast cancer incidence rate. Note that for both the breast cancer incidence rate and for the non-leukemia cancer mortality rate, $-1 < \hat{\beta}_2/\sqrt{Var(\hat{\beta}_2)} < 0$, and for the latter $0 < \hat{\beta}_1/\sqrt{Var(\hat{\beta}_1)} < 1$ as well. Finally,
1) The distributions of the observations, [$\underline{y}$, X] - and of their (initial) weights, $\hat{V}^{-1}$ - which comprise the LSS samples are quite bizarre - even for a, "non-experiment". See Figure 1f and Appendix I. The nature and degree of the idiosyncracies of size, shape, etc., of these distributions have required that several "data-instigated hypotheses" (Leamer, 1978), ad hockeries and maneuvers be introduced, post hoc, into the analysis as "salvage" operations in order that even plausible estimates of excess risk can be obtained from these data by regression models. ("Plausible estimates", $\hat{\underline{y}}$ and $\hat{\underline{\beta}}$, are those for which the sign, size, etc., are consistent with a priori information.) Several of these hypotheses and maneuvers are discussed in 2) - 10) below and in Appendices I-III.
2) The Hiroshima and Nagasaki data must be "pooled" in order to obtain

even plausible estimates of $\underline{\beta}$ for all models of all radiation-induced disease rates. This maneuver is required by the high degree of collinearity in the city-specific joint distributions of the gamma, D_γ, and neutron, D_n, doses. See Figures $\underline{1a}$ and $\underline{1f}$. The presence of the collinearity has characteristic effects on $\hat{\beta}$ and $Var(\hat{\beta})$: Both are inflated. In addition, one or more of the elements of $\hat{\beta}$ has the "wrong" sign. See Tables $\underline{1}$ and $\underline{2}$. The high degree of correlation of D_γ and D_n is, in part, inherent in the respective natures and constructions of the two radiation sources and in part due to aggregation of the observations into dose-groups for regression analysis (and to reduce the effects on $\hat{\beta}$ of random errors of measurement? vide supra). It is useful to represent the estimate $\hat{\beta}$ obtained from the pooled data $[y_1, X_1]$ (Hiroshima) and $[y_2, X_2]$ (Nagasaki) as the matrix-weighted average (See Leamer 1978 and Lecture $\underline{5}$):

$$\hat{\beta} = [(X_1'\hat{V}_1^{-1}X_1) + (X_2'\hat{V}_2^{-1}X_2)]^{-1} [X_1'\hat{V}_1^{-1}X_1\hat{\beta}_1 + X_2'\hat{V}_2^{-1}X_2\hat{\beta}_2] \quad (12)$$

$$Var(\hat{\beta}) = [X_1'\hat{V}_1^{-1}X_1 + X_2'\hat{V}_2^{-1}X_2]^{-1} \quad (13)$$

It is obvious that for the pooling maneuver to be effective in this case, the model matrices, X_1 and X_2 must have quite different correlations structures - as indeed they have for the Hiroshima and Nagasaki data. It is both interesting and useful to note that these pooled estimates of $\underline{\beta}$ can be represented as mixed estimates in which the Hiroshima observations (n = 8) provide the sample information, $\hat{\beta}_H$ and $Var(\hat{\beta}_H)$, on $\underline{\beta}$ and the Nagasaki observations (n = 8) provide the a priori information: $\underline{r} = R\underline{\beta} + \underline{v}$, $E(\underline{v}) = 0$, $Var(\underline{v}) = \Psi$. In this case, $\underline{r} = \hat{\beta}_N$, $R = I$, a (4x4) identity matrix, and $\Psi = R'Var(\hat{\beta}_N)R$ where $\underline{\beta}_N$ and $Var(\hat{\beta}_N)$ are the Nagasaki sample information on $\underline{\beta}$. For the LQ-L model, the compatibility coefficient for the Hiroshima and Nagasaki information is $\gamma = 2.24 \sim \chi^2(4)$ and the proportion of Nagasaki information included in the pooled estimates, $\hat{\beta}$ and $Var(\hat{\beta})$, is $\Theta_p = 0.48$ - as might be anticipated. See Table $\underline{2}$.

Reducing the effects of collinearity on the sample estimates $\hat{\beta}$ and $Var(\hat{\beta})$ by augmenting the sample with additional data selected to, say, maximize $|X'X|$ (and thus minimize $Var(\hat{\beta})$) is the most direct and obvious remedy. Such a maneuver is not uncommon as a post hoc, "salvage operation" (Gaylor & Merrill 1968; Dykstra 1971; Belsley, et al 1980. See Lecture $\underline{13}$). Although typically there are no guarantees that the new data available will be either (inherently) consistent with the original data or have the required correlation structure, where they are appropriate, such a remedy is preferable to the alternative of biased estimation of $\underline{\beta}$ as, say, in Ridge Regression (Belsley, et al 1980; Montgomery & Peck 1983). However, one notes that the respective radiation sources (as well as the population at risk) at the two sites were quite distinct: Hiroshima (gun-type; uranium core), Nagasaki (implosion-type; plutonium core). Thus the compatibility, or consistency, of the evidence on induced response provided by such (possibly) disparate "non-experiments" must be assessed. There is no evidence in the BEIR III report that this was done. However, we have done so using the methods of mixed estimation as described above. Although the evidence on $\underline{\beta}$ from the two samples is consistent this may be due only to the fact that the data from each sample includes so much "noise" that any characteristic differences are obscured - "buried". See Appendix I.
3) Examination of the regression diagnostics discloses that the

Table 1. Pooled (Hiroshima + Nagasaki) Sample Estimates of $\bar{\beta}$ for the Models of Choice. Poisson Regression. Iterative Re-Weighted Least Squares Estimates (Generalized Least Squares (GLS)).

a) Leukemia Incidence Rate. 10^5PY. n = 16. <u>LQ-L</u>.
  y = 3.345 + 0.100D_γ + 8.448*$10^{-4}D_\gamma^2$ + 2.730D_n - 0.228C
      (2.772) (1.043)   (1.460)   (3.550) (-2.130*10^{-2})
  RSS = 10.646 = $\chi_c^2(11)$. $P(\chi^2 > \chi_c^2| 11) = 0.473$.
  $\hat{\beta}_2/\hat{\beta}_1$ = <u>8.45*10^{-3}</u>. $\hat{\beta}_3/\hat{\beta}_1$ = <u>27.30</u>.

b) Cancer (σ leukemia) Mortality Rate. 10^5PY. n = 16. <u>LQ-L</u>.
  y = 240.760 + 0.210D_γ - 3.639*$10^{-4}D_\gamma^2$ + 6.062D_n - 3.186C
     (24.380) (0.517)   (-0.190)   (2.612)   (-0.295)
  RSS = 13.908 = $\chi_c^2(11)$. $P(\chi^2 > \chi_c^2| 11) = 0.238$.
  BEIR III Constraint: $\underline{r}$ = R$\hat{\beta}$. Ψ = [0] (Non-sample Information on $\underline{\beta}$)
  $\underline{r} = \begin{bmatrix}0\\0\end{bmatrix}$, R = $\begin{bmatrix}0, & 8.45*10^{-3}, & -1, & 0, & 0\\0, & 2.73*10^1, & 0, & -1, & 0\end{bmatrix}$

c) Breast Cancer Incidence Rate. 10^5PY. n = 20. <u>LQ-L</u> and <u>L-L</u>.
  1) y = 16.843 + 0.247D_γ - 1.190*$10^{-4}D_\gamma^2$ + 0.319D_n + 7.310C. (LQ-L)
        (6.446) (2.424)   (-0.427)   (0.658)   (2.409)
     RSS = 8.176 = $\chi_c^2(15)$. $P(\chi^2 > \chi_c^2| 15) = 0.916$.

  2) y = 16.988 + 0.218D_γ + 0.311D_n + 7.272C. (L-L)
        (6.568) (3.180)   (0.634)   (2.394)
     RSS = 8.402 = $\chi_c^2(16)$. $P(\chi^2 > \chi_c^2| 16) = 0.06$
     BEIR III Constraint: $\underline{r}$ = R$\underline{\beta}$, Ψ = [0] (Non-sample Information on $\underline{\beta}$).
     $\underline{r}$ = (0), R = (0, 1, -1, 0); or, r = 0.218, R = (0, 0, 1, 0)
The models of choice according to the BEIR III Report are LQ-L, $\overline{\text{LQ-L}}$ and L-L, respectively.

Table 2. LQ-L Model of Leukemia Incidence. City-Specific, Pooled, Matrix-Weighted Average and Mixed Estimates of $\underline{\beta}$.

a) Hiroshima. n_1 = 8, m = 4.
   y = 3.230 - 1.505D_γ - 4.330*$10^{-3}D_\gamma^2$ + 18.17D_n
      (4.907)(-0.947)   (-0.875)   (1.199)
   RSS = 3.288 = $\chi_c^2(4)$.
b) Nagasaki. n_2 = 8, m = 4.
   y = 4.025 - 0.050D_γ - 1.251*$10^{-3}D_\gamma^2$ + 75.950D_n
      (2.892)(-0.318)   (-0.570)   (1.096)
   RSS = 4.394 = $\chi_c^2(4)$.
c) Matrix-Weighted Average.
   y = 3.360 + 0.084D_γ + 8.570*$10^{-4}D_\gamma^2$ + 2.724D_n
      (5.807) (0.978)   (1.582)   (3.770)
d) Hiroshima + Nagasaki. $n_1 + n_2$ = 16, m = 5.
   y = 3.345 + 0.099D_γ + 8.448*$10^{-4}D_\gamma^2$ + 2.730D_n - 0.28C
      (2.772) (1.043)   (1.460)   (3.550) (-0.021)
   RSS = 10.646 = $\chi_c^2(11)$.
e) Mixed Estimation. Sample (Hiroshima). A Priori (Nagasaki)
   y = 3.360 + 0.084D_γ + 8.570*$10^{-4}D_\gamma^2$ + 2.724D_n
      (5.804) (0.978)   (1.581)   (3.766)
   γ = 2.240 $\sim \chi^2(4)$. Θ_p = 0.48.

estimates, $\hat{\beta}$, Var($\hat{\beta}$), $\hat{y}$, etc., for all models of all disease rates are strongly affected by "small parts of the data" (Welsch, 1984). In particular, they are <u>dominated</u> by the observations at zero radiation dose (e.g., the "background" leukemia incidence rate at Hiroshima [y; D_γ, D_n]$_i$ = (3.69; 0, 0)$_1$). See Table <u>3</u> and Appendix II. The number of diagnostics which exceeded the respective size-adjusted cut-offs for all four rival models (LQ-L, L-L, Q-L, L'-L) is given for each <u>observation</u> in Table <u>3a</u> (Belsley, et al, 1980). The total number of diagnostics which exceeded the respective size-adjusted cut-offs is given for each <u>model</u> in Table <u>3b</u>. Two features of these Tables are quite striking. First, observation #1 <u>dominates</u> the model estimates, $\hat{\beta}$, Var($\hat{\beta}$) and $\hat{y}$ for the leukemia incidence sample (Table <u>3ai</u>). Second the estimate, $\hat{\beta}$, for the LQ-L model is most affected by "small parts of the data" (Welsch, 1984) and that for the L'L model (vide infra) is the least affected (Table <u>3aii</u>).

4) The prior odds for the Q-L model, with respect to the L-L model, must exceed 5:3 in order to dominate the evidence of the (pooled) LSS leukemia incidence sample for the latter. See Table <u>3b</u>. Table <u>3b</u> also discloses that the prior odds in favor of the LQ-L model, with respect to the L'-L model, the <u>linear spline</u> (vide infra) must be nearly 5:1 in order to dominate the evidence of the (pooled) LSS leukemia incidence sample for the latter model (which was regarded as a "non-starter" in the BEIR III Report). The question must be asked: Is the <u>prior</u> evidence for the LQ-L and Q-L models really so cogent? (This well illustrates the difference between the two definitions of scientific truth of a model: Truth as determined by <u>concordance</u> or Truth as determined by <u>consensus</u> (Kuhn, 1970). It is evident that a "threshold" model - the linear spline- is the more <u>concordant</u> with the data. Compare Figures <u>3a-d</u>. Appendix II. However, the non-threshold model is the more <u>consistent</u> with consensus: "... in its estimates of low-dose risk the Committee chose not to include the class of functions with a threshold, i.e., functions in which the cancer risk is zero up to some positive value of the dose-scale" (NRC/NAS 1980).)

5) The 0.90 confidence limits on the response, $\hat{y}_i$, the "inverse estimate", $\hat{D}_\gamma$ (y_i), and the neutron RBE are <u>enormous</u> for all models of all disease rates: 1) For the LQ-L model the 0.90 confidence limits on the dose, D_γ (y_i), required to elicit a leukemia incidence rate 1.5 times, "background", are 7 and 55 (rads). The size(s) of the confidence limits are, as remarked by Park and Snee (1983) "... inversely proportional to the quality of the data used to make the estimate and directly proportional to the amount of extrapolation involved." It is evident that in most cases in which estimates of excess radiation risk are of concern, either estimates of response, $\hat{y}_i(D_\gamma)$, or inverse estimates, $\hat{D}_\gamma(y_i)$, are required. But, these estimates are the <u>extrapolations</u> for which $D_\gamma > 0$, $D_n = 0$. See Figures <u>1a</u> and <u>5b</u>. Thus, only part of the ambiguity in the BEIR III forecasts is due to the (very) poor quality of the construction samples; part is a consequence of the irrelevance to the question at issue (low LET radiation risk) of the presence of observations on D_n in the sample (This is one aspect of the so-called <u>scaling</u> problem.) and in part to <u>specification error</u>: The addition of D_γ^2 to the L-L model or D_γ to the Q-L model (to achieve the LQ-L model) does <u>not</u> improve the fit "significantly". However, either addition <u>inflates</u> the variance of the estimates, $\hat{y}_i$, $\hat{\beta}$, $\hat{D}_\gamma(y_i)$, etc. See Tables <u>3</u> and <u>4</u>. Compare Figure <u>5b</u> with Figures <u>5c</u> and <u>5d</u>. If one can consider the linear spline model, L'-L, as a viable alternative candidate to LQ-L then Tables <u>3</u> and <u>4</u>

and Figure 6b disclose some enormous effects of mis-specification of the model.

The 0.95 confidence limits (Fieller's Theorem. (Finney 1971)) on neutron RBE ($\hat{\beta}_3/\hat{\beta}_1 = 11.53$) are enormous: 3.14 and 38.72 for the L-L model of leukemia incidence. (In the BEIR III Report this ratio is included in the constraint matrix, R, for the L-L model of non-leukemia cancer mortality!)

The BEIR III Report, with "accurate ambiguity", (Leamer, 1978) has described the L-L and Q-L models as upper and lower confidence limits on the LQ-L model but fails to specify the level of confidence ($1-\alpha = ?$). Over most of the range of $D\gamma$ the curves for the L-L and Q-L models of the leukemia incidence rate can be shown to coincide (approximately) with the 0.60 confidence limits on the estimates of y_i obtained from the LQ-L model. See Figure 5a, Appendix III.

6) Of the three alternative polynomial representations (LQ-L, L-L and Q-L) of the non-threshold hypothesis of leukemia incidence rate, the L-L is the model of choice according to current regression criteria (Akaike/Atchison, PRESS, regression diagnostics, size of set estimates) - as well as the (classical) chi-squared test. It is a "useful" model. See Tables 4 and 5 and Figures 5b, 5c and 5d.

7) A polygon model (of the rival, "threshold", hypothesis), the linear spline, L'L, is a still more useful model of the leukemia incidence data (Wold, 1974). See Tables 4 and 5 and Figures 2a, 6a and 6b. In our notation the L'-L model is written as the polygon

$$y = \beta_0 + \beta_1(D\gamma - D_0)_+ + \beta_3 D_n \quad (14)$$

where the spline function (Montgomery & Peck, 1982; Wold, 1974) is

$$(D\gamma - D_0)_+ = \begin{cases} D\gamma - D_0, & D\gamma > D_0 \\ 0, & D\gamma \leq D_0 \end{cases} \quad (15)$$

The "knot", D_0, in the spline may be interpreted as an estimated -no-observed-effect-level (E-NOEL) (Park & Snee, 1983) for radiation leukemogenesis since the 0.90 CL on D_0 do not include zero (Montgomery & Peck 1982). The spline model is described in Table 6 and Figures 6a and 6b, Appendix III.

Comparison of the index plots of e_i^* for the L-L (Figure 3b) and Q-L (Figure 3c) with that for the LQ-L model (Figure 3a) suggests that the estimates of β for all three models may be dominated by the (Nagasaki) observations #12 and #14. (See Figures 3h, i and j.) Examination of Figure 3d, the index plot of e_i^* for the L'-L model, discloses that the estimates of β for the linear spline will probably not be sensitive to any observation. In the event, the estimates of β for the L'-L model are affected by observation #14 - as might be expected from examination of Figure 3c. However, it is evident from Figures 3a - d that the polygon model is more concordant with those than any of the three polynomial models. (If the a priori evidence against the L'-L model is strong enough to exclude it from consideration altogether then, of course, it is possible that some version of robust regression (See Lecture 4 and Montgomery & Peck 1982) might be used to obtain more appropriate estimates of β for the LQ-L, L-L and Q-L models. Such methods would reduce the influence of observations such as #12 and #14 by an appropriate "down-weighting". See Table 7.

The L'-L model represents leukemia incidence as a process in which the first derivative of the response, $\partial y/\partial D\gamma = \beta_1$, is a discontinuous function of the dose, $D\gamma$: $D\gamma \leq D_0$, $\beta_1 = 0$. $D\gamma > D_0$, $\beta_1 \neq 0.35$. (The Q-L model represents leukemia incidence as a process in

Table 3a. Model-Sample Interactions for Leukemia Incidence.

i) Frequency of Diagnostics[a]
   Exceeding Size-Adjusted Cut-off For Each Observation
   Index #. 1 2 3 4 5 6 7 8 9 10 11 12 13 14 15 16[b]
   Freq.   10 1 1 1 0 0 0 2 2 1 1 2 1 7 0 4
            Hiroshima (#1 - #8) Nagasaki (#9 - #16)

ii) Frequency of Diagnostics[a]
    Exceeding Size-Adjusted Cut-off[c] For Each Model
    Model. LQ-L   L-L   Q-L   L'-L
    Freq.   10   9   8   7

---
[a] h_i, RSTUDENT, COVRATIO, DFFITS and DFBETAS ($\hat{\beta}_1$, $\hat{\beta}_2$ and $\hat{\beta}_3$ only).
[b] The index #'s are simply the row numbers of the observation matrix [y, X]. They are also called case #'s. [c]See Appendix II.

Table 3b. Comparison of Rival Hypotheses. Leukemia Incidence Rate.
(n = 16) Classical and Bayesian.

	Sampling Theory		Bayesian Theory	
Null Hypothesis	RSS[df]	Hypotheses	Prior odds for H_0 to Give Equal Posterior Odds	
LQ-L[a]	10.64[11]	H_0: LQ-L H_1: L-L	1.80	
L-L	11.76[12]	H_0: LQ-L H_1: Q-L	1.07	
Q-L	12.54[12]	H_0: L-L H_1: Q-L	0.60	
L'-L	8.78[11]	H_0: LQ-L H_1: L'-L	4.65	

---
[a] LQ-L Model does not differ significantly from L-L or Q-L: ΔRSS(max) = 1.90. $(k_1-k_0) = 1$. $P(\chi^2 < \text{RSS}| 1) = 0.83$ (< 0.95).

Table 4. Evaluation of Rival Models of Leukemia Incidence. LSS Sample.
(n = 15)

Model, M_i[a]	RSS(df)	R_f^{2c}	$\bar{R}_f^2$	PRESS	R_p^2
LQ-L	10.27(11)[b]	0.860	0.822	17.39	0.764
L-L	11.36(12)	0.850	0.825	18.04	0.762
Q-L	12.40(12)	0.834	0.806	18.91	0.747
L'-L	8.52(12)	0.881	0.861	12.49	0.825

---
[a] These models do not include the indicator variable, C, since the analyses disclose that for the leukemia incidence rates (and also non-leukemia cancer mortality rates) we have $\hat{\beta}_4/\sqrt{\text{Var}(\hat{\beta}_4)} \ll 1.0$.
[b] ΔRSS = 11.36 - 10.27 = 1.09. RSS $\sim \chi^2(1)$ $P(\chi^2 < \Delta\text{RSS}| 1) = 0.70$. Hence the addition of the quadratic term, $D\gamma^2$, to the L-L model does not improve the goodness-of-fit "significantly". (Similarly, for Q-L)
[c] R_f^2 - Coefficient of determination. $\bar{R}_f^2$ - Adjusted coefficient of determination ("shrunken R^2"). R_p^2 - Predictive coefficient of determination. R_f^2 and $\bar{R}_f^2$ are functions of RSS. R_p^2 is a function of PRESS. See Montgomery & Peck 1982.

Table 5. Akaike Information Criterion. Leukemia Incidence Models. (n = 16).

Model, M_i [a]	RSS	df_i	AIC_i
LQ-L	10.646	11	-11.354
L-L	11.765	12	-12.354
Q-L	12.542	12	-11.458
L'-L	8.778	11[b]	-13.222[c]

[a] These models do include the indicator variable, C, despite the fact that the analyses have disclosed that it is "insignificant" since it is included in the models discussed in the BEIR III Report.
[b] Including 1 df for "knot", D_0.
[c] MAICE. Minimum AIC. The selection criterion.

Table 6. Comparison of Rival Models of Leukemia Incidence Rate.

a. L'-L. (Linear Spline. $D_0 = 38.8$)

$y = 3.69 + 0.35(D_\gamma - D_0)_+ + 2.95D_n - 0.21C$
 $(3.42)\ (3.83)\qquad\qquad (4.31)\ (-0.17)$
RSS = 8.78 $\sim \chi^2(11)$

b. L-L.

$y = 2.96 + 0.22D_\gamma + 2.52D_n + 0.21C$
 $(2.59)\ (3.65)\qquad (3.32)\quad (0.16)$
RSS = 11.76 $\sim \chi^2(11)$

Table 7. Sensitivity of Estimates of β in LQ-L Model of Leukemia Incidence to Deletion of a Single Observation. (See Index Plots of DFBETAS, Figures 3e, f, g.)

a. n = 16. k = 5.

$y = 3.345 + 0.099D_\gamma + 8.448*10^{-4}D_\gamma^2 + 2.730D_n - 2.847*10^{-2}C.$
$\hat{\beta}_2/\hat{\beta}_1 = \underline{8.533*10^{-3}}$, $\hat{\beta}_3/\hat{\beta}_1 = \underline{27.576}$.

b. n = 15 (obs. #12 deleted). k = 5.[a]

$y = 3.312 + 0.178D_\gamma + 5.255*10^{-4}D_\gamma^2 + 2.399D_n - 6.518*10^{-2}C.$
$\hat{\beta}_2/\hat{\beta}_1 = \underline{2.952*10^{-3}}$, $\hat{\beta}_3/\hat{\beta}_1 = \underline{13.478}$.

c. n = 15 (obs #14 deleted). k = 5.[a]

$y = 3.574 + 4.566*10^{-2}D_\gamma + 7.879*10^{-4}D_\gamma^2 + 3.153D_n - 24.413*10^{-2}C.$
$\hat{\beta}_2/\hat{\beta}_1 = \underline{17.2561*10^{-3}}$, $\hat{\beta}_3/\hat{\beta}_1 = \underline{69.054}$.

[a] Note the profound change on the ratios, $\hat{\beta}_j/\hat{\beta}_1$, j=2, 3, that are introduced as constraints, $\underline{r} = R\underline{\beta}$, $\Psi = [0]$, on $\underline{\hat{\beta}}$ for the LQ-L model of non-leukemia cancer mortality. Observations #12 and #14 clearly dominate these latter estimates.

which the first derivative, $\partial y/\partial D\gamma = 2\beta_2 D\gamma$, is a <u>continuous</u> function of the dose, $D\gamma$.) However, the spline function, like the other rival models, LQ-L, L-L, and Q-L, etc. represents a model of the cancer process in which the response itself (the zeroth derivative) is a discontinuous function of the <u>site</u>, C, at $D\gamma = D_n = 0$: $y = \beta_0 + \beta_4$ (Hiroshima). $y = \beta_0$ (Nagasaki). However, $\hat{\beta}_4/\sqrt{Var(\hat{\beta}_4)} > 1$ only for breast cancer incidence. See also Figure <u>5a</u>, Appendix III and Lecture <u>17</u>.

Spline models were not considered in the BEIR III Report. But, "The existence of carcinogenic thresholds can, in general, be neither proved nor disproved by the conventional bioassay; however, the concept of a threshold has worked well in the past for many toxicological responses. The E-NOEL and its uncertainty should be reported along with other estimates of potential risk when the model is shown to give an adequate fit to the data. Risk managers should be made aware when the existence of a threshold is consistent with the data" (Park & Snee, 1983). It has also been noted that, "... there is a tension between honesty and prudence. Probabilistic reports about adverse consequences to health are very often slanted to be conservative. I am arguing that it is better to report honestly, and that prudence should be appropriately represented in the evaluation process, not in the assessment process" (Raiffa, 1982).

8) The pooling maneuver described above (augmentation of the Hiroshima data with those of Nagasaki in order to reduce the degree and hence the effects on $\hat{\beta}$ and $Var(\hat{\beta})$ of the correlation between $D\gamma$ and D_n) <u>did not</u> adequately stabilize the estimates, $\hat{\beta}$, of the LQ-L, L-L and Q-L models of cancer ($\bar{s}$ leukemia) <u>mortality</u> rates as it <u>did</u> for the cognate models of the leukemia <u>incidence</u> rates. Although the joint distributions of the model matrix, X, and person-years, nT, are (nearly) identical for the two disease rates (See Appendix I) the maneuver failed to stabilize the estimates of β for the models of the mortality rate. The LSS observations on non-leukemia cancer mortality rates provide much "weaker" sample evidence for the β's of the respective models than do those for leukemia incidence because the <u>levels</u> of the rate, y, for cancer mortality exceed those of the leukemia incidence rate and, of course, $Var(\hat{\beta}) = (X'\hat{V}^{-1}X)^{-1}$, $\hat{V} = Diag[\hat{y}_i/n_i T_i]$, for all $\hat{\beta}$. See Figure <u>1d</u> and Appendices I and II. Therefore, (or so it would seem) in a maneuver reminiscent of the data augmentation maneuver by which the other cause of the weak data (multicollinearity in city-specific observations) was reduced, the BEIR III Committee "pooled" the leukemia incidence information on $\underline{\beta}$ with cancer ($\bar{s}$ leukemia) mortality information, $\hat{\beta}$, by introducing the former as linear <u>constraints</u>, $r = R\hat{\beta}$, on the estimates $\hat{\beta}$, of the corresponding model of the data from the latter sample in order to achieve the required stability. For example, for the LQ-L model we have

$$r = \begin{pmatrix} 0 \\ 0 \end{pmatrix}, \quad R = \begin{pmatrix} 0, & \hat{\beta}_2/\hat{\beta}_1, & -1, & 0, & 0 \\ 0, & \hat{\beta}_3/\hat{\beta}_1, & 0, & -1, & 0 \end{pmatrix}, \quad \psi = [0]. \tag{16}$$

where the ratios, $\hat{\beta}_2/\hat{\beta}_1 = 8.533 \times 10^{-3}$ and $\hat{\beta}_3/\hat{\beta}_1 = 27.576$, are obtained from $\hat{\beta}$ of the LQ-L model of the leukemia incidence rates. See Table <u>1</u>. (Although the motivations for these two maneuvers are quite similar - a <u>"salvage"</u> of weak data - the respective justifications are not equally well-founded: "pooling" information on such disparate, (complementary) responses as leukemia incidence and <u>non</u>- leukemia cancer mortality would seem to require some suspension of disbelief in order to carry it out.) This gave constrained estimates of β - the models $\overline{LQ-L}$, $\overline{L-L}$ and $\overline{Q-L}$ - in which the respective estimates of β can be represented as the matrix-weighted averages of sample and non-sample information, i.e., mixed estimates: $\hat{\beta} = [X'\hat{V}^{-1}X + R'\Psi^{-1}R]^{-1} (X'\hat{V}^{-1}\underline{y} + R\Psi^{-1}\underline{r})$ (Theil 1971). Ψ is the dispersion matrix of the (stochastic) constraint, $\underline{r} = R\underline{\beta} + \underline{v}$, $E(\underline{v}) = 0$, $Var(\underline{v}) = \Psi$. It can be shown that the mixed estimates approach the constrained estimates, asymptotically, as $\Psi \longrightarrow [0]$, a null matrix. Examination of the regression diagnostics for the constrained models, $\overline{LQ-L}$, $\overline{L-L}$ and $\overline{Q-L}$, of the cancer (s leukemia) mortality rate discloses that the respective estimates $\hat{\beta}^{**}$ and $Var(\hat{\beta}^{**})$ are dominated by the pseudo-observations, $[Q\underline{r}, QR]$, which represent the a priori information obtained from the cognate models of the leukemia incidence rate - "a small part of the prior knowledge" (Welsch, 1984). This evidence from the diagnostics is consistent with the estimates of the proportions, Θ_p, of non-sample (leukemia incidence) information in the posterior estimates, $\hat{\beta}^{**}$; $\Theta_p = 0.41$, 0.25 and 0.27, respectively, for LQ-L, L-L and Q-L. See Table 8. For example, in the plot of the hat matrix diagonals in Figure 4d of Appendix II for the $\overline{LQ-L}$ model on non-leukemia cancer mortality it can be seen that for the pseudo-observations, $h_{17} \simeq h_{18} \simeq 1.00$. This implies that (in some co-ordinate system) these two observations almost completely determine two of the five elements of $\hat{\beta}^{**}$. Or, about 40% of the information on β for non-leukemia cancer mortality that is represented by $\hat{\beta}^{**}$ does not come from the cognate sample, i.e., $\Theta = 0.41$ (Belsley, et al, 1980).

Thus, although the observations on the leukemia response were excluded, ab initio, from the cancer mortality sample, considerable information on that response must be reintroduced, a priori, as the "data-instigated" hypothesis, $\underline{r} = R\hat{\beta}$, into the estimates, $\hat{\beta}^{**}$, in order to reduce $Var(\hat{\beta}^{**})$ to more seemly levels - and hence confer a (rather specious?) "reality" on the estimates of β for each of the models of non-leukemia cancer mortality rates. The constraint matrices, $\underline{r}$ and R, are constructed from the cognate equations for leukemia incidence. The dispersion matrix, Ψ, of the constraint is, in principle, obtained as $\Psi = R'Var(\hat{\beta})R$, where $Var(\hat{\beta}) \longrightarrow [0]$, a null matrix, is the stipulated dispersion matrix for the estimates, $\hat{\beta}$, for the cognate model of the leukemia incidence rate. This reduction in variance is achieved because $Var(\hat{\beta}^{**}) = [X'\hat{V}^{-1}X + R\Psi^{-1}R']^{-1}$.

As noted previously, the plots of DFBETAS and COVRATIO in Figures 3h-3k disclose that the estimates, $\hat{\beta}$, and $Var(\hat{\beta})$ of the leukemia models themselves are in turn dominated by a few of the observations, especially the (Hiroshima) observation at zero dose $[y; D_\gamma, D_n]_i = [3.69; 0, 0]_1$ and $[nT]_1 = 559266$. Here a comment on "multiple-row effects" is in order (Belsley, et al 1980). Although the value of h_1 for the LQ-L model of the pooled (n=16) sample is large - exceeds the size-adjusted cutoff $2k/n = 0.63$ - the index plots of the deletion diagnostics DFBETAS, DFFITS, and COVRATIO, do not suggest that deletion of this observation will substantially alter the estimates $\hat{\beta}$, $Var(\hat{\beta})$, for the LQ-L model. However, examination of Figure 3e discloses the reason for this apparent anomaly: the observations at zero dose, #1 (Hiroshima) and #9 (Nagasaki), are included in a sub-set of several other potentially influential observations at similar positions (values of (D_γ, D_n)) and with similar weights, (n_iT_i/y_i). For example, in Figure 3a the effects of the observation with index #2, $[y, D_\gamma, D_n]_2 = [2.57, 1.7, 0.2]$ and $[nT]_2 = 262511$ are masked by the observation at index #1. Deletion of the sub-set of observations of

Table 8. Proportion, Θ_p, of Posterior Information on the Coefficient Vector of the Hybrid Models $\overline{LQ-L}$, $\overline{L-L}$ and $\overline{Q-L}$ of Cancer Mortality ($\overline{s}$ leukemia) that is Provided by the Leukemia Incidence Model (Prior Information)[a]

$$\Theta_p = (p+1)^{-1} \text{Trace}\{R^T\Psi^{-1}R[X^T\hat{V}^{-1}X + R^T\Psi^{-1}R]^{-1}\}$$

Hybrid Model	Constrained Estimates, Θ_p	Mixed Estimates, Θ_p
$\overline{LQ-L}$	0.406	0.366
$\overline{L-L}$	0.250	0.228
$\overline{Q-L}$	0.265	0.221
$\overline{L'-L}$	0.333	0.299

[a] The proportion, Θ_s, of the posterior information on the coefficient vector that is provided by the <u>sample</u> (the Ca Mortality ($\overline{s}$ Leukemia) data is, obviously, $\Theta_s = (1-\Theta_p)$.

Table 9. Comparison of Constrained[a] and Mixed Estimation for $\overline{LQ-L}$ Model of Non-Leukemia Cancer Mortality.

a) Constrained Estimation. $\Psi = [0]$.

	1	D_γ	D_γ^2	D_n	C
$\hat{\beta}_j$	238.82	0.14	$1.16*10^{-3}$	3.75	0.44
$\hat{\beta}_j/\sqrt{\text{Var}(\hat{\beta}_j)}$	27.41	4.04	4.04	4.04	0.00
γ	$2.15[\sim\chi^2(2)]$				
Θ_p	0.406				
$\hat{\beta}_2/\hat{\beta}_1$	$8.46*10^{-3}$				
$\hat{\beta}_3/\hat{\beta}_1$	27.40				

b) Mixed Estimation. $\Psi \neq [0]$.

	1	D_γ	D_γ^2	D_n	C
$\hat{\beta}_j$	239.00	0.14	$1.0*10^{-3}$	4.06	0.02
$\hat{\beta}_j/\sqrt{\text{Var}(\hat{\beta}_j)}$	27.18	1.42	1.72	3.59	0.04
γ	$1.90[\sim\chi^2(2)]$				
Θ_p	0.366				
$\hat{\beta}_2/\hat{\beta}_1$	$6.93*10^{-3}$				
$\hat{\beta}_3/\hat{\beta}_1$	28.07				

[a] Note that in Figure <u>4d</u>, $h_{17} = h_{18} \simeq 1.0$. These refer to the pseudo-observation matrix, [Qr, QR], where $Q'Q = \Psi^{-1}$, and r, R and Ψ represent the constraints obtained from $\hat{\beta}$ for the LQ-L model of Leukemia incidence. The Figure can be interpreted as showing that these two "observations" <u>completely</u> determine 2 out of 5 (40%) of the elements of $\hat{\beta}^{**}$ for the $\overline{LQ-L}$ model (Belsley, et al 1980). Figures <u>4d</u> and <u>3f</u> demonstrate that the <u>case</u> statistics of a model will disclose when the resultant regression may depend heavily, "on a small part of the prior knowledge", or, "on a small part of the data" (Welsch 1984).

index #s 1, 2 and 9 will profoundly alter the estimates $\hat{\beta}$ and Var($\hat{\beta}$) for the LQ-L model; it is an influential subset. (But, "... we must mention an inherent problem in delimiting influential subsets of the data, namely, when to stop - with subsets of size two, three or more?" (Belsley, et al, 1980)).

Also the (Nagasaki) observations [0; 38.8, 0.1]$_{12}$ and [57.4; 131.1, 0.9]$_{14}$ profoundly influence the estimates of $\hat{\beta}$ for the models, LQ-L and $\overline{\text{LQ-L}}$, respectively, of the leukemia incidence rate and non-leukemia cancer mortality rate. For example, the constraints on the ratios of parameter estimates, $\hat{\beta}_2/\hat{\beta}_1$ and $\hat{\beta}_3/\hat{\beta}_1$, for the LQ-L model of non-leukemia cancer mortality are obtained from the cognate estimates of the LQ-L model of leukemia incidence. For the full sample of observations on leukemia incidence (n = 16) these are $\hat{\beta}_2/\hat{\beta}_1$ = 8.533*10^{-3} and $\hat{\beta}_3/\hat{\beta}_1$ = 27.576. For the reduced sample (n=15, obs. #12 deleted) these ratios are deflated by a factor of about 0.5: $\hat{\beta}_2/\hat{\beta}_1$ = 2.952*10^{-3} and $\hat{\beta}_3/\hat{\beta}_1$ = 13.478. For the reduced sample (n=15, obs. #14 deleted) these ratios are inflated by a factor of about 2.0: $\hat{\beta}_2/\hat{\beta}_1$ = 17.256*10^{-3} and $\hat{\beta}_3/\hat{\beta}_1$ = 69.054. (These estimates are approximations.) See Table 7. This would seem to introduce some (hitherto unremarked) ambiguity into the $\overline{\text{LQ-L}}$, $\overline{\text{L-L}}$ and $\overline{\text{Q-L}}$ models of excess risk of non-leukemia cancer mortality (Table V-19 through V-21 of the BEIR III Report). Thus, in the BEIR III specification of the constraint matrix R, "... the Committee preferred to rely on human data ... it was agreed that the leukemia experience might provide a reasonable, if arbitrary guide" (NRC/NAS 1980). This would seem to be an "arbitrary guide", indeed.

Table V-11 of the BEIR III Report overstates the precision of the estimates, $\hat{\beta}_j$**, j = 1, 2, 3, for the low-LET components for each of these models by factors of between 1.25 to 2.50. The information obtained from the cognate models of the leukemia incidence rates that is used to stabilize the estimates, $\hat{\beta}_j$**, for the non-leukemia cancer mortality rates is just not as good as it would appear to be from Table V-11: Var($\hat{\beta}$) for the cognate model of leukemia incidence and thus ψ - does not approach a null matrix (as may be inferred from casual examination of Table V-8 in the BEIR III report itself which discloses that the 0.90 confidence limits on the denominator, $\hat{\beta}_1$, of the constraint, $\hat{\beta}_2/\hat{\beta}_1$ = 8.45*10^{-3} include zero!).

Since the (joint) distributions of radiation dose [D γ, D$_n$] and person-years at risk, nT, are coincident for the two responses, leukemia incidence and non-leukemia cancer mortality (See Appendix I), it would seem that the stability of the estimates of $\hat{\beta}$ for models $\overline{\text{LQ-L}}$, $\overline{\text{L-L}}$ and $\overline{\text{Q-L}}$ of the latter response is achieved by a kind of "double-count" of the same sample: once as non-leukemia cancer mortality to give the matrix [Py, PX] and again as leukemia incidence to give the matrix [Qr, QR] where P'P = $\hat{V}^{-1}$ and Q'Q = ψ^{-1}. We shall demonstrate this presently, but first it will be useful to compare the method of constrained estimation presented in the BEIR III Report with the (more appropriate) method of mixed estimation. The $\overline{\text{LQ-L}}$ model of non-leukemia cancer mortality is used as an example. The information, $\hat{\beta}$, Var($\hat{\beta}$), in the respective samples for leukemia incidence and non-leukemia cancer mortality as conveyed by the respective linear-quadratic models are given in Tables 1a and 1b.

In the present context, it is postulated that the estimates, $\hat{\beta}$ and Var($\hat{\beta}$), for leukemia incidence represent the non-sample, or a priori, information on radiation-induced non-leukemia cancer mortality. Two further conjectures are 1) that part of the information on $\hat{\beta}$ for cancer mortality has been "captured" in the cognate model of leukemia incidence and 2) that it can be adequately rendered by the linear form, $\underline{r} = R\beta + \underline{v}$, where $E(\underline{v}) = 0$, $Var(\underline{v}) = \Psi$ and, again

$$\underline{r} = \begin{pmatrix} 0 \\ 0 \end{pmatrix}, \quad R = \begin{pmatrix} 0, & \hat{\beta}_2/\hat{\beta}_1, & -1, & 0, & 0 \\ 0, & \hat{\beta}_3/\hat{\beta}_1, & 0, & -1, & 0 \end{pmatrix} \quad (17)$$

The matrices ($\underline{r}$, R) are common to both procedures, constrained as well as mixed estimation, and $\hat{\beta}_2/\hat{\beta}_1 = 8.45*10^{-3}$ and $\beta_3/\hat{\beta}_1 = 27.31$. The difference between the two procedures lies in the characteristic assumptions of the precision with which the a priori, or non-sample, information on β is known. This information, of course, is contained in the sample estimates, $\hat{\beta}$ and $Var(\hat{\beta})$, for the cognate models of leukemia incidence. These assumptions are represented by the characteristic dispersion matrices, Ψ, of the form. The BEIR III Report stipulates a high degree of precision, say $\Psi = kI = RVar(\hat{\beta})R'$ where $k=10^{-10}$ and I is the (2x2) identity matrix, i.e., the set and point estimates of β for leukemia incidence are coincident, $Var(\hat{\beta}) = [0]$, a null matrix. If this were in fact the case, constrained estimation is indeed appropriate. This is the method that was used to construct the estimates in Table V-11 of the BEIR III Report. However, it is quite evident from Table 1a that this stipulated covariance matrix, $Var(\hat{\beta}) = [0]$, for leukemia incidence rather overstates the true precision of the information on β that is conveyed by the LQ-L model of the leukemia incidence data. Instead, the correct dispersion matrix for the constraint, $\underline{r} = R\beta$, is

$$\Psi = R[Var(\hat{\beta})]R' = \begin{pmatrix} 1.704*10^{-3}, & 3.609*10^{-3} \\ 3.609*10^{-3}, & 9.331*10^{0} \end{pmatrix} \quad (18)$$

$Var(\hat{\beta}) = (X'\hat{V}^{-1}X)^{-1}$ is the correct, non-null, covariance matrix of β for leukemia incidence. As remarked above, the use of constrained estimation requires that $Var(\hat{\beta}) = [0]$, the null matrix. The consequences of the two assumptions, for permissable inferences on β are quite distinct. We present in Table 9 the ratios of the estimates of the coefficients to the respective standard errors for each assumption: $\hat{\beta}_j/\sqrt{Var(\hat{\beta}_j)}$

It has been remarked above that the proposition that it is permissable to pool information on leukemia incidence rates with that on non-leukemia cancer mortality rates in order to achieve more seemly, as well as stable, estimates of β for models of the latter would seem dubious to some. The novelty of this proposition may, perhaps, be made more vivid by a demonstration that it is tantamount to ignoring the latency, etc. distinctions between the two responses, leukemia incidence and non-leukemia cancer mortality, and simply pooling the respective samples of n=16 observations each. Each of the rival models then includes a second dichotomous indicator variable, A, to distinguish the natural ($D\gamma = D_n = 0$) levels of response. This is quite similar to pooling the Hiroshima and Nagasaki samples in order to reduce the collinearity in the variables $D\gamma$ and D_n and thus achieve more seemly and stable estimates of β for the rival models of both leukemia incidence and non-leukemia cancer mortality (as well as breast cancer incidence). It will be recalled that a dichotomous indicator variable, C, was included in each of the rival models in order to distinguish between the natural ($D\gamma = D_n = 0$) levels of response at the two sites, Hiroshima and Nagasaki (vide supra).

The GLS estimate of β for the LQ-L model obtained from the pooled samples (n=32) is presented in Table 10. It is seen to be consistent with the mixed estimate of β. Moreover, both the point estimates of β_1, β_2 and β_3, as well as the respective precisions, $\hat{\beta}_j/\sqrt{Var(\hat{\beta}_j)}$, are

Table 10. LQ-L Models of Non-Leukemia Cancer Mortality Rates.

a. LQ-L. Pooled Samples. I. (Hiroshima plus Nagasaki <u>Leukemia Incidence</u> Rate). <u>n = 16</u>.

$y = 3.34 + 0.10D_\gamma + 0.84*10^{-3}D_\gamma^2 + 2.73D_n - 0.03C$
$(2.77)\ (1.04)(1.46)\phantom{*10^{-3}D_\gamma^2\ +\ }(3.55)\ (-0.02)$
RSS = 10.65 ~ $\chi^2(11)$. $\hat\beta_2/\hat\beta_1$ = <u>$8.41*10^{-3}$</u>, $\beta_3/\hat\beta_1$ = <u>27.30</u>

b. LQ-L. Pooled Samples. II. (Hiroshima plus Nagasaki <u>Non-leukemia Cancer Mortality</u> Rate). <u>n = 16</u>.

$y = 240.76 + 021D_\gamma - 0.36*10^{-3}D_\gamma^2 - 6.06D_n - 3.19C$
$(24.38)\ (0.52)\ (-0.19)\phantom{*10^{-3}D_\gamma^2\ }(2.61)\ (-0.30)$
RSS = 13.91 ~ $\chi^2(11)$

c. LQ-L. Mixed Estimation (Leukemia Incidence Constraints). <u>n = 16</u>.

$y = 239.00 + 0.14D_\gamma + 1.00*10^{-3}D_\gamma^2 + 4.06D_n + 0.02C$
$(27.18)\ (1.42)(1.72)\phantom{*10^{-3}D_\gamma^2\ +\ }(3.59)\ (0.04)$
$\gamma = 1.90$ ~ $\chi^2(2)$. $\hat\beta_2/\hat\beta_1$ = <u>$7.14*10^{-3}$</u>, $\hat\beta_3/\hat\beta_1$ = <u>29.00</u>
$\Theta_p = 0.366$

d. LQ-L. Pooled Samples II. (<u>Leukemia Incidence</u> plus <u>Non-leukemia Cancer Mortality</u> Rates). <u>n = 32</u>.

$y = 3.38 + 0.10D_\gamma + 0.78*10^{-3}D_\gamma^2 + 3.16D_n - 0.15C + 237.19A$
$(2.81)\ (1.05)(1.44)\phantom{*10^{-3}D_\gamma^2\ +\ }(4.29)\ (-0.12)(55.24)$
RSS = 28.68 ~ $\chi^2(27)$. $\hat\beta_2/\hat\beta_1$ = <u>$7.80*10^{-3}$</u>, $\beta_3/\hat\beta_1$ = <u>31.60</u>
C = 1 - Hiroshima. C = 0 - Nagasaki
A = 1 - Non-leukemia Cancer Mortality. A = 0 - Leukemia Incidence

---

Table 11. Evaluation of Rival Models of Breast Cancer Incidence. LSS Sample. (n = 20)

Model, M_i[a]	RSS(df)	R_f^2	$\bar R_f^2$	PRESS	R_p^2
LQ-L	8.28(15)	0.781	0.723	13.56	0.638
L-L	8.40(16)	0.775	0.732	12.17	0.674
Q-L	17.03(16)	0.578	0.499	29.41	0.272
L	8.93(17)	0.765	0.738	11.03	0.710
L*	8.44(17)	0.774	0.748	10.38	0.723

[a] These models <u>do</u> include the indicator variable, <u>C, since</u> the analyses have disclosed that it is "significant": $\hat\beta_4/\sqrt{Var(\hat\beta_4)} \geq 2.00$ for all models.

seen to be quite consistent with the cognate estimates for leukemia incidence rate - as would be expected. The two ratios of the coefficient estimates, $\hat{\beta}_2/\hat{\beta}_1$ and $\hat{\beta}_3/\hat{\beta}_1$ ("neutron RBE") for the mixed estimates (n = 16) and pooled estimates (n = 32) for non-leukemia cancer mortality are also consistent with the cognate ratios for leukemia incidence, $8.40*10^{-3}$ and 27.30, respectively, stipulated as constraints on the estimates $\underline{\hat{\beta}}$ for the LQ-L model of non-leukemia cancer mortality in the BEIR III Report. Thus, it would seem that the increased precision in the estimate of $\underline{\beta}$ for the LQ-L model of non-leukemia cancer mortality that is achieved by the constrained estimation methods ($\overline{\text{LQ-L}}$ model) of the BEIR III Report is tantamount to a kind of "double-count" of a single sample: once to obtain the observation matrix, [y, X], for non-leukemia cancer mortality data and a second time to obtain the non-stochastic ($\psi \longrightarrow [0]$) constraint matrix, [r, R], from the leukemia incidence data. That is, the sample and non-sample, or a priori, information are derived from the same set of data to give the augmented "observation" matrix

$$\begin{bmatrix} Py, & PX \\ Qr, & QR \end{bmatrix}$$

where $P'P = \hat{V}^{-1}$ and $Q'Q = \psi^{-1}$. See Belsley, Kuh and Welsch 1980.

9) The BEIR III Report offers the L-L model of breast cancer incidence as the, "model of choice". However, the estimate of the coefficient of the neutron dose, D_n, for both the LQ-L and L-L models of breast cancer incidence is much less than its standard error: $\hat{\beta}_3/\sqrt{Var(\hat{\beta}_3)}$ = 0.63, for the L-L model. See Table 1. The Q-L model can be excluded as an alternative by the PRESS statistic, although this was not done in the BEIR III Report. See Tables 11 and 12. Thus, while it is true, as remarked in the BEIR III Report, that for the L-L model the estimated coefficient for the neutron dose does not differ significantly from the estimated coefficient for the gamma dose, implying that $\beta_3 = \beta_1$, (which is consistent with an RBE = 1) it seems important to note also that the estimated coefficient for the neutron dose does not differ significantly from zero. See Figure 8b. (This suggests that perhaps the neutron dose is over-estimated in the T65D dosimetry?) The usual model selection criteria of regression analysis suggests that the term in D_n should be deleted from the model since the increase in bias in the estimates, $\underline{\hat{\beta}}$, $\hat{y}$, etc. is less than the concomitant decrease in variance which results from deleting D_n. Thus, the mean squared error of estimate is decreased by omitting D_n, an important desideratum (Montgomery & Peck, 1982). A novel alternative is retain the term in D_n by "pooling" it with that in D_γ to form the pseudo-dose, $D^* = (D_\gamma + D_n)$. This is equivalent to the conclusion that $\beta_3/\beta_1 = 1$ as noted in the BEIR III Report. But, such an inference is inadmissable unless $\hat{\beta}_j/\sqrt{Var(\hat{\beta}_j)}$ for both variables are well in excess of unity i.e., both differ significantly from zero, and seems wholly contrary to any reported practice in regression analysis. (But it is perhaps consistent with the practice of "pooling" information on leukemia incidence with that on non-leukemia cancer mortality?) This is apparently the maneuver adopted in the BEIR III Report. However, this procedure also over-states the true precision with which the effect of D_n on the incidence rate, y, represented by the weight, β_3, can be estimated from the LSS data. This precision is correctly estimated by the ratio $\hat{\beta}_3/\sqrt{Var(\hat{\beta}_3)} = 0.63$. If it is thought necessary (say, on the basis of prior information) to retain D_n in the model, then it must be weighted by the "non-significant" coefficient estimate, $\hat{\beta}_3$, provided by LSS data or else a quasi-Bayesian procedure

must be adopted, such as mixed estimation, by which the weak data may be "strengthened" by non-sample information - or conjecture. (For instance, the conjecture, H_0: $\beta_1 \equiv \beta_3$, can be represented as the constraint, $r = R\underline{\beta}$, on the L-L model where $\underline{r} = 0$, $R = (0, 1, -1, 0)$, $\Psi \equiv [0]$. The conjecture, H_0: $\beta_3 \equiv 0$, can be represented as the constraint $\underline{r} = R\underline{\beta}$ where $\underline{r} = 0$, $R = (0, 0, 1, 0)$, $\Psi \equiv [0]$.) The rival models L-L, L* and L are described in Table 13.

10) It is not without interest to note that the regression methods used in the BEIR III Report consistently underestimate the size of the dispersion matrix, $Var(\hat{\underline{\beta}})$, for all models of all responses. In the BEIR III Report $Var(\hat{\underline{\beta}}) = h(X'\hat{V}^{-1}X)^{-1}$ where $h = RSS/(n-k)$ is the dispersion factor (McCullagh & Nelder, 1983). The correct estimate is $Var(\hat{\underline{\beta}}) = (X'V^{-1}X)^{-1}$. For the L-L models of leukemia incidence, non-leukemia cancer mortality and breast cancer incidence we find that $h = 0.97$, 1.26, 0.54, respectively. The effect of the error on subsequent inferences on the vectors, $\underline{\beta}$, of the rival models of the respective responses is trivial for the first response but not for the second and third. (One assumes that $h = 1$ if the model "fits" the data.)

11) It can also be shown that, owing to the peculiar size and shape of the joint distribution of $D\gamma$ and D_n in the LSS sample, (See Figure 1a) the Least Squares estimates of the coefficients of any linear model of response are quite sensitive to the presence of small random errors of measurement of the predictor variables. Thus, the presence of random errors of measurement of dose of the size suggested by Jablon (1971) may inflate the coefficient of $D\gamma$ and deflate the coefficient of D_n (in the L-L model) so that the lack of "significance" of the GLS estimate of the latter in the case of breast cancer incidence may be due to random errors in the T65D dose estimates as well as to the systematic errors recently disclosed by the studies of Loewe, Mendelsohn, (1981) and others. We have obtained heuristic estimates of $\underline{\beta}$ for the L-L model which have been corrected (to terms of first order) for the bias induced by random errors of measurement of dose by the methods described by Theil (1971) and Seber (1977). Standard errors of measurement, s_1 and s_3, of $D\gamma$ and D_n, respectively, of the order of 8 rads each, may induce biases, $\underline{b} = (n-k)(X'X)^{-1}\hat{D}\hat{\underline{\beta}}$, vide infra, of (approximately) + 50% and - 100% in the respective estimated weights, $\hat{\beta}_1$ and $\hat{\beta}_3$, for the L-L model of leukemia incidence. See Tables 14a, 14b and 14c. Note the inflation of the Perturbation Index as s_j, $1 \leq j \leq 3$, increases. Only for $s_1 = s_3 = 2$, is $PI < 0.10$, and hence b_j is negligible. Note that the sign of the bias, b_j, differs for $\hat{\beta}_1$ and $\hat{\beta}_3$. Hence, the effect of the random errors of measurement, s_j, may be either to inflate or deflate the estimate, $\hat{\beta}_j$. Note that the magnitude of the effect of the random errors of measurement of a given size, $s_1 = s_2$, depends upon the respective ratios, $s_1/\sqrt{Var(D\gamma)}$ and $s_3/\sqrt{Var(D_n)}$. See Table 14c and the Perturbation Index, Table 14b. It must be emphasized that these Errors-in-Variables estimates are the heuristic approximations given by the transformation, $\hat{\underline{\beta}}_E = [X'X-(n-k)\hat{D}]^{-1}(X'X)\hat{\underline{\beta}}$ (Theil 1971) where $\hat{\underline{\beta}} = (X'X)^{-1}X'\underline{y}$ (i.e., unweighted observations) and $\hat{D}$ is the (non-sample) estimate of the dispersion matrix of the random errors of measurement. See also Lecture 13.

As is the case with Ridge Regression methods (Belsley, et al, 1980; Montgomery & Peck, 1982) for (partially) defeating the effects of collinearity in the predictor variables on the estimates $\hat{\underline{\beta}}$ and $Var(\hat{\underline{\beta}})$, the EIV methods have not yet been developed for Poisson models. Thus, the approximate degrees of bias induced in $\hat{\underline{\beta}}$ by random errors of measurement of dose were estimated for the cognate models of

Table 12. Akaike Information Criterion.[a] Breast Cancer Incidence
         Models, M_i. (n = 20)

Models, M_i*	RSS	df_i	AIC_i
LQ-L	8.176	15	-21.824
L-L	8.402	16	-23.598
Q-L	17.032	16	-14.968
L*	8.444	17	-25.556[b]
L	8.930	17	-25.070

[a] These models <u>do</u> include the indicator variable, C, since the analyses have disclosed that it is "significant": $\hat{\beta}_4/\sqrt{Var(\hat{\beta}_4)} \geq 2.00$ for all models.
[b] MAICE

Table 13. Comparison of Rival Models of Breast Cancer Incidence.
         (n = 20)

a. L-L

   $y = 16.99 + 0.22 D_\gamma + 0.31 D_n + 7.27 C$[a]
       (6.57) (3.18)   (0.63)   (2.39)
   RSS = 8.40 $\sim \chi^2(16)$. $P(\chi^2 < 8.40| 16) = \underline{0.06}$

b. L*. $D* = D_\gamma + D_n$. (Constraint: $\beta_3 = \beta_1$)

   $y = 16.89 + 0.23 D* + 7.40 C$
       (6.70) (4.94)[b] (2.51)
   RSS = 8.44 $\sim \chi^2(17)$. $P(\chi^2 < 8.44| 17) = \underline{0.04}$[c]

c. L. $D_n = 0$. (Constraint: $\beta_3 = 0$)

   $y = 16.66 + 0.25 D_\gamma + 7.72 C$
       (6.61) (4.93)   (2.62)
   RSS = 8.93 $\sim \chi^2(17)$. $P(\chi^2 < 8.93| 17) = \underline{0.06}$

[a] Note that the estimate, $\hat{\beta}_4$, of the coefficient of the indicator variable, C, differs significantly from zero for <u>all</u> of the rival models of the breast cancer incidence rate (including LQ-L and Q-L). This is <u>not</u> the case for <u>any</u> of the models of either the leukemia incidence rate for the non-leukemia cancer mortality rate.
[b] This implies a precision of estimate $\hat{\beta}_3/\sqrt{Var(\hat{\beta}_3)} = 4.94$ for the coefficient, β_3, of D_n.
[c] It is the practice to <u>reject</u> models for which $P(\chi^2 < RSS| \nu) < 0.05$ on the (quite reasonable) grounds that the sample "fits" the model too well to have arisen by <u>random sampling</u> from the population described by the model. See Finney 1971 and Lecture 13. The <u>inflation</u> of the precision of estimate of β_3 (4.94) over that of the sample (0.63) is achieved by the addition of information on β_3 (constraint: $\beta_3 = \beta_1$). However, this additional information was obtained from the sample itself - a double-count - hence the RSS appears suspiciously <u>low</u>.

Table 14a. Errors-in-Variables Estimates. L-L Model. Leukemia Incidence Rates. n = 16.

1. Poisson Regression. Generalized Least Squares. [Py, PX]

$$y = 2.963 + 0.223D_\gamma + 2.523D_n + 0.205C$$
$$(2.588)\ (3.646)\ \ \ \ (3.324)\ \ \ \ (0.158)$$
$$RSS = 11.765 \sim \chi^2(12).\ R^2 = 0.807$$

2. Least Squares. [y, X]. (Surrogate)

a) $s_1 = s_3 = 0^a$
$$y = -0.086 + 0.278D\gamma + 2.661D_n + 0.502C$$
$$(-0.01)\ \ (4.808)\ \ \ \ (4.845)\ \ \ (0.05)$$
$$RSS = 2607.29.\ \bar{R}^2 = 0.907$$

b) $s_1 = s_3 = 2$
Bias:
$$\underline{b} = (0.922,\ -0.011,\ 0.161,\ -1.897)^b$$
Bias-Corrected Estimates:
$$y = 1.372 + 0.260D_\gamma + 2.919D_n - 2.528C^c$$

---

a The standard deviation of the random errors of measurement of D_γ and D_n are s_1 and s_3, respectively.
b $E(\hat{\underline{\beta}}) = \underline{\beta} - \underline{b}.\ \underline{b} \simeq n(X'X)^{-1}D\hat{\underline{\beta}}.$ (Seber 1977)
c $\hat{\underline{\beta}}_{EV} \simeq [\bar{X}'X - n\bar{D}]^{-1}(X'X)\hat{\underline{\beta}}.$ (Theil 1971)

$$D = \begin{matrix} 0 & 0 & 0 & 0 \\ 0 & s_1^2 & 0 & 0 \\ 0 & 0 & s_3^2 & 0 \\ 0 & 0 & 0 & 0 \end{matrix}$$

Table 14c. Mean and Standard Deviation of Distributions of D_γ and D_n. Leukemia Incidence Rates.

	Mean	$\sqrt{\text{Variance}}$
D_γ	81.76	88.79
D_n	5.89	10.73

Table 14b. Errors-in-Variables Estimates. L-L Model. Leukemia Incidence Rates. n = 16. Least Squares Regression.

(s_1, s_3)[a]	PI[d]	b_j[b]		$b_j/\hat{\beta}_j$[c]		$b_j/\sqrt{Var(\hat{\beta}_j)}$	
		(0, 0)		(0, 0)		(0, 0)	
(0, 0)	0	(-0.012,	0.161)	(-4.25*10^{-2},	6.065*10^{-2})	(-0.204,	0.294)
(2, 2)	0.09	(-0.070,	1.009)	(-0.252,	0.379)	(-1.210,	1.837)
(5, 5)	0.56	(-0.179,	2.582)	(-0.644,	0.970)	(-3.102,	4.703)
(8, 8)	1.44						

[a] The standard deviation of the random errors of measurement of D_γ and D_n are s_1 and s_3, respectively.
[b] $\underline{b} \cong n(X'X)^{-1}\hat{D}\hat{\underline{\beta}}$
[c] $\underline{\hat{\beta}} = (\hat{\beta}_0, \hat{\beta}_1, \hat{\beta}_2, \hat{\beta}_3)$ is the Least Squares estimate of $\underline{\beta}$ on $[\underline{y}, X]$.
[d] Perturbation Index, $PI = Tr[(X'X)^{-1}D]$ (Beaton, et al 1976).

the [y, X] observation matrix – rather than for the [Py, PX] matrix. Such developments of EIV and RR methods are an area of current research (See Carroll, 1984 and Schaeffer, 1984). Our heuristic, linear model, estimates are included in an effort to motivate further research into the problems of determining the size of the random errors of measurement in the LSS data and the methods by which the effects of such errors on the estimates of β may be reduced.

4. SUMMARY

The paper has been written in the <u>descriptive</u> and <u>normative</u> modes: An account – or, more often, because of limitations of space, a suggestion – of what <u>is</u> included in the BEIR III Report as well as an account of what <u>ought</u> (according to standard usages) to have been included in that report to assure a well-informed deployment of the recommendations therein by the target group to whom it is addressed: "It [BEIR III] provides the scientific bases upon which [radiation protection] standards may be decided after non-scientific social values have been taken into account" (NRC/NAS 1980).

We have demonstrated by standard methods and criteria that the "model of choice", LQ-L, that was selected in the BEIR III Report to convey the evidence of the lifetime cancer experience of those persons present in Hiroshima and Nagasaki in early August 1945 is probably <u>less</u> suitable, for that purpose, on several criteria, than are the rival models, L-L, Q-L and L'-L.

Some few additional remarks concerning the "threshold" L'-L model are appropriate. First, let us recur to the index plots of RSTUDENT, Figures <u>3a</u>-<u>3d</u> in the context of the recent comments of Cook & Wang (1983): "Outliers and influential observations, for example, are always judged relative to some <u>model</u>, either implicit or explicit" ... "The selection of a transformation can be properly viewed as <u>model selection</u>" ... "An outlying or influential case in the original scale, for example, may conform in the transformed scale. This could be because the transformed scale is more appropriate than the original, or because the selection of the transformation was <u>controlled</u> by the case in question. In either situation, it is surely important to find if the evidence for the selected transformation is spread evenly throughout the data or rests only within a few cases." [italics added]. The introduction of the spline function may be viewed quite properly as <u>transformation</u> of D_γ: $D_\gamma \longrightarrow (D_\gamma - D_0)_+$. This is quite analogous to the transformations, $D_\gamma \longrightarrow \log_{10} D_\gamma$, or, more appropriately, $D_\gamma \longrightarrow D_\gamma^2$ as for the Q-L model.

If one compares <u>3c</u> (Q-L) with <u>3b</u> (L-L) it would seem that the transformation, $D_\gamma \longrightarrow D_\gamma^2$, <u>is controlled</u> by case #12 – since e_{14}^* is <u>inflated</u> by the transformation. However, comparison of <u>3d</u> (L'-L) with <u>3b</u> (L-L suggests that, "... the transformed scale $[(D_\gamma - \overline{D_0})_+]$ is more appropriate than the original."

The presence (or absence) of a "threshold" in the response function apparently raises ontological issues for some. We offer no comment on these issues. Although our introduction of the L'-L model of the leukemia incidence rates antedates (Herbert, 1983) the Park and Snee paper (1983), it may be regarded as merely being responsive to the latter's recommendations of good practice: "Risk managers should be made aware when the existence of a threshold is consistent with the data." However, both the <u>plausibility</u> and <u>utility</u> of spline functions are fully discussed in the literature cited (Montgomery & Peck, 1982;

Wold, 1974)). A brief exposition of these aspects is also presented in Herbert, 1983 and Lecture 17.

We have also shown that the idiosyncracies of the LSS data required that several "salvage operations" be performed in order to achieve plausible estimates of β for each of the rival models. All of these may be usefully subsumed under the rubric of "pooling maneuvers" to implement "data-instigated hypotheses" (Leamer 1978).

1. It is necessary to pool the city-specific samples, Hiroshima and Nagasaki, owing to the high level of correlation of $D\gamma$ and D_n within each, in order to make the sign, size and statistical significance of the estimates $\hat{\beta}_j$ consistent with prior information for all responses: leukemia incidence rate, breast cancer incidence rate and non-leukemia cancer mortality. See Figures 1a and 1f.

2. It is necessary to "pool" the disease-specific samples, leukemia incidence rate and non-leukemia cancer mortality rate, owing to the weakness of the information, $w_i = n_i T_i/y_i$, in the latter in order to make the sign, size and statistical significance of the estimates $\hat{\beta}_j$ consistent with prior information. Although the number of person-years at risk, $n_i T_i$, and dose level, $(D\gamma, D_n)_i$, $1 \leq i \leq 16$, are identical for the leukemia incidence sample and the non-leukemia cancer mortality sample, the level of response, y_i, $1 \leq i \leq 16$, for the latter exceeds that for the former by as much as two orders of magnitude. Hence the non-leukemia cancer mortality data are much "weaker" than the leukemia incidence data. See Figures 1b, 1c, 1d and 1e.

As remarked earlier the increase in precision achieved in the estimates of β for the LQ-L model, is a kind of "double-count" of a single sample: once to obtain the matrix of sample information, [y, X], and a second time to obtain the matrix of a priori information, [r, R]. See Table 10d.

3. It is necessary to "pool" the radiation doses, $D^* = D\gamma + D_n$, for breast cancer incidence in order to make the size and significance of the estimate, $\hat{\beta}_3$, (for D_n) consistent with prior information.

One recalls Chargaff's aphorism: "One of the most insidious and nefarious properties of scientific models is their tendency to take over, and sometimes supplant, reality" (Chargaff, 1963).

It is useful to observe that each of these "salvage operations" (or, in a more felicitous locution, "data-instigated hypotheses" (Leamer, 1978)) can be implemented by the methods of mixed estimation by which the sample information, $\hat{\beta}$ and $Var(\hat{\beta})$, on the (unknown) parameter, β, of the model, $y = X\beta + \varepsilon$, can be combined with non-sample, or apriori, information, $r = R\beta + v$, $E(v) = 0$, $Var(v) = \Psi$, to give the posterior estimates, β^{**}, $Var(\beta^{**})$.

For 1) the sample information is $\hat{\beta}_H$, $Var(\hat{\beta}_H)$ and the non-sample information is $r = \hat{\beta}_N$, $R = I$, $\Psi = RVar(\hat{\beta}_N)R'$ where the subscripts H and N denote the city-specific samples for Hiroshima and Nagasaki, respectively, and I is a 4X4 identity matrix.

For 2) the sample information is $\hat{\beta}$, $Var(\hat{\beta})$ for the pooled (Hiroshima + Nagasaki) observations on non-leukemia cancer mortality rates and the a priori information is

$$r = \begin{pmatrix} 0 \\ 0 \end{pmatrix} \quad R = \begin{pmatrix} 0, & \hat{\beta}_2/\hat{\beta}_1, & -1, & 0, & 0 \\ 0, & \hat{\beta}_3/\hat{\beta}_1, & 0, & -1, & 0 \end{pmatrix}, \quad \Psi = R\,Var(\hat{\beta})R'$$

where $\hat{\beta}$ and $Var(\hat{\beta})$ are the estimates for the cognate model of pooled observations on leukemia incidence. The BEIR III Report stipulates that for the leukemia incidence rates, $Var(\hat{\beta}) \equiv [0]$, a null matrix, a choice which is obviously an overstatement of the precision obtainable from the LSS data.

For 3) the sample information is $\hat{\underline{\beta}}$, $\text{Var}(\hat{\underline{\beta}})$ for the L-L model of the pooled observations on breast cancer incidence and the a priori information is

$\underline{r} = (\hat{\beta}_1)$, $R = (0, 0, 1, 0)$, $\Psi = R \text{Var}(\hat{\underline{\beta}})R' = \text{Var}(\hat{\beta}_3)$

where $\hat{\beta}_1$ is the estimate of the coefficient of D_γ for the same model (L-L) of the same data (breast cancer incidence) and $\overline{\text{Var}(\hat{\underline{\beta}})}$ is the corresponding variance-covariance matrix which is stipulated in the BEIR III Report to be $\text{Var}(\hat{\underline{\beta}}) \equiv [0]$. The procedure is clearly a "double-count" of the data since $[\underline{y}, X]$ and $[\underline{r}, R]$ are obtained from the same set of observations - thus L^* fits too well (See Table 13b).

Although the motivation for the "salvage operations" which yielded the risk estimates of the BEIR III Report - the retrieval of plausible estimates, $\hat{\underline{\beta}}$, for a given model from weak data - is unexceptionable, the justification for them does not seem to be equally well founded in each instance.

The BEIR III models of radiation response can neither be fully appreciated nor effectively deployed in risk estimation unless the idiosyncracies of the LSS data which led to the selection of the respective "constrained" models to convey the data evidence are clearly identified and their effects on the estimates $\hat{\underline{\beta}}$ and $\text{Var}(\hat{\underline{\beta}})$ understood. Or, to repeat the comments of Welsch which introduced this account of our examination of the BEIR III Report: "It is important to know when the resultant regression depends heavily on a small part of the prior knowledge, on a small part of the data or on the exact choice of model or fitting process" (Welsch, 1984).

This lecture is an expansion of a lecture presented at the Fourth Annual Coolfont Conference on Radiation and Health, July 8-13, 1984 (Herbert, 1984) and of an earlier lecture presented at the Sixteenth Midyear Topical Symposium of the Health Physics Society (Herbert, 1983).

ACKNOWLEDGEMENT

The author wishes to acknowledge, with thanks, the patience and dedication, as well as expertise, of Marie Woodall in the preparation of this manuscript.

Much of this work was accomplished during the time the author was a consultant to the Nuclear Regulatory Commission, Office of Standards Development. The author wishes to acknowledge his debt to Robert Alexander, Assistant Director, for proposing the problem which led to this work and for his subsequent support and criticism, and to Dr. Allen Brodsky, Senior Scientist, NRC, for his interest, criticisms and recommendations over the past several years.

REFERENCES

AKAIKE, H., 1974, A New Look at the Statistical Model Identification. IEEE. Trans. on Augo. Cont. 19:(6) 711-723.
AKAIKE, H., 1977, On Entropy Maximization Principle In Applications In Statistics. P. R. Krishnaiah, Ed. North-Holland Pub. 27-41.
ALLEN, D. M., 1974, The Relationship Between Variable Selection and Data Augmentation and a Method for Prediction. Technometrics. 16: (1). 125-127.
ATKINSON, A. C., 1981, Likelihood Ratios, Posterior Odds and Information Criteria. J. Economet. 16: 15-20.

ATKINSON, A.C., 1980, A Note on the Generalized Information Criterion for Choice of a Model. Biometrika. 67:(2) 413-418.

BEATON, A. E., RUBIN D. B. and BARONE, J. L., 1976, The Acceptability of Regression Solutions: Another Look at Computational Accuracy, Journal American Statistical Association, 71: pp

BELSLEY, D. A., KUH, E. and WELSCH, R. E., 1980, Regression Diagnostics: Identifying Influential Data and Sources of Collinearity. John Wiley & Sons. N.Y.

BOX, G.E.P., 1979, Robustness in the Strategy of Scientific Model Building in Robustness in Statistics. R. L. Launer & G. N. Wilkinson, Eds. Academic Press. N.Y. 201-236.

BROWN, W., SORHUS, C., CHOU-YANG, B. and RICHARDS, J., 1981, A Note of Caution on the Use of Individual Observations for Estimating Outdoor Recreational Demand Functions. Amer. J. of Agri. Econ. 65:(1) 154-157.

BROWNLEE, K. A., 1965, Statistical Theory and Methodology in Science and Engineering. 2nd ed. John Wiley, N.Y.

CARROLL, R. J., SPEIGELMAN, C. H., LAN, K. K. G., BAILEY, K. T. and ABBOTT, R. D., 1984, On Errors-in-Variables for Binary Regression Models. Biometrika. 71: 19-25.

COOK, R. D. and WEISBERG, S., 1982, Residuals and Influence in Regression. Chapman and Hall. N.Y.

COOK, R. D. and WANG, P. C., 1983, Transformations and Influential Cases in Regression. Technometrics, 25:(4) 337-343.

DUNCAN, G. T., 1978, An Empirical Study of Jackknife-Constructed Confidence Regions in Nonlinear Regression. Technometrics. 20:(2) 123-129.

DYKSTRA, O., 1971, Augmentation of Experimental Data to Maximize X'X. Technometrics. 13:(3) 682-688.

ELLETT, W., 1984, Unpublished Communication. National Research Council Commission on Life Sciences, Washington, D.C. 20418.

FABRIKANT, J. I., 1979, The Effects on populations of Exposure to Low Levels of Ionizing Radiation, Proceedings of the Delaware Valley Region of the American Association of Physicists in Medicine Conference "Known Effects of Low-Level Radiation Exposure, April p. 96.

FILLIBEN, J. J., 1975, The Probability Plot Correlation Coefficient Test for Normality, Technometrics, 17:(1) 111-17.

FINNEY, D. J., 1971. Statistical Method in Biological Assay. 2nd ed. Griffin. London.

FROME, E. L., 1983, The Analysis of Rates Using Poisson Regression Models. Biometrics. 39:(3) 665-675.

FROME, E. and HUDSON, D., 1980, Interval Estimation for X Predictions when the Dependent Variable Follows the Poisson Distribution. Proc. Amer. Statis. Comp. Sec. 294-296.

FROME, E., KUTNER, M. H. and BEAUCHAMP, J. J., 1973, Regression Analysis of Poisson-Distributed Data. J. Amer. Statis. Assn. 68: 935-940.

GAYLOR, D. W. and MERRILL, J. A., 1968, Augmenting Existing Data in Multiple Regression. Technometrics. 10:(1) 73-81.

HERBERT, D., 1983, Model or Metaphor? More Comments on the BEIR III Report. Proceedings of the Health Physics Society 16th Midyear Topical Meeting: Epidemiology Applied to Health Physics. Albuquerque, N.M. 357-90.

HERBERT, D., 1984, Application of Poisson Regression Analysis to BEIR III Data on Radiocarcinogenesis, Coolfont IV. Proceedings of The

ASA Conference on Radiation and Health, Berkley Springs, W.V.
HOCKING. R., 1983, Developments in Linear Regression Methodology: 1959-1983. Technometrics 25(3) 219-230.
JABLON, S., 1971, Atomic Bomb Radiation Dose Estimation at ABCC. Tech. Report 23-71. Atomic Bomb Casualty Commission, National Research Council, National Academy of Science, Washington, D.C.
JOHNSON, N. L. and KOTZ, S., 1969, Discrete Distributions, Houghton Mifflin Co., Boston.
KUHN, T., 1970a, The Structure of Scientific Revolutions. Univ. of Chicago Press, Chicago.
KUHN, T., 1970b, Logic of Discovery or Psychology of Research? in Criticism and the Growth of Knowledge, I. Lakatos and A. Musgrave, eds., Cambridge University Press, Cambridge, 1-23.
LAND, C., 1982, Private Communication, NCI, Environ. Epidem. Branch, Landow Bldg. 3C16, Bethesda, MD.
LAND, C., 1981, Biological Models in Epidemiology: Radiation Carcinogenesis, Environmental Science Research 21: Proceedings of the 13th Rochester International Conference on Environmental Toxicology (Measurements of Risk), Ed. by G. Berg and H. D. Maille N.Y. Plenum.
LEAMER, E. E., 1978, Specification Searches: Ad Hoc Inference With Nonexperimental Data. John Wiley, N.Y.
LOEWE, W. E. and MENDELSOHN, E., 1981, Revised Dose Estimates at Hiroshima and Nagasaki. Health Physics. 41:(4) 663-666.
MCCULLAGH, P. and NELDER, J. A., 1983, Generalized Linear Models, Chapman & Hall. N.Y.
MOLINA, E. C., 1942, Poisson's Exponential Binomial Limit, D. Van Nostrand Co., Princeton, N.Y.
MONTGOMERY, D. and PECK, E., 1982, Introduction to Linear Regression Analysis. John Wiley, N.Y.
MOSTELLAR, F. and TUKEY, J. W., 1977, Data Analysis and Regression. A Second Course in Statistics. Addison-Wesley. Reading, MA.
NETER, J. and WASSERMAN, W., 1974, Applied Linear Statistical Models, Richard D. Irwin, Inc., Homewood, IL.
NRC/NAS80 The Effects on Populations of Exposure to Low Levels of Ionizing Radiation:1980. National Research Council. National Academy of Science. Washington, D.C. (BEIR III).
PARK, C. N. and SNEE, R. D., 1983, Quantitative Risk Assessment: State-of-the-Art for Carcinogenesis. Amer. Statis. 37:(4) 427-441.
PREGIBON, D., 1981, Logistic Regression Diagnostics. Annals of Statis. 9: 705-724.
RAIFFA, H., 1982, Science and Policy: Their Separation and Intergration in Risk Analysis. Amer. Statis. 36:(3) 224-231.
SCHAEFER, R. L., ROI, L. D. and WOLFE, R. A., 1984, A Ridge Logistic Estimator. Commun. Statis. - Theor. Meth. 13:(1) 99-113.
SEBER, G. A. F., 1977, Linear Regression Analysis. John Wiley, N.Y.
SMITH, P. L., 1979, Splines as a Useful Convenient Statistical Tool, American Statistician, 33:(2) 57-62.
SNEE, R. E., 1977, Validation of Regression Models: Methods and Examples, Technometrics, 19:(4) 415-428.
STONE, M., 1974, Cross-Validation Choice and Assessment of Statistical Predictions. J. Royal Statis. Soc. Series B, 36: 111-147.
THEIL, H., 1971, Principles of Econometrics. John Wiley, N.Y.
VELLEMAN, P. F. and WELSCH, R. E., 1981, Effecient Computing of Regression Diagnostics. Amer. Statis. 35(4) 234-242.
WATKINS, J., 1974, Against "Normal Science", in Criticism and the

Growth of Knowledge. Ed. by I. Lakatos and A. Musgrave. Cambridge. University Press. 25-37.
WELSCH, R. E., 1984, An Introduction to Regression Diagnostics. Proceedings of the AAPM First Midyear Topical Symposium: Multiple Regression Analysis: Applications in the Health Sciences. Amer. Inst. of Phys. N.Y.
WOLD, S., 1974, Spline Functions in Data Analysis. Technometrics. 16: (1). 1-11.
ZWEIFEL, J. R., 1966, Use of the Likelihood Principle for the Determination of Carcinogenic Activity in Pulmonary Tumor Assays, Journal National Cancer Inst., 36:(5) 937-46.

APPENDIX I. DATA

Several of the features of the LSS data which are crucial to informed exploitation of the models presented in the BEIR III Report are presented in Figures 1 and 2. These plots are not included in the BEIR III Report itself.

Examination of these Figures recalls Brownlee's well-known astringent iconoclasm: "The justification sometimes advanced that a multiple regression analysis on observational data can be relied upon if there is an adequate theoretical background is utterly specious and disregards the unlimited capability of the human intellect for producing plausible explanations by the car-load lot" (Brownlee, 1965).

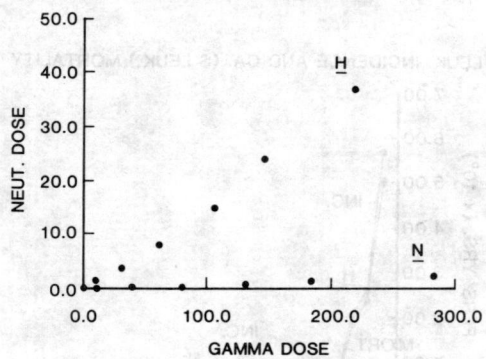

Figure 1a. Scattergram of the distribution of dose D_γ and D_n for the pooled (Hiroshima + Nagasaki) samples for the leukemia incidence rate. n = 16 (8 from each city). The Figure presents quite vividly the high degree of collinearity that encumbers the city-specific estimates of β with consequences that are disclosed in Table 2. The cognate distributions for the non-leukemia cancer mortality rate are nearly coincident with these and those for the breast cancer incidence rate are quite similar. Thus, it is necessary to pool the city specific samples in each case - or else use biased estimation methods (Montgomery & Peck 1982).

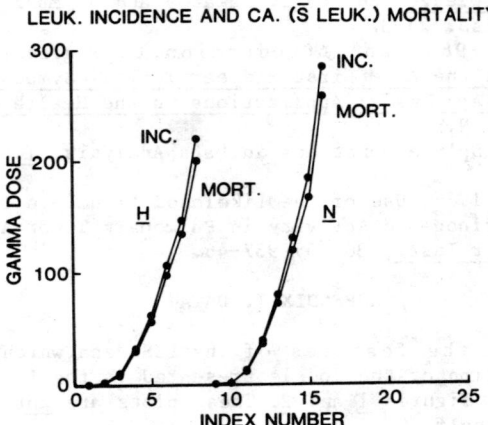

Figure 1b. Index plot of D_γ for the leukemia incidence rate and the non-leukemia cancer mortality rate. The first eight observations comprise the Hiroshima sample (H). The last eight observations comprise the Nagasaki sample (N). The abscissae, the index #'s, are simply the row numbers, $i = 1, \ldots, n = 16$ of the observation matrix, $[\underline{y}, X]$. The observations #8 and #16 refer to the groups at the hypocenters of the respective radiation sources at the two cities.

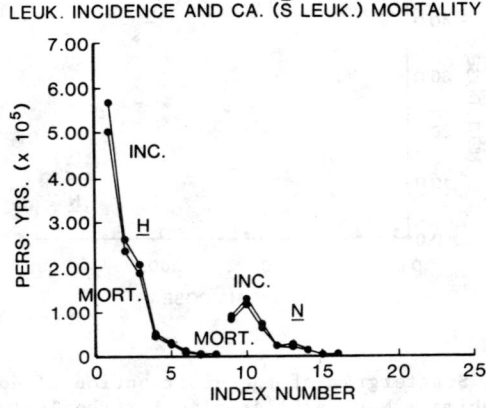

Figure 1c. Index plot of person years at risk for the leukemia incidence rate and the non-leukemia cancer mortality rate. From Figure 1b it is evident that the observations at lower doses will be "stronger" than those at higher doses.

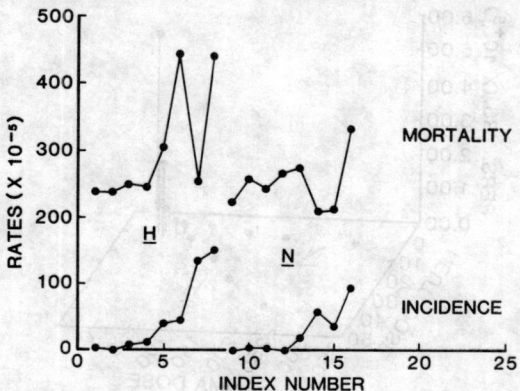

Figure 1d. Index plot of observed rates, y, for the leukemia incidence rate and the non-leukemia cancer mortality rate. The ratio of the averages of the two sets of observations is 7.16

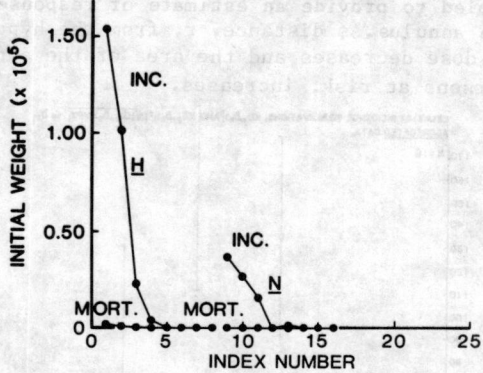

Figure 1e. Index plot of the initial weights, $w_i = n_i T_i / \hat{y}_i$, for the observations on the leukemia incidence rate and the non-leukemia cancer mortality rate. The ratio of the averages of the two sets of observations is 63.50. The data on leukemia incidence are "stronger" than those on non-leukemia cancer mortality by more than an order of magnitude. This feature of the LSS data seems to provide most of the empirical evidence for the LQ-L model of the latter response (a "data-instigated" hypothesis?. See Leamer 1978).

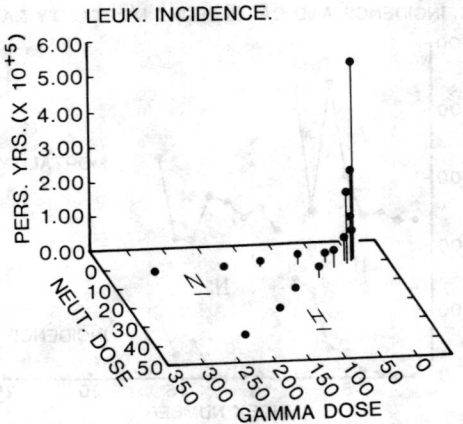

Figure 1f. Three-dimensional plot of person years at risk, nT, vs gamma dose, D_γ and neutron dose, D_n, for the leukemia incidence rate. The large correlation of D_γ and D_n within each city is characteristic of the respective radiation sources. The large values of nT at the lower doses - the largest value is at $D_\gamma = D_n = 0$ - is a result of the spherical symmetry of the "experimental design": The intensity of the unfiltered radiations at each city can be described, approximately as, $I = I_0 r^{-2} e^{-kr}$ where k^{-1} is a characteristic relaxation length. The disease experience of all persons within an annulus with center at the hypocenter is pooled to provide an estimate of response at the average dose within the annulus. As distance, r, from the hypocenter is increased the average dose decreases and the area of the annulus, and hence the number of persons at risk, increases.

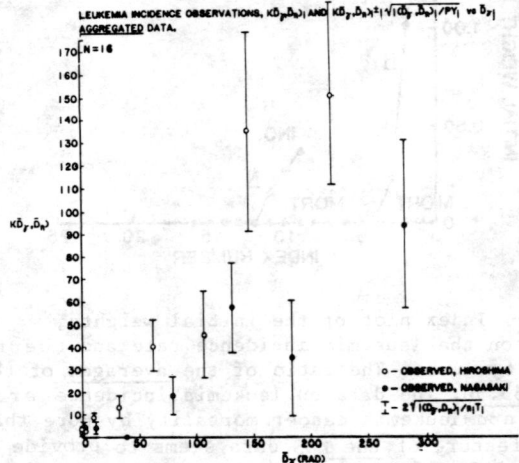

Figure 2a. Marginal distribution ($D_n = 0$) of observed response, $\underline{y} = I(D_\gamma, D_n)$, over D_γ for the leukemia incidence rate in the LSS sample. Note that in Figures 2a-c, reading from left to right, the Hiroshima observations correspond to index numbers 1-8 and the Nagasaki observations to index numbers 9-16. Note y=0 at D_γ = 38.8 rad.

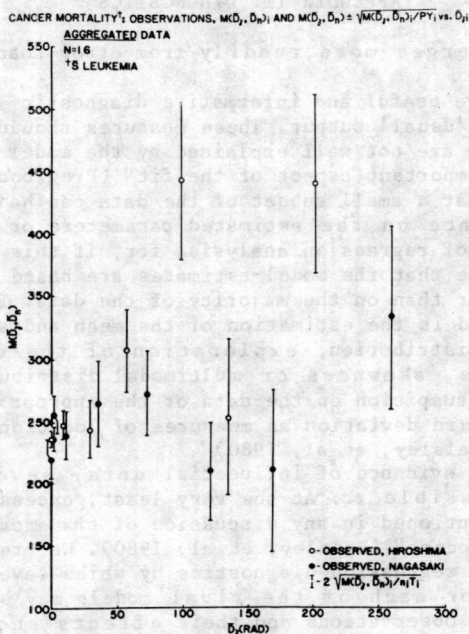

Figure 2b. Marginal distribution ($D_n = 0$) of the observed response, $y = M(D_\gamma, D_n)$, over D_γ for the non-leukemia cancer mortality rate in the LSS sample.

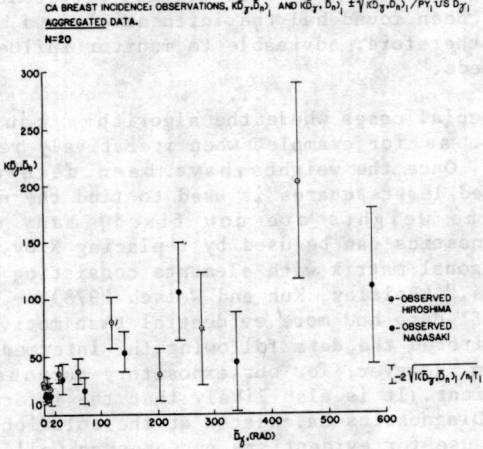

Figure 2c. Marginal distribution ($D_n = 0$) of the observed response, $y = I(D_\gamma, D_n)$, over D_γ for the breast cancer incidence rate in the LSS sample.

APPENDIX II. DIAGNOSTICS

"Truth emerges more readily from error than from confusion."
(Bacon 1613)

"We want to derive useful and informative diagnostic measures ... to supplement the 'usual' output. These measures should readily identify observations that are not well explained by the model as well as those dominating some important aspect of the fit" (Pregibon 1981).

"The fact that a small subset of the data can have a disproportionate influence on the estimated parameters or predictions is of concern to users of regression analysis, for, if this is the case, it is quite possible that the model-estimates are based primarily on this data subset rather than on the majority of the data. If, for example, the task at hand is the estimation of the mean and standard deviation of a univariate distribution, exploration of the data will often reveal outliers, skewness or multimodal distributions. Any one of these might cast suspicion on the data or the appropriateness of the mean and standard deviation as measures of location and variability, respectively." (Belsley, et al, 1980).

"If there is evidence of influential data, several corrective actions are possible ... At the very least, excessively influential data should be mentioned in any discussion of the model fitting and estimation on process" (Belsley, et al, 1980). We present in Figures 3 and 4 some of the regression diagnostics by which several influential observations for each of the rival models may be recognized. The presence of these observations and their effects should be kept in mind when using any of these models to interpret the evidence of the LSS data. See Belsley, et al 1980 and Pregibon 1981.

Since the estimates $\hat{\beta}$ and $Var(\hat{\beta})$ for the Poisson regression model are obtained by iterative re-weighted least squares methods (Frome 1973) we should note that, "There is one fundamental problem with non-linear regression diagnostics calculated only at the solution: data that are influential during specific iterations can cause the minimization algorithm to find a local minimum different from the one that would have been found had the influential data been modified or set aside. It is, therefore, advisable to monitor influential data as an algorithm proceeds."

...

"There are special cases where the algorithms can be monitored quite easily ... as for example, when iteratively re-weighted least squares is used ... Once the weights have been determined at each iteration, weighted least squares is used to find the next approximate solution. Since the weights are now fixed, many of the linear regression diagnostics can be used by replacing X by TX and y by Ty, where T is the diagonal matrix with elements consisting of the square root of the weights." (Belsley, Kuh and Welsch 1978).

If the BEIR III data had more evidential than motivational value we would have monitored the data following the intermediate as well as the final iteration. However, for our expository purposes the latter is quite sufficient.(It is also likely that the information provided by the Regression Diagnostics calculated at the solution will prove to be of considerable use for evidentiary purposes as well.)

Note that a <u>size-adjusted cut-off</u> such as 2k/n for the hat matrix diagonals will identify approximately the same proportion of potentially influential observations regardless of sample size n. (Belsley, Kuh and Welsch 1980)

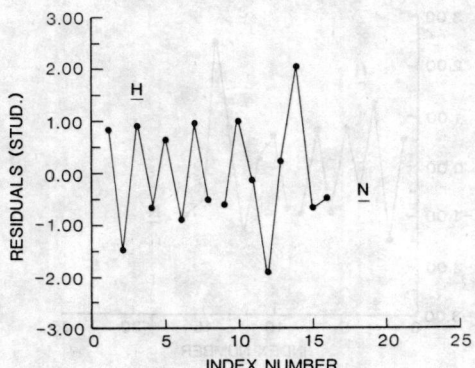

Figure 3a. Index plot of e_i^* (RSTUDENT) for the LQ-L model of the leukemia incidence rate. The e_i^* are scaled residuals, $e_i = (y_i - \hat{y}_i)$. To a first approximation, we may say that the ith response is not well explained by the model if $e_i^* > 2.0$. Thus, it seems that the (Nagasaki) observations #12 and #14 are outliers. These two observations will dominate the estimates of β_1, β_2 and β_3 in the model. See Figures 3h, i and j. There is evidence of negative correlation between the residuals in the Hiroshima data, presumably due to the method of grouping of the observations (Ellett 1984). In general the Hiroshima data seems to be better "explained" by the model.

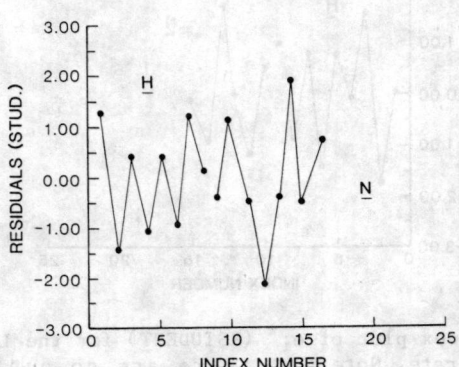

Figure 3b. Index plot of e_i^* (RSTUDENT) for the L-L model of the leukemia incidence rate. Note that the (Nagasaki) observations #12 and #14 are outliers - as is also the case for the LQ-L model. The effects of those observations on the estimates of β for the L-L model are found to be the same as for the LQ-L model. Refer to Figures 3a and 3h, i and j.

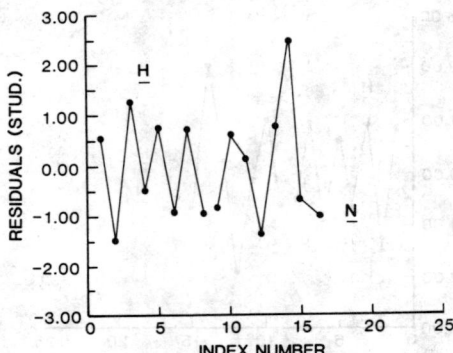

LEUKEMIA INCIDENCE. Q-L.

Figure 3c. Index plot of e_i^* (RSTUDENT) for the Q-L model of the leukemia incidence rate. Note that the (Nagasaki) observations #12 and #14 are <u>outliers</u> - as is also the case for the LQ-L and L-L models. The effects of these observations on the estimates of $\underline{\beta}$ for the Q-L model are found to be the same as for the LQ-L and L-L models. Refer to Figures 3a and 3h, i and j.

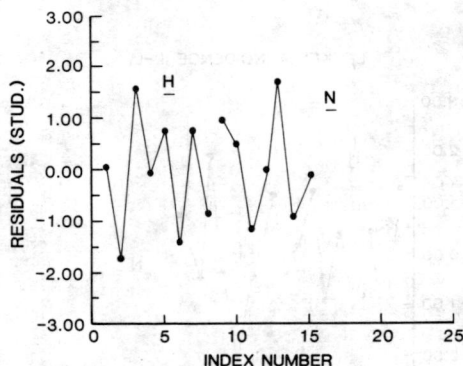

LEUKEMIA INCIDENCE. L'-L.

Figure 3d. Index plot of e_i^* (RSTUDENT) for the L'-L model of the leukemia incidence rate. Note that there are <u>no</u> <u>outliers</u> for this model of these data. That is, there are <u>no</u>, "... observations which are not well explained by the model ...". Thus, this "threshold" model seems to have mapped these data into a (binomial) "white-noise sequence". See Lecture 13.

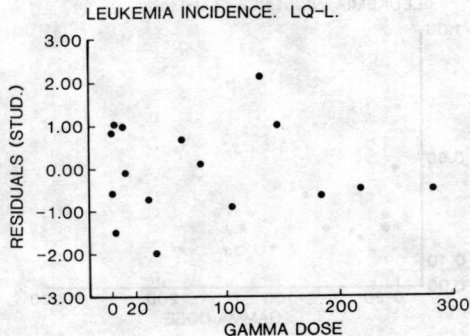

Figure 3e. Scattergram of e_i^* vs D_γ for the LQ-L model of the leukemia incidence rate. Note that <u>large</u> values of e_i^* typically occur <u>near</u> to the <u>centroid</u> of the distribution of X (Belsley, et al. 1980).

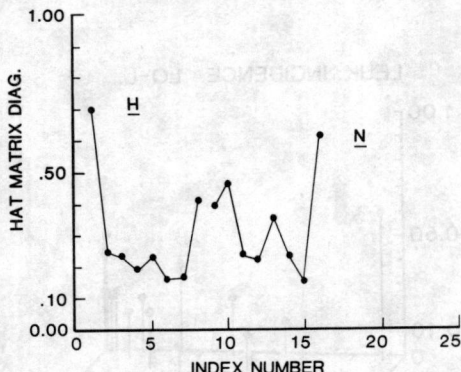

Figure 3f. Index plot of h_i, the hat matrix diagonal for the LQ-L model of the leukemia incidence rate. The hat matrix diagonal, h_i, can be considered to be a measure of the influence of y_i on $\hat{y}_i$: $h_i = \partial \hat{y}_i / \partial y_i$ (Velleman & Welsch, 1981). Alternatively, $1/h_i$ is the <u>effective</u> number of observations which determine $\hat{y}_i$ (Belsley, et al, 1980). In another sense, h_i is a measure of the (Mahalanobis) distance of the ith observation from the centroid of the data and/or a measure of the <u>weight</u> associated with the ith observation. The size-adjusted cut-off is $2k/n = 0.63$. Thus, the (Hiroshima) observation #1 and the (Nagasaki) observation #16 are high leverage observations and may have considerable effect on the estimates $\hat{y}$, $\hat{\beta}$, $Var(\hat{\beta})$, RSS, PRESS, etc. for the models.

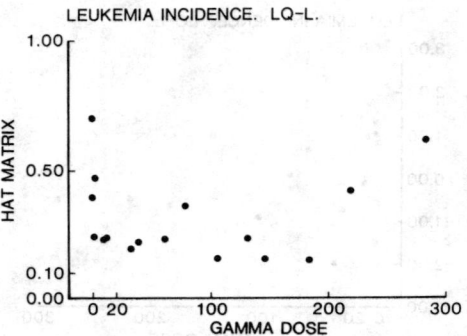

Figure 3g. Scattergrams of h_i vs D_γ for the LQ-L model of the leukemia incidence rate. Note that the large value of h_1 is due to the large value of the weight, $w_1 = n_1 T_1/\hat{y}_1 = 1.717$ while the large value of h_{16} is mostly due to the fact that it is relatively remote $(D_\gamma, D_n) = (283.7, 2.7)$ from the centroid of the set of observations, $(\overline{D}_\gamma, \overline{D}_n) = (3.34, 0.20)$, since w_{16} is only $6.529*10^{-4}$. Note also that observation #1 is masking the effects of the other observations at low doses which also carry large weights. Note that large values of h_i typically occur remote from the centroid of the distribution of X (Belsley, et al, 1980).

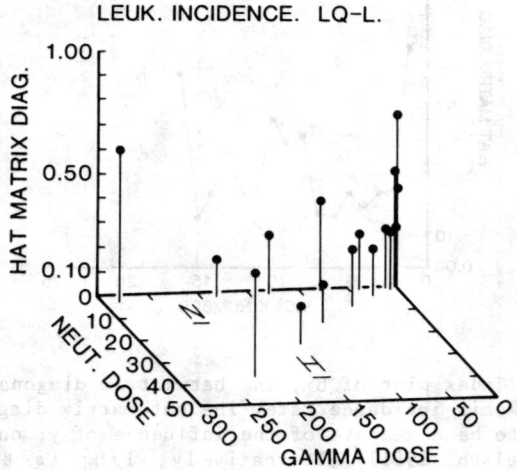

The lower figure presents a three dimensional scattergram of h_i vs D_γ and D_n for the LQ-L model of the leukemia incidence rate. Note that h_1 is in the Hiroshima (H) sample while h_{16} is in the Nagasaki (N) sample.

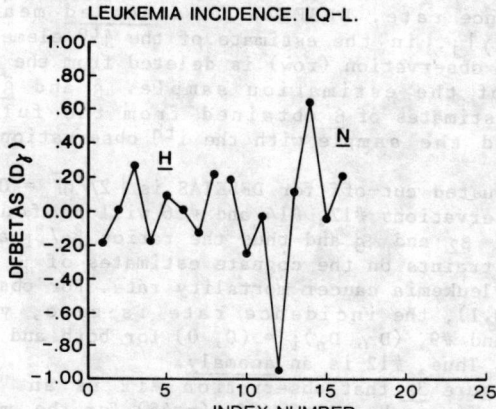

Figure 3h.

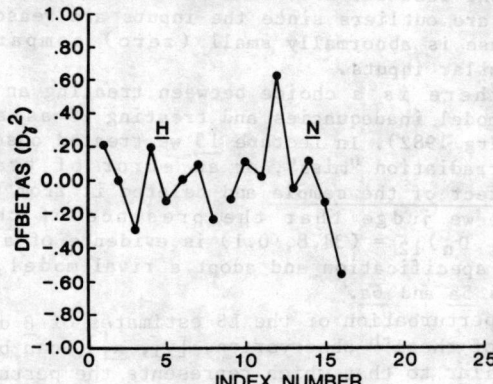

Figure 3i.

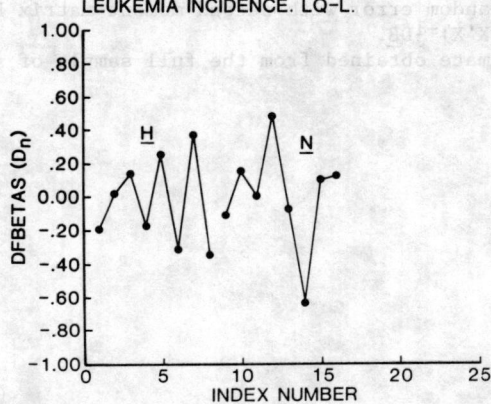

Figure 3j.

Figure 3h, i and j. Index plots of DFBETAS for the LQ-L model of the leukemia incidence rate. DFBETAS is a scaled measure of the change, $[\hat{\beta} - \hat{\beta}(i)]_j$, in the estimate of the jth element of $\hat{\beta}$ which occurs when the ith observation (row) is deleted from the observation matrix $[y, X]$ of the estimation sample. $\hat{\beta}$ and $\hat{\beta}(i)$ denote, respectively, the estimates of β obtained from the full sample (n observations) and the sample with the ith observation deleted (n-1 observations).

The size-adjusted cut-off for DFBETAS is $|2/\sqrt{n}| = 0.50$. Thus, it is evident that observations #12, #14 and #16 will profoundly affect the estimates of β_2 and β_3 and thus the ratios $\hat{\beta}_2/\hat{\beta}_1$ and $\hat{\beta}_3/\hat{\beta}_1$ that are imposed as constraints on the cognate estimates of β in the LQ-L model of the non-leukemia cancer mortality rate. For observation #12, $(D_\gamma, D_n) = (38.8, 0.1)$, the incidence rate is zero, $y_{12} = 0$. For observations #1 and #9, $(D_\gamma, D_n)_i = (0, 0)$ for both and $y_i = 3.69$ and 2.24, respectively. Thus, #12 is an anomaly.

We noted in Figure 3a that observation #12 is an "outlier" in quite the same sense as observation #1 (n=45) for the empirical model of ablation of head and neck cancer in Lecture 13. See Figures 4b, 4d and 5 in that lecture. The first two figures are cognate to Figures 3a and 3l of the present lecture. Both observations #12 (Lecture 14) and #1 (Lecture 13) are outliers since the inputs are reasonable in each case but the response is abnormally small (zero) compared to other observations at similar inputs.

Now, "... there is a choice between treating an observation as informative about model inadequacies and treating it as an outlier." (Cook and Weissberg 1982). In Lecture 13 we treated observation #1 as an "outlier", an irradiation "miss", or an error of transcription, etc. , i.e., a defect of the sample and deleted it from the sample. In the present lecture we judge that the presence of the anomalous response at $(D_\gamma, D_n)_{12} = (38.8, 0.1)$ is evidence of a defect in the model, an error of specification and adopt a rival model, the linear spline. See Figures 5a and 6a.

Note that the perturbation of the LS estimates of β due to the deletion (absence) of the ith observation, $[y_i, x_i']$, can be represented by a form quite similar to that which represents the perturbation due to the presence of random error in X:
a) Deletion of $[y_i, x_i']$
$\hat{\beta}(i) = \hat{\beta} - [e_i/(1-h_i)](X'X)^{-1}x_i$.
b) Presence of random error with second moment matrix D
$E(\hat{\beta}) = \beta - n(X'X)^{-1}\hat{D}\beta$.
$\hat{\beta}$ is the LS estimate obtained from the full sample of size n.

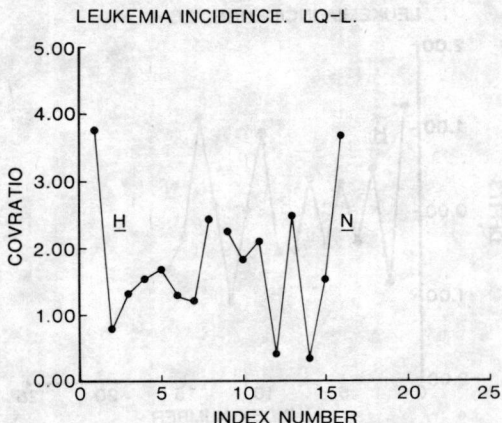

Figure 3k. An index plot of COVRATIO for the LQ-L model of the leukemia incidence rate. "The effects that particular data rows may have on the efficiency of estimation are most effectively conveyed by COVRATIO, a ratio of the determinants of the estimated variance-covariance matrices of the parameters, the numerator having been row-deleted" (Belsley, et al, 1980). Note that $|(X_{(i)}'\hat{V}_{(i)}^{-1}X_{(i)})^{-1}|/|(X'\hat{V}^{-1}X)^{-1}| \cong (1-h_i)^{-1}$ where the numerator is the determinant of $Var(\hat{\underline{\beta}}_{(i)})$. (If $\hat{V}^{-1} = I$, the (nxn) identity matrix, then the ratio is exactly equal to $(1-h_i)^{-1}$.)

"A value of COVRATIO greater than one indicates that the absence of the associated observation impairs efficiency, while a value of less than one indicates the reverse" (Belsley, et al, 1980). The size-adjusted cut-offs are $1 \pm 3k/n = 0.06, 1.94$. Thus, observations #1 and #16 clearly dominate the efficiency, $Var(\hat{\underline{\beta}})$, with which $\hat{\underline{\beta}}$ is estimated for the LQ-L model of leukemia incidence. It is also evident that observation #1, at $D_\gamma = D_n = 0$, is the more dominant of these two, though only slightly so, of course.

"The regression coefficients are, of course, a fundamental element in any structural analysis of a regression equation. Their estimated values, as well as the precision, or reliability, with which they are estimated, are of central importance. Thus, those diagnostics, such as the DFBETAS or the collinearity analysis, [VIF_i, etc.], that point to characteristics of the data to which the coefficient estimates or their estimated standard errors are particularly sensitive [COVRATIO, etc.], are especially useful for examining the suitability of the data for structural estimation." (Belsley, et al, 1980)

(Note that deletion of observations #12 and #14 would greatly decrease the volume of the confidence ellipsoid on $\underline{\beta}$.

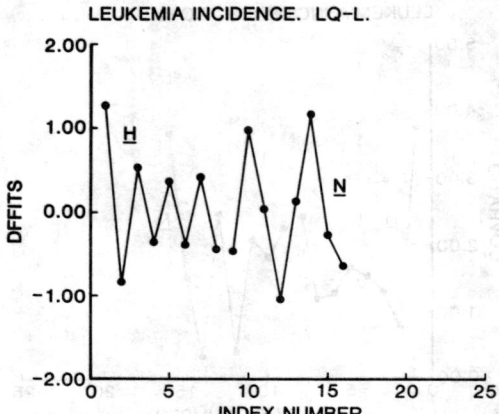

Figure 31. An index plot of DFFITS for the LQ-L model of the leukemia incidence rate. DFFITS is a scaled measure of the change in predicted response, $\hat{y}_i - \hat{y}_i(i) = x_i'[\hat{\beta} - \hat{\beta}(i)] = h_i e_i/(1-h_i)$, when the ith observation is deleted. "... DFFITS is an effective measure of the combined influence of a specific row on all coefficients taken together. DFFITS joins the two key diagnostics, e_i^* and h_i, in an economical way that serves to flag circumstances where either $|e_i^*|$ or h_i is large, or both are moderate but act jointly, and hence points to those rows that have the largest effect on predictions" (Belsley, et al, 1980). DFFITS is cognate to Cooks's D (Belsley, et al 1980; Cook and Weisberg 1982). Note that the individual elements of PRESS, $e_i^2/(1-h_i)^2$, are proportional to the unscaled values of DFFITS: $h_i e_i/(1-h_i)$. The size-adjusted cut-off is $2\sqrt{k/n} = 1.12$. Thus, the plot suggests that the presence of observations #1, #12, and #14 will degrade the predictive performance of the LQ-L model of leukemia incidence in new data.

"In prediction, and in the estimation of predictive error, less interest is often attached to the estimates of individual coefficients than to their combined effect. Diagnostics such as DFFITS are aimed at pinpointing sources of such overall sensitivity. Some of the diagnostic measures, such as the hat-matrix diagonals ... appear to be useful in both the predictive and structural context." (Belsley, et al, 1980).

Figures 3a - 31 emphasize the importance of constructing the case statistics, e_i^*, h_i, DFBETAS$_j$, etc., as well as the aggregate statistics, $\hat{\beta}$, Var($\hat{\beta}$), RSS' CDF(e_i^*), etc., for any regression model of a given set of data. (Note that the aggregates are functions of all n observations in the sample and are constructed on the assumption that the model is correct while the case statistics are functions of individual observations and disclose the presence of "problems" with that assumption.)

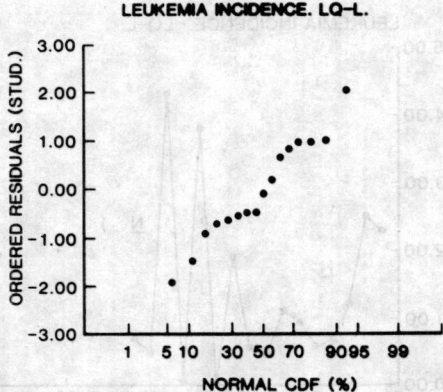

Figure 3m. Normal probability plot of e_i^* (RSTUDENT) for the LQ-L model of the leukemia incidence rate (n=16). The description of Daniel and Wood (1971) anent Normal probability plots of random normal deviates is strikingly apt: "Since samples of cumulative distributions will contain random errors, we need to acquire some feeling for normal departures from normality. ... Sets of 16 show shocking wobbles; ..." However, the Filliben probability plot correlation coefficient (Filliben 1975) is $r_F = 0.981$. For n=16 this suggests that the residuals have a Normal distribution. (See Lecture 13.) Similarly, for the models L-L, Q-L and L'-L the values of r_F are 0.992, 0.963 and 0.989, respectively, suggesting that each has a Normal distribution.

After examining Figures 1 - 3 it is surely interesting (to say the least) to recall that the information on radiocarcinogenesis contained in these LSS data and conveyed by the LQ-L model thereof that provides much of the empirical content (e.g., the radioepidemiological tables) of the current procedures for adjudication of "toxic tort" cases brought by individuals presenting with a confirmed diagnosis of cancer and a docmented history of antecedent exposure to ionizing radiation. The theoretical infrastructure is provided by Bayes' Theorem (i.e., the Probability of Causation). But so far as we know the suitability of these data for estimating a specific linear regression model by GLS has not been examined hitherto. ("Hell is truth seen too late" J. Locke.)

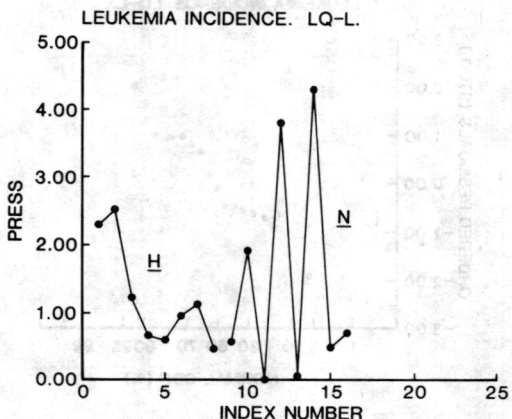

Figure 3n. Index plot of the elements of PRESS for the LQ-L model of the leukemia incidence rates. It is evident that the presence of the (Nagasaki) observations #12 and #14 will degrade the predictive performance of the model in new data. vide supra.

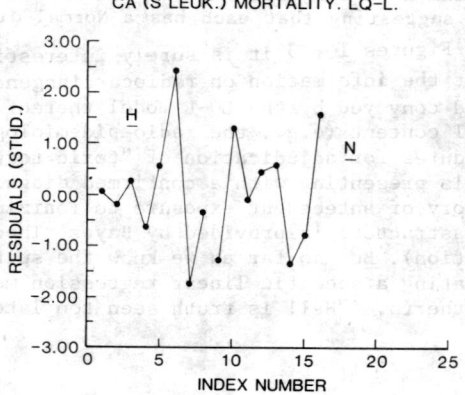

Figure 4a. Index plot of the Studentized residuals, e_i^*, for the LQ-L model of the non-leukemia cancer mortality rates. There is a single outlier.

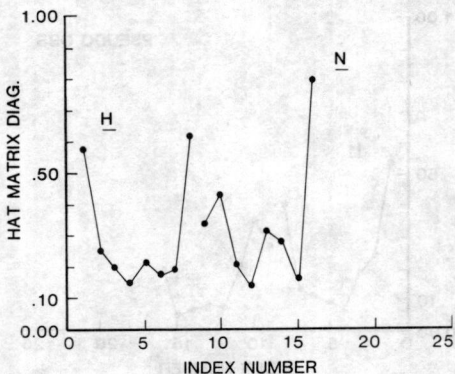

Figure 4b. Index plot of the hat matrix diagonals for the LQ-L model of the non-leukemia cancer mortality rates.

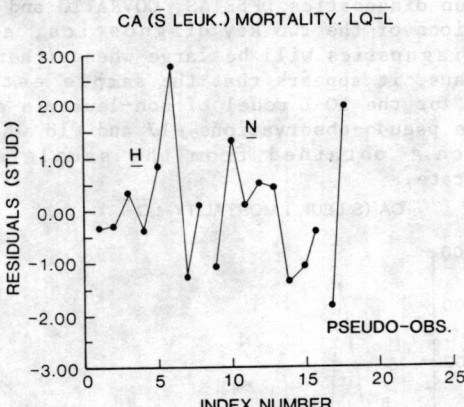

Figure 4c. Index plot of the Studentized residuals, e_i^*, for the $\overline{\text{LQ-L}}$ model of the non-leukemia cancer mortality rates. The pseudo-observations are the rows of the $[Qr, QR]$ matrix where $Q'Q = \psi^{-1}$ and $\psi = R\text{Var}(\hat{\underline{\beta}})R'$ where $\text{Var}(\hat{\underline{\beta}})$ is the covariance matrix of $\underline{\beta}$ for the LQ-L model of the leukemia incidence rates. Note that there are now two outliers. The imposition of the constraint, $\underline{r} = R\underline{\beta}$, inflates the residual sum of squares: $\text{RSS} = (\underline{y}-X\hat{\underline{\beta}})'\hat{V}^{-1}(\underline{y}-X\hat{\underline{\beta}}) + (\hat{\underline{\beta}}_{CE}-\hat{\underline{\beta}})'(X'\hat{V}^{-1}X)(\hat{\underline{\beta}}_{CE}-\hat{\underline{\beta}})$. The increment of RSS is given by the compatibility statistic, γ^*:

$\text{RSS}' = \text{RSS} + \gamma = 13.91 + 2.15 = 16.06$.

RSS' is the residual sum of squares for the $\overline{\text{LQ-L}}$ model and RSS is the residual sum of squares for the LQ-L model of the non-leukemia cancer mortality rate. See Table 9a.

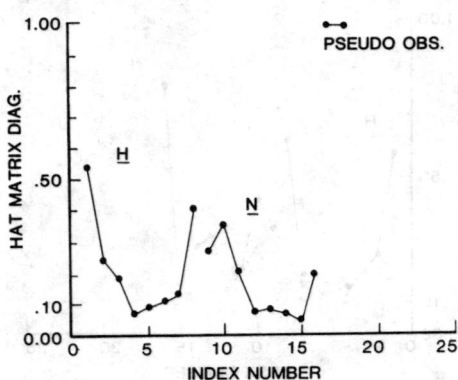

Figure 4d. Index plot of the hat matrix diagonals, h_i, for the $\overline{LQ-L}$ model of the non-leukemia cancer mortality rate. The size-adjusted cut-off is $2k/n = 0.56$. Thus, the pseudo-observations are influential points: $h_{17} = 0.999$, $h_{18} = 0.999$. Note that the other single-row deletion diagnostics DFBETAS, COVRATIO and DFFITS, as well as PRESS, are functions of the two key diagnostics, e_i^* and h_i. In general, these diagnostics will be large when either e_i^* or h_i - or both - are large. Thus, it appears that the sample estimates, $\hat{y}$, $\hat{\beta}$, $Var(\hat{\beta})$ and PRESS for the LQ-L model of non-leukemia cancer mortality are dominated by the pseudo-observations #17 and #18 which represent the information on β obtained from the sample estimates of the leukemia incidence rate.

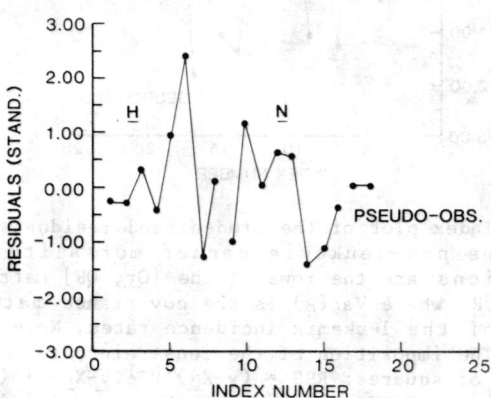

Figure 4e. Index plot of the residuals, $e_i = (y_i - \hat{y}_i)$. Note that when $h_i = 1.0$ we have $e_i = 0$, $i = 17, 18$. This is equivalent to saying that in some co-ordinate system, one parameter is determined completely by y_i. (Belsley, et al 1980) In the present instance, two out of five, or 40% of the elements of $\hat{\beta}^{**}$ for the $\overline{LQ-L}$ model are determined by the prior information on β. This is in close agreement with the estimate $\theta_p = 0.40$ in Table 8. See Table 9a.

APPENDIX III. POINT AND SET ESTIMATES

Interval estimates of $y_i(D\gamma)$ and $D\gamma(y_i)$ were not included in the BEIR III Report for any of the rival models of radiocarcinogenesis. We have presented a few of them in Figures 5 to 8. Examination of these Figures recalls the observations of Park and Snee cited above: "The size of the confidence limits is inversely proportional to the quality of the data used to make the estimate and directly proportional to the amount of extrapolation involved. This important information is lost if the confidence limits and best estimates are not routinely reported. The width of the confidence interval is one of the best measures risk assessors, and risk managers, have to evaluate the quality of the estimates of potential risks. It is important to distinguish between those situations in which the risk is precisely estimated and those in which it is not" (Park & Snee, 1983).

Interval Estimates of Response Rates.

The estimates response, $\hat{y}_i$, at a specified value of the vector of exogenous variables $\underline{x}_i' = (1, D_\gamma, D_n, ...)$ is $\hat{y}_i = \underline{x}_i \underline{\hat{\beta}}$. $\hat{y}_i$ is a point estimate of y_i. Interval estimates of response may be obtained if the sampling distribution of y_i is known: $y_i \sim N(X\beta; V)$. y_i is distributed asymptotically normal with $E(y_i) = E(\underline{x}_i \hat{\beta}) = E(\underline{x}_i' \hat{\beta}) = \underline{x}_i \hat{\beta}$ and $Var(\hat{y}_i) = \underline{x}_i'(X'V^{-1}X)^{-1}\underline{x}_i'$. The ratio, $t = (\underline{x}_i \hat{\beta} - \underline{x}_i'\hat{\beta})/\sqrt{\underline{x}_i^+(X'V^{-1}X)^{-1}\underline{x}_i^+}$, is distributed as Students' t with (n-k) degrees of freedom. The $(1-\alpha)$ confidence interval for the conditional response at $\underline{x}_i'$ is the region curcumscribed by $\hat{y}_{i-}, \hat{y}_{i+} = \underline{x}_i'\underline{\hat{\beta}} \pm t(n-k; 1-\alpha/2)/\sqrt{\underline{x}_i^+(\overline{X}'V^{-1}X)^{-1}\underline{x}_i^+}$.

The interval estimates of the level of dose, D_γ^*, that is required to educe a response rate, $\hat{y}_i^*$, which represents either a specified multiple of, or is at a specified increment to, the "background" or spontaneous, response rate would seem to be of some interest as an objective measure of a "maximum permissable dose" that would serve as, ".. a scientific basis for the development of suitable radiation protection standards", (NRC/NAS 1980), and that is also consistent with current practice in environmental toxicology. This proposition has, apparently, not been discussed before. Let us obtain an interval estimate of the gamma dose, D_γ^*, at which the induced incidence rate, y_i^*, is a multiple, n, of the background rate, y_0. We assume that the neutron dose is zero when the gamma dose is D_γ^* and follow the methods of Frome, et al (1973, 1980 and 1983). The difference, $d = (ny_0 - y)$, is distributed with $E(d) = 0$ and $Var(d) = n^2 Var(y_0) + Var(y_i)$. The ratio $d/\sqrt{Var(d)} \sim N(0, 1)$, asymptotically. The limits, $D_\gamma^*{}_-$ and $D_\gamma^*{}_+$, of the $(1-\alpha)$ confidence interval on D_γ^* are obtained as solutions to the equation, $d^2 = z_{(1-\alpha)}^2 Var(d)$, where $z = x_{(1-\alpha)}$ is the unit Normal deviate.

If we write the GLS estimate of the LQ-L equation for $D_n = 0$ as $y = a_0 + a_1 D_\gamma + a_2 D_\gamma^2$ then the interval estimates of D_γ^* for the LQ-L model of response ($D_n = 0$) are obtained as the non-negative real roots of the equation: $A_0 + A_1 D_\gamma + A_2 D_\gamma^2 + A_3 D_\gamma^3 + A_4 D_\gamma^4 = 0$ where

$$A_0 = z_{1-\alpha}^2[n^2(y_0/n_0) + Var(a_0)] + 2na_0 y_0 - n^2 y_0^2 - a_0^2$$

$$A_1 = 2z_{1-\alpha}^2 Cov(a_0, a_1) + 2na_1 y_0 - 2a_0 a_1$$

$$A_2 = z_{1-\alpha}^2[Var(a_1) + 2Cov(a_0, a_2)] + 2na_2 y_0 - (a_1^2 + 2a_0 a_2)$$

$$A_3 = 2z_{1-\alpha}^2 Cov(a_1, a_2) - 2a_1 a_2$$

$$A_4 = z_{1-\alpha}^2 \text{Var}(a_2) - a_2^2$$

N_0 is the sample size from which y_0 is determined. The $(1-\alpha)$ interval estimates of D_γ^* for the L-L and Q-L models ($D_n = 0$) are obtained as the non-negative real roots of cognate equations. (Frome & Dufrain 1979; Frome & Hudson 1980).

We have constructed confidence limits on both the <u>direct</u>, $\hat{y}_i$, and <u>inverse</u>, D_γ^2, estimates for all three models of the conditional leukemia incidence rate as well as for the conditional <u>excess</u>, ($y_i - y_0$) incidence rate. Some of these are now discussed in the context of the recommendations of the BEIR III Report. See Figures <u>5</u> - <u>7</u>.

The Report describes a rather novel method for setting confidence limits to the levels of conditional response predicted by the several models considered, LQ-L, L-L and Q-L: "These members would prefer to regard the linear (L or L-L) model not as central, but rather as one extreme on which credible upper bounds (in the form of confidence limits) could be based; the other extreme would be provided by the pure quadratic (Q or Q-L) model, on which credible lower bounds could be based." (NRC/NAS 1980). Note first that the forecasts of the L-L and Q-L models are not to be circumscribed by any confidence limits whatever. Note also that the level of confidence, $(1-\alpha)$, that is to be associated with the region of LQ-L forecasts that is circumscribed above and below by the L-L and Q-L forecasts, respectively, is not specified. Finally, note that since neither the data, the zeroth order elements in the estimates, $\hat{\beta}$, and Var($\hat{\beta}$) nor the off-diagonal elements in the estimates, Var($\hat{\beta}$), for the several models are presented in the Report, one cannot construct any estimates of the region of response, y_i, predicted by the LQ-L model that is to be circumscribed by the forecasts of the L-L and Q-L models.

Figure <u>5a</u> presents a superposition of the estimates of the conditional ($D_\gamma \geq 0$, $D_n = 0$) leukemia incidence rate of the three models, LQ-L, L-L and Q-L together with the 0.95 confidence limits on the former, which are seen to be enormous. It can be shown that over most of the range of D_γ the region circumscribed by the forecasts from the L-L and Q-L models coincides (approximately) with the 0.60 confidence limits on LQ-L forecasts. (In the region of greatest interest, the region of lower gamma dose, the level of confidence falls to zero, of course.) The Figure also presents a superposition of the estimates of the conditional spontaneous ($D_\gamma = 0$, $D_n = 0$) incidence rates for four ($=2^2$) response strata specified by the dichotomous endogenous variables U = (sex, site). The m = 2^2 strata define 4 levels of U, coded as, say, (0, 0), (0, 1), (1, 0) and (1, 1). It is evident that the respective variations in the level of conditional response, y_i, that are produced by variations in the endogenous, U, and exogenous, X, variables may be comparable.

<u>Further Remarks on Spline Functions and the L'-L Model.</u>

Let us return to the most general expression for a conditional response rate, $y = y(U, X)$, where U denotes a vector of <u>endogenous</u> variables (sex, site, etc.) and X denotes a vector of <u>exogenous</u> variables (dose, dose-squared, etc.). The conditional response is typically represented by the linear form $y = X\beta + \varepsilon$ (in the usual notation for Normal theory). The dependence on X is explicit, that on U is implicit in β: $\beta = \beta(U)$. For instance, for the LSS data we have considered it may be shown that the intercept, β_0, of the conditional incidence rate for breast cancer is a function of (bomb) site. (As is also evident in Figure 5a for the leukemia incidence rate where it appears that the effects on the level of the conditional response rate, y, of such changes in U may be comparable to the effects of changes in $D\gamma$ of one or two orders of magnitude – depending on the model.) Moreover, it is a discontinuous function, $y = \beta_0 + \beta_1 D\gamma + \beta_2 D_\gamma^2 + \beta_4 C$, where C is the dichotomous site variable C = 0 (Nagasaki) and C = 1 (Hiroshima); The intercept changes discontinuously from β_0 (Nagasaki) to $\beta_0 + \beta_4$ (Hiroshima) at $D\gamma = 0$.

We may contemplate, as a general proposition, that the mechanism of conditional biological response to low-LET radiation may change discontinuously between <u>regions</u> of the exogenous variables, X, as well as between <u>strata</u> defined by the endogenous variables, U. Levels of X which delimit these regions define "thresholds" in these variables. These changes in mechanism are disclosed by changes in the coefficient vector, $\underline{\beta}$, of linear models. that is, we have $\underline{\beta} = \underline{\beta}(U, X)$, in general. Thus, the intercept, β_0, slope, β_1, etc., of a dose-response equation may change discontinuously at one or more levels of the dose. For example, we have $y = \beta_0 + \beta_1 D\gamma + \beta_2 D\gamma^2 + \beta_4 C$, for $D\gamma \leq D_0$ and $y = (\beta_0 + \gamma) + \beta_1 D\gamma + \beta_2 D\gamma^2 + \beta_4 C$ for $D\gamma > D_0$: the intercept changes discontinuously, by amount β_4, at $D\gamma = D_0 > 0$.

The regression estimates of the coefficients in such equations, where $\underline{\beta} = \underline{\beta}(X)$, may be implemented by the introduction of indicator variables, just as in the case of equations in which $\underline{\beta} = \underline{\beta}(U)$, (the site variable C=0, 1 is an indicator variable) in a <u>piece-wise regression</u>. Alternatively, the estimation may be implemented by the introduction of <u>spline functions.</u> In general, splines are piece-wise polynomaials of order k. The join points of the pieces are called knots. The values of the function and its first (k-1) derivatives agree at the knots. That is, the spline is a continuous function with (k-1) continuous derivatives. A spline function with knot k_i is defined as $(x-k_i)_+ = (x-k_i)$, for $(x-k_i) \geq 0$, and $(x-k_i)_+ = 0$ for $(x-k_i) < 0$.

As well as representing characteristic structural changes in the mechanism of a process the spline functions also provide basis functions for empirical approximations that are a useful adjunct to the polynomials in the better-known Taylor series approximations. For instance, it is not infrequently the case that a low-order polynomial provides but a poor fit to the data and a modest increase in the degree of the polynomial dose not sufficiently improve it. In such cases, a Taylor series will <u>not</u> provide an adequate approximation. A spline function may (Wold 1974).

Moreover, there is the question of <u>local</u> vs <u>global</u> behaviour of the two classes of approximation functions. For polynomials – as well as for most other mathematical functions – the behaviour of these functions in a <u>small</u> region of the argument determines their behaviour <u>everywhere</u>. Although this property of "rigidity" makes such functions good filters (for smoothing and interpolation over the region of observations) it may also introduce bias into the estimates of response in some other region of the treatment variables (e.g., the Q-L model). A spline function may not. (Wold 1974).

It will have been noted that although the LQ-L, L-L and Q-L models have been proposed only on theoretical arguments as representative of alternative mechanisms of conditional response they also represent 3 of the possible $2^6 = 64$ alternative Taylor Series approximations to the conditional response – to terms of second order in $D\gamma$ and D_n. In other words, these three rival polynomial models may be considered either as equations that describe the mechanism which generated the population from which the sample was selected or as simply empirical graduations of the data. From either perspective, the spline function is a contender.

We are immediately concerned with the linear spline, $k = 1$, for which only one discontinuity or knot, at $x = k_1 = x_0$, occurs in the region of interest (Montgomery & Peck 1982; Smith 1979; Wold 1974). Thus we write, in an obvious notation, $y = \beta_0 + \beta_1 x + \beta_2(x-x_0)_+$. Although in principle the positions of the knots are free parameters of the model that enter in a non-linear manner it is permissible to regard the determination of knot positions as corresponding to the choice of a functional type in ordinary curve-fitting practice. Therefore, the determination of the position of a knot does not require one degree of freedom. Furthermore, just as to each functional type there may correspond a transformation of the variables x or y, e.g., $x \longrightarrow \log x$, the spline function corresponds to the transformation $x \longrightarrow (x-k_1)_+$.

We have fitted a linear spline to the pooled leukemia incidence data ($\beta_4 = 0$): $y = \beta_0 + \beta_1 D_\gamma + \beta_2 (D_\gamma - D_0)_+ + \beta_3 D_n + \epsilon$ where $(D_\gamma - D_0)_+ = (D_\gamma - D_0)$ for $D_\gamma - D_0 \geq 0$ and $(D_\gamma - D_0)_+ = 0$, for $D_\gamma - D_0 < 0$. We found that the term in D_γ could be neglected, giving the L'-L model: $y = \beta_0 + \beta_2(D_\gamma - D_0)_+ + \beta_3 D_n$.

We have obtained Maximum Likelihood estimates of the position, D_0, of the knot with upper and lower confidence limits, D_{0+} and D_{0-}, respectively, by a variation of the Box-Cox procedure for determining empirically the optimal value of the exponent of the dependent variables in a power-law transformation in linear regression (Montgomery & Peck 1982). This appears to be the same method used by Zweifel (1966) to determine the respective threshold doses for chemical carcinogens for murine pulmonary tumors (although that author does not refer to the Box-Cox procedure nor does he estimate the confidence limits). It is worth noting that the deletion diagnostics for the L'-L model of the leukemia incidence rate, disclosed <u>no discordant</u> observations (RSTUDENT < 2.0).

Further Remarks on Mixed Estimation and the $\overline{L-L}$ Model

We have described in the paper how the regression diagnostics for the $\overline{LQ-L}$ model of non-leukemia cancer mortality were obtained from the augmented matrix (See Belsley, et al 1980):

$$\begin{bmatrix} P\underline{y}, & PX \\ Q\underline{r}, & QX \end{bmatrix}$$

Here $[\underline{y}, X]$ is the <u>observation</u> matrix for the model of non-leukemia cancer mortality with estimated dispersion matrix, $\hat{V}$, and $[\underline{r}, R]$ is the matrix of a priori information on $\underline{\beta}$, or <u>constraint</u> matrix, with dispersion matrix, Ψ, and where $PP' = \hat{V}^{-1}$ and $QQ' = \Psi^{-1}$. $[\underline{r}, R]$ and Ψ are obtained from the cognate model of the leukemia incidence rate. However, we did not illustrate the procedure.

It would perhaps be useful to indicate briefly the elements of the procedure using the $\overline{L-L}$ model of the non-leukemia cancer mortality rate. For this model the constraint, $\underline{r} = R\underline{\beta}$, is represented graphically by the line, $\hat{\alpha}_1 = \hat{\beta}_1/11.53$, in Figure <u>8a</u> which also presents the point and interval estimate of $(\alpha_1, \beta_1) = (\beta_1, \beta_3)$ for the model of the leukemia incidence rate. In this figure the estimates, $(\beta_1, \beta_3) = (0.22, 2.55)$, refer to the model of the pooled <u>sample</u> with the site (city) variable <u>omitted</u>, $\beta_4 \equiv 0$, since $\hat{\beta}_4/\sqrt{Var(\hat{\beta}_4)} \ll 1.0$ for leukemia incidence (and also for non-leukemia cancer mortality). Thus, the ratio $\hat{\beta}_3/\hat{\beta}_1$ = 11.50 rather than 11.30 as when the model includes the site variable and $(\hat{\beta}_1, \hat{\beta}_3)$ = (0.22, 2.52) as it does in the BEIR III Report.

For both constrained and mixed estimates of $\underline{\beta}$ = $(\beta_0, \beta_1, \beta_3, \beta_4)$ for the L-L model of non-leukemia cancer mortality we have $\underline{r}$ = 0 and R = [0, 11.30, -1, 0]. For both estimates Ψ = R Var$(\hat{\underline{\beta}})$R' where $\hat{\underline{\beta}}$ is the estimate of $\underline{\beta}$ for the L-L model of the leukemia incidence rate. For mixed estimation we take Var$(\hat{\underline{\beta}})$ = $(X'\hat{V}^{-1}X)^{-1}$. (Recall that the confidence ellipsoid is described by the quadratic form $(\underline{\beta} - \hat{\underline{\beta}})'(X'\hat{V}^{-1}X)(\underline{\beta} - \hat{\underline{\beta}})$ = Q.) For constrained estimation we take Var$(\hat{\underline{\beta}})$ = kI where I is the 4x4 identity matrix and k = 10^{-10}. Thus, ψ = 10^{-8}. Then for the constrained estimates, we have Q$\underline{r}$ = $\psi^{-\frac{1}{2}}\underline{r}$ = 0 and QR = $\psi^{-\frac{1}{2}}$R = [0, 11.30*10^4, -10^4, 0]. The regression diagnostics are obtained from the augmented matrix described above.

†Doll, R., Payne, P., and Waterhouse, J.: Cancer Incidence in Five Continents. International Union Against Cancer. Springer-Verlag. N.Y. 1966

Further Remarks on the BEIR III Model of Breast Cancer Incidence.
The received model (BEIR III) of the radiation-induced incidence rate of breast cancer is presented in Table 13 as L*: y = β_0 + β_1D* + β_4C where D* = $(D_\gamma + D_n)$. As noted therein this model is inconsistent with several of the normative practices of regression analysis: 1) The value of RSS suggests that the fit of the model to the sample is "too good" for the latter to have been generated by a random process from the population described by the model (Finney 1971). (Since the coefficient of D_n was chosen to equal that of D_γ this result is not altogether unexpected.) 2) Forming a linear combination of two predictor variables such as D* = $D_\gamma + D_n$ is consistent with standard practice in only two circumstances: a) When the sample estimates of the coefficients of each predictor variable, i.e., D_γ and D_n, are both significantly larger than their respective standard errors and the difference between the two coefficients is significantly smaller than the standard error of their difference or b) When the two predictor variables are highly correlated. In the latter case, the above model is just a regression on the first principal component, D* = $(D_\gamma + D_n)/\sqrt{2}$ (Montgomery and Peck 1982). But for these breast cancer incidence data neither of these circumstances obtain: a) $\hat{\beta}_1/\sqrt{Var(\hat{\beta}_1)}$ = 3.16 > $\hat{\beta}_3/\sqrt{Var(\hat{\beta}_3)}$ = 0.63 (See Table 13). b) The collinearity of D_γ and D_n, although large in both the Hiroshima and Nagasaki samples, has been reduced by pooling these two samples - data augmentation. (See Figures 1a and 8b and Table 2.) Thus, the received model of breast cancer incidence would seem to represent largely a result of some rather inappropriate practices of regression analysis - the right answers to the wrong questions or a Type III error (See Lecture 19).

"It is apparent that the crucial elements in risk analysis are a) the numbers and b) the methodology. It is on them that critical scrutiny must concentrate" (Hoos 1980). Recent concentrations of critical scrutiny (Loewe, et al 1981) on the T65D estimates of gamma and neutron doses by which, in August of 1945, the extant populations of Hiroshima and Nagasaki were converted to the respective LSS samples - disclosed some major problems with some of the "numbers" in the BEIR III Report. It is obvious from Tables 8, 9, 12 and 13 (and Figure 8b) that the BEIR III "methodology" (as well as other "numbers") bears looking into also (see also Appendices I-III).

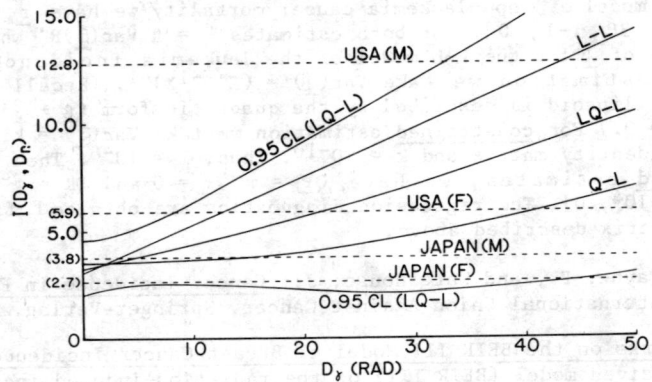

Figure 5a. The marginal ($D_n = 0$) dose-response curves for the rival models of the leukemia incidence rate $y = I(D_\gamma, D_n)$. The 0.95 confidence limits (CL) for the LQ-L model are super-imposed. The spontaneous, or "background", ($D_\gamma = D_n = 0$) leukemia incidence rates for males (M) and females (F) for the US (Connecticut) and Japan (Miyagi prefecture) are also superimposed for comparison.

The 0.95 interval estimates are obviously enormous. The "confidence limits" proposed in the BIER III Report - L-L (upper) and Q-L (lower) - are obviously unsatisfactory.

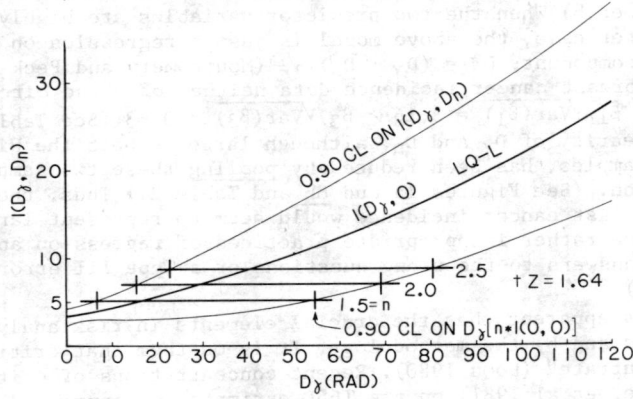

Figure 5b. The marginal ($D_n = 0$) dose-response curve for the LQ-L model of the leukemia incidence rate $y = I(D_\gamma, D_n)$ together with the 0.90 confidence limits (CL) are presented. The inverse estimates, $\hat{D}_\gamma$ (y), of the dose required to educe the incidence rates, $y = ny_0$, where y_0 is the spontaneous or "background" incidence rate and $n = 1.5, 2.0, 2.5$ are also presented. The interval inverse estimates are enormous.

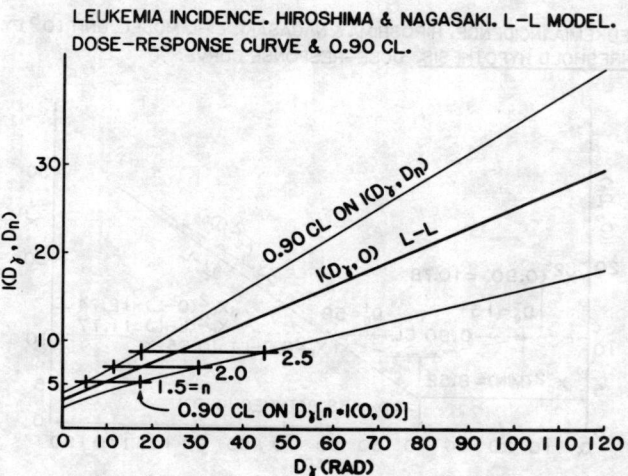

Figure 5c. The marginal ($D_n = 0$) dose-response for the L-L model of the leukemia incidence rate $y = I(D_\gamma, D_n)$ together with the 0.90 confidence limits (CL) are presented. The <u>inverse</u> estimates, $\hat{D}_\gamma(y)$, of the dose required to educe the incidence rates, $y = ny_0$, where y_0 is the spontaneous or "background" incidence rate and $n = 1.5, 2.0, 2.5$ are also presented. The <u>interval</u> inverse estimates are <u>enormous</u>.

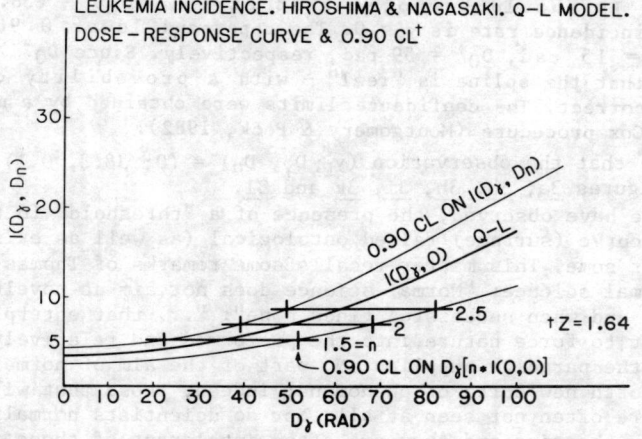

Figure 5d. The marginal ($D_n = 0$) dose-response curve for the Q-L model of the leukemia incidence rate $y = I(D_\gamma, D_n)$ together with the 0.90 confidence limits (CL) are presented. The <u>inverse</u> estimates, $D_\gamma(y)$, of the dose required to educe the incidence rates, $y = ny_0$, where y_0 is the spontaneous or "background" incidence rate and $n = 1.5, 2.0, 2.5$ are also presented. The <u>interval</u> inverse estimates are <u>enormous</u>.

It is evident from Figures 5b-d that the set estimates for the L-L model are the narrowest. Note that Figures 5a-d refer to models for which the indicator variable C has been <u>omitted</u>. Since the analyses have disclosed that the $\hat{\beta}_4/\sqrt{\text{Var}(\hat{\beta}_4)} < 1.0$, retaining it would inflate the <u>variance</u> of estimate more than omitting it would inflate the <u>bias</u> of estimate.

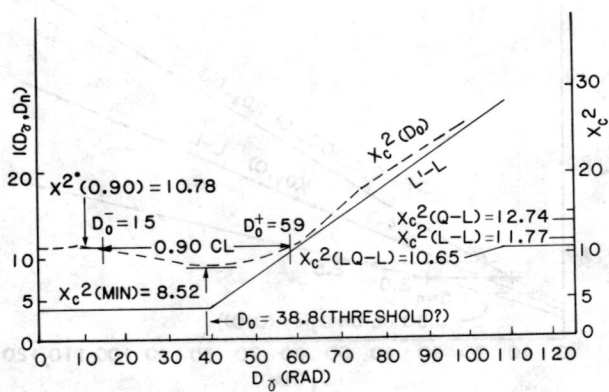

Figure 6a. The marginal ($D_n = 0$) dose-response curve (solid) for the L'-L model (linear spline) of the leukemia incidence rate, $y = I(D_\gamma, D_n)$. The indicator variable, C, is omitted from the model. The dashed curve is a plot of the RSS = $\chi_c^2(D_0)$ for the model as a function of the position of the spline "knot", D_0. The value of $D_0 = 38.8$ defines an E-NOEL and (perhaps) identifies a "threshold" in the leukemogenesis process. It should be noted that at $(D_\gamma, D_n) = (38.8, 0.1)$ the observed incidence rate is $y = 0$. The upper and lower 0.90 CL on D_0 are $D_0^- = 15$ rad, $D_0^+ = 59$ rad, respectively. Since $D_0^- > 0$, one may conclude that the spline is "real" - with a probability of 0.10 of being incorrect. The confidence limits were obtained by a modification of a Box-Cox procedure (Montgomery & Peck, 1982).

Note that the observation $(y_i; D_\gamma, D_n) = (0; 38.8, 0.1)$ is index #12 in Figures 3a, 3b, 3h, 3i, 3k and 3l.

As we have observed, the presence of a "threshold" in the dose-response curve (surface) raised ontological (as well as existential?) issues for some. This matter recalls some remarks of Thomas Kuhn(1970a) anent normal science: "Normal science does not aim at novelties of fact or theory and when successful finds none": "... that enterprise seems an attempt to force nature into the preformed and relatively inflexible box that the paradigm supplies. No part of the aim of normal science is to call forth new sorts of phenomena; indeed, those that will not fit the box are often not seen at all. Nor do dcientists normally aim to invent new theories and they are often intolerant of those invented by others. Instead, normal-scientific research is directed to the articulation of those phenomena and theories that the paradigm already supplies." (Kuhn's comments are equally appropriate to the BEIR III version of the L-L model of breast cancer incidence.)

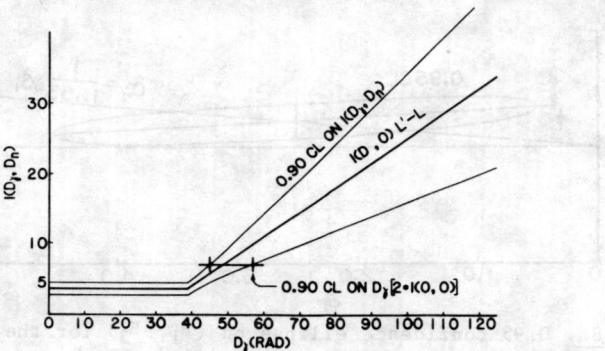

Figure 6b. The marginal ($D_n = 0$) dose-response curve for the L'-L model of the leukemia incidence rate, $y = I(D\gamma, D_n)$, together with the 0.90 confidence limits (CL). The <u>inverse</u> estimate $D\gamma(y)$ of the dose required to educe the incidence rate $y = 2y_0$, where y_0 is the spontaneous or "background" incidence rate is also presented. The <u>interval</u> inverse estimate is quite narrow. Comparison of the 0.90 confidence limits on the L'-L model in Figure 6b with the cognate limits on the LQ-L model of Figure 5b suggests that most of the width of the latter in the region 0-50 rad may be largely the effect of <u>specification error</u>.

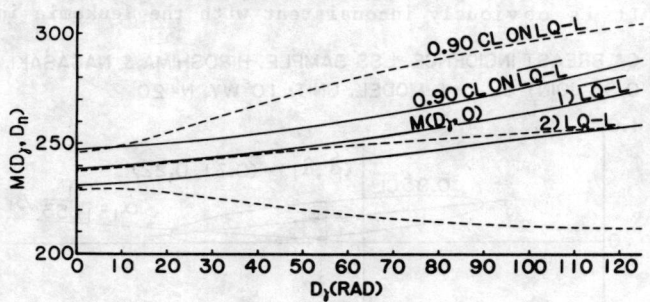

Figure 7. The marginal ($D_n = 0$) dose-response curves of the $\overline{LQ}$-L and LQ-L models of the non-leukemia cancer mortality rate, $y = M(D, D_n)$. Both models include the indicator variable, C. The respective 0.90 confidence limits are super-imposed. Note that the inverse estimates, $D\gamma(y)$ are enormous even for the $\overline{LQ}$-L model. For example, the upper and lower 0.90 CL on the estimate of the dose required to educe a mortality rate only slightly above "background" - say $250/10^5 PY$ - are, respectively, about 18 rads and about 88 rads.

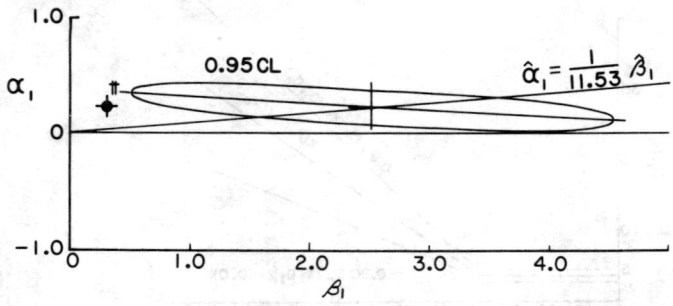

Figure 8a. 0.95 confidence ellipse on (β_1, β_3) for the L-L model of the leukemia incidence rate. (The axes are labelled according to the BEIR III style where $\alpha_1 = \beta_1$ and $\beta_1 = \beta_3$.) The line through the origin, $\alpha_1 = \alpha_1/11.53$ represents the constraint, $r = R\beta$, described in the BEIR III Report. Here, $r = 0$, $R = [0, 11.50, -1, 0]$, $\Psi = [0]$. The Figure shows that the stipulation, $\Psi = [0]$, is a considerable overstatement of the precision with which the elements of r and R can be estimated from the leukemia incidence rate data since the estimate of Ψ is $R\text{Var}(\hat{\beta})R'$ and the volume of the confidence ellipsoid is proportional to the determinant, $|\text{Var}(\hat{\beta})| = |(X'\hat{V}^{-1}X)|$. It is evident from the Figure that the pooling maneuver was successful in reducing the collinearity of $D\gamma$ and D_n. However, the confidence limits on the estimates of β_1 and β_3 are still quite large. Broadly speaking, joint confidence regions circumscribe the region of parameter space within which one can believe what one pleases about the parameter vector, with a specified probability, $(1-\alpha)$, of being wrong. In Figure 8a $(1-\alpha) = 0.05$. The solid circle at $(\beta_1, \beta_3) = (0.22, 0.31)$ represents the point estimate $(\hat{\beta}_1, \hat{\beta}_3)$ for the L-L model of breast cancer incidence. It is obviously inconsistent with the leukemia incidence estimates.

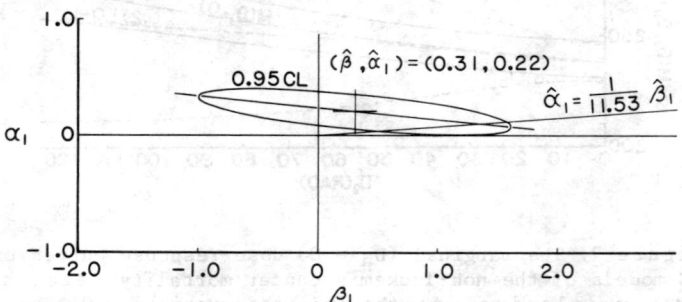

Figure 8b. 0.95 confidence ellipse on $(\hat{\beta}_1, \hat{\beta}_3)$ for the L-L model of the breast cancer incidence rate. (The axes are labelled according to the BEIR III style where $\alpha_1 = \beta_1$ and $\beta_1 = \beta_3$.) The ellipse cuts the $\beta_1 = 0$ axis. Thus, the data are consistent with the conjectures that either $D\gamma > 0$, $D_n = 0$ in the T65D dosimetry or that these neutrons were very inefficient carcinogens. (The data may be consistent with other conjectures as well, of course.)

STATISTICAL ANALYSIS OF ROC DATA
IN EVALUATING DIAGNOSTIC PERFORMANCE

Charles E. Metz
The University of Chicago, Chicago, IL 60637

ABSTRACT

Receiver Operating Characteristic (ROC) analysis is now widely recognized as the most meaningful approach to the problem of measuring and specifying diagnostic performance in medicine. An ROC curve plots the probability of true detection against the probability of false detection as some "confidence threshold" is varied. In the evaluation of medical imaging systems, ROC curves are measured by requiring each observer (eg., radiologist) to report a categorical "confidence rating" for each image to represent his level of certainty concerning his diagnosis. The resulting rating data can be used to calculate estimates of a set of points on the ROC curve for the particular observer-modality combination in question. ROC rating data arise from multinomial distributions that can be related to the parameters of an underlying model in terms of signal detection theory. This paper describes the statistical properties of ROC rating data sets, appropriate procedures for ROC curve-fitting, and tests that can be used to evaluate the statistical significance of apparent differences between measured ROC curves.

INTRODUCTION

Receiver Operating Characteristic (ROC) analysis originated in statistical decision theory[1] and was developed in the 1950's and 1960's to evaluate the detectability of signals in radar[2,3] and sensory psychology.[4,5] The potential usefulness of ROC analysis in evaluating medical diagnoses was first noted by Lusted,[6,7] and subsequently it has been applied fruitfully to a broad variety of tasks in medical diagnosis[8,9] and particularly in diagnostic medical imaging.[10-17] These medical applications have stimulated several methodological developments[18-21], notably in the statistical analysis of ROC data.[17,22-26]

Essentially, an "ROC curve" describes the continuum of trade-offs that are available between the "sensitivity" and the "specificity" (ie., between the "hit rate" and the "false alarm rate" or between the "true positive fraction" and the "false positive fraction") of a diagnostic test as the "threshold of abnormality" or "critical confidence level" is varied.[13] ROC analysis is now widely recognized as the most meaningful approach to the problem of measuring and specifying diagnostic performance in medicine because it separates two aspects of decision performance that are confounded by other methods[17]: (1) the inherent discriminability of a diagnostic test, and (2) the particular balance between sensitivity and specificity that a decision-maker chooses (consciously or unconsciously) to adopt.

Limitations of Other Methods

For many years diagnostic performance was reported as a kind of "batting average": the percentage of diagnostic decisions that turned out to be correct. This "percent correct" measure, sometimes called "accuracy", suffers from several grave limitations. First, the value obtained depends strongly on the prevalence of the disease in question in the population studied [eg., if the prevalence of a disease is 5%, a totally worthless diagnostic test can be "95% accurate" simply by diagnosing all patients as not having the disease], and no statistical procedure can correct meaningfully for this dependence. Second, the "percent correct" measure does not reveal the relative frequencies of "false positive" and "false negative" diagnostic errors, which can have very different consequences; thus two diagnostic tests yielding equal percentages of correct decisions may not be of equal value if the errors of one are mostly "false negative" while those of the other are mostly "false positive." Finally, the usefulness of this measure is limited by its dependence on the "threshold of abnormality" employed with the the diagnostic test, which we discuss in greater detail below.

An alternative approach, which seems superficially more attractive, is to measure diagnostic performance in terms of a pair of indices: "sensitivity" (the probability of deciding that the disease in question is present when it is, in fact, present) and "specificity" (the probability of deciding that the disease is absent when it is, in fact, absent). Since "sensitivity" is the same thing as "hit rate" or "true positive fraction" and since [1-"specificity"] is the same as "false alarm rate" or "false positive fraction", a pair of these alternative measures is equivalent to a "sensitivity, specificity" pair.[13] In effect, the "accuracies" of the decision process for actually positive and actually negative patients or images are measured separately. A pair of indices of this kind does not depend on the prevalence of disease in the population studied as long as decisions are made in the same way, and false positive (Type I) and false negative (Type II) errors are accounted for separately. The most obvious problem in using sensitivity and specificity (or, equivalently, true and false positive fractions) arises in comparing the disease detection performance of two diagnostic systems (eg., imaging modalities and/or observers) when one system is found to be superior in terms of sensitivity while the other is better in terms of specificity.[13] This dilemma can be resolved by recognizing the additional problem that both the sensitivity and the specificity of a diagnostic system change if a different "threshold of abnormality" or "critical confidence level" is employed in the decision process.[13] Then by deliberately varying the threshold of abnormality or critical confidence level, one can generate a curve that describes the trade-offs available between sensitivity and specificity. Essentially this is an ROC curve, although ROC curves usually display true-positive fraction (ie., sensitivity) as a function of false positive fraction (ie., 1-specificity). From the perspective of statistical hypothesis testing, an ROC curve is conceptually equivalent to a curve showing the relationship between statistical power and the probability of Type I error as the "critical value" of the test statistic is varied. Theoretical aspects of ROC analysis are discussed in the next section.

We mention in passing two other approaches to the evaluation of diagnostic performance in medical imaging. For many years, the "detail visibility" provided by x-ray imaging systems has been reported in terms of "Contrast-Detail Curves" (or "C-D Diagrams") which plot the minimum detectable contrast of a test object (such as a disk or square patch) as a function of object size. Although the original proposal of this methodology[27] included an attempt to account for the possibility of "false positive" detection, the approach is almost invariably used by simply asking the observer to report the lowest contrast object with fixed size (or smallest object with fixed contrast) that he "can see" in a rectangular array of objects such that contrast decreases in one direction and size decreases in the other. Thus, at best, the results of this method are subject to variations in the minimum level of "certainty" that each observer requires to "see" an object. At worst it is subject to "wishful thinking" since the observer knows the positions of the objects that he cannot, in fact, "see." In effect, Contrast-Detail curves attempt to measure sensitivity without any control for possible changes in specificity across observers or imaging conditions. The limitations of this approach are usually confounded by the use of only one image of the object array to measure the curve, allowing potentially large variations in the results due to statistical fluctuations in image noise outcomes.[28]

Finally, we mention the "Two-Alternative Forced Choice" (2-AFC) paradigm[4,23] in which an observer views pairs of images, one of which actually contains the visual signal, and is required to state the image that includes the signal (eg., lesion or disease). If the actually "positive" image is varied randomly (from left to right, say) with equal probability, then the fraction of correct decisions in this task can range from 0.5 (indicating chance performance) to 1.0 (indicating perfect performance). With this paradigm, the observer does not need to adopt any "confidence threshold" against which to compare his impression of each image; instead, his impressions of the two images (his "decision variables") are compared to each other. It can be shown that the expected fraction of correct decisions in this 2-AFC experiment equals the expected area under the ROC curve that would be measured with the same images viewed one-at-a-time.[4,23] As we mention later, the area under an ROC curve is often used as an index to summarize the "quality" of the curve. Thus if only the ROC area (A_Z) is of interest, it can be measured directly by the 2-AFC paradigm with some apparent saving in experimental effort. The chief disadvantage of the 2-AFC approach lies in the fact that the trading relationship between sensitivity and specificity (ie., the ROC curve) is never determined, so the relative advantages of alternative diagnostic systems at different false positive rates are never known. Further, the apparent efficiency of the 2-AFC method is somewhat deceptive, since greater statistical precision in measuring the area under an ROC curve can be gained with a given number of images if rating-scale data (sufficient to determine the ROC, as we show below) are obtained from the observer.[23]

Signal Detection Theory and ROC Analysis

The conceptual basis for ROC analysis lies in signal detection theory.[4,29] Consider the situation in which a simple diagnostic test yields a single numerical value as its result, and imagine the probability density functions for test results in actually normal and actually abnormal patients. Without loss of generality, assume that both densities are bell-shaped, with the density for actually abnormal patients centered on a higher value (ie., to the right). The test result value constitutes a "decision variable" (x) on which interpretation of the test (ie., a diagnosis) is based. In this situation the test is usually interpreted by adopting some "threshold of abnormality" (t) and by deciding that the test is "Normal" if $x < t$ and "Abnormal" if $x \geq t$. Since the "true positive fraction" (ie., sensitivity) of the test is given by the area under the "actually abnormal" density to the right of t, and since the "false positive fraction" (ie., 1-specificity) is given by the area under the "actually normal" density to the right of t, one can see clearly that these decision fractions increase or decrease together as the "decision criterion" t is moved to the left or right. If true positive fraction (TPF) is plotted as a function of false positive fraction (FPF) as t is moved from $+\infty$ to $-\infty$, then a curve is swept out in the unit square, beginning at (0,0) and ending at (1,1). This is the ROC curve associated with the diagnostic test (and, strictly speaking, with the decision strategy that compares some monotonic function of the numerical test result against a variable cut-off value).

This notion can be generalized to include decision tasks that involve perceptual processes by assuming, rather generally, that in such situations an observer formulates (by some unknown rule) a quantity on which he will base his decision, selecting one alternative if the quantity exceeds a critical value and selecting the other alternative if it doesn't. Then changing the critical value will change both TPF and FPF, in general, and an ROC curve will be swept out.

Although signal detection theory can be used to address the question of how an <u>ideal</u> decision-maker would combine all available information to formulate a "decision variable" to achieve optimal discrimination capacity in a two-alternative decision task [he should, it happens, formulate any monotonic transformation of the data likelihood ratio], the general conceptual model described here can be used as a basis for interpreting any empirical ROC curve measured by varying some "threshold of abnormality" or "confidence threshold" on a "decision-variable axis." In decision tasks involving sensory information and/or subjective judgements, the decision variable is thought of as the observer's subjective assessment of his confidence in a positive decision, given the information he has available. Although different (ie., lower and possibly higher) ROC curves can be achieved by the observer if he uses different strategies for combining available evidence, one may reasonably assume that in repeating a given two-alternative decision task he <u>attempts</u> to adopt some fixed algorithm for formulating a decision variable and some fixed confidence threshold. Variations in the observer's algorithm and/or confidence threshold can be thought of--and sometimes modeled--as an additional random statistical component in the decision variable.[30,31] One should note that since an ROC curve depends only on the <u>rank order</u> of decision variable outcomes, it does not change with monotonic transformations of the decision-variable[29,32]; thus an observer's ROC curve does not depend on whether his subjective assessment of confidence in a positive decision (ie., his decision variable) is thought of as an estimate of likelihood ratio, or log-likelihood ratio, or log posterior odds, etc.

In effect, an ROC <u>curve</u> is an empirical description of the ability a diagnostic system to discriminate between two states of the world, while each <u>point</u> on the curve represents a different compromise between true positive fraction and false positive fraction that can be achieved by adopting a different "threshold of abnormality" or "critical confidence level" in the decision process. A "strict" criterion (eg., one that calls patients "positive" only when the evidence of disease is very strong) yields a low false positive fraction but a relatively small true positive fraction also--ie., a point on the lower left portion of the ROC curve. Progressively less strict decision criteria yield larger fractions of both kinds--ie., points higher and to the right on the ROC curve. If the ROC curves associated with two diagnostic systems do not cross, the system with the higher ROC clearly provides superior discrimination performance. If the curves cross, then one can rank the systems for the range of false positive or true positive fractions of interest in a diagnostic task, taking into account the costs and benefits of alternative diagnoses if that is desired.[13]

A variety of indices has been proposed to specify and/or summarize empirical ROC curves.[4,15,17,29] An index or set of indices <u>specifies</u> an ROC curve if the entire ROC curve can be reconstituted from a known value (or values) of the index (or indices). In general, an ROC curve is specified by assuming that it follows some particular functional form with one or more adjustable parameters and then indicating the parameter values that provide the best fit to available data, in some sense. The "binormal" functional form for the ROC curve is used most widely and has been shown to provide good fits to empirical ROC curves measured in a wide variety of situations. This form is expressed most easily by the pair of equations

$$FPF(t) = \Phi(-t) \tag{1}$$

and

$$TPF(t) = \Phi(a-bt) \tag{2}$$

where Φ is the cumulative standard normal distribution, where the parameters "a" and "b" determine the ROC <u>curve</u>, and where the parameter "t" determines a particular <u>point</u> on the ROC curve. These equations show that the ROC curve is given explicitly by

$$TPF(FPF) = \Phi[a + b \cdot \Phi^{-1}(FPF)] . \tag{3}$$

On "normal deviate axes" the ROC curve is given by

$$\Phi^{-1}(TPF) = a + b \cdot \Phi^{-1}(FPF) \tag{4}$$

where $\Phi^{-1}(FPF)$ and $\Phi^{-1}(TPF)$ represent the "normal deviates" corresponding to the probabilities FPF and TPF, respectively. Thus a "binormal" ROC curve plots as a straight line on normal-deviate axes, with ordinal intercept "a" and slope "b." From Eqs. (1) and (2), a binormal ROC curve can be <u>interpreted</u> in terms of an <u>effective</u> decision variable "<u>x</u>" arising from two Gaussian densities such that

$$f(x|n) = \exp[-x^2/2]/(2\pi)^{1/2} \tag{5}$$

for actually negative trials (eg., actually "normal" patients) and

$$f(x|s) = b \cdot \exp[-(bx-a)^2/2]/(2\pi)^{1/2} \tag{6}$$

for actually positive trials (eg., actually "abnormal" patients). In this interpretation, the "actually negative" density has zero mean and unit standard deviation; the "actually positive" density has mean (a/b) and standard deviation (1/b); and t represents the cut-off employed on the effective decision variable (<u>x</u>) axis. We emphasize that an ROC curve does not uniquely define the underlying distributions, however.[29,32]

Though the binormal form specifies a wide variety of empirical ROC curves by means of its two adjustable parameters "a" and "b", no functional form with only a single adjustable parameter has proved generally adequate. Thus one can state that, in general, empirical ROC curves must be <u>specified</u> by two parameters. In some situations it is desirable to <u>summarize</u> an ROC curve by a single index value, however, so that several diagnostic systems can be ranked. Although a large number of such summary indices is available,[4,15,17,29] the most meaningful candidates appear to be TPF(FPF_0), the true positive fraction at a reference value of false positive fraction,[17,25,26] and A_Z, the area under the ROC curve.[15,17,23-25] For ROC curves with binormal form, these summary indices are given by:

$$TPF(FPF_0) = \Phi [a + b \cdot \Phi^{-1} (FPF_0)] \qquad (7)$$

and

$$A_Z = \Phi [a/(1+b^2)^{1/2}] \qquad (8)$$

respectively.

CONFIDENCE RATING DATA

Two experimental approaches can be used to measure conventional ROC curves for imaging systems.

In the first approach, sometimes called the "Yes/No Method," the observer views sequentially a series of images, some of which are "actually positive" and some of which are "actually negative," and he is required to give a <u>binary</u> (eg., "Positive" or "Negative") response for each image. The series is then re-read on several subsequent occasions, with the observer motivated to use a "stricter" or "more relaxed" confidence threshold on each occasion. The observer's responses from each reading session can be used to calculate different "sensitivity, specificity" or "TPF, FPF" pairs, which are plotted as points in the unit square. Vertical and horizontal error bars can then be calculated for each point on the basis of binomial statistics,[4] and a smooth curve can be drawn through the points. This approach follows directly from the conceptual basis for ROC analysis, but it is experimentally inefficient, requiring that the series of images be read M times to generate estimates of M points on the ROC curve.

The "Rating Method" is more efficient and is almost always used in practice. In this approach,[4,11,13,17,33] the observer is required to select one of several <u>ratings</u> (ie., categories of confidence) to represent his impression that the image arose from one or the other state of truth. The use of K categories provides estimates of (K-1) operating points on a conventional ROC curve [in addition to the (0,0) and (1,1) points] from a single set of rating data. Usually five or six categories are employed.

Relationship to Decision Variable Outcomes

Confidence rating data are interpreted in terms of the following model. With I categories, the observer is assumed to establish (I-1) cut-points, t_i, on the decision variable axis, partitioning that axis into I intervals. The probability of the rating "i" is then equal to the probability that the decision variable outcome lies between t_{i-1} and t_i, where $t_0 = -\infty$ and $t_I = +\infty$. Thus for actually negative images the probability of a rating "i" is given by

$$p_i = \int_{x=t_{i-1}}^{t_i} f(x|n) \, dx \tag{9}$$

where $f(x|n)$ is the "actually negative" decision-variable probability density, while for actually positive images the corresponding probability is

$$\pi_i = \int_{x=t_{i-1}}^{t_i} f(x|s) \, dx \tag{10}$$

where $f(x|s)$ is the "actually positive" density.

Calculation of ROC Point Estimates

Since the FPF associated with a cut-off value t_i on the decision variable axis represents the probability that the decision variable $\underline{x}$ has an outcome $\geq t_i$ for an actually negative trial, we have

$$FPF(t_i) = \int_{x=t_i}^{\infty} f(x|n) \, dx \tag{11}$$

$$= \sum_{j=i+1}^{I} p_j \tag{12}$$

from Eq. (9), where $1 \leq i \leq I-1$ and I is the number of categories. Similarly, the TPF associated with the cut-off t_i is

$$TPF(t_i) = \int_{x=t_i}^{\infty} f(x|s) \, dx \tag{13}$$

$$= \sum_{j=i+1}^{I} \pi_j \tag{14}$$

These relationships provide the theoretical basis for calculating estimates of ROC curve points from a set of rating data by the method described next. A less formal description of this procedure with a simple example has been published elsewhere.[33]

Suppose that rating data in I categories are generated from M_n independent "actually negative" trials and M_s independent "actually positive" trials. The data will consist of I numbers $\underline{k_i}$ ($1 \leq i \leq I$) representing the number of actually negative trials assigned the rating "i", and I additional numbers $\underline{\ell_i}$ ($1 \leq i \leq I$) representing the number of actually positive trials assigned the rating "i". Clearly,

$$\sum_{i=1}^{I} \underline{k}_i = M_n \tag{15}$$

and

$$\sum_{i=1}^{I} \underline{\ell}_i = M_s. \tag{16}$$

If the trials are independent, then the set of random variables $\{\underline{k}_i : 1 \leq i \leq I\}$ and the set $\{\underline{\ell}_i : 1 \leq i \leq I\}$ follow multinomial distributions with categorical probabilities p_i and π_i, respectively.

The partial sum

$$\mathbf{K}_{>i} \equiv \sum_{j=i+1}^{I} \underline{k}_j \tag{17}$$

represents the number of actually negative trials for which a rating <u>greater than i</u> was obtained. Hence, for M_n such trials, this partial sum follows a binomial distribution with expected value $M_n \cdot \text{FPF}(t_i)$, so for $1 \leq i \leq I-1$:

$$\widehat{\text{FPF}}(t_i) = \mathbf{K}_{>i} / M_n \tag{18}$$

provides an unbiased estimate of the FPF value associated with the observer's ith cut-off value, t_i, however he may choose it. According to binomial statistics, the standard deviation of this estimate is

$$\sigma_{\widehat{\text{FPF}}} = [\text{FPF}(1-\text{FPF})/M_n]^{1/2}. \tag{19}$$

Similarly, the partial sum

$$\mathbf{L}_{>i} \equiv \sum_{j=i+1}^{I} \underline{\ell}_j \tag{20}$$

represents the number of actually positive trials for which a rating <u>greater than i</u> was obtained. For M_s such trials, an unbiased estimate of the TPF corresponding to the ith cut-off value, t_i, is:

$$\widehat{\text{TPF}}(t_i) = \mathbf{L}_{>i} / M_s \tag{21}$$

with standard deviation

$$\sigma_{\widehat{\text{TPF}}} = [\text{TPF}(1-\text{TPF})/M_s]^{1/2} \tag{22}$$

The above procedure can be carried out for $1 \leq i \leq I-1$ to obtain estimates of (I-1) coordinate pairs (FPF_i, TPF_i) on the ROC curve. These (I-1) coordinate pairs, which proceed from lower left to upper right in the ROC space as i decreases from (I-1) to 1, correspond to the (I-1) cutoff values that the observer adopts in defining the I categories of confidence he employs.

One should note that the data $\{k_j, \ell_j : i+1 \leq j \leq I\}$ used to calculate the ith coordinate pair are included in the calculation of the (i-1)st, (i-2)nd, etc. coordinate pair estimates also. Thus the estimates of the (I-1) points on the ROC curve obtained in this way are correlated. Further, Eqs. (19) and (22) show that the ROC point estimates are subject to uncertainties in both the horizontal and vertical directions, with magnitudes that depend both on the coordinate values and the numbers of trials of each kind.

FITTING ROC CURVES TO RATING DATA: MLE

The preceding discussion of the statistical properties of ROC point estimates obtained by cumulating rating data did not make any assumptions regarding the functional form of the ROC curve. If one chooses not to make any such assumptions, then the (I-1) ROC coordinate pairs calculated directly from the rating data can be plotted in the unit square with horizontal and vertical error bars obtained from Eqs. (19) and (22), and a smooth free-hand curve can be drawn near the points and through the (0,0) and (1,1) corners [which represent the (FPF, TPF) pairs corresponding to $t_I = +\infty$ and $t_0 = -\infty$, respectively].

If a more objective and reliable method for curve-fitting is desired, then some assumption must be made concerning the functional form of the ROC curve, and some principle must be selected to define the "best fit" of a curve with adjustable parameters. We mentioned earlier that the "binormal" form expressed by Eqs. (1)-(3) is usually assumed. With this model, Eqs. (5), (6), (9), and (10) show that the probabilities of obtaining a rating "i" from an actually negative and an actually positive trial are given by

$$p_i = \Phi(t_i) - \Phi(t_{i-1}) \tag{23}$$

and

$$\pi_i = \Phi(bt_i - a) - \Phi(bt_{i-1} - a) \tag{24}$$

respectively, where $t_0 = -\infty$ and $t_I = +\infty$. Thus with the binormal model [or any model involving two adjustable curve parameters] and rating data in I categories, one must determine the I+1 adjustable parameter values $\{a, b, t_i : 1 \leq i \leq I-1\}$ that produce a "best fit" of the ROC curve to the rating data in some sense. Since rating data in I categories for both actually negative and actually positive trials generally contain 2(I-1) degrees of freedom, an exact (but unreliable) fit can (almost) always be obtained when I=3. For more than three categories (ie., I>3), some statistical procedure must be adopted to take advantage of the redundancy available in that "overdetermined" situation.

Conventional least-squares methods are not appropriate for a variety of reasons mentioned earlier: (1) the ROC point estimates are correlated; (2) they follow binomial rather than Gaussian statistics, with precision that varies appreciably over the range of the data; and

(3) the ROC point estimates are uncertain in both the vertical and horizontal directions. Instead, a maximum likelihood estimation (MLE) algorithm should be used.

MLE algorithms for fitting binormal ROC curves to rating data are available in the literature[17,34,35] and provide estimates not only of the parameters of the best-fit ROC curve (in a maximum likelihood sense), but also estimates of the uncertainties in those parameters. We sketch briefly here the theoretical basis for such algorithms, which usually employ the "method of scoring"[36,37] to converge iteratively on the MLE solution.

Since a set of rating data in I categories from M_n independent "actually negative" trials and M_s independent "actually positive" trials is drawn from a joint multinomial distribution, the probability of obtaining a particular observed set of data $\{k_m, \ell_m: 1 \leq m \leq I\}$ given a set of parameter values $\{a, b, t_i: 1 \leq i \leq I-1\}$ is

$$P = (M_n)!(M_s)! \prod_{m=1}^{I} p_m^{k_m} \pi_m^{\ell_m} \Big/ (k_m)!(\ell_m)! . \qquad (25)$$

If the binormal model is assumed, then this probability is related to the model parameters through Eqs. (23) and (24). The task of maximum likelihood estimation is to find the set of parameter values $\{a, b, t_i: 1 \leq i \leq I-1\}$ that maximizes this probability.

Maximizing P is equivalent to maximizing its natural logarithm after deleting any multiplicative factors that do not depend on the adjustable parameters. Thus our task becomes the maximization of

$$L = \sum_{m=1}^{I} k_m \ln p_m + \sum_{m=1}^{I} \ell_m \ln \pi_m \qquad (26)$$

through appropriate choice of the parameters $\{a, b, t_i: 1 \leq i \leq I-1\}$, which determine L through Eqs. (23), (24), and (26). A necessary condition is that $\partial L/\partial a = 0$, $\partial L/\partial b = 0$, and $\partial L/\partial t_i = 0$ for $1 \leq i \leq I-1$. The task, then, becomes solution of these (I+1) non-linear equations.

The Newton-Raphson method for solving (I+1) equations with (I+1) unknowns of the form $h_i(\theta_j: 1 \leq j \leq I+1) = 0$ can be expressed in matrix form as

$$\vec{\theta}(n+1) = \vec{\theta}(n) + [G(n)]^{-1} \vec{h}(n) \qquad (27)$$

where $\vec{\theta}(n)$ is a vector representing the (I+1) solution estimates at the nth iteration; $\vec{h}(n)$ is a vector representing the values of the functions h_i obtained with the solution estimates $\vec{\theta}(n)$; and $G^{(n)}$ is an (I+1)x(I+1) matrix with elements

$$g_{ij}^{(n)} = -\partial h_i / \partial \theta_j \qquad (28)$$

evaluated at $\vec{\theta} = \vec{\theta}(n)$. This iterative scheme can be used to obtain maximum likelihood estimates of binormal ROC curves by setting $h_i = \partial L/\partial \theta_i$, where $\{\theta_i: 1 \leq i \leq I+1\}$ represents the (I+1) adjustable parameters $\{a, b, t_i: 1 \leq i \leq I-1\}$, and by suitably differentiating Eq. (26) with the aid of Eqs. (23) and (24).

As mentioned above, the "Method of Scoring"[36,37] is usually employed to carry out the MLE calculations for ROC curve fitting. This method can be thought of as an approximation to the Newton-Raphson method. If one writes out an expression for the (i,j)th element in the G matrix of the Newton-Raphson method [given by Eq. (28) with $h_i = \partial L/\partial \theta_i$ and with L given by Eq. (26)], one finds that it consists of four sums: two involve products of first-order partial derivatives of p_m or π_m, while the other two involve second-order partial derivatives of p_m or π_m. One can show that the two sums involving second-order partial derivatives equal zero if the data are equal to the expected data [ie., if $k_m = M_n p_m$ and $\ell_m = M_s \pi_m$], and more generally, these sums are usually small in the neighborhood of the solution vector. The distinction between the Newton-Raphson method and the Method of Scoring lies in the fact that the latter method evaluates the elements of G at the expected data so that the sums involving second-order partial derivatives of p_m and π_m become zero (i.e., are omitted). This substantially decreases the computational load, and since the basic Newton-Raphson scheme is essentially "self-correcting" in the neighborhood of the solution, the simpler Method of Scoring is quite robust.[38]

An important practical benefit of the Method of Scoring is that it automatically assesses the precision of the parameter estimates it produces; this is possible due to the multinomial nature of the input data. One can show that after the solution is achieved, the elements of $[G]^{-1}$ represent estimates of the variances of -- and covariances among -- the parameter estimates[36,37]; specifically:

$$(g^{-1})_{ij}^{\text{final}} \cong \text{Cov}(\hat{\theta}_i, \hat{\theta}_j) . \qquad (29)$$

Further, it is known that the parameter estimates follow a multivariate normal distribution in the limit of large numbers of trials.[36,37] Thus, if the numbers of trials are sufficiently large, the parameter estimates can be assumed to follow a multivariate normal distribution around the estimated values with a variance-covariance structure calculated by means of Eqs. (28) and (29). Extensive simulation studies have shown the parameter estimate distributions to be reasonably normal with as few as 50 trials of each kind, and very close to normal with a few hundred trials of each kind.[22]

STATISTICAL TESTS FOR DIFFERENCES
BETWEEN INDEPENDENT ESTIMATES OF ROC CURVES

Often the statistical significance of an apparent difference between measured ROC curves is of interest. Several tests are possible, depending on the null hypothesis to be tested--that is, depending on the sense in which the "similarity" or "difference" of the measured ROC curves is to be judged. In general, one can approach this problem by: (i) choosing a null hypothesis that can be related to the parameters of the ROC curves; (ii) estimating the relevant parameters of the two ROC curves in question from sets of rating data; (iii) estimating the uncertainties and correlations in those parameter estimates; (iv) forming a test statistic that should follow some standard distribution if the null hypothesis is true; and (v) calculating the probability ("p-value") that an outcome of the test statistic at least as extreme as that found could arise from the assumed distribution.

Three diffferent null hypotheses and their associated statistical tests are sketched below. The parameters of the ROC curves implied by the rating data sets in question can be estimated by means of the binormal model and the Method of Scoring. Further, when the rating data sets for estimation of two ROC curves can be assumed statistically independent from each other, the Method of Scoring provides all of required estimates of parameter variance and covariance. [A procedure applicable to correlated rating data sets is described later.] Formulation of a test statistic with a known distribution is more problematic, since the exact joint distribution of maximum-likelihood estimates of ROC parameters is generally unknown, and thus the distribution of any test statistic is generally unknown also. However, since MLE parameter estimates are known to follow a jointly normal distribution in the limit of large numbers of trials,[36,37] one can assume this property, form appropriate test statistics, and then investigate the small-sample performance of the resulting statistical tests by computer simulation. In our experience, all of the statistical tests described here perform adequately with rating data sets that include as few as 50 actually positive and 50 actually negative trials.

Bivariate Chi-Square Parameter Test[22]

Consider two diagnostic systems, "x" and "y", and suppose that a set of confidence-rating data is available for estimation of each system's ROC curve. If the binormal form is assumed to describe each of these ROC curves, they can be specified by the parameter pairs (a_x, b_x) and (a_y, b_y). A null hypothesis that the two rating data sets arose from a single common ROC curve is equivalent to the hypothesis that $a_x = a_y$ and $b_x = b_y$. Suppose that this null hypothesis is true and that the parameter estimates $\hat{a}_x$, $\hat{b}_x$, $\hat{a}_y$, and $\hat{b}_y$ are jointly normal. Then the test statistic $v = \delta\, W^{-1}\, \delta'$ should follow a Chi-square distribution with two degrees of freedom if δ is the row vector $(\hat{a}_x - \hat{a}_y, \hat{b}_x - \hat{b}_y)$ and W is the 2x2 covariance matrix with elements

$$w_{11} = \text{Var}(\hat{a}_x) + \text{Var}(\hat{a}_y) - 2\,\text{Cov}(\hat{a}_x, \hat{a}_y) \tag{30}$$

$$w_{22} = \text{Var}(\hat{b}_x) + \text{Var}(\hat{b}_y) - 2\,\text{Cov}(\hat{b}_x, \hat{b}_y) \tag{31}$$

and

$$w_{12} = w_{21} = \text{Cov}(\hat{a}_x, \hat{b}_x) + \text{Cov}(\hat{a}_y, \hat{b}_y)$$
$$- \text{Cov}(\hat{a}_x, \hat{b}_y) - \text{Cov}(\hat{a}_y, \hat{b}_x) \; . \tag{32}$$

When the sets of rating data for estimation of the two ROC curves are statistically independent, then all cross-curve covariance terms [specifically, the last terms in Eqs. (30) and (31) and the last two terms in Eq. (32)] are necessarily zero. The remaining terms are generally non-zero, but all of these can be estimated by using the Method of Scoring separately with the two ROC curve data sets. Simulation studies have shown this test to be reliable for independent data sets containing as few as 50 actually positive and 50 actually negative trials.[22]

True Positive Fraction (TPF) Test[25,26,39]

In some applied situations, the question of practical importance may be not whether two diagnostic systems yield identical ROC curves, but rather whether the two systems yield ROC curves with the same True Positive Fractions (TPF's) at a particular False Positive Fraction (FPF_o). Thus for two systems "x" and "y", the relevant null hypothesis is that $TPF_x(FPF_o) = TPF_y(FPF_o)$. One should note that this null hypothesis is quite distinct from that of the bivariate Chi-square test: since two ROC curves can cross, the null hypothesis of the TPF test can be true at some FPF_o when that of the bivariate test is false.

If each of the ROC curves in question can be specified by the binormal model, then Eq. (4) can be used to recast the null hypothesis for the TPF test in the form

$$a_x + b_x \Phi^{-1}(FPF_o) = a_y + b_x \Phi^{-1}(FPF_o), \qquad (33)$$

or with $t_o \equiv \Phi^{-1}(1-FPF_o) = -\Phi^{-1}(FPF_o)$:

$$(b_x-b_y)t_o - (a_x-a_y) = 0 . \qquad (34)$$

Thus if the parameter estimates $\hat{a}_x$, $\hat{b}_x$, $\hat{a}_y$, and $\hat{b}_y$ are jointly normal and the null hypothesis is true, the random variable $\underline{v} = (\hat{b}_x-\hat{b}_y)t_o - (\hat{a}_x-\hat{a}_y)$ follows a normal distribution with zero mean and standard deviation $\sigma_v = [w_{11} - 2t_o w_{12} + t_o^2 w_{22}]^{1/2}$, where the w_{ij}'s are given by Eqs. (30)-(32). Again, the cross-curve covariance terms in those expressions must equal zero if the two rating data sets are independent, and the remaining terms in the expressions can be estimated by the Method of Scoring. In our experience this TPF test is reliable for FPF_o values greater than about 0.02 if the rating data sets contain at least 50 trials of each kind.

Area (A_Z) Test[17]

A third possible approach to the difference between measured ROC curves involves the "Area Index", A_Z, which summarizes each ROC curve in terms of the area beneath it in the unit square. Here, the relevant null hypothesis is that the two rating data sets in question arose from ROC curves with equal areas beneath them. One should note that this test is distinct from the two tests described above: since ROC curves can cross, the null hypothesis of the area test can be true when that of the bivariate test is false and that of the TPF test is false except at a single FPF_o value.

Equation (8) expresses the A_Z index in terms of the two parameters of a binormal ROC curve. For sufficiently large numbers of trials, the relative uncertainties in the parameter estimates for two ROC curves ($\hat{a}_x, \hat{b}_x, \hat{a}_y$, and $\hat{b}_y$) become small, and these estimates approach a jointly normal distribution. Thus if the null hypothesis is true, the difference between the A_Z indices for two systems "x" and "y":

$$\underline{v} = \Phi(\hat{a}_x/[1+\hat{b}_x^2]^{1/2}) - \Phi(\hat{a}_y/[1+\hat{b}_y^2]^{1/2}) \qquad (35)$$

becomes an approximately normal random variable with zero mean and variance

$$\sigma_v^2 \cong \sum_{i=1}^{4} \sum_{j=1}^{4} (\partial v/\partial \theta_i)(\partial v/\partial \theta_j) \text{Cov}(\hat{\theta}_i, \hat{\theta}_j) \qquad (36)$$

where $\{\theta_i: i=1, 2, 3, 4\} = \{a_x, b_x, a_y, b_y\}$ represents the set of four parameters of the two ROC curves. Again, if the rating data sets used to estimate the ROC curves are statistically independent, the cross-curve covariance terms in Eq. (36) vanish, and the remaining terms can be estimated with the Method of Scoring. We have found this test to be reliable for data sets containing as few as 50 trials of each kind.

A non-parametric test for differences in the A_z index has been proposed by Hanley and McNeil.[23] Based on Wilcoxon statistics, this alternative test is only slightly less powerful than the parametric test described here. Hanley and McNeil provide tables for estimation of the numbers of trials needed to achieve acceptable statistical power with their test. Since the powers of the two A_z tests are similar, their table can be used to obtain conservative estimates of the power of the parametric test also. A more direct approach for estimating the statistical power of the A_z test (and of the other two parametric tests) described here is sketched in the last section below.

EXTENTION TO CORRELATED ESTIMATES OF ROC CURVES

Several common situations in medical image evaluation produce <u>conditionally correlated</u> rating data -- that is, pairs of ratings that are correlated even when ratings from "actually negative" or "actually positive" images are analyzed separately. For example, if two images of each patient in a clinical ROC experiment are made with different imaging systems, then the ratings from the two images of each patient will tend to agree, even when the data from "actually negative" and "actually positive" patients are analyzed separately, because any variation in the patient that is <u>shared</u> by the two images (such as lesion size, confusing background structure, etc.) will tend to cause the ratings to vary together across patients. Similarly, when a single set of images is displayed in two different ways (for example, with and without digital enhancement), the two ratings of each image (from a given state of truth) will tend to covary due to the shared image data. In such situations, the probability of obtaining a particular <u>pair</u> of ratings ("i" and "j") for each patient or image is not simply equal to the product of the marginal probabilities p_i and p_j or π_i and π_j, and some generalized model is needed to serve as a basis for analyzing the correlated rating data.

This need is particularly important in testing the significance of apparent differences between ROC curves, because intuition (correctly) suggests that statistical power in an observer performance experiment should be increased if the patients imaged by two systems are matched, for example, or if the same noisy images are viewed in processed and unprocessed form. In these situations, the sampling variation <u>shared</u> by paired ratings causes the two estimated ROC curves to tend to vary together (eg., to be atypically high or atypically low). Thus, when rating data sets for the measurement of two ROC curves are (positively) correlated, a given difference between the curves should be interpreted as more significant than if the rating data sets were independent. Hence, to properly assess the significance of an apparent difference between ROC curves measured from the same patient sample or the same image sample, the effect of <u>curve covariance</u> on the variance of the difference must be estimated and incorporated into the test.

The way that this curve covariance can be incorporated into the three tests sketched above should be clear already. For independent rating data sets, we argued that the cross-curve covariance terms $\text{Cov}(\hat{a}_x,\hat{a}_y)$, $\text{Cov}(\hat{a}_x,\hat{b}_y)$, $\text{Cov}(\hat{a}_y,\hat{b}_x)$ and $\text{Cov}(\hat{b}_x,\hat{b}_y)$ in Eqs. (30)-(32) and Eq. (36) can be set equal to zero. For correlated rating data, those terms are generally non-zero, however, and they must be estimated by some method. Conventional paired statistics are inadequate for that purpose, in general, so new approaches are required. One such approach, which is useful for testing differences in the area index A_z (only), employs paired Wilcoxon statistics together with certain fairly rough but empirically fruitful assumptions.[24] An alternative approach, based on generalization of the binormal model, is sketched next.

The Bivariate Binormal Model[25]

In the conventional binormal model described earlier, the decision variable is assumed, in effect, to arise from one of two univariate normal probability densities, $f(x|n)$ and $f(x|s)$, corresponding to "actually negative" and "actually positive" trials. The probability of a rating "i" from either state of truth is then given by integration of the appropriate density between the category boundaries t_{i-1} and t_i on the decision-variable axis. This point of view can be generalized to include <u>two</u> possibly correlated decision variables, <u>x</u> and <u>y</u>, which, in effect, arise from one of two <u>bivariate</u> normal densities, $f(x,y|n)$ and $f(x,y|s)$. Each of these bivariate densities has generally different means and standard deviations in the x and y directions, and each is characterized by its own correlation coefficient, r_n and r_s. The two decision variables <u>x</u> and <u>y</u> can be those due to readings of the same image by the same observer under different conditions, readings of different images of the same patient by the same or different observers, etc.

The probability of a <u>pair</u> of ratings "i" and "j" from an "actually negative" or "actually positive" image, p_{ij} or π_{ij} respectively, is now given by an integral of the appropriate bivariate density in a rectangular region of the x-y plane. Each such region is defined by the boundaries t_i and t_{i-1} used to categorize the decision variable <u>x</u> and by the boundaries u_j and u_{j-1} used to categorize the decision variable <u>y</u>.

This model reduces to the conventional binormal model for a single set of rating data if either of the decision variables is considered alone. Thus, for consistency with the conventional univariate binormal model, the mean of the "actually negative" density is taken to be (0,0), and the marginal standard deviations of that density in both directions are taken to be unity. Similarly, the x and y coordinates of the mean of the "actually positive" density are taken to be (a_x/b_x) and (a_y/b_y), respectively, and the marginal standard deviations are taken to be $(1/b_x)$ and $(1/b_y)$. Here (a_x,b_x) and (a_y,b_y) can be interpreted as the parameter pairs that specify each of the ROC curves individually. These relationships imply that the probability of the rating pair "i" and "j" from an "actually negative" trial is given by

$$p_{ij} = L(t_i, u_j, r_n) + L(t_{i-1}, u_{j-1}, r_n)$$
$$- L(t_{i-1}, u_j, r_n) - L(t_i, u_{j-1}, r_n) \qquad (37)$$

where $L(x,y,r)$ is the cumulative bivariate normal distribution function.[40] Similarly, the probability of the same rating pair from an "actually positive" trial is given by

$$\pi_{ij} = L(b_x t_i - a_x, b_y u_j - a_y, r_s) + L(b_x t_{i-1} - a_x, b_y u_{j-1} - a_y, r_s)$$
$$- L(b_x t_{i-1} - a_x, b_y u_j - a_y, r_s) - L(b_x t_i - a_x, b_y u_{j-1} - a_y, r_s). \quad (38)$$

The Method of Scoring can be used to design an algorithm for maximum likelihood estimation of the parameters of this bivariate binormal model from _paired_ rating data. The crucial feature of this approach is that it provides the cross-curve parameter covariance estimates needed for the three statistical tests described in the previous section when they are applied to correlated rating data. A FORTRAN program to implement this MLE algorithm and to perform the various statistical tests is available from the author.[25]

Statistical Tests for Differences between
ROC Curves Estimated from Correlated Rating Data

With the estimates of cross-curve parameter covariance provided by the Method of Scoring when it is applied to the bivariate binormal model, each of the three statistical tests described earlier can be carried out in a straightforward way. Although the parameter variances and the intra-curve parameter covariances can be estimated from the marginal (ie., unpaired) rating data, the same quantities are estimated automatically in the bivariate binormal MLE calculation. We recommend the latter estimates for use in the statistical tests, both because they are consistent with the cross-curve covariance estimates and because they tend to be slightly more precise due to the "sharing" of information between the correlated marginal data sets. In practice, differences in the parameter variance and intra-curve covariance estimates obtained by the two methods are usually very small.

We have used Monte Carlo simulation methods to evaluate the performance of all three statistical tests for differences between ROC curves measured with correlated rating data.[25] Although in principle these tests are exact only in the limit of large numbers of trials, we have found them to perform well for as few as 50 trials of each kind in each marginal data set. For the worst case studied (the area [A_Z] test with $a_x = a_y = 2.0$, $b_x = b_y = 0.85$, $r_n = 0.5$, $r_s = 0.85$, and 50 trials of each kind for each marginal data set), the type I error rate at the 95% confidence level was about 3% (instead of the ideal 5%). Generally, we have found all three tests to be "conservative" the sense that, for small data sets, they yield Type I errors somewhat _less_ frequently than expected. For 250 or more trials of each kind, the Type I error rates for all three tests were very nearly equal to α for all conditions studied.

Power Estimation

The "power" of a statistical test is the probability that the test will show an _actual_ difference to be statistically significant when the test is used with a particular critical significance level (usually $\alpha = 0.05$). In designing experiments, one usually seeks at least 80% power.

At the risk of raising a potentially confusing point -- but in the hope of stimulating a resonance of understanding -- we note that the relationship between the power of a statistical test and the significance level used for that test (a "Power Curve") is strictly analogous to the relationship between the True Positive Fraction of a diagnostic system and the False Positive Fraction of that system (an ROC curve). This analogy reflects the origins of ROC analysis in statistical decision theory.[1] Here we discuss the power of statistical tests for differences between ROC curves. Thus, in effect, we have a delightful (or distressing) opportunity to discuss ROC curve analysis of ROC curve analysis. For the purpose of this discussion we may assume that the significance level (α) used for the statistical test is held constant at some conventional value (eg., 0.05), however.

Generally, the power of a given statistical test depends on: (i) the magnitude of the actual difference to be demonstrated; (ii) the number of "trials" in the experiment; and (iii) the amount of correlation between the data sets [eg., the extent to which extraneous factors have been controlled.] In ROC analysis, statistical power depends also on: (iv) the particular ROC curves to be compared; (v) the balance between the numbers of "actually negative" and "actually positive" trials used in the experiment; and (vi) the number and distribution of "operating points" measured on each ROC curve [ie., the number of categories in the rating data and how those categories are used by the observer(s)].

Often the question is asked: "How many images are required in an observer performance experiment to demonstrate the significance of an assumed difference between two ROC curves?" The answer depends on the factors listed above. To be precise, however, one should note that this question is improperly phrased, because one can never be certain that any finite number of trials (eg., images) will ensure the statistical significance of an actual difference. Statistical hypothesis testing is statistical, and the same sampling variability that can fall in one direction to cause false rejection of the null hypothesis (Type I error) can fall in the other direction to cause false retention of the null hypothesis (Type II error). More properly, the question should be: "How many images are required to provide an acceptable probability of demonstrating an assumed difference between two ROC curves?" In other words: "How many images are required to achieve adequate statistical power?"

This question can be addressed by noting that the matrix G defined in Eq. (28) gives, through the elements of its inverse in Eq. (29), a relationship between the expected values of the rating data in an ROC experiment, on one hand, and the variability and correlation of ROC curve parameter estimates calculated from samples of those rating data, on the other. With that relationship and with quantitative assumptions concerning the expected rating data (ie., concerning the true parameter values of the underlying bivariate binormal model), one can predict the variability of the test statistic of each of the three statistical tests described earlier in this paper. In the limit of large numbers of trials, the test statistics of the univariate TPF and A_Z tests follow a non-central normal distribution when an actual difference exists, and the statistic of the bivariate Chi-square test follows a non-central Chi-square distribution with two degrees of freedom.[40] Thus statistical power can be predicted from assumptions regarding the true parameter values of the bivariate binormal model by integrating these non-central distributions between appropriate limits that are related to the significance level (α) to be used with the tests. To predict statistical power for tests involving independent data sets, the bivariate binormal parameters r_n and r_s are simply assumed equal to zero.

Because the statistical power of each of the three tests depends in a complex way on all six factors listed earlier in this section, no simple "rule of thumb" or table can be formulated to relate statistical power to the set of factors associated with a particular experimental design. A FORTRAN computer program can be used to predict the power of all three tests as a function of the total number of trials for any assumed set of factors, however.[25] Monte Carlo simulation studies involving various pairings of different ROC curves have shown that this program provides reliable power estimates for all three tests. Overall, the fractions of comparisons found significant agreed with predicted power to within a few percent for most pairs of curves and to within 10% for all.

THE HORIZON

Swets and Pickett[17] identify three components of variability in estimates of ROC curves due to "within-reader" variation (random reading-to-reading fluctuations in the human cognitive process), "between-reader" variation (associated with different levels of observer skill), and "case sample" variation (associated with the spectra of "actually negative" and "actually positive" stimuli employed in the experiment). Though their analysis of ROC curve differences focuses specifically on the area (A_Z) index, it indicates clearly in a general way the gains in statistical power that can be achieved by matching readers and cases (ie., from increasing correlations) and by replicating readings within and across observers. Further, it prescribes and illustrates methods that can be used to enhance statistical power when replicated readings are available. McNeil and colleagues[41] have published an interesting case study that employs aspects of the Swets/Pickett methodology in combination with other techniques.

The approach described by Swets and Pickett and the approach described here have different strengths and weaknesses. Swets and Pickett emphasize replication of readings within and across observers, and they show how such replication can be used to enhance statistical power. Perhaps the major practical shortcoming of their technique is that correlations between ROC index estimates must be obtained by product-moment methods, which may be imprecise when only a few observers and/or replications are available. The methodology described here, on the other hand, is capable of producing relatively stable estimates of index estimate correlation from only two readings of each image, but it has not been developed to allow replicated estimates of each ROC curve to be pooled in a simple way that enhances power. Combining the strengths of these two methodologies in a new, unified approach to test differences between ROC curves seems a worthwhile task for the future.

ACKNOWLEDGMENT

This work was supported by Contract No. DE-AC02-82ER60033 from the U.S. Department of Energy.

REFERENCES

[1] A. Wald, *Statistical Decision Functions* (Wiley, New York, 1950).
[2] W.W. Peterson, T.G. Birdsall, and W.G. Fox, IRE Trans. *PGIT-4*, 171 (1954).
[3] W.P. Tanner, Jr. and J.A. Swets, Psych. Rev. $\underline{61}$: 401 (1954).
[4] D.M. Green and J.A. Swets, *Signal Detection Theory and Psychophysics* (Wiley, New York, 1966; Krieger, Huntington NY, 1974).
[5] J.A. Swets, Science $\underline{182}$, 990 (1973).
[6] L.B. Lusted, Radiology $\underline{74}$, 178 (1960).
[7] L.B. Lusted, *Introduction to Medical Decision Making* (Thomas, Springfield IL, 1968).
[8] B.J. McNeil, E. Keeler and S.J. Adelstein, N. Engl. J. Med. $\underline{17}$, 163 (1975).
[9] E.A. Robertson, M.H. Zweig and A.C. Van Steirtghem, Am. J. Clin. Path. $\underline{79}$, 78 (1982).
[10] D.J. Goodenough, K. Rossmann and L.B. Lusted, Radiology $\underline{105}$, 199, (1972).
[11] D.J. Goodenough, K. Rossmann and L.B. Lusted, Radiology $\underline{110}$, 89, (1974).
[12] W.S. Andrus and K.T. Bird, Chest $\underline{67}$, 378 (1975).
[13] C.E. Metz, Sem. Nucl. Med. $\underline{8}$, 283 (1978).
[14] B.J. McNeil and H.Z. Mellins, Radiol. Clinics No. America $\underline{17}$, 175 (1979).
[15] J.A. Swets, Invest. Radiol. $\underline{14}$, 109 (1979).
[16] J.A. Swets, R.M. Pickett, S.F. Whitehead, D.J. Getty, J.A. Schnur, J.B. Swets, and B.A. Freeman, Science $\underline{205}$, 753 (1979).
[17] J.A. Swets and R.M. Pickett, *Evaluation of Diagnostic Systems: Methods from Signal Detection Theory* (Academic Press, New York, 1982).
[18] S.J. Starr, C.E. Metz, L.B. Lusted and D.J. Goodenough, Radiology $\underline{116}$, 533 (1975).
[19] C.E. Metz, S.J. Starr and L.B. Lusted, Radiology $\underline{121}$, 337, (1976).
[20] International Atomic Energy Agency, in *Medical Radionuclide Imaging*, Vol. I (IAEA, Vienna, 1977), pp. 585-615.
[21] P.C. Bunch, J.F. Hamilton, G.K. Sanderson and A.H. Simmons, J. Appl. Photog. Eng. $\underline{4}$, 166 (1978).
[22] C.E. Metz and H.B. Kronman, J. Math. Psych. $\underline{22}$, 218 (1980).
[23] J.A. Hanley and B.J. McNeil, Radiology $\underline{143}$, 29 (1982).
[24] J.A. Hanley and B.J. McNeil, Radiology $\underline{148}$, 839 (1983).
[25] C.E. Metz, P.-L. Wang and H.B. Kronman, presented at the VIIIth Conference on Information Processing in Medical Imaging, Academic Hospital, The Free University of Brussels, Belgium (1983). To be published in the proceedings, edited by F. Deconinck (Martinus Nijhoff, The Hague, in press).
[26] B.J. McNeil and J.A. Hanley, Med. Dec. Making $\underline{4}$, No. 2 (1984) (in press).
[27] G.C.E. Burger, Acta Radiologica $\underline{31}$: 193 (1949).
[28] L.-N. Loo, K. Doi, M. Ishida, C.E. Metz, H.-P. Chan, Y. Higashida and Y. Kodera, Proc. SPIE $\underline{419}$: 68 (1983).
[29] J.P. Egan, *Signal Detection Theory and ROC Analysis* (Academic Press, New York, 1975).

30. W.A. Wickelgren, J. Math. Psych. 5: 102 (1968).
31. D.J. Goodenough and C.E. Metz, in *Information Processing in Scintigraphy*, edited by C. Raynaud and A.E. Todd-Pokropek (CEA, Orsay, France, 1975), pp. 400-419.
32. J.A. Swets, W.P. Tanner, Jr. and T.G. Birdsall, Psych. Rev. 68: 301 (1961).
33. C.E. Metz, in *The Physics of Medical Imaging: Recording System Measurements and Techniques*, edited by A.G. Haus (AIP, New York, 1979), pp. 546-572.
34. D.D. Dorfman and E. Alf, J. Math. Psych. 6: 487 (1969).
35. D.R. Grey and B.J.T. Morgan, J. Math. Psych. 9: 128 (1972).
36. C.R. Rao, *Advanced Statistical Methods in Biometric Research* (Hafner, Darien CT, 1970).
37. M. Kendall and A. Stuart, *The Advanced Theory of Statistics*, 4th Ed. (MacMillan, New York, 1979), Vol. 2, Chap. 18.
38. D.D. Dorfman, L.L. Beavers, and C. Saslow, Bull. Psychon. Soc. 1: 207 (1973).
39. J.A. Hanley and B.J. McNeil, Med. Dec. Making 2: 371 (1982).
40. M. Zelen and N.C. Severo, Chapter 26 in *Handbook of Mathematical Functions*, edited by M. Abramowitz and I.A. Stegun (Nat. Bureau Stds., Washington DC, 1968).
41. B.J. McNeil, J.A. Hanley, H.H. Funkenstein and J. Wallman, Radiology 149: 75 (1983).

CLINICAL DOSE-RESPONSE MODELS II. PROBIT AND LOGIT MODELS OF EXPERIMENTAL AND NON-EXPERIMENTAL DATA

Donald E. Herbert, Ph.D.
University of South Alabama, College of Medicine
Dept. of Radiology, Mobile, Alabama 36688

ABSTRACT

The ethical, medical-legal, socio-economic, logistic, etc., constraints which frustrate the application of optimal experimental designs to obtain information on the radiation responses of clinical interest in the target system - cancer patients - and which form the basis of the problems of estimation for regression models of these responses constructed from data obtained in non-experiments do not apply to the most common surrogate system, mice, the radiation responses of which are accessible to elucidation by designed experiments. Other constraints are in force, however, with the result that the accessible region of the treatment variables for radiation experiments on the surrogate system often do not coincide with the region of clinical interest for the target system. This lack of congruence of the regions of accessibility and of clinical interest in the treatment variables together with the inherent differences in the surrogate and target system comprise the (well-known) problems of interpretation - the so-called "mouse-to-man" problems.

In the present lecture we anatomize a recently published designed experiment in small animal radiobiology which is an exemplar of many of the features of its genre. We compare it to a designed experiment in toxicology which exhibits most of the classic features of a well-designed experiment: orthogonality in the design matrix, X, large samples at each level of the treatment variables, absence of extreme levels of treatment and outlying responses, co-incidence of the regions of interest and accessibility, etc. Both experiments have for their immediate purpose the construction of regression models of a binary response. The evidence from the radiobiological experiment is also used to discriminate between an empirical and a mechanistic model of the response in the original paper.

The anatomization and comparison are based on the methods of regression diagnostics and validation described in Lectures 1, 3, 7 and 13.

1. INTRODUCTION

"The study of fallacy in concrete examples ought to play a greater part in our educational curriculum. Certain works have a permanent value in this respect" (Pearson 1911).

"If the design of an experiment is faulty, any method of interpretation which makes it out to be decisive must be faulty too" (Fisher 1966).

"There are two kinds of scientist, those who are interested in dissecting and reducing to order the external world and those who are mainly interested in understanding their own minds" (Kendall 1976).

"All models are wrong but some are useful" (Box 1979).

Multiple regression methods were developed to obtain optimal (usually in a Least Squares or Maximum Likelihood sense) estimates, say

$\hat{\beta} = (X'X)^{-1}X'\underline{y}$, $Var(\hat{\underline{\beta}}) = \hat{\sigma}^2(X'X)^{-1}$ and $\hat{\sigma}^2 = (\underline{y}-X\hat{\underline{\beta}})(\underline{y}-X\hat{\underline{\beta}})/(n-k)$ of the unknowns $\underline{\beta}$ and σ^2 in a linear model say, $\underline{y} = X\underline{\beta} + \underline{\varepsilon}$, from a set of observations, [$\underline{y}$, X], obtained from a <u>designed experiment</u>. (By way of introduction only, we prefer to instance the so-called Normal theory model wherein the response, y_i, is <u>continuous</u> and $-\infty < y_i < \infty$ (see Myers, Lectures <u>1</u> and <u>2</u>). However, the principal interest in the paper is the model for which the response is <u>discrete</u> and $y_i = 0$ or 1 (see Frome, Lecture <u>7</u> and also Lecture <u>13</u>). [$\underline{y}$, X] is the n x k <u>observation matrix</u> where k=m-1. In such experiments, the location, scale and shape of the joint distribution of an optimal number, n, of observations, [$\underline{y}$, X], can be selected to provide optimal (usually in a $\min |(X'X)^{-1}|$ sense) estimates of the <u>unobservables</u> $\underline{\beta}$ and σ^2, where $\underline{\beta}$ is the (mx1) vector and σ^2 is a scalar. For example, the levels of the treatment variables may be selected to make the spectrum of eigenvalues, λ_j, of the (pxp) <u>correlation matrix</u>, $X_0'X_0$, <u>uniform</u>, $\lambda_1 = \ldots = \lambda_p = 1$, and the set of diagonal elements, [h_j], of the (nxn) hat matrix, $X(X'X)^{-1}X'$ <u>uniform</u>, $h_1 = \ldots = h_n = p/n$. See Box and Lucas (1959). However, for biological systems of clinical interest a properly designed experiment usually represents a dubious practice and so the observations, [$\underline{y}$, X], must often (usually?) be made on a <u>surrogate</u> system say S_2, rather than on the system of immediate interest, the <u>target</u> system, say S_1, since legal, ethical, logistical, financial, etc., constraints must be imposed on the joint distribution of [$\underline{y}$, X] which are inconsistent with the requirements for an optimal distribution of observations, [$\underline{y}$, X], for estimation of the unknowns $\underline{\beta}$ and σ^2 by Least Squares or Maximum Likelihood methods.

Thus, although the estimates, $\hat{\underline{\beta}}$, and $\hat{\underline{y}} = X\hat{\underline{\beta}}$ obtained from the surrogate system may be precise, (small $Var(\hat{\underline{\beta}})$, $Var(\hat{\underline{y}})$), their meaning for the system of interest is usually ambiguous; the estimates on the surrogate must be "scaled" to provide a description of the system of interest which can be exploited, say, for either prediction or control of the response, $\underline{y}$. The "mouse to man" problem (Schneiderman, Mantel and Brown 1975) is the biological locution. The scaling problem in radiation biology is exacerbated by the common practice of "overdriving" the surrogate system at extreme levels of the treatment variables in order to elicit an extreme - and hence unambiguous - response in a short time. (See Ang, van der Kogel, van der Scheuren 1983). Thus, frequently one is trying to scale between <u>different responses</u>, as well as between <u>different systems</u>. The scaling problem for a biological system is non-trivial; it was correctly described long ago by Schneiderman (1967) as, "... a leap made largely in the dark." Because of the scaling problem the results of designed experiments are frequently - and often accurately - described as "mouse-bound". As well as non-trivial, (and frequently intractable) the scaling problem is also non-statistical; it must be addressed on the basis of knowledge of the subject matter. Of course, some of the <u>implementation</u> of the solution may be a statistical problem.

An ethically acceptable alternative to experimental studies of the target system are the non-experimental studies described at some length in Lecture <u>13</u>. However, observations obtained in non-experimental studies of the system of immediate interest usually have distributions which are less-than-optimal for regression methods. The same legal, ethical, etc., constraints which require that experiments be run on surrogate systems assure the presence of a high degree of collinearity, in the columns of X, as well as the presence of extreme - and hence perhaps, "influential" - rows, $\underline{x}_i'$, of X and of "outlying" responses (rows of) y, or, in Pregibon's (1981) locution: "... observations which are not well explained by the model or which dominate some aspects of the fit." These features of the joint distribution of [y, X] in non-experimental studies give rise to sample estimates of response, $\hat{y}_i$, Var($\hat{y}_i$) and parameters $\hat{\beta}$, Var($\hat{\beta}$) which are unstable, often inconsistent with prior information - or perhaps even implausible - on the respective sizes, signs, statistical significance, etc.

Lecture 13 adumbrated, with examples, a schedule of post hoc salvage operations on the data obtained in a non-experimental study by which the respective effects of these model-sample interactions on $\hat{y}$, $\hat{\beta}$, Var($\hat{\beta}$), etc., could be identified, diagnosed and perhaps ameliorated. (Lectures 1-5 provide better-founded and more detailed expositions.) Thus, while the scaling problem for the estimates of β and y_i that are obtained from a multiple regression analysis of non-experimental observations on the system of interest may be minimal - and can often be ameliorated by randomization, etc. - it is often the case that the estimates $\hat{\beta}$ and $\hat{y}$ cannot be immediately exploited for model discrimination, or for control or predication of response because of the effects of the model-sample interactions previously described. There is obviously a kind of "uncertainty principle" operating here: The more precisely the parameters, β, and response, y, of a linear model of radiation response can be estimated, the less meaningful these estimates become for answering the insistent questions of patient management, radiation risk, etc. For a non-experiment, the problem is estimation of $\hat{\beta}$ and $\hat{y}$. For an experiment the problem is interpretation of $\hat{\beta}$ and $\hat{y}$.

It would seem obvious that an experiment on a surrogate system must be conceived, designed, executed and evaluated by methods that will ensure that the estimates, $\hat{y}_i$, Var($\hat{y}_i$) and $\hat{\beta}$, Var($\hat{\beta}$) are not encumbered by the effects of such features as collinearity, outliers, etc., of the joint distribution of [y, X] which are characteristic of non-experimental studies, i.e., by methods which minimize the model-sample interactions characteristic of non-experimental estimates of y_i and β. The presence of one or more of these features may lead to estimates $\hat{\beta}$, Var($\hat{\beta}$) which provide such strong evidence for rival hypotheses that the confidence which can be placed in the investigators' explanation of what the experimental data "mean" - even for the surrogate system - is considerably diminished. The inferences from such experiments are beclouded by both the scaling problems (interpretations) characteristic of classical experiments and the estimation problems which encumber non-experimental studies - in other words the very worst of both worlds.

By way of motivation, let us begin (as we did in Lecture 13) by assuming that we have at hand a specific instance of a quite general enterprise, namely, a clinical investigator who was interested in obtaining a regression model that 1) predicts well the level of a specified radiation response to specified adjustable inputs within the region of interest, say R_2, of a target system, say S_1, in a unique state specified by a set of covariates, and 2) provides information about the roles of each of the inputs in evoking, and of each of the covariates in modulating, the response of interest. As in Lecture 13, the target system, S_1, is the head and neck cancer patient in a state specified (for purposes of exposition) by a binary covariate, the T-stage. The response of interest is P(S) = P(E_1 and $\bar{E}_2$) where S is the binary event, success, E_1 is the binary event, tumor ablation and E_2 is the binary event, normal tissue complication. The inputs are the adjustable variables of treatment, total radiation dose D, number of treatments or fractions, N, total duration of treatment, T, dose per fraction, (D/N). Because of the presence of various ethical, medical-legal, socio-economic, etc., constraints, which conflict with the desiderata of good design, e.g., maximum Det[X'X], etc. (See Lecture 13 and Myers 1971) a regression model <u>cannot</u> be constructed from observations obtained in optimally designed experiments on S_1 in R_2. However, <u>non</u>-experimental observations on S_1 in a region R_3 are available. But, the presence of the same constraints which render the designed experiments infeasible have resulted in distributions of the sample non-experimental observations in which estimates, $\hat{\beta}$, of the parameter vector, β, in the regression model are encumbered by the effects of <u>model-sample interactions</u> such as multicollinearity, outlying responses and extreme levels of the inputs or treatment variables (see Belsley, Kuh and Welsch 1980; Pregibon 1981 and Cook and Weissberg 1981). Moreover, in non-experimental studies the allocation of patients to treatment levels is non-random. (But, see Stewart and Jackson 1975). Thus, the principal difficulty in the construction of a regression model of the response from non-experimental data is that of <u>estimation</u> of β.

The constraints which frustrate the attainment of a good experimental design for the response of the target system S_1 in the region of interest R_2 hold with much less force - some are event irrelevant - for a different response in a <u>surrogate</u> system S_2 - mice - in a region R_4. The latter may be referred to as the experimentally <u>accessible</u> region. However, other desiderata require that the location of <u>this</u> region be more or less remote from the region of interest R_4. For example, "... most biological investigations have been carried out with fraction sizes which are considerably larger than those usually employed in the clinic." ... "It is however, very difficult to use small fraction sizes in radiobiological studies" (Ang, et al 1983). Since the requirements of optimal design can be met in the accessible region, R_4, good estimates of the parameter vector of the regression model can be readily obtained. However, the meaning of these estimates and this model of <u>this</u> response, for the response of the target system S_1 in the region R_2 is questionable. Thus, there are two characteristic problems with the construction of regression models of the response on <u>surrogate</u> systems: <u>interpretation</u> and <u>extrapolation</u> of the estimates of the parameter vectors.

The foregoing discussion is illustrated in Figures <u>1a</u>, <u>1b</u> and <u>1c</u>, which are projections of the regions, R_1, R_2, R_3 and R_4 onto the (D, N), (N, T) and (D, D/N) planes. Figure <u>1</u> is identical to Figure <u>1</u> in Lecture 13. Region R_3 refers to the non-experimental data on the head and neck cancer patients (<u>target</u> system, S_1) of that lecture. Region R_4 refers to the region of the experiments on the <u>surrogate</u> systems, S_2, to be described in the present lecture. Region R_1 refers to the region of the input variables covered by the current received wisdom as summarized by the empirical models of Ellis (1969) and Supe, et al (1983).

In the present lecture we present a textual criticism of a recently published analysis of a designed experiment on a surrogate system (mouse) which had for its purpose the estimation of the parameters β for two rival models of an acute radiation response in normal tissues. This experiment together with its analysis seems typical of much of the current published literature in the field of clinicl radiation biology. The anatomization of the paper is done by

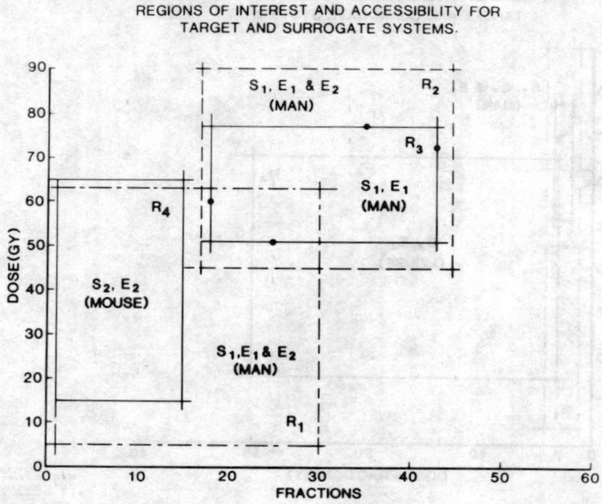

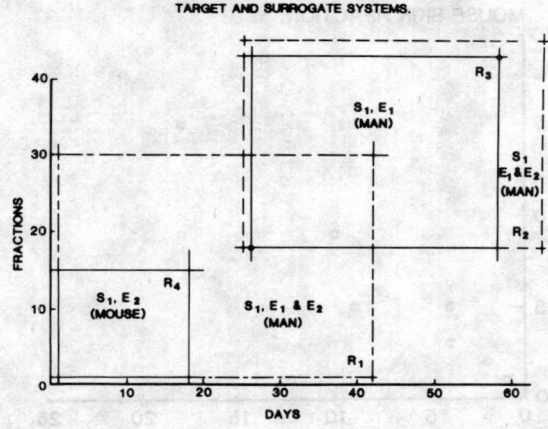

Figure 1a. Regions of a priori information, R_1, (clinical) interest, R_2, feasibility, T_3, and accessibility, R_4, for target, S_1, and surrogate, S_2, systems. D-N plane.

Figure 1b. Regions of a priori information, R_1, (clinical) interest, R_2, feasibility, R_3, and accessibility, R_4, for target, S_1, and surrogate, S_2, systems. N-T plane.

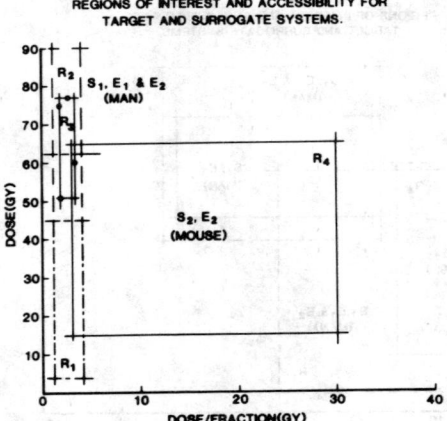

Figure 1c. Regions of a priori information, R_1, (clinical) interest, R_2, feasibility, R_3, and accessibility, R_4, for target, S_1, and surrogate, S_2, systems. D-D/N plane.

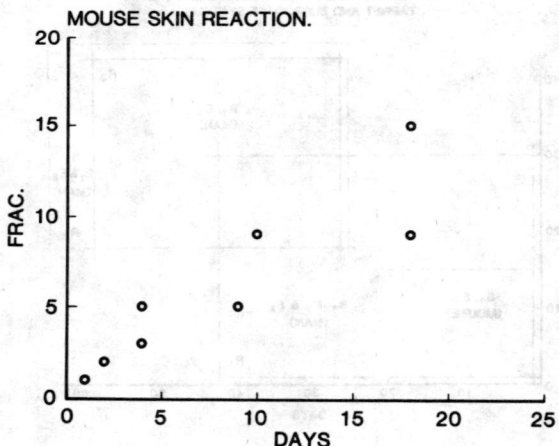

Figure 2a. Scattergram of the distribution of treatment regimens for skin reaction experiment in N-T plane. There is strong multicollinearity present: r = 0.907.

comparing the radiobiology experiment with the non-experimental study of Lecture 13 and with a "text-book case" in experimental design in bioassay taken from Finney (1971).

The purpose of the present lecture is not, of course, to pillory the work of any investigator. Rather we use the Herring (1980) paper as typical of its genre in order to vivify a discussion of what seem to us to be some characteristic difficulties in design and analysis of this type of investigation with the object of, "improving the breed"- so to speak. (See Note 1 Added in Proof)

2. DATA. MOUSE SKIN REACTION #2

"Models are not the source of information, however; data are. And it is the case that the data can be unequally informative about different parametrizations" (Thisted 1980).

The radiobiological data are taken from the 1980 paper by Herring which presents, "... an approach to fitting models to radiation therapy data in order to extract dose-response curves for tumor local control and normal tissue damage." In this paper two rival models of normal tissue time-dose response were compared by fitting two different logistic regression models to a single set of observations on the conditional binary radiation responses elicited in a designed and controlled experiment on the surrogate system, S_2, over the region R_4. See Figure 1. The specific response was "... mouse foot skin reactions," of specified severity. The severity of observed reaction was scored on an ordinal scale (of 29 levels) devised by Denekamp (1973). In the 1980 paper the scores on the response of the individual mice were collapsed to a binary (0, 1) scale, by scoring a "response" (1) only if the severity of the reaction exceeded a specified level (2 on the Denekamp scale) and a "non-response" (0) otherwise. (This practice discards information, of course. The standard error of a quantal or binary estimate is large when compared with the standard error of the more quantitative, ordinal, estimates. There are, of course, regression models which are appropriate to ordinal response data. One is described in Nelder and McCullough (1983). However, since the original observations are unavailable to us, we could not make use of these models for comparison.) The observations on the binary response were included in the paper and so we could re-construct the two alternative models of those data. The treatment variables were total radiation dose, D, duration of completed irradiation schedule, T, and number of treatments, or radiation fractions, N. The experiment required the irradiation of n = 293 mice. These animals were distributed over five levels of dose, D, at each of eight levels of fractions and duration (N, T). There were thus forty levels of treatment (D, N, T) with n_i = 293/40 ≃ 7 mice irradiated at each level.

The data presented in tabular form in Herring (1980) were N and T and the number of responders, r, in n animals exposed at each level of (N, T). We estimated the corresponding values of dose, D, from the graphs of the dose-response curves at each of the eight levels of (N, T) presented in the paper. Comparison of the estimates, $\hat{y}_i$, $\hat{\beta}$, $\text{Var}(\hat{\beta})$, ln likelihood, etc., we obtained from the data acquired in this manner with the estimates presented in the paper cited suggested that the data were "captured" with an adequate fidelity: 1) Our estimates of the respective maximum log likelihoods for models M_1 and M_2 and the respective ratios, $\hat{\beta}_j/\sqrt{\text{Var}(\hat{\beta}_j)}$, for each of these models (M_1: j = 1,

..., 3; M_2: j = 1, 2) were about 5% <u>larger</u> for our version of the data than were cognate estimates presented in the paper cited. 2) The correlation coefficient, r, for the estimated proportions of response, $\hat{P}_i$, i = 1, ..., 40, from the model M_1 of our version of the data and from which the estimates presented in the paper were derived was r = 0.999. The mean difference in the estimated values of P_i was 0.014.

The fact that the maximum log likelihood for the probit models was higher (< 5%) than for the cognate logistic models indicated that the estimated, $\hat{y}_i$, and observed, y_i, responses are more concordant for the probit models. But this is quite consistent with the fact that at 33 of the 40 treatment levels in the experiment, the response was either 0 or 1: "For most purposes the logistic and normal agree closely over the whole range. The only exception is when special interest attaches to the regions where the probability of success is either very small, or very near one. Then the normal curve approaches its limit more rapidly than the logistic" (Cox 1970). (Because there may be some <u>small</u> differences in the values of D between these data and those of the original experiment, it may be appropriate to consider the data presented in this paper to be that obtained in a <u>replicate</u> of the experiment described in Herring (1980). See <u>Note 2 Added in Proof</u>.

3. MODELS OF MOUSE-SKIN REACTION

"One use of a priori information is the initial selection of predictor variables and specification of the model. Injudicious use of a priori information at this stage can lead to (a) bias if important predictor variables are not included or are mis-specified in the model or (b) multicollinearities if redundant predictor variables are included or (c) increased variance of estimate if irrelevant predictor variables are included" (Gunst 1980).

One of the rival models, M_1, is <u>empirical</u>: It is the Taylor series model which we have already described in Lecture <u>13</u>.

$$M_1: z = \beta_0 + \beta_1 x_1 + \beta_2 x_2 + \beta_3 x_3. \qquad (1)$$

$x_1 = \log_{10} D$, $x_2 = \log_{10} T$, $x_3 = \log_{10} N$ and z is the logit transform of the probability, π, of the conditional binary response: $z = \log_e[\pi/(1-\pi)]$. As previously noted, empirical models describe response in terms of the adjustable variables of treatment, here D, T and N. The empirical model is not altogether without theoretical justification. As remarked in Lecture <u>13</u>, the Weber-Fechner law states that many responses of biological systems are proportional to the <u>logarithm</u> of the stimulus. See Lecture <u>13</u>.

The second model, M_2, is <u>mechanistic</u>, the so-called LQ model:

$$M_2: z = \beta_0 + \beta_1 D + \beta_2 D^2/N. \qquad (2)$$

It is based on the not implausible radiobiological conjecture that the likelihood of occurrence of a binary response in a tissue is determined in part by the surviving fraction of cells, S, in the irradiated tissue, which, in turn, is determined by the level of treatment (D, N, T) and, in particular by the dose per fraction, D/N. The model is S = $\exp\{-N[\alpha(D/N) + \beta(C/N)^2]\} = \exp\{-[\alpha D + \beta D^2/N]\}$ (Douglas and Fowler 1976).(But see <u>Note 3 Added in Proof</u>)

As anticipated by Box (1979) both models are <u>wrong</u> - for the reasons set forth by Gunst (1980): Both are mis-specified with respect to the variable, T, which represents the effect on the likelihood of occurrence of the event, E_2, of the proliferation of tissue induced by irradiation. The fact of such proliferation is fundamental in the received wisdom of radiobiology; indeed, it is the second of the "four R's" (Repair, Repopulation, Redistribution, Reoxygenation). In the empirical model, M_1, it is assumed that the response is affected by T at <u>all</u> levels of treatment included in the sample. In the mechanistic model, M_2, it is assumed that the response is affected by T at <u>none</u> of the levels of treatment included in the sample. Actually, of course, "The Truth lies somewhere in between" (probably), since in the reference (Douglas and Fowler 1976) cited in Herring (1980) one finds that "... within 8 days; ... proliferation is negligible." Thus, we could anticipate, a priori, that the likelihood of response will be noticibly affected by T at some of the levels of treatment within the sample, those for which $T > T_0$, and not at all at others — those for which $T \leq T_0$. Therefore, the <u>prior</u> odds, $P(M_1)/P(M_2)$, for M_1 (in which the effect of T is mis-specified) should exceed those for M_2 (in which the effect of T is omitted altogether (See Leamer 1978) although it was remarked in the Herring (1980) paper that the sample evidence (as measured by the maximum log likelihood) for M_2 was stronger than for M_1. Moreover, from Lecture <u>13</u> we may anticipate that the estimates of β_0, β_1 and β_2 for M_2 will (probably) be <u>aliased</u> by an amount depending on both the design matrix of the experiment and the form and strength of the association of the response with the omitted treatment variable (See Draper and Smith 1981).

The empirical model is not altogether without theoretical justification. As remarked in Lecture <u>13</u>, the Weber-Fechner law states that many responses of biological systems are proportional to the <u>logarithm</u> of the stimulus. See Finney (1971). However, it is perhaps both fruitful, as well as accurate, to regard the empirical model as descriptive of a "black-box". See Lecture <u>13</u>.

We note that both the empirical and mechanistic models are in fact polynomials in the treatment variables or their transforms. The former is a polynomial of first degree in x_j, $1 \leq j \leq 3$. The latter is of third degree since it includes the interaction term, $N^{-1}D^2$. Marquardt's (1980) remarks on polynomial models will prove to be useful to our subsequent analyses: "... most polynomials, whether linear, quadratic or higher order should be viewed as being in the nature of a low-order power series expansion of the response function about the centroid of the experimental region."

Estimates, $\hat{\underline{\beta}}$ and $\text{Var}(\hat{\underline{\beta}})$ for the rival models of the binary mouse-skin reaction (Herring, 1980) and the single empirical model of the T Castaneum mortality (Finney 1971) were obtained by the Maximum Likelihood methods described in Lecture <u>13</u>. For purposes of the present exposition each of these can be represented as the generalized least squares estimates (Johnston 1972, Theil 1971):

$$\hat{\underline{\beta}} = (X'\hat{V}^{-1}X)^{-1}X'\hat{V}^{-1}z^* \text{ and } \text{Var}(\hat{\underline{\beta}}) = \sigma^2(X'\hat{V}^{-1}X)^{-1}. \tag{3}$$
$$\text{RSS} = \Sigma(y_i - n_i\hat{\pi}_i)^2/n_i\hat{\pi}_i(1-\hat{\pi}_i). \tag{4}$$

z^* is the <u>pseudo</u>-observation (See Lecture <u>13</u>) and $\hat{V}^{-1}$ is an (nxn) diagonal matrix. σ^2 is a dispersion factor. In general, $\sigma^2 = 1$, unless there is evidence of under- or over-dispersion. In these cases σ^2 = RSS /(n-k) where RSS is the Pearson chi-squared statistic (See McCullagh and Nelder 1983). However, a dispersion factor was not mentioned in the analysis for either model presented in Herring (1980). In fact, the concordance of either of the rival models with the data was only assessed "by eye". It was said to be, for each model "very good indeed". (The exact locutions were: "The fit of equation (1) [the <u>empirical</u>

model] to the data is very good indeed, as shown in Figure 2," and, "The fit of this model [the mechanistic model] to the data is shown in Figure 3. Again, the fit is very good indeed.") The residual sum of squares, RSS, (distributed asymptotically as a chi-squared statistic) were not computed for either model in Herring (1980).

Moreover, there was little or no discussion of the validation of either M_1 or M_2. But, "Most will agree, ... that before a model is used, some check of its validity should be made." ... "Methods to determine the validity of regression models include comparison of model predictions and coefficients with theory, collection of new data to check model predictions, comparison of results with theoretical model calculations, and data splitting or cross-validation in which a portion of the data is used to estimate the model coefficients, and the remainder of the data is used to measure the predictive accuracy of the model" (Snee 1977).

In Table 1 are presented various measures of the interaction of a specified model, M_j, with a given sample of observations of size n. These are of two kinds: 1) Aggregate statistics such as RSS (sum of squares of residuals) which gives an overall measure of concordance of the model and sample, or, the predictive performance of model on initial sample, PRESS (weighted sum of squares of residuals) which gives an overall measure of predictive performance of model on new sample and AIC which provides an alternative measure of predictive performance of model on new sample. 2) Case statistics such as t_i, h_i and D_i^1, $1 < i < n$ which have a value for each case included in the sample and which give a measure of the role of each case in the estimates $\hat{y}_i$, $\hat{\beta}$, $Var(\hat{y}_i)$, and $Var(\hat{\beta})$.

Table 1a presents the deletion diagnostics together with the respective critical values. (See Belsley, Kuh and Welsch 1980 and Cook and Weissberg 1982.) We have found that values of $D_i^1 > 0.20$ indicate that the ith observation has considerable influence on $\hat{\beta}$, since this one-step approximation underestimates D_i. (Lecture 13 and Pregibon 1981).

Table 1c presents the various model selection or identification criteria. RSS (residual sum of squares) is a measure of concordance of the model, M_j, with the sample. On the null hypothesis that M_j is the "true" model, RSS is distributed, asymptotically, as chi-squared with f = (n-k) degrees of freedom. Thus, RSS should be (approximately) equal to f. The RSS may be "inflated" by the presence of several different errors of specification in the null hypothesis, H_0, such as when the systematic part of the "true" model differs from that specified by H_0 because the latter failed to include an important predictor variable or when the random part of the "true" model differs from that specified by H_0 because the form of the error distribution is different - for example, that the distribution is not binomial. (The "true" model is that which describes the population from which the sample is obtained.)

The PRESS statistic is a weighted RSS. The weight of the ith observations is $(1-h_i)^{-1/2}$. PRESS thus assigns greatest weight to those treatment levels which are "extreme" - large h_i - or at which the response is an "outlier" - large e_i. PRESS is also a cross-validation measure which describes the sum of squared predictive residuals achieved by successive division of the sample (size n) into a "construction" subsample of size (n-1) and a "validation" subsample of size 1 in all n possible ways. PRESS is an alternative to cross-validation or "data-splitting" (Snee 1977) when the sample size is too small to split into a "construction" subsample of size $n_1 \gg 1$ and a "validation" subsample of size $n_2 \gg 1$. (Snee (1977) has concluded, "... that data splitting is an effective method of model validation when it is not practical to collect new data to test the model.") In the selection of one model from several alternatives, the rule is to select that model for which PRESS is <u>least.</u> See Cook and Weisberg, 1982.

The AIC statistic is an information - theoretic measure. See Akaike (1977). The first factor on the RHS is a measure of the concordance of the model and sample: If the fit is good - 2lnL is small. The second factor on the RHS penalizes models which achieve a good fit by the use of many parameters. (See Crocker 1972.) In the selection of one model from several alternatives the rule is to select that model for which AIC is <u>least</u>. The AIC measure thus explicitly "rewards parsimony". See Box, Hunter and Hunter (1978). AIC is an alternative to Mallow's C_p statistic. See Lectures <u>1</u> and <u>2</u> and Montgomery and Peck (1982). Table <u>1b</u> describes the collinearity diagnostics, VIF and $\Sigma\lambda^{-1}$. The VIFs are the diagonal elements of the correlation matrix, $(X_0X_0)^{-1}$. "The VIF for each term in the model measures the combined effect of the dependencies among the regressors on the variance of that term. One or more large VIFs indicate multicollinearity. Practical experience indicates that if any of the VIFs exceed 5 or 10 it is an indication that the associated regression coefficients are poorly estimated because of multicollinearity" (Montgomery and Peck 1982).

The λ_j are the eigenvalues, or characteristic roots, of the correlation matrix, X_0X_0. "... the presence of small characteristic roots indicates collinearity. Besides the individual roots we also look at the sum of the reciprocals of the characteristic roots. If any of the individual characteristic roots are less than 0.01, or the sum of the reciprocals of the roots is greater than, say five times the number of explanatory variables in the problem, then we say that the variables are collinear. If the above conditions do not hold, the variables are regarded as non-collinear." (Chatterjee and Price 1977).

4. THE FINNEY BIOASSAY EXPERIMENT

The data described below as the Finney toxicological experiment were taken from p. 162 of Finney 1971. These observations are taken to describe the paradigm for multifactor dose-response, or bioassay, experiments. In this experiment there are two treatment variables x_1 and x_2 (vide infra). Observations on response (vide infra) were made at 3 levels of x_1 and 4 levels of x_2. We remark in passing that for the toxicology experiment there is no requirement for a <u>surrogate</u> system, S_2, since there are no constraints on the levels of treatment for the system of interest, S_1, that would compromise the criteria of good experimental design. Therefore, the region of interest, R_2, is completely overlapped by the region of <u>experimental</u> accessibility, R_4, and thus, <u>extrapolation</u> is not required. Thus, an <u>empirical</u> model is quite adequate. The 3 x 4 = 12 levels of 2 treatment variables, or factors (x_1, x_2), comprise a (3 x 4) <u>factorial design</u>. Note that since at least 3 levels of each factor, $\underline{X_1}$ and $\underline{X_2}$, are included in the design matrix, terms of second order can, in principle, be estimated:

$$z = \beta_0 + \beta_1 x_1 + \beta_2 x_2 + \beta_{12} x_1 x_2 + \beta_{11} x_1^2 + \beta_{22} x_2^2. \tag{5}$$

This important class of designs is characterized by the fact that the effect on the response of a change of level in each factor or treatment variable can be assessed independently of the presence of the other variables. This is achieved by including in the design each of the possible combinations of the levels of each treatment variable. The

Table 1. Model-Sample Interaction Measures. Definitions.

a) Deletion Diagnostics. $i = 1, \ldots, n$.

1. Standardized Residuals, t_i
$t_i = (y_i - n_i \hat{\Pi}_i / \sqrt{n_i \hat{\Pi}_i (1-\hat{\Pi}_i)})$. $y_i = r_i$
Critical Value: $|t_i| > 2.0$ (Absolute cut-off)

2. Hat Matrix Diagonals, h_i
$h_i = (V^{-1/2} X (X'V^{-1}X)^{-1} X' V^{-1/2})_{ii}$
$v^{ii} = n_i f_i^2 / \hat{\Pi}_i (1-\hat{\Pi}_i)$.
Critical Value: $h_i > 2k/n$. $k = p+1$. (Size-adjusted cut-off)

3. Beta Shift, D_i^1 (One-step approximation)
$D_i^1 = D_i^1(X'V^{-1}X, m) = [t_i/(1-h_i)]^2 (h_i/m)$
Critical Value: $D_i^1 > 0.20$. See Lecture 13.

b) Variance Inflation Factors, VIF_j. $j = 1, \ldots, p$

1. $VIF_j = \sum\limits_{q=1}^{k} v_{jq}^2 \lambda_q^{-1}$.
Critical Value: $VIF_j > 5.0$.
λ_q, V_q - eigenvalue, eigenvector of (pxp) correlation matrix, $X_0'X_0$.
v_{jq} - projection of jth variable on V_q.

2. $\sum\limits_{i=1}^{p} \lambda_j^{-1}$.
Critical Value: $\Sigma \lambda_j^{-1} > 5p$.

c) Model Selection Criteria.

1. $RSS = \Sigma t_i^2$
$RSS \sim \chi^2(n-k)$. **Asymptotic** on null hypothesis.
$k = p+1$

2. $PRESS = \Sigma t_i^2/(1-h_i)$. ($t_i/\sqrt{1-h_i}$ is a **predictive residual**).[a]

3. $AIC = -2\ln L + 2p$
L - Maximum Likelihood function for model, M_j.
p - Number of treatment variables in M_j.

---

[a] This definition of the PRESS statistic is a Studentized version of that given by Montgomery and Peck (1982) for the so-called Normal theory model in which, $PRESS = \Sigma e_{(i)}^2$ where $e_{(i)} = e_i/(1-h_i)$ is the so-called predictive residual. Using this as a model selection criterion will result in models that fit relatively well at remote rows of X. To correct for this effect, the Studentized version of $e_{(i)}$ is $e_{(i)} = e_i/\hat{\sigma}\sqrt{1-h_i}$ where $\hat{\sigma}$ is the estimated residual mean squared error (for the Normal theory model). For the probit (and logit) model the Studentized version of the predictive residual is given above (Cook and Weisberg 1982).

order in which the n_i observations on the response are obtained at each of the twelve treatment levels is determined from a table of random numbers (a randomized factorial design). (See Myers 1971) The nominal value of n_i was 10. Each exposure was replicated three times. Thus, the effective number at risk at each treatment level (D, N, T) was approximately 30: n_i = 27.5. (See p 446).

It is instructive to demonstrate that the combinations of treatment levels in a factorial experiment can be described by a symmetric array termed the design matrix. (See Myers 1971) To this end the treatment levels can be coded as follows: For x_1 we have the four levels, $x_{1i}' = 6(x_{1i}-\overline{x}_1)/d_1$, i = 1, 2, 3, 4. For x_2 we have the three levels, $x_{2i}' = 2(x_{2i}-\overline{x}_2)/d_2$, i = 1, 2, 3. $\overline{x}_j$, d_j are the mean and range, respectively, of the jth variable. The design matrix, D, for the Finney experiment is thus described as:

$$x_1 \begin{array}{c} x_2 \\ (-3, -1), (-3, 0), (-3, +1) \\ (-1, -1), (-1, 0), (-1, +1) \\ (+1, -1), (+1, 0), (+1, +1) \\ (+3, -1), (+3, 0), (+3, +1) \end{array}$$

Here, x_1 is $\log_{10}$ concentration (mg/ml) and x_2 is $\log_{10}$ deposit (mg/cm^2) of a film of the toxic agent pyrethrum and the binary response y, is mortality (alive/dead) of the (flour)beetle, T. Castaneum. The probit model of the response is

$$z = 0.403 + 3.990x_1 + 1.186x_2 \qquad (6)$$
$$(2.911)(10.534) \quad (2.967)$$
$$RSS = 4.438. \quad P(\chi^2 > RSS| 9) = 0.880 \qquad (7)$$

The numbers in parentheses beneath each estimate, $\hat{\beta}_i$, are the ratios of the Maximum Likelihood estimates of $\hat{\beta}_j$ to the respective standard errors: $\hat{\beta}_j/\sqrt{\text{Var}(\hat{\beta}_j)}$.

Since RSS is much less than the number of degrees of freedom, f=9, and the estimates, $\hat{\beta}_j$, exceed the respective estimates $\sqrt{\text{Var}(\hat{\beta}_j)}$, the model seems to be a "useful" one. However, we shall want to examine more closely several aspects of the "fit" described by case statistics.

The collinearity diagnostics for the first-order empirical model of the Finney data, $z = \beta_0 + \beta_1 x_1 + \beta_2 x_2$, are presented in Table 2.

5. EVALUATION OF THE MOUSE SKIN EXPERIMENT

The experimental data on the mouse skin reaction are summarized in Table 1 and Figures 2 and 3 of Herring (1980) and we do not repeat them here. These Figures also present the dose-response curves at each of the eight levels of (N, T) that were estimated from the empirical model.

There are several striking features of the distributions of treatment levels, number at risk at each level, and observed response in these data on mouse-skin reaction. First, there seems to be no evidence in the distribution of treatment levels in the mouse-skin experiment that the design included consideration of any of the common criteria, for instance, maximizing Det[(X'X)] which will minimize the maximum Var($\hat{y}_i$) over the region of observations (the so-called D-optimality), or minimizing Trace [(X'X)$^{-1}$]. See Seber (1977) or Montgomery and Peck (1982).

Rather, it is evident in Table 1 and Figure 2 of Herring (1980) that N and T are highly correlated. A scattergram of the eight levels (N, T) is presented in Figure 2a: r_{NT} = 0.907. A scattergram of the eight levels of X_1 and X_2 on a log scale is presented in Figure 2c; r_{12} = 0.960. These Figures should be compared to Figures 2b and 2d, respectively, which are constructed from the toxicology data of Finney (1971). Figures 2b and 2d describe a proper, orthogonal factorial, experiment. Note that the degree of collinearity is unchanged by a monotonic variable transformation in the orthogonal design. However, in the mouse-skin experiment the log transformation increases the large correlation of N and T already present in the sample data. The increase is important to the subsequent estimation of β_1, β_2 and β_3 for the empirical model: The respective VIFs are inflated by the following factors: 1.149, 2.233, 2.120. As we shall see, the effect of the log transformation, "exaggerates a difficulty into an impossibility": the VIFs for the terms in N and T have been inflated from only somewhat over 5 to well over 10. See Table 3 and Montgomery and Peck (1982). (It is perhaps of some interest to note that the, "lazy V", configuration of the distribution of the levels of the treatment variables N and T achieved by the experimental design of Figure 2a occurs also in the (uncontrolled) distribution of the levels of the treatment variables, gamma dose, $D\gamma$ and neutron dose, D_n, in the non-experimental LSS data on radiation carcinogenesis that is discussed in Lecture 14. In the latter study the distribution is achieved by data augmentation - pooling of the Hiroshima and Nagasaki data. This post-hoc maneuver was required in order to reduce the effects of multicollinearity on the estimates of β in the dose-response equation for radiocarcinogenesis. The residual multicollinearity achieved by this "salvage operation" is much less than that in the designed experiment described in Figures 1a and 1b.)

In the present analysis, as in any other, it will be rewarding to examine index plots of the various regression diagnostics, t_i, h_i, D_i^1, etc. (Cook and Weisberg 1982 and Pregibon 1981). It will be recalled that in an index, or case plot, the abscissae are the row numbers of the observation matrix, [y, X]. Where possible, it is usually fruitful to order these rows according to some qualitative or quantitative feature of the observations, e.g., stage of disease, dose, etc. Also, it will be of assistance in understanding, evaluating, and interpreting both the model and the sample, if these index plots of t_i, h_i, D_i^1, etc., are accompanied by index plots of the variables of treatment, x_1, x_2, ...

Index plots of x_1 and x_2 for the mouse-skin experiment are presented in Figure 3a, 3b and 3c. Index plots of x_1 and x_2 for the Finney toxicology experiment are presented in Figure 3d and 3e for comparison. The lack of orthogonality between X_1 and X_2 for the mouse-skin experiment is evident: both x_1 and x_2 are increasing functions of the index number.

Scattergrams of the joint distributions of X_1 and X_2 and of X_1 and X_3 for the mouse-skin experiment are presented in Figures 4a, 4b. The scattergram of D and D/N is presented in 4c. It is evident from those Figures that all the treatment variables D, N and T are highly correlated in this design and the correlation between D and (D/N) is only somewhat less. The collinearity diagnostics for the model M_1, X = (1, X_1, X_2, X_3) of the mouse-skin experiment data are presented in Table 3. We recall from Lectures 1-5 that the presence of collinearity in the joint distribution of (D, N, T) will inflate both the estimates, $\hat{\beta}$, and their variance-covariance matrix, Var($\hat{\beta}$). We recall from Lecture

Table 2. Collinearity Diagnostics. T. Casteneum Mortality. Empirical Model. $X = (x_1, x_2)$

1) Correlation Matrix: 1.0   0
                                   1.0
2) Eigenvalue, Eigenvector:
   $\lambda_1 = 1.0$, $V_1 = (0.707, 0.707)$
   $\lambda_2 = 1.0$, $V_2 = (-0.707, 0.707)$
3) Condition Number, $\kappa = \lambda_1/\lambda_2 = 1.0$
4) Proportion of Variance, $\lambda_1/\Sigma\lambda_j = 0.500$
5) $\Sigma\lambda_j^{-1} = 2.00 < 5p = 10$
6) Variance Inflation Factors, VIF: (1.0, 1.0)

Table 3. Collinearity Diagnostics. Mouse Skin Reaction #2. A. Empirical Model M_1. $X = (x_1, x_2, x_3)$

1) Correlation Matrix: 1.0, 0.836, 0.849
                               1.0, 0.960
                                       1.0

2) Eigenvalue, Eigenvector
   $\lambda_1 = 2.765$, $V_1 = (0.560, 0.587, 0.585)$
   $\lambda_2 = 0.196$, $V_2 = (-0.828, 0.361, 0.430)$
   $\lambda_3 = 3.963 \times 10^{-2}$, $V_3 = (0.042, -0.724, 0.688)$

3) Condition Number, $\kappa = \lambda_1/\lambda_p = 69.763$.
4) Proportion of Variance, $\lambda_1/\Sigma\lambda_i = 0.922$
5) $\Sigma\lambda_j^{-1} = 30.70 > 5p = 15$
6) Variance Inflation Factor, VIF: (3.655, 14.038, 12.016)
7) Inflation of Joint Confidence Region for β: [a]
   $100[(\Pi\lambda_i)^{-1/2} - 1] = 582.6 \approx 600\%$
   Determinant of $X_0'X_0 = \Pi\lambda_i = 2.14 \times 10^{-2}$

---

[a] Selection of the design to achieve a small volume of the confidence ellipsoid is appropriate if the goal of the experiment is the estimation of the parameters in a specified model to a specified level of precision. Of course, there may be other experimental goals:
1) The prediction of the levels of certain responses which depend upon some unknown parameters.
2) The selection of one of several rival models to be most consistent with the "true nature of things."
3) The determination of a course of action in a situation in which the optimal action depends on a) what the correct model is and b) what the values of the parameters are (See Bard 1974). Each of the goals has its own design criteria. All of these criteria may be derived from information theory (Bard 1974). It is not altogether clear from the Herring paper which, if any, of these goals was the "object of the exercise."

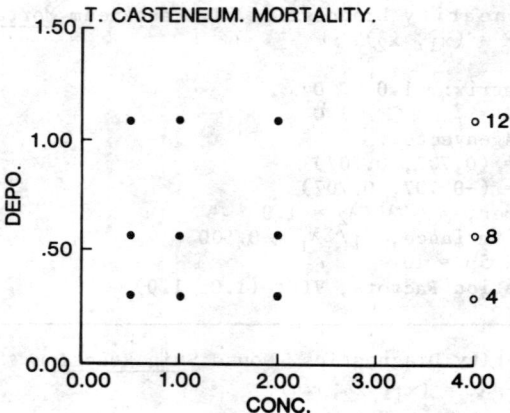

Figure 2b. Scattergram of the distribution of treatment regimens for mortality experiment in Deposit-Concentration plane. The variables are orthogonal: r = 0. The solid circles denote regimens for which $0 < r_i/n_i < 1.0$. The open circles denote regimens for which $r_i/n_i = 1.0$.

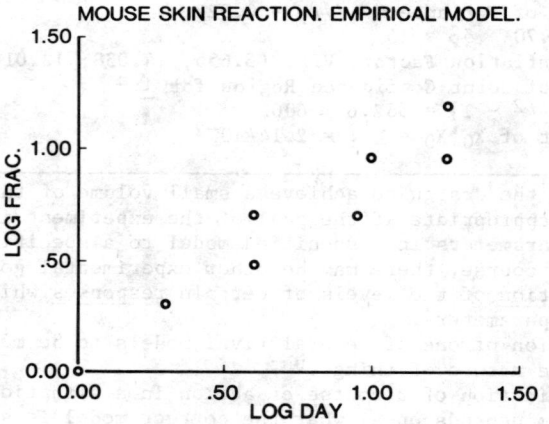

Figure 2c. Scattergram of the distribution of treatment regimens for skin reaction experiment in log N - log T plane. The log-transformation has enhanced the multicollinearity: r = 0.960. Compare with Figure 5b in Lecture 13 for non-experimental data in which r=0.978.

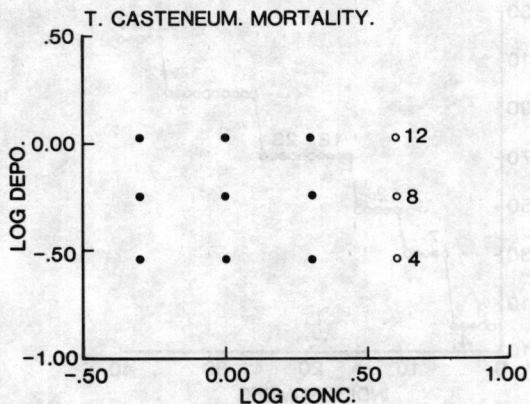

Figure 2d. Scattergram of the distribution of treatment regimens for mortality experiment in log Depo. - log Conc. plane. The log-transformation has no effect on the degree of orthogonality of the treatment variables.

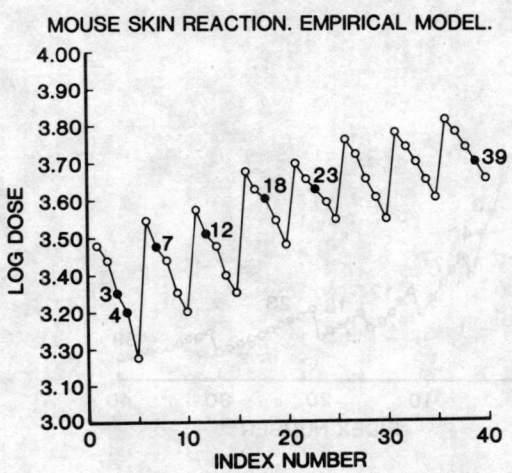

Figure 3a. Index plot of the distribution of log D_i in treatment regimens for skin reaction experiment. The solid circles denote regimens for which $0 < r_i/n_i < 1.0$. The open circles denote regimens for which $r_i/n_i = 0$ or 1.0. The regimens #1 - #5 are single-dose experiments. Regimens #6 - #40 are fractionated dose experiments. There is a qualitative distinction.

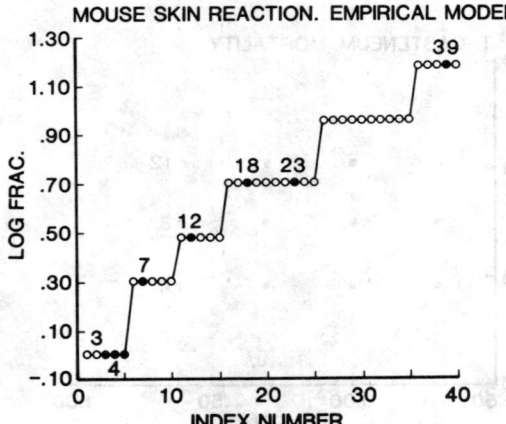

Figure 3b. Index plot of the distribution of log N_i in treatment regimens for skin reaction experiment. Comparison with Figure 3a shows that the variables are correlated: $r = 0.836$.

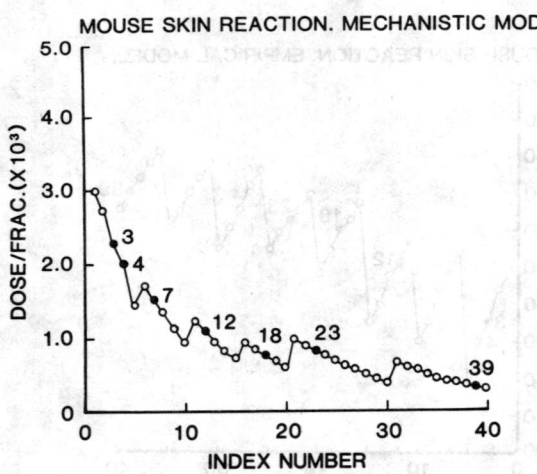

Figure 3c. Index plot of the distribution of $(D/N)_i$ for skin reaction experiment (D/N) is the key variable in the radiobiological model of skin reaction. (It is, of course, also an adjustable variable of treatment in a response surface model.) Note that the levels of dose per fraction in the single dose experiment (#1 - #5) exceed those for clinical practice by about an order of magnitude.

13 that it will also exacerbate the effects of errors of measurement of dose on the estimates, $\hat{\beta}$. The marginal bivariate distributions can be adequately summarized, respectively, by the equations for the three consraints: $N = 0.70T$ ($R^2 = 0.82$), $D = 2210T^{0.30}$ ($R^2 = 0.70$) and $D = 2150N^{0.35}$ ($R^2 = 0.72$). (Each equation is a simple Least Squares fit to the data.) The (socio-economic) relation between N and T is that for a treatment schedule of 5X/week (5/7 = 0.71). The (medical-legal) relation between D and T is that described by the Strandvist "cube root law": $D \propto T^{0.33}$. (See Lecture 13). We shall have reason to note later that the relation between D and N can be described equally well by a "square root law": $D = 1726N^{0.50}$. ($R^2 = 0.72$). We find that the constraints imposed on the treatment variables in the designed experiment are precisely those of the non-experimental study described in Lecture 13. Thus, the difficulties of interpretation, characteristic of experimental studies on surrogate systems, (mice), have been compounded by the difficulties of estimation, characteristic of non-experimental studies on the target systems (men).

The second curious feature of the mouse-skin experiment is the size of the sample allocated to each level of treatment, (D, N, T). An index plot of the sample size is presented in Figure 5a. The average size is $\bar{n}_i = 7.32$ with a coefficient of variation of 25.19%. An index plot of the sample size for the Finney experiment is presented in Figure 5b. The average size is $\bar{n}_i = 27.50$ and the coefficient of variations is CV = 12.01%. The mouse-skin experiment has only about 0.25x the number at each treatment level of the Finney experiment. The standardized variation in this number is 2x that of the Finney experiment.

We next notice in Figure 2 of Herring (1980) that there are only 7 of the 40 treatment levels (D, N, T) at which the proportion of responders, $\hat{\pi}_i = r_i/n_i$, differs from 0 or 1. The variance of the estimate of z_i obtained from a sample of size n_i of which r_i response is $Var(z_i) = \hat{\pi}_i(1-\hat{\pi}_i)/n_i f_i^2$. Therefore, the amount of information on z_i is the weight, $w_i = n_i f_i^2/\hat{\pi}_i(1-\hat{\pi}_i)$. Here f_i is the Normal density function at z_i. We have $w_i \longrightarrow 0$ as $\pi_i \longrightarrow 0$ or 1. Thus, in only 7 of the 40 observations on n_i individuals is the information on the probit of response more reliable than if only a single individual were irradiated at that level. (See Finney 1971). The evidence that there are such a modest number of "effective" observations included in the sample recalls Crocker's (1972) comments: "The expectation of R^2 [the multiple correlation coefficient] in the null case was shown ... to be $E(R^2) = p/(n-1)$ where p is the number of predictors and n is the sample size. Values of R near 1.0 are thus easily obtained by chance as the number of predictors approaches (one less than) the sample size - a common problem where data are limited." The 7 regimens for which $0 < r_i/n_i < 1.0$ are labeled by the respective index numbers in the Figures for the mouse-skin experiment: #3, #4, #7, #12, #18, #23 and #39. The respective levels of response, r_i/n_i are: 0.67(#3), 0.08(#4), 0.83(#7), 0.50(#12), 0.375(#18), 0.33(#23), 0.50(#39). Note that there are 3 out of 12 regimens for which r/n = 1.0 in the Finney experiment. Those have index numbers #4, #8 and #12 in Figures 2 to 5.

The reader will no doubt recall several papers from the current radiobiological literature, in which the experimental data from which dose-response curves (and surfaces) are constructed include a large number of treatment regimens in which the majority of responses are extreme: r/n = 0 or 1. Thus, the Herring (1980) data are not at all atypical in any aspect of the design; they differ more in degree than in kind from much of the data described - or invoked - in the current literature. (And that difference is <u>trivial</u>.)

But prominent writers in the field of bioassay have vividly described the difficulties which encumber the use of bioassay observations in which $r_i/n_i = 0$ or 1. They have also provided empirical - and often Draconian - remedies: "It is the considered opinion of the present writer that observations with zero or 100 percent response should not be used at all - that when they occur another experiment should be performed with different dosages at values where observations of zero or 100 percent are very unlikely - but this is probably an extreme position. I may point out, however, that in one widely advocated Karber ... method of estimate of L.D.50, if at two successive doses an observation of zero percent mortality is made, only the larger of the two doses is used, and similarly only the smallest of several consecutive doses which show 100 percent mortality is considered in making the calculation. ... My suggestion, then, is to use at most one such observation at each extremity."
...
"The definitive procedure now for dealing with zero's and 100 percent observations ... is therefore to use the rule of substituting $1/2n$ for zero and $(2n-1)/2n$ for 100 percent observations." (See Berkson 1953).

For the mouse-skin data, neither of the two latter remedies seem appropriate. If the (empirical) Karber method of deletion is followed, there will remain 1 level of (N, T) with 4 levels of dose, 5 levels of (N, T) with 3 levels of dose and 2 levels of (N, T) with only 2 levels of dose. In other words, the number of observations on the principal treatment variable, radiation dose will have been reduced from 40 to 19 (but see below). If the (empirical) substitution method of Berkson is followed, the bias thereby introduced into the initial estimates of response will be very large - because of the small sample size, n_i. For example, for $n_i = 6$, $r_i/n_i = 0$ will be replaced by $1/2n_i = 0.08$ and $r_i/n_i = 1$ will be replaced by 0.91. Therefore, we have eschewed both of these standard empirical remedies; they are appropriate only in data in which 1) there are only a few extreme responses and 2) the sample sizes are large, say, $n_i = 25$; in such a case $1/2n_i = 0.02$ and $(2n_i-1)/2n_i = 0.98$. But see <u>Note 4 Added in Proof</u>.

An index plot of the initial weights, w_i, for the mouse-skin experiment is presented in Figure <u>6a</u>. The amount of information on the probit transform, z_i, contributed by a sample obviously depends not only on the sample size, n_i, but also the level of observed response, $r_i/n_i = \hat{\pi}_i$, the estimate of the parameter π of a binomial distribution. A plot of the <u>unit</u> weights, $I_i/n_i = f_i^2/P_iQ_i$ and $J_i/n_i = 1/P_iQ_i$ for quantiles and proportions, respectively, vs the proportion, $P_i = \Pi_i$, over the range $0 \leq P_i \leq 1.0$ is presented as Figure <u>6b</u>. Here f_i is the density function of the standard Normal distribution, $f_i = e^{-u^2/2}/\sqrt{2\pi}$, and $Q_i = 1-P_i$. Figure 6b discloses that for the extreme expected values of response, $\hat{\pi}_i \longrightarrow 0$ and $\hat{\pi}_i \longrightarrow 1.0$, the <u>standardized</u> residuals, $t_i = (r_i - n_i\hat{\pi}_i)/\sqrt{n_i\hat{\pi}_i(1-\hat{\pi}_i)}$, may be quite large even if the simple residuals $(r_i-n_i\hat{\pi}_i)$ are small. Figure <u>6b</u> also discloses that the weight of observations at extreme levels response is much less than that at intermediate levels and thus, the latter will tend to dominate the estimates, $\hat{z}_i$, $Var(\hat{z}_i)$, $\underline{\hat{\beta}}$ and $Var(\underline{\hat{\beta}})$ for any model. Since the weight of the residual increases faster than the weight of the observation decreases as $\hat{\pi}_i \longrightarrow 0$ or 1, it is possible for an observation to have large values of h_i and $|t_i|$ simultaneously. See Figure 7a.

The presence of only 7 regimens for which $0 < r_i/n_i < 1$ may simply

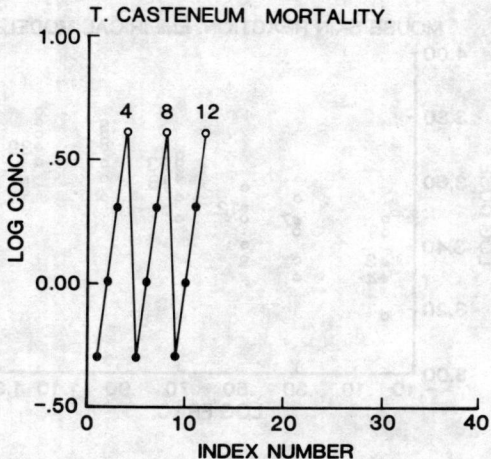

Figure 3d. Index plot of the distribution of Concentration$_i$ for mortality experiment. Compare Figure 3a.

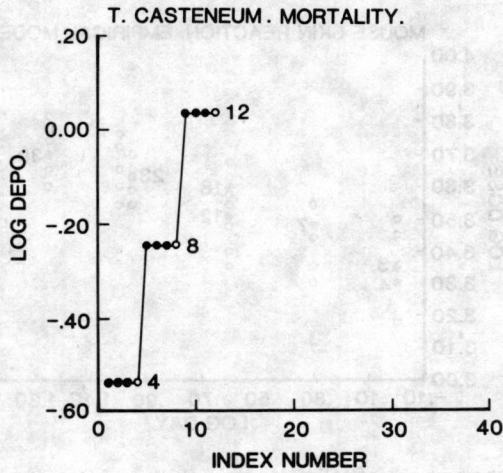

Figure 3e. Index plot of the distribution of Deposit$_i$ for mortality experiment.

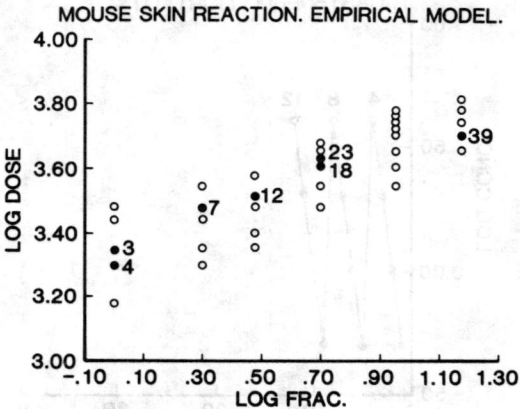

Figure 4a. Scattergram of the distribution of treatment regimens for skin reaction experiment in log D - log N plane. The distribution consists of a family of distributions of D indexed by values of N = 1, 2, 3, 5, 9 and 15.

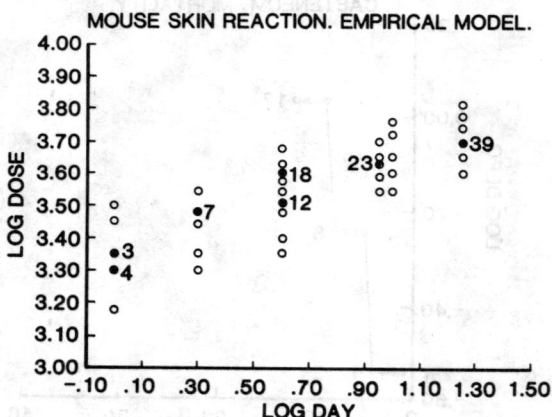

Figure 4b. Scattergram of the distribution of treatment regimens for skin reaction experiment in logD - logT plane. The distribution consists of a family of distributions of D indexed by values of T = 1, 2, 4, 9, 10 and 18 days. The excessive number (33/40) of extreme responses, $r_i = 0$ or n_i, may be due in part to an infelicitous choice of conditional ranges of D and T (or N) such that the former produces "overwhelming" and the latter "trivial" effects on the response. See text.

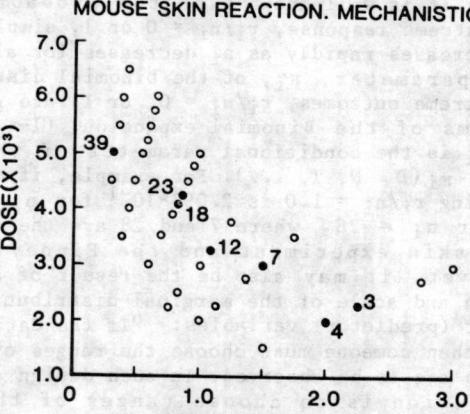

Figure 4c. Scattergram of the distribution of treatment regimens for skin reaction experiment in D-(D/N) plane. The distribution consists of a family of distributions of (D, D/N) indexed by N=1,2,3,5,9 and 15.

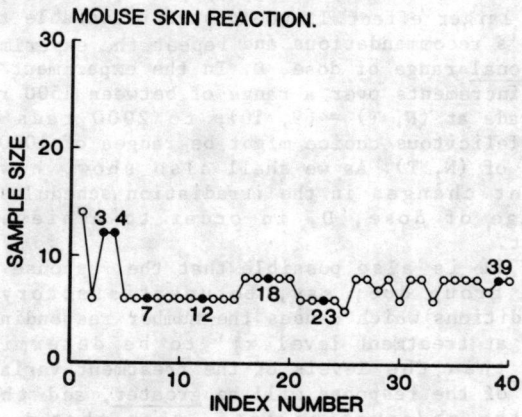

Figure 5a. Index plot of the distribution of sample size, n_i, in treatment regimens for skin reaction experiment. Note that the excessive number (33/40) of extreme responses, r_i = 0 or n_i, may be due in part to the small numbers at risk ($n_i \approx 7$) at each treatment level. Note that the largest levels of n_i are included in the single-dose experiments.

be the consequence of allocating too small sample size to each of the 40 treatment regimens: it is well-known that the probability of the occurrence of an extreme response, $r_i/n_i = 0$ or 1, simply as a result of random sampling increases rapidly as n_i decreases for all values of the (conditional) parameter, π_i, of the binomial distribution. The probability of the extreme outcomes, $r_i/n_i = 0$, or 1, are given by the first and last terms of the Binomial expansion, $(1-\pi_i)^{n_i}$ and $\pi_i^{n_i}$, respectively. Here, π_i is the conditional parameter of the Bernoulli distribution, $\pi_i = \pi_i(D, N, T, ...)$. For example, if $\pi_i = 0.80$, the probability of observing $r_i/n_i = 1.0$ is $2.097*10^{-1}$ for $n_1 = 7$, but is only $1.934*10^{-3}$ for $n_i = 28$, where 7 and 28 are the average sample sizes for the mouse-skin experiment and the Finney experiment, respectively. However, it may also be the result of an unfortunate choice of the location and scale of the marginal distribution of one or more of the treatment (predictor) variables: "If the data come from an experimental design, then someone must choose the ranges over which the predictor variables are to be observed. In such design situations, an implicit guiding criterion is to choose ranges of the predictor variables that will produce neither trivial nor overwhelming effects on the response variables, that is, effects that are of comparable magnitude a priori." (Marquardt 1980) [and also minimize $Det[(X'X)^{-1}]$]. The treatments at which $0 < r_i/n_i < 1$ occur in the mid-range of D at both extremes of (N, T). Observations #3 and #4 are at (N, T) = (1, 1) and observation #39 is at (N, T) = (15, 18). See Figures 3c and 4a and 4b. Thus, it would seem likely that the conditional ranges of D that were chosen at each level (N, T) have produced, "... overwhelming effects on the response variables." (We note that in the Finney experiment, the three extreme ($r_i/n_i = 1.0$) responses occurred at the largest value of $x_1 = \log_{10}(conc.)$.) Moreover, since for $T < T_0$ there is no effect of proliferation on response, the range of $(T-T_0)$ - from 0 to 10 has a "trivial" effect on the response, while N, which varies from 1 to 15 has a larger effect. It would seem reasonable to follow the first of Berkson's recommendations and repeat the experiment, but using a narrower <u>conditional</u> range of dose, D. In the experiment described, D was varied in 5 increments over a range of between 1500 rads at (N, T) = (1, 1) to 2250 rads at (N, T) = (9, 10), to 2000 rads at (N, T) = (15, 18). A more felicitous choice might be ranges of 1000-1200 rads at each set of values of (N, T). As we shall also show, however, there must be important changes in the irradiation schedules, N and T, as well as in the range of dose, D, in order to achieve a properly designed experiment.

Of course, it is also possible that the response is <u>correlated</u> within each dose group due, say, to unsatisfactory control of experimental conditions which causes the number responding, r_i, within a group of size n_i at treatment level $\underline{x_i}'$ to be determined by some covariate other than the levels of the treatment variables. In such cases the variance of the response will be <u>greater</u>, and the weight to be attached to the observation <u>less</u>, than that described by the binomial distribution. "The extreme situation is that, in every batch tested, either all members respond or all fail to respond, so that the evidence from a batch is no more reliable than that from an individual" (Finney 1971) - such as has apparently occured in 33 of the 40 "batches" in the mouse-skin experiment. It is possible that the dichotomization of the scores on the ordinal scale of response has induced an effective correlation within-groups. Whatever the cause, one immediate effect of the evidence of the correlation in the responses,

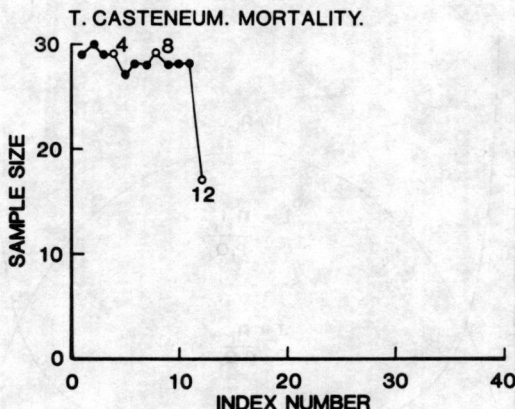

Figure 5b. Index plot of the distribution of sample size, n_i, in treatment regimens for <u>mortality</u> experiment.

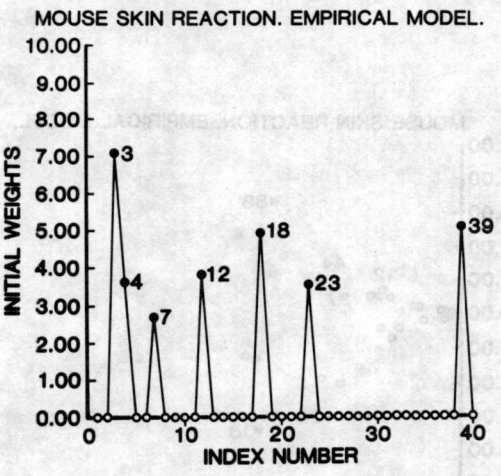

Figure 6a. Index plot of the distribution of initial (probit) weights $w_i = n_i f_i^2 / \pi_i (1-\pi_i)$, in treatment regimens for <u>skin reaction</u> experiment. Only the regimens for i = 3, 4, 7, 12, 18, 23 and 39 have nonzero weight. These are the regimens for which $0 < r_i/n_i < 1.0$. For the remainder, $r_i/n_i = 0$ or 1.0.

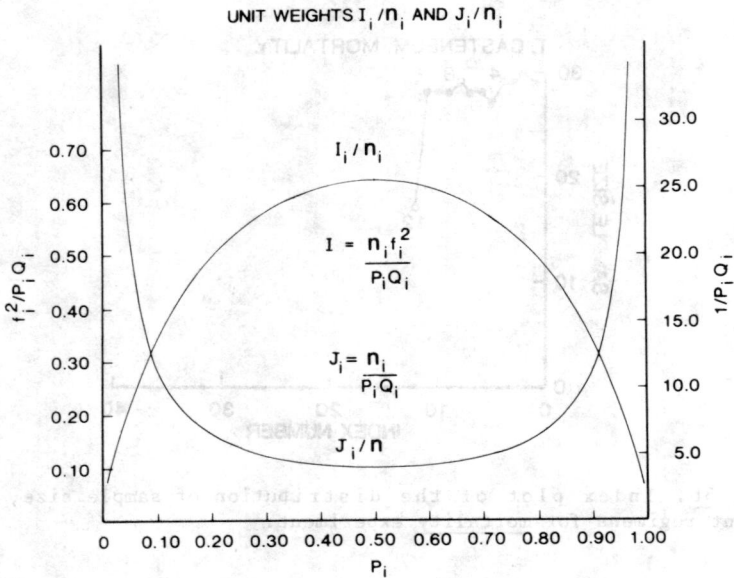

Figure 6b. Plot of unit weights ($n_i = 1$) for proportions and percentiles as a function of $P_i = \pi_i$. ($Q_i = 1 - \pi_i$).

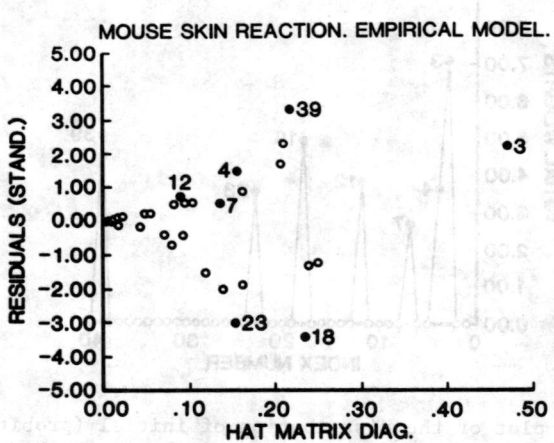

Figure 7a. Scattergram of the distribution of key diagnostics, t_i and h_i, for (first-order) response surface model, M_1, of <u>skin reaction</u> experiment. Note that $|t_i|$ and h_i have a strong positive correlation, $r = 0.78$: large values of $|t_i|$ and h_i tend to occur together.

y_i, is the implication that the estimated variance-covariance matrix $\text{Var}(\hat{\beta})$ for the empirical model, $X = (\underline{1}, \underline{X}_1, \underline{X}_2, \underline{X}_3)$ should be <u>inflated</u> by the (over-) dispersion factor, $\sigma^2 = \text{RSS}/(n-k)$: $\text{Var}(\hat{\beta}) = \sigma^2(X'\hat{V}^{-1}X)^{-1}$. This was <u>not</u> done in the Herring (1980) analysis of the mouse-skin experiment. Failure to do so, as we shall show below, strongly affects the validity of the inferences made <u>in that paper</u> for this model. The factor σ^2 is sometimes called the heterogeneity factor (Finney 1971). If RSS exceeds the 0.95 quantile of $\chi^2(n-k)$ the data are said to be <u>heterogeneous</u>. (It is assumed here that the increased dispersion of response is random. However, if the lack of concordance is systematic rather than random, due to a mis-specification error, say, to the failure to include higher order terms in the model, such as x_1^2, then it is not appropriate to include the overdispersion factor (vide infra). See Finney 1971.

In order to develop an appreciation of the Herring (1980) data, it is useful to fit the simple bioassay model, $z = \beta_0 + \beta_1 x_1$, to the set of 5 levels of D at each of the 8 levels of (N, T) where $x_1 = \log D$. The pattern in these estimates, $\hat{\beta}$, $\text{Var}(\hat{\beta})$, will disclose the nature and strength of the dependence of the slope, β_1, and <u>intercept</u>, β_0, on the covariates N and T, $\beta_j = \beta_j(N, T)$, $j = 0, 1$, within this family. However, we find at once that the estimates, $\hat{\beta}_0$, $\hat{\beta}_1$, could only be obtained at $(N, T) = (1, 1)$, since the (iterative) Maximum Likelihood methods do <u>not</u> converge for the remaining 7 levels (see Figure <u>2a</u>). One obvious reason for failure of the standard methods is that the sample observations on the conditional dose-response process at these levels simply do not contain enough information. For most, $r_i/n_i = 0$ or 1. $1 \leq i \leq 5$ (see Figure <u>3a</u>). This is a common experience: "The convergence ... is rarely a problem, unless one or more of the elements of $\underline{\beta}$ are infinite. This can occur, for example, when the data are sparse and, for some observations $y_i = 0$ or $y_i = n_i$" (McCullagh and Nelder 1983). (In the notation of this lecture $y_i \equiv r_i$.)

We next fit the model $z = \beta_0 + \beta_1 x_1 + \beta_2 N + \beta_3 T = \beta_0' + \beta_1 x_1$ where $\beta_0' = [\beta_0 + \beta_2 N + \beta_3 T]$. The iterative maximum likelihood procedure now converges, but for this candidate, RSS = 180.68 which greatly exceeds the number of degrees of freedom, f = 36; the model is obviously inconsistent with the data. (This model assumes that the effects of N and T on the response only affect the <u>intercept</u> of the dose-response curve. In order to represent the effects of N and T on the <u>slope</u> of the dose-response curve, <u>interaction</u> terms such as $x_1 T$ or $x_1 NT$ must be included in the model. See Lecture <u>13</u>.)

6. THE EMPIRICAL MODEL

Let us next consider the empirical, <u>response surface</u>, model, $z = \beta_0 + \beta_1 x_1 + \beta_2 x_2 + \beta_3 x_3$. There is some theoretical basis for choosing the transforms, $D \longrightarrow \log D$, etc., in the Weber-Fechner law (Finney 1971) which states that the level of response is proportional to the logarithm of the level of stimulus. We note at once the small a priori probability, $P(M_1)$, for this model (See Leamer 1978). M_1 is a "non-starter" (or, "strawman"?) since it is manifestly incompatible with the a priori information on β_3 that was cited in Herring (1980): "... over this time, [0-8 days] proliferation was negligible." (Douglas and Fowler 1976). One could not anticipate detecting <u>any</u> effect of T on the observed response at about half of the treatment levels included in the sample since proliferation is "negligible" for $T < 8$. Since M_1 describes the received wisdom on S_1 in R_1 (See Lecture <u>13</u>) one recalls

Table 4. Empirical Models of Mouse Skin Reaction #2. (Probit)

Model M_1. (Linear in x_1, x_2, x_3)

$z = -109.93 + 32.28x_1 - 7.44x_2 - 2.62x_3$
   (-8.81) (8.78) (-4.96) (-2.47)
RSS(36) = 63.76. PRESS = 82.104. AIC = 140.03
$P(\chi^2 > RSS | 36) = 0.003$. (Reject H_0)

Model M_{1i}. (Quadratic in x_1)[a, b]

$z = 511.92 - 328.86x_1 - 9.78x_2 - 3.27x_3 + 52.52x_1^2$
   (4.07) (-4.39) (-4.84) (-2.47) (4.71)
$z = -1.23 + 7.03x_1^* - 3.57x_2^* - 1.40x_3^* + 1.22x_1^{**2}$
   (4.71) (7.36) (-4.84) (-2.47) (4.71)
RSS(35) = 40.56. PRESS = 56.59. AIC = 106.96
$P(\chi^2 > RSS | 35) = 0.238$. (Accept H_0)

Model M_{1ii}. (Quadratic in x_3)

$z = -151.64 + 45.44x_1 - 8.90x_2 - 12.23x_3 + 5.55x_3^2$
   (-7.83) (7.82) (-4.84) (-5.02) (4.74)
RSS(35) = 45.45. PRESS = 60.40. AIC = 108.84
$P(\chi^2 > RSS | 35) = 0.111$. (Accept H_0)

Model M_{1iii}. (Linear Spline in x_3. $T_0 = 9$)[c]

$z = -137.19 + 40.98x_1 - 15.21x_2 + 7.96x_3'$
   (-7.80) (7.79) (7.28) (4.30)
RSS(36) = 47.06. PRESS = 64.43. AIC = 122.88
$P(\chi^2 > RSS | 36) = 0.103$. (Accept H_0)

Model M_{1iv}. (Linear Spline in x_3. $T_0 = 4$)[c]

$z = -149.55 + 44.64x_1 - 15.57x_2 + 9.86x_3' - 4.77x_0'$
   (-7.38) (7.37) (-6.76) (4.54) (-4.54)
RSS(35) = 40.00. PRESS = 61.52. AIC = 117.23
$P(\chi^2 > RSS | 35) = 0.258$. (Accept H_0)

Model M_{1v}. (Spline in x_3, Quadratic in x_1)[a,b]

$z = 8.44 + 43.31x_1^*s_1 - 15.11x_2^* + 4.88x_3' + 39.14x_1^{**2}$
   (6.47) (7.14) (-6.68) (2.30) (3.23)
RSS(35) = 38.58. PRESS = 60.35. AIC = 114.28
$P(\chi^2 > RSS | 35) = 0.311$. (Accept H_0)

---

[a] $x_j^* = (x_j - \bar{x}_j)/s_j$, $1 \leq j \leq 3$
[b] $x_1^{**2} = (x_1 - d)^2$. d - Dykstra Statistic
[c] $x_3' = \begin{cases} \log(T/T_0). & T \geq T_0 \\ 0. & T < T_0 \end{cases}$

$x_0' = \begin{cases} 1 & T \geq T_0 \\ 0 & T < T_0 \end{cases}$

the oft-cited hazard of extrapolation of any, empirical or methanistic model: "... the mechanism may change ...".(Box, Hunter and Hunter 1978)

Note that for fixed values of x_2 and x_3 the empirical model is $z = [\beta_0 + \beta_2 x_2 + \beta_3 x_3] + \beta_1 x_1$. The results of fitting this model to the 40 levels of (D, N, T) are presented in Table 4. The iterative Maximum Likelihood procedure also converges for this model. However, the first thing that one notices is that although $\hat{\beta}_j/\sqrt{Var(\hat{\beta}_j)} > 2$ for $j = 0, \ldots, k$, a comparison of the chi-squared measure of concordance, RSS, with the number of degrees of freedom, f, discloses that this model does not fit these data. (It will be recalled that Herring (1980) described the fit as, "very good indeed".) Note that this situation is the reverse of that described in Lecture 13 for the empirical (Taylor series) model of response E_1 in the non-experimental study — there the concordance was satisfactory — RSS $\sim$ f — but the precision of estimate was poor: $\hat{\beta}_j/\sqrt{Var(\hat{\beta}_j)} \leq 2$, $j = 0, \ldots, k$.

At this point let us note with Finney (1971) the effect of small expected frequencies upon the sampling distribution of the Pearson statistic: If none of the expected frequencies, $n_i P_i$, $n_i Q_i$, are "too small", (where $\hat{\pi}_i = P_i$) the sampling distribution of the Pearson statistic, RSS, is adequately approximated by a chi-squared distribution in which the parameter is equal to the number of degrees of freedom, f, i.e., the expectation or mean of the distribution is f, the variance is 2f, etc. When, however, some of the expected frequencies are small, the sampling distribution of RSS will differ from that of chi-squared, in part because the former becomes noticibly discontinuous and, in part, because the moments of the distribution of RSS beyond the first, f, may be somewhat different from those of chi-squared; in particular, the variance may frequently exceed 2f." It has been recently, and repeatedly, demonstrated in some Monte Carlo experiments (Roscoe and Byars 1971) that in many cases the familiar requirement for the validity of the approximation by chi-squared, namely, that the expected frequencies each exceed some arbitrary minimum value, say 1 or 5 or 10 ("pooling" of the observations in adjacent "cells" may be required in order to achieve a recommended, albeit arbitrary, minimum, see,for example,Finney 1971 and Snedecor and Cochran 1967) may be too conservative. For instance, for some comparisons it has been shown that a requirement based upon a minimum expected frequency for every cell may be replaced by a requirement of a minimum average expected frequency over all cells, e.g., the chi-squared distribution is a valid approximation to the sampling distribution of RSS at the 0.05 level of significance if the average expected frequency exceeds 2. The average expected frequency for the mouse-skin data exceeds 3.5 for all the models which we examined. It should be repeated that no problem (with the approximation of the sampling distribution of the Pearson statistic by the chi-squared distribution) arises if RSS $\simeq$ f the number of degrees of freedom.

As remarked above, if this model is to be the basis of any inferences on the conditional response of the surrogate system, S_2, then the sample covariance matrix, $Var(\hat{\beta})$, should include the over-dispersion factor, $\hat{\sigma}^2 = RSS/(n-k)$. That is, the estimated covariance matrix is inflated by $\hat{\sigma}^2$: $Var(\hat{\beta}) = \hat{\sigma}^2 (X'\hat{V}^{-1}X)^{-1}$. Moreover, in constructing confidence limits, the $(1-\alpha/2)$ quantile of the t-distribution rather than that of the Normal distribution is the appropriate factor. (See Finney 1971; McCullagh and Nelder 1983) However, the lack of concordance of model and sample suggested by the value of the heterogeneity (Finney 1971) or over-dispersion (McCullagh and Nelder 1983) factor, σ^2, which is estimated as $\hat{\sigma}^2 = RSS/(n-k) = 1.771$, may be a consequence of either the presence of a variation in the response variable that is in excess of that given by the binomial distribution - an experimental error - or the absence of an important treatment variable from the response equation - a specification error. The former is strongly suggested by the information plot, Figure 6a, in which at only 7 of the 40 treatment levels do we find $0 < r_i/n_i < 1.0$, $1 \leq i \leq 40$. (See Finey 1971). The latter is supported by the evidence of the residuals plot of Figure 8a. Of course, these events are not mutually exclusive; it is entirely possible that RSS is inflated by the joint failure of both of these a priori assumptions: 1) $Var(y_i) = \pi_i(1-\pi_i)$ and 2) the model matrix, X, is correctly specified. Indeed this is the most likely case for these data, given the evidence of Figures 6a and 8a. However, since we cannot untangle these factors, we shall simply estimate the overdispersion factor for this model as if the inflation of RSS were due solely to the failure of the first assumption. Then $\hat{\sigma}^2 = 1.771$.

The inferences which can be made (correctly, now) from the isoeffect equation are very different indeed from those presented in the Herring paper in which the exponents of N and T in the derived isoeffect equation are 0.209 ± 0.065 and 0.103 ± 0.042, respectively, and the results of the analysis are summarized as, "At this point one might be tempted to conclude that equation (1) [the empirical model] is correct and that overall treatment time is important for these data." However, the correct estimates (Fieller's theorem, Finney 1971) of these exponents from the probit model, together with the appropriate 0.95 CL are now $N = 0.230$, $(0.124, 0.332)$ and $T = 0.081$, $(-0.008, 0.171)$. These estimates include the inflations of $Var(\hat{\beta})$ due to the over-dispersion factor, $\hat{\sigma}^2$, and to the use of the quantile of the Student's t distribution as a factor. Since the 0.95 CL on the exponent of T include 0, the correct inference is that on the evidence of this model of these data, the Ellis equation is not correct - overall treatment time is not important. This (contrary) conclusion is demonstrated still more directly and vividly by examining the 0.95 CL on $\hat{\beta}_3$ in the full dose-response equation: $\hat{\beta}_3 = -2.616$ $(0.243, -5.475)$. Thus, a closer examination of the data suggests that two important inferences made in the Herring paper may be questionable: 1) the concordance of the empirical model and data is really rather poor, rather than "very good indeed." 2) if that model is used, then the coefficient of the term in T - x_3 - appears not to differ significantly from zero, if $Var(\hat{\beta})$ is inflated by the dispersion factor. However, as we shall see presently, the empirical model is simply inappropriate for these data. It represents a misspecification of the prior information on the dimensions of the model matrix, X, since the effect of T on response is simply misstated: "... for if it is known a priori [and here it is] that a response to a stimulus x tails off beyond a certain level which is well within the range of x in the data, then a linear term only in x will not be adequate for the model." (McCullagh and Nelder 1983).

As well as a priori misspecification of the effect of T, there are also errors in the design of the experiment which will cause these data to misrepresent the effect of T on the response. One reason that T appears to have no notable or possibly no statistically significant effect on the response is, in part, the result of an improper design of the experiment which is disclosed by the collinearity diagnostics for this model of these data. These are presented in Table 3. There is, obviously, a high degree of correlation between all three pairs of variables. Therefore, $\text{Det}[(X'X)]$ is far from a maximum. The largest correlation is that between x_2 and x_3. This pair also has the largest projections on the eigenvector, V_3, of the smallest eigenvalue, λ_3. The VIFs for β_2 and β_3 each exceed 10 indicating that the estimates of these coefficients may be significantly affected by the collinearity. See Snee 1973. In particular, the multicollinearity is large enough to significantly inflate the variance of $\hat{\beta}$ for the empirical model.

We recall that the data in the non-experimental study described in Lecture 13 was similarly afflicted by the strong multicollinearity in N and T (or, x_2 and x_3), owing to the presence of the (socio-economic) constraint, $7N-5T = 0$, for system S_1 in region R_2. However, there seems to be no good reason for that constraint to have been incorporated into the design of the present experiment. One can only speculate on possible motivations for this curious feature of the design.

Let us next examine the regression diagnostics, t_i, h_i and D_i^1 for this empirical, response surface, model. The scattergram of the joint distribution of the key diagnostics t_i and h_i is presented in Figure 7a. Contrary to the cognate plot for the empirical model of the non-experimental data in Lecture 13 we see that a strong __association__ between t_i and h_i is present: there is a tendency for large (absolute) values of t_i to be associated with large values of h_i; the correlation coefficient of $|t_i|$ and h_i is $r = 0.77$. This will lead to large values of D_i^1. (See Table 1) The cognate plot for the Finney experiment in Figure 7b shows __no__ correlation between t_i and h_i.

The index plot of the residuals t_i is presented in Figure 8a. There are 6 outliers - observations for which $|t_i| > 2.0$. These are #s 3, 18, 23, 28, 38, 39. Four of these - #s 3, 18, 23 and 39 - are those observations for which $0 < r_i/n_i < 1.0$. The index plot of t_i for the Finney experiment is presented in Figure 8b; no "outlying" responses are disclosed.

It is evident in the Figure 8a that the residuals, t_i, are __serially correlated__. That is, responses for which $t_i > 0$, $i = 1, \ldots, 40$, tend to occur together. Likewise, responses for which $t_i < 0$ tend to occur together, so that $E(t_i t_j) \neq 0$. The presence of correlation in the residuals means that the sample estimates, $\text{Var}(\hat{\beta}_j)$, are __deflated__ (and hence the precision of the sample estimates, $\hat{\beta}_j$, are __overstated__). See Montgomery and Peck 1982. It is also suggested that there is additional information on the response that is not included in the model - as a result of an unidentified but important treatment variable being omitted from the model matrix, X. The parabolic pattern of the index plot of residuals suggests that a second-order term in $\underline{X}_1$, $\underline{X}_2$ or $\underline{X}_3$ could be usefully included in the model matrix, X, since each of these variables increases with index number. See Figures 3a and 3b.

Figure 8a provides a splendid instance of sample evidence (in the pattern of the residuals) for a model - here M_1 - that failed to "capture" __all__ of the information on the conditional response that was present in the sample. Some of that information has obviously "leaked" into the residuals; M_1 has __failed__ to map the data into a "white noise" vector, $\underline{\varepsilon}$ (See Box, Hunter and Hunter 1978, and Lecture 13.)

We recall that the Herring (1980) paper stated that the mechanistic model in which the dose D was included in a quadratic, D^2/N, as well as a linear, D, term, fit the data better than did the empirical model in which there were no terms of second degree. Thus, the most likely candidate is x_1^2. (But see below.) The Figure 8a also suggests that the lack of fit of M_1 is systematic as well as random and

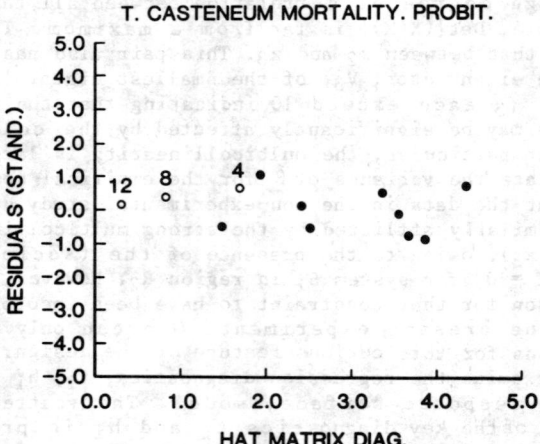

Figure 7b. Scattergram of the distribution of key diagnostics, t_i and h_i, for first-order response surface model of <u>mortality</u> experiment. Note that t_i and h_i have a small correlation, $r = -0.22$.

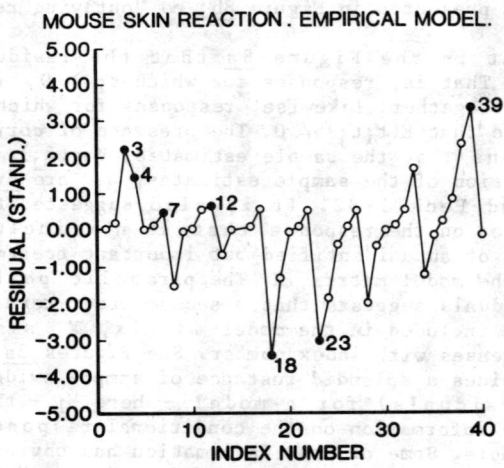

Figure 8a. Index plot of the distribution of residuals, t_i, for response surface model, M_1, of <u>skin reaction</u> experiment. Note the <u>presence</u> of a (second-order) <u>pattern</u> and evidence of (positive) <u>serial correlation</u>. A <u>runs test</u> of serial correlation in the residuals (Draper and Smith 1981) gives $z_0 = -1.48$ where z_0 is a unit Normal deviate: $P(z \leq z_0) = 0.07$.

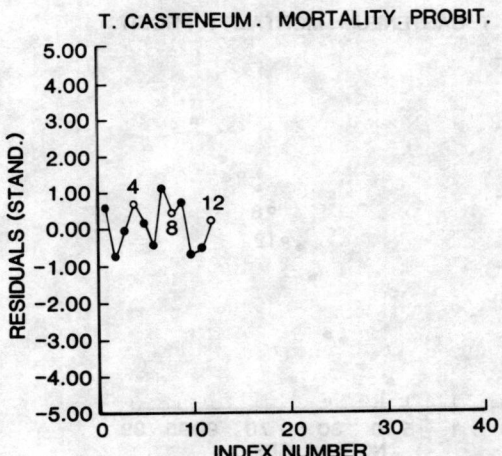

Figure 8b. Index plot of the distribution of the residuals, t_i, for response surface model of <u>mortality</u> experiment. Note the <u>absence</u> of any <u>pattern</u> or evidence of <u>serial correlation</u>.

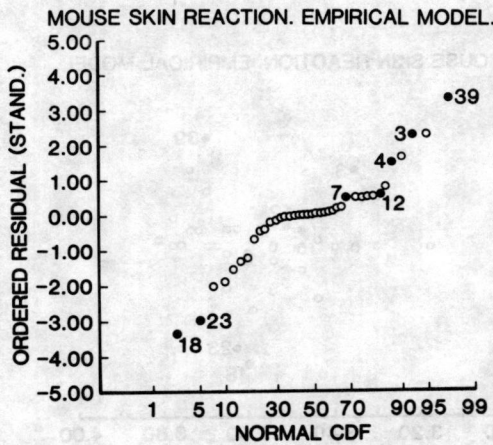

Figure 9a. Normal probability plot of residuals, t_i, for response surface model, M_1, of <u>skin reaction</u> experiment. The distribution is obviously <u>non-Normal</u>.

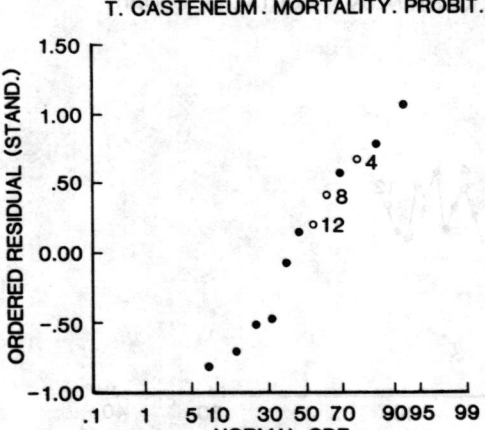

Figure 9b. Normal probability plot of residuals, t_i, for response surface model of <u>mortality</u> experiment. The distribution is obviously Normal.

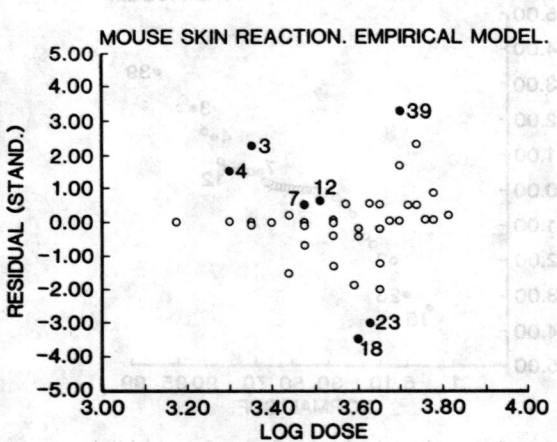

Figure 10a. Scattergram of the distribution of residuals, t_i, over $\log D_i$ for model M_1 of <u>skin reaction</u> experiment. There is evidence of a (second-order) <u>pattern</u> dominated by those 7 regimens for which $0 \leq r_i/n_i < 1.0$. (See <u>Note 5 Added in Proof</u>.)

thus the previous correction for over-dispersion may be questioned. However, the presence of so many extreme (r_i/n_i = 0 or 1) responses - 33 out of 40 - strongly suggests that the variance of the response exceeds that of a "pure" binomial distribution. We shall return to this below.

The cumulative distribution of t_i for the mouse-skin experiment is presented in Figure 9a. The plot is consistent with a heterogeneous distribution - say, a super-position of two Normal distributions with different location parameters but similar dispersions. Alternatively, it is also consistent with a homogeneous but "light-tailed" distribution, such as the Cauchy distribution. The 6 outliers are in the tails of the respective cumulative distribution. The cumulative distribution of t_i for the toxicology experiment is presented in Figure 9b. The plot is consistent with that for a single, homogeneous, Normal distribution.

The respective Filliben correlation coefficients, r_F, (Filliben 1975; See also Lecture 13) are 0.955 and 0.984. On the basis of the sampling distributions for r_F for n = 40 and n = 12, we reject H_0 for the model of the mouse-skin experiment and accept it for the toxicology experiment. (The use of r_F as an adjunct to the probability plot is recommended whenever n < 50. (See Seber 1977))

The residuals, t_i, for the empirical model of the mouse-skin experiment are plotted against x_1 in Figure 10a. There is clearly a quadratic pattern in the distribution, suggesting that a model which includes second degree term in dose, say x_1^2, should be considered. The residuals, t_i, for the Finney experiment are plotted against x_1, in Figure 10b. There is the suggestion of a similar quadratic pattern in the distribution, but since there are no outliers - $|t_i| \leq 2.0$, i = 1, ..., 12 - there is no compelling evidence for considering the addition of a second-order term to the model.

Plots of the residuals, t_i, against the estimated response, $\hat{z}_i$, for the mouse-skin and Finney experiments are presented in Figure 11a and 11b, respectively. The 7 observations for which $0 \leq r_i/n_i \leq 1.0$ in the mouse-skin experiment are superposed on the double-bow pattern which is typical for a binary response. See Lecture 13. Note again that the largest (absolute value) residuals occur at intermediate levels of response since the variance of a proportion, r/n, is greater in the vicinity of r/n = 0.5 than in the vicinity of r/n = 0 or 1. (See Figure 6b.)

An index plot of the hat matrix diagonals, h_i, for the empirical model of the mouse-skin experiment is presented in Figure 12a. There are 7 "high leverage" - $h_i \geq$ 2m/n = 0.20 - observations. These are observation #s 3, 18, 19, 33, 34, 38 and 39. Three of these, #s 3, 18 and 39 are among the 7 observations for which $0 < r_i/n_i < 1.0$. An index plot of the hat matrix diagonals, h_i, for the Finney experiment is presented in Figure 12b. There are no "high leverage" observations present: $h_i < 2m/n = 0.50$, i = 1, .., 12. But it is of interest to note that the h_i have a bimodal distribution in Figure 12b. The larger h_i are for extreme treatment levels, x_i'. The smaller h_i are for treatment levels near the centroid of the model matrix or for those levels for which r_i/n_i = 1.0. See Figures 2b and 2d.

Note also in Figure 12a that several h_i = 0, although for Normal theory models we have $1/n < h_i < 1/c$ for models which include a constant term. c is the number of times x_i' is replicated. In this model h_i is a measure of both the weight of the observation as well as its distance from the centroid of X; the lower bound is zero since the

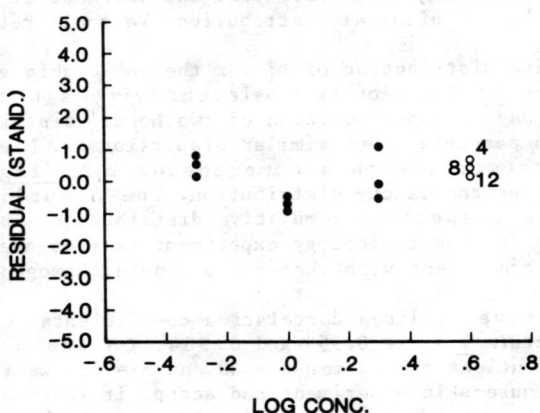

Figure 10b. Scattergram of the distribution of residuals, t_i, over logConc$_i$ for mortality experiment. There is no convincing evidence of a pattern.

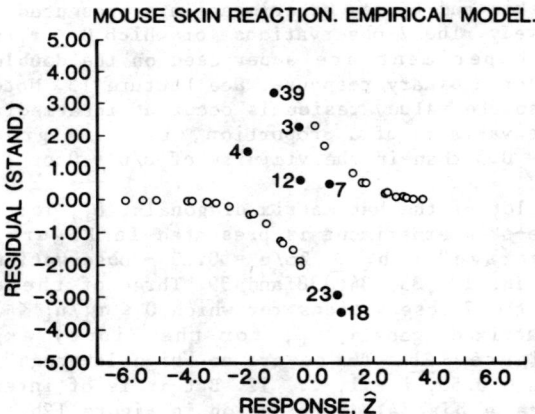

Figure 11a. Scattergram of the distribution of residuals, t_i, over estimated response, $\hat{z}_i$, for model M_1 of skin reaction experiment. Note that the residuals are largest in the vicinity of $\hat{z}_i = 0$ ($\pi_i = 0.50$) as is characteristic of a Binomial distribution.

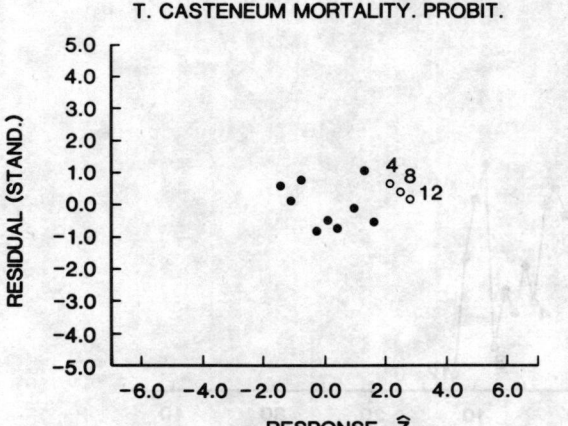

Figure 11b. Scattergram of the distribution of residuals, t_i, over estimated response, $\hat{z}_i$, for response surface model of mortality experiment. Note that the residuals are largest in the vicinity of $\hat{z}_i = 0$ ($\pi_i = 0.50$) as is characteristic of a Binomial distribution.

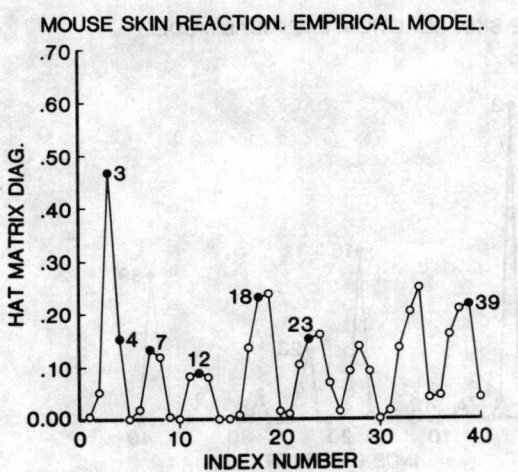

Figure 12a. Index plot of the distribution of hat matrix diagonals, h_i, for model M_1 of skin reaction experiment. Note that at 7 of the regimens $h_i > 2k/n = 0.20$. This is inconsistent with one of the desiderata of good experimental design. Note that since the single dose experiment includes both an outlying response (large t_3) and an extreme $\underline{x}_3'$ (large h_i) it will dominate $\hat{\underline{\beta}}$.

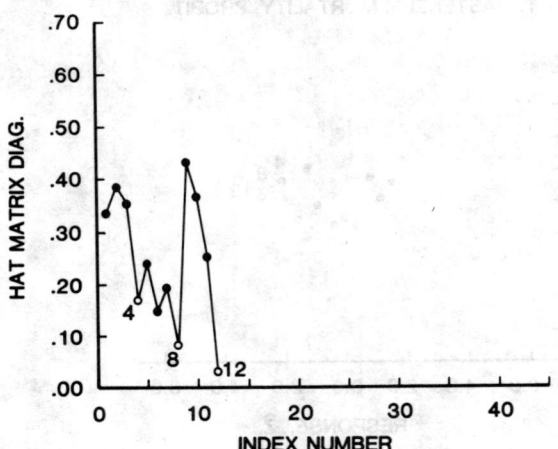

Figure 12b. Index plot of the distribution of hat matrix diagonals, h_i, for response surface model of mortality experiment. Note that at all of the regimens $h_i \leq 2k/n = 0.50$, as required for a well-designed experiment. See Table 8.

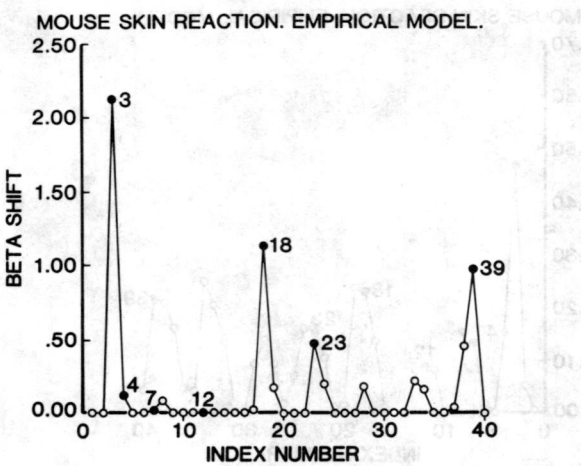

Figure 13a. Index plot of the distribution of $Q(\hat{\beta} - \hat{\beta}_{(i)}) = D_i^1$ (one-step approximation to Cook's D_i) for model M_1 of skin reaction experiment. Note that estimates of β are dominated by treatment regimens #3, #18, #23 and #39. $Q(.)$ denotes a quadratic form.

weight of extreme response is zero.

Consider now the β-shift diagnostics, $D_1{}^i(X'\hat{V}^{-1}X, m)$. It will be recalled that this diagnostic is a measure of the perturbation, $\hat{\underline{\beta}} - \hat{\underline{\beta}}(i)$, in the estimates, $\hat{\underline{\beta}}$, which is produced by the deletion of the ith case to give the estimates, $\hat{\underline{\beta}}(i)$. Since $D_1{}^i = t_i{}^2/m[h_i/(1-h_i)^2]$, both "outliers" and "high leverage" points and affect the estimates, $\hat{\underline{\beta}}$. In Figure 13a there are three observations for which the beta shift exceeds 1.0 ! These are observations #s 3, 18 and 39. Referring to Figures 8a and 12a we conclude that observation #s 3, 18 and 39 shift $\hat{\underline{\beta}}$ because they are "outliers" - large t_i. However, observations #s 18 and 39 are also "high leverage" points - large h_i. This is an unusual circumstance in which an observation is both an "outlier" and a "high leverage" point owing to the association described in Figure 7a.

Note also in Figures 8a and 12a that $t_3 > t_4$ and $h_3 > h_4$, respectively. The result of this is that $D_3{}^1 \gg D_4{}^1$. This is an example of the dominance, or "masking", of the effect on the sample estimate, $\hat{\underline{\beta}}$, of one observation, here #4, by the presence of another in close proximity, here #3. See Figures 4a and 4b. (See Belsley, Kuh and Welsch 1980; Cook and Weisberg 1982.)

Figure 13b presents the index plot of $D_i{}^1$ for the Finney experiment. We see that $D_i{}^1 \ll 1.0$ for all values of i. It should be noted in Figure 13a that of the 40 observations in the mouse-skin experiment, 6 have $D_i{}^1 = 0$. Thirteen more have $D_i{}^1 \leq 10^{-3}$. Thus, nearly half of the observations have little influence on the estimates, $\hat{\underline{\beta}}$, for this model! It is therefore evident that the estimates $\hat{\underline{\beta}}$ for the empirical model are dominated by the observation #s 3, 18 and 39.

On the evidence of the regression diagnostics, especially Figures 8a and 10a, we are led to entertain a modification of the empirical model that was not considered in Herring (1980): We include a term in $x_1{}^2$. Thus, we consider the model, $z = \beta_0 + \beta_1 x_1 + \beta_2 x_2 + \beta_3 x_3 + \beta_4 x_1{}^2$. The results of fitting this model are presented in Table 4. The concordance is much better: RSS = 40.56 $\simeq$ f = 35. Moreover, $\Delta\overline{RSS}$ = RSS (36) - RSS(35) $\gg$ 3.82, $\Delta f = 1$, indicating that the additional term is highly "significant". However, we also note that the signs of $\hat{\beta}_0$ and $\hat{\beta}_1$ are reversed from those of the previous empirical model. Note as well that $\hat{\beta}_4 > 0$, implying that the trace of the quadratic form in the z-x_1 plane is concave up: $\partial^2 z/\partial x_1{}^2 = 2\beta_4 > 0$. That is, as would be expected from the data, the response increases, "faster than linear", with the radiation dose. In each of the first order response surface models hitherto formulated, the coefficients, β_0 and β_i, i = 1, 2, 3 are the intercept and slopes, respectively, at the origin: $x_1 = x_2 = x_3 = 0$, which is remote from the centroid, $\overline{x}_1$, $\overline{x}_2$, $\overline{x}_3$, of the sample observations - especially $x_1 \gg 0$. Thus, $\overline{x}_1$ and $x_1{}^2$ are highly correlated. (Snee 1973) Therefore, in the case of polynomial models it is fruitful to transform the variables. The transformation is linear: $x_{ij} \longrightarrow (x_{ij} - \overline{x}_i)/s_i$; i = 1, 2, 3 and j = 1, ..., n = 40. $\overline{x}_i$ and s_i are the mean and standard deviation, respectively, of x_i. That is, the variables are standardized: centered and scaled. The transformation reduces the degree of correlation between $\hat{\beta}_0$ and $\hat{\beta}_1$ and between $\hat{\beta}_1$ and $\hat{\beta}_4$. The latter effect is shown dramatically in Figure 14. Figure 14a shows the correlation between x_1 and $x_1{}^2$. For these data the correlation coefficient is r = 0.99. Figure 14b shows the correlation between the centered variables, $(x_1 - \overline{x}_1)$ and $(x_1 - \overline{x}_1)^2$ where $\overline{x}_1$ = 3.566: r = -0.40. Note that the non-zero correlation coefficient r = -0.40, is due entirely to the extreme position of treatment level, #5, under the transformation, $x_1 \longrightarrow (x_1 - \overline{x}_1)$. If the distribution of x_1 were badly

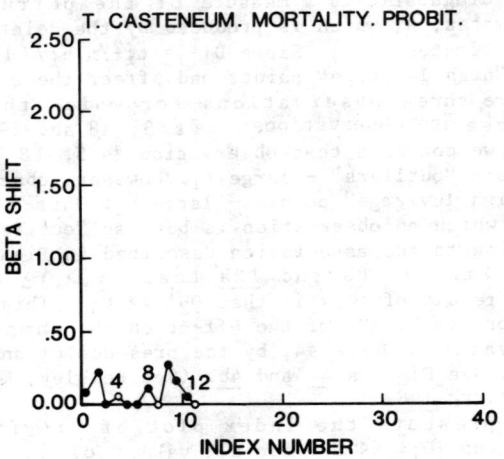

Figure 13b. Index plot of the distribution of $Q(\hat{\underline{\beta}} - \hat{\underline{\beta}}_{(i)}) = D_i^1$ for response surface model of <u>mortality</u> experiment. Note the absence of any dominant observations.

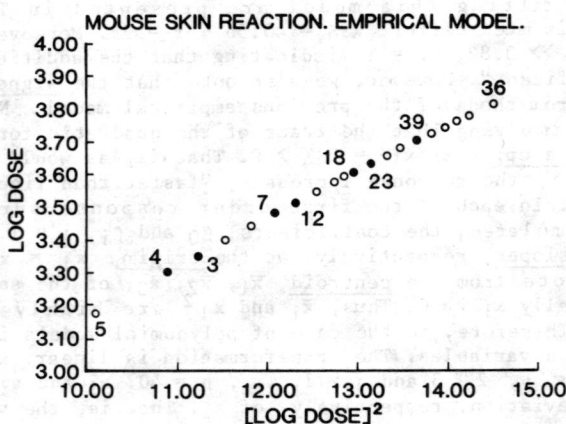

Figure 14a. Scattergram of the distribution of treatment regimens for <u>skin reaction</u> experiment in $\log D - (\log D)^2$ plane. Note the <u>strong</u> collinearity, r = <u>0.99</u>.

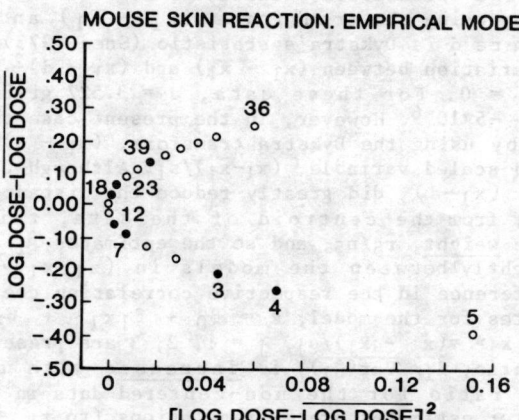

Figure 14b. Scattergram of the distribution of treatment regimens for skin reaction experiment in $(\log D - m) - (\log D - m)^2$ plane where $m = \overline{\log D}$. Note the weak collinearity for the centered variables, $r = -0.40$.

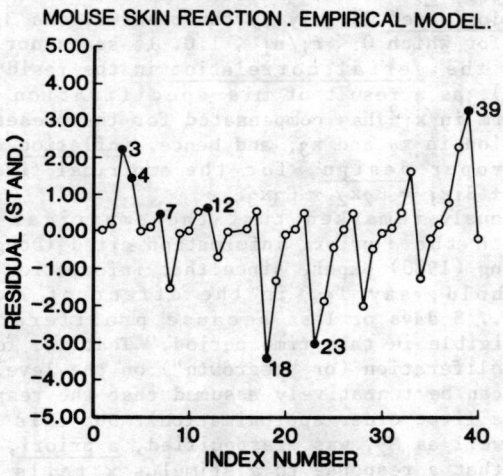

Figure 15a. Index plot of the distribution of residuals, t_i, for response surface model, M_1, of skin reaction experiment. Note the presence of a (second-order) pattern and evidence of (positive) serial correlation.

skewed, then the correlation between the first and second order terms can be reduced by using the transforms $(x_1 - \bar{x}_1)$ and $(x_1 - d)^2$ as variables, where d is Dykstra's statistic (Snee 1973) and is selected to make the covariation between $(x_1 - \bar{x}_1)$ and $(x_1 - d)^2$ zero: $\Sigma(x_{1i} - \bar{x}_1)(x_{1i} - d)^2 = 0$. For these data, d = 3.527 giving a correlation coefficient, r = $-5*10^{-6}$. However, in the present case there is little to be gained by using the Dykstra transform, $(x_1 - d)$, and so we used the centered and scaled variable, $(x_1-\bar{x}_1)/s_1$. Although the transformation, $x_1 \longrightarrow (x_1-d)$, did greatly reduce the distance of the extreme observation, #5, from the centroid of the data, this observation carries little weight, r_5/n_5, and so the estimates $\hat{\beta}_1$ and $\hat{\beta}_1/\sqrt{Var(\hat{\beta}_1)}$, differ only slightly between the models in $(x_1-\bar{x}_1)^2$ and $(x_1-d)^2$, despite the difference in the respective correlation coefficients.

The estimates for the model, $z = \beta_0 + \beta_1 x_1^* + \beta_2 x_2^* + \beta_3 x_3^* + \beta_4 x_1^{*2}$, where $x_j^* = (x_j - \bar{x}_j)/s_j$, j = 1, 2, 3 are presented in Table 4. Note that the ratio, $\hat{\beta}_1/\sqrt{Var(\hat{\beta}_1)}$, is increased with respect to the corresponding ratio for the non-centered data in Table 4. This is because the latter estimates are extrapolations (to $x_1 = x_2 = x_3 = 0$) whereas, those for x_1^* are obtained at the centroid, $(\bar{x}_1, \bar{x}_2, \bar{x}_3)$ of the data and hence, greater precision of estimation is achieved thereby (Snee 1973).

Plots of the regression diagnostics, t_i, h_i and D_i^1 for this second-order model are presented in Figures 15b, 15c and 15d. Figure 15a is the index plot of the residuals for the model matrix, X = ($\underline{1}$, $\underline{X}_1$, $\underline{X}_2$, $\underline{X}_3$), in Table 4. Note that the second-order pattern (and thus the serial correlation) of index plot of the residuals is much diminished between Figures 15a and 15b. (However, one could make a case for the presence of a "cubic" pattern in the residuals.) Moreover, the coefficient estimates, $\hat{\beta}$, for the quadratic model as for the linear model, are still dominated largely by the observations 3, 4, 7, 12, 18, 23 and 39 - those for which $0 < r_i/n_i < 1.0$. It seems not unlikely that the presence of the serial correlation in the residuals, and hence deflation of $Var(\hat{\beta})$, as a result of mis-specification of the model (absence of a term in x_1^2) has compensated for the presence of the high degree of correlation in x_2 and x_3, and hence, inflation of $Var(\hat{\beta})$, as a result of improper design, for the empirical first-order Taylor series M_1: $z = \beta_0 + \beta_1 x_1 + \beta_2 x_2 + \beta_3 x_3$.

We have previously remarked that the empirical model M_1 was inconsistent with the a priori information cited (Douglas and Fowler 1976) in the Herring (1980) paper, since that information describes an effective threshold, say T_0, in the effect of T on the observed response. T_0 is "... 8 days or less because proliferation has been shown to be negligible in this time period." That is, for $T < T_0$ there is no effect of proliferation (or "regrowth") on the level of response z. For $T \geq T_0$ it can be tentatively assumed that the response is linear in T (at least to a first-order approximation). But this implies that the model M_1, as well as M_2, was misspecified, a priori, for "... if it is known a priori that a response to a stimulus x tails off beyond a certain level which is well within the range of x in the data, then a linear term only in x will not be adequate for the model." (McCullagh and Nelder 1983) (M_2 was mis-specified a priori since it omitted the effect of T altogether. vide supra.) From the foregoing observations and remarks, one might expect that the adequacy of the model M_1 can be improved by including a quadratic, as well as linear, term in x_3. There is some evidence in the sample for this expansion of the model. We note that from the design, T as well as D, increases with index number.

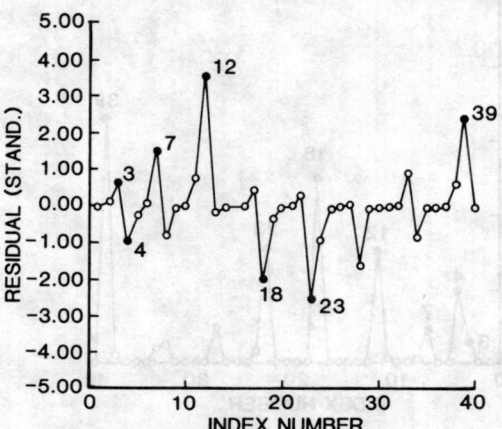

Figure 15b. Index plot of the distribution of residuals, t_i, for response surface model, M_{1i}, of skin reaction experiment. Note the absence of any pattern and of any evidence for serial correlation. However, there are still several "outliers", $t_i > 2.00$.

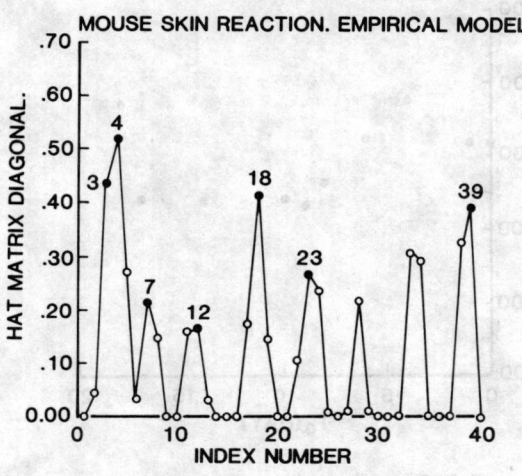

Figure 15c. Index plot of the distribution of hat matrix diagonals, h_i, for response surface model, M_{1i}, of skin reaction experiment. there are 8 treatment regimens for which $h_i > 2k/n = 0.25$. This is inconsistent with one of the desiderata of good experimental design.

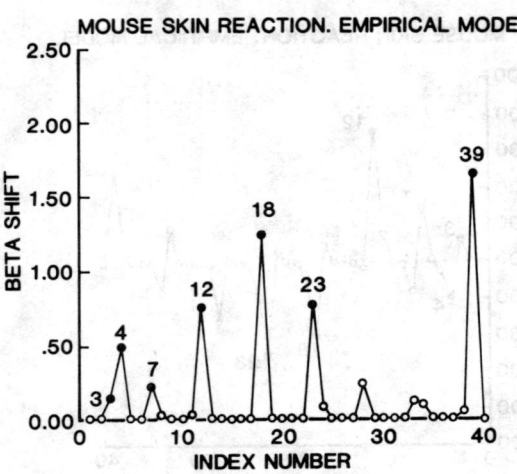

Figure 15d. Index plot of the distribution of $Q(\hat{\underline{\beta}} - \hat{\underline{\beta}}(i)) = D_i^1$ for response surface model, M_{1i}, of skin reaction experiment. Note that the estimates of $\underline{\beta}$ are dominated by 6 of the treatment regimens.

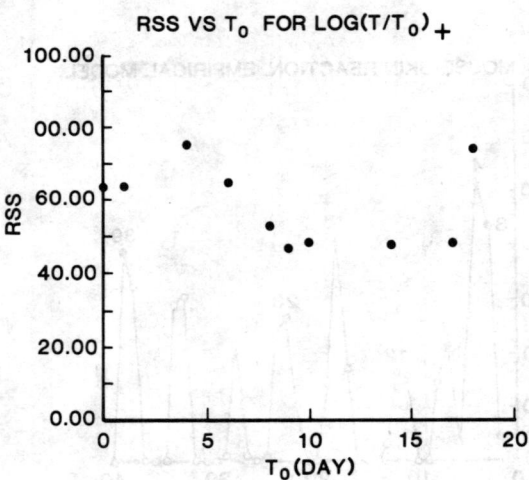

Figure 16. Plot of RSS vs "knot", T_0, for linear spline model, M_{1iv}, of skin reaction experiment. The minimum RSS occurs at $T_0 = 9$.

Moreover, T increases <u>monotonically</u> with the index number. See Figures <u>3a</u> and <u>4b</u>. Thus, the quadratic pattern in the residuals for M_1 presented in Figure <u>15a</u> could be interpreted as a corroboration of the a priori evidence of Douglas and Fowler (1976) for a response that includes both linear and quadratic terms in T (or x_3).

Therefore, we examine the model, $z = \beta_0 + \beta_1 x_1 + \beta_2 x_2 + \beta_3 x_3 + \beta_4 x_3^2$. The results are presented in Table <u>4</u>. Since the centroid, $\bar{x}_3 = 0.743$, is near the origin, $x_3 = 0$, the data are not centered. The addition of the term in x_3^2 seems to have improved the "fit" significantly: RSS = 45.45, f = 35; for M_1, RSS = 63.76, f = 36, ΔRSS = 18.31, Δf = 1. Since ΔRSS > 3.84, the improvement is significant at P = 0.95.

However, there is another interpretation of the a priori evidence of Douglas and Fowler (1976) and the argument of McCullagh and Nelder (1983). The a priori information that proliferation is negligible for $T < T_0 \sim 8$ days can be interpreted as evidence for the presence of an inherent, hitherto ignored, <u>biological</u>, constraint in the system S_2 in the region R_4.

We have previously remarked that the design of the mouse-skin experiment includes the (socio-economic) constraint on <u>treatment variables</u> N and T that was described in the non-experimental clinical data of Lecture 13: 7N - 5T = 0 ($R^2 = 0.82$). It also includes a version of the biological constraint on the <u>treatment variables</u> D and T, D = $2210 T^{0.30}$, ($R^2 = 0.70$), that was described for the latter study. (This has the force of a <u>medical-legal</u> constraint since it is believed by some to describe the relation between levels of D and T required to evoke "tolerance" in the connective tissues of the irradiated patient.) However, the present biological constraint is on the <u>continuity</u> of the <u>response</u>, z, and its first <u>derivative</u>, with respect to T (or x_3):
$$\partial z / \partial T = 0, \ T < T_0; \ \partial z / \partial T \neq 0, \ T \geq T_0 \qquad (8)$$
The response is <u>continuous</u>, the first derivative is <u>discontinuous</u>, at $T = T_0$. Note that for the model that includes the quadratic term, x_3^2, the first derivative is <u>continuous</u> at $T = T_0$ - and everywhere else. (This is characteristic of <u>polynomials</u>, of course.)

As we observed earlier, model M_1 in which the response and each of its first derivatives are everywhere <u>continuous</u> represents a consensus. It is the received wisdom derived from a <u>clinical</u> experience with system S_1 over region R_1. The inherent constraints on the derivatives of the response surface are $\beta_1 > 0$, $-1 < \beta_j / \beta_1 < 0$. That the mechanism of response apparently "<u>shifts</u>" between two different systems, S_1 and S_2, in two different regions, R_1 and R_4, should not be altogether unexpected, especially when one region, R_4, is essentially a boundary region for two of the variables (N and T). ("What is often overlooked in the application of regression models is their great dependence on the <u>validity</u> of the model employed, generally one of a linear nature." (Mantel 1983). Italics added.

The postulated "shift" in the mechanism in this case is expressed by the discontinuous change in β_3 at $T = T_0$, that is, $\beta_3 = \beta_3(T)$, i.e., a constant that is not "constant". Such parameter shifts are not without precedent. For instance it is a commonplace in chemical kinetics, that rate "constants" are functions of such treatment variables as concentration and such covariates as temperature and catalyst.

Thus, we are led to consider one member of the class of empirical basis functions known as spline functions - the <u>linear</u> spline, or "broken-stick" function. Although this function is also discussed in

Lecture 14, it is worth-while to repeat in the present context some of those more general remarks anent spline functions. It should be recalled from Lecture 13, that the inherent constraints on the range of the parameter, θ, of the binomial distribution, $0 \leq \theta \leq 1$, led us to introduce the probit transform, z, of the response, y (See Figure 1 of Lecture 13). Here, inherent constraints on the continuity of that transform and its first derivative lead us to a transform on T – or x_3.

Now for those general remarks:

1) "Data which appear to behave according to two different straight-line relationships on opposite sides of an undertermined joint point occur frequently in a wide variety of contexts." (Watts and Bacon 1974).

2) "In many situations, approximations of a set of data by a polygonal curve is more advantagous than approximation by a polynomial." (Bellman and Roth 1969).

3) "... ordinary polynomials are inadequate in many situations. This is particularly the case when one approximates functions which arise from the physical world rather than the mathematical world. Functions which express physical relationships are frequently of a disjointed or disassociated nature. That is to say that their behaviour in one region may be totally unrelated to their behaviour in another region. Polynomials along with most other mathematical functions have just the opposite property. Namely, their behaviour in a small region determines their behaviour everywhere. Splines do not suffer this handicap since they are defined piecewise ..." (Wold 1974).

4) "Splines are generally defined to be piecewise polynomials of degree n whose function values and first n-1 derivatives agree at the points where they join. The abscissas of these join points are called knots. Polynomials may be considered a special case of splines with no knots ..." (Smith 1979).

The general problem which we consider is that the form of the regression is fixed but one or more components of the parameter vector may change discontinuously between two (connected) regions of the space of treatment variables. A simple example will illustrate the problem. Let us assume that the model is $y = \beta_0 + \beta_1 x$ for $x < x_0$ and $y = \beta_0^* + \beta_1^* x$ for $x \geq x_0$, $\beta_j^* \neq \beta_j$, $j = 0, 1$. Thus, the parameter vector is a discontinuous function of the treatment variable x: $\underline{\beta} = \underline{\beta}(x)$. (We recall in Lecture 13 that the parameter vector may be a discontinuous function of the covariate(s), U: $\underline{\beta} = \underline{\beta}(U)$. That is, $\underline{\beta}$ changes discontinuously between prognostic strata.) If $\beta_0^* = \beta_0$ and $\beta_1^* \neq \beta_1$, then the function is continuous and the first derivative, $dy/dx = \beta_1$ is discontinuous at $x = x_0$.

This may be conveniently described in terms of the linear (n=1) spline functions, $(x-x_0)_+$ and $(x-x_0)_+^0$:

$$y = \beta_0 + \beta_1(x-x_0)_+ + \beta_{10}(x-x_0)_+^0. \tag{9}$$

where

$$(x-x_0)_+ = \begin{cases} x-x_0 & x \geq x_0 \\ 0 & x < x_0 \end{cases} \tag{10}$$

and

$$(x-x_0)_+^0 = \begin{cases} 1 & x \geq x_0 \\ 0 & x < x_0 \end{cases} \tag{11}$$

If $\beta_1 \neq 0$ and $\beta_{10} = 0$, then the function is continuous and the first derivative is discontinuous at $x = x_0$. Of course, $\underline{\beta}$ changes discontinuously. (See Lecture 13 for the discussion of the regression of z on the dichotomous, S = 0, 1, variable of stage of disease.)

We may represent the linear spline as a transformation of the variable x, x ⟶ $(x-x_0)_+$, that is quite analogous to the transformations, x ⟶ logx, or, x ⟶ x^2.

For the present mouse-skin experiment we consider the model, z = $\beta_0 + \beta_1 x_1 + \beta_2 x_2 + \beta_3 x_3' + \beta_4 x_0'$, where

$$x_3' = \log(T/T_0)_+ = \begin{cases} \log(T/T_0) & T \geq T_0 \\ 0 & T < T_0 \end{cases} \quad (12)$$

and

$$x_0' = \begin{cases} \log 10 & T \geq T_0 \\ \log 1 & T < T_0 \end{cases} \quad (13)$$

The results for T_0 = 4 and 9 are presented in Table 4.

We find that for T_0 = 4, both the response and its first derivative, β_3, are discontinuous at T_0, that is, both $\hat{\beta}_3$ and $\hat{\beta}_4$ are significantly greater than their respective standard errors. However, for T_0 = 9, only $\hat{\beta}_3$ exceeds its standard error by a significant amount. Although, the spline with T_0 = 4 fits these data much better than that for T_0 = 9, the latter seems to represent with greater fidelity the inherent biological continuity constraint that appears to be described in the Dogulas and Fowler (1976) paper. It is, of course, possible that more than one constraint on the continuity of the response and its derivatives should be included in the model. However, the ill-conditioning of the design matrix increases with the number of "knots" (Smith, 1979). It is also possible that the discontinuity in the function as well as in the first derivative is inherent rather than contingent. However, inherent, discontinuities in the level of response seem much less likely, a priori, than discontinuities is its rate of change. On the other hand discontinuities in the level of response are quite often a consequence of changes - "slippage" - in the external conditions of an experiment and we have remarked earlier some of the peculiarities in the distribution of response in the present experiment, which are consistent with - if not suggestive of - some "slippage". Again, the reader is reminded that we are only exploiting the heuristic value of the Herring (1980) and thus we beg further questions in this matter.

Although an a priori estimate of the approximate position of the knot is available, it was estimated from the sample to be T_0 = 9 by plotting the RSS of the model M_{1ii} against T_0. This plot is presented in Figure 16: The minimum RSS is at T_0 = 9. (Recall that in Lecture 14 the position of the "threshold" gamma dose, D_0, for radioleukemogenesis was estimated by a similar procedure. The procedure used there to determine confidence limits on D_0 could also obviously be similarly employed here but it is an unnecessary exercise.) Note that T_0 = 0, RSS = 63.76, corresponds to X = (1, X_1, X_2, X_3) and T_0 = 18, RSS = 74.46, corresponds to X = (1, X_1, X_2), i.e., deletion of X_3.

The model z = $\beta_0 + \beta_1 x_1 + \beta_2 x_2 + \beta_3 x_3'$ was fit to the data. The results are presented in Table 4. The transformation, $x_3 \longrightarrow x_3'$, has improved on M_1 somewhat more than did the addition of the quadratic term x_3^2. That is, the aggregate statistics, RSS, PRESS and AIC are nearly as good for model M_{1iii} as for M_{1ii} and much better than for M_1 for which f = 36 also. Thus, the sample evidence concurs in the a priori evidence for the "broken-stick" function. It is also favored by considerations of "parsimony". (See Lecture 13).

The end-result of our model building on the a priori evidence of the Weber-Fechner law (Finney 1971) the empirical clinical models of Ellis (1969) and the Douglas Fowler (1976) paper is the model, M_{1iv}: z = $\beta_1(x_1-\overline{x}_1) + \beta_2 x_2 + \beta_3 x_3' + \beta_4(x_1-d)^2$. Note that this model describes a polygon in x_3 as well as a polynomial in x_1. (Belleman and Roth 1969).

The results of fitting this model to the experimental data are presented in Table 4. It fits these data more closely than any of the rivals considered so far (including the radiobiological model): RSS = 38.58. f = 35. Our construction of M_{1v} is an example of the "motivated iteration" described by Box 1976. See also Cook and Weisberg 1982.

However, there are still several responses which are not well explained by this model and several observations which dominate some aspects of the fit. Moreover, the sign of the estimated coefficient of the spline transform, $\log(T/T_0)_+$, in M_{1iii} differs from that of logT in M_1. For the former $\hat{\beta}_3 > 0$; for the latter, $\hat{\beta}_3 < 0$. The latter is consistent with a priori information on β_3. However, since the respective values of the statistic, $\hat{\beta}_3/\sqrt{Var(\hat{\beta}_3)}$, for M_1 and M_{1iii}, are -2.47 and 4.30, and since the estimates $\hat{\beta}_j$, $0 \leq j \leq 3$ are <u>aliased</u> owing to the <u>omission</u> of a second-order term (probably x_1^2), we do <u>not</u> view the sign of $\hat{\beta}_3$ for $\log(T/T_0)_+$ as a mere numerical artifact.

The knowledge that T has little or no effect on the response for T $< T_0 = 9$ helps to explain the <u>fact</u> that in the simple empirical model, M_1, Table 3a, we find that $\hat{\beta}_3/\sqrt{Var(\hat{\beta}_3)} = 0.5\ \hat{\beta}_2/\sqrt{Var(\hat{\beta}_2)}$, that is, the precision with which the weight for N is estimated is twice that with which the weight for T is estimated since the <u>effective</u> ranges for T is that of $T - T_0$: $0 < T-T_0 < 9$. However, the effective range for N is $1 < N < 15$. Or, $0 < \log(T/T_0) < 0.3010$ and $0 < \log N < 1.176$ a factor of nearly 4. (One recalls that, "The analyst should always look at the range $(x_{max} - x_{min})$ of the variables in physical units to see whether an insignificant regression coefficient may have resulted from the variable being varied over a small range" (Snee 1973) - leading to "trivial" effects on the level of response. (Marquardt 1980.) vide supra)

7. THE MECHANISTIC MODEL

Let us now examine the mechanistic or, radiobiological, model of these data, M_2: $z = \beta_0 + \beta_1 D + \beta_2 D^2/N$. It is appropriate to recall once again that there are <u>two</u> distinct kinds of regression models: 1) empirical, or predictive, models (response surface models) and 2) mechanistic, or "causal", models. In the construction of the latter we are requiring that the sample information increase our theoretical knowledge of the process that generated the sample as well as enhance our performance in forecasting a specific response. Now, "Mechanistic models, when appropriately employed, have several advantages over empirical models. They provide greater scientific insight, a better basis for extrapolation, and usually more parsimonious representation." (Box, Hunter and Hunter 1978). The conditional clause in the first sentence is of course, crucial, since it is often difficult to decide a priori when a given mechanistic model refers to the same population as does the sample; this is one aspect of the so-called "scaling" problem (or, "mouse to man" problem) discussed earlier.

Of course, if the mechanistic model is <u>incorrect</u>, that is, if it describes a population which is different from the one from which the sample of observations was drawn, then the aforesaid advantages disappear and a simple response surface model, a Taylor-series in the inputs (here the adjustable variables of treatment) is much superior.

"The set of variables that will be most appropriate for a causal regression equation is therefore not necessarily the same set that will be best for estimation [prediction] purposes." (Seber 1977) However, it is frequently the case that one or more variables selected for a <u>causal</u>

Table 5. Collinearity Diagnostics. Mouse Skin Reaction #2.
B. Radiobiological Model. $X = (D, D^2/N)$

a) Correlation Matrix: 1.0, -0.106
   1.0

b) Eigenvalue, Eigenvector
   $\lambda_1 = 1.106$, $V_1 = (-0.707, 0.707)$
   $\lambda_2 = 0.894$, $V_2 = (0.707, 0.707)$

c) Condition Number, $\kappa = \lambda_1/\lambda_2 = 1.237$
d) Proportion of Variance, $\lambda_1/\Sigma\lambda_j = 0.553$
e) $\Sigma\lambda_j^{-1} = 2.02 < 5p = 10$
f) Variance Inflation Factors, VIF: (1.011, 1.011)
g) Inflation of Joint Confidence Region for β:
   $100 \ [(\Pi\lambda_i)^{-1/2} - 1] = 0.567$ 0.6%
   Determinant of $X_0'X_0 = \Pi\lambda_i = 0.99$

Table 6. Mechanistic Model (Linear in $D, D^2/N$) of Mouse Skin Reaction #2. Probit.

a) $z = -13.04 + 1.82*10^{-3}D + 1.81*10^{-6}(D^2/N)$
   (-7.82) (7.44)   (7.57)
   RSS(37) = 45.85. PRESS = 54.332. AIC = 112.72
   $P(\chi^2 > RSS|37) = 0.151$. (Accept H_0)

b) $z = -14.34 + 1.75*10^{-3}D + 2.17*10^{-6}(D^2/N) + 0.15(T-9)_+$
   (-7.69) (6.80)   (7.43)   (2.74)
   RSS(36) = 33.89. PRESS = 45.34. AIC = 107.02.
   $P(\chi^2 > RSS|36) = 0.569$. (Accept H_0)

Table 7. Semi-Empirical Models of Mouse Skin Reaction #2. Probit. Model M_{1v}. (Quadratic in D/N)

a) $z = -25.017 + 4.137*10^{-3}D + 1.173*10^{-2}(D/N) - 2.011*10^{-6}(D/N)^2$
   (-7.132) (7.169)   (5.842)   (-3.845)

b) $z = -15.063 + 4.137*10^{-3}D + 5.781*10^{-3}(D/N)* - 2.011*10^{-6}(D/N)**2$
   (-7.168) (7.169)   (6.974)   (-3.845)
where
$(D/N)* = [(D/N) - m]$. $(D/N)**1 = [(D/N) - d]^2$.
   m - Mean(D/N). d - Dykstra's d
   RSS(36) = 32.95. PRESS = 41.54. AIC = 100.83.
   $P(\chi^2 > RSS|36) = 0.614$. (Accept H_0)

Model M_{1vi}. (Spline in T. Quadratic in D/N)

c) $z = -27.585 + 4.285*10^{-3}D + 1.458*10^{-2}(D/N) + 0.117(T-T_0)_+$
   (-6.69) (7.06)   (5.09)   (1.55)
   $- 1.849*10^{-6}(D/N)^2$
   (-3.73)
   RSS(35) = 28.71. PRESS = 41.07. AIC = 100.51.
   $P(\chi^2 > RSS|35) = 0.765$ (Accept H_0)

model on the basis of theory or conjecture can be usefully included in an _empirical_ model. We shall find this to be the case for these data.

However, we note that the _mechanistic_ model, M_2, is _inconsistent_, a priori, with a large body of _general_ knowledge on models of physiochemical mechanisms: "...usually interactions should not be included without main effects, nor higher degree terms without their lower degree relations." (McCullagh and Nelder 1983) However, the radiobiological model includes the interaction term, $D(D/N)$, but not a term representing the main effect of either, N, or, the dose per fraction, (D/N), although the main effect of dose, D, is included. And, of course, as remarked earlier, M_2 does not include a term in T which would represent the effect of proliferation, or "regrowth", on the observed response. Thus, the model is _inconsistent_ with both general and specific prior biological and clinical information. (Thus, it would seem that the prior probability, $P(M_2)$ is quite small, as noted.)

The joint distribution of the model matrix, $X = (D, D^2/N)$ is presented in Figure 17a. The number of fractions, N, is an index of the family of six diverging straight lines which describe the distribution of treatment regimens; counter-clockwise from the lower right N = 1, 2, 3, 5, 9 and 15. The number of levels of treatment is the same as for the empirical model: n = 40. (vide infra). The 7 treatment regimens for which $0 < r_i/n_i < 1$ are labelled by index number.

Note that the largest sample size, $n_1 = 14$, (see Figure 4a) is at the treatment regimen at the extreme lower right - an extreme level (large h_i) of the matrix X'X (although, as we shall see, not of the matrix, $X'\hat{V}^{-1}X$, in which it is greatly "downweighted"). The estimated equation together with the chi-squared measure of concordance are presented in Table 6. The "fit" of this model seems good, RSS = 45.85 f = 37. Moreover, each of the coefficient estimates is about 8X the respective standard error.

We were careful to point out that Table 6 is derived from the same sample as Table 4. This is a non trivial distinction between our analysis and that of Herring (1980) for a careful examination of the latter paper discloses that the analysis therein belies to some degree the previously stated intent to present "... two different time-dose models and their fit to the same set of experimental data ...". Instead, that paper compares the respective fits of the empirical model, M_1, and the mechanistic model, M_2, on two rather different sets of data. The estimates, $\hat{\beta}$, $\hat{y}$, RSS for M_2 were obtained from a sample of 36 treatment regimens that was derived from the initial sample of 40 treatment regimens (from which were obtained the cognate estimates for M_1) by pooling the responses, r_i/n_i, for common levels of (D_i, N_i), ignoring T_i: Four of the regimens for which T = 4 were pooled with 4 for which T = 9. The following example describes the method by which the respective responses were "pooled". For observation #17 (D, N, T) = (4250, 5, 4) and for observation #23, (D, N, T) = (4250, 5, 9). We find $r_{17}/n_{17} = 8/8$ and $r_{23}/n_{23} = 2/6$. In the paper, the pooled observation was (D, N, r/n) = (4250, 5, 10/14). In other words, for each of the two regimens defined by common values of (D, N) we have $r/n = (r_4 + r_9)/(n_4 + n_9)$ where r_i/n_i, i = 4, 9 are the observed responses at T = 5 and D = 9, respectively. But, as a general practice, before two sample estimates can be "pooled" it is necessary to assure their "compatibility", that is, that they are estimates of the parameter of the same population (See the statistics τ and γ in Lecture 13). If they are compatible, then they are pooled as a _weighted_ average (or a _matrix-weighted_ average if appropriate. See Lecture 13) in which the respective weights are the reciprocals of the variances of the separate

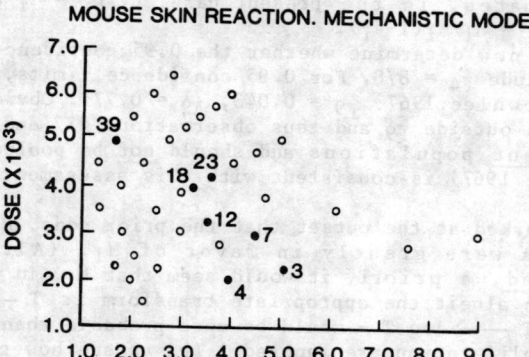

Figure 17a. Scattergram of the distribution of treatment regimens for
<u>skin reaction</u> experiment in D − (D^2/N) plane. The distribution
consists of a family of distributions indexed by values of N = 1, 2,
3, 5, 9 and 15.

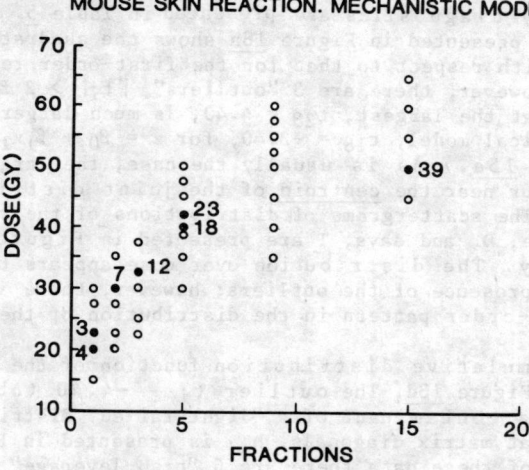

Figure 17b. Scattergram of the distribution of treatment regimens for
<u>skin reaction</u> experiment in D-N plane. The distribution consists in a
family of distributions of D indexed by values of N = 1, 2, 3, 5 and
15. Compare with Figure <u>2d</u>. It is obvious that a proper design for a
dose-response <u>surface</u> is <u>not</u> achieved by a kind of paratactic aggregation of designs for dose-response <u>curves</u>.

sample estimates. In the present case, $r_i/n_i = \hat{\pi}_i$ an estimate of $\hat{\pi}_i$ with weight $w_i = n_i/\hat{\pi}_i(1-\hat{\pi}_i)$.

Let us now determine whether the 0.95 confidence limits on π for $\hat{\pi}_9 = 2/6$ include $\hat{\pi}_4 = 8/8$. For 0.95 confidence limits, $\underline{\hat{\pi}}_9$ and $\overline{\hat{\pi}}_9$ on $\hat{\pi}_9$ we have (Brownlee 1967) $\underline{\hat{\pi}}_9 = 0.043$, $\overline{\hat{\pi}}_9 = 0.777$. Obviously, $\pi_4 = 8/8 = 1.0$, lies well outside $\overline{\hat{\pi}}_9$ and thus observations #17 and #23 are samples from different populations and should not be pooled. Fisher's exact test (Brownlee 1967) is consistent with this assessment ($P = 0.015$ on $H_0: \pi_4 = \pi_9$).

We remarked at the outset that the prior odds, $P(M_1)/P(M_2)$ of the rival models were greatly in favor of M_1. (Although both are mis-specified, a priori, it would seem that M_1, in which the effects are included - albeit the appropriate transform is $T \longrightarrow \log(T/T_0)_+$ rather than $T \longrightarrow \log T$ - would be more probable than M_2, in which the effects of proliferation are ignored.) Let us see how great they must be in order to dominate the sample evidence for M_2. (See Leamer 1978) The Bayes factor, $B = P(y|M_1)/P(y|M_2) = (RSS_2/RSS_1)^{n/2} n^{(k_2-k_1)/2} = 2.162*10^{-4}$. Then, the posterior odds are $P(M_1|y)/P(M_2|y) = BP(M_1)/P(M_2)$. In order to dominate the sample evidence for M_2, the prior odds for M_1 must exceed 4606:1.

Let us now consider the model, M_{1iii} in which $\log T$ is replaced by the "broken stick" function $\log(T/T_0)_+$ where $T_0 = 9$. For M_{1iii} we have RSS = 47.06, k* = 4 and $B = 9.391*10^{-2}$. In order to dominate the sample evidence for M_2, the prior odds for M_{1iii} need only be 11:1 - a quite plausible ratio.

We next examine the nature and degree of the model-sample interaction as disclosed by the respective regression diagnostics. The collinearity diagnostics are presented in Table 5. The index plot of the residuals presented in Figure 18a shows the quadratic pattern to be diminished with respect to that for the first-order response surface in Figure 15a. However, there are 3 "outliers", $|t_i| > 2$ for i = 4, 23 and 39. Note that the largest, $t_{23} = 4.40$, is much larger than the largest for the empirical model, $t_{18} = -3.40$, for $z = \beta_0 + \beta_1 x_1 + \beta_2 x_2 + \beta_3 x_3$. (See Figure 15a.) As is usually the case, the largest values of the residuals occur near the centroid of the joint distribution of X in Figure 17. The scattergrams of distributions of the residuals over the variables dose, D, and days, T are presented in Figures 18b and 18c, respectively. The distribution over dose appears to be featureless, save for the presence of the outliers; however, there is a suggestion of a second-order pattern in the distribution of the residuals over T (days).

The cumulative distribution function of the residuals, t_i, is presented in Figure 18d. The outlier $t_i = -4.40$ (observation #23) appears as a contaminant of a "light-tailed" distribution. The index plot of the hat matrix diagonals, h_i, is presented in Figure 18e. For this model of these data there are 6 "high-leverage" points ($h_i \geq 2k/n = 0.15$) for i = 3, 4, 7, 34, 38 and 39. Note that 4 of these are from the set of 7 points with non-zero initial weights - that is, for $0 < r_i/n_i < 1.0$, i = 3, 4, 7 and 39. As is usually the case the largest values of h_i occur at the extremes of the joint distribution of X in Figure 17. Note that the presence of #3 "masks" the effect of #4, again.

The index plot of D_i^1 for the mechanistic model, M_2, is presented in Figure 18f. There are four observations which strongly affect $\hat{\beta}$, whereas for the empirical model, there are six which do so. For the mechanistic model the largest values of D_i^1 occur at observations #23 and 39. Note that both of these belong to the set of 7 observations for

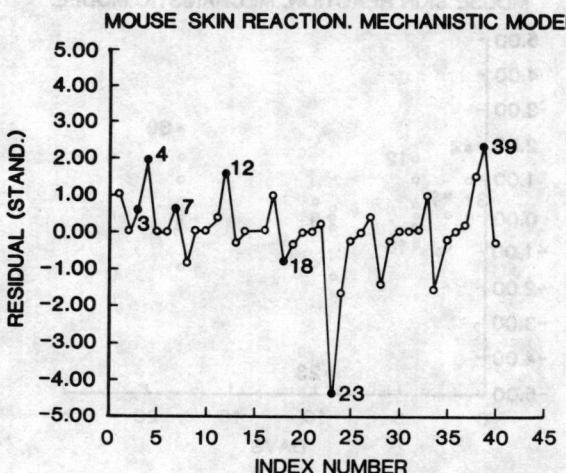

Figure 18a. Index plot of residuals, t_i, for radiobiology model, M_2, of skin reaction experiment. There are 3 "outliers", $t_i \geq 2.00$.

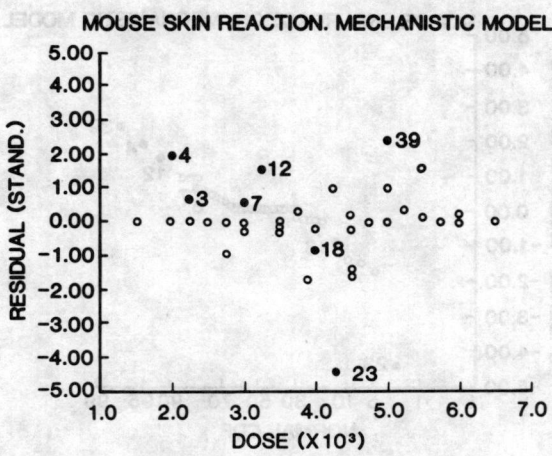

Figure 18b. Scattergram of the distribution of residuals, t_i, over dose, D_i, for model M_2 of skin reaction experiment. There is no evidence of a pattern.

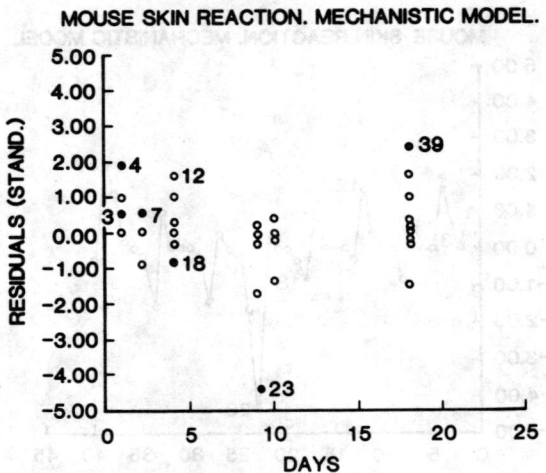

Figure 18c. Scattergram of the distribution of residuals, t_i, over time, T_i, for model M_2 of <u>skin reaction</u> experiment. There is evidence of a (second-order) <u>pattern</u> dominated by those 7 regimens for which $0 < r_i/n_i < 1.0$.

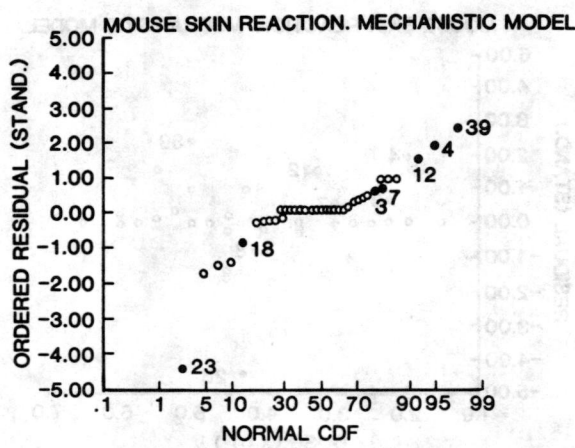

Figure 18d. Normal probability plot of residuals, t_i, for model M_2 of <u>skin reaction</u> experiment. The plot is consistent with that of a "<u>contaminated</u>" Normal distribution. the "contaminant" is treatment regimen #23.

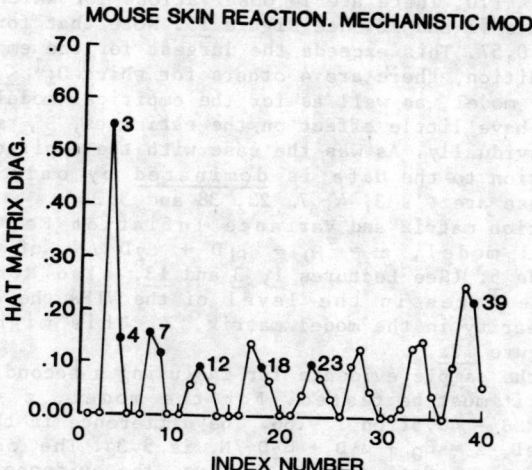

Figure 18e. Index plot of hat matrix diagonals, h_i, for model M_2 of skin reaction experiment. there are 5 regimens of which $h_i > 2k/n = 0.15$. This is inconsistent with the desiderata of a good experimental design.

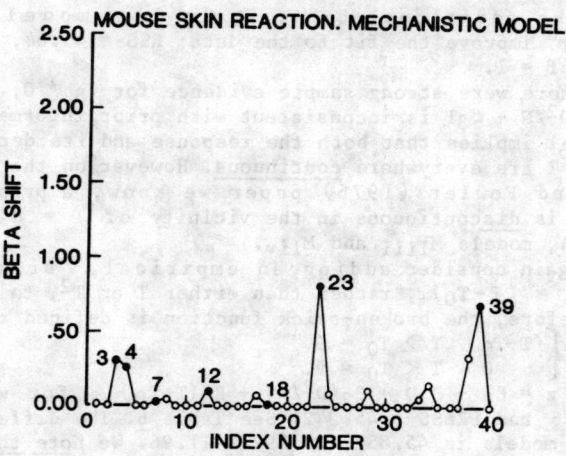

Figure 18f. Index plot of $Q(\hat{\underline{\beta}} - \hat{\underline{\beta}}_{(i)}) = D_i^1$ for model M_2 of skin reaction experiment. The estimates of $\underline{\beta}$ will be dominated by treatment regimens #23 and #39.

which $0 < r_i/n_i < 1.0$. There are 14 observations for which $D_i{}^1 = 0$. Two of these are points #1 and #2. See Figure 15. Note that for the largest of these, $h_3 = 0.57$. This exceeds the largest for the empirical model, $h_3 = 0.47$. In addition, there are 4 others for which $D_i{}^1 \le 1.0 * 10^{-3}$. Thus, for this model, as well as for the empirical model, nearly half the observations have little effect on the estimates, $\hat{\beta}$, and $\hat{z}_i$, when considered individually. As was the case with the empirical model, the fit of this equation to the data is <u>dominated</u> by only a few - 6 - observations. These are #'s 3, 4, 7, 23, 38 and 39.

The correlation matrix and Variance Inflation Factors for the radiobiological model, $z = \beta_0 + \beta_1 D + \beta_2 D^2/N$, of those data are presented in Table 5. (See Lectures 1, 3 and 13, also Montgomery and Peck 1982.) One notes in the level of the VIFs the absence of any effect of collinearity in the model matrix, X. This might have been expected from Figure 17a.

Although the sample evidence for including a second order term in T is not strong, it must be tested. For the model, $z = \beta_0 + \beta_1 D + \beta_2 D^2/N + \beta_3 T^2$, RSS = 40.51 on f = 36. The difference in the RSS between this and the model, $z = \beta_0 + \beta_1 D + \beta_2 D^2/N$, is 5.34. The corresponding difference in degrees of freedom is 1. Thus, the evidence of the sample suggests that the second-order term in T may be "real", since ΔRSS = 5.34 > 3.82, $\Delta f = 1$.

However, we recall yet again that a model in which T appears only as a term of second degree <u>may</u> not make sense, "... higher degree terms [should not be included] without their lower degree relations." (McCullagh and Nelder, 1983). However, there is no evidence for a linear term in T in the plots of the residuals. Moreover, the collinearity diagnostics show that the degree of correlation in D and T in the sample distribution is large enough to be troublesome: $r = 0.806$. However, the model, $z = \beta_0 + \beta_1 D + \beta_2 D^2/N + \beta_3 T$ was fitted to the sample. The results were consistent with anticipations: the presence of the additional term, T, in the model does <u>not</u> "significantly" improve the fit to the data: RSS = 45.64, f = 36. ΔRSS = 0.21 < 3.82. $\Delta f = 1$.

Even if there were strong sample evidence for $\beta_3 \ne 0$, the model, $z = \beta_0 + \beta_1 D + \beta_2 D^2/N + \beta_3 T$ is inconsistent with prior information on $\underline{\beta}$. The above model implies that both the response and its derivative, β_3, with respect to T are everywhere <u>continuous</u>. However on the evidence of the Douglas and Fowler (1976) paper we know, a priori, that the derivative, β_3, is <u>discontinuous</u> in the vicinity of T = 8 or 9 days. See also Table 4, models M_{1iii} and M_{1iv}.

Thus, we again consider adding an empirical, "broken-stick", function, $T' = (T-T_0)_+$, rather than either T or T^2, to the original model, M_2. As before, the broken-stick function is defined as

$$(T-T_0)_+ = \begin{cases} T-T_0 & T \ge T_0 = 9 \\ 0 & T < T_0 = 9. \end{cases} \qquad (14)$$

The equation is $z = \beta_0 + \beta_1 D + \beta_2(D^2/N) + \beta_3(T-T_0)_+$, for which RSS = 33.892, f = 36 and PRESS = 45.342. See Table 6. The difference in RSS between the two models is 45.85 - 33.892 = 11.96. We <u>note that</u> $\beta_3 > 0$; it is <u>positive</u> as is the case for β_1 and β_2. Also $\hat{\beta}_3/\sqrt{\text{Var}(\hat{\beta}_3)} = 2.736$.

The mechanistic model which we have examined was presented in Herring (1980) as derived from prior radiobiological experiments and conjectures. Included in this speculation was the postulate that the dose per fraction, D/N, was a useful transformation of the treatment variables. It can be regarded as a <u>second-order interaction term</u>, DN^{-1}, in a Taylor series expansion of the probit or logit transform of response. In general, we can view it as including a transformation of the variable $N \longrightarrow N^{-1}$, quite similar in principle to the "broken-stick" transformation which produces the linear spline function, $T \longrightarrow (T-T_0)_+$, (or $\log T \longrightarrow \log(T/T_0)_+$), or the logarithmic transformation $D \longrightarrow \log_{10} D$.

In Herring (1980) the form of the mechanistic model was based on two hypotheses that

1) $S = e^{-[\alpha(D/N) + \beta(D/N)^2]N}$ (15)

and

2) $z = a + b(\log S)$ (16)

where S is the surviving fraction of cells in the tissue at risk of occurrence of the binary event (skin reaction > 2 on the Denekamp scale) and z is the logit transform of P, the probability of occurrence of the binary event. The a priori evidence - cell-survival experiments and clinical experience - for this particular form of a causal or mechanistic model of response also suggests a rival, causal or mechanistic, model on one variable in still another polynomial model: It seems quite plausible that the dose per fraction, D/N, is both an important factor in a causal model of radiation response as well, of course, as another adjustable variable of treatment in a predictive model. The index plot of the variable (D/N) is presented in Figure 3c and the scattergram of the variables D and (D/N) is presented in Figure 4c. Note that the latter plot now frequently appears in the current discussions of the so-called linear-quadratic model in the radiobiological literature. (Fowler and Stern 1963 ; Tucker 1984; Fowler 1984) In these plots, sets of isoeffects are graduated by the equation, $D^{-1} = a_0 + a_1(DN^{-1})$. (Since the dose appears on both LHS and RHS the expression is more of an identity than an equation.) Different "effects" (different responses in different tissues (Withers, et al 1977)) are said to be distinguished by characteristic values of $a_0/a_1 = \alpha/\beta$ where α and β are, respectively, the coefficients of the first- and second-order terms in $d = D/N$ in the cell-survival equation, $S = e^{-(\alpha d + \beta d^2)N}$. (Douglas and Fowler 1976)

Thus, it is of interest to examine a semi-empirical model, the Taylor series approximation to terms of second-order:

$z = \beta_0 + \beta_1 D + \beta_2(D/N) + \beta_3(D/N)^2$ (17)

Note that this model now includes the main effects of both treatment variable, D, and (D/N), as well as the homogeneous second-order term in the latter - in place of the interaction term D(D/N) of the radiobiological model, M_2. The results are presented in Table 7 for both raw and standardized forms of (D/N). In this model we have once more assumed that the only adjustable variables of treatment are D and (D/N) - an error of specification.

It will be noted at once that this semi-empirical model fits the data well: RSS = 32.95, f = 36. In fact, it fits significantly better than does the radiobiological model, M_1: ΔRSS = 12.90, $\Delta f = 1$. Indeed, it is more concordant with the sample than any of the rival models we have hitherto considered. Furthermore, for this model the measures of mean square predictive error - PRESS and AIC - are less than for any of the alternative models. (See Lecture 1 and 13, also Montgomery and Peck 1982.) However, the diagnostics for this model disclose the presence of outlying responses, e.g., $t_{12} = 3.07$, $t_{23} = 3.70$, very extreme levels of treatment, e.g., $h_3 = 0.82$, $h_{39} = 0.45$ and correspondingly large beta-shifts, e.g., $D_{12}{}^1 = 0.58$, $D_{23}{}^1 = 0.83$, $D_{39}{}^1 = 0.91$. Moreover, since this model also does not include a function of T, the sample estimates, $\hat{\beta}$, are biased (aliased) (See Gunst 1980 and Draper and Smith

1981). In order to reduce the bias in the sample estimates of β we include the "broken stick" function $(T-9)_+$ to give the semi-empirical model M_{1vi}.

The model, M_{1vi}, is linear in D, quadratic in (D/N) and linear in the piece-wise continuous function of T: $z = \beta_0 + \beta_1 D + \beta_2(D/N) + \beta_3 (T-T_0)_+ + \beta_4(D/N)^2$. The analysis is presented in Table 7. The index plots of t_i, h_i and d_i^1 are presented in Figures 19a, 19b and 19c. Note that the larger values of t_i occur near the centroid of the sample and the smaller values are remote from the centroid. The converse is true for h_i. See Figure 4c. The RSS statistic suggests that this model fits this sample very well indeed - in fact better than any of its rivals. It is noteworthy that the residual sum of squares for the initial radiobiological model M_2 exceeds that of this semi-empirical model by a factor of 1.60, or $\Delta RSS = 45.85 - 28.71 = 17.14$, a highly significant difference. Moreover, the latter is more consistent with the prior information and the PRESS and AIC statistics for M_{1vi} suggest that this model is quite "portable". However, Figures 19a, 19b, and 19c disclose that there remain several observations which are not well explained by the model ($t_{12} > 3.0$, $t_{23} < -2.5$) and several which dominate some important aspects of the fit ($D_{12}^1 > 0.50$, $D_{23}^1 > 0.90$, $D_{34}^1 > 1.30$, $D_{39}^1 > 0.80$). See Pregibon 1981. We note that the sign and significance of the estimated coefficient of $(T-9)_+$ are rather anomalous: $\hat{\beta}_3 > 0$, $\hat{\beta}_3/\sqrt{Var(\hat{\beta}_3)} < 2.0$. These are idiosyncracies which we ascribe to the presence of some degree of collinearity between D and $(T-9)_+$ (We note that both D and $(T-9)_+$ have large components on V_3, the eigenvector of λ_3 where $\lambda_1 > \lambda_2 > \lambda_3 > \lambda_4$ are the eigenvalues of the appropriate correlation matrix, $X_0 X_0$) and to the narrow range of $(T-9)_+$ in the sample. The small value of RSS implies a large R^2 which we assume may be due in some measure to the fact that we have a five-term equation and only about seven effective observations (See Crocker 1972).

The results of our analyses suggest that the estimates, RSS, $\hat{\beta}$, $Var(\hat{\beta})$, etc., for any model will be dominated by those cases for which $0 < r_i/n_i < 1$: $i = 3, 4, 7, 12, 18, 23, 39$. More precisely, these are the observations for which the response will not be well explained by the model or which will dominate some aspect of the fit. In general, those cases near the centroid of the sample will have small values of h_i and large values of t_i: $i = 18$ and 23. The converse will be true for those cases remote from the centroid: $i = 3, 4$, and 39.

We have only a few remaining comments to make. They deal largely with the sample evidence for the possible roles for proliferation in modulating the response to a given radiation dose.

First, let us examine the design of this experiment according to still another criterion. In Lecture 13 we discussed the so-called small eigenvalue problem of non-experimental studies. That is, it is the presence of small eigenvalues, λ_j, $0 < j < p$, of the correlation matrix, $X_0'X_0$, that inflates the sample estimates, $\hat{\beta}$ and $Var(\hat{\beta})$, for a regression model. It is also the presence of small eigenvalues, μ_j, of the centered covariance matrix, $_0X'_0X$, that inflates the hat matrix diagonals h_i. (See Cook and Weisberg 1982.) We have seen that one desideratum for a designed experiment is orthogonality of the treatment variables. This feature of the sample distribution of X may be specified in terms of the shape - uniform or nonuniform - of the spectrum of eigenvalues, λ_j, $j = 1, \ldots p$, of the correlation matrix, $X_0'X_0$: If the treatment variables are orthogonal there are no "small" eigenvalues; the eigenvalue spectrum is uniform: $\lambda_1 = \ldots = \lambda_p = 1.0$.

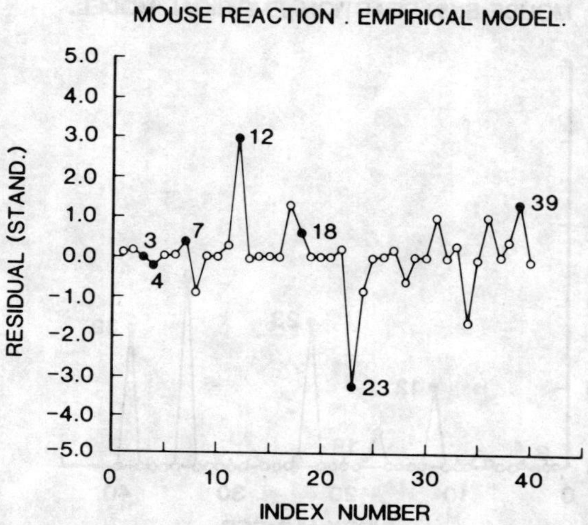

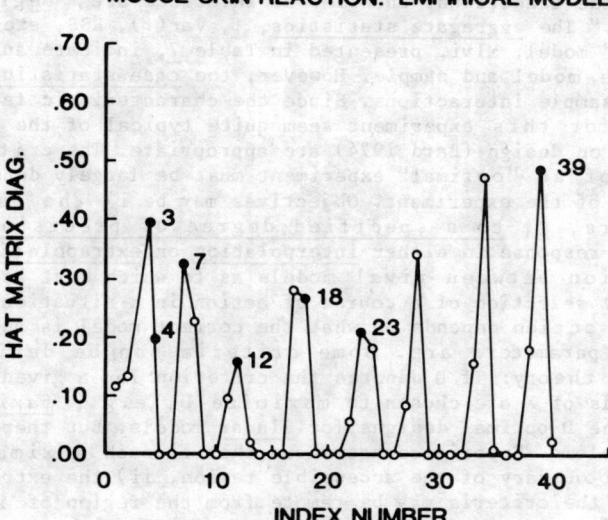

Figure 19a. Index plot of residuals, t_i, for the semi-empirical model M_{lvi}, of the skin reaction experiment. The model obviously fits these data better than any of the alternatives considered so far. However, 2 "outliers" remain at regimen #12 and #23. Note that the t_i are larger the closer is $\underline{x}_i{}'$ to the centroid of the distribution of X.

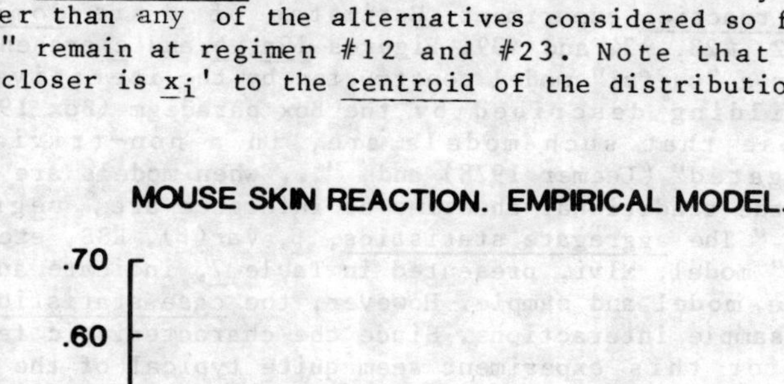

Figure 19b. Index plot of the hat matrix diagonals, h_i, for model M_{lvi} of skin reaction experiment. There are 7 treatment regimens for which $h_i \geq 2k/n = 0.25$. Note that the h_i are larger, the more remote is $\underline{x}_i{}'$ from the centroid of the distribution of X.

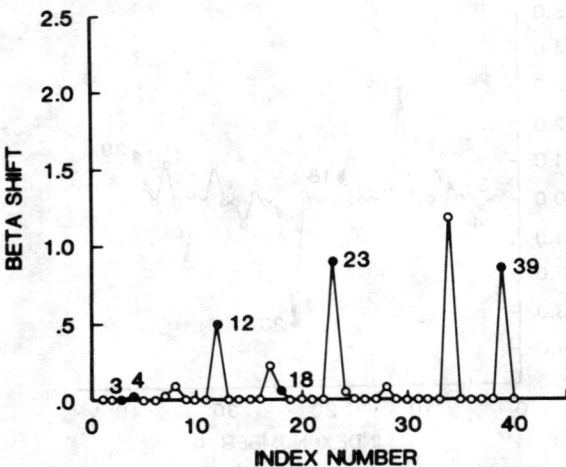

Figure 19c. Index plot of beta shifts, $Q(\hat{\beta}-\hat{\beta}_{(i)}) = D_i^1$, for model, Mlvi, of skin reaction experiment. Estimates of β are dominated by regimens #12, #23, #34 and #39. Figures 19a, b and c present the case statistics for a "useful" model constructed by the iterative process of model building described by the Box paradigm (Box 1979). It is often the case that such models are, in a non-trivial sense, "data-instigated" (Leamer 1978) and, "... when models are instigated by the data, the traditional theories of inference are, regrettably, invalidated." The aggregate statistics, $\hat{\beta}$, $Var(\hat{\beta})$, RSS, etc., for the process "exit" model, Mlvi, presented in Table 7, indicate an adequate "fit" of the model and sample. However, the case statistics disclose strong model-sample interactions. Since the characteristic features of the design for this experiment seem quite typical of the genre some brief remarks on design (Bard 1974) are appropriate. The criteria for the design of an "optimal" experiment must be largely determined by the objective of the experiment. Objectives may be i) the estimation of parameters, β, to a specified degree of precision; ii) the estimation of response in either interpolation or extrapolation; iii) discrimination between rival models as to which best accords with "reality"; iv) selection of a course of action in a situation in which the optimal action depends on what the correct model is and what the values of the parameters are. Some criteria can be derived from information theory: If D denotes the criterion for a given objective then the levels of X are chosen to maximize D, e.g., maximize D = det{X'X} - the D-optimal designs for linear models. But there are some practical problems. Experience has shown that i) such maxima usually lie at the boundary of the accessible region, ii) the extreme values prescribed by the criteria may be remote from the region of interest, iii) the properties of the system/process differ between the center and the boundary of the accessible region, and iv) for non-linear models the criteria cannot be computed without initial estimates $\hat{\beta}_0$ and $Var(\hat{\beta}_0)$ since the matrix cognate to X'X, say F'F, has elements which are functions of β (Box and Lucas 1959).

A nonuniform eigenvalue spectrum is evidence of undue sensitivity of the estimates, $\hat{\beta}$ and $\hat{z}_i$, to errors of measurement in the treatment variables as well as the possible inconsistencies between the sample and prior information on the sign, size and statistical significance of one or more elements of the coefficient vector, β, or of the size and sign of the estimates, $\hat{y}_i$, remarked above. From these earlier remarks it follows that a second criterion for a well-designed experiment is the uniformity of the spectrum of the hat matrix diagonals, h_i, $i = 1, \ldots, n$: $h_i = \bar{h} = p/n$, $i = 1, \ldots, n$: "For a D-optimal design all observations have the same influence and $[h_i] = p/n$" (Atkinson 1982). It will be recalled that for a D-optimal design, $\text{Det}[X'X]$ is maximized (See Montgomery and Peck 1982). This requirement will minimize the effects on the estimates of β of a small proportion of outliers in the response. (Box and Draper 1975) For purposes of comparison of the spectra of h_i for the Finney experiment, the two models of the mouse-skin data and the empirical model of the non-experimental observations in Lecture 13, the degree of uniformity will be specified by both the ratio of the order statistics, $h_{(n)}/h_{(1)}$, and by the coefficient of variation of the respective distributions of h_i. This is presented in Table 8a. It is evident that the mouse-skin response experiment resembles that of the non-experimental study more closely than that of a proper experiment.

The observation matrix, $[y, X]$, for a proper experiment for estimation of the parameter vector, β, for the (first-order) empirical model M1a is presented as the 2^k factorial design in Table 8b. The cognate experiment for the (quadratic) radiobiological model M2 is a 3^k factorial design (Myers 1971). Note that a proper design requires a large number $n_i \geq 30$ of animals at risk at each of the 2^k treatment levels (Compare Figure 5b). These levels of X must be chosen so that i) the variables are orthogonal (See Figure 2d) and ii) the range of each is such as to produce "... neither trivial nor overwhelming effects ..." on the level of response (Marquardt 1980). Obviously, both the level of the treatment variables and the number of animals at risk jointly determine the likelihood of observing an extreme ($r_i = 0$ or n_i) level of a binary response. Extreme levels of treatment also are "influential" observations (large h_i, $D_i{}^1$). Thus, it would seem desirable to exclude single-dose (N=T=1) observations since these are not only qualitatively distinct from fractionated irradiation (and hence of little clinical interest) but tend to be made at extreme levels (high D, low N and T) and thus, dominate the estimates $\hat{\beta}$, $\text{Var}(\hat{\beta})$, RSS, etc. for any model.

There is a high degree of collinearity present in the matrix X for the empirical model, M_1, (Table 3) and a very low degree in the cognate matrix for the mechanistic model, M_2 (Table 5). The transformation required for M_1, $(D, N, T) \longrightarrow (\log_{10}D, \log_{10}N, \log_{10}T)$, further inflated the colinearity already present in the initial design (Figures 2a, c). However, the transformation required by M_2, $(D, N, T) \longrightarrow (D, D^2/N)$ has deflated the collinearity present in the initial design (Figure 4) by two means: 1) deletion of T (Hence the sample estimate of β for M_2 is biased. See Draper and Smith 1981) and 2) because of an artefact of the design, D and D^2/N are nearly orthogonal. Some remarks of Montgomery and Peck (1982) are illuminating and appropriate: "Multicollinearity is often caused by the choice of model, such as when two highly correlated regressors are used in the regression equation. In these situations some respecification of the regression equation may lessen the impact of multicollinearity. One approach to model

Table 8a. Uniformity of Distributions of $\lambda_j{}^a$ and h_i.

	$\lambda_1/\lambda_p{}^a$	$h(n)/h(1)$	$\bar{h}_i$	$Var(h_i)$	Coefficient Variation
1. T. Casteneum Mortality $X = (x_1, x_2)$	1.0	$1.383*10^1$	0.250	$1.64*10^{-1}$	51.26
2. Mouse Skin Reaction #2					
a) $X = (x_1, x_2, x_3)$	69.763	$6.550*10^6$	0.100	$9.64*10^{-3}$	98.19
b) $X = (D, D^2/N)$	1.238	$1.595*10^7$	$7.5*10^{-1}$	$1.04*10^{-1}$	135.88
3) Sq. Ca. Ablation[b] $X = (x_1, x_2, x_3)$	105.839	$2.605*10^2$	$8.89*10^{-2}$	$5.41*10^{-3}$	82.73

[a] λ_1 and λ_p are eigenvalues of $X_0'X_0$. [b] See Lecture 13.

Table 8b. Observation Matrix, [y, X], for Estimation of $\underline{\beta}$ (See Myers 1971). First Order Response Surface. $2k$ Factorial[a]. $z = \beta_0 + \beta_1 x_1 + \beta_2 x_2 + \beta_3 x_3 + \epsilon$ (Model M1a).

i) Model Matrix (Coded Levels. See p 397)

X_0	X_1	X_2	X_3
1	-1	-1	-1
1	1	-1	-1
1	-1	1	-1
$X = 1$	1	1	-1
1	-1	-1	1
1	1	-1	1
1	-1	1	1
1	1	1	1

ii) ith Row of Response Matrix (3 replications) $(r_{i1}/n_i1, r_{i2}/n_{i2}, r_{i3}/n_{i3})$ $1 \leq i \leq 2^k = 8$. $y_i = r_i./n_i. = \Sigma r_{ij}/\Sigma n_{ij}$. $n_{ij} \cdot n_{ij} \geq 10$, $0 \leq r_{ij} \leq n_{ij}$. The 2^k treatments are run in random order within each of the 3 replicates. Replication provides data to discriminate between a) "lack of fit" of model and b) overdispersion relative to assumed binomial variation within treatments when RSS is "significantly" large. If b) is the case then $Var(\underline{\hat{\beta}})$ must be multiplied by heterogeneity factor, $\sigma^2 = RSS/(n-4)$ (Finney 1971, McCullagh and Nelder 1983).

[a] "... an important feature of the 2^k factorial experimental designs (and more generally the class of orthogonal designs) is that the estimates of the model coefficients are uncorrelated with one another." (Myers 1971) The mechanistic model, M2, requires a 3^k factorial design: "... 3^k factorial experimental designs can be useful when the regression model is best represented by a second order relationship. ... Like the 2^k factorial, the 3^k factorial design falls into the class of orthogonal designs ..." (Myers 1971). It is important to note that the proper designs for a dose-response surface are not simply paratactic aggregations of the proper designs for dose-response curves. Compare Figures <u>4a</u> and <u>4b</u> with Figure <u>2d</u>.

respecification is to redefine the regressors. For example, if x_1, x_2 and x_3 are nearly linearly dependent it may be possible to find some function such as $x = (x_1 + x_2)/x_3$ or $x = x_1 x_2 x_3$ that preserves the information content in the original regressors but reduces the ill-conditioning." The sample correlation coefficients of (D, N) and (D, D^2) are 0.811 and 0.986, respectively. The correlation coefficient for (D, D^2/N) is -0.106. Thus, the transformation (D, D^2, N) $\longrightarrow$ (D, D_2/N) is an orthogonal transformation for this sample. It must be emphasized that this transformation is an artifact of the (peculiar) design of this experiment only.

It will be recalled that we remarked earlier that the sample distribution of D and N for the mouse-skin experiment could be equally well graduated by either of two equations: 1) $D = 2054 N^{0.35}$ ($R^2 = 0.71$) or $D = 1726 N^{0.50}$ ($R^2 = 0.72$). It is obvious, that for the latter D^2/N is nearly constant - and hence (nearly) uncorrelated with D. This is quite evident in Figure 17b.

Let us look at the situation from still another perspective. We note the obvious similarity between the pairs of Figures 13a and 13b and 17b and 17a. The first pair represents the orthogonal transformation: $(x_1, x_1^2) \longrightarrow [(x_1 - \bar{x}_1), (x_1-d)^2]$, where d is the Dykstra statistic (See Daniel and Wood 1971). The second pair represents the orthogonal transformation described above: (D, D^2, N) $\longrightarrow$ (D, D^2/N). For the first pair of Figures, the transformation is quite general: $\bar{x}_1$ is the sample mean and d is estimated from the sample itself under the constraint that the correlation between $(x_1-\bar{x}_1)$ and $(x_1-d)^2$ be exactly zero. For the second pair the transformation is an artefact of the design. However, there exists a general method by which regressor varibles which are orthogonal linear combinations of the initial treatment variables can be obtained from any sample. These combinations are the eigenvectors of $X'X$ (or $X'\hat{V}^{-1}X$) or $X_0'X_0$. The method is known as principal components regression. It is an alternative to Ridge regression. See Montgomery, Lecture 4 and Montgomery and Peck 1982. The method can be regarded as an application of the spectral decomposition of a (pxp) symmetric matrix: $X_0'X_0 = \lambda_1 \underline{V}_1 \underline{V}_1' + \lambda_2 \underline{V}_2 \underline{V}_2' + \ldots + \lambda_p \underline{V}_p \underline{V}_p'$, where λ_j is the jth eigenvalue of the correlation matrix, $\underline{V}_j$ is the (pxl) corresponding eigenvector and the (pxp) matrix, $\underline{V}_j \underline{V}_j'$, is the dyadic, or outer product of V_j. It will be recalled that the inner product is the scalar, $\underline{V}_j'\underline{V}_j$, which is the norm of $\underline{V}_j$ (See Rao 1973). The $\underline{V}_j$ are the principal components pf $X_0'X_0$.

In the present case, the orthogonal transforms (D, D^2/N) have more a substantive interpretation than is generally the case for the principal components. However, we must emphasize that the (near) orthogonality ($r = -0.10$) is an accident of the design for this particular experiment, whereas the orthogonality of the $\underline{V}_j$ is a general property of the principal components. Moreover, the orthogonality achieved by the principal components transformation of X conserves the sample information on conditional response since no treatment variables are deleted; the eigenvectors, V_j, of $X_0'X_0$ are, of course, linear combinations of all the treatment variables.

Thus, the design of this experiment assures that the data are more informative for the radiobiological model, M_2, (treatment variables, D, D^2/N) than they are for M_1 (treatment variables logD, logN, logT). We recall Thisted's (1980) remarks quoted earlier: "Models are not the source of information, however; data are. And it is the case that the data can be unequally informative about different parametrizations."

It is a characteristic requirement of designed experiments that the levels of all factors that might influence the response between groups are either closely controlled at specified levels or randomly allocated (See Box, Hunter and Hunter 1978). It follows that experimental results are vulnerable to failure to achieve either effective control and/or randomization. One cue that such a failure, (sometimes known as "slippage") has occurred - and there may be multiple such failures, of course - is evidence of a time shift or trend in the response between groups when these are ordered with respect to time. Such failure is more insidious in "longitudinal" experiments in which elapsed time is a variable of treatment than in "cross-sectional" experiments in which all treatment groups could be exposed in a very short time. The Finney toxicological experiment is an example of the latter.

We have previously included the spline functions as variables of treatment in our models in order to account for the sample evidence of discontinuities in time in the response and its first derivative, in terms of intrinsic changes in the pharmacokinetics of the process which generated the sample. However, we have noted the similarity of the two phenomena, $\beta = \beta(X)$ and $\beta = \beta(U)$, in which the parameter vector is a discontinuous function of the variables of treatment, X, as well as of the covariates, U, which uniquely specify the state of the system, S_m, $m = 1, 2$ (See the discussion of Lecture 13).

It is useful to examine an alternative representation of the former in which time is treated as a covariate which necessarily differs between groups. Changes in response between groups may be partly due to changes in experimental conditions between groups (failure of the ceteris paribus qualification) that are distinguished by the different values of elapsed time in the experiment. Changes in experimental conditions may, of course, be treated as changes in the state of the system, S_1. Let us examine the method of piecewise linear regression (See Neter and Wasserman 1974). In this method we introduce binary variables to distinguish differences in elapsed time, the intervals ($T > T_0$ and $T \leq T_0$) in precisely the same manner as we used them to distinguish differences in stage of disease T_2 and T_3 (which, of course, also distinguishes differences in elapsed time albeit of greater duration, say, months rather than days).

Thus, the representation of the discontinuity at T_0 in the response and its first derivative with respect to time can be included in the response surface model by introducing the binary variable x_4: $z = \beta_0 + \beta_1 \log D + \beta_2 \log N + \beta_3 \log(T/T_0)x_4 + \beta_4 x_4$ where

$$x_4 = \begin{cases} 1 & T \geq T_0 \\ 0 & T < T_0 \end{cases} \tag{18}$$

The main effect of x_4 represents the change in the response; the interaction term, $x_4 \log(T/T_0)$, represents the change in the first derivative. See Table 4. However, as remarked previously, when we introduced the spline functions, $\log(T/T_0)_+$ and $(T-T_0)_+$ there were too few points on either side of the "knot" at T_0 to provide firm support for estimates of β (4 to 5 points on either side of T_0, are recommended (Wold 1974) but there are only 6 distinct levels of T in the sample).

Recall that in the non-experimental study of Lecture 13 the range of T exceeds that of N (32 vs 15). However, in the present experiment, the effective range of T is that of T-9 = 8, whereas that of N is 14. Thus, the sample evidence for the association of T with response would be expected, a priori, to be weaker in the sample of Lecture 13 than for the present sample. (Note also that the range of both variables is greater in the non-experimental study than in the designed experiment.) However, it is not only the narrow effective range of T, but, also the scarcity of observations within that range that frustrates obtaining any convincing assessment of the role of T in the response from the data of this sample.

ACKNOWLEDGEMENT

We acknowledge, with thanks, the dedication, as well as expertise, of Marie Woodall in the preparation of the manuscript.

REFERENCES

AKAIKE, H. (1977), On Entropy Maximization Principle, in Applications of Statistics, P. R. Krishnaiah, ed., North-Holland, 27-41.
ANG, K. K., VAN DER KOGEL, A. J. and VAN DER SCHUEREN, E. (1983), The Effect of Small Radiation Doses on the Rat Spinal Cord: The Concept of Partial Tolerance, International Journal Radiation Oncology, Biology, Physics, 9: 1487-1491.
ATKINSON, A. C. (1982), Robust and Dignostic Regression Analyses, Commun. in Statistics - Theory & Methods. 11(22): 2559-2571.
BARD, Y. (1974), Nonlinear Parameter Estimation, Academic Press, N.Y.
BELLMAN, R. and ROTH, R. (1969), Curve Fitting by Segmented Straight Lines, Journal of the American Statistical Assn. 64: 1079-1084.
BELSLEY, D. A., KUH, E. and WELSCH, R. E. (1980), Regression Diagnostics: Identifying Influential Data and Sources of Collinearity, John Wiley & Sons, N.Y.
BERKSON, J. (1953), A Statistically Precise and Relatively Simple Method of Estimating the Bioassay with Quantal Response, Based on the Logistic Function. Journal of the American Statistical Assn. 565-599.
BOX, G.E.P. (1976) Science and Statistics. Journal of the American Statistical Association. 71: 791-799.
BOX, G.E.P. (1979) Robustness in the Strategy of Scientific Model Building, in Robustness in Statistics, R.Launer and G. N. Wilkinson eds., Academic Press, N.Y.
BOX, G.E.P. and DRAPER, N. R. (1975), Robust Designs, Biometrika, 62(2) 347-352.
BOX, G.E.P., HUNTER, W. G. and HUNTER, J. S. (1978), Statistics for Experimenters, John Wiley & Sons, N.Y.
BOX, G.E.P. and LUCAS, H. L. (1959), Design of Experiments in Nonlinear Situations. Biometrika, 46: 77-90.
BROWNLEE, K. A. (1965) Statistical Theory and Methodology in Science and Engineering, 2nd ed., John Wiley & Sons, N.Y.
CHATTERJEE, S. and PRICE, B. (1977), Regression Analysis by Example, John Wiley & Sons, N.Y.
COOK, R. D. and WEISBERG, S. (1982), Residuals and Influence in Regression, Chapman and Hall, N.Y.
COX, D. R. (1970), The Analysis of Binary Data, Methuen & Co. Ltd., London.
CROCKER, D. C. (1972) Some Interpretations of the Multiple Correlation Coefficient, The American Statistician, 31-33 (April).
DANIELS, C. and WOOD, F. S. (1971), Fitting Equations to Data: Computer Analysis of Multifactor Data for Scientists and Engineers, Wiley-Interscience. N.Y.
DOUGLAS, B. G. and FOWLER, J. F. (1976) The Effect of Multiple Small

Doses of X Rays on Skin Reactions in the Mouse and a Basic Interpretation, Radiation Research, 66: 401-426.
DRAPER, N. R. and SMITH, H. (1981) Applied Regression Analysis, 2nd ed. John Wiley & Sons, N.Y.
ELLIS, F. (1969), Dose, Time and Fractionation: A Clinical Hypothesis, Clinical Radiology, 20: 1-7.
FILLIBEN, J. J. (1975) The Probability Plot Correlation Coefficient Test for Normality, Technometrics, 17(1): 111-117.
FINNEY, D. J. (1971), Probit Analysis, 3rd ed., University Press, Cambridge.
FISHER, R. A. (1958) Statistical Methods for Research Workers, 13th ed. Oliver & Boyd, London.
FOWLER, J. F. (1984) Fractionated Radiation Therapy After Standqvist, ACTA Radiologica, Oncology, 23: 209-216.
FOWLER, J. F. and STERN, B. E. (1963) II. Dose-Time Relationships in Radiotherapy and the Validity of Cell Survival Curve Models, British Journal of Radiology, 36: 163-173.
FROME, E., Lecture 7, Regression Methods for Binomial and Poisson Distributed Data, See This Volume.
GUNST, R. F. (1980) Comment on G. Smith & F. Campbell paper, A Critique of Some Ridge Regression Methods, Journal of the American Statistical Assn., 75: 98-100.
HERBERT, D. E., Lecture 13, Clinical Dose Response Models, I. Regression Diagnostics and Biased Estimation, See This Volume.
HERBERT, D. E., Lecture 14, Clinical Radiocarcinogenesis Applications of Regression Diagnostics and Bayesian Methods to Poisson Regression Models, See This Volume.
HERRING, D. F. (1980), Methods for Extracting Dose Response Curves from Radiation Therapy Data, I: A Unified Approach, International Journal of Radiation Oncology, Biology, Physics, 6: 225-232.
JOHNSTON, J. (1972) Econometric Methods, 2nd ed., McGraw-Hill, N.Y.
KENDALL, M. (1976), Statisticians - Production and Consumption, American Statistician, 30(2): 49-53.
LEAMER, E. E. (1978) Specification Searches: Ad Hoc Inference with Nonexperimental Data, Wiley-Interscience, N.Y.
LEAMER, E. E. Lecture 5, Bayesian Regression and Sensitivity Analysis, See This Volume.
MANTEL, N. (1983), Cautions on the Use of Medical Databases, Statistics In Medicine, 2: 355-362.
MARQUARDT, D. W. (1980), Comment on A Critique of Some Ridge Regression Methods, Journal of the American Statistical Assn., 75: 87-91.
MCCULLAGH, P. and NELDER, J. (1983), Generalized Linear Models, Chapman and Hall, N.Y.
MONTGOMERY, D., Lecture 4, Biased Estimation and Robust Regression, See This Volume.
MONTGOMERY, D. and PECK, E. (1982) Introduction to Linear Regression Analysis, John Wiley & Sons, N.Y.
MYERS, R. H. (1971) Response Surface Methodology, Edwards Brothers Inc., Distributor, Ann Arbor, MI.
MYERS, R. H. Lecture 1 and 2, Regression Analysis - Basics and Current Frontiers, See This Volume.
NETER, J. and WASSERMAN, W. (1974), Applied Linear Statistical Models: Regression, Analysis of Variance and Experimental Designs, Richard Irwin, Inc. Homewood, IL.
PEARSON, K. (1957) The Grammar of Science, Meridian Books, Inc., N.Y.
PREGIBON, D. (1981) Logistic Regression Diagnostics, Annals of Statis-

tics, 9(4): 705-724.
RAO, C. R. (1965), <u>Linear Statistical Inference and Its Applications</u>, 2nd ed., John Wiley & Sons, N.Y.
ROSCOE, J. T. and BYARS, J. A. (1971) An Investigation of the Restraints with Respect to Sample Size Commonly Imposed on the Use of the Chi-Square Statistic, <u>Journal of the American Statistical Assn.</u>, 66: 755-759.
SCHNEIDERMAN, M. A., MANTEL, N. and BROWN, C. C. (1975), From Mouse to Man - or How to Get from the Laboratory to Park Avenue and 59th St., <u>Annals of the New York Academy of Sciences</u>, 246:
SEBER, G.A.F. (1977), <u>Linear Regression Analysis</u>, John Wiley & Sons, N.Y.
SMITH, P. L. (1979) Splines as a Useful and Convenient Statistical Tool <u>American Statistician</u>, 33(2): 57-62.
SNEDECOR, G. W. and COCHRAN, W. G. (1967) <u>Statistical Methods</u>, 6th ed., Iowa State University Press, Ames, IA.
SNEE, R. D. (1973) Some Aspects of Nonorthogonal Data Analysis: Part I Developing Prediction Equations, <u>Journal of Quality Technology</u>, 5(2): 67-79.
SNEE, R. D. (1973) Some Aspects of Nonorthogonal Data Analysis: Part I. Comparison of Means, <u>Journal of Quality Technologu</u>, 5(3): 109-122.
SNEE, R. D. (1977) Validation of Regression Models: Methods and Example, <u>Technometrics</u>, 19(4): 415-428.
STEWART, J. G. and JACKSON, A. W. (1975) The Steepness of the Dose Response Curve Both for Tumor Cure and Normal Tissue Injury, <u>The Laryngoscope</u>, LXXXV: (7) 1107-1111.
SUPE, S. J., NAGALAXMI, K. V. and MEENAKSI, L. (1983) Tumor Significant Dose, <u>Medical Physics</u>, 10(1): 51-56.
THEIL, H. (1971) <u>Principles of Econometrics</u>, John Wiley & Sons, N.Y.
THISTED, R. A. (1980) Comment on Smith and Campbell paper, A Critique of Some Ridge Regression Methods, <u>Journal of the American Statistical Assn.</u>, 75: 81-86.
TUCKER, S. L. (1984), Tests for the Fit of the Linear-Quadratic Model to Radiation Isoeffect Data, <u>International Journal of Radiation Oncology, Biology, Physics</u>, 10: 1933-1939.
WATTS, D. G. and BACON, D. W. (1974) Using An Hyperbola as a Transition Model to Fit Two-Regime Straight-Line Data, <u>Technometrics</u>, 16(3): 369-373.
WELSCH, R., Lecture 3, Introduction to Regression Diagnostics, See This Volume.
WITHERS, H. R., THAMES, H. D. and PETERS, L. J. (1983), A New Isoeffect Curve for Change in Dose Per Fraction, <u>Radiotherapy and Oncology</u>, 1: 187-191.
WOLD, S. (1974) Spline Functions in Data Analysis, <u>Techonometrics</u>, 16(1): 1-11.

NOTES ADDED IN PROOF

1. As remarked in the <u>Preface</u> to this volume (Altman 1980) "When a paper containing incorrect results (not necesarily through statistical mistakes) is published there may be serious consequences, although surprisingly this does not seem to be generally appreciated: ... (v) If the results go unchallenged the researcher(s) involved may use the same substandard statistical methods again in subsequent work and others may copy them ... the above points make the misuse of statistics very much an ethical, as well as scientific, issue." The Herring (1980) paper has been cited 5 times since publication - but never before challenged.

2. "These benefits [of either replication or secondary analysis of published studies] include the verification and refinement of original findings and the refutation of them." (Hedrick 1985)

3. The mechanistic model, M_2, is based on the so-called "F_e Concept" (Douglas and Fowler 1976). Three of the (four) basic assumptions of this concept are as follows: a) "That repair occurs after single doses of X rays ... Equal amounts of repair occur after dose fractions of equal size." b) "That the measured radiation effect in the system under investigation is correlated with the surviving number of critical cells in that system, i.e., with the surviving fraction of cells, for all fractionation schedules ... ." c) "That the effect of a dose of radiation is dependent on the magnitude of that dose, and not significantly on its position in a series of doses of radiation." It is further stipulated that the surviving fraction of cells irradiated by a single dose of radiation of size d is $S = e^{-[\alpha d + \beta d^2]}$. Then the surviving fraction, S_n, of cells in a tissue following n fractions of radiation of size d is $S = e^{-n[\alpha d + \beta d^2]} = e^{-n[\alpha(D/n) + \beta(D/n)]} = e^{-\alpha D + \beta D^2/n}$. This is (obviously) an exponential decay process of n steps in which the "decay constant", say ε, is a function of dose: $\varepsilon = \alpha d + \beta d^2$. Here $D = nd$, the total dose.

Let us examine the jth step in this process:
$$(N_j - N_{j-1}) = -\varepsilon_j N_{j-1}. \quad 0 \leq \varepsilon_j \leq 1. \tag{19a}$$
N_j is the number of surviving cells at the jth step. This may be rewritten as
$$N_j = (1-\varepsilon_j)N_{j-1} \tag{19b}$$
After n steps
$$N_n = N_0 \prod_n (1-\varepsilon_j) \tag{20}$$
where N_0 is the initial number of cells.

According to the "F_e concept" $\varepsilon_j = \varepsilon$. $1 \leq j \leq n$. $0 \leq \varepsilon \leq 1$. Thus,
$$(N_n/N_0) = (1-\varepsilon)^n. \tag{21}$$
Then we must have, at the lim $n \to \infty$
$$S_n = (N_n/N_0) = (1-n\varepsilon/n)^n = e^{-n\varepsilon} \tag{22a}$$
Or
$$S_n = e^{-n[\alpha d + \beta d^2]} \tag{22b}$$
Note that this assumes that the overall probability of "death" at each of the j steps is constant: $\varepsilon_j = \varepsilon$. While this may be plausible for cells <u>in vitro</u> it seems to us that an alternative, random, process is more appropriate to the destruction of tissues. The rival process is well-known in the breakage of brittle solids and also in the growth of biological systems as the Theory of Proportionate Effects (Aitchison and Brown 1966, Bury 1975, Epstein 1947).

Let us recur to Equation (20):
$$N_n = N_0 \prod_n (1-\varepsilon_j). \quad 0 \leq \varepsilon_j \leq 1.$$
This may be rewritten as
$$S_n = \prod X_j. \quad 0 \leq X_j \leq 1.0.$$
Then by a multiplicative analogue of the central limit theorem, "If $\{X_j\}$ is a sequence of independent positive variates having the same probability distribution and such that
$$E\{\log X_j\} = \mu$$
and $D^2\{\log X_j\} = \sigma^2$
both exist, then the product $\prod^n X_j$ is asymptotically distributed as $\Lambda(n\mu, n\sigma^2)$." (Aitchison and Brown 1966). $\Lambda(.)$ is the two-parameter log normal distribution function. (In the present case $\log X_j < 0$.) Thus,
$$E\{S_n\} = e^{n[\mu + 0.5\sigma^2]}$$
$$D^2\{S_n\} = e^{2n[\mu + 0.5\sigma^2]}(e^{n\sigma^2}-1)$$

The mode and median are, respectively, $e^{n[\mu-\sigma^2]}$ and $e^{n\mu}$. (See also Hald 1965).

4. We recall that the iterative Maximum Likelihood methods of estimation of $\underline{\beta} = (\beta_0, \beta_1)$ for probit model, $z = \beta_0 + \beta_1 x_1$, of the dose-response data at 7 of the 8 levels of (N, T) failed to converge because of the excessive number of responses $r_i/n_i = 0$ or 1, $1 \leq i \leq 5$ at each level. However, these methods did converge for the model $z = \beta_0 + \beta_1 x_1 + \beta_2 x_2 + \beta_3 x_3$ of the pooled data from the eight levels of (N, T). Thus, the experiment may be viewed as a salvage operation - data augmentation - required by the presence of extreme levels of response in each of these 7 samples. The same kind of operation was required in Lecture 13 because of the presence of extreme levels of treatment in the samples for Stages T_2 and T_3 and in Lecture 14 because of the presence of multicollinearity in the Hiroshima and Nagasaki samples. (Note that the "salvage operation" recommended by Berkson (1953) for dealing with extreme ($r_i=0$ or n_i) responses (substituting $r_i^* = 1/2$ or $n_i - 1/2$, respectively) is also data augmentation - with pseudodata.

5. We remarked in Lecture 13 that if the time of observations, t^*, differs from the time for end-point response, t_1, then the dose-response equation must include a second-order term in log dose, say x_1^2. We note that the empirical model M_{1i} (Table 4) fits the data (rather) "significantly" better than the mechanistic model M_2 (Table 6): $P(\chi^2 > 5.29 | 2) = 0.071$. Thus, there would seem to be some question as to whether the interaction term, D^2/N, required by M_2 described an inherent feature of the radiation response of the system (mouse foot skin) or is perhaps more descriptive of an artifact of the experiment.

ADDITIONAL REFERENCES

ALTMAN, D.G. (1982) Statistics in Medical Journals, Statis. In Med. I:

AITCHISON, J. and BROWN, J.A.C. (1966) The LogNormal Distribution. Cambridge Univ. Press. Cambridge, MA.

BURY, K.V. (1975) Statistical Models in Applied Science. John Wiley & Sons. N.Y.

DOUGLAS, B.G. and FOWLER, J.F. (1976) The Effect of Multiple Small Doses of X Rays on Skin Reactions in the Mouse and a Basic Interpretation. Radiation Research. 66: 401-26.

EPSTEIN, B. (1947) The Mathematical Description of Certain Breakage Mechanisma Leading to the Logarithmico-Normal Distribution. J. of the Franklin Inst. 224: 471-77.

HALD, A. (1965) Statistical Theory with Engineering Applications. John Wiley & Sons. N.Y.

HEDRICK, T.E. (1985) Justifications for and Obstacles to Data Sharing, In Sharing Research Data, Fienberg, S., Martin, M. and Straf, M., eds. National Academy Press. Washington, D.C., 123-147.

SAMPLED MEAN RADIATION TREATMENT DESIGN: Physical and Biophysical Parameters

Larry D. Simpson, Ph.D.
Medical Physics Division
University of Rochester Cancer Center
601 Elmwood Ave. -- Box 647
Rochester, N.Y. 14642
(716) - 275-5261

I. INTRODUCTION

The radiation oncologist prescribes a spatially-conscribed, invisible beam of energy to intersect a non-visible tumor volume inside the patient. The intersection is prescribed to be maintained for a few minutes while a certain prescribed dose is delivered to the tumor. Actually, a sequentially-set-up group of beams are prescribed, all intersecting the tumor, but just arriving from different directions. This prescription may be repeated 1-3 times per day for up to 4-7 weeks. The radiation oncologist's goal is to complete the dose delivery, accurate to within $\pm$ 5%.

This is a physically-taxing, unique form of medical treatment. These treatment executions require the presence of and challenge the medical radiation physicist and the radiation biophysicist.

The responsibility of the medical radiation physicist has been to, under standard laboratory conditions, assure the calibration of the beam geometry and activity, assure that once a beam direction is chosen for a beam as part of a treatment, that that beam does intersect the prescribed tumor volume, assure that all this can be reproduced for the many fractions of the patient's treatment, in short to minimize the influence of the physical part of the treatment delivery on the therapeutic benefit of the prescribed energy. The primary constraint is that normal, often radiosensitive tissues must not be overdosed, while delivering the prescribed dose to the targeted tumor volume.

Often, little recognition is explicitly given to random or systematic uncertainties inherent in this fractionated physical treatment delivery. Clinical and laboratory experience demonstrate that 3-5% accuracy is necessary for some diseases and desirable for most. A new method is proposed in the second section of this presentation, which addresses these points and could provide needed documentation of the effects of estimates of these uncertainties in the physical dose delivery chain for the patient's treatment record -- the sampled mean treatment plan.

The radiation biophysicist assists in describing biological consequences of the stochastic absorption of the prescribed energy to the tumor and the normal tissues. Regression techniques are used to describe or model past tumor control probabilites and the probability of particular acute and late normal tissue damage for particular dose fraction magnitude-fraction time interval-total number of fractions, for the radiation oncologists. The most frequently utilized model is the TDF model of Orton and Ellis (1973), even given the frequent rational criticisms directed its way. It is, even to this day, frequently utilized in professional consultations by medical physicists and radiation oncologists for medical malpractice cases. An alternative formulation, the partial tolerance fraction (PTF), is presented for the first time in the third section of this presentation.

The problem of combining the "clinical effect" models with the sampled mean treatment plan is raised in the final section of this presentation, together with other applications.

Radiation oncologists and participant allied health professionals respect the frequency distribution of uncomplicated outcome as a function of time after the initiation of treatment. They acquire this respect from their personal experience and the peer-reviewed published clinical scientific papers (as well, from anecdotal experiential conversations with their peers and basic scientists).

However, these same health professionals, who presume some biological variability in their outcome of the proposed treatments, too frequently assume the treatment is precisely and accurately delivered as they have prescribed. And why shouldn't they -- they are never really shown anything different than the "prescription" and some numerical documentation (with no error estimators) in the patient's treatment record.

The medical oncologist prescribes and delivers drugs intravenously, intrathecally, intraperitoneally, or orally. This is repeated in time, perhaps with different dosages, and perhaps at irregular time intervals. The surgical oncologist physically removes diseased tissues while permitting healing of residual normal tissues/functions. This is most frequently not repeated in time.

Note, that medical oncologists and surgical oncologists presume to control relatively large potential errors, for example; "the wrong drug", "5 cc rather than 10 cc", "leaving an adequate surgical margin", "using the wrong anesthetic", etc. One does not hear them talking of "3-5%" of anything.

What does that mean, compared to the type of documentation currently placed in the patient's radiation treatment records? What do statement-of-accuracy demands on the radiation oncology team mean relative to other oncology specialties?

II. SAMPLED MEAN RADIATION TREATMENT DESIGN

Medical radiation physicists should focus on creating radiation treatment simulation programs which will handle "fields treated" as well as the more conventional "fields prescribed".

For example, consider a three-"prescribed field" plan performed nominally on a computerized treatment planning workstation to correspond to the "perfect" dose delivery in one perfect set-up of each field. In actual fact, this plan, when executed, will be composed of, say, 30 fractions of 200 rad/fraction, prescribed to a tumor volume -- in other words it is a 90-field plan with 66.7 rad (+ or - the random and systematic uncertainties of the calibration and dose computational algorithm) delivered to a tumor volume per field. The assigned tumor volume and surrounding normal tissues, including the skin surface, are in fact positioned 90 times, and there is a "wander" in pointing each field for its treatment time, for each fraction.

Conventional planning creates an erroneous prediction of the perfectly delivered doses with perfect execution of the prescribed patient set-up for each daily treatment. Well-defined, precisely placed fields superimposed upon precisely-defined, precisely-positioned skin, organs, heterogeneities, and tumor volumes lead, then to a set of isodoses which often tempt the physician, dosimetrist, medical physicist, and technologist to make unrealistic changes in field parameters or patient-position parameters.

Comments such as these are frequently heard during the treatment planning process:

"A 1 cm change in this field's width and a 0.5 cm change in the isocentric set-up point in the patient, here, should do it!"

"Say, Chris, we need to recut these blocks and reposition them 3 mm to the left."

"The radiation field should coincide with the light field within +/- 2 mm."

"Well-executed SSD set-ups are just as good as isocentric set-ups."

"The parameters for this optimized plan are: Field 1, 10.5 x 7.5 cm, gantry angle 39 degrees, weight 1.25....."

These and other similar statements remain untestable given the current state-of-the-art in treatment planning programs.

Tools should be created to test them over a range of uncertainties and hence, force a more scientific basis to the physical aspects of the treatment design and computation process currently being followed in the clinic.

Goitein (1982) has reviewed this problem most recently. He also identifies a major problem in the statistical insignificance imparted to the treatment planning process by our putting isodose lines on paper, optimized, and CT-corrected for heterogeneities via the latest model, without providing the physician a distribution of uncertainties for the plan. He suggests as one simple possibility, giving the physician three plans -- a normal plan, lower-bound plan, and an upper-bound plan:

It is proposed that these extrema plans can be better represented by another technique. Gaussian-distributed set-up points in the patients as well as a similar treatment of other treatment parameters which can randomly wander about some mean parameter (the prescribed parameter, presumably) will be relatively represented as varying source orientation and spatial locations. Random normal deviates can then be generated to permit all 90 fields (from the example given above) to be planned but sampled from assumed Gaussian distributions of beam locations and parameters. The variance assigned to a distribution specifies the mean deviation of that parameter.

An example of the type of effect which can be incorporated into our treatment planning process is shown in Figure 1 and Table 1. The primary effect demonstrated is the effective broadening of the penumbra, as one would expect for this single 10 MV, 10 x 10 cm field, gantry angle 0, collimator angle 0, sampled from a Gaussian distribution of set-up points with a mean of 12 cm and a mean deviation of 4 mm.

Obtaining estimates of the mean deviations in treatment parameters is equally important and a substantial unaccomplished task; according to Goitein (1982), these will depend on the "site of interest, the condition of the patient, and the techniques and policies of the institution".

These parameters will most likely specifically vary with each patient treated. The physical treatment plan process in the future should include the following sequential steps:

a) do initial, conventional, perfect plan, satisfying dose constraints of a normal dose optimization routine (whether manual or computer-assisted) to selected points,

b) iteratively maximize the acceptable mean deviations in each parameter which still permits satisfaction of plan (a), where "uncertainty constraints" have been placed on each point used °in the perfect plan process generating plan (a) and then display plan (b), now defined as plan (a) with these uncertainties displayed,

c) verify that this plan is satisfied in the actual treatments for that patient -- i.e., that the patient is treated within acceptable windows of the various critical treatment parameters for that particular plan.

It is felt that the interactions of step (a) and (b) must be studied for a range of routine treatment planning problems. Then, let the knowledge gained here specify the criteria necessary to begin looking in the future at the various cost-effective technologies available to be brought to bear on step (c).

Current treatment planning computer programs, modified to incorporate these effects, will thus assist further in also investigating mechanisms of improved treatment planning techniques, their clinical and physical significance, and constitute the first of many steps necessary for incorporating clinical-effect models into the radiation physical treatment- and clinical-design process.

III. THE PARTIAL TOLERANCE FRACTION (PTF) CONCEPT

The TDF model of Orton and Ellis (1973) has been clearly useful as a teaching tool for recognizing the effects of dose fraction size, number of fractions, and interfraction time interval on late effects on normal connective tissues. Its application has been, however, criticized, when directed at the prediction of effects in normal tissues other than normal connective tissues, when directed at the prediction of effects for small numbers of fractions, when directed at the prediction of effects in tumors, and for not distinguishing between acute and late effects in the normal connective tissues. Alternative models, the linear quadratic model, and the cell kinetic model, have been presented but have not received widespread incorporation into the clinic to date. This is changing rapidly as the availability of parameters for these alternative models, usable in the clinic, and the education process of the limitations of the TDF concept and its abuses, both, increase.

Perhaps, this author's definition of the PTF will bridge the gap by, at the very least, answering the criticism above, that the TDF model is not applicable for predicting tolerance dose schedules for tissues other than normal connective tissues.

The PTF for a particular normal tissue is defined as:

1) $$PTF = \frac{n}{N}$$

where N is the number of fractions which would lead to a 5% chance of the late complications scored by Cohen (1983) for the particular fractionation schedule in question, and n is the actual number of fractions which have been delivered or will be delivered by that same particular fractionation schedule.

Thus,

2) $$PTF = \frac{nd(1-tex-nex)^{-1}_x - tex(1-tex-nex)^{-1}}{NSD(1-tex-nex)^{-1}}$$

where d is the dose fraction size, tex is the exponent on overall time (T), nex is the exponent on N, x is a variable depending on the number of fractions delivered per week and, the day of the week that the treatments are started (on the average x is 5.49, 3.07, 2.21, 1.55, 1.39, and 0.70 for 1 fraction per week, 2, 3, 4, 5, and 10, respectively), and NSD is the equivalent single dose intercept, all for 100 cm field sizes.

The estimates of pertinent parameters are provided by Cohen (1983) for best fits from clinical data for the model

3) $D = (NSD)VT^{tex}N^{nex}$

where D is the total dose at which there is a 5% probability of incurring the late complication scored, and V is a field size correction factor given by

4) $V = (\frac{100}{A})^Y$

where A is the field area (cm^2) actually used in treatment and Y is the volume effect exponent, also given by Cohen (1983). Y is given values .24, .24, .145, .24, .22, and .13, for normal connective tissue, brain, spinal cord, lung, gut, and kidney, respectively.

Tables 2-7 contain PTF for the various normal tissues for a range of fraction sizes, number of fractions, for the 100 cm^2 field size.

Figures 2-7 are graphs of N versus d.

PTF's, like the TDF's are additive for that tissue. They are most valid for conventional fraction numbers. The single fraction PTF is given only for the convenience of using the tables. They are most valid, only when used to predict a 5% probability of the late effect scored. Arriving at a cumulative PTF of 1.0 for some conbination of schedules for a particular tissue is tolerance. Split-course treatments may be treated in the conventional way, with a decay factor on the earlier schedule given by

5. $DF = (\frac{T}{T+R})^{tex}$

for that particular tissue where T, here, is the treatment elapsed time for the course under consideration and R is the gap time before picking up the last part of the split course.

The PTF tables have been very well received in the clinic and they have removed some of the conflicts which evolve during discussions between medical physicists (the one often requested to "give me the TDF on this, and how much further can I go with this new schedule") and the radiation oncologist who is summarizing past treatments and planning new treatment schedules with the singularly abused TDF tables.

The PTF tables answer only one of the criticisms addressed by Withers, Thames, and Peters (1983), that is, using the TDF for other than normal connective tissue late effects, but not the other primary concern, that is addressing the problem of the prediction of dose schedules on the basis of some anticipated early, acute effect. Their model is based on the linear-quadratic survival model and is being studied carefully by our group for inclusion in our treatment planning systems, as well.

IV. APPLICATIONS OF SAMPLED MEAN PLANS AND PTF

Computer programs are being written to take advantage, clinically of the two concepts discussed in this presentation. Daily, physical dose distributions are to be simulated by random sampling from the normally-distributed set of possible prescribed treatment fields' parameters and the resultant sampled mean daily PTF distribution for the normal tissues in that treatment volume of the patient calculated. The cumulative prescribed, "perfect" physical and PTF distribution for the total prescription can be generated as well as the respective cumulative sampled mean plans. It is this author's hypothesis that these plans will provide a means to conduct more meaningful discussions with the radiation oncologists in the daily treatment planning decision-making process, and thus answer, better, some of the questions posed in the Introduction. Additionally, criteria for immobilization requirements can be better stated on an individual basis, especially when optimized plans are generated, utilizing tumor response functions liberated from clinical data with the techniques discussed during this conference.

REFERENCES

1. Withers, H., Thames, H., Peters, L., "A new isoeffect curve for change in dose per fraction", Radiotherapy and Oncology, 1:187 (1983).

2. Orton, C., Ellis, F., "A simplification in the use of the NSD concept in practical radiotherapy", Br. J. Radiology, 46:529 (1973).

3. Cohen, L., Creditor, M., "Iso-effect tables for tolerance of irradiated normal human tissues", Int. J. Radiation Oncology Biol. Phys., 9:233 (1983).

4. Goitein, M., Med. Phys., 9:580 (1982).

TABLE 1

Sampled Mean Plan

Single Field
10 x 10 cm
100 cm SAD, isoc. depth 12 cm
Phantom Thickness 24 cm
Distribution: 54%, 0 mm displacement
               35%, 4 mm displacement
               11%, 8 mm displacement

Penumbra

Off - Axis	Perfect	Sampled	Percentage Points Difference (Sampled - Perfect)
+ 0	100%	100%	0
3.6	97.2	96.6	- .6
3.8	96.0	94.8	-1.2
4.0	94.7	92.6	-2.1
4.2	91.0	87.9	-3.1
4.4	86.3	82.2	-4.1
4.6	74.8	71.6	-3.2
4.8	60.6	59.5	-1.1
5.0	43.6	45.4	+1.8
5.2	30.4	33.8	+3.4
5.4	20.9	24.5	+3.6
5.6	15.0	18.0	+3.0
5.8	12.8	14.6	+1.8
6.0	11.0	12.0	+1.0
6.2	10.1	10.6	+0.5
6.4	9.3	9.5	+0.2

TABLE 2A

LATE EFFECTS IN CONNECTIVE TISSUES/SKIN (ELLIS)
PARTIAL TOLERANCE FRACTION TABLES
compiled by Larry Simpson--01/31/84

the number frac per week is : 1
the exponent on N is : 0.240
the exponent on T is : 0.110
the NSD is : 1768

Dose per Fraction, d	Number of Fractions, n ---->								
	1	5	10	15	20	25	30	35	40
50	0.004	0.02	0.04	0.05	0.07	0.08	0.10	0.11	0.13
100	0.010	0.05	0.10	0.14	0.19	0.23	0.28	0.32	0.37
150	0.017	0.09	0.17	0.26	0.34	0.43	0.51	0.59	0.68
200	0.027	0.14	0.27	0.40	0.53	0.66	0.79	0.92	1.05
250	0.037	0.19	0.37	0.56	0.74	0.93	1.11	1.30	1.48
300	0.049	0.25	0.49	0.74	0.98	1.23	1.47	1.72	1.96
350	0.063	0.32	0.63	0.94	1.25	1.56	1.87	2.18	2.49
400	0.077	0.39	0.77	1.15	1.53	1.91	2.29	2.67	...
450	0.092	0.46	0.92	1.37	1.83	2.29	2.74	...	...
500	0.108	0.54	1.08	1.62	2.15	2.69	...	...	...
550	0.125	0.63	1.25	1.87	2.49	...	...	...	...
600	0.143	0.72	1.43	2.14	2.85	...	...	...	...
650	0.161	0.81	1.61	2.42	...	...	...	...	...
700	0.181	0.91	1.81	2.71	...	...	...	...	...
750	0.201	1.01	2.01	...	...	...	...	...	...
800	0.222	1.11	2.22	...	...	...	...	...	...
900	0.266	1.33	2.66	...	...	...	...	...	...

TABLE 2B

LATE EFFECTS IN CONNECTIVE TISSUES/SKIN (ELLIS)
PARTIAL TOLERANCE FRACTION TABLES
compiled by Larry Simpson--01/31/84

the number frac per week is : 2
the exponent on N is : 0.240
the exponent on T is : 0.110
the NSD is : 1768

Dose per Fraction, d	Number of Fractions, n ---->								
	1	5	10	15	20	25	30	35	40
50	0.004	0.02	0.04	0.06	0.07	0.09	0.11	0.13	0.14
100	0.010	0.05	0.10	0.15	0.20	0.25	0.30	0.35	0.40
150	0.019	0.10	0.19	0.28	0.38	0.47	0.56	0.66	0.75
200	0.029	0.15	0.29	0.44	0.58	0.73	0.87	1.02	1.16
250	0.041	0.21	0.41	0.62	0.82	1.02	1.23	1.43	1.64
300	0.054	0.27	0.54	0.81	1.08	1.35	1.62	1.89	2.16
350	0.069	0.35	0.69	1.03	1.37	1.72	2.06	2.40	2.74
400	0.085	0.43	0.85	1.27	1.69	2.11	2.53	2.95	...
450	0.101	0.51	1.01	1.52	2.02	2.52	...	...	...
500	0.119	0.60	1.19	1.78	2.37	2.97	...	...	...
550	0.138	0.69	1.38	2.06	2.75	...	...	...	...
600	0.157	0.79	1.57	2.36	...	...	...	...	...
650	0.178	0.89	1.78	2.67	...	...	...	...	...
700	0.199	1.00	1.99	2.99	...	...	...	...	...
750	0.222	1.11	2.22	...	...	...	...	...	...
800	0.245	1.23	2.45	...	...	...	...	...	...
900	0.293	1.47	2.93	...	...	...	...	...	...

LATE EFFECTS IN
CONNECTIVE TISSUES/SKIN (ELLIS)   TABLE 2C

PARTIAL TOLERANCE FRACTION TABLES
compiled by Larry Simpson--01/31/84

the number frac per week is : 3
the exponent on N is : 0.240
the exponent on T is : 0.110
the NSD is : 1768

Dose per Fraction, d   Number of Fractions, n ----->

Dose	1	5	10	15	20	25	30	35	40
50	0.004	0.02	0.04	0.06	0.08	0.10	0.11	0.13	0.15
100	0.011	0.06	0.11	0.16	0.22	0.27	0.32	0.37	0.43
150	0.020	0.10	0.20	0.30	0.40	0.50	0.59	0.69	0.79
200	0.031	0.16	0.31	0.46	0.62	0.77	0.92	1.08	1.23
250	0.044	0.22	0.44	0.65	0.87	1.08	1.30	1.51	1.73
300	0.058	0.29	0.58	0.86	1.15	1.43	1.72	2.00	2.29
350	0.073	0.37	0.73	1.09	1.45	1.81	2.18	2.54	2.90
400	0.089	0.45	0.89	1.34	1.78	2.23	2.67	...	...
450	0.107	0.54	1.07	1.60	2.14	2.67	...	...	...
500	0.126	0.63	1.26	1.88	2.51	...	...	...	...
550	0.146	0.73	1.46	2.18	2.91	...	...	...	...
600	0.166	0.83	1.66	2.49	...	...	...	...	...
650	0.188	0.94	1.88	2.82	...	...	...	...	...
700	0.211	1.06	2.11	...	...	...	...	...	...
750	0.234	1.17	2.34	...	...	...	...	...	...
800	0.259	1.30	2.59	...	...	...	...	...	...
900	0.310	1.55	...	...	...	...	...	...	...

LATE EFFECTS IN
CONNECTIVE TISSUES/SKIN (ELLIS)   TABLE 2D

PARTIAL TOLERANCE FRACTION TABLES
compiled by Larry Simpson--01/31/84

the number frac per week is : 4
the exponent on N is : 0.240
the exponent on T is : 0.110
the NSD is : 1768

Dose per Fraction, d   Number of Fractions, n ----->

Dose	1	5	10	15	20	25	30	35	40
50	0.004	0.02	0.04	0.06	0.08	0.10	0.12	0.14	0.16
100	0.012	0.06	0.12	0.17	0.23	0.28	0.34	0.40	0.45
150	0.021	0.11	0.21	0.32	0.42	0.53	0.63	0.74	0.84
200	0.033	0.17	0.33	0.49	0.65	0.82	0.98	1.14	1.30
250	0.046	0.23	0.46	0.69	0.92	1.15	1.38	1.61	1.84
300	0.061	0.31	0.61	0.91	1.22	1.52	1.82	2.13	2.43
350	0.077	0.39	0.77	1.16	1.54	1.93	2.31	2.69	...
400	0.095	0.48	0.95	1.42	1.89	2.36	2.84	...	...
450	0.114	0.57	1.14	1.70	2.27	2.83	...	...	...
500	0.134	0.67	1.34	2.00	2.67	...	...	...	...
550	0.155	0.78	1.55	2.32	...	...	...	...	...
600	0.177	0.89	1.77	2.65	...	...	...	...	...
650	0.200	1.00	2.00	2.99	...	...	...	...	...
700	0.224	1.12	2.24	...	...	...	...	...	...
750	0.249	1.25	2.49	...	...	...	...	...	...
800	0.275	1.38	2.75	...	...	...	...	...	...
900	0.329	1.65	...	...	...	...	...	...	...

LATE EFFECTS IN
CONNECTIVE TISSUES/SKIN (ELLIS) TABLE 2E

PARTIAL TOLERANCE FRACTION TABLES
compiled by Larry Simpson--01/31/84

the number frac per week is : 5
the exponent on N is : 0.240
the exponent on T is : 0.110
the NSD is : 1768

Dose per Fraction, d   Number of Fractions, n ---->

Dose per Fraction, d	1	5	10	15	20	25	30	35	40
50	0.004	0.02	0.04	0.06	0.08	0.10	0.12	0.14	0.16
100	0.012	0.06	0.12	0.18	0.23	0.29	0.35	0.40	0.46
150	0.022	0.11	0.22	0.32	0.43	0.54	0.64	0.75	0.86
200	0.034	0.17	0.34	0.50	0.67	0.83	1.00	1.16	1.33
250	0.047	0.24	0.47	0.70	0.94	1.17	1.40	1.64	1.87
300	0.062	0.31	0.62	0.93	1.24	1.55	1.86	2.17	2.47
350	0.079	0.40	0.79	1.18	1.57	1.96	2.35	2.74	...
400	0.097	0.49	0.97	1.45	1.93	2.41	2.89	...	...
450	0.116	0.58	1.16	1.73	2.31	2.89	...	...	...
500	0.136	0.68	1.36	2.04	2.71	...	...	...	...
550	0.157	0.79	1.57	2.36	...	...	...	...	...
600	0.180	0.90	1.80	2.70	...	...	...	...	...
650	0.203	1.02	2.03	...	...	...	...	...	...
700	0.228	1.14	2.28	...	...	...	...	...	...
750	0.253	1.27	2.53	...	...	...	...	...	...
800	0.280	1.40	2.80	...	...	...	...	...	...
900	0.335	1.68	...	...	...	...	...	...	...

LATE EFFECTS IN
BRAIN (COHEN 1983)
TABLE 3A

PARTIAL TOLERANCE FRACTION TABLES
compiled by Larry Simpson--01/31/84

the number frac per week is : 1
the exponent on N is : 0.560
the exponent on T is : 0.030
the NSD is : 769

Dose per Fraction, d | Number of Fractions, n ----->

Dose per Fraction, d	1	5	10	15	20	25	30	35	40
50	0.002	0.01	0.02	0.02	0.03	0.03	0.04	0.04	0.05
100	0.007	0.04	0.07	0.10	0.13	0.16	0.19	0.22	0.25
150	0.017	0.09	0.17	0.25	0.33	0.41	0.50	0.58	0.66
200	0.034	0.17	0.34	0.50	0.67	0.83	1.00	1.16	1.33
250	0.057	0.29	0.57	0.86	1.14	1.43	1.71	2.00	2.28
300	0.089	0.45	0.89	1.34	1.78	2.23	2.67	...	...
350	0.130	0.65	1.30	1.95	2.59	...	...	...	...
400	0.180	0.90	1.80	2.69	...	...	...	...	...
450	0.239	1.20	2.39	...	...	...	...	...	...
500	0.309	1.55	...	...	...	...	...	...	...
550	0.390	1.95	...	...	...	...	...	...	...
600	0.482	2.41	...	...	...	...	...	...	...
650	0.586	2.93	...	...	...	...	...	...	...
700	0.702	...	...	...	...	...	...	...	...
750	0.831	...	...	...	...	...	...	...	...
800	0.973	...	...	...	...	...	...	...	...
900	1.296	...	...	...	...	...	...	...	...

LATE EFFECTS IN
BRAIN (COHEN 1983)
TABLE 3B

PARTIAL TOLERANCE FRACTION TABLES
compiled by Larry Simpson--01/31/84

the number frac per week is : 2
the exponent on N is : 0.560
the exponent on T is : 0.030
the NSD is : 769

Dose per Fraction, d | Number of Fractions, n ----->

Dose per Fraction, d	1	5	10	15	20	25	30	35	40
50	0.002	0.01	0.02	0.02	0.03	0.03	0.04	0.05	0.05
100	0.007	0.04	0.07	0.10	0.13	0.16	0.20	0.23	0.26
150	0.018	0.09	0.18	0.26	0.35	0.43	0.52	0.60	0.69
200	0.035	0.18	0.35	0.52	0.69	0.87	1.04	1.21	1.38
250	0.060	0.30	0.60	0.90	1.19	1.49	1.79	2.09	2.38
300	0.093	0.47	0.93	1.40	1.86	2.32	2.79	...	...
350	0.136	0.68	1.36	2.03	2.71	...	...	...	...
400	0.188	0.94	1.88	2.81	...	...	...	...	...
450	0.250	1.25	2.50	...	...	...	...	...	...
500	0.323	1.62	...	...	...	...	...	...	...
550	0.407	2.04	...	...	...	...	...	...	...
600	0.503	2.52	...	...	...	...	...	...	...
650	0.612	...	...	...	...	...	...	...	...
700	0.733	...	...	...	...	...	...	...	...
750	0.867	...	...	...	...	...	...	...	...
800	1.015	...	...	...	...	...	...	...	...
900	1.353	...	...	...	...	...	...	...	...

LATE EFFECTS IN
BRAIN (COHEN 1983)

TABLE 3C

PARTIAL TOLERANCE FRACTION TABLES
compiled by Larry Simpson--01/31/84

the number frac per week is : 3
the exponent on N is : 0.560
the exponent on T is : 0.030
the NSD is : 769

Dose per Fraction, d   Number of Fractions, n ----->

Dose per Fraction, d	1	5	10	15	20	25	30	35	40
50	0.002	0.01	0.02	0.02	0.03	0.04	0.04	0.05	0.05
100	0.007	0.04	0.07	0.10	0.14	0.17	0.20	0.23	0.27
150	0.018	0.09	0.18	0.27	0.36	0.44	0.53	0.62	0.71
200	0.036	0.18	0.36	0.54	0.71	0.89	1.07	1.24	1.42
250	0.061	0.31	0.61	0.92	1.22	1.53	1.83	2.14	2.44
300	0.095	0.48	0.95	1.43	1.90	2.38	2.85	...	...
350	0.139	0.70	1.39	2.08	2.77	...	...	...	...
400	0.192	0.96	1.92	2.88	...	...	...	...	...
450	0.256	1.28	2.56	...	...	...	...	...	...
500	0.331	1.66	...	...	...	...	...	...	...
550	0.417	2.09	...	...	...	...	...	...	...
600	0.516	2.58	...	...	...	...	...	...	...
650	0.627	...	...	...	...	...	...	...	...
700	0.751	...	...	...	...	...	...	...	...
750	0.888	...	...	...	...	...	...	...	...
800	1.040	...	...	...	...	...	...	...	...
900	1.385	...	...	...	...	...	...	...	...

LATE EFFECTS IN
BRAIN (COHEN 1983)

TABLE 3D

PARTIAL TOLERANCE FRACTION TABLES
compiled by Larry Simpson--01/31/84

the number frac per week is : 4
the exponent on N is : 0.560
the exponent on T is : 0.030
the NSD is : 769

Dose per Fraction, d   Number of Fractions, n ----->

Dose per Fraction, d	1	5	10	15	20	25	30	35	40
50	0.002	0.01	0.02	0.02	0.03	0.04	0.04	0.05	0.05
100	0.007	0.04	0.07	0.11	0.14	0.17	0.21	0.24	0.27
150	0.018	0.09	0.18	0.27	0.36	0.45	0.54	0.63	0.72
200	0.037	0.19	0.37	0.55	0.73	0.91	1.09	1.27	1.46
250	0.063	0.32	0.63	0.94	1.25	1.57	1.88	2.19	2.50
300	0.098	0.49	0.98	1.47	1.95	2.44	2.93	...	...
350	0.142	0.71	1.42	2.13	2.84	...	...	...	...
400	0.197	0.99	1.97	2.95	...	...	...	...	...
450	0.263	1.32	2.63	...	...	...	...	...	...
500	0.339	1.70	...	...	...	...	...	...	...
550	0.428	2.14	...	...	...	...	...	...	...
600	0.529	2.65	...	...	...	...	...	...	...
650	0.643	...	...	...	...	...	...	...	...
700	0.771	...	...	...	...	...	...	...	...
750	0.912	...	...	...	...	...	...	...	...
800	1.067	...	...	...	...	...	...	...	...
900	1.422	...	...	...	...	...	...	...	...

LATE EFFECTS IN
BRAIN (COHEN 1983)

TABLE 3E

PARTIAL TOLERANCE FRACTION TABLES
compiled by Larry Simpson--01/31/84

the number frac per week is : 5
the exponent on N is : 0.560
the exponent on T is : 0.030
the NSD is : 769

Number of Fractions, n ----->

Dose per Fraction, d	1	5	10	15	20	25	30	35	40
50	0.002	0.01	0.02	0.02	0.03	0.04	0.04	0.05	0.05
100	0.007	0.04	0.07	0.11	0.14	0.17	0.21	0.24	0.27
150	0.019	0.10	0.19	0.28	0.37	0.46	0.55	0.64	0.73
200	0.037	0.19	0.37	0.55	0.74	0.92	1.10	1.28	1.47
250	0.063	0.32	0.63	0.95	1.26	1.58	1.89	2.21	2.52
300	0.099	0.50	0.99	1.48	1.97	2.46	2.95	...	...
350	0.144	0.72	1.44	2.15	2.87	...	...	...	...
400	0.199	1.00	1.99	2.98	...	...	...	...	...
450	0.265	1.33	2.65	...	...	...	...	...	...
500	0.342	1.71	...	...	...	...	...	...	...
550	0.432	2.16	...	...	...	...	...	...	...
600	0.533	2.67	...	...	...	...	...	...	...
650	0.648	...	...	...	...	...	...	...	...
700	0.777	...	...	...	...	...	...	...	...
750	0.919	...	...	...	...	...	...	...	...
800	1.075	...	...	...	...	...	...	...	...
900	1.433	...	...	...	...	...	...	...	...

LATE EFFECTS IN
SPINAL CORD (COHEN 1983)   TABLE 4A

PARTIAL TOLERANCE FRACTION TABLES
compiled by Larry Simpson--01/31/84

the number frac per week is : 1
the exponent on N is : 0.420
the exponent on T is : 0.060
the NSD is : 1023

Dose per
Fraction, d   Number of Fractions, n ----->

	1	5	10	15	20	25	30	35	40
	1	0.02	0.03	0.04	0.05	0.07	0.08	0.09	0.10
50	0.003	0.02	0.03	0.04	0.05	0.07	0.08	0.09	0.10
100	0.010	0.05	0.10	0.15	0.19	0.24	0.29	0.33	0.38
150	0.021	0.11	0.21	0.31	0.41	0.52	0.62	0.72	0.82
200	0.036	0.18	0.36	0.54	0.72	0.90	1.07	1.25	1.43
250	0.055	0.28	0.55	0.83	1.10	1.37	1.65	1.92	2.19
300	0.078	0.39	0.78	1.17	1.56	1.95	2.33	2.72	...
350	0.105	0.53	1.05	1.57	2.09	2.62	...	...	...
400	0.136	0.68	1.36	2.03	2.71	...	...	...	...
450	0.170	0.85	1.70	2.55	...	...	...	...	...
500	0.208	1.04	2.08	...	...	...	...	...	...
550	0.250	1.25	2.50	...	...	...	...	...	...
600	0.295	1.48	2.95	...	...	...	...	...	...
650	0.344	1.72	...	...	...	...	...	...	...
700	0.397	1.99	...	...	...	...	...	...	...
750	0.453	2.27	...	...	...	...	...	...	...
800	0.513	2.57	...	...	...	...	...	...	...
900	0.643	...	...	...	...	...	...	...	...

LATE EFFECTS IN
SPINAL CORD (COHEN 1983)   TABLE 4B

PARTIAL TOLERANCE FRACTION TABLES
compiled by Larry Simpson--01/31/84

the number frac per week is : 2
the exponent on N is : 0.420
the exponent on T is : 0.060
the NSD is : 1023

Dose per
Fraction, d   Number of Fractions, n ----->

	1	5	10	15	20	25	30	35	40
		0.02	0.03	0.04	0.06	0.07	0.08	0.10	0.11
50	0.003	0.02	0.03	0.04	0.06	0.07	0.08	0.10	0.11
100	0.011	0.06	0.11	0.16	0.21	0.26	0.31	0.36	0.41
150	0.022	0.11	0.22	0.33	0.44	0.55	0.66	0.77	0.88
200	0.039	0.20	0.39	0.58	0.77	0.96	1.15	1.34	1.53
250	0.059	0.30	0.59	0.88	1.17	1.47	1.76	2.05	2.34
300	0.084	0.42	0.84	1.25	1.67	2.08	2.50	2.91	...
350	0.112	0.56	1.12	1.68	2.24	2.80	...	...	...
400	0.145	0.73	1.45	2.17	2.89	...	...	...	...
450	0.182	0.91	1.82	2.72	...	...	...	...	...
500	0.222	1.11	2.22	...	...	...	...	...	...
550	0.267	1.34	2.67	...	...	...	...	...	...
600	0.315	1.58	...	...	...	...	...	...	...
650	0.368	1.84	...	...	...	...	...	...	...
700	0.424	2.12	...	...	...	...	...	...	...
750	0.484	2.42	...	...	...	...	...	...	...
800	0.548	2.74	...	...	...	...	...	...	...
900	0.687	...	...	...	...	...	...	...	...

LATE EFFECTS IN
SPINAL CORD (COHEN 1983)   TABLE 4C

PARTIAL TOLERANCE FRACTION TABLES
compiled by Larry Simpson--01/31/84

the number frac per week is : 3
the exponent on N is : 0.420
the exponent on T is : 0.060
the NSD is : 1023

Dose per Fraction, d   Number of Fractions, n ---->

d	1	5	10	15	20	25	30	35	40
50	0.003	0.02	0.03	0.05	0.06	0.07	0.09	0.10	0.11
100	0.011	0.06	0.11	0.16	0.21	0.27	0.32	0.37	0.42
150	0.023	0.12	0.23	0.35	0.46	0.57	0.69	0.80	0.91
200	0.040	0.20	0.40	0.60	0.80	0.99	1.19	1.39	1.59
250	0.061	0.31	0.61	0.92	1.22	1.52	1.83	2.13	2.43
300	0.087	0.44	0.87	1.30	1.73	2.16	2.59	...	...
350	0.117	0.59	1.17	1.75	2.33	2.91	...	...	...
400	0.150	0.75	1.50	2.25	3.00	...	...	...	...
450	0.189	0.95	1.89	2.83	...	...	...	...	...
500	0.231	1.16	2.31	...	...	...	...	...	...
550	0.277	1.39	2.77	...	...	...	...	...	...
600	0.328	1.64	...	...	...	...	...	...	...
650	0.382	1.91	...	...	...	...	...	...	...
700	0.440	2.20	...	...	...	...	...	...	...
750	0.503	2.52	...	...	...	...	...	...	...
800	0.569	2.85	...	...	...	...	...	...	...
900	0.714	...	...	...	...	...	...	...	...

LATE EFFECTS IN
SPINAL CORD (COHEN 1983)   TABLE 4D

PARTIAL TOLERANCE FRACTION TABLES
compiled by Larry Simpson--01/31/84

the number frac per week is : 4
the exponent on N is : 0.420
the exponent on T is : 0.060
the NSD is : 1023

Dose per Fraction, d   Number of Fractions, n ---->

d	1	5	10	15	20	25	30	35	40
50	0.003	0.02	0.03	0.05	0.06	0.08	0.09	0.11	0.12
100	0.011	0.06	0.11	0.17	0.22	0.28	0.33	0.39	0.44
150	0.024	0.12	0.24	0.36	0.48	0.60	0.72	0.83	0.95
200	0.042	0.21	0.42	0.62	0.83	1.03	1.24	1.45	1.65
250	0.064	0.32	0.64	0.95	1.27	1.59	1.90	2.22	2.54
300	0.090	0.45	0.90	1.35	1.80	2.25	2.70	...	...
350	0.121	0.61	1.21	1.82	2.42	...	...	...	...
400	0.157	0.79	1.57	2.35	...	...	...	...	...
450	0.196	0.98	1.96	2.94	...	...	...	...	...
500	0.240	1.20	2.40	...	...	...	...	...	...
550	0.289	1.45	2.89	...	...	...	...	...	...
600	0.341	1.71	...	...	...	...	...	...	...
650	0.398	1.99	...	...	...	...	...	...	...
700	0.459	2.30	...	...	...	...	...	...	...
750	0.524	2.62	...	...	...	...	...	...	...
800	0.593	2.97	...	...	...	...	...	...	...
900	0.744	...	...	...	...	...	...	...	...

LATE EFFECTS IN
SPINAL CORD (COHEN 1983)   TABLE 4E

PARTIAL TOLERANCE FRACTION TABLES
compiled by Larry Simpson--01/31/84

the number frac per week is : 5
the exponent on N is : 0.420
the exponent on T is : 0.060
the NSD is : 1023

Number of Fractions, n ----->

Dose per Fraction, d	1	5	10	15	20	25	30	35	40
50	0.003	0.02	0.03	0.05	0.06	0.08	0.09	0.11	0.12
100	0.012	0.06	0.12	0.17	0.23	0.28	0.34	0.39	0.45
150	0.024	0.12	0.24	0.36	0.48	0.60	0.72	0.84	0.96
200	0.042	0.21	0.42	0.63	0.84	1.05	1.26	1.47	1.67
250	0.065	0.33	0.65	0.97	1.29	1.61	1.93	2.25	2.57
300	0.091	0.46	0.91	1.37	1.82	2.28	2.73	...	...
350	0.123	0.62	1.23	1.84	2.45	...	...	...	...
400	0.159	0.80	1.59	2.38	...	...	...	...	...
450	0.199	1.00	1.99	2.98	...	...	...	...	...
500	0.243	1.22	2.43	...	...	...	...	...	...
550	0.292	1.46	2.92	...	...	...	...	...	...
600	0.346	1.73	...	...	...	...	...	...	...
650	0.403	2.02	...	...	...	...	...	...	...
700	0.465	2.33	...	...	...	...	...	...	...
750	0.530	2.65	...	...	...	...	...	...	...
800	0.600	3.00	...	...	...	...	...	...	...
900	0.753	...	...	...	...	...	...	...	...

LATE EFFECTS IN
LUNG (COHEN 1983)   TABLE 5A
       PARTIAL TOLERANCE FRACTION TABLES
     compiled by Larry Simpson--01/31/84

the number frac per week is : 1
the exponent on N is : 0.250
the exponent on T is : 0.040
the NSD is : 1116
Dose per   Number of Fractions, n ----->
Fraction, d

	1	5	10	15	20	25	30	35	40
50	0.012	0.06	0.12	0.18	0.23	0.29	0.35	0.41	0.46
100	0.031	0.16	0.31	0.46	0.61	0.76	0.92	1.07	1.22
150	0.054	0.27	0.54	0.81	1.08	1.35	1.62	1.89	2.16
200	0.081	0.41	0.81	1.22	1.62	2.02	2.43	2.83	...
250	0.111	0.56	1.11	1.66	2.21	2.77	...	...	...
300	0.143	0.72	1.43	2.15	2.86	...	...	...	...
350	0.178	0.89	1.78	2.67	...	...	...	...	...
400	0.215	1.08	2.15	...	...	...	...	...	...
450	0.253	1.27	2.53	...	...	...	...	...	...
500	0.294	1.47	2.94	...	...	...	...	...	...
550	0.336	1.68	...	...	...	...	...	...	...
600	0.380	1.90	...	...	...	...	...	...	...
650	0.425	2.13	...	...	...	...	...	...	...
700	0.472	2.36	...	...	...	...	...	...	...
750	0.520	2.60	...	...	...	...	...	...	...
800	0.569	2.85	...	...	...	...	...	...	...
900	0.672	...	...	...	...	...	...	...	...

LATE EFFECTS IN
LUNG (COHEN 1983)   TABLE 5B
       PARTIAL TOLERANCE FRACTION TABLES
     compiled by Larry Simpson--01/31/84

the number frac per week is : 2
the exponent on N is : 0.250
the exponent on T is : 0.040
the NSD is : 1116
Dose per   Number of Fractions, n ----->
Fraction, d

	1	5	10	15	20	25	30	35	40
50	0.012	0.06	0.12	0.18	0.24	0.30	0.36	0.42	0.48
100	0.032	0.16	0.32	0.48	0.63	0.79	0.95	1.10	1.26
150	0.056	0.28	0.56	0.84	1.12	1.39	1.67	1.95	2.23
200	0.084	0.42	0.84	1.26	1.67	2.09	2.51	2.92	...
250	0.115	0.58	1.15	1.72	2.29	2.86	...	...	...
300	0.148	0.74	1.48	2.22	2.96	...	...	...	...
350	0.184	0.92	1.84	2.76	...	...	...	...	...
400	0.222	1.11	2.22	...	...	...	...	...	...
450	0.262	1.31	2.62	...	...	...	...	...	...
500	0.303	1.52	...	...	...	...	...	...	...
550	0.347	1.74	...	...	...	...	...	...	...
600	0.392	1.96	...	...	...	...	...	...	...
650	0.439	2.20	...	...	...	...	...	...	...
700	0.487	2.44	...	...	...	...	...	...	...
750	0.537	2.69	...	...	...	...	...	...	...
800	0.588	2.94	...	...	...	...	...	...	...
900	0.694	...	...	...	...	...	...	...	...

TABLE 5C

LATE EFFECTS IN
LUNG (COHEN 1983)

PARTIAL TOLERANCE FRACTION TABLES
compiled by Larry Simpson--01/31/84

the number frac per week is : 3
the exponent on N is : 0.250
the exponent on T is : 0.040
the NSD is : 1116

Dose per Fraction, d   Number of Fractions, n ----->

Dose per Fraction, d	1	5	10	15	20	25	30	35	40
50	0.013	0.07	0.13	0.19	0.25	0.31	0.37	0.43	0.49
100	0.032	0.16	0.32	0.48	0.64	0.80	0.96	1.12	1.28
150	0.057	0.29	0.57	0.85	1.14	1.42	1.70	1.99	2.27
200	0.085	0.43	0.85	1.28	1.70	2.13	2.55	2.98	...
250	0.117	0.59	1.17	1.75	2.33	2.91	...	...	...
300	0.151	0.76	1.51	2.26	...	...	...	...	...
350	0.187	0.94	1.87	2.81	...	...	...	...	...
400	0.226	1.13	2.26	...	...	...	...	...	...
450	0.267	1.34	2.67	...	...	...	...	...	...
500	0.309	1.55	...	...	...	...	...	...	...
550	0.354	1.77	...	...	...	...	...	...	...
600	0.400	2.00	...	...	...	...	...	...	...
650	0.447	2.24	...	...	...	...	...	...	...
700	0.496	2.48	...	...	...	...	...	...	...
750	0.547	2.74	...	...	...	...	...	...	...
800	0.599	3.00	...	...	...	...	...	...	...
900	0.707	...	...	...	...	...	...	...	...

TABLE 5D

LATE EFFECTS IN
LUNG (COHEN 1983)

PARTIAL TOLERANCE FRACTION TABLES
compiled by Larry Simpson--01/31/84

the number frac per week is : 4
the exponent on N is : 0.250
the exponent on T is : 0.040
the NSD is : 1116

Dose per Fraction, d   Number of Fractions, n ----->

Dose per Fraction, d	1	5	10	15	20	25	30	35	40
50	0.013	0.07	0.13	0.19	0.25	0.31	0.37	0.44	0.50
100	0.033	0.17	0.33	0.49	0.66	0.82	0.98	1.15	1.31
150	0.058	0.29	0.58	0.87	1.16	1.45	1.74	2.03	2.32
200	0.087	0.44	0.87	1.30	1.74	2.17	2.60	...	...
250	0.119	0.60	1.19	1.78	2.38	2.97	...	...	...
300	0.154	0.77	1.54	2.31	...	...	...	...	...
350	0.191	0.96	1.91	2.86	...	...	...	...	...
400	0.230	1.15	2.30	...	...	...	...	...	...
450	0.272	1.36	2.72	...	...	...	...	...	...
500	0.315	1.58	...	...	...	...	...	...	...
550	0.361	1.81	...	...	...	...	...	...	...
600	0.408	2.04	...	...	...	...	...	...	...
650	0.456	2.28	...	...	...	...	...	...	...
700	0.506	2.53	...	...	...	...	...	...	...
750	0.558	2.79	...	...	...	...	...	...	...
800	0.611	...	...	...	...	...	...	...	...
900	0.721	...	...	...	...	...	...	...	...

LATE EFFECTS IN
LUNG (COHEN 1983)   TABLE 5E

PARTIAL TOLERANCE FRACTION TABLES
compiled by Larry Simpson--01/31/84

the number frac per week is : 5
the exponent on N is : 0.250
the exponent on T is : 0.040
the NSD is : 1116

Dose per   Number of Fractions, n ----->
Fraction, d

d	1	5	10	15	20	25	30	35	40
50	0.013	0.07	0.13	0.19	0.25	0.31	0.38	0.44	0.50
100	0.033	0.17	0.33	0.50	0.66	0.83	0.99	1.15	1.32
150	0.059	0.30	0.59	0.88	1.17	1.46	1.75	2.04	2.33
200	0.088	0.44	0.88	1.31	1.75	2.18	2.62	...	...
250	0.120	0.60	1.20	1.80	2.39	2.99	...	...	...
300	0.155	0.78	1.55	2.32	...	...	...	...	...
350	0.192	0.96	1.92	2.88	...	...	...	...	...
400	0.232	1.16	2.32	...	...	...	...	...	...
450	0.274	1.37	2.74	...	...	...	...	...	...
500	0.317	1.59	...	...	...	...	...	...	...
550	0.363	1.82	...	...	...	...	...	...	...
600	0.410	2.05	...	...	...	...	...	...	...
650	0.459	2.30	...	...	...	...	...	...	...
700	0.509	2.55	...	...	...	...	...	...	...
750	0.561	2.81	...	...	...	...	...	...	...
800	0.615	...	...	...	...	...	...	...	...
900	0.726	...	...	...	...	...	...	...	...

LATE EFFECTS IN GUT (COHEN 1983) TABLE 6A

PARTIAL TOLERANCE FRACTION TABLES
compiled by Larry Simpson--01/31/84

the number frac per week is : 1
the exponent on N is : 0.290
the exponent on T is : 0.080
the NSD is : 1528

Dose per Fraction, d — Number of Fractions, n ---->

d	1	5	10	15	20	25	30	35	40
50	0.004	0.02	0.04	0.06	0.08	0.09	0.11	0.13	0.15
100	0.011	0.06	0.11	0.16	0.22	0.27	0.32	0.38	0.43
150	0.021	0.11	0.21	0.31	0.41	0.51	0.61	0.71	0.81
200	0.032	0.16	0.32	0.48	0.64	0.80	0.96	1.12	1.28
250	0.046	0.23	0.46	0.69	0.92	1.14	1.37	1.60	1.83
300	0.061	0.31	0.61	0.92	1.22	1.52	1.83	2.13	2.44
350	0.078	0.39	0.78	1.17	1.56	1.95	2.33	2.72	...
400	0.096	0.48	0.96	1.44	1.92	2.40	2.88	...	...
450	0.116	0.58	1.16	1.74	2.32	2.90	...	...	...
500	0.137	0.69	1.37	2.06	2.74	...	...	...	...
550	0.160	0.80	1.60	2.39	...	...	...	...	...
600	0.183	0.92	1.83	2.75	...	...	...	...	...
650	0.208	1.04	2.08	...	...	...	...	...	...
700	0.234	1.17	2.34	...	...	...	...	...	...
750	0.261	1.31	2.61	...	...	...	...	...	...
800	0.289	1.45	2.89	...	...	...	...	...	...
900	0.348	1.74	...	...	...	...	...	...	...

LATE EFFECTS IN GUT (COHEN 1983) TABLE 6B

PARTIAL TOLERANCE FRACTION TABLES
compiled by Larry Simpson--01/31/84

the number frac per week is : 2
the exponent on N is : 0.290
the exponent on T is : 0.080
the NSD is : 1528

Dose per Fraction, d — Number of Fractions, n ---->

d	1	5	10	15	20	25	30	35	40
50	0.004	0.02	0.04	0.06	0.08	0.10	0.12	0.14	0.16
100	0.012	0.06	0.12	0.18	0.23	0.29	0.35	0.41	0.46
150	0.022	0.11	0.22	0.33	0.44	0.55	0.66	0.77	0.88
200	0.035	0.18	0.35	0.52	0.69	0.86	1.04	1.21	1.38
250	0.050	0.25	0.50	0.74	0.99	1.23	1.48	1.72	1.97
300	0.066	0.33	0.66	0.99	1.31	1.64	1.97	2.30	2.62
350	0.084	0.42	0.84	1.26	1.68	2.09	2.51	2.93	...
400	0.104	0.52	1.04	1.55	2.07	2.59	...	...	...
450	0.125	0.63	1.25	1.87	2.50	...	...	...	...
500	0.148	0.74	1.48	2.21	2.95	...	...	...	...
550	0.172	0.86	1.72	2.57	...	...	...	...	...
600	0.197	0.99	1.97	2.96	...	...	...	...	...
650	0.224	1.12	2.24	...	...	...	...	...	...
700	0.252	1.26	2.52	...	...	...	...	...	...
750	0.281	1.41	2.81	...	...	...	...	...	...
800	0.311	1.56	...	...	...	...	...	...	...
900	0.375	1.88	...	...	...	...	...	...	...

LATE EFFECTS IN
GUT (COHEN 1983)   TABLE 6C

PARTIAL TOLERANCE FRACTION TABLES
compiled by Larry Simpson--01/31/84

the number frac per week is : 3
the exponent on N is : 0.290
the exponent on T is : 0.080
the NSD is : 1528

Dose per Fraction, d | Number of Fractions, n ---->

Dose per Fraction, d	1	5	10	15	20	25	30	35	40
50	0.004	0.02	0.04	0.06	0.08	0.10	0.12	0.14	0.16
100	0.012	0.06	0.12	0.18	0.24	0.30	0.36	0.42	0.48
150	0.023	0.12	0.23	0.35	0.46	0.57	0.69	0.80	0.91
200	0.036	0.18	0.36	0.54	0.72	0.90	1.08	1.26	1.44
250	0.052	0.26	0.52	0.77	1.03	1.28	1.54	1.79	2.05
300	0.069	0.35	0.69	1.03	1.37	1.71	2.05	2.39	2.73
350	0.088	0.44	0.88	1.31	1.75	2.18	2.62	...	...
400	0.108	0.54	1.08	1.62	2.16	2.70	...	...	...
450	0.130	0.65	1.30	1.95	2.60	...	...	...	...
500	0.154	0.77	1.54	2.31	...	...	...	...	...
550	0.179	0.90	1.79	2.68	...	...	...	...	...
600	0.206	1.03	2.06	...	...	...	...	...	...
650	0.233	1.17	2.33	...	...	...	...	...	...
700	0.262	1.31	2.62	...	...	...	...	...	...
750	0.293	1.47	2.93	...	...	...	...	...	...
800	0.324	1.62	...	...	...	...	...	...	...
900	0.391	1.96	...	...	...	...	...	...	...

LATE EFFECTS IN
GUT (COHEN 1983)   TABLE 6D

PARTIAL TOLERANCE FRACTION TABLES
compiled by Larry Simpson--01/31/84

the number frac per week is : 4
the exponent on N is : 0.290
the exponent on T is : 0.080
the NSD is : 1528

Dose per Fraction, d | Number of Fractions, n ---->

Dose per Fraction, d	1	5	10	15	20	25	30	35	40
50	0.005	0.03	0.05	0.07	0.09	0.11	0.13	0.15	0.17
100	0.013	0.07	0.13	0.19	0.25	0.32	0.38	0.44	0.50
150	0.024	0.12	0.24	0.36	0.48	0.60	0.72	0.84	0.96
200	0.038	0.19	0.38	0.57	0.76	0.94	1.13	1.32	1.51
250	0.054	0.27	0.54	0.81	1.07	1.34	1.61	1.88	2.14
300	0.072	0.36	0.72	1.08	1.43	1.79	2.15	2.50	2.86
350	0.092	0.46	0.92	1.37	1.83	2.28	2.74	...	...
400	0.113	0.57	1.13	1.70	2.26	2.82	...	...	...
450	0.136	0.68	1.36	2.04	2.72	...	...	...	...
500	0.161	0.81	1.61	2.41	...	...	...	...	...
550	0.187	0.94	1.87	2.81	...	...	...	...	...
600	0.215	1.08	2.15	...	...	...	...	...	...
650	0.244	1.22	2.44	...	...	...	...	...	...
700	0.274	1.37	2.74	...	...	...	...	...	...
750	0.306	1.53	...	...	...	...	...	...	...
800	0.339	1.70	...	...	...	...	...	...	...
900	0.409	2.05	...	...	...	...	...	...	...

LATE EFFECTS IN
GUT (COHEN 1983)   TABLE 6E

PARTIAL TOLERANCE FRACTION TABLES
compiled by Larry Simpson--01/31/84

the number frac per week is : 5
the exponent on N is : 0.290
the exponent on T is : 0.080
the NSD is : 1528

Dose per   Number of Fractions, n ----->
Fraction, d

	1	5	10	15	20	25	30	35	40
50	0.005	0.03	0.05	0.07	0.09	0.11	0.13	0.15	0.17
100	0.013	0.07	0.13	0.19	0.26	0.32	0.38	0.45	0.51
150	0.025	0.13	0.25	0.37	0.49	0.61	0.73	0.85	0.97
200	0.039	0.20	0.39	0.58	0.77	0.96	1.15	1.34	1.53
250	0.055	0.28	0.55	0.82	1.09	1.36	1.63	1.90	2.17
300	0.073	0.37	0.73	1.09	1.45	1.81	2.18	2.54	2.90
350	0.093	0.47	0.93	1.39	1.85	2.32	2.78	...	...
400	0.115	0.58	1.15	1.72	2.29	2.86	...	...	...
450	0.138	0.69	1.38	2.07	2.76	...	...	...	...
500	0.163	0.82	1.63	2.45	...	...	...	...	...
550	0.190	0.95	1.90	2.85	...	...	...	...	...
600	0.218	1.09	2.18	...	...	...	...	...	...
650	0.247	1.24	2.47	...	...	...	...	...	...
700	0.278	1.39	2.78	...	...	...	...	...	...
750	0.310	1.55	...	...	...	...	...	...	...
800	0.344	1.72	...	...	...	...	...	...	...
900	0.414	2.07	...	...	...	...	...	...	...

LATE EFFECTS IN
KIDNEY (COHEN 1983)   TABLE 7A

PARTIAL TOLERANCE FRACTION TABLES
compiled by Larry Simpson--01/31/84

the number frac per week is : 1
the exponent on N is : 0.250
the exponent on T is : 0.190
the NSD is : 465

Dose per Fraction, d | Number of Fractions, n ----->

Dose per Fraction, d	1	5	10	15	20	25	30	35	40
50	0.011	0.06	0.11	0.16	0.21	0.27	0.32	0.37	0.42
100	0.037	0.19	0.37	0.55	0.73	0.91	1.09	1.27	1.45
150	0.075	0.38	0.75	1.12	1.49	1.87	2.24	2.61	2.98
200	0.125	0.63	1.25	1.87	2.49	...	...	...	...
250	0.186	0.93	1.86	2.78	...	...	...	...	...
300	0.257	1.29	2.57	...	...	...	...	...	...
350	0.338	1.69	...	...	...	...	...	...	...
400	0.429	2.15	...	...	...	...	...	...	...
450	0.530	2.65	...	...	...	...	...	...	...
500	0.639	...	...	...	...	...	...	...	...
550	0.758	...	...	...	...	...	...	...	...
600	0.885	...	...	...	...	...	...	...	...
650	1.021	...	...	...	...	...	...	...	...
700	1.165	...	...	...	...	...	...	...	...
750	1.318	...	...	...	...	...	...	...	...
800	1.479	...	...	...	...	...	...	...	...
900	1.825	...	...	...	...	...	...	...	...

LATE EFFECTS IN
KIDNEY (COHEN 1983)   TABLE 7B

PARTIAL TOLERANCE FRACTION TABLES
compiled by Larry Simpson--01/31/84

the number frac per week is : 2
the exponent on N is : 0.250
the exponent on T is : 0.190
the NSD is : 465

Dose per Fraction, d | Number of Fractions, n ----->

Dose per Fraction, d	1	5	10	15	20	25	30	35	40
50	0.013	0.07	0.13	0.20	0.26	0.32	0.39	0.45	0.51
100	0.044	0.22	0.44	0.66	0.88	1.10	1.32	1.54	1.76
150	0.091	0.46	0.91	1.36	1.82	2.27	2.72	...	...
200	0.152	0.76	1.52	2.28	...	...	...	...	...
250	0.226	1.13	2.26	...	...	...	...	...	...
300	0.313	1.57	...	...	...	...	...	...	...
350	0.412	2.06	...	...	...	...	...	...	...
400	0.523	2.62	...	...	...	...	...	...	...
450	0.645	...	...	...	...	...	...	...	...
500	0.779	...	...	...	...	...	...	...	...
550	0.923	...	...	...	...	...	...	...	...
600	1.078	...	...	...	...	...	...	...	...
650	1.244	...	...	...	...	...	...	...	...
700	1.419	...	...	...	...	...	...	...	...
750	1.605	...	...	...	...	...	...	...	...
800	1.801	...	...	...	...	...	...	...	...
900	2.223	...	...	...	...	...	...	...	...

LATE EFFECTS IN
KIDNEY (COHEN 1983)   TABLE 7C

PARTIAL TOLERANCE FRACTION TABLES
compiled by Larry Simpson--01/31/84

the number frac per week is : 3
the exponent on N is : 0.250
the exponent on T is : 0.190
the NSD is : 465

Dose per Fraction, d   Number of Fractions, n ----->

d	1	5	10	15	20	25	30	35	40
50	0.015	0.08	0.15	0.22	0.29	0.36	0.43	0.50	0.57
100	0.050	0.25	0.50	0.74	0.99	1.23	1.48	1.72	1.97
150	0.102	0.51	1.02	1.52	2.03	2.54	...	...	...
200	0.170	0.85	1.70	2.55	...	...	...	...	...
250	0.253	1.27	2.53	...	...	...	...	...	...
300	0.350	1.75	...	...	...	...	...	...	...
350	0.461	2.31	...	...	...	...	...	...	...
400	0.584	2.92	...	...	...	...	...	...	...
450	0.721	...	...	...	...	...	...	...	...
500	0.870	...	...	...	...	...	...	...	...
550	1.032	...	...	...	...	...	...	...	...
600	1.205	...	...	...	...	...	...	...	...
650	1.390	...	...	...	...	...	...	...	...
700	1.587	...	...	...	...	...	...	...	...
750	1.795	...	...	...	...	...	...	...	...
800	2.014	...	...	...	...	...	...	...	...
900	2.485	...	...	...	...	...	...	...	...

LATE EFFECTS IN
KIDNEY (COHEN 1983)   TABLE 7D

PARTIAL TOLERANCE FRACTION TABLES
compiled by Larry Simpson--01/31/84

the number frac per week is : 4
the exponent on N is : 0.250
the exponent on T is : 0.190
the NSD is : 465

Dose per Fraction, d   Number of Fractions, n ----->

d	1	5	10	15	20	25	30	35	40
50	0.017	0.09	0.17	0.25	0.33	0.41	0.49	0.57	0.65
100	0.056	0.28	0.56	0.84	1.11	1.39	1.67	1.94	2.22
150	0.115	0.58	1.15	1.72	2.29	2.86	...	...	...
200	0.192	0.96	1.92	2.87	...	...	...	...	...
250	0.285	1.43	2.85	...	...	...	...	...	...
300	0.395	1.98	...	...	...	...	...	...	...
350	0.519	2.60	...	...	...	...	...	...	...
400	0.659	...	...	...	...	...	...	...	...
450	0.813	...	...	...	...	...	...	...	...
500	0.982	...	...	...	...	...	...	...	...
550	1.164	...	...	...	...	...	...	...	...
600	1.359	...	...	...	...	...	...	...	...
650	1.568	...	...	...	...	...	...	...	...
700	1.790	...	...	...	...	...	...	...	...
750	2.024	...	...	...	...	...	...	...	...
800	2.271	...	...	...	...	...	...	...	...
900	2.803	...	...	...	...	...	...	...	...

LATE EFFECTS IN　　　　　　　　TABLE 7E
KIDNEY (COHEN 1983)

PARTIAL TOLERANCE FRACTION TABLES
compiled by Larry Simpson--01/31/84

the number frac per week is : 5
the exponent on N is : 0.250
the exponent on T is : 0.190
the NSD is : 465

Number of Fractions, n ---->

Dose per Fraction, d	1	5	10	15	20	25	30	35	40
50	0.017	0.09	0.17	0.26	0.34	0.42	0.51	0.59	0.67
100	0.058	0.29	0.58	0.87	1.15	1.44	1.73	2.02	2.30
150	0.119	0.60	1.19	1.78	2.38	2.97	...	...	...
200	0.199	1.00	1.99	2.98	...	...	...	...	...
250	0.296	1.48	2.96	...	...	...	...	...	...
300	0.409	2.05	...	...	...	...	...	...	...
350	0.539	2.70	...	...	...	...	...	...	...
400	0.684	...	...	...	...	...	...	...	...
450	0.844	...	...	...	...	...	...	...	...
500	1.019	...	...	...	...	...	...	...	...
550	1.207	...	...	...	...	...	...	...	...
600	1.410	...	...	...	...	...	...	...	...
650	1.627	...	...	...	...	...	...	...	...
700	1.857	...	...	...	...	...	...	...	...
750	2.100	...	...	...	...	...	...	...	...
800	2.357	...	...	...	...	...	...	...	...
900	2.909	...	...	...	...	...	...	...	...

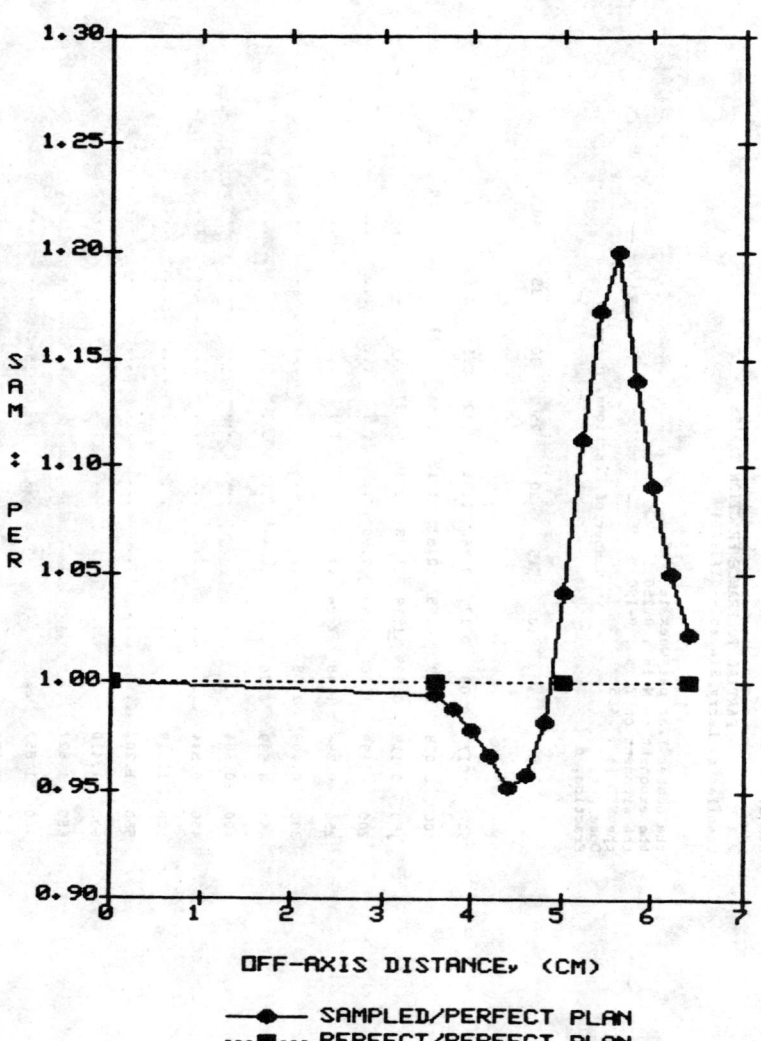

FIGURE 1
APPROXIMATION TO THE SAMPLED MEAN PLAN

FIGURE 2

LATE EFFECTS TOLERANCE FRACTION NUMBER, N, --1 FX/WK

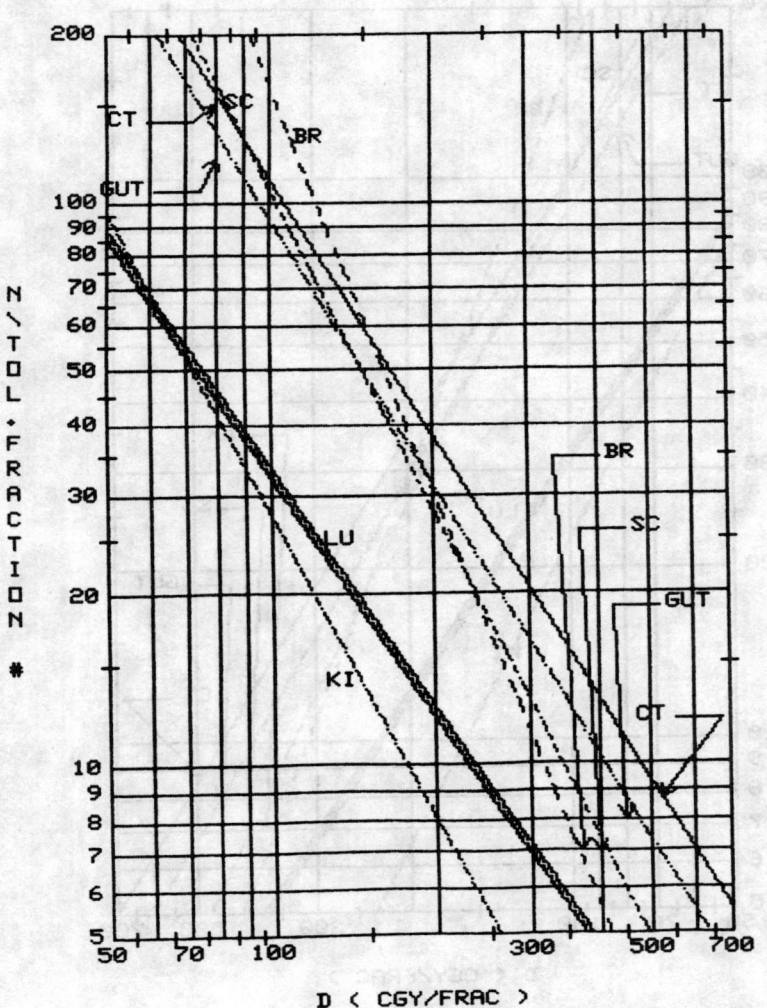

FIGURE 3

LATE EFFECTS TOLERANCE FRACTION NUMBER, N, --2 FX/WK

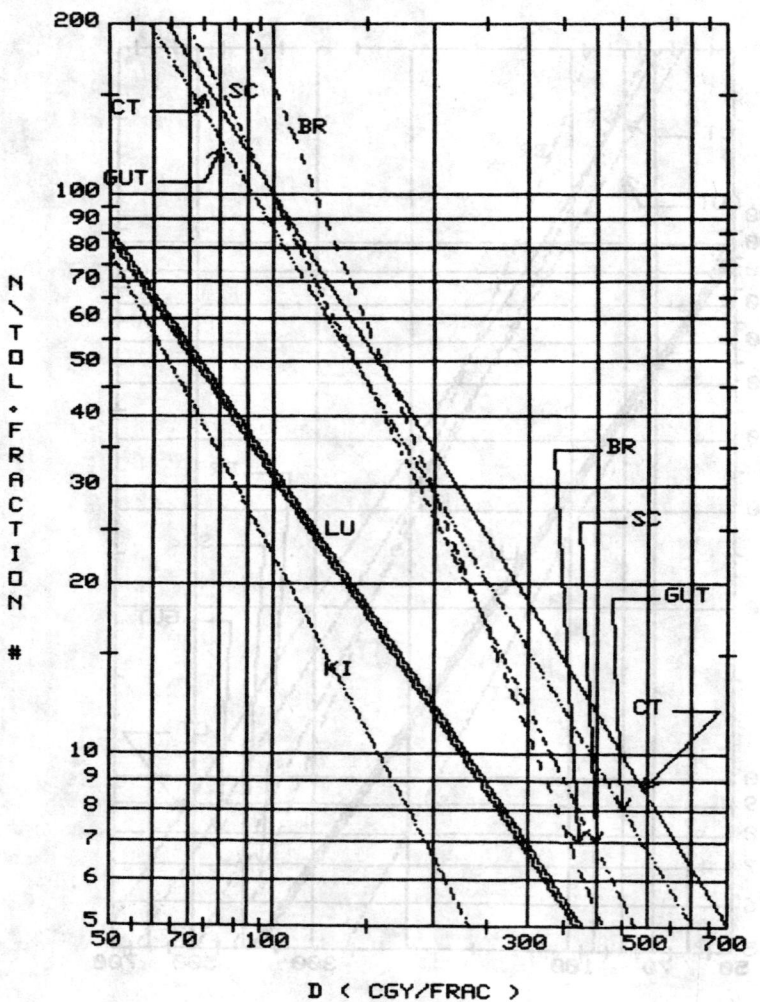

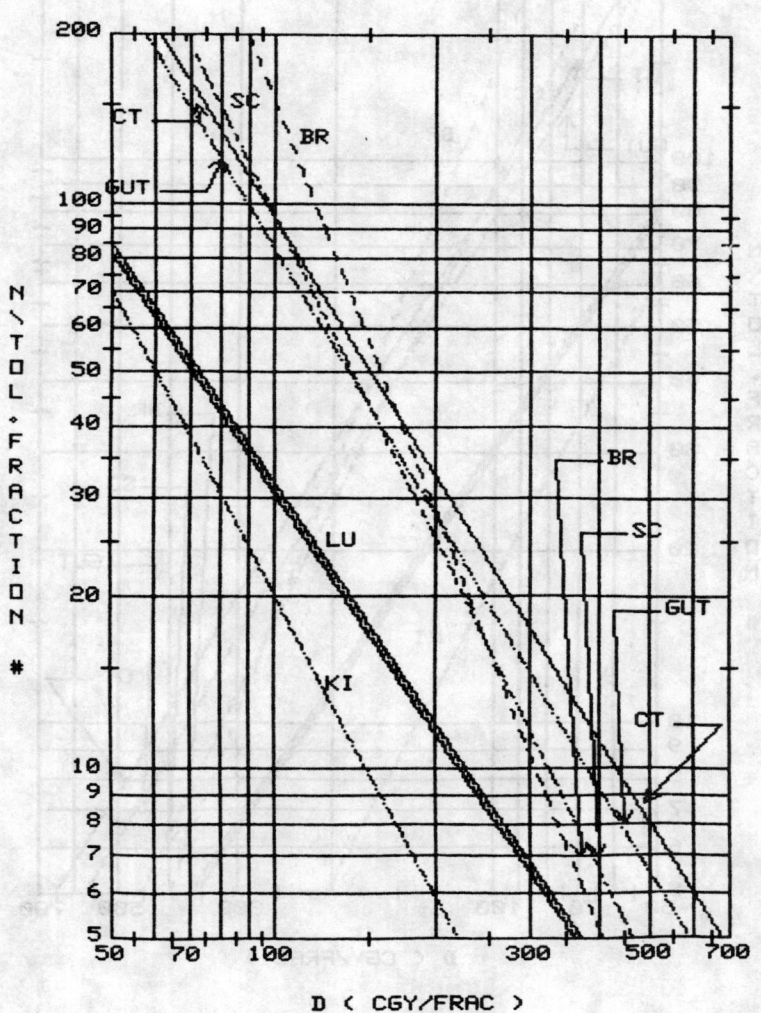

FIGURE 4

LATE EFFECTS TOLERANCE FRACTION NUMBER, N, --3 FX/WK

FIGURE 5

LATE EFFECTS TOLERANCE FRACTION NUMBER, N, --4 FX/WK

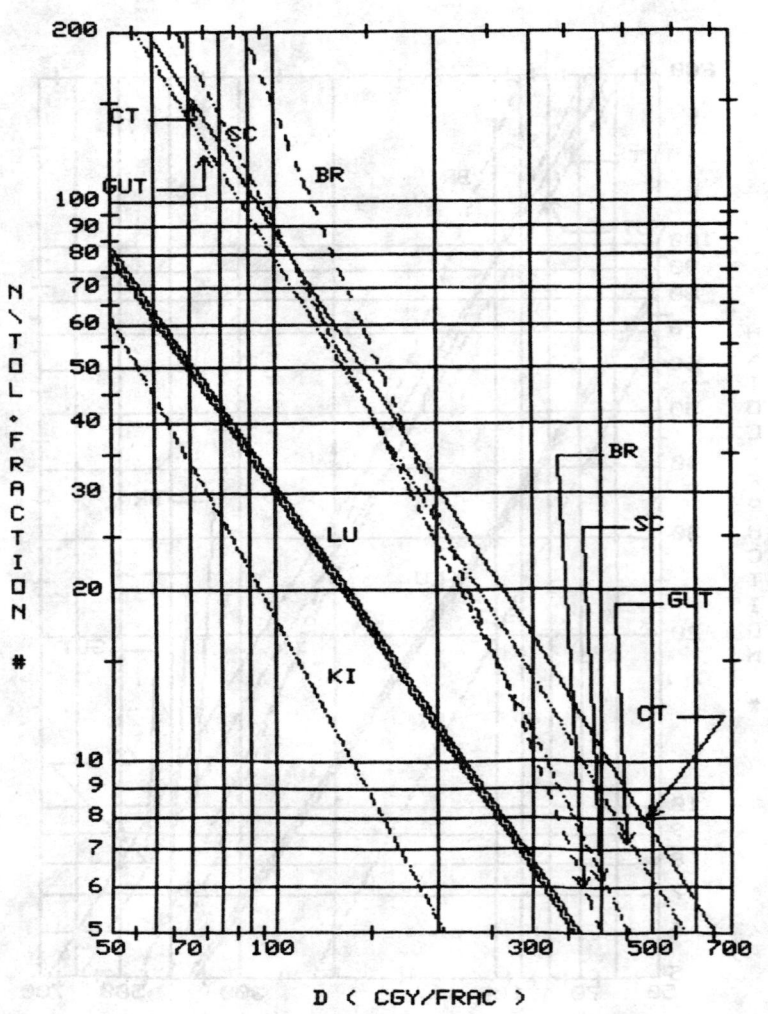

FIGURE 6

LATE EFFECTS TOLERANCE FRACTION NUMBER, N, --5 FX/WK

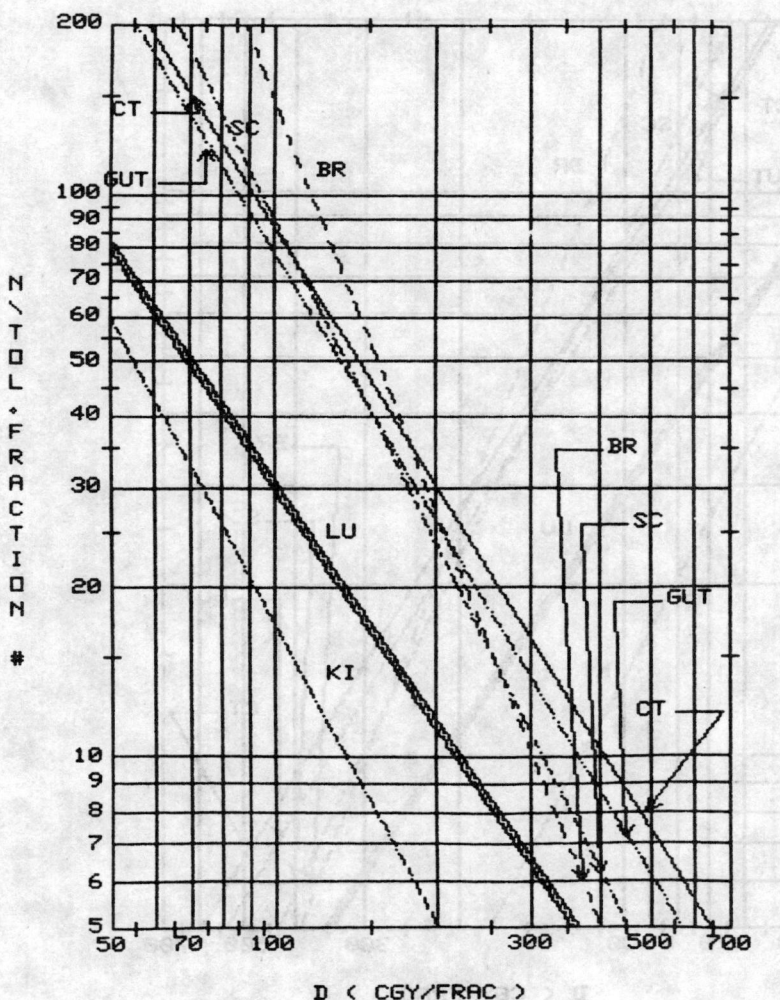

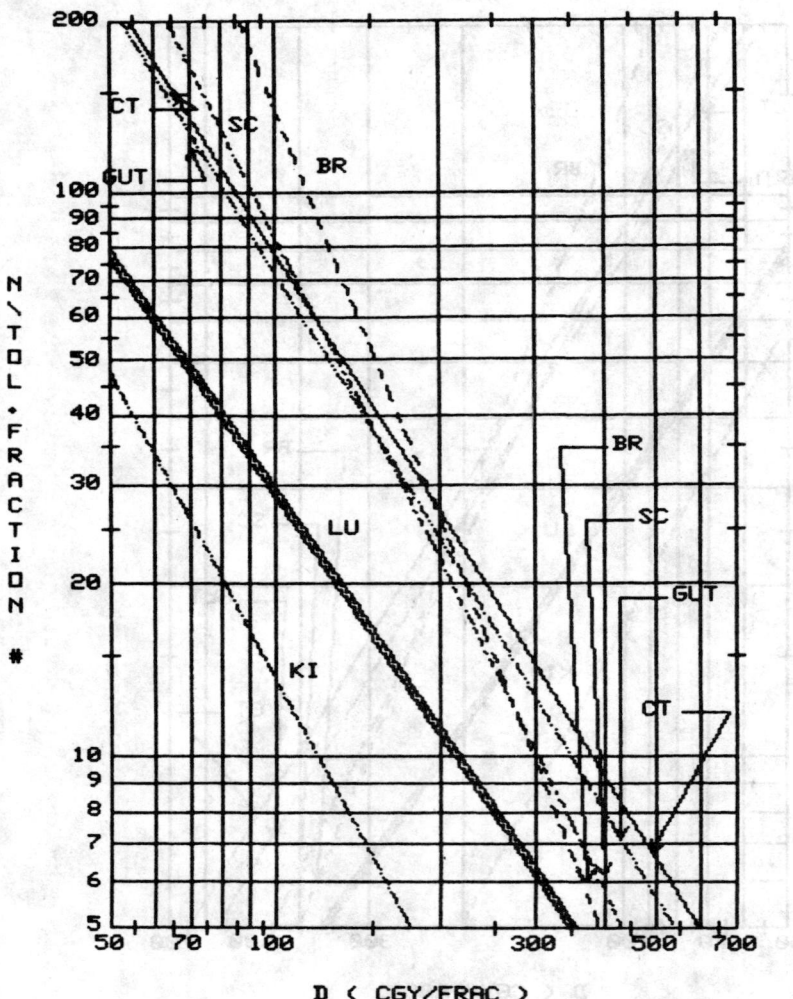

FIGURE 7
LATE EFFECTS TOLERANCE FRACTION NUMBER, N, -10 FX/WK

CLINICAL DIAGNOSTIC MODELS. DISCRIMINANT ANALYSIS OF IN VITRO
NMR MEASUREMENTS ON NORMAL AND MALIGNANT TISSUES

Donald E. Herbert, Ph.D.
University of South Alabama, College of Medicine,
Dept. of Radiology, Mobile, Alabama 36688

ABSTRACT

There is currently no consensus in the received wisdom concerning the clinical value of NMR measurements (in vivo and in vitro) of proton T_1 and T_2 relaxation times in discriminating between normal ($\overline{Ca}$) and malignant (Ca) tissues. There is rather a sharp dichotomy which may be discerned in the "mutually impenetrable" views of the groups represented by Damadian, et al and Hollis, et al. However, the data which has been adduced in support of either view has little evidential value: In the first place there is little information contained in the small samples which are used to obtain point estimates of the misclassification rates achieved by the use of the T_1 and T_2 measurements. In the second place, the statistical methods by which the (small) information available is exploited for diagnosis are inappropriate: The Malignancy Index is an _improper_ linear model.

It is shown that the correct methods for taking qualitative decisions on the basis of quantitative data in an optimal manner are those of discriminant analysis - a _proper_ linear model. The data of Damadian, et al, are used to illustrate several aspects of the more insistent of the problems in classification and identification. These include the minimization of the frequency of occurrence of Type I, II and III errors, the statistical significance of the discriminant function, the effect on Type I and Type II errors of adding (or deleting) from the discriminant function a specified subgroup of diagnostic tests, the effects on the estimated discriminant function of the idiosyncracies (including measurement errors in the predictors) of the sample _distribution_ of measurements (as well as the sample _size_), interval estimates of the misclassification rate for a discriminant function, measures of the "resemblance" of an individual to a given group, and methods for the generalization of discriminant analysis to k > 2 measurements on individuals from g > 2 mutually exclusive and exhaustive categories or groups.

1. INTRODUCTION

"The problem of discrimination and classification is insistent in sciences" (R. C. Bose and S. N. Roy 1938).
"One of the difficulties in treating patients with cancer is the fact that the neoplasm is often discovered late. Recent statistics emphasize the need for early detection." ... "... the purpose of the study was to test the capabilities of NMR to distinguish normal and malignant tissue ..." "... the data presented here indicate that NMR is capable of distinguishing normal tissue from malignant pathology with a high degree of consistency" (Koutcher, Goldsmith and Damadian 1978).
"Koutcher et al introduced the concept of a malignancy index in the course of these investigations. This combines two or more NMR parameters to yield a figure which gives improved cancer discrimination. It may well be that a similar index will be of value in a medical imaging system, when the results of a number of scans could be combined

to yield a map of malignancy index rather than simply of spin density or relaxation time."(Mansfield and Morris, 1982)

"Statistics is a form of social control over the professional behaviour of researchers" (Harris 1975).

In 1978 there was published an account of a function by which the reported differences in the conditional frequency distributions of proton T_1 and T_2 measurements on biopsy specimens could be fruitfully exploited in the diagnosis of malignancy in the breast, colon and lung (Koutcher, Goldsmith and Damadian 1978). This function is the Malignancy Index. The motivation for this function was described by Goldsmith et al: "Since some degree of overlap is evident [in the conditional distributions of T_1 and T_2 on the two conditions: disease free ($\overline{Ca}$) and diseased (Ca).] ... we hoped that a combined malignancy index would be more reliable than any single parameter in discriminating normal from malignant specimens. We decided to try a normalized sum of the relaxation constants T_1 and T_2, since the former is generally 10X greater than the latter. Thus, we defined a 'malignancy index' for each specimen by substitution into the following 'separation algorithm': Malignancy Index, $\Lambda = (T_1)_i/(\overline{T_2})\text{normal} + (T_2)_i/(\overline{T_2})$ normal where $(T_1)_i$ and $(T_2)_i$ are T_1 and T_2 of the ith specimen, and $(\overline{T_1})$ normal and $(\overline{T_2})$ normal are the mean values of T_1 and T_2 for the normal population" (Goldsmith, Koutcher and Damadian, 1977). It is evident that the respective weights of the relaxation times in the Malignancy Index, Λ, are only scaling factors, chosen so that Λ is dimensionless and its two components are commensurable.[a] Dawes (Judgement Under Uncertainty ,1982) has described devices such as Λ as improper linear predictive models[b] (vide infra).

A tissue was classified as benign ($\overline{Ca}$) or malignant (Ca) according as $\Lambda_i \leq \Lambda_o$ or $\Lambda_i > \Lambda_o$ where Λ_o is an arbitrary "threshold" value of Λ. In the literature of the Malignancy Index, Λ_o is chosen to be either the upper extreme of the sample frequency distribution of Λ on $\overline{Ca}$ or the lower extreme of the sample frequency distribution on Ca. The basis for either choice of Λ_o is not given in the literature, however, it is readily shown that these choices of Λ_o confer a (speciously) lower value on their sample estimates of the missclassification rates achieved by the use of Λ than do more standard choices.

Assume that because of the overlap of the conditional ($\overline{Ca}$ and Ca) joint distributions of the T_1 and T_2 measurements, the above classification rule results in the incorrect identification of r_1 of n_1 $\overline{Ca}$ patients and r_2 of n_2 Ca patients. The correct sample estimates of the respective error rates are $P_1 = r_1/n_1$ (False Positive) and $P_2 = r_2/n_2$ (False Negative). The correct sample estimate of the total error rate is simply the sum: $P_1+P_2 = (r_1/n_1) + (r_2/n_2)$ (See Armitage 1977). This (standard) definition provides a distinction between False Positive and False Negative errors; these have substantially different social and clinical implications. The relative sizes of the two error rates, say P_1/P_2, depends upon the choice of Λ_0. One optimal rule is to select Λ_0 to minimize the total error rate, $P_1 + P_2$.

In Koutcher, et al (1978), the position of Λ_0 is chosen to minimize the total error rate which is however, defined as $(r_1+r_2)/(n_1+n_2)$. Minimization is achieved by selection of Λ_0 so that either $r_1 = 0$ or $r_2 = 0$.

It can be shown, from estimates of the confidence limits on percentiles (See Eisenhart, et al 1947; Hald 1965; Elvebeck and Taylor 1969) that because of the small sample size, n_i, of each of the conditional distributions of T_1 and T_2 ($7 \leq n_i \leq 20$. i=1, $\overline{Ca}$; i=2, Ca) at each anatomical site (breast, colon and lung) the 0.90 confidence limits, $\underline{P_i}$ and $\overline{P_i}$, on the proportions, P_1 and P_2, of the respective $\underline{\text{populations}}$, $\overline{Ca}$ and Ca, that had values of Λ beyond the threshold, $\Lambda \geq \Lambda_0$ and $\Lambda < \Lambda_0$, respectively, were $\underline{\text{large}}$; typically, $\underline{P_i} \sim 0$, $\overline{P_i} \sim 0.20$. (Herbert 1986). From this it would surely appear that the value of Λ in clinical diagnosis cannot be even roughly assessed from the Koutcher, et al (1977) studies.[c] It can also be shown, on the basis of certain testable assumptions, that a convincing assessment of the value of T_1 and T_2 biopsy measurements in discrimination between normal and malignant conditions can only be obtained from estimates of the location and scale parameters of the conditional, Normal, frequency distributions which are determined from sample sizes of the order of $n_1 = n_2 = 100$ $(1=\overline{Ca}, 2=Ca)$.[d]

We will now show that the respective sample sizes are not the only, or even the most troublesome, deficiency of the previous studies (Koutcher, et al 1977): As well as the $\underline{\text{small samples}}$, the novel $\underline{\text{methods}}$ by which the information contained therein is exploited, preclude a convincing assessment of the diagnostic utility of T_1 and T_2 measurements on the basis of these data. (Increasing the sample size cannot, of course, correct inherent technical deficiencies in methodology.) But, although the sample sizes of the Koutcher, et al (1977), study are $\underline{\text{inadequate}}$[e] to support a convincing assessment of the diagnostic utility of T_1 and T_2 measurements on biopsy specimens, these data are useful both to motivate the study of the construction, and to illustrate the characteristic features, of a proper linear predictive model, the linear discriminant model. (See Lachenbruch 1975 and Lecture $\underline{10}$; Dawes 1982) Therefore, we proceed to present $\underline{\text{heuristic}}$ discriminant analyses of the Koutcher et al (1978) data in order to demonstrate this model which is (evidently) not as widely known as the usefullness of both its insights and methods warrants.

Moreover, since it can be shown that the two key statistics of the discriminant model, the Mahalanobis distance, D^2, and the discriminant weights, $\underline{L}^T$, are $\underline{\text{invariant}}$ under affine transformations (Rao 1973 and Lachenbruch 19$\overline{75}$), $\underline{U} = AX + \underline{C}$, where $\underline{X}$, $\underline{U}$ and $\underline{C}$ are (kx1) and A is (kxk), such analyses of $\underline{\text{in vitro}}$ data also demonstrate a (possible) methodology for assessing the diagnostic utility of the proton T_1 and T_2 measurements in $\underline{\text{in vivo}}$ studies such as NMR imaging, for which $\underline{X} = (T_1, T_2)$.

It is worth noting at the outset that the recommendations on sample size that were reported above, namely, $n_1 = n_2 \geq 100$, will be supported by the heuristic studies to be described.

We discuss problems of classification for (only) two mutually exclusive and exhaustive disease groups. This is the problem most widely treated in the rich literature on the classification problem and, of course, the only case for which there are, as yet, any NMR data to which we had access. The classification problem for three or more diagnostic groups is treated in the references to be cited later.

2. BACKGROUND

The ambiguity that encumbers taking a qualitative, binary $\underline{\text{decision}}$, on the state, or condition, of a system, e.g., disease present, disease absent, on the basis of the information contained in a quantitative $\underline{\text{measurement}}$ on that system is well known. Its proximate cause was correctly perceived in the Koutcher, et al studies (1978) and the Goldsmith, et al studies (1977); it resides in the "overlap" of the respective conditional frequency distributions of the measurement on the two conditions of the system. The present paper is motivated by the evident immediate interest in exploiting methods which will reduce this "overlap", and hence the level of ambiguity in the clinical significance of T_1 and T_2 measurements on biopsied tissues, and in the more obvious implications of such studies for differential diagnoses based on the NMR images obtained from patients.

Intuitively, the proposition that some synthesis of the diagnostic information included in each of two (or more) single measurements or tests (for example, a <u>linear form</u> on the measurements is one common method for achieving such a synthesis) which have frequency distributions for which the location, scale, etc., parameters are characteristic of the conditions, or categories, to be diagnosed may provide a less ambiguous criterion than any of the single tests in determining the identification or classification of an individual (that is, his assignment to one of several <u>mutually exclusive</u> and <u>exhaustive</u> - complementary - categories) is very attractive. It is certainly plausible. However, several questions of concept and practice concerning its implementation immediately arise: 1) How shall the component tests be selected? (Number? Kind?) 2) How shall the respective test scores be combined? (Additively? Multiplicatively?) 3) By what criterion shall the respective weights (Coefficients? Exponents?) of the component tests be determined? (Requiring the coefficients to be "scale factors" - as in the Malignancy Index - is an example of one such criterion (Koutcher, Goldsmith and Damadian 1978). We shall find that there are still better ones.) 4) How shall the "threshold" of the combination be determined; i.e. that value which "optimally" divides the scores of one group from those of the complementary group(s) and hence determines the relative sizes of the False Positive and False Negative error rates? 5) How shall <u>point</u> and <u>interval</u> estimates of the diagnostic performance of the combination be specified and estimated? 6) How shall the diagnostic performance of the combination in both present and future samples be assessed? 7) How do the <u>idiosyncracies</u> of the frequency distribution of the particular sample - outliers, collinear variables, etc. - as well as the sample <u>size</u> (both relative and absolute) from which these estimates are obtained affect these estimates? 8) Although the problem of optimal assignment of an individual is set in the context of decision between <u>known</u> mutually exclusive and exhaustive disease groups (vide supra) it is frequently the case in medical diagnosis that the external, or a priori, evidence that the individual in question belongs to one of the alternatives considered is equivocal or absent. How does the "optimal" procedure change when it is not known whether the <u>known</u> alternatives are mutually exclusive and exhaustive for a given <u>individual</u>? 9) How does the problem change when the apriori probabilities of the known alternative conditions, although known, are of vastly different sizes (as in screening)? These are some of the questions which we will address in this paper and for which we will exploit the data of Koutcher, et al (1978), in <u>motivating</u> - as well as illustrating - the (correct) answers (See Lachenbruch, Lecture <u>10</u> and Lachenbruch 1975).

3. CONCEPTS AND METHODS

There has been an augmenting increase in the literature relating to the principal methodology of classification - <u>discriminant analysis</u> (Lachenbruch 1975). Between the mid 1930's and the early 1980's there were well over 1000 articles on the classification problem published in books, journals, and proceedings (P. Lachenbruch, Personal Communication). Therefore, this paper will describe and illustrate only briefly the motivation and the implementation of several of the many solutions to the different aspects of the classification problem for those individuals who are to be assigned, on the basis of two measurements (k=2) to one of two groups (g=2). (However, we will indicate the procedures for generalization of these solutions to circumstances in which $g > 2$ and $k > 2$.)

These are the solutions first associated with the names of 1) Tiedeman, et al (1953 and 1954) (measure of "resemblance" of an individual to a specified group on the basis of $k \geq 2$ tests) 2) Fisher (1936) and Welch (1939)("optimal" allocation − in the long run − of individuals between g = 2 complementary groups on the basis of $k \geq 2$ tests), 3) Bryan (1951) and Rao (1948) ("optimal" allocation − in the long run − of individuals between $g > 2$ groups on $k \geq 2$ tests), Cornfield (1965) (estimation of "risk" of an individual for developing one of two complementary conditions, given "scores" on $k \geq 2$ tests) and Lachenbruch (1967 & 1968) (point and interval estimates of proportion of errors incurred − in the long run − in the allocation of individuals between g=2 groups on the basis of $k \geq 2$ tests). Obviously, there has been considerable subsequent elaboration of, and innovation in, each of these solutions by these and other scientists. (Lachenbruch has been especially prolific of solutions to these problems.)

As noted, it is often the case in science that one must make qualitative decisions on the basis of quantitative data. For instance, one may be required to estimate an unknown level of a qualitative variable on the basis of known levels of one or more quantitative variables which are associated with it. Establishing the sex of a skeleton on the basis of a set of anthropometric measurements is one such example in which the qualitative variable is binary: male or female (Ashton 1967 and Rao 1952). Establishing the differential diagnosis for a patient on the basis of one or more laboratory tests is another − and for the present paper, more apposite − example in which the qualitative (binary) variable is disease status: viral hepatitis or obstructive jaundice (Rao, 1952). In both examples, the association which obtains between the qualitative and quantitative variables is that the location (and perhaps other) parameter and form of the frequency distribution of the latter is conditional on the former, that is, the quantitative variable has a conditional frequency distribution on the qualitative variable. Let us assume that the qualitative variable, U, is two-dimensional − binary − describing the two states denoted as $\bar{E}$ and E, which have local prior probabilities, $P(\bar{E})$ and $P(E)$. Moreover, the states or conditions denoted by E and $\bar{E}$ are complementary, i.e., they are mutually exclusive and exhaustive: $P(\bar{E}$ and $E) = 0$ and $P(\bar{E}$ or $E) = P(\bar{E}) + P(E) = 1.0$, respectively. The quantitative variable is the (kxl) vector, $\underline{X}$, of measurements for the ith individual. It will be convenient in the sequel to represent this problem as follows: It is assumed that the alternative conditions are mutually exclusive, $E \cap \bar{E} = \emptyset$, and exhaustive, $\overline{E \cup E} = \emptyset$. Here $\emptyset$ is the empty, or null, set. Then, if it is known from non-sample, or a priori, evidence that $\underline{X} \varepsilon \, E \cup \bar{E}$ with known apriori probabilities of $\bar{E}$ and E, now denoted as q_1 and q_2, respectively, the practical classification problem devolves into one of deciding, with minimum probability of error, "Which", of the two complementary states or conditions gave rise to an observed vector, $\underline{X}$. However, it is frequently the case in medical practice that the non-sample evidence for $X \in \bar{E} \cup E$ is either weak or that the prior probability for one condition greatly exceeds that of the other: $q_i \gg q_j$. In this case, one must depend upon the internal evidence provided by $\underline{X}$ itself and the classification problem devolves from (the Bayesian) one of estimation of the (posterior) <u>probability of a hypothesis</u> into (the Sampling Theory) one of <u>hypothesis testing</u>. We have assumed that the quantitative variable, $\underline{X}$, is k-dimensional, that is, there are k measurements (or "tests") $\underline{X} = (x_1, x_2, ..., x_k)$ on the individual in question which have a joint frequency distribution which is k-variate Normal: $\underline{X} \sim N_k(\underline{\mu}, \Sigma).^f$ $\underline{\mu}$ is (kx1) and Σ is (kxk). Then the association between $\underline{X}$ and U can be described as $\underline{\mu} = \underline{\mu}(U)$, $\Sigma = \Sigma(U)$. Since U is binary, the parameters take only the values $(\underline{\mu}_1, \Sigma_1)$ for condition $\bar{E}$ and $(\underline{\mu}_2, \Sigma_2)$ for condition E. In this context the examples described above may be reduced to the following kind. Given that an individual is completely specified by the set of variables (U, $\underline{X}$) and that for a given individual only the measurements $\underline{X} = (x_1, x_2, ..., x_k)$ are known, what is the value of U, or better, since U is a <u>binary</u> variable, <u>which</u> of the two possible values of U is present? Note that the (generalized) <u>distance</u> in X-space between the two points which represent the individuals' set of measurements, $\underline{X}$, and the parameter, $\underline{\mu}_j$, which describes the <u>centroid</u> of the jth group is given by $C_j = (\underline{X} - \underline{\mu}_j)^T \Sigma_j^{-1} (\underline{X} - \underline{\mu}_j)$. $j = 1, 2$. Since $\underline{X} \sim N_k(\underline{\mu}_j, \Sigma_j)$, the statistic, C_j, is distributed as chi-square with k degrees of freedom. C_j is a measure of the "<u>resemblance</u>" of $\underline{X}$ to group j (Tiedeman 1953 and 1954).

In such use of $\underline{X}$ to infer on the value of U it is possible to commit errors because the two conditional distributions of $\underline{X}$ <u>overlap</u>. (vide supra) The probability of commiting such errors increases as the degree of overlap increases. There are two kinds of errors. One is the assignment of an individual of group 1 to group 2. The second is the assignment of an individual of group 2 to group 1. Let us denote the probability of the first kind of error, a Type I error, by P_1 and the probability of the second kind, a Type II error, by P_2. P_1 is the probability of a False Positive and P_2 is the probability of a False Negative, usually denoted as α and β, respectively (Armitage, 1977). Table <u>1a</u> defines the possible outcomes to taking the decision.

Let f_1 and f_2 be the conditional, Normal, density functions of the set of measurements X on conditions $\bar{E}$ and E, respectively. Then

$$P_1 = \int_{R_2} f_1(\underline{X}; \Theta_1) d\underline{X} \tag{1}$$

and

$$P_2 = \int_{R_1} f_2(X; \Theta_2) dX \tag{2}$$

where $\Theta_j = (\underline{\mu}_j, \Sigma_j)$, $j = 1, 2$.
R_1 is the region of X-space in which one would assign the ith individual on the evidence of $\underline{X}$, to group 1. R_2 is the region of X-space in which one would assign him (her) to group 2. We assume that $R_1 \cap R_2 = \emptyset$ and $\overline{R_1 \cup R_2} = \emptyset$, where $\emptyset$ is the null set.

The problem devolves into specifying the regions, R_j, such that the average probability, P, of misclassification is a <u>minimum</u>, where
$$P = q_1 P_1 + q_2 P_2. \tag{3}$$
This quantity is minimized if the boundary between R_1 and R_2 is chosen so that
$$q_2 f_2(\underline{X}; \Theta_2) - q_1 f_1(\underline{X}; \Theta_1) < 0 \tag{4}$$
for all points in R_1. This gives rise to the optimal rule: Assign $\underline{X}$

to group 1 if $f_1(\underline{X}; \Theta_1)/f_2(\underline{X}; \Theta_2) > q_2/q_1$ and to group 2 otherwise (Lachenbruch 1975 and Welch 1939).

If $\underline{X} \sim N_k(\underline{\mu}_j, \Sigma_j)$ we have $f_j(\underline{X}; \Theta_j) = (2\pi)^{-p/2} |\Sigma_j|^{-1/2} \exp(-0.5 (\underline{X}-\underline{\mu}_j)^T \Sigma_j^{-1}(\underline{x}_j' - \underline{\mu}_j))$. $j = 1, 2$. An important special case is when $\Sigma_1 = \Sigma_2 = \Sigma$; $\underline{\mu}_1 \neq \underline{\mu}_2$. Then

$$f_1(\underline{X}; \Theta_1)/f_2(\underline{X}; \Theta_2) = \exp[\underline{L}^T\underline{X} - 0.5\underline{L}^T(\underline{\mu}_1 + \underline{\mu}_2)] \tag{5}$$

where $\underline{L}^T = (\underline{\mu}_1 - \underline{\mu}_2)^T S^{-1}$. Taking logarithms, the optimal rule is to assign $\underline{X}$ to group 1 if $\underline{L}^T[\underline{X}-0.5(\underline{\mu}_1 + \underline{\mu}_2)] > \ln q_2/q_1$, otherwise to group 2.

The linear discriminant function may be written as $z = \underline{L}^T\underline{X} - 0.5\underline{L}^T(\underline{\mu}_1 + \underline{\mu}_2)$ $(= L(\underline{X}) = \text{LDF})$. The difference in the function for $\overline{X} = \underline{\mu}_1$ and $\overline{X} = \underline{\mu}_2$ is $z_1 - z_2 = \underline{L}^T\underline{\mu}_1 - \underline{L}^T\underline{\mu}_2 = (\underline{\mu}_1 - \underline{\mu}_2)^T\Sigma^{-1}(\underline{\mu}_1 - \underline{\mu}_2)$. This (generalized distance between the respective centroids of the two groups is the Mahalanobis distance, $\Delta^2 = (\underline{\mu}_1 - \underline{\mu}_2)^T\Sigma^{-1}(\underline{\mu}_1 - \underline{\mu}_2)$. Note as well that since $\text{Var}(\underline{X}) = \Sigma$ we have $\text{Var}(z) = \underline{L}^T\Sigma\underline{L} = (\underline{\mu}_1 - \underline{\mu}_2)^T\Sigma^{-1} \Sigma\Sigma^{-1}(\underline{\mu}_1 - \underline{\mu}_2) = \Delta^2$ (Lachenbruch 1975). (Δ^2 is a measure of the "resemblance" of the two groups to the degree that a group can be characterized by $\underline{\mu}_j$ (vide supra).

It is or both interest and importance to note that the sample point estimate of Δ^2, which is $D^2 = (\overline{X}_2 - \overline{X}_1)^T S^{-1}(\overline{X}_2-\overline{X}_1)$, is a biased estimate: $E(D^2) > \Delta^2$, where $E(\ )$ denotes an expected value. The sample estimates of the mean vector, $\underline{\mu}_j$, and covariance matrix, Σ_j, are $\overline{X}_j$ and S_j, respectively. The size of the bias depends on both the number of variables, k, and the sample sizes, n_1 and n_2 (Lachenbruch 1975 and Chakravati 1967).

An unbiased point estimate of Δ^2 is given by δ^2 where (Chakravati, et al 1967)

$$\delta^2 = \frac{n_1+n_2 - k-3}{n_1+n_2 - 2} D^2 - \frac{(n_1+n_2)k}{n_1 n_2} \tag{6}$$

The difference $(D^2-\delta^2) > 0$, describes the "shrinkage" of the Mahalanobis distance that will be observed when the estimated discriminant function, $\underline{L}^T$, is used for the assignment of future individuals drawn from the same population. Note that the shrinkage increases with k, the number of predictor variables. This shrinkage of the Mahalanobis distance is cognate to the "shrinkage" of the multiple correlation coefficient, R^2, observed in multiple regression analysis (vide infra) which is described by the so-called adjusted multiple correlation coefficient, R^2, where

$\overline{R}^2 = 1 - \left(\frac{n-1}{n-k}\right)(1-R^2)$ and $n = n_1 + n_2$ (Lachenbruch 1968 and Montgomery and Peck 1982).

It is readily shown that Δ^2 is invariant under nonsingular, linear transformations of $\underline{X}$, such as, $\underline{X}^* = A\underline{X} + \underline{C}$. That is, $\Delta^2(\underline{X}^*) = \Delta^2(\underline{X})$ (Lachenbruch 1975). Thus, the level of diagnostic information present in the invivo measurements of T_1 and T_2 that is represented in NMR images may be anticipated to some degree by an assessment of degree of the diagnostic information present in invitro measurements of T_1 and T_2 provided, of course, that the invivo and invitro measurements are related by a non-singular linear, or affine, transformation, as seems, indeed, to be the case at present (1984).

It is evidently important to distinguish those cases in which the between-group differences in the conditional joint distributions of $\underline{X}$ disclosed by the sample is "real" from those in which it is not. An obvious method for this is to determine whether Δ^2 is significantly different from zero. The hypothesis, $H_0: \Delta^2 = 0$, may be tested by the statistic (Lachenbruch 1975).

$$F_c = \frac{n_1 n_2}{n_1+n_2} \frac{(n_1+n_2-k-1)}{(n_1+n_2-2)k} D^2 \qquad (7)$$

which is distributed as F with k and (n_1+n_2-k-1) degrees of freedom on the null hypothesis $H_0: \underline{\mu}_1 = \underline{\mu}_2$, $\Sigma_1 = \Sigma_2 = \Sigma$. This is equivalent to a multivariate generalization of the Student's t-test, namely, the Hotelling's T^2 test since $T^2 = n_1 n_2 D^2/(n_1+n_2)$ (Herbert 1986; Lachenbruch 1975; Morrison 1967).

If the observed between-group difference, $\underline{\mu}_1 - \underline{\mu}_2$ (for $\Sigma_1 = \Sigma_2$), is greater than would be expected to occur in random sampling from a population described by the null hypothesis, H_0, then it is next of interest to determine whether <u>any</u> of the observed difference may be the result of sampling fluctuations in the values of one or more of the measurements included in $\underline{X}$. In other words, can a subset, (k-m), of the measurements be eliminated from the discriminant function without "significantly" changing the Mahalanobis distance between the two groups? A variance ratio test of the hypothesis, $H_0: \Delta_k^2 = \Delta^2_{k-m}$, is required. The test statistic is

$$F_c = \frac{n_1+n_2-k-1}{(k-m)} \frac{n_1 n_2 (D_k^2 - D_m^2)}{(n_1+n_2)(n_1+n_2-2) + n_1 n_2 D_m^2} \qquad (8)$$

which is distributed as F with (k-m) and (n_1+n_2-k-1) degrees of freedom (Rao 1973 and Lachenbruch 1975).

The procedure of examining the effects upon D^2 of the <u>deletion</u> of one (or more) of the <u>columns</u> of X, eg. $\underline{X}_j \longrightarrow \underline{0}$, where $\underline{X}_j$ is the jth column vector and $\underline{0}$ is the (nx1) null vector, is analogous to the <u>transformation</u> of one (or more) of the columns of X, eg., $\underline{X}_j \longrightarrow \log_{10}(\underline{X}_j)$. Both procedures should be used in achieving the most "useful" specification of the X matrix, which is a compromise between redundancy and parsimony in the set of predictors.

Let us briefly consider an alternative, Decision Theory, development of the discriminant analysis model of the classification problem. This is based on minimizing the average cost of misclassification, 1, (Rao 1973 and Anderson 1951). Table <u>1b</u> defines the possible losses incurred in taking the decisions on membership. If l_i is the loss incurred in the misclassification of a member of group i as a member of group j then it is required to find regions R_1 and R_2 for which l is a minimum where

$$l = l_1 q_1 P_1 + l_2 q_2 P_2. \qquad (9)$$

l is a minimum if R_1 is defined by

$$l_2 q_2 f_2(\underline{X}; \theta_2) < l_1 q_1 f_1(\underline{X}; \theta_1). \qquad (10)$$

This gives rise to the optimal rule: Assign $\underline{X}$ to group 1 if

$$f_1(\underline{X}; \theta_1)/f_2(\underline{X}; \theta_2) > l_2 q_2/l_1 q_1 \qquad (11)$$

For $\underline{X} \sim N_k(\underline{\mu}_i, \Sigma)$ i = 1, 2 we have the following rule: Assign $\underline{X}$ to group 2 if

$$\underline{L}^T[\underline{X} - 0.5(\underline{\mu}_1+\underline{\mu}_2)] > l_2 q_2/l_1 q_1 \qquad (12)$$

otherwise to group 2. It is usually the case, however, that the loss ratio, l_2/l_1, is difficult to estimate. Therefore, it is rarely introduced <u>explicitly</u> in practice, although it is usually taken into account in the final decision - informally.

Both of the above rules are mass assignment rules; their use will minimize either the loss incurred, or the frequency of misclassification, over the long run. It is important to have some evaluation of the diagnostic performance of a given set of tests under these rules. The performance is most usefully specified in terms of error rates (Lachenbruch 1968; Dunn 1966 and 1971; Johns 1961; Hills 1966). There are actually three, not just two, types of errors incurred in the classification problem which must be considered (Herbert 1986). However, it is useful to first more clearly distinguish the discriminant model from the hypothesis testing model of the classification problem (Rao 1952 and 1973). In the case of the latter (Sampling Theory), there is a well-specified null hypothesis, H_0: $\underline{X} \in \overline{E}$, and one (or perhaps more), rather vague, alternative hypothesis, H_1. On the basis of the sampling distribution of a specified statistic, say, $C_j = (\underline{X} - \underline{\mu}_1)^T \Sigma^{-1} (\underline{X} - \underline{\mu}_1)$ distributed as chi-squared on k df, H_0 may be accepted or rejected with a specified probability of occurrence of a Type I error, α. In those (less frequent) cases in which the sampling distribution (eg., the non-central distributions) of the statistic on H_1 is known, the probability, β, of a Type II error may also be specified. The emphasis however, is usually on the null hypothesis, H_0 (Rao 1952).

In the discriminant model of the problem of classification it is a question of selecting, on the basis of the statistic, $z = \underline{L}^T [\underline{X} - 0.5(\underline{\mu}_1 + \underline{\mu}_2)]$, one of several (in the present case two) alternative hypotheses, H_1: $\underline{X} \in E$, H_2: $\underline{X} \in E$, with a minimum weighted average of Type I and Type II errors; $q_1 P_1 + q_2 P_2$. The weights are the respective apriori probabilities of the two conditions — q_1 and q_2. Note the distinction between the two intervals described by the respective Type I and Type II errors. In the hypothesis-testing model the interval refers to that of the statistic, C_j, a function of the population values of the diagnostic tests, whereas in the discriminant model the interval refers to the population values of the diagnostic tests themselves and not some aggregate of them. See the distinction between tolerance intervals and confidence intervals in Mood, Graybill and Boes (1974). The concept of the probability of a hypothesis, say $q_j = P(H_j)$ (prior), or $P(H_j | \underline{X})$ (posterior), has no place in the hypothesis testing model because of the requirement of a repetitive element in the definition of probability in Sampling Theory. However, this concept is essential to the discriminant function model of the classification problem. In the absence of estimates of the prior probabilities, the discriminant solution to the classification problem devolves into a hypothesis test, or a sequence of such tests, as we shall describe in a later section of the present paper.

In the case where $l_1 = l_2$, $q_1 \neq q_2$, $\underline{X} \sim N_k(\underline{\mu}_j, \Sigma)$, the sample point estimates of the probabilities of misclassification (Type I and Type II error rates) are given (Lachenbruch 1975) by

$$P_1 = \Phi \left(\frac{\ln\left(\frac{1-q_1}{q_1}\right) - \frac{D^2}{2}}{D} \right) \qquad (13)$$

and

$$P_2 = \Phi \left(-\frac{\ln\left(\frac{1-q_1}{q_1}\right) + \frac{D^2}{2}}{D} \right) \qquad (14)$$

Where $\Phi( )$ denotes the cumulative distribution function of the standard Normal distribution. For $l_1 = l_2$, $q_1 = q_2$, these point estimates devolve into a familiar expression: $P_1 = \Phi(-D/2) = P_2$ (Lachenbruch 1968 and 1975; Cornfield 1965). Note that these are biased estimates since

$E(D^2) > \Delta^2$.

The decision rule described above will minimize the <u>total</u> error rate, $P_1 + P_2$. By varying the position of the threshold, say by varying the ratio, q_2/q_1, the relative sizes of the error rates P_1 and P_2 will change. A plot of P_1 vs P_2 (or, P_1 vs $1-P_2$) will describe the Operating Characteristic Curve of the linear discriminant function (vide infra) (Lusted 1975).

0.90 <u>interval</u> estimates for the error rates, P_j, can be obtained from the tables prepared (by Monte Carlo methods) by Dunn and Varady for several sizes of sample and for the constraint, $n_1 = n_2$ (Dunn 1966 and 1971). An alternative, heuristic method for obtaining interval estimates is described and illustrated in another section of this paper. Note that this method is biased.(See Eisenhart et al 1947).

The methods of obtaining <u>point</u> and <u>interval</u> estimates of the misclassification rates so far described depend on the validity of the assumption that the observations have (conditional) Normal distributions on $\bar{E}$ and E. These estimates are thus <u>asymptotic</u>; large samples, n_1, $n_2 \longrightarrow \infty$, are required for their validity.

We shall discuss below still another method of obtaining point and interval estimates of the misclassification rates that is valid for <u>small</u> samples and is based on the Bernoulli distribution. This is the Lachenbruch "leaving-one-out" method (Lachenbruch 1967 and 1975). It provides (nearly) <u>unbiased</u> estimates of the error rates.

In all cases it is found that to determine either <u>point</u> or <u>interval</u> (sample) estimates of the discriminant function and the respective error rates for two-group discrimination it is most profitable to divide the sample <u>equally</u> between the two groups: $n_1 = n_2$ (Rao 1952 and 1973).

The optimal rule given above has been derived on the assumption that the variance-covariance matrices are equal: $\Sigma_1 = \Sigma_2 = \Sigma$. This assumption is not always justified. In particular, when one of the groups, say $\bar{E}$, is comprised of non-diseased persons, and the second group, say E, is comprised of diseased persons, it is usually found that $\Sigma_2 > \Sigma_1$ (Lachenbruch 1973 and 1980). The heterogeneity of variance may be due to the presence of subgroups of disease (Lachenbruch 1980). For example, for $\bar{E}$ we may have $\underline{X} \sim N_k(\mu_1, \Sigma)$ and for E we have $\underline{X} \sim \sum_{g=1}^{w} \gamma_g N_k(\mu_{2j}, \Sigma)$ where $\sum_{g=1}^{w} \gamma_g = 1$. γ_g are the weights which represent the proportions of the various <u>subgroups</u> of E (Lachenbruch 1973 and 1980). On the other hand, the heterogeneity of variance, $\Sigma_2 > \Sigma_1$, may describe the derangement of homeostasis in each of the individuals within a single disease group (Elvebeck and Taylor 1969).

If the two matrices, Σ_1 and Σ_2, are not greatly different, the <u>linear discriminant function</u> is still <u>optimal</u>. However, if they differ considerably, <u>but</u> both conditional distributions are still Normal, then the <u>quadratic discriminant function</u> should be considered (Lachenbruch 1975, 1977 and 1980; Marks 1974; Geisser 1964; Smith 1947; Gilbert 1969). In this case, the optimal rule is: Assign to group 1 if

$$Q(\underline{X}) = \ln[f_1(\underline{X}; \Theta_1)/f_2(\underline{X}; \Theta_2)] > \ln \frac{l_2 q_2}{l_1 q_1}, \qquad (15)$$

otherwise assign to group 2. $Q(\underline{X})$ is the quadratic discriminant function (Rao 1973):

$$Q(\underline{X}) = (\underline{\mu}_2^T \Sigma_2^{-1} - \underline{\mu}_1^T \Sigma_1^{-1})\underline{X} + \tfrac{1}{2}\underline{X}^T(\Sigma_1^{-1} - \Sigma_2^{-1})\underline{X} + \tfrac{1}{2}[\underline{\mu}_1^T \Sigma_1^{-1}\underline{\mu}_1$$
$$- \underline{\mu}_2^T \Sigma_2^{-1}\underline{\mu}_2 - \log(|\Sigma_2|/|\Sigma_1|)] \tag{16}$$

Note that $Q(\underline{X}) \longrightarrow L(\underline{X})$ as $\Sigma_1^{-1} - \Sigma_2^{-1} \longrightarrow 0$ where 0 is the (kxk) null matrix. It has been found that $Q(X)$ is not optimal if the conditional distributions deviate much from multivariate Normal; i.e., the procedure is not robust with respect to non-normality (Lachenbruch 1977; Gilbert 1969).

Some discussions of the classification problem do not place enough emphasis on the verification of the statistical assumptions on which the validity of the various models and procedures depends.g We shall now examine two of that body of methods which can help to determine whether two of the assumptions underlying a given model for discriminant analysis are indeed satisfied by the sample - either the raw data or their transforms, eg. their logarithms.

i.) The hypothesis of Normality of the conditional joint distributions of $\underline{X}$, H_0: $\underline{X} \sim N_k(\underline{\mu}_i, \Sigma_i)$, can be assessed by several procedures. A Normal probability plot of the marginal cumulative distributions of each component of $\underline{X}$, together with the sample estimate of the Filliben probability plot correlation coefficient (Filliben 1975) is only an approximate method, since Normality of the marginal distributions of the components, x_i, $i = 1, \ldots, k$, is a necessary but not sufficient condition for Normality of the joint distribution of $\underline{X} = (x_1, \ldots, x_k)$. See Hald 1965.

Higher dimensional probability plots provide graphical appreciation of the nature of the distribution (Gnanadesikan 1977) although there are no statistics cognate to that of Filliben's Coefficient for the univariate probability plot (This is an area of current research. Filliben, Private Communication).

For bivariate observations, $\underline{X} = (x_1, x_2)$, we have noted that $C_j = (\underline{X}-\underline{\bar{X}}_j)S_j^{-1}(\underline{X}-\underline{\bar{X}}_j)$ is distributed as chi-square with k=2 degrees of freedom if $\underline{X}$ is bivariate Normal, $\underline{X} \sim N_2(\underline{\mu}_j, \Sigma_j)$: $C_j \sim \chi^2(2)$. The cumulative distribution function is (rewriting $C_j \equiv C$):
$$P(\chi^2 < C|2) = P(C) = 1 - e^{-0.5C} \tag{17}$$
or
$$\log_{10}P(C) = -0.217C \tag{18}$$
For the order statistics $C_{(1)}, \ldots, C_{(n_j)}$ of the n_j observations, the corresponding estimates of $1-P(C_{(i)})$ are $(n_j - i + 0.5)/n_j, \ldots, 1/2n_j$. Therefore, $C_{(i)}$, and $(n_j - i + 0.5)/n_j$ when plotted on semi-logarithmic paper will be distributed at random about a straight line through the point (0, 1) with slope -0.217. (It is not known how much uncertainty the use of the estimates, $\underline{\bar{X}}_i$, S_i, in place of $\underline{\mu}_i$, Σ_i introduces (Hald 1965).) Another method for bivariate data is to plot the contour ellipse for a specified value of P and compare the observed and theoretical numbers of observations which lie within the ellipse (Hald 1965).

ii.) The hypothesis, H_0: $\Sigma_1 = \Sigma_2$ can be readily tested by a generalization of the Bartlett test for the homogeneity of g variances in which the determinants of the sample covariance matrices, $|S_i|$, assume the role of generalized variances (Morrison 1967). Box has shown that the quantity MC^{-1} is distributed approximately as chi-square with $0.5(g-1)k(k+1)$ as $n_i \longrightarrow \infty$. Here

$$M = \Sigma(n_1-1)\ln|S| - \sum_{i=1}^{g}(n_i-1)\ln|S| \tag{19}$$
and

$$c^{-1} = 1 - \frac{2k^2 + 3k - 1}{6(k+1)(k-1)} \left[\sum_{i=1}^{g} \frac{1}{(n_i - 1)} - \frac{1}{\Sigma(n_i - 1)} \right] \qquad (20)$$

for $n_i = n$, all i.

We shall consider, in another section, still other methods for assessing the degree to which the sample data are consistent with the a priori information on which the validity of the discriminant model depends, e.g. Regression Diagnostics (Belsley et al 1980).

One aspect of the classification problem concerns the concept and methods for achieving a linear combination, $z = \underline{X}^T \underline{V}$, of a set of k measurements $\underline{X}^T = (x_1, x_2, \ldots x_k)$ such that the "overlap" of the conditional distributions of this combination for two (or more) groups is a minimum. A useful measure of overlap is the ratio of the within- to the between-groups sum of squares. Fisher (1936) was the first to propose this ratio as a two-group discriminant criterion. If B' is the between-groups sum of squares of z and W' is the within-groups sum of squares of z then the problem of minimizing overlap devolves into finding the vector $\underline{V}^T = (v_1, v_2, \ldots, v_k)$ such that the ratio, λ, is a maximum where

$$\lambda = B'/W' = \underline{V}^T B \underline{V} / \underline{V}^T W \underline{V} \qquad (21)$$

and B and W are the (kxk) between- and within-groups sums of squares matrices of $\underline{X}$ (Bryan 1951; Rao 1948; Press 1972; Lindeman 1980). We consider the problem for g = 2 groups.

If the conditional distributions of $\underline{X}$ on groups 1 and 2 are, respectively, $\underline{X} \sim N_k(\underline{\mu}_1, \Sigma_1)$ and $\underline{X} \sim N_k(\underline{\mu}_2, \Sigma_2)$ and $\underline{\mu}_j$ and Σ_j are estimated from samples of size n_j by $\overline{X}_j$ and S_j, respectively, then B is the (kxk) matrix

$$B = n_1 * n_2 / (n_1 + n_2) \, (\overline{\underline{X}}_2 - \overline{\underline{X}}_1)(\overline{\underline{X}}_2 - \overline{\underline{X}}_1)^T \qquad (22)$$

and W is the (kxk) matrix

$$W = (n_1 - 1) S_1 + (n_2 - 1) S_2 \qquad (23)$$

The elements of the vector, $\underline{V}$, can be found as the solution of the vector differential equation[h]

$$\partial \lambda / \partial \underline{V} = 0 \qquad (24)$$

It can be shown that the solutions can be found in terms of the eigenvectors of the square, asymmetric, matrix $W^{-1}B$:

$$(W^{-1}B - \lambda I)\underline{V} = 0. \qquad (25)$$

It can be shown also that the determinantal equation, $|W^{-1}B - \lambda I| = 0$, describes a sufficient as well as necessary condition for maximizing the discriminant criterion, λ (Lindeman 1980).

By considering the rank of the matrix $W^{-1}B$ it may be shown that the number, r, of non-zero eigenvalues λ_j, $j = 1, \ldots, r$, is the smaller of the two integers g-1 and k where g is the number of groups; here, g=2. (r is the number of non-zero eigenvalues of a square matrix, $W^{-1}B$. The rank of a product of two matrices cannot exceed the smaller of the ranks of the two factor matrices, here B and W^{-1}. Since W^{-1} is non-singular it is of rank k. There are only (g-1) independent rows of B, therefore, B is of rank (g-1). Thus, the rank of $W^{-1}B$ is r, the smaller of the two integers (g-1) and k. For g=2 there is only 1 non-zero eigenvalue of $W^{-1}B$.)

It can be shown that $\underline{V}$ is the right eigenvector, $\underline{P}_2$, of the matrix, $W^{-1}B$. That is, for g=2 ($\lambda_1 \equiv 0$) the spectral decomposition of $W^{-1}B$ is given by $W^{-1}B = (\lambda_2 / \underline{P}_2^T \underline{Q}_2) \underline{Q}_2 \underline{P}_2^T$ where $\underline{Q}_2$ is the left eigenvector of $W^{-1}B$. (For the asymmetric matrix A, the determinantal equation, $|A - \lambda I| = 0$, has $m > 0$ roots some of which may be complex even if A is real. Corresponding to the root λ_j there are two vectors, $\underline{P}_j$ and $\underline{Q}_j$, the right and left eigenvectors, respectively, where $A\underline{P}_j =$

$\lambda_j \underline{P}_j$ and $A^T \underline{Q}_j = \lambda_j \underline{Q}_j$. See Rao 1973.)

The relations between the statistics, λ_2, $\underline{V}_2$, of this model and the statistics, D^2, $\underline{L}^T$, of the earlier model are described by the simple proportions (Rao 1973):

$$D^2 = \frac{(n_1+n_2)(n_1+n_2-2)}{n_1 * n_2} \lambda_2 \qquad (26)$$

and

$$\underline{L}^T = (\overline{X}_2 - \overline{X}_1)^T S^{-1} = c\underline{P}_2^T = c\underline{V}_2 \qquad (27)$$

where c is a multiplicative constant.

It can be shown that the quantity $[n_1+n_2 - (g+k)/2] \ln(1+\lambda_m)$ is distributed asymptotically as chi-squared with g+k-2m degrees of freedom, where m is the <u>order</u> of λ_m, i.e., for the largest eigenvalue, m=1, for the next largest, m=2, etc. Thus, the <u>significance</u> of λ_m and hence that of the corresponding discriminant function may be assessed. For $r > 1$, Bartlett's statistic, V, must be computed where V is the scalar

$$V = [n_1+n_2-1-(g+k)/2] \sum_{m=1}^{r} \ln(1+\lambda_m). \qquad (28)$$

V is distributed approximately as chi-squared with k(g-1) degrees of freedom and the residuals computed (Harris 1975).

In the interpretation of the discriminant function coefficients it is of interest to know, among other things, the <u>relative</u> contribution of each of the k variables to the discrimination achieved. This can be determined by comparison of the standardized discriminant coefficient, $v_{mi}^* = \sqrt{w_{ii}} \, v_{mi}$, where v_{mi} is the ith component of the eigenvector, $\underline{V}_m$, and w_{ii} is the ith diagonal element of W, the within-groups sum of squares matrix (Tatsuoka 1971).

The eigenvector-eigenvalue methods just discussed are readily generalized to the classification problem for three or more groups, $g > 2$. See Rao 1948 and Bryan 1951. (But see also the methods of Lindeman, 1980 and Tatsuoka 1971 (vide infra) for $g > 2$.)

One fundamental aspect of the classification problem concerns the concept and measure of the "<u>resemblance</u>" of the individual in question to each of one or more (here two) distinct groups. It is an aspect which becomes more important as the non-sample evidence for $\underline{X} \in E \cup \overline{E}$ becomes weaker and the <u>classification problem</u> devolves into one of <u>hypothesis testing</u> (Rao 1952).

Tiedeman, et al (1953), chose to measure resemblance in terms of dissimilarity and selected the C_j statistic as a measure of the <u>dissimilarity</u> of an individual to a given group on the basis of a specified set of measures: Given the individual with a set of scores $\underline{X}$. With respect to group j, with parameters $\underline{\mu}_j$, Σ_j we have, $C_j = (\underline{X} - \underline{\mu}_j)^T \Sigma_j^{-1}(\underline{X} - \underline{\mu}_j)$ (Tiedeman 1953 and 1954). This seems to be an appropriate choice since the larger is C_j with respect to a given group the greater is the (generalized) distance of the point which represents his scores from the point which represents the centroid of the scores of the group and hence the less that individual resembles the "average member" of that group.[i] Moreover, if the conditional joint frequency distribution of the k measurements for the reference group is k-variate normal, $\underline{X} \sim N_k(\underline{\mu}_j, \Sigma_j)$, then C_j is distributed as chi-squared with k degrees of freedom and the C_j value of the set of scores, $\underline{X}$, of an individual provides an estimate of the <u>proportion</u> of the group that are at distances from the centroid, $\underline{\mu}_j$, which exceed (or are exceeded by) the distance of the individual in question. Geometrically this proportion can be represented by the contour ellipsoid (vide infra) on which the given value of $\underline{X}$ lies (Hald 1965; Lindeman 1980).

A rather different aspect of the classification problem has to do with the assessment of the risk, for an individual with measurements $\underline{X}$, of the transition to the condition E from an initial condition $\overline{E}$ (Lachenbruch 1975 and 1980; Cornfield 1965; Halperin, et al, 1971; Truett, et al, 1967; Press 1978). This risk can be expressed as the posterior probability $P(E|\underline{X})$. The risk can in turn be related to the apriori probability of E, $P(E)$, and the conditional probability of $\underline{X}$ given E, $P(\underline{X}|E)$, by Bayes' Theorem which can be described by the simple relations which obtain between the marginal, joint and conditional probabilities of E and X (Lachenbruch 1975; Rao 1952 and 1973):

$$P(E|\underline{X}) = P(\underline{X}\&E)/P(\underline{X}) = P(\underline{X}|E)P(E)/P(\underline{X}) \tag{29}$$

where

$$P(\underline{X}) = P(\underline{X}\&\overline{E}) + P(\underline{X}\&E) = P(\underline{X}|\overline{E})P(\overline{E}) + P(\underline{X}|E)P(E) \tag{30}$$

since $\overline{E}$ and E are exhaustive as well as (mutually) exclusive for $\underline{X}$.
Then

$$P(\overline{E}) = q_1, \quad P(E) = (1-q_1) = q_2 \tag{31}$$

and the respective likelihoods of the measurements, $\underline{X}$, are

$$P(\underline{X}|\overline{E}) = f_1(\underline{X}; \Theta_1) \text{ and } P(\underline{X}|E) = f_2(\underline{X}; \Theta_2) \tag{32}$$

Therefore, we may write

$$P(E|\underline{X}) = \frac{1}{1+\left(\frac{1-q_1}{q_1}\right)\frac{f_1(\underline{X}; \Theta_2)}{f_2(\underline{X}; \Theta_2)}} \tag{33}$$

If $\underline{X} \sim N_k(\underline{\mu}_1, \Sigma)$ on $\overline{E}$ and $\underline{X} \sim N_k(\underline{\mu}_2, \Sigma)$ on E, and $\underline{\mu}_1 \neq \underline{\mu}_2$, $\Sigma_1 = \Sigma_2$, then the risk can be described by a logistic function:

$$P(E|\underline{X}) = \frac{1}{1+\frac{1-q_1}{q_1} e^{-L^T\underline{X} + L^T(\overline{X}_1+\overline{X}_2)/2}} \tag{34}$$

where
$$L^T = (\overline{X}_2-\overline{X}_1)^T \Sigma^{-1}$$

In this case we may estimate the risk of E, given $\underline{X}$, in terms of the linear discriminant function, $L(\underline{X})$:

$$P(E|\underline{X}) = \frac{1}{1+\left(\frac{1-q_1}{q_1}\right) e^{-L(\underline{X})}} \tag{35}$$

If $\underline{X} \sim N_k(\underline{\mu}_1, \Sigma_1)$ on $\overline{E}$ and $\underline{X} \sim N_k(\underline{\mu}_2, \Sigma_2)$ on E, and $\underline{\mu}_1 \neq \mu_2$, $\Sigma_1 \neq \Sigma_2$, then the risk of E given $\underline{X}$ may be described in terms of the quadratic discriminant function, $Q(\overline{X})$:

$$P(E|X) = \frac{1}{1 + \frac{1-q_1}{q_1} e^{-Q(\underline{X})}} \tag{36}$$

If we recall the definition, $\text{logit } P = \log[(1-P)/P]$, we find that the logit of risk is a linear (or quadratic) function of $\underline{X}$ in which the coefficients are estimated by the linear (or quadratic) discriminant function. In the linear case we have

$$\log\left(\frac{1-P(E|\underline{X})}{P(E|\underline{X})}\right) = \beta_0 + \sum_{i=1}^{k} \beta_i x_i \tag{37}$$

where

$$\beta_0 = \log\left(\frac{1-q}{q}\right) - 0.5\underline{L}^T(\overline{\underline{X}}_1 + \overline{\underline{X}}_2) \tag{38}$$

and

$$\beta_i = (\underline{L}^T)_i, \quad 1 < i < k. \tag{39}$$

For $P(E|\underline{X}) = P$, a constant, $0 \leq P \leq 1$, and $k > 2$, the logit equation describes an <u>isorisk surface</u>. For $k=2$, the logit equation describes an <u>isorisk curve</u>.

Still another Bayesian approach to the classification problem was developed by Geisser (1964) and Press (1972) who describe a predictive odds ratio, $P(\underline{X} \in E_i|\underline{X})/P(\underline{X} \in E_j|\underline{X})$, for classifying $\underline{X}$ into group E_i as compared to group E_j. $i, j = 1, \ldots, g \geq 2$. The ratio devolves into the ratio of the associated multivariate Student-t densities. In this approach there is no complicated distribution theory, the sample sizes need not be large, the covariance matrices, Σ_i, need not be equal, and it is readily extended to $g > 2$. (This is a procedure which does not seem to be as widely used as perhaps it should be.)

One useful feature of the discriminant model of the classification problem is the analogy between <u>multiple regression analysis</u> on a binary response variable and the two-group ($g=2$) discrimination problem (Lachenbruch 1975; Kleinbaum 1978). In the binary regression model of discrimination, the relation between the response variable, y_i, which describes the status, or condition, of the individuals which comprise the two complementary groups, $\overline{E}$ and E, (say $y_i = 0$ in $\overline{E}$ and $y_i = 1$ in E) and the predictor variables $\underline{X} = (x_1, x_2, \ldots x_k)$ is, in a sense, the inverse of that which obtains in discriminant model. In <u>discriminant model</u>, the values of the parameters, Θ, (and perhaps the functional form, $f(\underline{X}; \Theta)$, as well) of the frequency distributions of $\underline{X}$ are conditional on U, e.g., $\underline{X} \sim N_k(\underline{\mu}(U), \Sigma(U))$. In the <u>binary regression model</u>, the value of the response variable, y, is conditional on $\underline{X}$; it is a linear <u>function</u> of X: $\underline{y} = y(\underline{X})$. If we denote the coefficient vector of regression analysis by $\underline{\alpha} = (\alpha_0, \alpha_1, \ldots, \alpha_k)$ then $y_i = \alpha_0 + \alpha_1 x_1 = \ldots + \alpha_k x_k + u = \underline{X}^T \underline{\alpha} + u$, where $\underline{X}^T$ is $1*(k+1)$ and $\underline{\alpha}$ is $(k+1)*1$.

It will be useful in the sequel to introduce now the standard notation of the multiple regression model. (See Lectures 1 and 2 and Montgomery & Peck 1982; Belsley, et al 1980.) If $\underline{y}$ is nx1, $\underline{X}$ is nx(k+1), $\underline{\alpha}$ is (k+1)x1, we have estimates of the coefficient vector, $\underline{\alpha}^*$, and response, $\underline{y}$:

$$\underline{\hat{\alpha}} = (X^T X)^{-1} X^T \underline{y} \text{ and } \underline{\hat{y}} = X\underline{\hat{\alpha}} = X(X^T X)^{-1} X^T \underline{y} = H\underline{y}. \tag{40}$$

H is the so called (nxn) hat matrix. It describes the projection of $\underline{y}$ onto the X subspace (Montgomery & Peck 1982). Also,

$$\text{Var}(\underline{\hat{\alpha}}) = s^2 (X^T X)^{-1} \text{ where } s^2 = (\underline{\hat{y}} - \underline{y})^T (\underline{\hat{y}} - \underline{y})/(n-k-1). \tag{41}$$

Then, $\text{Var}(\underline{\hat{y}}) = s^2 H$

Since $\underline{y}$ is binary, the members of the two complementary conditions, $\overline{E}$ and E, may be coded as any two distinct values, say $y_i = a_1$ and $y_i = a_2$ for individuals in conditions $\overline{E}$ and E, respectively. However, it is simplest to code the response variable as $y_i = -n_2/(n_1+n_2)$ for all n_1 individuals in condition $\overline{E}$ and $y_i = n_1/(n_1+n_2)$ for all n_2 individuals in condition E (Fisher 1936; Kleinbaum & Kupper 1978). Then the mean of y for the combined groups is $\overline{y} = 0$.

If $\underline{\beta} = (\beta_0, \beta_1, \ldots, \beta_k)$ is the coefficient vector for the two-group discriminant function we have $\beta_0 = \frac{1-q_1}{q_1} - 0.5*\underline{L}^T(\overline{\underline{X}}_1 + \overline{\underline{X}}_2)$, and

$\beta_i = (L^T)_i, 1 < i < k$. It must be recalled however, that in two-group discriminant analysis, the coefficients of the discriminant function are <u>not</u> parameters in the usual sense and hence can only be determined up to a scale factor (Rao 1970 and 1973). However, the asymptotic variance of the coefficient, β_i, is estimated by

$$\text{Var}(\beta_i) = s^{ii}\left(\frac{1}{n_1} + \frac{1}{n_2}\right)$$

where s^{ii} is the ith element of the matrix S^{-1}. Then $\underline{\beta} = \underline{\alpha}*/c$, where

$$c = \frac{n_1*n_2}{n_1+n_2} \bigg/ \left[(n_1 + n_2 - 2) + \frac{n_1*n_2}{n_1+n_2} D^2\right] \tag{42}$$

Here $D^2 = (\bar{X}_2-\bar{X}_1)^T S^{-1}(\bar{X}_2-\bar{X}_1)$ is the sample estimate of the Mahalanobis distance between the two groups. For other y codes, c may be different, of course. It follows at once that $\text{Var}(\underline{\beta}) = \text{Var}(\underline{\alpha}*)/c^2$. This has been found to agree fairly well with the asymptotic expression given above (Kleinbaum and Kupper 1978).

There is also an interesting relationship between the squared multiple correlation coefficient, R^2, of the binary regression model and the estimates, D^2, of the Mahalanobis distance for the discriminant model (Lachenbruch 1968, 1975; Kleinbaum and Kupper 1978):

$$D^2 = \frac{(n_1+n_2)(n_1+n_2-2)}{n_1 + n_2} \frac{R^2}{1-R^2} \tag{43}$$

Here

$$R^2 = 1-\text{RSS}/\text{SSY} \tag{44}$$

where $\text{RSS} = \sum_{i=1}^{n} (\hat{y}_i-y_i)^2 = \sum_{i=1}^{n} e_i^2$ and $\text{SSY} = \sum_{i=1}^{n} (y_i-\bar{y})^2$.

The utility of the relation between the discriminant analysis on two groups and multiple regression analysis with a binary response variable is now not so much the fact that its existence permits the implementation of the discriminant model by means of the more familiar multiple regression analysis algorithms (which are implemented by more widely available computer programs). Rather, it is that the existence of the analogy provides a basis for some cogent recommendations for sample size, methods for <u>validation</u> of the estimated discriminant function - point-estimates of its predictive performance in <u>future</u> samples - and assessment of the nature and degree of the dependence of the sample estimates of the coefficients, etc. of the discriminant function upon the idiosyncracies of the sample: The nature (intrinsic or contingent) and degree of collinearity of variables, the identification of responses not well explained by the model and cases that dominate some aspect of the "fit" of the model. The last is especially important for, "It is known by now that the proportion of <u>gross errors</u> in data, depending on circumstances, is normally between 0.1% and 10% with several percent being the rule rather than the exception" (Hampel 1974). (italics added).

For one example of the fruitfulness of the binary regression analogy, consider the requirements for sample size. It is generally accepted that in multiple regression analysis, on k predictor variables, the sample sizes $n(=n_1+n_2)$ should not be <u>less</u> than 100 or 20k whichever is the <u>greater</u> (Lindeman 1980). Thus, <u>taking</u> into account Rao's (1952) recommendation that discriminant functions be estimated on the basis of <u>equal</u> numbers in each group we can conclude that the <u>minimum</u> sample required to estimate a discriminant function is $n_1 = 50$, $n_2 = 50$ observations. (It is of perhaps more than historical interest to note that the sample sizes from which Fisher wrote his 1936 paper were $n_1 = n_2 = n_3 = 50$.) Because of the dependence of the degree of shrinkage, $(D^2-\Delta^2)$ on k, the smallest <u>useful</u> sample size may well exceed this minimum.

For another example, although it is (quite properly) recommended that a second sample of size n' of observations on individuals of known status ($\bar{E}$ or E) be used to validate any estimated discriminant function (Lindeman 1980), it is usually the case that additional cases of the required kind and in the required number are difficult to obtain for logistic, fiscal, ethical, etc., reasons. For similar reasons the initial sample is often too small to provide for "data-splitting" or cross-validation (Montgomery and Peck 1982). A similar prescription for validation obtains for regression models and there also, the required "second look" at the estimated model either from a new sample or a "split" sample is often very difficult to secure. In the latter case, the so-called PRESS statistic may be constructed as a simulation of "data-splitting", in order to provide an estimate of the degree to which a multiple regression equation will "hold" its predictive power between the initial and future samples (Montgomery & Peck 1982; Allen 1974). The method achieves cross-validation of the model when the initial sample is too small for conventional "data splitting" methods.

This statistic is defined by Allen (1974) as follows:

$$\text{PRESS} = \sum_{i=1}^{N} (y_i - \hat{y}_{i(j)})^2 = \sum_{i=1}^{N} [e_i/(1-h_i)]^2 \qquad (45)$$

where $\hat{y}_{i(j)}$ is the estimated value of y_i obtained when $\underline{\alpha}$ is estimated from the sample of size $(n-1)$ in which the jth observation, $j = 1, \ldots, n = n_1 + n_2$, has been deleted. The construction of the PRESS statistic provides n cross-validations in which each "training sample" is of size $(n-1)$ and each "validation sample" is of size 1 (Montgomery & Peck 1982). A squared multiple correlation coefficient, R'^2, which provides a more familiar assessment of predictive performance of the equation in future samples can be constructed from PRESS: $R'^2 = 1-\text{PRESS}/\text{SSY}$. $(R^2 - R'^2)$ is a measure of the degree of "shrinkage" of the multiple correlation coefficient, R^2, described by Lachenbruch.

h_i is the ith diagonal element of the nxn, so-called, "hat matrix", $H = X(X^TX)^{-1}X^T$. It is a measure of the distance of the (row) vector $\underline{x}_i^T$ from the centroid of the distribution of the pooled groups. Note that $\partial \hat{y}_i / \partial y_i = h_i$. e_i is the ith residual: $y_1 - \hat{y}_i$. These key diagnostics, e_i and h_i, identify cases for which the response is not well explained by the model or which dominate some aspects of the fit of the model to the data. Thus, PRESS measures the degree to which the estimates of the model are peculiar to the sample in hand. It is a summary measure of the nature and degree of model-sample interaction. In using PRESS to discriminate between rival regression models, M_g, on the basis of a single sample the rule is to choose that model for which PRESS is least.

An estimate, D'^2, of the Mahalanobis distance can be obtained from R'^2 by the transformation given above, and then the probabilities of misclassification, $P'_j = \phi(-D'/2)$, $j = 1, 2$, to be achieved by use of $L(\underline{X})$ in future samples, can be estimated. It has been found, empirically, that $D'^2 \simeq \delta^2$, the unbiased estimate of Δ^2. (This is, of course, not unexpected.) This use of the PRESS statistic is based on the hypothesis, $H_0: \underline{X} \sim N_k(\underline{\mu}_j, \Sigma)$, and hence, (only) valid for large samples. It is similar in both concept and method to the Lachenbruch "leaving-one-out" method, which, however, is based on the Bernoulli distribution and hence, valid for small samples (Lachenbruch 1967 and 1975).

It is, of course, well-known that the observed dispersion of the conditional distribution of a biological meaasurement, x, summarizes both the inherent biological variability and the random error in the measurement: x = x' + u, where x' is the true value and u is the random error of measurement. It is assumed that x' is distributed conditionally Normal: $x' \sim N(\mu_b, \sigma_b^2)$. It is assumed that $E(u) = 0$ and $Var(u) = \sigma_m^2$. Then x is distributed conditionally Normal: $x \sim N(\mu, \sigma^2)$ where $\sigma^2 = \sigma_b^2 + \sigma_m^2 = \sigma_b^2(1 + \sigma_m^2/\sigma_b^2)$. (See Hald 1965) The presence of measurement errors in two variables x_1 and x_2 <u>deflates</u> the sample estimate, r_{12}, of the <u>true</u> correlation coefficient, r_{12}': $r_{12}' = r_{12}'/\sqrt{(1+\sigma_{m_1}^2/\sigma_{b_1})(1+\sigma_{m_2}^2/\sigma_{b_2})}$. (See Lark, Craven and Bosworth 1968.)

The problems for the inferences made from Normal theory <u>regression</u> models, $y_i = \underline{x_i}'\underline{\beta} + \varepsilon_i - \infty < y_i < \infty$, which arise from the presence of errors of measurement, u_i, <u>in the predictors</u>, x_i, are also well-known (Theil 1971). Such measurement errors will <u>bias</u> the least squares estimates, $\hat{\underline{\beta}} = (X'X)^{-1}X'\underline{y}$, that is, $E(\underline{\beta}) \neq \underline{\beta}$. <u>Unbiased</u> estimates of $\underline{\beta}$, say $\underline{\beta}^*$, can be obtained if the dispersion matrix, E, of the errors of measurement of X is known, a priori. A useful approximation is (Theil 1971) $\hat{\underline{\beta}}^* \cong (\overline{X'X} - nE)^{-1}X'\underline{y} = (X'X-nE)^{-1}X'X\hat{\underline{\beta}}$. $n = n_1+n_2$ is the total sample size. Usually only the variances of the measurement errors are known, so that E is a diagonal matrix. Moreover, E must be estimated from <u>non-sample</u> evidence. Comparison of the cognate error-in-variables estimates, $\hat{\underline{\alpha}}^*$ with the least squares estimates, $\underline{\alpha}$, of the <u>binary</u> regression model, $y_i = \underline{x_i}'\underline{\alpha} + \underline{u}$, $y_i = 0, 1$, of the discrimination problem as described above, will provide an assessment of the effects of such errors on the estimated discriminant function coefficients, $\underline{L}^T$.

Before examining the Lachenbruch method let us condider a test of the utility of the classification procedure based on the discriminant model. One method for assessing the value of an estimated discriminant function is to test whether the assignments of the members of the training sample between complementary groups by the <u>optimal partition</u> of the measurement space that is achieved by requiring the total error to be a minimum differs significantly from the assignments achieved by a purely <u>random partition</u> (Press 1972).

Accordingly, we define the Q statistic, which is distributed as chi-square with one degree of freedom on the null hypothesis (random partition). For g groups we have

$$Q = \frac{(n-e)^2}{e} + \frac{(\overline{n}-\overline{e})^2}{\overline{e}} \qquad (46)$$

where n and $\overline{n}$ denote the number of correct and incorrect assignments made by the discriminant function and e and $\overline{e}$ denote the expected numbers of correct and incorrect assignments on the basis of random partition. Then, let $N = \sum_{i=1}^{g} n_i$, denote the total size of the training sample. The probability of correct random classification is $1/g$. Then we have

$$n = \sum_{i=1}^{g} n_{ii}, \quad \overline{n} = N-n \qquad (47)$$

$$e = N/g, \quad \overline{e} = N-N/g \qquad (48)$$

Q may then be rewritten as

$$Q = (N-ng)^2/N(g-1) \qquad (49)$$

Use of this method with the initial sample, <u>underestimates</u> the number of misclassifications to be expected in use of the discriminant function in subsequent samples. For <u>small</u> samples, the size of this bias is large.

Let us now consider an unbiased method for obtaining both point and interval estimates of the misclassification rates P_1 and P_2 for the two-group discrimination which is based on the binomial distribution. If we denote the sample estimates P_1 and P_2 obtained from a sample of size $n = n_1+n_2$ as $P_1(n_1, n_2)$ and $P_2(n_1, n_2)$, the method to be described yields estimates of $P_1(n_1-1, n_2)$ and $P_2(n_1, n_2-1)$. If n_1 and n_2 are not too small these approximations to P_1 and P_2 should be satisfactory. The method is due to Lachenbruch (1967 and 1975). It is known as the Lachenbruch, "Leaving-one-out" method.

The motivation as well as the implementation of this method are cognate to those of the PRESS statistic for multiple regression analysis, namely, to achieve cross-validation of the model when the size of the initial sample is too small for conventional "data-splitting" methods. The procedure is as follows: Estimate the discriminant function for each of the n samples of size n_1+n_2-1 that are obtained by the successive deletions of one observation from the initial sample and record for each whether the deleted observation is misclassified. If r_1 observations on $\overline{E}$ are misclassified and r_2 observations on E are misclassified by this procedure, then the sample point estimates of the respective misclassification rates are $P_1 = r_1/n_2$ and $P_2 = r_2/n_2$. These are unbiased estimates.

The respective $(1-\alpha)$ interval estimates are obtined as $\underline{P}_j$ and $\overline{P}_j$, $j = 1, 2$ where (Brownlee 1967)

$$\underline{P}_j = \frac{r_j}{(n_j-r_j+1)(F(1-\alpha/2; 2(n_j+r_j+1) 2r) + r} \qquad (50)$$

and

$$\overline{P}_j = \frac{(r_j+1)F(1-\alpha/2; 2(r_j+1), 2(n_j-r_j))}{(n_j+r_j) + (r_j+1) F(1-\alpha/2; 2(r_j+1, 2(n_j-r_j))} \qquad (51)$$

For $r_j = 0$, these reduce to
$$\overline{P}_j = 1-(\alpha/2)^{1/n_j} \qquad (52)$$
$$\underline{P}_j = 0 \qquad (53)$$

These are also unbiased estimates. They are, of course, not dependent on the form of the distribution of $\underline{X}$.

We now examine the Lachenbruch procedure and that of Allen (the PRESS statistic) (1974) from a more general point of view. In each of these procedures the sample data were perturbed by the deletion, seriatim, of a single case and the effect of each individual perturbation as measured by the discrepancy between the observed and predicted values of the identity and response, respectively, were summed to give the statistics, r_i/n_i ($i = 1, 2$) and PRESS of Lachenbruch (1975) and Allen (1974), respectively. These are summary measures of the effect of the presence of anomalies or idiosyncracies in the sample and hence of the degree to which the sample estimates for a specified model, depend on the sample at hand, that is, the degree to which they are sample-specific. However, it is also fruitful to examine profiles of the effects on the sample estimates of the model of the deletion of individual cases as well as the sum of the seriatim deletions. (As we observed earlier, it is known that the proportion of gross errors in data, depending on circumstances, is normally between 0.1% and 10% with several percent being the rule rather than the exception (Hampel 1974; Ahmed and Lachenbruch 1977).) The presence of gross errors is most readily detected from the profiles rather than the sums.) Moreover, although large values of the sample size, $n = n_1+n_2$, are required to reduce the variance of estimates of model parameters and model predictions it is also the case that the range of values of

$\underline{X}$, within a sample increases with sample size, n, since the difference between the modes of the distributions of both extreme values (largest and smallest values of $\underline{X}$) increases with the sample size, since $\underline{X} \sim N_k$ ($\underline{\mu}_j, \Sigma_j$). Therefore, the probability of the model estimates being sample-specific may sometimes increase with sample size because of the increased likelihood of the presence of values of $\underline{X}$, that are quite remote from the centroid, $\underline{\mu}_j$, j = 1, 2, even in a homogeneous sample. It is also the case that the degree of contamination of the sample may increase with the sample size, so that large samples may be more heterogeneous - and hence more biased - than smaller samples, because of a larger proportion of more atypical members in the former. (But, of course with larger samples the effect of any single case on the sample estimates tends to decrease.)

In order to assess the effect of sample idiosyncracies on the estimated discriminant function one might wish to compare estimates of the Mahalanobis distance, Δ^2, and of the coefficient vector, $\underline{\beta}$, obtained with and without the ith case. These may be described as $D^2 - D^2(i)$ and $\underline{\hat{\beta}} - \underline{\hat{\beta}}(i)$, respectively. The latter is related to the Cook's D and DFFITS statistics (Belsley, Kuh and Welsch 1978; Cook and Weisberg 1982). A related measure of interest is the comparison of estimates of the variance-covariance matrix, S_j, j = 1, 2, with and without, the ith case: $S_j - S(i)_j$. Another related measure is the comparison of estimates of the Normality of the marginal distributions with and without the ith case: $r_j - r(i)_j$, j = 1, 2, where r_j is Filliben's probability plot correlation coefficient for the jth conditional distribution (Filliben 1975). It is often the case, of course, that the ith case will strongly influence all such measures.

This method - deletion of single cases - of assessing the effects of the anomalies or idiosyncracies of the sample in hand (model-sample interactions) upon the estimated discriminant function, $L(\underline{X})$, and its related measures, such as D^2, is a special case of a set of more general procedures known as Regression Diagnostics that have been recently developed for similar purposes for multiple regression analysis (Belsley, et al 1980; Cook and Weisberg 1982).

Regression Diagnostics provide quantitative assessments of the degree to which the results of a regression analysis depend on the idiosyncracies of the particular sample of data in hand (Belsley, et al 1980; Krasker 1980; Ramsey, et al 1980; Hymans 1980). They provide (some) protection against the universal tendency to confuse those features of the study population that are summarized by sample estimates which are unique to the particular sample at hand, with true features of the population. At present, however, these methods are more descriptive than prescriptive: While they are useful in the detection, classification and measurement of the effects of idiosyncracies in the sample distribution on the sample estimates of the model they do not provide explicit recommendations or procedures for ameliorating these. (See, however, Welsch's bounded influence estimators (1980).)

It must be emphasized that the assessment of the interaction between model and sample which is provided by Regression Diagnostics is rather different from that provided by an examination of the more familiar interval estimates of the coefficient vector, $\hat{\alpha}$, predicted response, $\underline{y}$, etc. Interval estimates circumscribe - at a given level of uncertainty - the range of point estimates which may be observed in future samples in which the distribution of $\underline{X}$ - described by the design matrix, X, of the experimental studies for which

Table 1a. Classification Matrix

	Condition	
	$\underline{X} \in \overline{E}$	$\underline{X} \in E$
Decision $\underline{X} \in \overline{E}$	Correct	Type I
Decision $\underline{X} \in E$	Type II	Correct

Table 1b. Loss Matrix.

	Condition	
	$\underline{X} \in \overline{E}$	$\underline{X} \in E$
Decision $\underline{X} \in \overline{E}$	0	1_1
Decision $\underline{X} \in E$	1_2	0

Table 2a. Contingency Table for $H_1: X \in \overline{E} \cup E$. Type III Error.

	$P(\chi^2 < E_0| k-1) > (1-\alpha)$	$P(\chi^2 < E_0| k-1) \leq (1-\alpha)$
Decision	$X \in (\overline{E} \cup E)^*$	$X \in (\overline{E} \cup E)^*$

Table 2b. Contingency Table when $H_1: X \in \overline{E} \cup E$ is Rejected.

	$P(\chi^2 < E_1| 1) > (1-\alpha)$	$P(\chi^2 < E_1| 1) \leq (1-\alpha)$
$P(\chi^2 < E_2| 1) > (1-\alpha)$	$X \in \overline{E} \cup E^{**}$	$X \in \overline{E}$
$P(\chi^2 < E_2| 1) \leq (1-\alpha)$	$\underline{X} \in E$	?

Table 3. Sample sizes of the conditional distributions of proton T_1 and T_2 for biopsy specimens.

	Colon[a]	Breast[a]	Lung[a]
$n_1(Ca)$	16	11	22
$n_2(Ca)$	20	12	7
$n_1 + n_2$	36	23	29

[a] See Koutcher, et al (1978)

Table 4. Test of Equality of Covariance Matrices. $H_0: \Sigma_1 = \Sigma_2 = \Sigma$.

	Colon	Breast	Lung
MC^{-1}a= χ_c^2(df)	7.380(3)	2.673(3)	15.492(3)
$P(\chi^2 < \chi_c^2| df)$	0.939[b]	0.555[b]	0.999[c]
$|S_1|/|S_2|$ [d]	2.279[e]	0.244	0.052

[a] Box statistic (Morrison, 1967). [b] Accept H_0. [c] Reject H_0.
[d] Ratio of determinants of sample covariance matrices.
[e] Unusual. It is more commonly the case that $|S_1|/|S_2| < 1.0$, when group 1 is Normal and group 2 is Disease.

multiple regression methods were initially developed - is always the same; i.e., the distribution of $\underline{X}$, described by the rows (and columns) of X is <u>invariant</u>. However, Regression Diagnostics evaluates the effect upon the regression estimates, of changes in the distribution of $\underline{X}$, namely, the effect of the <u>absence</u> of each of the observations - the rows of $\underline{X}$ - an occurrence which is most likely in future samples in the <u>non-experimental</u> studies for which these diagnostic methods were developed.

In the foregoing discussions, procedures for assigning an individual between alternative groups, $\overline{E}$ and E, with minimum total loss or with minimum total frequency of misclassification, that is, minimum number of Type I plus Type II errors incurred <u>in the long run</u>, have been presented. The validity of these procedures depends on the validity of the <u>a priori</u> information for the individual in question that the two groups are: 1) mutually exclusive, 2) exhaustive, and 3) that the <u>ratio</u> - as well as the <u>sum</u> - of the a priori probabilities of the respective groups is <u>known</u>. This prior knowledge is <u>external</u>, or non-sample, evidence. It may be summarized as follows in terms of the prior probabilities of the complementary groups, $\overline{E}$ and E:
1) Mututally exclusive
   $P(\overline{E} \text{ and } E) = 0$
2) Mutually exhaustive (complementary)
   $P(\overline{E} \text{ or } E) = P(\overline{E}) + P(E) = q_1 + q_2 = 1.0$
3) <u>Known</u> ratio of prior probabilities $P(E)/P(\overline{E}) = q_2/q_1$

When external evidence of this sort is weak or equivocal the assignment of an individual to one of the two groups may incur another kind of error, namely, the assumption that he belongs to one or the other when, in fact, he belongs to <u>neither</u>. Instead he belongs to some <u>third group</u> for which both the distribution function and a priori probability are unknown. ("In statistical analysis we have to take into account the possibility that he might belong to an unknown group whose existence has yet to be established" Rao 1973.) In the absence of unambiguous <u>external</u> evidence concerning the existence and the nature of the third group, it is necessary to examine by means of the <u>internal</u> evidence of the measurements, $\underline{X}$, of the individual, whether he belongs to the larger group comprised of the two (complementary) sub-groups, $\overline{E}$ and E. In this case, in the <u>absence</u> of the <u>a priori</u> information, the classification problem has devolved from the <u>probability of a hypothesis</u> into a <u>test of hypothesis</u> based on functions of the two statistics, D^2 and L^T, of the discriminant analysis:
$H_o: \underline{X} \in \overline{E} \cup E. \ (\overline{E} \cap E = \emptyset)^{11}$ $H_1: \underline{X} \in \overline{E} \cup E / \emptyset$
The test statistic, E_0, is distributed <u>asymptotically</u> as chi-squared with (k-1) degrees of freedom (Rao 1973).

$E_0 = (\underline{X}-\overline{\underline{X}}_1)^T S^{-1}(\underline{X}-\overline{\underline{X}}_1) - [L^T(\underline{X}-\underline{X}_1)]^2/D^2$

Note that the first term on the right hand side is the (chi-squared) measure of "resemblance" of the individual in question to the group, $\overline{E}$ (Tiedeman 1953, 1954). (The statistic E_0 is also related to the statistic, h_i, since the latter is a linear function of the (Mahalanobis) distance, d_i^2, of the ith observation from the <u>centroid</u> of the distribution of the remaining $(n_1 + n_2 - 1)$ observations: $d_i^2 = h_i n(n-2)/(1-h_i)(n-1)$ See Velleman and Welsch 1981. As remarked previously, if $\underline{X} \sim N_k(\mu_j, \Sigma_j)$ then $(\underline{X}-\overline{\underline{X}}_1)^T S^{-1}(\underline{X}-\overline{\underline{X}}_1)$ is distributed as <u>chi-squared with k degrees of freedom</u>. The numerator of the second term on the right hand side is just the projection on the axis of the discriminant function (the right eigenvector of $W^{-1}B$) of the vector difference $(\underline{X}-\overline{X}_1)$. Thus, the difference between the first and second terms can be interpreted as a squared distance between the observation and its projection on the discriminant axis, (the axis of the marginal distribution of LDF) i.e., it is a measure of the distance of the observation "away" from (or perpendicular to) the discriminant axis. If $\underline{X} \sim N_k(\underline{\mu}_j, \Sigma_j)$ then $[\underline{L}^T(\underline{X}-\overline{X}_1)]^2/D^2$ is distributed as chi-squared with 1 degree of freedom. Hence, E_0 is distributed as chi-squared with $(k-1)$ degrees of freedom.

If E_0 is not significant then the internal evidence suggests that the case, $\underline{X}$, may belong to one of the two groups $\overline{E}$ or E, i.e., $\overline{E}$ and E are exhaustive of the possibilities for this case. But, if E_0 is significant, than there is internal evidence that this case belongs to a third, unknown, group, $(\overline{E} \cup E)*$, for which the mean is not simply related to the means of the two known groups: it is "away" from (in a direction perpendicular to) the line joining μ_1 and μ_2. In this case, the individual cannot be classified by the discriminant function, $\underline{L}^T$; to use the discriminant function to assign (optimally) the individual to one of the two specified groups, $\overline{E}$ or E, would give the right answer to the wrong question - a Type III error (Kimball 1957). The question "Which?" (of the two alternative groups) has no meaning! The possible alternative inferences are described in Table 2a.

If E_0 is not significant then membership in one of the two groups can in some cases be inferred by a further test of hypothesis. Consider the two statistics $E_1 = [\underline{L}^T(\underline{X}-\overline{X}_1)]^2/D^2$ and $E_2 = [\underline{L}^T(\underline{X}-\overline{X}_2)]^2/D^2$. Each is distributed as chi-squared with 1 degree of freedom (vide supra). The possible alternative inferences are described in Table 2b.

If E_1 and E_2 are both significantly large, then it is possible that the individual in question belongs to a third unknown group, $(E \cup E)**$, with centroid collinear with - "along" - the mean vectors, μ_1 and μ_2. If neither E_1 nor E_2 are significant, then the internal evidence for membership is quite ambiguous as well, and the individual (again) cannot be classified by the discriminant function, $\underline{L}^T$. The remaining two alternatives will assign the individual, as is shown, to either group 1 or group 2, with probability of Type I error equal to α.[k] This method of assignment devolves into a procedure quite similar to that described by Tiedeman (1953).

There are other sets of circumstances, in which, although some estimates of the apriori probabilities of the complementary conditions are available, and hence a discriminant model of the classification problem is appropriate and could be constructed from the sample in hand, there are more optimal procedures. One such circumstance arises when the only estimates of the respective a priori probabilities of $\overline{E}$ and E are the sizes of the "available" samples, n_1 and n_2, from which the coefficients of the discriminant function are estimated. It is rarely the case that the available numbers provide valid estimates of the required a priori probabilities, q_1 and q_2, for the target population in a given case and unless they do, they should not be used. (As noted above, in order to estimate the coefficients of the discriminant function, Rao (1953) recommends that the samplesize be chosen so that $n_1 = n_2$.

The second set of circumstances arises when, although these a priori probabilities are known, one is much greater than the other, say $q_2/q_1 = 10^{-3}$. In both cases, the classification problem devolves into the test of hypothesis, H_0: $X \varepsilon E$ described previously. Assignments are made only on the criterion of resemblance and the a priori information is ignored. Resemblance of the ith individual - to either group - is measured by the Mahalanobis distances for the individual: $(\underline{X}-\underline{\bar{X}}_j)^T S_j^{-1}$ $(\underline{X}-\underline{\bar{X}}_j)$, j = 1, 2. $\underline{X} \sim N(\underline{\mu}_j, \Sigma_j)$, then the criterion is distributed as chi-squared with k degrees of freedom (Tatsuoka 1971; Tiedeman 1953 & 1954).

4. DATA

The data are taken from Koutcher, et al (1978), and will be used to illustrate the concepts and methods discussed earlier. The sample sizes are such that the motivational (inspirational?) and heuristic values of these samples exceeds their evidential value in the question of the diagnostic value of NMR measurements on biopsy specimens. The respective sample sizes are presented in Table 3.

The conditional ($\overline{Ca}$ and Ca) distributions of the measurements of T_1 and T_2 from biopsies of normal and malignant tissues in colon, breast and lung are presented in the scattergrams of Figures 1, 2 and 3, respectively, and in the probability plots of Figures 4, 5 and 6, respectively. Note the appearance of extreme heterogeneity in each of the conditional distributions of the lung measurements. Those (extreme) cases for which the corresponding diagonal element, h_i, $1 \leq i \leq n$, of the respective hat matrix, H, for the cognate binary regression model exceeded the size-adjusted cut-off - $h_i \geq 2m/n$ - are designated by "?" (Belsley, et al 1980 in Figures 1-6). These are (possibly) "influential" observations whose presence in the sample may affect the consistency of the sample data with the basic assumptions of the linear discriminant model for two groups - conditional multivariate normality of the distributions, equality of variance - covariance matrices, S_j, etc. - as well as the degree of overlap, D^2, of the two conditional distributions.

Since the linear discriminant model for g = 2 groups has a surrogate in the multiple regression model with a binary response variable, y, the sample estimate of the coefficients of the discriminant function, like the α-vector of the cognate binary regression equation, $y = \underline{X}^T \alpha$, may be expected to be sensitive to the presence in the sample of poorly fit responses, or outliers, and extreme levels of $\underline{X}$. Such cases identify the nature and degree of model-sample interaction that is present in a given estimate, $\hat{\underline{\alpha}}$; that is, the extent to which the regression results depend on the particular sample at hand. The methods of regression diagnostics have been developed to systematically identify and measure these aspects of the regression model-sample interaction (Belsley, Kuh and Welsch 1978; Cook and Weisberg 1982). The case statistics for the surrogate binary regression model have some diagnostic value for the discriminant model as well. Characteristic case plots of the Studentized residuals, e_i^*, hat matrix diagonals, h_i, and Cook's D_i are presented in Figures 1b, 1c and 1d, respectively, for the regression model of the colon data. Cook's D_i is a function of the two key diagnostics e_i and h_i: $D_i = e_i^2 h_i / m(1-h_i)^2$, where m = k+1 (Montgomery and Peck 1982). It is a measure of the squared distance from $\hat{\underline{\alpha}}$ to $\hat{\underline{\alpha}}_{(i)}$ relative to the fixed geometry of $X^T X$, where $\hat{\underline{\alpha}}_{(i)}$ is the sample estimate of $\underline{\alpha}$ with the ith case deleted. See Lecture 13. The joint distribution of h_i and e_i^* is described by the scattergram presented in Figure 1e. Critical values of e_i^* and h_i, are ± 2.0 and $2m/n = 0.167$, respectively, where $n = n_1 + n_2$, m = k+1. Thus, there are three "outliers" (case #14, 15, and 34) and three extreme "design" points (case #8, #30, and #34). Two explanations may be appropriate. One, the fact that case #34, has both

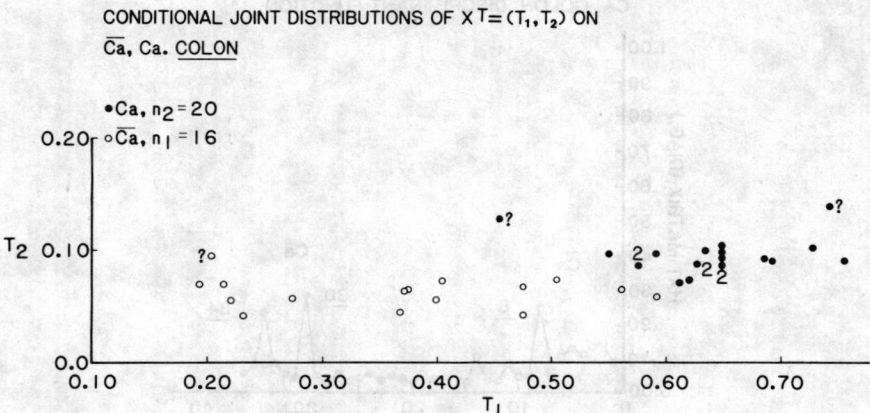

Figure 1a. Superposition of the scattergrams of conditional ($\overline{Ca}$ and Ca) joint distributions of proton T_1 and T_2 measurements of biopsies of <u>Colon</u>. Case #8, #30 and #34 have $h_i > 2k/n = 0.17$ (Figure 1c) and also appear as anomalies on the chi-squared probability plots of Figure <u>4</u> where they are denoted by "?".

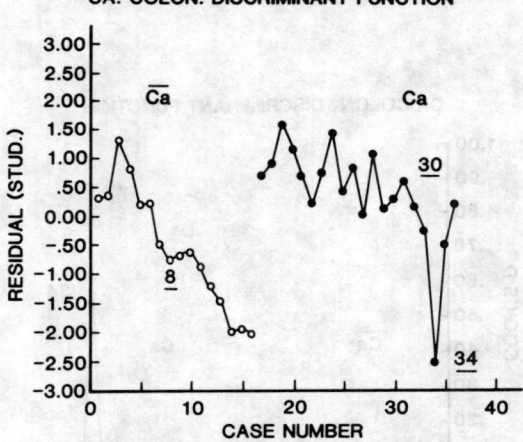

Figure 1b. Case plot of Studentized residuals, e_i^*, for binary regression model of discriminant function for g = 2 groups. <u>Ca Colon</u>. There are two outliers: #16 and #34. There are 3 outliers in the $\overline{Ca}$ group (#'s 13, 14, 15) and 1 outlier in the Ca group (#34).

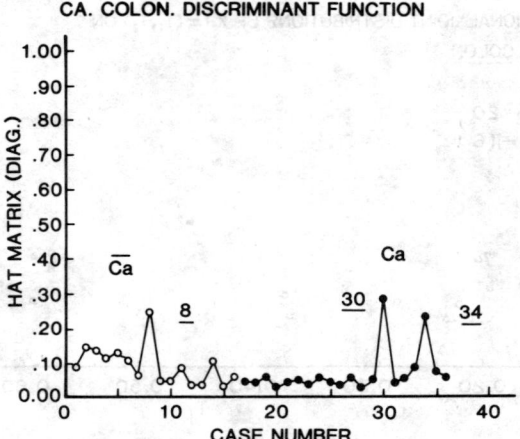

Figure 1c. Case plot of hat matrix diagonal, h_i, for the binary regression model of the linear discriminant function for $g = 2$ groups. Ca Colon. There are 3 possible high leverage points ($h_i > 2k/n = 0.17$): #8, #30, #34.

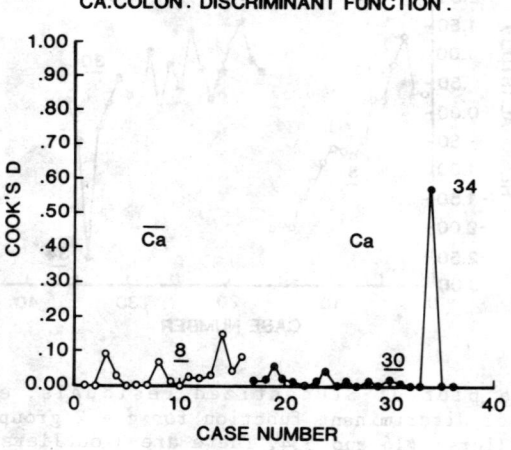

Figure 1d. Case plot of Cook's D_i for the binary regression model of the linear discriminant function for $g = 2$ groups, Ca Colon. Case #34 will dominate the estimates of the discriminant function. Note that only for #34 are both e_i^* and h_i large. See Figure 1e.

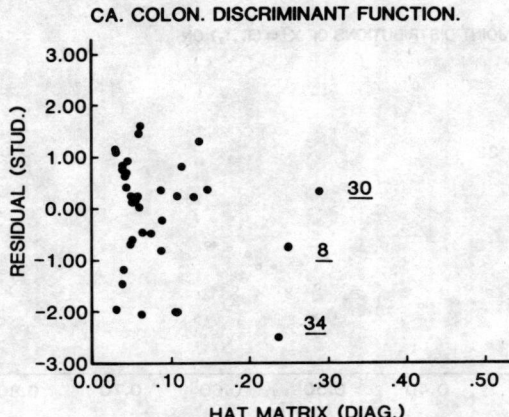

Figure 1e. Scattergram of key diagnostics, e_i^* and h_i, for binary regression model of the linear discriminant function for $g = 2$ groups, Ca Colon.

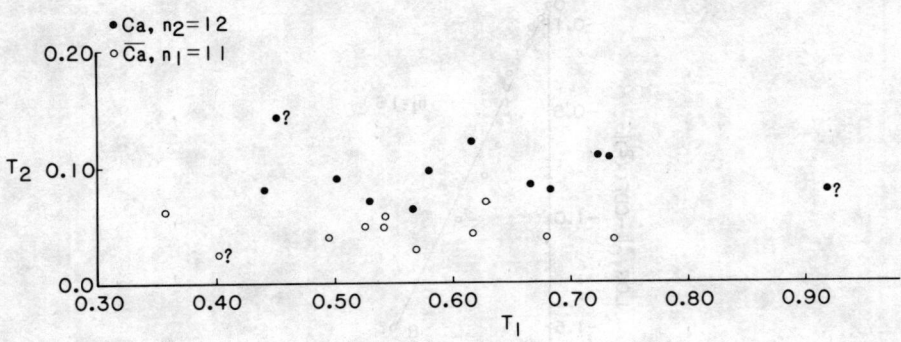

Figure 2. Superposition of the scattergrams of conditional ($\overline{Ca}$ and Ca) joint distributions of proton T_1 and T_2 measurements on biopsies of Breast. "?" appears as an anomaly in the plots of Figure 5.

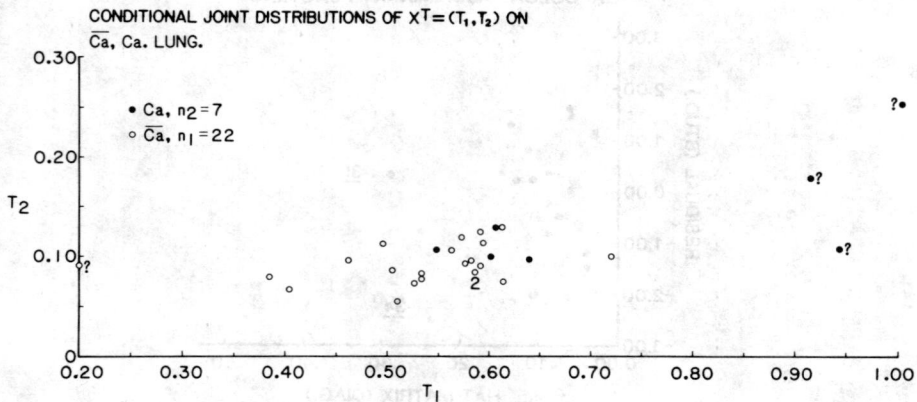

Figure 3. Superposition of the scattergrams of conditional ($\overline{Ca}$ and Ca) joint distributions of proton T_1 and T_2 measurements on biopsies of Lung. "?" appears as a anomaly in the plots of Figure 6.

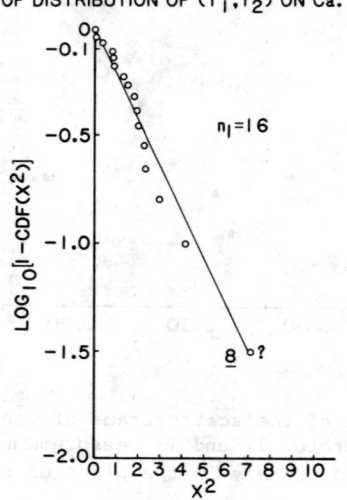

Figure 4a. Chi-squared probability plot of joint distribution of proton T_1 and T_2 measurements on biopsies of normal Colon. See Hald 1965 and Gnanadeskikan 1977.

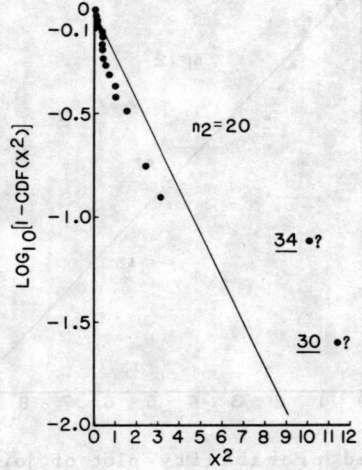

Figure 4b. Chi-squared probability plot of joint distribution of proton T_1 and T_2 measurements on biopsies of Ca Colon.

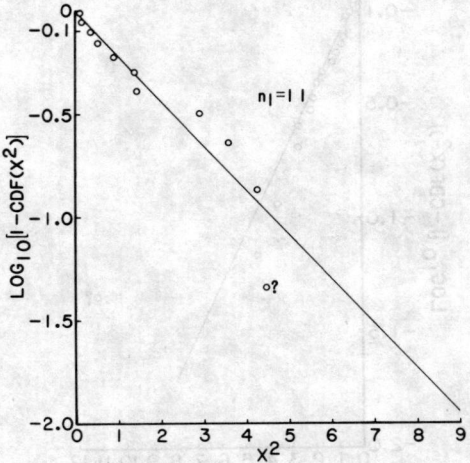

Figure 5a. Chi-squared probability plot of joint distribution of proton T_1 and T_2 measurements on biopsies of normal Breast. See Hald 1965 and Gnanadesikan 1977.

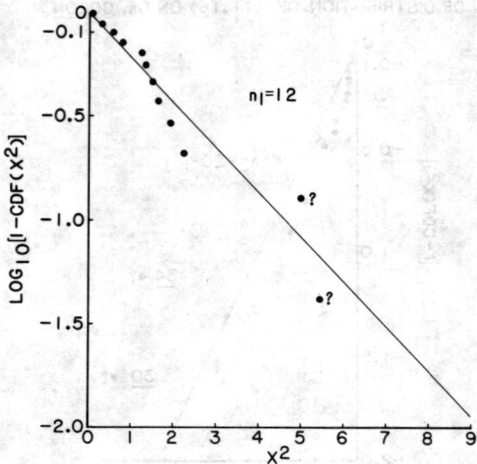

Figure 5b. Chi-squared probability plot of joint distribution of proton T_1 and T_2 measurements on biopsies of Ca Breast.

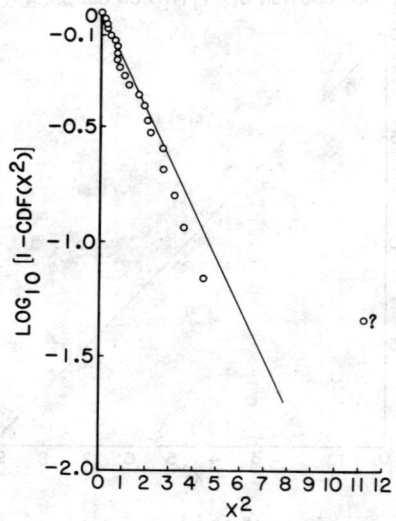

Figure 6a. Chi-squared probability plot of joint distribution of proton T_1 and T_2 measurements on biopsies of normal Lung. See Hald 1965 and Gnanadesikan 1977.

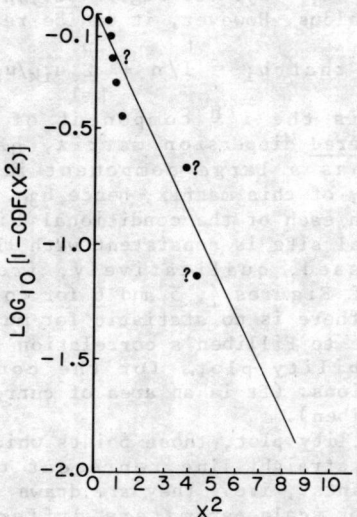

Figure 6b. Chi-squared probability plot of joint distribution of proton T_1 and T_2 measurements on biopsies of Ca Lung.

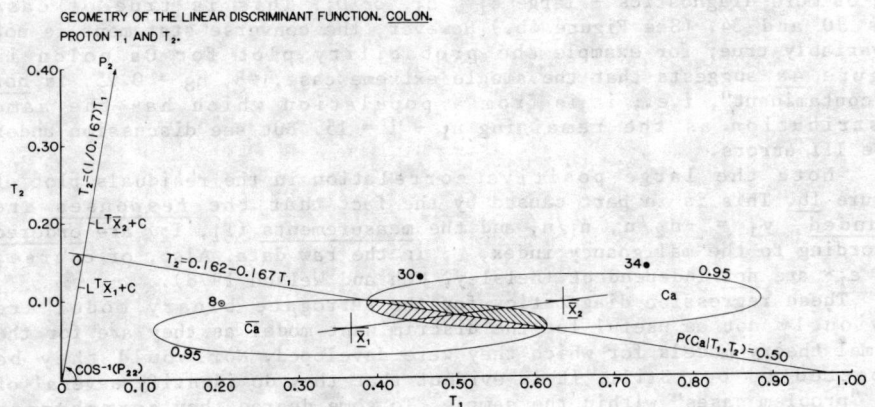

Figure 7. Superposition of the 0.95 contour ellipses for conditional (C̄a and Ca) joint distributions of proton T_1 and T_2 measurements on biopsies of colon. The right eigenvector, P_{22}, of the non-zero eigenvalue, λ_2, of the $W^{-1}B$ matrix and the discriminant curve (0.50 isorisk curve) are also presented.

$e_{34} < -2.0$ and $h_{34} > 2m/n$ follows from the dichotomy of y_{34}. Two, the case #30, for which $(T_1, T_2) = (0.455, 0.130)$ is in the vicinity of the centroid $(\overline{T}_1, \overline{T}_2) = (0.516, 0.082)$ of the pooled ($\overline{Ca}$ and Ca) joint distribution of size $n = n_1 + n_2$. See Figure 1a. Thus, the large value of h_{30} may seem anomalous. However, it can be readily shown (see Cook and Weisberg, 1982) that $h_i = 1/n + \sum_{k=1}^{p} u_{ik}^2/\mu_k$ where μ_k is the kth eigenvalue and u_{ik} is the ith component of the corresponding eigenvector of the centered dispersion matrix, $_0X_0X$. For the colon data, the case #30 has a large component in the direction of the smallest eigenvalue, μ_1, of this matrix, hence h_{30} is large.

The degree to which each of the conditional bivariate distributions at each anatomical site is consistent with the hypothesis H_0: $X \sim N_2(\mu_j, \Sigma_j)$ can be assessed, qualitatively, from the chi-squared probability plots of Figures 4, 5 and 6 for colon, breast and lung, respectively. However, there is no statistic for multivariate distributions that is cognate to Filliben's correlation coefficient (Filliben 1975) for normal probability plots for the conditional marginal (univariate) distributions. (It is an area of current research. Private communication from Filliben).

For any probability plot, those points which depart appreciably from the characteristic straight line represent cases which may be regarded as "contaminants", i.e., they are drawn from a population for which the location and/or scale parameters differ appreciably from those of the population from which the rest of the sample was drawn. (As remarked above, it is known by now that the proportion of gross errors in data, depending upon circumstances, is normally between 0.1% and 10% with several percent being the rule rather than the exception (Hampel 1974).)

It is evident that these "contaminants" may have large values of one or more diagnostics - large e_i^*, h_i, or D_i. This is true of case #'s 30 and 34. (See Figure 4b.) However, the converse statement is not invariably true; for example the probability plot for $\overline{Ca}$ colon in Figure 4a suggests that the single extreme case, #8, $h_8 = 0.25$, is not a "contaminant", i.e., it is from a population which has the same distribution as the remaining $n_1 - 1 = 15$. But see discussion under Type III errors.

Note the large positive correlation in the residuals plot of Figure 1b. This is in part caused by the fact that the responses are bounded, $y_i = -n_2/n$, n_1/n, and the measurements (T_1, T_2) are ordered according to the malignancy index, Λ, in the raw data. Also, of course, the e_i^* are not independent (Belsley, Kuh and Welsch 1978).

These regression diagnostics for the surrogate binary model are obviously not as useful for the discriminant model as they are for the Normal theory models for which they were developed. Nor would they be expected to be. Still, it is evident that they do identify several of the "problem cases" within the sample. To some degree they corroborate the evidence of the multivariate probability plots, hitherto the principal "diagnostics" for a discriminant analysis. Clearly both may be overread. However, it would seem that it is better to include than to omit them from any analysis.

5. HEURISTIC DISCRIMINANT ANALYSES OF THE KOUTCHER, ET AL (1978) DATA

The ratios $|S_1|/|S_2|$ and the results of the test of the hypothesis,

H_o: $\Sigma_1 = \Sigma_2$ for the conditional distributions at each anatomical site - colon, breast and lung - are presented in Table 4 (Morrison 1967). The validity of this test is questionable for breast and lung since $n_j \leq 20$, j=1, 2 for each of the conditional distributions at both sites. The results of the test determine the validity of the linear discriminant model of these data. Note that $|S_1|/|S_2| > 1$ for colon. This is rather atypical; it is more often the case that the dispersion of measurements in the diseased condition exceeds that of the normal condition. This is partly because in a disease condition there is usually a much stronger time-dependence in the value of any physiological observation than is the case in normal conditions and any sample of suitable size invariably includes patients at different "ages" of the disease condition in question. Moreover, the time-dependence is usually rather complex, with cyclic, "seasonal" and "trend" components present as well as "noise". Also, of course, there is greater likelihood of diet and/or drug perturbation - "interference" - of the measurements obtained on tissues in disease conditions.

Table 5 presents the estimates, D^2, of the Mahalanobis distance between the centroids of the bivariate distributions of T_1 and T_2 for the two conditions, $\overline{Ca}$ and Ca, for each of the $2^2-1 = 3$ discriminant functions possible at each site. The first column is the Mahalanobis distance between the joint distributions of T_1 and T_2, columns two and three are the distances for the respective marginal distributions. Those values of D^2 marked with an asterisk are those of the best linear discriminant. Thus, for colon both T_1 and T_2 are required for the best discrimination, while for breast, the addition of the set of T_1 measurements to the set of T_2 measurements does not significantly improve the discrimination achieved by T_2 alone, i.e., $D^2(T_1, T_2) = 7.298$ does not differ significantly ($P(F > F_c | 1, 20) = 0.12$) from $D^2(T_2) = 6.706$. Similarly, adding T_2 measurements to T_1 measurements for lung does not significantly improve the discrimination achieved by T_1 alone, i.e., $D^2(T_1, T_2) = 3.108$ does not differ significantly ($P(F > F_c | 1, 26) = 0.21$) from $D^2(T_1) = 2.710$.

Note that for colon and breast $D^2(T_1, T_2) \simeq D^2(T_1) + D^2(T_2)$, whereas for lung, $\overline{D^2(T_1, T_2) < D^2(T_1) + D^2(T_2)}$. The difference is due to the difference in the correlation coefficients, $r(T_1, T_2)$, of the joint distributions: In colon and breast, $r(T_1, T_2) \simeq 0$, whereas in lung, $r(T_1, T_2) > 0$.

The effects of transforming the data - $(T_1, T_2) \longrightarrow (\log T_1, \log T_2)$ - upon the estimates, D^2, as well as upon the validity of the hypotheses, H_o: $\underline{X} \sim N_2(\mu_i, \Sigma_i)$, and H_0: $\Sigma_1 = \Sigma_2(=\Sigma)$, were also examined. It was found that the raw data were more consistent with the latter requirements than were the transformed data.

Table 5 presents the effects of the deletion of a single column, the variable T_1 or T_2, from the matrix of observations, $[\underline{y}, X]$, on the estimates, D^2, of the Mahalanobis distances. It will also be useful to examine the effects of the deletion of a single row, an entire observation, from the matrix $[\underline{y}, X]$. The rows to be deleted are the cases identified as "extreme", i.e., those for which $h_i \geq 2m/n$, $n = n_1 + n_2$, in the binary regression model, $\underline{y} = X\underline{\alpha} + \underline{\varepsilon}$. Examination of these deletion diagnostics will permit an assessment of the degree to which the estimates of the discriminant model coefficients depend upon the particular sample at hand; more precisely, the degree to which the idiosyncracies of the particular sample affect the estimates of the error rates which can be achieved by the linear discriminant function, which in turn will determine the degree to which these estimates are

Table 5. Estimates, D^2, of the Mahalanobis Distance for Conditional Distributions of Proton T_1 and T_2 for Biopsy Specimens.

	$D^2(T_1, T_2)$	$D^2(T_2)$	$D^2(T_1)$
Colon	12.616*[a]	4.912	7.010
Breast	7.298[a]	6.706*	0.255
Lung	3.208[b]	2.271	2.720*

* This is D^2 of the best linear discriminant, LDF, for these data. The corresponding discriminant functions are:
1) Colon, $z = -26.517 + 163.415 T_2 + 27.290 T_1$
The relative contributions of each measurement to the discrimimant function are similar. The standarized discriminant weights are $8.35*10^{-2}$ and $9.79*10^{-2}$ for T_2 and T_1, respectively.
2) Breast, $z = -9.543 + 136.194 T_2$
3) Lung, $z = -9.130 + 12.401 T_1$
The respective Q-Statistics are:
1) Colon, $Q = 36.00$
2) Breast, $Q = 19.17$
3) Lung, $Q = 12.45$
All exceed 3.84, the 0.95 quantile of the chi-squared distribution with one degree of freedom.

---

[a] The correlation of T_1 and T_2 in the groups Ca and $\overline{Ca}$ is very **small**, hence $D^2(T_1, T_2) \simeq D^2(T_1) + D^2(T_2)$
[b] The correlation of T_1 and T_2 in the groups Ca and $\overline{Ca}$ is very **large**, hence $D^2(T_1, T_2) < D^2(T_1) + D^2(T_2)$

Table 7. Point Estimates of Error Rates of LDF(T_1, T_2) Based on **Sample** Estimates of Prior Odds Ratio $(1-\pi_1)/\pi_1$ and Δ^2. Normal Distribution.

	P_1[a]	P_2[b]	$q_1 P_1 + q_2 P_2$[c]
Ca Colon			
$q_1 = 16/36$ $D^2 = 12.626$	0.0436	0.0330	0.0377
Ca Breast			
$q_1 = 11/23$ $D^2 = 7.298$	0.0951	0.0838	0.0892
Ca Lung			
$q_1 = 22/29$ $D^2 = 3.208$	0.0618	0.603	0.192

---

[a] $P_1 = \Phi\left[\dfrac{\ln\{(1-q_1)/q_1\} - D^2/2}{D}\right]$
[b] $P_2 = \Phi\left[-\dfrac{\ln\{(1-q_1)/q_1\} + D^2/2}{D}\right]$
[c] $q_2 = 1 - q_1$

Table 6. Deletion Diagnostics. Colon. $|S_1||S_2| = 2.28$, $MC-1 = 7.38$, $P(\chi^2 < MC-1 | 3) = 0.939$, $D_0^2 = 12.626$. $(-D_0/2) = 0.0375$.

i^a	h_i	e_i^*	C_1	C_2	MC^{-1}	$|S_1|/|S_2|$	$P(\chi^2 < MC^{-1}|3)^b$	$D_0^2-D_i^2$	$\Phi(-D-i/2)$
8(Ca)	0.249	-0.769	6.81 (0.970)^c	18.17 (0.995)	9.33	1.30	0.975	-0.211	0.039
30(Ca)	0.289	0.311	22.04 (0.995)	8.20 (0.983)	14.64	5.56	0.998	-0.413	0.040
34(Ca)	0.238	-2.518	43.49 (0.995)	10.66 (0.995)	8.26	4.64	0.959	2.703^d	0.025

h_i - ith diagonal element of hat matrix, $H = X(X^TX)^{-1}X^T$, where X is (nx3) and $n = n_1+n_2$ (=36).
Size-adjusted cut-off: $h_0 = 2m/n = 0.167$. $m = k+1 = 3$.
e_i^* - ith Studentized residual
e_i^* and h_i are regression diagnostics obtained from the binary regression model.
C_i - Chi-squared statistic for ith observation with respect to jth group: $C_j = (X_i-\bar{X}_j)^T S_j^{-1}(X_i-\bar{X}_j)$, $j = 1, 2$.
D_{-i}^2 - Mahalanobis distance between group centroids with ith row deleted: $D_{-i}^2 = [(\underline{X}_2-\underline{X}_1)^T S^{-1}(\underline{X}_2-\underline{X}_1)]_{(i)}$

---

a See Figure 7.
b The Statistic MC^{-1} is distributed as chi-squared with 3df on $H_0: \Sigma_1 = \Sigma_2$.
c The figure in parentheses is $P(\chi^2 < C_i|2)$, $i = 1, 2$; e.g., $C_i = 5.99$ defines points on the 0.95 contour ellipse (Figure 7) since $P(\chi^2 < 5.99|2) = 0.95$
d "The greatest increase in D^2 correponds to the deletion of observations whose scores are furtherest (sic) from that for the mean of the other species (group)." See Campbell (1978).

susceptible of generalization to the (target) population of interest. It must be emphasized that this assessment is rather different from that provided by an examination of the more customary <u>interval</u> estimates of the coefficient vector, predicted response, etc., vide supra.

Some deletion diagnostics for Colon are presented in Table 6. There are three "extreme" cases, $h_i \geq 2m/n = 0.167$, these are case #'s 8, 30, and 34 (See Figure 7). Note that the greatest change (increase) in D^2 corresponds to the deletion of (Ca) case # 34, the case which is most remote from the <u>complementary</u> ($\overline{Ca}$) group. This seems to be a general effect which is due to the fact that the deletion of the case does not reduce the between-groups sum of squares, B, as much as it reduces the within-groups sum of squares, W (Campbell 1978). Note that the deletion of these points also strongly affects the validity of the hypothesis, $H_o: \Sigma_1 = \Sigma_2 = \Sigma$, and hence the validity of the <u>linear</u> discriminant model. It is evident from Figures 4a and 4b that deletion of case #'s <u>8</u>, <u>30</u> and <u>34</u> may also affect the validity of the hypothesis, $H_o: \underline{X} \sim N_2(\underline{\mu}_j, \Sigma_j)$, $j=1, 2$.

Figure 7 offers a collage of several features of the linear discriminant model previously discussed under <u>Concepts and Methods</u> in terms of the $\overline{Ca}$ and Ca cases for the Colon site. 1) The ellipses are the respective 0.95 contour ellipses for each group (Hald 1965). The equation of the contour ellipse is

$$c^2 = \frac{1}{(1-r^2)} \left[\left(\frac{x_1 - \overline{x}_1}{s_1}\right)^2 - 2r \left(\frac{x_1 - \overline{x}_1}{s_1}\right) \left(\frac{x_2 - \overline{x}_2}{s_2}\right) + \left(\frac{x_2 - \overline{x}_2}{s_2}\right)^2 \right] \quad (54)$$

where $x_j = T_j$ and $\overline{x}_j$, s_j, $j = 1, 2$ are the mean and standard deviation, r is the correlation coefficient and c^2 is a constant. The probability that (x_1, x_2) lies within the ellipse is $P(\chi^2 < c^2 | 2)$. Recall that this is just the Tiedeman (1953, 1954) measure of dissimilarity, $(\underline{X} - \overline{\underline{X}})^T S^{-1} (\underline{X} - \overline{\underline{X}})$, vide supra. The superpositions of the respective 0.95 contour ellipses illustrate the nature and degree of the overlap of the two conditional <u>joint</u> distributions. 2) Orthogonal projections of the cross-hatched region onto the T_1 and T_2 axes give corresponding measures of the size of the overlap of the respective <u>marginal</u> distributions. 3) The extreme cases identified by the case statistics of the binary regression model are case #'s <u>8</u>, <u>30</u> and <u>34</u>, which lie <u>outside</u> the respective 0.95 contour ellipses.

Note that on the hypothesis, $H_o: \underline{X} \sim N_2(\mu_1, \Sigma_1)$, $0.95*n_1 = 15$ observations should lie within the 0.95 contour ellipse for the condition $\overline{Ca}$. This is indeed what is observed in the Figure. This evidence for H_0 is also consistent with the evidence of the chi-squared probability plot of Figure 4a.

Figure 7 discloses that the $P(Ca| T_1, T_2) = 0.50$ isorisk line obtained from the logistic risk model passes through the two intersections of the contour ellipses, as we should expect. The isorisk line is orthogonal to the <u>right</u> eigenvector, $\underline{P}_2$, of the $W^{-1}B$ matrix, which describes the direction in T_1-T_2 space of the linear combination, $L^T \underline{X}$, of measurements which defines the linear discriminant function (Rao 1973). This direction is, by definition, the direction in which the overlap of the marginal distributions of the measurements - measured by the corresponding eigenvalue, $\lambda_2 = \lambda_2(D^2)$ - is a <u>minimum</u>. This is also quite evident in the Figure.

Figure 7 also presents an important feature of the linear discriminant model for g=2 groups which is not evident in the usual development of the classification rule - that of Welch (1939). It will be recalled that in that development the X-space was divided, optimally, into two regions, R_1 and R_2, chosen such that if $\underline{X} \in R_j$ the individual was assigned to group j. The $P(Ca|\underline{X}) = 0.50$ isorisk line defines the optimal division. However, it is evident from the Figure that the discriminant function can be used to decide on the identity of only those individuals for whom $\underline{X} \in \overline{Ca} \cup Ca$, i.e., for those individuals with values of $\underline{X}$ lying (roughly) within the region of the T_1-T_2 space that is "covered" by the two contour ellipses. The prescribed optimal partition of the measurement space which is achieved by the Welch rule and represented by the 0.50 isorisk line is not sufficiently unambiguous. It is necessary (to minimize Type I and Type II errors); it is not sufficient (to avoid Type III errors). This feature will be discussed in more detail below. (See also the earlier discussion of Type III errors.)

Figures 8a and 8b present the Normal probability plots of the conditional frequency distribution of the linear discriminant function, LDF, for the colon. For small samples, such plots may not be very informative; Seber (1977) in citing the examples of such plots presented in Daniels and Wood (1971) commments that "... samples of size 8 tell us almost nothing about normality; sets of 16 show shocking wobbles; sets of 32 are visibly better behaved; sets of 64 nearly always appear straight in their central regions but fluctuate at their ends; sets of 384 seem very stable except for their few lowest and highest points. It is recommended that n [the sample size] be at least 20 and preferably greater than 50." Therefore, a proper statistical test of the null hypothesis is appropriate. Filliben has described an appropriate statistic (the Filliben probability plot correlation coefficient, r_F) and prepared tables of its sampling distribution for sample sizes $3 \leq n \leq 100$. r_F for the distribution of the LDF on $\overline{Ca}$ colon (Figure 8a) is $r_F = 0.982$. For $n_1 = 16$ this lies between the 0.50 and 0.75 quantiles of the sampling distribution, hence, the null hypothesis, normality, is accepted. For the distribution of the LDF on Ca colon (Figure 8b), $r_F = 0.929$. For $n_2 = 20$, this lies between the 0.01 and 0.025 quantiles of the sampling distribution. Hence the null hypothesis is rejected. The anomalous position of case #34 is evidence of "contamination". However, note that, as prescribed by the Central Limit Theorem the distributions of linear combination of the variables T_1 and T_2 are more nearly "Normal" than the distribution of the single variables themselves (Hald 1965). Although case #34 appears as a "contaminant" in the plot of Ca colon, its deletion would have the result that a pooled estimate of the variance-covariance matrix, S, could not be used. Only if #34 is included, is the null hypothesis, H_0: $S_1=S_2=S$ valid. This would mean that the estimates, S_1 and S_2, respectively of the covariance matrices of the separate conditional distributions could not be so precisely determined. This would, in turn, inflate the interval estimates of the error rates, P_1 and P_2. Also, of course, a linear discriminant model would no longer be appropriate.

Figure 9a presents the superposition of the respective cumulative distribution functions, CDF, of the two conditional ($\overline{Ca}$ and Ca) distributions of the linear discriminant function, LDF, for the colon.[m] The Figure also presents the superposition of the 0.95 confidence limits on the quantiles of the respective distributions for two different sample sizes. The pth quantile, x_p, of a (univariate) Normal distribution, for which the estimates of the mean and variance

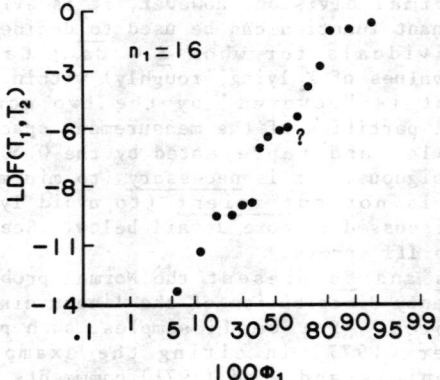

Figure 8a. Normal probability plot of distribution of LDF (T_1, T_2) on normal $(\overline{Ca})$ colon. Case #8 is denoted as "?".

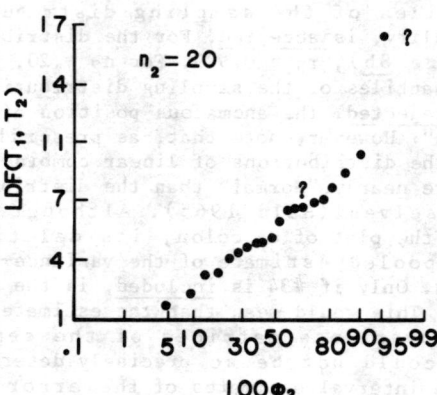

Figure 8b. Normal probability plot of the distribution of LDF (T_1, T_2) on Ca colon. Case #30 and #34 are denoted as "?".

obtained from a sample of size n are $\bar{x}$ and s^2, respectively, is $x_p = \bar{x} + z_p s$ where z_p is the standard normal deviate for the proportion P. The asymptotic estimates of the $(1-\gamma)$ confidence limits on x_p are $\bar{x} \pm z_0 s + t(\nu; \gamma/2) s \sqrt{\frac{1}{n} + \frac{z_p^2}{2(n-1)}}$ where $t(\nu; \gamma/2)$ is the $\gamma/2^{th}$ quantile of the Student's t distribution with $\nu = (n-1)$ degrees of freedom. The intersection of the two conditional CDFs provides a description of the point estimate $\phi(-D/2) = 0.0375$, of the misclassification rate. The two intersections, a-b, of the two limbs of the 0.95 confidence limits on the quantiles of the two conditional distributions provide a description of the $0.95^2 = 0.90$ interval estimate of the misclassification rate: $P[0 \leq P_1 \leq 0.12] = 0.90$. The interval, a - b = 0.12, corresponds to the average $\bar{n}_1 = \bar{n}_2 = 18$ of the available sample sizes of the two conditional distributions. The interval A-B = 0.05, corresponds to a hypothetical sample of $n_1 = n_2 = 100$.

Figure 9b presents the superposition of the ROC curves for T_1 and LDF (Lusted 1975). The minimum total error rates which can be achieved with each test are $P_1 = P_2 = 0.0928$ and $P_1 = P_2 = 0.0375$ for T_1 and LDF, respectively. Note that the latter is simply $\phi(-D/2)$. These are the points marked by the solid symbols. The respective indices of sensitivity (Green & Swets 1974) are d' = 2.65 and 3.55, respectively, where d' is the difference between the means of the conditional distributions (of T_1 or LDF) divided by the (common) standard deviation. (d' is obviously cognate to the Mahalanobis distance.) It is evident that the LDF has an ROC curve similar to that of a good screen-film system (Green & Swets 1974).

Table 7 presents point estimates of the error rates, P_1 and P_2, of the linear discriminant functions, LDF(T_1, T_2), for colon, breast and lung. These estimates assume that the respective samples provide unbiased estimates, D^2, and $(1-q_1)/q_1$, respectively, of the Mahalanobis distance, Δ^2, and prior odds ratio $(1-\pi_1)/\pi_1$.

Table 8 provides a comparison of two alternative procedures for obtaining unbiased point estimates of the predictive performance of LDF (T_1, T_2) in future samples. Both procedures are based on the assumption $X \sim N_2 (\underline{\mu}_j, \Sigma)$, $j = 1, 2$, that is, bivariate Normal with common variance-covariance matrix, Σ. The first procedure depends upon obtaining an unbiased estimate, δ^2, of the Mahalanobis distance parameter, from the sample estimate, D^2. The second procedure depends upon the relations between the PRESS statistic and the coefficient of determination, R^{*2} of the binary regression model and the biased sample estimate, D^2 of Δ^2. Note that the two estimates of the misclassification rate, P_1, to be achieved in future samples are quite comparable.

Table 9 presents a third method for obtaining unbiased point estimates of the respective misclassification rates, $\overline{P_1}$ and P_2. This method, which is based on the Bernoulli distribution, also provides interval estimates of P_i. This is the Lachenbruch "Leaving-One-Out" Method (Lachenbruch 1975). Both the motivation and implementation of the method are cognate to those of the PRESS statistic.

Table 10 presents a comparison of two procedures for obtaining interval estimates of the error rates: $P_1 = P_2 = P$. Both methods are based on the Normal distribution. One procedure, that of Dunn and Varady, is based on Monte Carlo methods. The second is a heuristic procedure described by Herbert (1986), and described in Figure 9a. See also Eisenhart (1947). The Table shows that the estimates obtained by the two procedures are comparable. The Table suggests as well that the sample sizes such that $n_1 = n_2 \cong 100$ are required to determine the

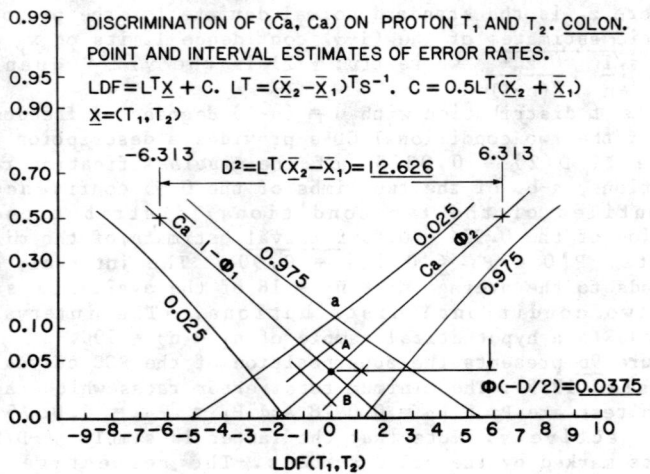

Figure 9a. Superposition of the conditional ($\overline{Ca}$ and Ca) distributions (CDF) of LDF (T_1, T_2). The 0.95 confidence limits on the probability of misclassification are shown for two different sample sizes: 1) Point a ($n_1 = n_2 = 20$) 2) Interval A-B ($n_1 = n_2 = 100$). Compare with Figure 7.

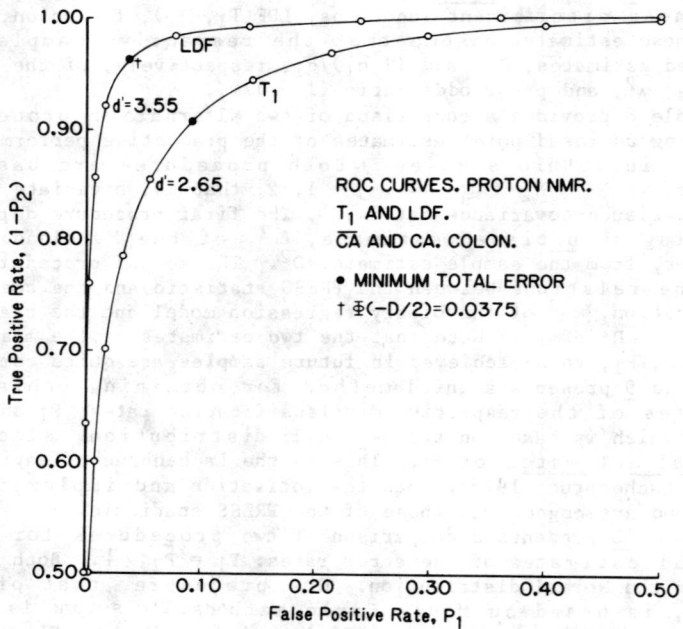

Figure 9b. Superposition of ROC curves for T_1 and LDF (T_1, T_2) for discrimination between $\overline{Ca}$ and Ca Colon.

Table 8. Comparison of Alternative Point Estimates of Error Rate, P_1, in Future Samples. "Shrunken Distance" Methods. Normal Distribution.

	Ca Colon (T_1, T_2) $(n_1 = 16, n_2 = 20)$	Ca Breast (T_2) $(n_1 = 11, n_2 = 12)$	Ca Lung (T_1) $(n_1 = 22, n_2 = 7)$
1) $\delta/2$[a]	1.680	1.214	0.764
$P_1 = \Phi(-\delta/2)$	0.045	0.112	0.223
2) $D*/2$[b]	1.549	1.109	0.688
$P_1 = \Phi(-D*/2)$	0.061	0.134	0.246

[a] D^2 is a biased estimate of Δ^2 ($E(D^2) > \Delta^2$).
$\delta^2 = (n_1+n_2 - p-3/n_1+n_2-2 \; D^2) - ((n_1+n_2)/n_1n_2)p$, is an unbiased estimate of Δ_2.
[b] $D*^2 = [(n_1+n_2)(n_1+n_2-2)/n_1n_2]R*^2/(1-R*^2)$ where $R*^2 = 1-\text{PRESS/SSY}$.
$\text{PRESS} = \sum_{i=1}^{n_1+n_2} \left(\frac{e_i}{1-h_{ii}}\right)^2$

Table 9. 0.95 Interval Estimates of Probabilities of Misclassification. $\overline{P}_i$ and $\underline{P}_i$ of LDF. Confidence Limits on Parameter of Binomial Distribution (P_i Estimated by Lachenbruch "Leaving-One-Out" Method.)

		Colon LDF(T_1, T_2)	Breast LDF(T_2)	Lung LDF(T_1)
Ca	n_1	16	11	22
	$\overline{P}_1$	0.302	0.285[a]	0.229
	P_1	0.063	0	0.046
	$\underline{P}_1$	0.002	0	0.001
Ca	n_2	20	12	7
	$\overline{P}_2$	0.168[a]	0.385	0.901
	P_2	0	0.083	0.571
	$\underline{P}_2$	0	0.002	0.184

[a] $\overline{P}_i = 1-(\alpha/2)^{1/n_i}$. $\alpha = 0.05$, $i = 1, 2$.

Table 10. Comparison of 0.90 Interval Estimates of P for LDF (T_1, T_2). Normal Distribution Methods. Colon.

	$\underline{P}$	$\overline{P}$	$\underline{P}$	$\overline{P}$
1) Dunn & Varady[a] (Monte Carlo) See Figure 9.	0.004	0.13	0.01	0.07
	($n_1 = n_2 = 25$)		($n_1 = n_2 = 100$)	
2) Herbert[a] (Student's t Approx.) See Figure 10.	0.004	0.12	0.02	0.07
	($n_1 = n_2 = 18$)		($n_1 = n_2 = 100$)	

[a] $H_0: \Sigma_1 = \Sigma_2 = \Sigma$

clinical utility of any combination of T_1 and T_2 measurements on biopsies of colon tissues.

It seems evident in the size of the confidence limits on P_1 and P_2 that the question addressed in the Koutcher, et al 1978 paper – whether the NMR measurements recommended therein have any clinical value in the discrimination between $\overline{Ca}$ and Ca tissues at these anatomic sites cannot be answered from the data presented in that paper.

Figures 10, 11 and 12, respectively, present a superposition of the family of isorisk curves, which can be derived from the logistic model of discriminant analysis, and the scattergrams for the sites colon, breast and lung (Cornfield 1965; Lachenbruch 1975). Each family of curves is derived on the assumption that the sample sizes, n_1 and n_2, do indeed provide accurate estimates of the respective a priori probabilities, i.e., that the sample estimates of the respective prior odds ratios are appropriate: $\pi_1/(1-\pi_1) = n_1/n_2$. (We have noted that, in general, this is not likely to be the case.)

Figure 10 also includes the scattergram of a distribution of T_1 and T_2 measurements on biopsies obtained from the non-diseased ($\overline{Ca}$) portion of the colon of eleven new patients with a diagnosis of colon cancer (Ca) (Koutcher, et al 1978). These are represented by the symbol "+" in the figure. Note that only two of the observations lie below the 0.50 isorisk line: $P(Ca| T_1, T_2) = 0.50$. Only these two have NMR measurements which are consistent with the histologic diagnosis. This apparent inconsistency is explained by Koutcher, et al (1978) as evidence of a "systemic effect" of cancer which is disclosed only to NMR measurements and not to histology. We have no comment on this explanation (at this time). Instead, we wish to exploit the heuristic value of this set of observations in the demonstration of the Type III error (Kimball 1957).

The Type III error was described briefly under Methods. It will be recalled that it may be defined (only half-facetiously) as giving the right answer to the wrong question (Kimball 1957). It is incurred (for example) when the linear discriminant function is used to assign, by a procedure which was developed to give minimum probability of incurring Type I and Type II errors, an individual between two known mutually exclusive groups (say, $\overline{Ca}$ and Ca) when in fact the individual in question belongs to neither of those groups but rather to some third, unknown, non-normal, non-malignant group. (See Rao 1973.) Although the two alternative groups considered are mutually exclusive ($\overline{E} \cap E = \emptyset$) for the individual in question they are not exhaustive of the possible groups to which the individual in question may belong: $\overline{E} \cup E \neq 0$, or $P(\overline{E}) + P(E) < 1.0$. In other words, before the discriminant function is used to classify an individual, the non-sample, or external, evidence for the validity of the model, that is, the hypothesis, H_0: $\underline{X} \sim \overline{Ca} \cup Ca$, or $P(\overline{Ca}) + P(Ca) = 1.0$, for that individual must be quite unambiguous. For those individuals for whom it is not the hypothesis, H_0, must be tested on the internal evidence of $\underline{X}$ itself. (As noted this is a circumstance which is not uncommon in medical diagnosis.) The alternative hypothesis is H_1: $X \in \overline{Ca} \cup Ca$. The $2^1 = 2$ possibilities are described in Table 2a. The test statistic, E_0, is distributed as chi-square on (k-1) degrees of freedom where

$$E_0 = (\underline{X}-\underline{\overline{X}}_1)^T S^{-1}(\underline{X}-\underline{\overline{X}}_1) - [\underline{L}^T(\underline{X}-\underline{\overline{X}}_1)]^2/D^2. \tag{55}$$

If E_0 is significant then H_0 is rejected and H_1 is accepted: the individual in question belongs to some unknown non-normal,

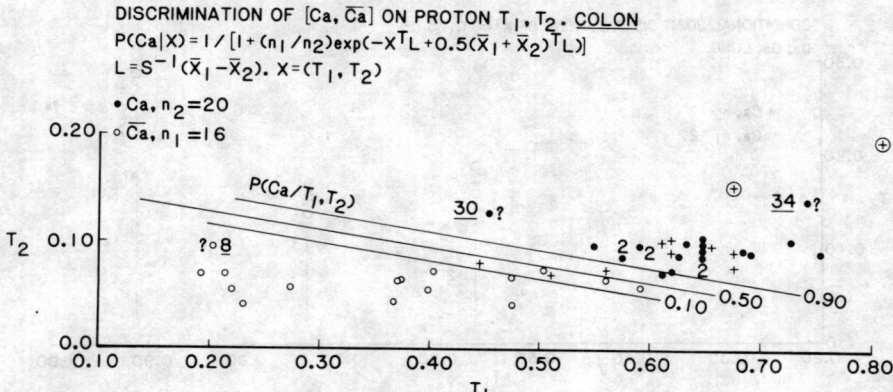

Figure 10. Superposition of isorisk lines from LDF (T_1, T_2) on scattergrams of the conditional ($\overline{Ca}$ and Ca) distributions of proton T_1 and T_2 measurements on biopsies of Colon.

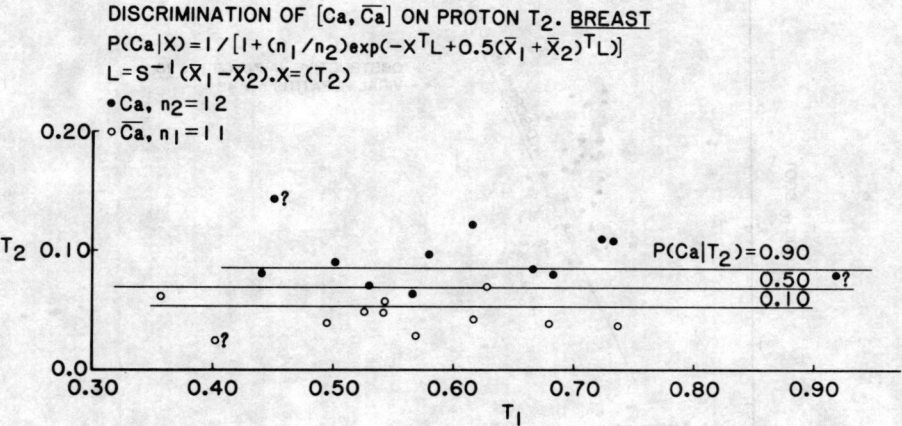

Figure 11. Superposition of isorisk lines from LDF (T_2) on scattergrams of the conditional ($\overline{Ca}$ and Ca) distributions of biopsies of Breast.

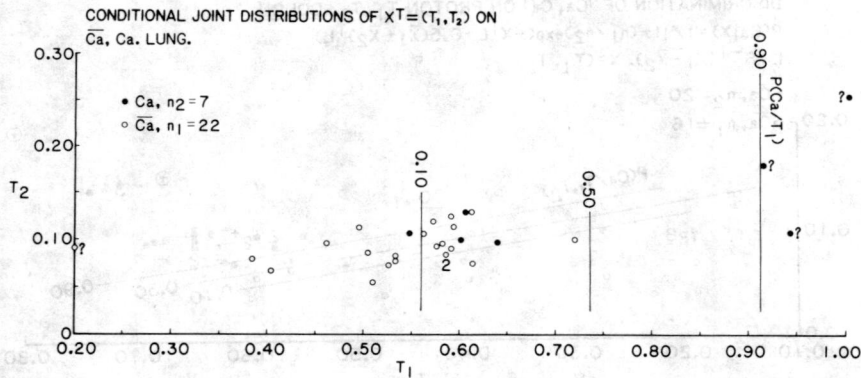

Figure 12. Superposition of isorisk lines from LDF (T_1) on scattergrams of the conditional ($\overline{Ca}$ and Ca) distributions of biopsies of Lung.

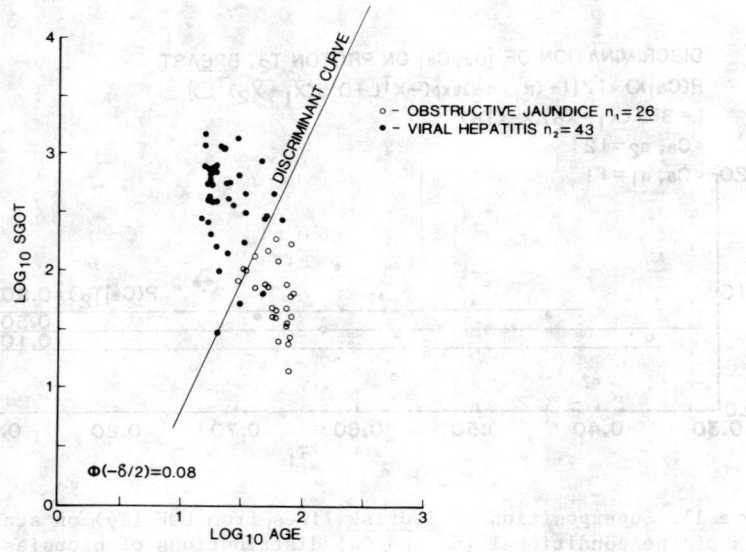

Figure 13a. Superposition of the scattergrams of the conditional (viral hepatitis and obstructive jaundice) distributions of age and SGOT.

non-malignant group, $\overline{Ca} \cup Ca$,* for which the mean is not simply related to the means of $\overline{Ca}$ and Ca. If E_0 is not significant then H_o may be accepted. Membership of the individual in question is then determined by the values of statistics E_1 and E_2, each distributed as chi-squared on one degree of freedom, where

$$E_1 = [\underline{L}^T(\underline{X}-\underline{X}_1)]^2/D^2 \tag{56}$$

$$E_2 = [\underline{L}^T(\underline{X}-\underline{X}_2)]^2/D^2. \tag{57}$$

The $2^2 = 4$ possibilities are described in Table 2b.

Tests of the hypothesis, H_0, for each of the individuals in the histologically benign ($\overline{Ca}$) group disclose that the two at the positions circled, "⊕", belong to a third, unknown group, with centroid of the joint distribution of T_1 and T_2 not simply related to those of the groups $\overline{Ca}$ and Ca, i.e., $\underline{X} \in \overline{(\overline{Ca} \cup Ca)}$**. The centroid of the unknown group lies in a direction "away" from, or perpendicular to, the line connecting the centroids of the $\overline{Ca}$ and Ca groups. See Figure 7.

The devolution of the problem of discrimination into the sequence of problems of hypothesis testing described above and that should occur in the absence of a priori information that the alternatives entertained are in fact exhaustive for the membership of the individual in question provided a simple instance of the more general problem described by LC Payne (1964). It is simply that a discriminant function, or, "A sympton, sign, or item of history is only of clinical significance to the extent that it discriminates between a set of possibilities: namely those that a clinician has thought of." ... "The greatest area of ignorance is that which we don't even know we don't know". (i.e, that we know not that we know not). This is the context in which Type III errors occur. The tests of membership developed by Rao (1973) that are implemented by the statistics, E_0, E_1 and E_2 which were illustrated in our example by the T_1 and T_2 measurements on normal colon tissues for Ca colon patients will minimize the likelihood of occurrence of Type III errors that arise when the possible memberships of the individual in question exceed, "... those that the clinician has thought of ...", at the time he uses the discriminant function.

The test statistics, E_0, E_1 and E_2 are constructed from the same sample information that is used to minimize the likelihood of occurrence of Type I and Type II errors, namely, L^T and D^2.

It is instructive to note that a test of the hypothesis H_0 above for each of the two (anomalous) cases #8 and #30 in Figure 7, discloses that both belong to a third unknown groups with centroid not simply related to those of $\overline{Ca}$ and Ca. For the (anomalous) case #34 it can be shown that it, too, belongs to a third, unknown, group. However, this group has a mean vector collinear with the mean vectors of the groups $\overline{Ca}$ and Ca. See Figure 7. Note in this Figure that the perpendicular distance from the line joining the centroids of the $\overline{Ca}$ and Ca groups is much greater for case #8 and case #30 than it is for case #34. Note in Figure 10 that this perpendicular distance for the two "new cases" (labeled "⊕") also exceeds that for case #34. These results are qualitatively consistent with the chi-squared measures of resemblance, C_j, of each of these observations to the proximal group which are presented in Table 6 (Tiedeman, et al 1953). Note that the Mahalanobis distance, d_i, for each of these cases are $d_8 = 3.21$, $d_{30} = 3.58$, and $d_{34} = 3.10$, far in excess of d_i for any other observation. For these three $h_i \simeq 12.8*d_i$.

These examples disclose the importance of the practice of determining the validity of the null hypothesis, H_0: $\underline{X} \in \overline{Ca} \cup Ca$, for the individual, or case, in question before assigning him on the basis of his discriminant score, $\underline{L}^T\underline{X}$, to either $\overline{Ca}$ or Ca. Note that the statistics, E_0, E_1, and E_2 provide measures of the <u>utility</u> of the LDF for assignment of <u>future</u> cases which is different from that of either the PRESS or the δ^2 statistics.

6. "THE LAW OF SMALL NUMBERS"

We now have a realizing sense of some of the difficulties, ambiguities, etc., that attend any attempt to assess the clinical value of a proposed diagnostic test on the basis of samples of the kind and <u>size</u> that frequently appear in the literature of these attempts. By way of emphasis of the problem of small samples we now quote liberally from one of the more recent and thoughtful papers addressing the problem (Kahneman, et al 1982).

"The law of large numbers guarantees that very large samples will indeed by highly representative of the population from which they are drawn. ... People's intuitions about random sampling appear to satisfy the law of small numbers, which asserts that the law of large numbers applies to small numbers as well."

"... the believer in the law of small numbers practices science as follows:
1) He gambles his hypotheses on small samples without realizing that the odds against him are unreasonbaly high.
2) He has undue confidence in early trends (e.g., the data of the first few subjects) and in the stability of observed patterns (e.g., the number and identify of significant results. <u>He over-estimates significance</u>.
3) In evaluating replications, his or others', he has unreasonably high expectations about the replicability of significant results. <u>He underestimates the breadth of confidence intervals</u>.
4) He rarely attributes a deviation of results from expectations to sampling variability, because he finds a causal 'explanation' for any discrepancy. Thus, he has little opportunity to recognize sampling variation in action. His belief in the law of small numbers, therefore, will forever remain intact."

7. PROPER AND IMPROPER LINEAR PREDICTIVE MODELS

We have shown that the data which have been offered in support of the rather optimistic pronouncements of the clinical diagnostic value of NMR measurements of T_1 and T_2 in biopsy specimens that were published by Koutcher, et al (1978), probably don't. Because of the samll samples $(n_1, n_2) < 25$ the <u>interval</u> estimates of the error rates, P_1 and P_2, of the linear discriminant function constructed to implement the diagnostic information contained in the conditional distributions of invitro measurements of T_1 and T_2 were too large to permit any convincing assessment - a result anticipated from the Law of Small Numbers.

It should be noted in the assessment of the Koutcher et al (1978) data that even such diagnostic information as was contained in the small samples from the conditional distributions of measurements of T_1 and T_2 on biopsies from the three anatomical sites, was not exploited by optimal methods in their solution - the linear combination of T_1 and T_2 called a Malignancy Index - to the classification problem which they addressed: discrimination between $\overline{Ca}$ and Ca. This is because the coefficients of the respective measurements in the Malignancy Index are simply chosen to be "scaling factors"; the object of this transformation was to achieve <u>commensurability</u> of the measurements which comprise $\underline{X}$.

We have described the standard linear discriminant model for exploiting, for diagnostic purposes, the difference in the mean vectors, ($\underline{\mu}_1 - \underline{\mu}_2$), of the conditional distributions of any set of biological measurements. This model has an alternative criterion for selecting the coefficients or weights of the respective measurements. They are chosen to <u>maximize</u> the strength of the association between the observed status ($\overline{Ca}$ or Ca) and the status estimated by the linear form, z, on T_1 and T_2, or, what amounts to the same thing, <u>minimization</u> of the frequency of misclassification achieved by use of z to assign individuals between complementary groups. Dawes (1982) (vide supra) has made a useful (albeit invidious) distinction between <u>proper</u> and <u>improper</u> linear predictive models: "A <u>proper linear model</u> is one in which the weights given the predictor variables are chosen in such a way as to optimize the relationship between the prediction and the criterion. [In the present paper these are denoted by the linear form, z, and the observation, y, respectively.] Simple regression analysis is the most common example ... Discriminant function analysis is another example of a proper linear model; weights are given to the predictor variables in such a way that the resulting linear composites maximize the discrepancy between two or more groups. ... An <u>improper linear model</u> is one in which the weights are chosen by some non optimal method. They may be chosen to be equal, they may be chosen on the basis of the intuition of the person making the prediction, or they may be chosen at random."

It is instructive to compare the <u>proper</u>, LDF, and <u>improper</u>, Λ, linear models with respect to the reduction of the degree of "overlap" of the conditional joint distributions of T_1 and T_2. The comparison may be made in terms of the sample estimates of the respective Mahalanobis distances D^2 for LDF and D_A^2 for Λ. The estimates of D_A^2 are obtained in the following way. Let $\underline{X} = (T_1, T_2)$ and $\underline{\Gamma}^T = (1/\overline{T}_1(\text{norm}), (1/\overline{T}_2(\text{norm}))$ so that $\Lambda = \underline{\Gamma}^T\underline{X}$. The analogue to the Mahalanobis distance is $D_A^2 = [\underline{\Gamma}^T(\underline{X}_1-\underline{X}_2)\underline{\Gamma}]^2/\underline{\Gamma}^T S \underline{\Gamma}$, where $\overline{X}_1$, $\overline{X}_2$, S are, respectively, the mean vectors for the groups $\overline{Ca}$ and Ca and the <u>pooled</u> estimate of the common variance-covariance matrix of the conditional distributions of measurements (Lachenbruch 1975). Table <u>11</u> presents the comparison.[n] Note that $D^2 \geq D_A^2$, as would be expected - although often <u>not by much</u>.

If we assume that the misclassification rates for Λ can be estimated as $\Phi(-D_A/2)$ then the effect on the misclassification rates achieved by the use of the LDF criterion to select the coefficients of the linear form on T_1 and T_2 does not greatly differ from that achieved by Λ. This is simply a specific instance of a general effect: for <u>small samples</u> the predictive power of improper linear models is not much different than that of proper linear models (See papers by Claudy 1972, Dawes 1980 and Einhorn and Hogarth 1975.) In fact, for small samples, it is frequently the case that an improper linear predictive model with <u>unit</u> coefficients, or weights, for the predictors, eg. $\underline{\Gamma}^T = (1, 1)$, will not differ much in predictive power from the optimal model (Claudy 1972). For instance, the values of $D_A/2$ for the <u>unit weight</u> model for Colon, Breast and Lung are 1.48, 0.45 and 0.87, respectively. The corresponding values of D/2 for the proper (LDF) model are 1.78, 1.35 and 0.90, respectively. The respective misclassification rates for the improper predictive model, $\Phi(-D_A/2)$, are 0.07, 0.33, and 0.19. The corresponding rates for the proper predictive model, $\Phi(-D/2)$, are 0.04, 0.09, and 0.18.[6] This "indifference" of small samples is obviously simply another aspect of the large interval estimates of the coefficients of the proper linear predictive models that are obtained from small numbers. See Table 12.

Note, however, that even with small numbers, it is possible to exploit the proper linear predictive models in determining the roles of a subset of the predictor variables; in the present case these subsets are T_1 and T_2. Thus, we note in Table 11 that the coefficient of T_1 in LDF (T_1, T_2) for Breast is not significantly greater than its standard error. Nor is the coefficient of T_2 in LDF (T_1, T_2) for Lung. Note further, that even with small numbers, the statistics, L^T and D^2, of the proper predictive model can be exploited to reduce the frequency of occurrence of Type III errors (in principle).

We have shown that the assessments of Koutcher et al, concerning the diagnostic value of T_1 and T_2 measurements could still not be supported by their data even when diagnostic decisions based on these measurements were implemented by optimal methods: The interval estimates of the error rates incurred in the use of this optimal methodology are still too large to provide a convincing assessment - a consequence of the small sizes of n_1 and n_2, which are less than 1/5 the minimum required: (n_1+n_2) must be at least 100 or 20k, whichever is larger (Lindeman 1980).

As well as an inadequate total size of sample, described as, say, the sum, $((n_1+n_2) \gg 100)$, the distribution of (n_1+n_2) over the complementary groups $\overline{Ca}$ and Ca described as, say, the ratio, (n_1/n_2), is also inadequate to achieve an "optimal" solution to the classification problem. Rao (1952) has shown that for estimation of the discriminant coefficient vector, L^T, this ratio should be approximately unity: $n_1/n_2 = 1.0$. Moreover, the non-sample, as well as sample information, is inadequate. The decision threshold, $\ln(q_2/q_1)$, for the discriminant function should be an unbiased estimate of the relative frequency $((1-\pi_1)/\pi_1)$ of the two groups in the target population, and therefore additional a priori, or non-sample - in this case epidemiologic - information is (usually) required rather than that provided by the sample, in the more general case.

We have shown that if the speculations of Mansfield and Morris (1982) anent the putative clinical value of a combination of "NMR parameters" in a medical imaging system are to be evaluated in a meaningful manner, then the respective coefficients, or weights, given to each of the parameters in the combination should be chosen to minimize the "overlap" of the conditional joint frequency distributions, as in a linear discriminant function (and not simply to make the measurements of the two parameters commensurable, as in the Malignancy Index of Koutcher, et al (1978).)

We have examined the conjectures of Koutcher et al and Mansfield, et al in the context of the more general (and insistent) "... problems of discrimination and classification." (Bose and Roy 1938) Although, we have illustrated the classification problem only for two tests, proton T_1 and T_2, and two groups, $\overline{Ca}$ and Ca, that is, for the special (and most common) case in which k=g=2, it is possible, and often fruitful, to use more than two tests, $k > 2$, in order to reduce the classification error rates still further. (The sample estimate of D^2 will never decrease - and will usually increase - as k is increased.) It is also possible to consider under the rubric of Discriminant Analysis more

Table 11. Comparison of the Mahalanobis distance function, D^2 and D_A^2, on T_1 and T_2 for Proper (LDF) and Improper (Λ) Linear Predictive Models.

	Colon	Breast	Lung
D^2	12.63	7.30	3.21
L^T	(27.29, 163.42)	(6.15, 148.28)	(8.81, 27.04)
$L'^{T a}$	(5.398, 4.135)	(1.069, 2.810)	(1.820, 1.276)
D_Λ^2	12.62	6.90	3.12
Γ^T	(2.73, 15.67)	(1.80, 21.70)	(1.87, 10.81)

[a] $L'^T = (\hat{\alpha}_1/\sqrt{Var(\hat{\alpha}_1)}, \hat{\alpha}_2/\sqrt{Var(\hat{\alpha}_2)})$ where $\hat{\alpha}_j$ is the estimated coefficient of T_j, $j = 1, 2$, in the regression model of the linear discriminant function, LDF. L'^T is the vector of standardized coefficients; each element is distributed (approximately) as Student's t with (n-3) degrees of freedom.

Table 12. Comparison of Misclassification Rates for Improper ("Unit Weight") and Proper (Discriminant Function) Predictive Models Estimated from Small Samples.

	Colon n = 36	Breast n = 23	Lung n = 29
Unit Weight Model. $\phi(-D*/2)$	0.07	0.33 [0.10][a]	0.19
Linear Discriminant Model. $\phi(-D/2)$	0.04	0.09	0.18

[a] "Weights" (1, 0) vs (1, 1)

than two complementary groups, $g > 2$, for the classification of an individual.

Some general remarks may be made concerning the theoretical and practical constraints on the kinds and number, k, of tests or measurements to be included in the vector, $\underline{X}$, in order for the methods previously discussed to be valid. In principle, only ratio or interval measurements with conditional Normal distributions^P on each subgroup should be included, however, it has been shown that nominal measurements such as the binary observations, $x_i = (0, 1)$, may be also included, as well as ordinal measurements; the discriminant function seems "robust" with respect to these two types of non-normality (Lachenbruch 1977; Gilbert 1969). It is also obviously better to include tests which are mutually uncorrelated, i.e., Σ_j should be a (nearly) diagonal matrix.

For $g = 2$ it has been shown that although the estimated error rates, P_1 and P_2, in the sample at hand decreases with k, the "shrinkage" of the Mahalanobis distance, D^2, and hence the inflation of the error rates in future samples, increases with k for a given sample size. Moreover, as we have remarked, the minimum required sample size is about 100 or 20k, whichever is the greater. Since, in general, it is difficult to obtain large homogeneous samples, as well as a large number, k, of nearly uncorrelated measurements, $\underline{X}$, these (and other considerations such as cost, risk, etc. of obtaining the measurement) limit the number of predictor variables which may be usefully included to (about) $k \leq 6$. (The important question, of course, is, 'Which 6?!') It might be fruitful to include five or six variables as seemingly disparate as possible; e.g., measurements of T_1 and T_2 combined with radiographic, histologic and biochemical observations. The nature - inherent (population) or contingent (sample) - and degree of mutual correlation within the set of k measurements must be determined for each choice.

8. GENERALIZATIONS OF LINEAR DISCRIMINANT MODEL

The linear discriminant model is readily generalized to the classification problem in which $k > 2$, $g = 2$. We will present only the "graphics" for a model for $k = 4$. The problem is a differential diagnosis of liver disease. The $g = 2$ complementary subgroups, $\overline{E}$ and E, of icteric patients are comprised of those with either viral hepatitis or obstructive jaundice. Liver disease seems to be one of the favorite examples in the literature of medical applications of discriminant analysis. (See Rao 1952.) This is perhaps because the alternatives require such disparate forms of treatment, here medicine or surgery, respectively. However, the loss ratio will be taken to be unity here.

The sample for the present example is taken from Herbert, Gydesen and Venzon (1972) and includes 26 patients with obstructive jaundice and 43 with viral hapatitis. The $k = 4$ tests are age (at diagnosis), white blood cell count, WBC, serum glutamic oxalacetic transaminase, SGOT and serum alkaline phosphatase, A/∅. Because of the small sample size this analysis also has only heuristic value.

Figure 13a is the scattergram of the 69 cases in the X-space defined by 1) age (an etiologic factor: viral hepatitis is a hazard that, "goes with the territory", occupied by the drug culture) and 2) serum levels of the enzyme SGOT (an elevated level is one effect of a mal de fois). Figure 13b presents the superpositions of the respective conditional joint (contour ellipses) and marginal (frequency curves)

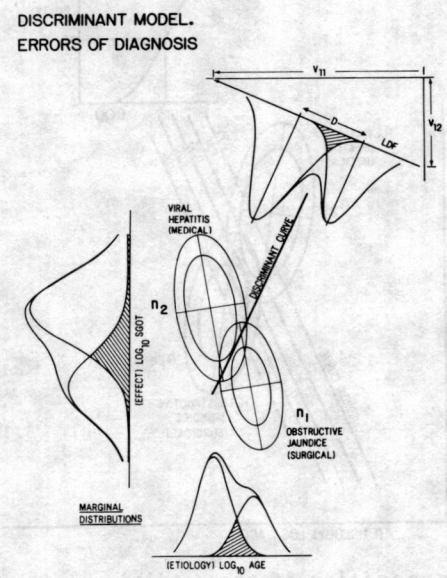

Figure 13b. Superpositions of the conditional *joint* and *marginal* distributions of $\underline{X}$ and $L(\underline{X})$.

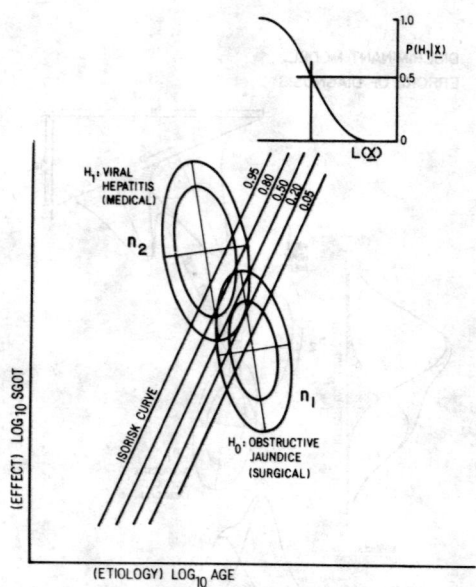

Figure 13c. Superposition of the conditional joint distributions of age and SGOT and the isorisk curves. Risk is specified in terms of the probability of the presence of viral hepatitis since surgery is the more traumatic (and irreversible) of the alternative maneuvers for management of the patient.

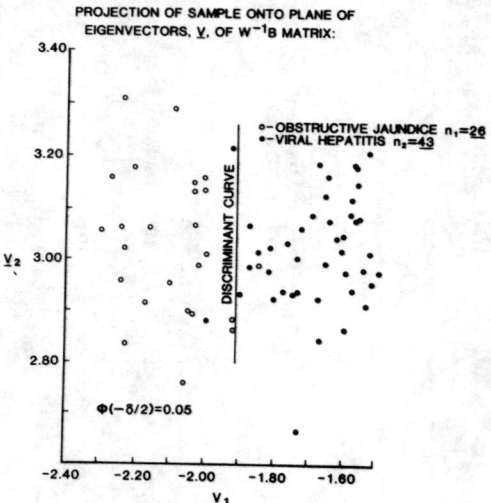

Figure 13d. Superposition of the scattergrams of the conditional joint distributions of the projections of the patient scores (age, SGOT, AØ and WBC) in the plane defined by the right eigenvectors of the first two eigenvalues of the $W^{-1}B$ matrix.

distributions of $\underline{X}$ and $L(\underline{X})$. The drawings are to scale and show vividly the decreased "overlap" in the conditional distribution that is achieved by forming the linear combination for which $\phi(-\delta/2) = 0.08$. The (right) eigenvector of $W^{-1}B$ is $\underline{V}_1 = (v_{11}, v_{12})$. See Bryan 1951 and Rao 1948.

Figure 13c presents the family of iso-risk curves which are readily derived from the logistic function in which $L(\underline{X})$ is the argument. See Cornfield 1962 and Cornfield, et al 1961. The sigmoid curve in the upper right is a section of the risk surface perpendicular to the discriminant curve (or parallel to the right eigenvector of $W^{-1}B$).

Figure 13d presents the scattergram of the sample in the plane defined by the $(4*1)$ right eigenvectors, $\underline{V}_1$ and V_2, of the first two eigenvalues, λ_1 and λ_2, of $W^{-1}B$. (There is only one non-zero eigenvalue since k = 4 and g = 2. The direction of the discriminant function, $L(\underline{X})$, is given by the right eigenvector of the first (non-zero) eigenvalue. The direction of the discriminant curve is given by the right eigenvector of second eigenvalue. Note that the eigenvectors of the asymmetric matrix $W^{-1}B$ are not orthogonal; we have orthogonalized $\underline{V}_1$ and $\underline{V}_2$ for display purposes. The Mahalanobis distance is proportional to λ_1. The estimated probability of misclassification is $\phi(-\delta/2) = 0.05$.

The eigenvalue-eigenvector methods of Bryan (1951) and Rao (1948) and the minimum chi-squared method of Tiedeman, et al (1954) and Tatsuoka (1971) are generalized, in an obvious manner, to the case of $g > 2$. However, the implementations of these generalizations to $g > 2$ are, perhaps, somewhat less straight-forward than that for $k > 2$ (g=2) and the reader should consult the literature cited before embarking on the use of these methods. Also, it should be recalled that the binary regression model of discriminant analysis is valid only in the case in which g = 2, and therefore the methods of Regression Diagnostics cannot apply directly for $g > 2$ (Kleimbaum and Kupper 1978).

ACKNOWLEDGEMENT

We acknowledge, with thanks, the dedication, as well as expertise, of Marie Woodall in the preparation of the manuscript.

FOOTNOTES

[a] The accepted statistical practice for achieving commensurability is to standardize the variables: $T_{ij} = (T_{ij}-\overline{T}_j)/s_j$, $j = 1, 2$; $i = 1, \ldots n_i$ (Armitage 1977).
[b] See Robyn Dawes, "The robust beauty of improper linear models in decision making" in Judgement Under Uncertainty: Heuristics and Biases, D. Kahneman, P. Slovic and A. Tversky, Eds. (1982). The present paper compares such improper models with proper linear predictive models such as the discriminant functions.
[c] The literature also includes a discussion of an application of the Malignancy Index to a still more exiguous sample: Ca cervix, $n_1 = 5$, $n_2 = 6$, described in a paper published in the journal, Gynecologic Oncology, in 1978. See Herbert 1986.
[d] Distribution-free methods will, in general, give interval estimates of P_i which are too large to be useful even for samples sizes of the order of $n_1=n_2=100$. It is necessary to reduce the variance of estimate by the general method of combining prior information, the parametric

form of the distribution (Gaussian), with that of the sample (the mean and variance). By this procedure the sample mean and variance are taken to be estimates of the parameters of a distribution for which the form is specified apriori (Herbert 1986). As in any procedure in which prior and sample information are combined it is necessary to examine their mutual compatibility. In this case it was a matter of testing the hypothesis that the data are consistent with a random sample of the same size from a Normal population by a probability plot and the Filliben statistic. See also the tests of bivariate Normality in the present paper.

e   This seems to be a defect which is common to many published studies: the size of the sample and the idiosyncracies (outlying and influential observations, collinear variables, etc.) of the distribution of observations therein, preclude any useful generalization of the results. Thus, these studies have only heuristic value - even as in the present case. This also seems to be a defect, however, of which many authors - and their reviewers - are unaware. Since the data in such studies are usually relevant to questions of no little inherent scientific interest as well as of great practical importance, their use to implement a demonstration of the appropriate methodology may stimulate the interest of the reader to examine such methods more closely. Thus, the motivational value of such data exceeds their evidential value. Such is the case for the Koutcher, et al (1978) data.

f   In the present exposition it is assumed that the form of the frequency distribution, eg. Normal, is invariant over the set of complementary conditions, $\bar{E}$ and E.

g   It is, in fact, the case that most expositions of the statistical aspects of any medical study do not place enough emphasis on the verification of the statistical assumptions on which the validity of the various models and procedures depends. Since these models vary greatly in their robustness with respect to departures of the data from these assumptions, it would seem that a great many questionable conclusions may have been drawn by the readers as well as the writers of these discussions.

h   Note the similarity of this procedure for determining the elements of V to the procedure for determining the position of the optimal cut-off or threshold, x_o, between two, overlapping, conditional frequency distributions. Both procedures are based on minimizing the overlap of two conditional distributions. See Lubin 1950.

i   This gives rise to a simple classification scheme which may be called the minimum chi-squared rule (Tatauoka 1971). This rule has the optimal property of minimizing the probability of misclassification when the two groups are multivariate Normal with equal covariance matrices and equal apriori probabilities. The rule is as follows: Compute the χ^2_{ij} statistic of the individual in question with respect to each of the alternative groups and assign him to that group with respect to which the value χ^2_{ij} is smaller. $j = 1, 2$.

When the covariance matrices are unequal, an adjusted chi-square statistic is computed:
$$\chi'^2_{ij} = \chi^2_{ij} + \ln|\Sigma_j|$$
The optimal rule is then to assign the individual in question, X_i, to the group for which χ^2_{ij} is a minimum.

When both the covariance matrices and the apriori probabilities, q_j, are unequal a further adjustment of χ^2_{ij} is required. Compute the statistic

$$\chi'^2_{ij} = \chi^2_{ij} + \ln|\Sigma_j| - 2\ln q_j.$$

The optimal rule is then to assign the individual in question to the group for which χ'^2 is the smaller.

The methods can obviously be generalized to the classification problem with three or more groups. $j = 1, \ldots, g$. $g > 2$.

j   In particular, numerical diagnostics are functions of the data whose values may detect <u>responses</u>, y_i, that are unusually large or small (<u>outliers</u>) and <u>predictors</u>, x_i', that are remote from most of the data (<u>extremes</u>) and may be influential ("high leverage" points). Here, y_i is the ith response and x_i' is the ith 1x(k+1) row vector.

k   In this event, it can be shown that the discriminant function constructed from measurements on the complementary groups $\overline{E}$ and E is sufficient for all third groups for which $\underline{X} \sim N_k(\alpha \underline{\mu}_1 + (1-\alpha)\underline{\mu}_2, \Sigma)$ where $0 \leq \alpha \leq 1$. Hence, $\overline{E} \cup E$, includes such third groups as well, in this hypothesis test. Note that we have supressed, for clarity of notation, the element index, i. That is, $X_i \longrightarrow X$, $E_{0i} \longrightarrow E_0$, etc. See Rao 1973.

l   This characteristic suggests one reason why the derangements of tissue <u>structure</u> produced by disease (and disclosed to histology) are <u>less</u> ambiguous than the derangements of tissue <u>function</u> which often precede them.

m   Figures <u>8a, b</u> and <u>9</u> are related to Figure <u>7</u> in the following way. The former represent the conditional, <u>marginal</u>, distributions obtained by <u>projection</u> of the conditional, <u>joint</u>, distributions onto the <u>right</u> eigenvector, P_2, of the $W^{-1}B$ matrix.

n   The reader will have remarked in Table <u>11</u> that for Colon, $D^2 = D_A^2$ and, $\underline{\Gamma}^T \cong 10(\overline{X}_2 - \overline{X}_1)^T S^{-1}$. This is an example of the fact that discriminant function coefficients are uniquely determined only up to a multiplicative constant: Any multiple, say h, of the coefficient vector, $(\overline{X}_2 - \overline{X}_1)^T S^{-1}$, will have the same value of the Mahalanobis distance, or, as remarked earlier, the LDF and D^2 are <u>invariant</u> under affine transformations. (The relation, $\underline{\Gamma}^T = 10(\overline{X}_2 - \overline{X}_1)S^{-1}$ is simply adventitious; the relation does not hold for the Malignancy Index at the other anatomical sites.)

o   The reader will have remarked that for Breast the missclassification rate achieved, 0.33, with the unit weight model, $\Gamma^T = (1, 1)$, greatly exceeds that achieved, 0.09, with the LDF model, $(\overline{X}_2 - \overline{X}_1)S^{-1} = (140.28, 6.15)$. This illustrates the consequences for a Bayesian regression model when the <u>location</u> of the prior distribution of the coefficient vector is <u>incorrect</u>. A moments reflection discloses that the unit weight model is, in fact, a Bayesian regression model in which the <u>prior</u> information on the coefficient vector dominates that obtained from the <u>sample</u> by estimation since it is stipulated <u>a priori</u> - the dispersion of the prior distribution is zero. It will be recalled that the Koutcher, et al (1978) paper stipulated that both T_1 and T_2 contributed to the discrimination between the conditions $\overline{Ca}$ and Ca for the Breast as well as for the other two sites and <u>further</u> stipulated, a priori, that the coefficient vector was $\underline{\Gamma}^T = (1/\overline{T}_1(normal), 1/\overline{T}_2(normal))$. (Thus, the Malignancy Index is a quasi-Bayesian regression model.) Let us now assume, a priori, that T_1 does <u>not</u> contribute to the required discrimination, but that T_2 does. This prior information may be represented by the coefficient vector, $\Gamma^T = (1, 0)$. For this vector we have $D_A/2 = 1.2948$ and $\Phi(-D_A/2) = 0.0977$ which differs by a negligible amount from the value of $\Phi(-D/2)$ for the discriminant model, $L(\underline{X})$, where $\underline{X} = T_2$. It is in fact the case that for <u>Small Numbers</u> if the prior information on the <u>sign</u> of the unit weight of a predictor,

eg. ± 1, is correct as well as that the variable in question is associated with the response, that is, has a non-zero weight, then the unit weight model is frequently nearly as good, or perhaps even slightly better a predictor, than is the cognate proper linear prediction model (Einhorn and Hogarth 1975; Claudy 1972).
P   Or measurements which can be "Normalized" by simple transformations; e.g., the log Normal (Galton) distribution (Armitage 1977).

REFERENCES

AHMED, S. W. and LACHENBRUCH, P. A. (1977) Discriminant Analysis When Scale Contamination if Present in the Initial Sample in Classification and Clustering. Academic Press. N.Y. pp. 331-353.

ALLEN, D. M. (1974) The Relationship Between Variable Selection and Data Augmentation and a Method for Prediction, Technometrics. 16: pp. 125-27.

ANDERSON, T. W. (1951) Classification by Multivariate Analysis, Psychometrika. 16: pp. 31-50

ARMITAGE, P. (1977) Statistical Methods in Medical Research. 4th Printing. Blackwell Scientific Pub. Oxford.

ASHTON, E. H., HEALY, M. J. R. and LIPTON, S. (1957) The Descriptive Use of Discriminant Functions in Physical Anthropology, Proc. Royal Soc. (Series B) 146: pp. 552-572.

BELSLEY, D. A., WELSCH, R. E. and KUH, E. (1980) Regression Diagnostics. Identifying Influential Data and Sources of Collinearity. John Wiley. N.Y.

BOSE, R. C. and ROY, S. N. (1938) The Distribution of the Studentized D^2 Statistic. Sankhya. 4: pp. 19-30.

BROWNLEE, K. A. (1967) Statistical Theory and Methodology. John Wiley. N.Y.

BRYAN, J. G. (1951) The Generalized Discriminant Function: Mathematical Foundation and Computational Routine. Harvard Educational Review XXI. pp. 90-05.

CAMPBELL, N. A. (1978) The Influence Function as an Aid in Outlier Detection on Discriminant Analysis. Appl. Statis. 27: pp. 251-258.

CHAKRAVATI, I. M., LAHA, R. G. and ROY, J. (1967) Handbook of Methods of Applied Statistics. I. John Wiley. N.Y.

CLAUDY, J. G. (1972) A Comparison of Five Variable Weighting Procedures. Ed. and Psy. Measurements. 32: pp. 311-322.

COOK, R. and WEISBERG, S. (1982) Residuals and Influence in Regression Chapman and Hall, N.Y.

CORNFIELD, J., GORDON, T. and SMITH, W. (1961) Quantal Response Curves for Experimentally Uncontrolled Varibles. Bulletin of the Int'l. Statis. Institute, 38: pp. 97-115.

CORNFIELD, J. (1962) Joint Dependence of Coronary Heart Disease on Seruum Cholesterol and Systolic Blood Pressure: A Discriminant Function Analysis. Fed. Proc. 21: pp. 58-61.

DANIEL, C. and WOOD, F. (1971) Fitting Equations to Data, Wiley-Interscience. N.Y.

DUNN, O. H. (1971) Some Expected Values of the Probabilities of Correct Classification in Discriminant Analysis. Techno. 13(2): pp. 345-53.

DUNN, O. J. and VARADY, P. D. (1966) Probabilities of Correct Classification in Discriminant Analysis. Biometrics. 22: pp. 908-24.

EINHORN, H. J. and HOGARTH, R. M. (1975) Unit Weighting Schemes for Decision Making. Organizational Behaviour and Human Performance.

13: pp 171-92.

EISENHART, C., HASTAY, M. and WALLIS, W. (1947) Selected Techniques of Statistical Analysis for Scientific and Industrial Research and Production and Management Engineering. Statistical Research Group, Columbia Univ. McGraw-Hill.

ELVEBECK, L. R. and TAYLOR, W. F. (1969) Statistical Methods of Estimating Percentiles. Annals of New York Academy of Sciences. 161: pp 538-48.

FILLIBEN, J. J. (1975) The Probability Plot Correlation Test for Normality. Techno. 17: pp. 111-117.

FISHER, R. A. (1936) The Use of Multiple Measurement in Taxonomic Problems. Annal. Eugen. 7: pp. 148-163.

GEISSER, S. (1964) Posterior Odds for Multivariate Normal Classification. J. of Royal Stat. Soc. 26: Series B. pp. 69-76.

GILBERT, E. S. (1969) The Effect of Unequal Variance-Covariance Matrices on Fisher's Linear Discriminant Function. Biometrics. 25: pp. 505-16.

GNANADESIKAN, R. (1977) Methods for Statistical Data Analysis on Multivariate Observations. John Wiley. N.Y.

GOLDSMITH, M., KOUTCHER, J. and DAMADIAN, R. (1977) Nuclear Magnetic Resonance in Cancer. XIII: Application of NMR Malignancy Index to Human Lung Tumours. Brit. J. Cancer. 36: pp. 235-242.

GREEN, D. M. and SWETS, J. A. (1974) Signal Detection Theory and Psychophysics. Robert E. Krieger Publishing CO., N.Y.

HALD, H. A. (1965) Statistical Theory with Engineering Applications. John Wiley. N.Y.

HALPERIN, M., BLACKWELDER, W. C. and VENTER, J. L. (1971) Estimation of the Multivariate Logistic Risk Function: A Comparison of the Discriminant Function and Maximum Likelihood Approaches. J. Chron. Dis. 24: pp. 125-158.

HAMPEL, F. R. (1974) The Influence Curve and its Role in Robust Estimation. JASA. 69: pp. 383-393.

HARRIS, R. J. (1975) A Primer of Multivariate Statistics. Academic Press. N.Y.

HERBERT, D., GYDESEN, F. and VENZON, D. (1972) The Identification of patients and the Categorization of diseases. Some Applications of Discriminant and Cluster Analyses to Medical Problems. Digest of The Third International Conference on Medical Physics, Including Medical Engineering, Edited by R. Kadefors, R. Magnusson and I. Petersen, Chalmers University of Technology, Goteborg, Sweden, (abstract) Section 43.4.

HERBERT, D. (1986) The Perceived Clinical Value of NMR Measurements on Biopsy Specimens. Part I. Magnetic Resonance Imaging. 4(3)

HILLS, M. (1966) Allocation Rules and Their Error Rates. J. Royal Statis. Soc. B28: pp. 1-31.

HYMANS, S. H. (1980) Some Comments on Papers by Dent and Geweke, Welsch and Kelejian in Evaluation of Econometric Models. Ed. by J. Kmenta and J. B. Ramsey. Academic Press. N. Y. pp. 219-222.

JOHNS, S. (1961) Errors in Discrimination. Annals Math. 32: pp. 1125-1144.

Judgement Under Uncertainty: Heuristics and Biases. (1982) Ed. by D. Kahneman, P. Slovie and A. Tversky. Cambridge Univ. Press.

KIMBALL, S. W. (1957) Errors of the Third Kind in Statistical Consulting. JASA. 57: pp. 133-138.

KLEINBAUM, D. and KUPPER, L. (1978) Applied Regression Analysis and

Other Multivariate Methods. Duxbury Press. North Scituate, MA.
KOUTCHER, J., GOLESMITH, M. and DAMADIAN, R. (1978) A Malignancy Index to Discriminant Normal and Cancerous Tissue. Cancer. 41: pp. 174-182.
KRASKER, W. S. (1980) Some Comments on the Papers by Welsch and Hill in Evaluation of Econometric Models. Ed. by J. Kmenta and J. B. Ramsey. Academic Press. N.Y. pp. 112-116.
LACHENBRUCH, P. A. (1967) An Almost Unbiased Method of Obtaining Confidence Intervals for Probability of Misclassification in Discriminant Analysis. Biometrics. 23: pp. 639-45.
LACHENBRUCH, P. A. (1968) On the Expected Probabilities of Misclassification in Discriminant Analysis Necessary Sample Size and a Relation with the Multiple Correlation Coefficient. Biometrics. 24: pp 823-34.
LACHENBRUCH, P. A. and MICKEY, M. R. (1968) Estimation of Error Rates in Discriminant Analysis. Techno. 10:(1) pp. 1-11.
LACHENBRUCH, P. A. and KUPPER, L. L. (1973) Discriminant Analysis When One Population is a Mixture of Normals. Biomed. J. Bd. 15: pp. 191-97.
LACHENBRUCH, P. A. (1975) Discriminant Analysis. Hafner Press. N.Y.
LACHENBRUCH, P. A. (1976) Some Unsolved Problems in Discriminant Analysis in Decision-Making and Medical Care. Can Information Science Help? Proc. of the IFIP Working Conference on Decision-Making and Medical Care. Ed. by F. T. de Dombal and F. Gremy. North-Holland Pub. Co., Amsterdam. pp 423-31.
LACHENBRUCH, P. A., CLARKE, W. R., BROFFIL, B. and LIN, L. (1977) The Effect of Non-Normality on the Quadratic Discriminant Function. MEDINFO. 77: Shires/Wolf, Eds. IFIP. North-Halland Publ. Co.
LACHENBRUCH, P. A. and BROFF, H. B. (1980) On Classifying Observations When One Population is a Mixture of Normals. Biomed. J. Bd. 22: pp. 295-301.
LACHENBRUCH, P. A. and CLARKE, W. R. (1980) Discriminant Analysis and Its Applications in Epidemiology. Methods of Information in Med. 19: pp. 220-26.
LACHENBRUCH, P. A. (1980) Note on Combining Risks Using the Logistic Discriminant Function Approach. Biom. J. 22: pp. 759-762.
LARK, P. D., CRAVEN, B. R. and BOSWORTH, R. C. (1968) The Handling of Chemical Data. Pergamon Press. N.Y.
LINDEMAN, R. H., MERENDA, P. and GOLD, R. (1980) Introduction to Bivariate and Multivariate Analysis. Scott, Foreman and Co. Glenview, IL. p. 215.
LUBIN, A. (1950) Linear and Non-Linear Discriminating Functions. Brit. Journal of Psychology. Stat. Sec. 3: pp 90-103.
LUSTED, L. (1975) Receiver Operating Characteristic Analysis and Its Significance in Interpretation of Radiologic Images. Current Concepts in Rad. 2: pp. 117-130.
MANSFIELD, P. and MORRIS, P. G. (1982) NMR Imaging in Biomedicine. Supplement 2. Advances in Magnetic Resonance. Academic Press. N.Y.
MARKS, S. and DUNN, O. (1974) Discriminant Functions When the Covariance Matrices are Unequal. JASA. 69: pp. 555-558.
METZ, C., GOODENOUGH, D. and ROSSMAN, K. (1973) Evaluation of Receiver Operating Characteristic Curve Data in Terms of Information Theory, with Applications in Radiography. Radiology. 109: pp. 297-303.
MONTGOMERY, D. and PECK, E. (1982) Introduction to Linear Regression Analysis. John Wiley. N.Y.
MOOD, A. M., GRAYBILL, F. A. and BOES, A. C. (1974) Introduction to

The Theory of Statistics. 3rd Ed. McGraw-Hill. N.Y.
MORRISON, I. (1967) Multivariate Statistical Methods. McGraw-Hill. N.Y.
PAYNE, L. C. (1964) The Role of the Computer in Refining Diagnosis. Lancet. pp. 32-35.
PRESS, S. J. (1972) Applied Multivariate Analysis. Holt, Rinehart and Winston, Inc. N.Y.
PRESS, S. J. and WILSON, S. (1978) Chossing Between Logistic Regression and Discriminant Analysis. JASA 73: pp. 699f.
RAMSEY, J. B. and KEMENTA, J. (1980) Problems and Issues in Evaluating Econometric Models in Evaluation of Econometric Models. Ed. by J. Kmenta and J. B. Ramsey. Academic Press. N.Y. pp. 1-11.
RAO, C. R. (1948) The Utilization of Multiple Measurements in Problems of Biological Classification. J. Royal Statis. Soc. (B) X: pp. 159-193.
RAO, C. R. (1952) Advanced Statistical Methods in Biometric Research. John Wiley. N.Y.
RAO, C. R. (1970) Inference on Discriminant Function Coefficients in Essays on Probability and Statistics. Ed. by E.C. Bose, et al. Chapel Hill Univ. of N. C. and Statistical Publishing Soc. pp. 587-602.
RAO, C. R. (1973) Linear Statistical Inference and Its Applications 2nd Ed. John Wiley. N.Y.
Seber, G.A.F. (1977) Linear Regression Analysis, John Wiley & Sons, N.Y.
SMITH, C. A. B. (1947) Some Examples of Discrimination. Ann. Eugneics. 13: pp. 272-282.
TATSUOKA, M. (1971) Multivariate Analysis: Techniques for Educational and Psychological Research. John Wiley. N.Y.
THEIL, H. (1971) Principles of Econometrics. John Wiley. N.Y.
TIEDEMAN, D. V., BRYAN, J. G. and RULON, P. (1953) The Utility of the Airman Classification Battery and Assignment of Airmen to Eight Airforce Specialties. Cambridge, MA. Educational Research Corp.
TIEDEMAN, D. V. and BRYAN, J. G. (1954) Prediction of College Field of Concentration. Harvard Educational Review XXIV. pp. 122-139.
TRUETT, J., CORNFIELD, J. and KARNEL, W. (1967) A Multivariate Analysis of the Risk of Coronary Heart Disease in Tramingham. J. Chron. Disc. 20: pp. 511-524.
VELLEMAN, P. F. and WELSCH, R. E. (1981) Efficient Computing of Regression Diagnostics. American Statistician. 35(4) 234-242.
WELCH, B. L. (1939) Note on Discriminant Functions. Biometrika. 31: pp. 218-20.
WELSCH, R. E. (1980) Regression Sensitivity Analysis and Bounded-Influence Estimation in Evlauation of Econometric Models. Ed. by J. Kmenta and J. B. Ramsey. Academic Press. N.Y. pp. 153-167.

NOTE ADDED IN PROOF

1) We note that the Koutcher et al (1978) paper has been cited approximately 36 times since its publication.

NONLINEAR REGRESSION ANALYSIS IN NUCLEAR MEDICINE

M. D. Harpen
University of South Alabama, Mobile, AL 36617

ABSTRACT

A review of the nonlinear regression procedures used in the field of nuclear medicine is presented. Special attention is given to the physiological models which are the basis of the functional forms used to fit data obtained in the most common radionuclide procedures. Specifically, three models are discussed in detail: the open-ended two-compartment model for the distribution and elimination of Iodine 131 Orthoiodohippurate (OIH) used in the determination of effective renal plasma flow (ERPF), the indicator dilution model for Technetium 99m labeled red blood cells used to correct the time-activity curves obtained in cardiac flow studies for recirculation, and the model for the antibody saturation curves used in radioimmunoassay.

INTRODUCTION

Nonlinear regression analysis is used in three general areas in nuclear medicine:

  i. Compartmental analysis of tracer distribution is used to determine the effective renal plasma flow (ERPF) and glomerular filtration rates (GFR) in renal studies[1-5], to determine the size of the iodide and pertechnetate traps in thyroid studies [6-7], to quantitate the regional and global ventilation of the lung in Xenon ventilation studies and to quantitate bone uptake and elimination of bone imaging agents [11,12]. In addition to diagnostic applications, compartmental analysis is used in the calculation of internally absorbed dose for practically all radiotracers used in nuclear medicine studies.

 ii. Nonlinear regression analysis has been used in indicator dilution analysis of blood flow [13-15]. Recently similar techniques have been employed in the analysis of blood flow using Technetium labeled compounds to determination of cardiac outputs and cardiac chamber volumes [16-18].

iii. In the in vitro nuclear medicine laboratory, nonlinear regression analysis is applied to antibody saturation curves used in radioimmunoassay [19-20].

What follows is a brief discussion of these three general areas of use of nonlinear regression analysis with some specific examples worked out in detail.

In the nonlinear regression approach, the dependent and independent variables are measured experimentally (X_i, Y_i). A model is proposed which gives a theoretical relation between these variables, $Y=f(X, \alpha_j)$ which contains several unknown parameters α_j. The goal is to determine the values of α_j which minimzes the discrepancies between Y_i and $f(X_i, \alpha_j)$. From the values of α_j the quantities of interest for example the ERPF are calculated.

Before the availability of computers, investigators would restrict themselves to linear relations between X and Y or functions which could be made linear either by a suitable transformation or non linear graph paper, for example log log or probit and then visually fit the best straight line through the data points.

With computers a more general and justifiable approach becomes available. One can specify a fitting parameter to be minimized, eg. the residual sum of the squares (i.e. least squares fit)

$$\Sigma_i (Y_i - f(X_i, \alpha_j))^2$$

or weighted sum of the squares (for example a chi-squared fit)

$$\Sigma_i w_i (Y_i - f(X_i, \alpha_j))^2$$

The function ($f(X_i, \alpha_j)$) is expanded in a Taylor series in α_j about the initial estimate of α_j and quadratic and higher terms are rejected. The fitting parameter then becomes that for a standard linear or multiple linear regression and a new estimate of the alphas are obtained. The function $f(X_i, \alpha_j,)$ is then re-expanded in a Taylor series about the new estimate for alpha and the process is repeated. The iterative procedure continues until successive fitting parameters differ by less than a set amount. Several initial estimates of the α_j may be tried to ensure that the final values are those producing the absolute minimum in the fitting parameter.

ERPF

Measurements of effective renal plasma flow are done routinely at most nuclear medicine facilities. The conventional approach is to administer approximately 300 uCi of Iodine 131 labeled Orthoiodohippurate (OIH) I.V. following the injection, approximately 8 blood samples are obtained at timed intervals. The blood samples are spun down in a centrifuge and the plasma fraction removed and counted in a deep well counter to determine this specific activity of the plasma or the whole blood is counted and the count rates corrected for hematocrit. This specific activity of the plasma is plotted versus time on semi-log paper and a biphasic curve results. Curved stripping is used and two sets of slopes (α_1, α_2) and intercepts (A_1, A_2) are obtained. The ERPF is calculated as

$$\text{ERPF} = I\alpha_1\alpha_2 / (\alpha_1 A_1 + \alpha_2 A_1) \qquad (1)$$

where I is the total injected activity.

Equation 1 results from the compartmental model of OIH distribution given in Fig. 1. Two compartments and the rates of transfer are involved. Volume 1 is the volume into which the injection is made and is interpreted loosely as the vascular space. Volume 2 is a volume that exchanges freely with V_1 and is interpreted loosely as the extravascular space. The rate constant k_3 is the logarithmic rate of removal of tracer from V_1 by renal clearance. The ERPF is given by the product of k_3 and V_1.

The rates of change of the activities in these two compartments are given by

$$d/dt\ A_1 = -(k_1 + k_3)A_1 + k_2 A_2$$

$$d/dt\ A_2 = k_1 A_1 - k_2 A_2$$

These two equations can be written in matrix form as:

$$d/dt \begin{vmatrix} A_1 \\ A_2 \end{vmatrix} = \underline{\Lambda} \begin{vmatrix} A_1 \\ A_2 \end{vmatrix} \qquad (2)$$

where Λ is given by

$$\underline{\Lambda} = \begin{vmatrix} -(k_1 + k_3) & k_2 \\ k_1 & -k_2 \end{vmatrix} \qquad (3)$$

A similarity transformation is found which will diagonalize the matrix

$$\underline{T}\underline{\Lambda}\underline{T}^{-1} = \begin{vmatrix} \gamma_+ & 0 \\ 0 & \gamma_- \end{vmatrix}$$

and uncouple the differential equations. In this case

$$\underline{T} = \begin{vmatrix} -(\alpha - \beta)/2\beta & 1/2\beta \\ (\alpha + \beta)/2\beta & -1/2\beta \end{vmatrix}, \quad \underline{T} = \begin{vmatrix} 1 & 1 \\ \alpha + \beta & \alpha - \beta \end{vmatrix} \qquad (4)$$

where

$$\alpha = (k_1 + k_3 - k_2)/2k_2, \quad \beta = ((k_1 + k_3 - k_2)^2 + 4k_1 k_2)^{\frac{1}{2}} \qquad (5)$$

and

$$\gamma_\pm = -k_2(1 + \alpha) \pm \beta k_2$$

Applying the matrix $\underline{T}$ to both sides of Eq. (2) yields

$$\underline{T}\, d/dt \begin{vmatrix} A_1 \\ A_2 \end{vmatrix} = \underline{T}\underline{\Lambda}\underline{T}^{-1}\, \underline{T} \begin{vmatrix} A_1 \\ A_2 \end{vmatrix} \tag{6}$$

$$d/dt \begin{vmatrix} A_+ \\ A_- \end{vmatrix} = \begin{vmatrix} \gamma_+ & 0 \\ 0 & \gamma_- \end{vmatrix} \begin{vmatrix} A_+ \\ A_- \end{vmatrix}$$

or

$$dA_+/dt = \gamma_+ A_+ \quad , \quad dA_- = \gamma_- A_- \tag{7}$$

with the familiar solutions

$$A_+ = A_{+_o} e^{\gamma_+ t} \quad , \quad A_- = A_{-_o} e^{\gamma_- t} \tag{8}$$

The solutions in terms of A1 and A2 are found by applying the matrix $\underline{T}^{-1}$ to $[A_+, A_-]$

$$\begin{vmatrix} A_1 \\ A_2 \end{vmatrix} = \underline{T}^{-1} \begin{vmatrix} A_+ \\ A_- \end{vmatrix} = \underline{T}^{-1} \begin{vmatrix} A_{+_o} e^{\gamma_+ t} \\ A_{-_o} e^{\gamma_- t} \end{vmatrix}$$

$$A_1 = A_{+_o} e^{\gamma_+ t} + A_{+_o} e^{\gamma_- t} \tag{9}$$

$$A_2 = (\alpha + \beta) A_{+_o} e^{\gamma_+ t} + (\alpha - \beta) A_{-_o} e^{\gamma_- t}$$

At T=0 A_1 is equal to the injected activity and A_2 is zero, thus:

$$A_{+_o} + A_{-_o} = I$$
$$(\alpha + \beta) A_{+_o} + (\alpha - \beta) A_{-_o} = 0 \tag{10}$$

FIG. 1. Schematic diagram of compartmental model for distribution and elimination of OIH.

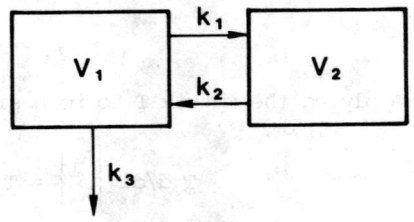

FIG. 2.

```
10      SUB ERPF (X(), Z(), K(), N)
20      REM X() BLOOD SAMPE TIMES
30      REM Z() FITTED VALUES OF SPECIFIC ACT.
40      REM K(1,2,3) COMPARTMENTAL RATE CONT.
50      REM K(0) SPECIFIC ACT AT T=0
60      REM N NUMBER OF BLOOD SAMPLES
70      REM M ITERATIONS BETWEEN SAMILE TIMES
80      REM W() ACT OF EXTRAVASCULAR SPACE
90      DIM W(100)
100     M = 10
110     Z(0) = K(0)
120     W(0) = 0
130     FOR I = 1 TO N
140     P = Z(I-1)
150     O = W(I-1)
160     FOR J = 1 TO M
170     P = P+ (P*(-K(1)-K(3)) + O*K(2) * (X(I)-X(I-1)/M
180     O = O + (P*(K(1) - O*K(2) * (X(I) - X(I-1)/M
190     NEXT J
200     Z(I) = P
210     W (I) = O
220     NEXT I
230     SUB END
```

Using Eqs. (5) and (10) and after a little algebra, the compartmental rate constant k_1, k_2 and k_3 can be expressed in terms of the parameters $(A_+, A_-, \gamma_+, \gamma_-)$.

$$k_1 = ((A_+^0 - A_-^0)/(A_+^0 + A_-^0) - (\gamma_+ + \gamma_-)/(\gamma_+ + \gamma_-))^{-1} \cdot$$
$$(1 - (A_+^0 - A_-^0)^2/(A_+^0 + A_-^0)^2) \, (\gamma_+ + \gamma_-)/2 \tag{11}$$

$$k_2 = ((A_+^0 - A_-^0)/(A_+^0 + A_-^0) - (\gamma_+ + \gamma_-)/(\gamma_+ + \gamma_-))(\gamma_+ + \gamma_-)/2 \tag{12}$$

$$k_3 = -I\gamma_+\gamma_-/(A_+^0\gamma_- + A_-^0\gamma_+) \tag{13}$$

When the parameters are obtained by fitting the two exponential function to the plasma specific activity-time curve, the above expression for k_3 has the dimensions of volume per time and represents the ERPF.

In the proceeding development, compartmental analysis was used to determine the functional form of the specific activity of the plasma. This function had four parameters which are adjusted to obtain the best fit for an individual data set. The ERPF is then calculated from the values of the four parameters. With the general availability of digital computers, a nonlinear regression analysis may be applied to the compartmental rate constants directly. Figure 2 lists a sub-program written in basic which returns the calculated values of A_1 for a given set of values of the rate constants k. When this sub-program is chained to a nonlinear regression program and data file for a specific patient, the rate constant k_3 and hence the ERPF may be determined. This is the general approach used in many popular compartmental analysis programs.[21]

CARDIAC OUTPUTS

The radionuclide determination of cardiac outputs involves an IV bolus injection of Technetium labeled red blood cells (Tc-RBC), followed by the acquisition of a dynamic scan of the passage of the bolus of activity through the left ventricle. Subsequent to the data acquisition, a time-activity curve for the left ventricle is constructed. Computer analysis of this curve along with a knowledge of the total injected activity as well as the specific activity of the whole blood after complete mixing of tracer with the blood pool is sufficient to determine the cardiac output.

The time-activity curves of the bolus transit, before recirculation of tracer, have been known to be described by the functional form known as the gamma variate[22-24].

$$t^\alpha e^{-t/\beta} / (\beta^{(\alpha+1)} \Gamma(\alpha+1)) \qquad (14)$$

Models for blood or tracer flow have been proposed to explain this functional form. Davenport[25] considered arterial or venous blood flow to be described by a series of dilution chambers with a constant flow rate (Q) from one chamber to the next. The time-activity curve for the ith chamber following a unit of activity introduced at t = 0 into the first chamber is given by

$$C_n(t) = C_o/(n-1)! \ (Qt/V)^{n-1} e^{-Qt/V} \qquad (15)$$

which is a form of Eq. (14).

An alternative approach is to consider the transport of tracer as being described by convective dispersion [26,27]. The one dimensional differential equation governing this phenomenon is

$$\frac{\partial}{\partial t} C(x,t) + \nu \frac{\partial}{\partial t} C(x,t) - \mu \frac{\partial^2}{\partial x^2} C(x,t) = 0 \qquad (16)$$

where ν is the averaged velocity in the vessel, μ is the coefficient of longitudinal dispersion and x is the distance along the vessel measured from the point of injection. If the boundary conditions are made such that c(x,o) is $\delta(x)$ then the solution to Eq. (8) is

$$C(x,t) = \frac{1}{2A\sqrt{\pi\mu}^{-\frac{1}{2}}} t^{-\frac{1}{2}} e^{-(x-\nu t)^2/(4\mu t)} \qquad (17)$$

For times on the order of the arrival time of the bolus (t∿x/v) the above expression may be written

$$C(t) = \text{cont.} \ t^\alpha e^{-t/\beta} \qquad (18)$$

where

$$\alpha = \frac{\nu x}{4\mu} - \frac{1}{2} \quad , \quad \beta = \frac{4\mu}{\nu^2}$$

If C(t) is the instantaneous activity of the left ventricle at time t then the rate of flow of activity out of the left ventricle at time t is kC(t) where k is the heart rate (HR) times the ejection fraction (EF). Assuming that all injected activity passes through the left ventricle, the total activity may be expressed as

$$A = k \int_0^\infty C(t) \, dt \qquad (19)$$

where C(t) is the left ventricular activity corrected for recirculation (Fig. 3). After complete mixing of the tracer with the blood pool, the equilibrium count rate observed over the left ventricle (C_∞) is A times the ratio of the left ventricular volume to the total blood volume, i.e.

$$V_V = V_{total} \, C_\infty / A \qquad (20)$$

Cardiac output is left ventricular volume times the product of heart rate and ejection fraction. Hence

$$\text{CARDIAC OUTPUT} = V_V (HR)(EF) = V_{total} \, HR \, EF \, C_\infty / A$$

$$= V_{total} \, C_\infty / \int_0^\infty C(t) \, dt \qquad (21)$$

V total is determined by dividing the total injected activity by the specific activity of the whole blood after mixing of tracer with the blood pool.

FIG. 3. Time activity curve for passage of bolus of labeled red cells through right ventricle.

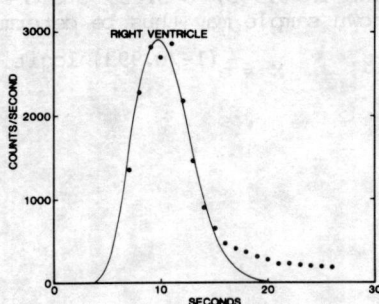

RADIOIMMUNOASSAY

Minute concentrations of a substance (ligand) are measured by performing a competitive binding study using a known concentration of antibody to the ligand and a known concentration of radiolabeled ligand. The uptake of radiolabeled ligand by the antibody is related to the unknown concentration of ligand.

The standard protocol for most radioimmunoassay is similar to that of the prolactin assay. Standard solutions of known concentration of prolactin (0,5,10,20,50,100,200 ng/ml) are incubated in antiprolactin--antibody coated tubes with tracer amounts of Iodine 125 labeled prolactin. Afterwards, the solution is poured out and the uptake of radiolabeled prolactin by the antibodies is measured. Assuming that for a given assay all binding sites of the antibody are saturated then the fraction of binding sites occupied by radio-labeled ligand (B/Bo) may be expressed as [28]

$$B/B_o = q^*/(q + q^*) \tag{22}$$

where B is the antibody uptake of radiolabeled ligand for a given assay, Bo is the uptake for the zero standard and q^* and q are the concentrations of labeled and unlabeled ligand. Eq.(27) may be rewritten as

$$\ln(B/B_o / (1 - B/B_o)) = -\ln q + \ln q^* \tag{23}$$

the log (B/Bo/(1-B/Bo)) term is written as logit (B/Bo). Thus a logit-log plot of the binding data should yield a straight line with a slope approximately equal to -1.

$$\text{logit}(B/Bo) = m \ln q + b \tag{24}$$

Figure 4 lists the recovered counts for a prolactin assay done recently at our institution. Figure 5 is the corresponding logit-log data. Linear regression analysis applied to the logit-log data yields m=-0.993, b=3.958 and r=-0.999. The concentration of an unknown sample may thus be determined by evaluation of

$$X = e^{(1-/0.993) \text{ logit }(B/Bo) + 3.958/0.993} \tag{25}$$

FIG. 4. Recovered counts for a prolactin assay (a) and corresponding logit-log data (b).

B	q	logit (B/Bo)	ln q
17651	0		
16014	5	2.2807	1.610
15060	10	1.7601	2.303
12801	20	0.9705	2.996
9406	50	0.1317	3.912
6074	100	−0.6450	4.605
3712	200	−1.3231	5.298

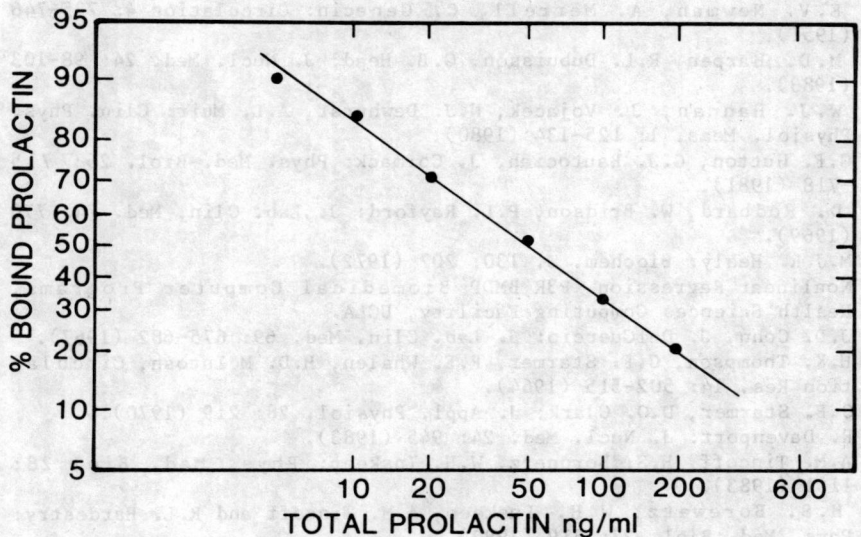

FIG. 5. Plot of B/Bo vs. Total prolactin on logit-log paper.

REFERENCES

1. W.N. Tauxe, F.T. Maher, W.F. Taylor: Mayo Clin. Proc. 46: 524-531 (1971).
2. W.N. Tauxe, E.V. Dubovsky, T. Kidd, F. Diaz: Eur. J. Nucl. Med. (1981).
3. W.N. Tauxe, M.K. Burbank, F.T. Maher, et al: Mayo Clin.Proc. 39: 761-766 (1964).
4. M.D. Blaufox, A. Cohen: Amer. J. Physiol. 218: 542 (1970).
5. C.M.E. Matthews: Phys. Med. Biol. 2: 36-53 (1957).
6. W.N.P. Lee, J.A. Siegel, M.D. Harpen, M.A. Greenfield: J. Clin. Endo. and Metab. 55(6): 1131-1137 (1982).
7. W.N.P. Lee, P. Mpanias, J.R. Wimmer, M.A. Greenfield, S.A. Kaplan: J. Nucl. Med. 19: 85 (1978).
8. M.T. Hays: J. Nucl. Med. 19: 789 (1978).
9. T.W. Van der Mark, R. Peset, H. Beekhuis, A. Kiers, A.E.C. Rookmaker, M.G. Waldring: J. Nucl. Med. 21: 1029 (1980).
10. H. Susskind, H.L. Atkins, S.H. Cohn: J. Nucl. Med. 18: 462 (1971).
11. N.D. Charkes, P.T. Makler Jr., C. Phillips: J. Nucl. Med. 19: 1301-1309 (1978).
12. N.D. Charkes, M. Brookes, P.T. Makler Jr.: J. Nucl. Med. 20: 1150-1157 (1979).
13. Y. Ishii, W.J. Mcintyre: Circulation 44: 37-46 (1971).
14. D.L. Kirch, C.E. Metz, P.P. Steele: Am. J. Cardiology 34: 711 (1974).
15. E.V. Newman, A. Merrell, C. Genecin: Circulation 4: 735-746 (1951).
16. M.D. Harpen, R.L. Dubuisson, G.B. Head: J. Nucl. Med. 24: 98-103 (1983).
17. W.J. Hannan, J. Vojacek, N.J. Dewhurst, A.L. Muir: Clin. Phys. Physiol. Meas. 1: 125-134 (1980).
18. G.F. Hutton, G.J. Bautociah, J. Cormack: Phys. Med. Biol. 26: 715-718 (1981).
19. D. Rodbard, W. Bridson, P.L. Rayford: J. Lab. Clin. Med. 74: 770 (1969).
20. M.J.R. Healy: Biochem. J. 130: 207 (1972).
21. Nonlinear Regression, P3R BMDP Biomedical Computer Programs, Health Sciences Computing Facility, UCLA.
22. J.D. Cohn, J. DelGuercio: J. Lab. Clin. Med. 69: 675-682 (1967).
23. H.K. Thompson, C.F. Starmer, R.E. Whalen, H.D. McIntosh, Circulation Res. 14: 502-515 (1964).
24. C.F. Starmer, D.O. Clark: J. Appl. Physiol. 28: 219 (1970).
25. R. Davenport: J. Nucl. Med. 24: 945 (1983).
26. A.M. Tincoff, H.S. Boronetz, W.H. Inskeep: Phys. Med. Biol 28: 1191 (1983).
27. H.S. Borewetz, W.H. Inskeep, A.M. Tincoff and R.L. Hardestry: Phys. Med. Biol. 27: 819 (1982).
28. E.P. Leonard and S.B. Jorgensen: Ann. Rev. Bioery 3: 293 (1974).

REGRESSION MODELS AND ACTUARIAL CURVES. I AND II

Richard S. Cox
Stanford University Medical Center, Stanford, CA. 94305

ABSTRACT

In the health sciences, survival analysis of clinical data is often required where the survival function is dependent upon a large number of explanatory variables. Recently, a powerful technique, known as the proportional hazards model, has been developed which allows the application of multivariate regression techniques to survival analysis. In the first section of this paper, an introduction to the mathematical definitions and techniques used in survival analysis are provided. In section two, the mathematical basis of the proportional hazards model is set forth. In section three, an example is given of the application of the proportional hazards model to a clinical study. Finally, in section four, some limitations and difficulties associated with the application of the model are discussed.

I. INTRODUCTION

In the health sciences and particularly in the study and treatment of cancer, actuarial analysis remains the principal method for quantitatively assessing the natural history of disease and the efficacy of treatment modalities. Typically, the value of a positive random variable, the time to failure, is observed along with associated explanatory variables or covariates. The "failure" point may be death, relapse, progression of disease or some other well-defined end point. The covariates may describe the condition of the patients at presentation (sex, age, extent of disease, blood counts, histology, etc.) or specify the treatment regimen (extent of radiotherapy, extent of chemotherapy, etc.). Under the assumption that the observations for each patient are independent, the actuarial problem consists of estimating the underlying failure-time distribution which gave rise to the observations and describing its dependence on the covariates.

In clinical investigations, the actuarial problem is usually complicated by the fact that observations are censored, i.e., that some patients may not have failed at the close of the study. Censored observations contain only partial information in that the failure time is greater than the time of observation, however, even partial information can improve our knowledge of the failure-time distribution. In medical applications, the type of censoring that arises is "random;" i.e., after initiation of the clinical study, patients are entered at different times and may be censored:
1) at the end of the study if the patient has not yet failed.
2) during the study due to loss to follow-up or some medical reason that requires the discontinuation of their participation in the study.

Fig. 1.1 illustrates the various possible results for the observation; some type 1 censoring is expected in almost all studies, even those with good follow-up. With random censoring, it is necessary to assume that the time to failure and time to censoring are independent in order to derive useful results. For most studies, this assumption is probably justified, however, when there is a large amount of type 2 censoring, it may not be.

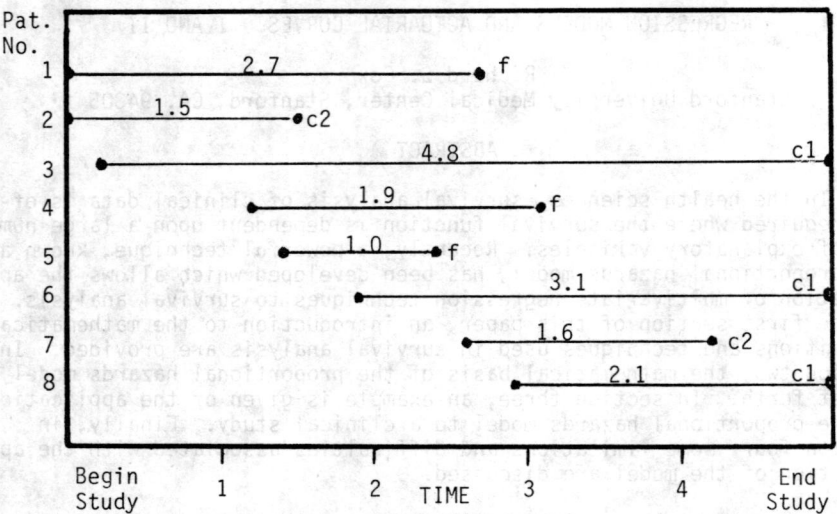

Figure 1.1. Hypothetical clinical study of eight patients over a five year time span showing the time of entry and outcome (f = failure, c1 = type 1 censoring, c2 = type 2 censoring).

Mathematically, the actuarial problem is defined by specifying the random variable $T \geq 0$ with density function $f(t;\vec{x})$ and distribution function $F(t;\vec{x})$. The second independent variable x represents the values of the covariates and may be thought of as an n-dimensional vector if n covariates are present. The survival function $S(t;\vec{x})$ for the patient group with covariate values $\vec{x}$ is the probability that the random variable T exceeds t,

$$S(t;\vec{x}) = \text{Prob}\{T > t; \vec{x}\} = 1 - F(t;\vec{x}). \quad (1.1)$$

The hazard rate $h(t;\vec{x})$ is defined as the conditional probability of failing in the time interval (t, t+dt) having survived past t

$$h(t;\vec{x}) = \text{Prob}\{t < T < t+dt \mid T > t; \vec{x}\}$$
$$= f(t;\vec{x}) / S(t;\vec{x}). \quad (1.2)$$

By integrating the hazard function, it can be shown that

$$S(t;\vec{x}) = \text{EXP}\left\{-\int_0^t h(u;\vec{x}) \, du\right\}, \quad (1.3)$$

where EXP is the exponential function. We have assumed that $f(t;\vec{x})$ and $F(t;\vec{x})$ are continuous, however, these concepts can be extended to the discrete case where $F(t;\vec{x})$ is discontinuous.

Methods used to estimate $S(t;\vec{x})$ fall into two broad categories, parametric and non-parametric. In the parametric method, a functional form for the time dependence of $S(t;\vec{x})$ that depends on the value of the regressor variable. One example is the exponential model:

$$S(t;\vec{x}) = \text{EXP}\{-\lambda t \, \text{EXP}(\vec{\beta}\cdot\vec{x})\} \quad (1.4)$$

where the components of $\vec{\beta}$ are the regression coefficients corresponding to the components of $\vec{x}$. The problem then consists of finding the values of $\vec{\beta}$ which best explain the observed failure-time data. In clinical studies, non-parametric estimation of $S(t;\vec{x})$ is more commonly used. Here, no underlying functional forms are assumed for $S(t;\vec{x})$, and must also be determined directly from the data.

For the description of the non-parametric estimators of $S(t;\vec{x})$, let us ignore for temporarily the dependence on the covariates, $\vec{x}$. The crudest non-parametric estimate of $S(t)$ is found by including only those patients, $s(t)$, who have survived for a time $T \geq t$ or those patients, $d(t)$, who have failed in a time $T \leq t$. This estimate, $\hat{S}_{rs}(t)$, called the reduced sample estimate, is given by

$$\hat{S}_{rs}(t) = s(t) / [s(t) + d(t)] = [n(t) - d(t)] / n(t), \quad (1.5)$$

where $\quad n(t) = s(t) + d(t) = $ the risk set at time t. $\quad (1.6)$

Clearly, the $\hat{S}_{rs}$ ignores any information provided by the patients censored before time t and, hence, is biased downward, however this estimator is still being used in the literature today. A better estimate of $S(t)$ is provided by utilizing an "actuarial" method[1] in which the overall survival probability is broken up into a product of k probabilities corresponding to k equal time intervals:

$$\hat{S}_a(t_k) = \hat{P}\{T>t_1\} \cdot \hat{P}\{T>t_2|T>t_1\} \cdot \ldots \cdot \hat{P}\{T>t|T>t_{k-1}\}. \quad (1.7)$$

The individual terms of the product, e.g., $\hat{P}\{T>t_2|T>t_1;\vec{x}\}$, are computed as follows:

Let $s(t_2)$ = all patients surviving for time $T > t_2$
$c(t_2)$ = all patients censored during time $t_1 \leq T < t_2$
$d(t_2)$ = all patients failing during time $t_1 < T \leq t_2$.

The risk set for time t_2 is defined as

$$n(t_2) = s(t_2) + d(t_2) + 0.5\, c(t_2), \quad (1.8)$$

and $\quad \hat{P}\{T>t_2|T>t_1\} = [\,n(t_2) - d(t_2)\,] / n(t_2). \quad (1.9)$

Thus, breaking up the total time interval into a series of shorter intervals allows the use of data on patients who were under observation for times less than the full term of the study and who would have been eliminated in the reduced sample calculation. The actuarial estimate of $S(t)$ contains less bias and is an improvement over the reduced sample estimate.

In 1958, Kaplan and Meier[2] derived the generalized maximum likelihood or product-limit estimator $\hat{S}_{pl}(t)$ of the survival function. $\hat{S}_{pl}(t)$ is similar to $\hat{S}_a(t)$ in that the total time interval is broken up into subintervals, however, here the t_k are variable and corresponding to the actual ordered observation times of the n patients. The calculation of

$$\hat{P}_i = \hat{P}\{T>t_i|T>t_{i-1}\} \quad (1.10)$$

goes as follows:

$n(t_i)$ = the risk set for time t_i = $s(t_i) + d(t_i)$
$d(t_i)$ = number failing at time t_i.
$\hat{P}_i$ = 1 for a failure or 0 for a censoring

$$\hat{p}_i = [n(t_i) - d(t_i)] / n(t_i) \qquad (1.11)$$

$$\hat{S}_{pl}(t) = \prod_{i=1}^{n} \hat{p}_i = \prod_{i \in U} \hat{p}_i . \qquad (1.12)$$

where U is the set of uncensored observations (since $\hat{p}_i = 1$ for a censored observation). These formulae assume that no ties occur in the time ordering, which is usually a good assumption in clinical studies with limited numbers of patients and accurate time data (to the day). For m tied uncensored observations it can be shown that the above formula for $\hat{p}_i$ holds with $d(t_i) = m$. For tied censored and uncensored observations, the uncensored observations are taken as having occurred first. If the last observation at time t_f is censored, $\hat{S}_{pl}(t_f)$ is > 0 and for $t > t_f$ it is best to think of $\hat{S}_{pl}(t)$ as undefined because there are no observations beyond this point. Since the product-limit estimator makes the maximum use of the time and censoring data over all time periods, it contains the least bias. A comparison of the estimators for $S(t)$ for the data of Fig. 1.1 is provided in Table 1.1 and Fig. 1.2. Although these calculations are tedious to carry out by hand for large clinical studies with many failures, a product-limit algorithm is easily coded for a computer and $\hat{S}_{pl}$ should be used whenever the results of clinical studies are being described.

Pat Num	Time	Outcome	Risk set	Time	$\hat{S}_{rs}$	$\hat{S}_a$	Time	$\hat{S}_{pl}$
5	1.0	f	8				1.0	0.88
2	1.5	c	7	1.0	0.88	0.88		
7	1.6	c	6					
4	1.9	f	5	2.0	0.67	0.73	1.9	0.70
8	2.1	c	4					
1	2.7	f	3	3.0	4.0	0.52	2.7	0.47
6	3.1	c	2					
3	4.8	c	1	4.0	0.25	0.52	4.8	0.47

Table 1.1. Comparison of the reduced sample, $\hat{S}_{rs}$, actuarial, $\hat{S}_a$, and product limit estimates of the survival function for the hypothetical clinical study diagrammed in Fig. 1.1.

When one or more covariates are present, an assessment of the significance of differences in $S(t;\vec{x})$ for different values of $\vec{x}$ is usually required. The various methods for making this assessment proceed roughly as follows. Assume that the null hypothesis, H0, holds (i.e., that there is no difference between the various $S(t;\vec{x})$). For this patient data, calculate a statistic that is a measure of the difference divided by the standard deviation of this statistic. If this estimator is normally distributed, we may then assign a p-value corresponding to the observed difference. If the p-value is below some arbitrary value (typically 0.05 for clinical studies), H0 is rejected and the difference is considered to arise because of the effect of the covariate rather than random chance.

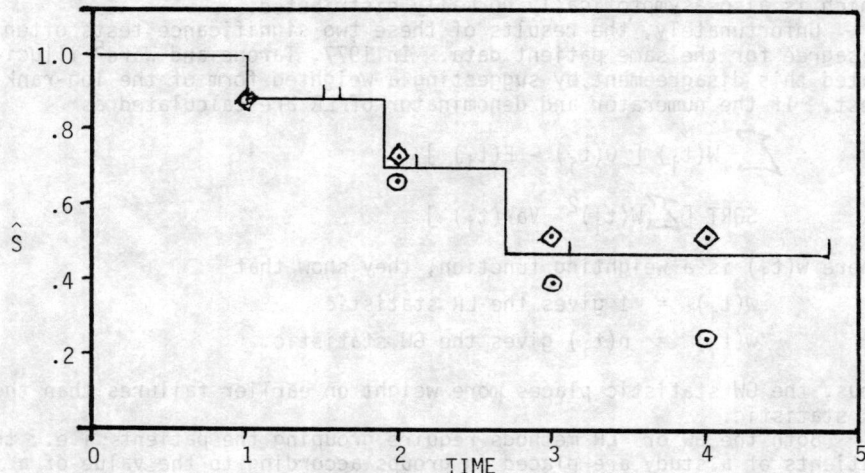

Figure 2.1. Plots of $\hat{S}(t)$ for the clinical study diagrammed in Figure 1.1 ($\hat{S}_{rs}$ = circles, $\hat{S}_a$ = diamonds, $\hat{S}_{pl}$ = broken line). The tick marks on $\hat{S}_{pl}$ indicate censorings and the drops indicate failures.

The first non-parametric tests developed were two-sample methods which were later generalized to k-samples (k > 2). In 1959, the log-rank test was introduced by Mantel and Haenszel.[3] Their method consists of time ordering the observations in the two patient groups (A and B) and constructing a 2 x 2 contingency table at each failure. From the ratio of patients remaining in the risk set at the time of each failure, the expectation ($E(t_i)$) that the failure at time t_i occurred in group A and the corresponding variance $Var(t_i)$ is computed. The log-rank statistic is taken as

$$LR = \sum_{i \in U} [O(t_i) - E(t_i)] / SQRT \sum_{i \in U} [Var(t_i)] \quad (1.13)$$

where $O(t_i) = 1$ if the failure occurred in group A and 0 otherwise. The sums extend over the failures. The LR statistic is asymptotically normally distributed so a P-value can be assigned. In 1965, Gehan[4] introduced a generalization of the Wilcoxon test appropriate to survival data containing censored observations. In this generalization, a number, N, is calculated by the following rules:

1. If patient i in group A is observed for a longer time than patient j in group B and patient j is a failure, add one to N.
2. If patient i in group A is a failure and is observed for a shorter time than patient j in in group B, subtract one from U.
3. Otherwise, leave N unchanged.

Under H0, the expected value of N is zero. Gehan then calculates the variance of N for the patient groups and forms the statistic

$$GW = N / SQRT[Var(N)] \quad (1.14)$$

which is also asymptotically normally distributed.

Unfortunately, the results of these two significance tests often disagree for the same patient data. In 1977, Tarone and Ware[5] illucidated this disagreement by suggesting a weighted form of the log-rank test. If the numerator and denominator of LR are calculated as

$$\sum W(t_i) [O(t_i) - E(t_i)]$$

$$SQRT [\sum W(t_i)^2 Var(t_i)]$$

where $w(t_i)$ is a weighting function, they show that

$w(t_i)$ = 1 gives the LR statistic

$w(t_i)$ = $n(t_i)$ gives the GW statistic.

Thus, the GW statistic places more weight on earlier failures than the LR statistic.

Both the GW or LR methods require grouping the patients; i.e., the patients of a study are placed in groups according to the value of a particular covariate and the test statistic determined between the groups. For many clinical studies these methods have been extremely useful in identifying the covariates correlated with failure; however, the need to split the patient population limits their ability to detect correlation between covariates. A covariate unrelated to survival may appear to influence survival if it is correlated with another covariate that does. If this correlation remains undetected, an incorrect inference may be drawn. Correlations among the covariates may be investigated by grouping the patients according to one covariate, then subgrouping the groups according to the second, etc., and applying the significance tests between appropriate subgroups. Clinical studies often have a relatively small number of patients for whom a large number of covariates may be specified. As the available patients are spread across larger numbers of subgroups, the accuracy with which differences in survival functions can be determined decreases rapidly.

II. THE COX PROPORTIONAL HAZARDS MODEL

In 1972, the British Statistician, D. R. Cox,[6] developed a very general model which allows the application of multivariate regression techniques to the actuarial problem. In this formulation, the hazard function is factored into a baseline hazard function, $h'(t)$, times a second function, $C(\vec{x},t)$, which carries the covariate information. Note that in the most general formulation of the model, the function C may carry time dependent parts as well. An important special case of this model is the so-called "proportional hazards" model which is the subject of this paper. In the proportional hazards model the hazard function is simplified to:

$$h(t,\vec{x}) = h'(t) EXP\{ \vec{\beta} \cdot \vec{x} \} \qquad (2.1)$$

where $h'(t)$ is the arbitrary baseline hazard function and $\vec{\beta}$ is the regression coefficients associated with the covariates represented by $\vec{x}$.

If we consider the special case of one covariate with two values (e.g., sex: x=0 for females, x=1 for males)

$$h(t,0) = h'(t) \, \text{EXP}\{ \beta \cdot 0 \} = h'(t)$$
$$h(t,1) = h'(t) \, \text{EXP}\{ \beta \cdot 1 \} = g \, [\, h'(t) \,]$$

where $g = \text{EXP}\{ \beta \}$

and, thus, the ratio of the hazard functions at one time is proportional to the ratio of the hazard functions at any other time. The constant g is called the relative risk. Defining the baseline survival function $S'(t)$:

$$S'(t) = \text{EXP}\{ - \int_0^t h'(u) \, du \} \qquad (2.2)$$

then $\quad S(t,0) = S'(t) \quad$ and $\quad S(t,1) = [\, S'(t) \,]^g \qquad (2.3)$

Initially, the baseline hazard function and regression coefficients are unknown and the actuarial problem involves estimating them. Since no specific functional form is assumed for the hazard function, the Cox method is often referred to as "non-parametric" even though a modeling process is employed. It would seem that the term "partially-parametric" is actually a better description of this method.

In fact, we are usually more concerned with determining the relative importance of the regression coefficients than we are with the exact baseline hazard function. To estimate the regression coefficients, Cox employs a maximum likelihood technique, analogous to the procedures used in parametric modeling. Let us assume that we have made observations $(t_k, \vec{x}_k)$ on n patients (k = 1 to n) where the variable t carries the time and censoring information and the variable $\vec{x}$ contains the covariate information. The full likelihood is just the conditional probability that the sequence of n events $(t_1, \vec{x}_1, \ldots\ldots t_n, \vec{x}_n)$ will occur given the values of the regression coefficients. Thus,

$$L_f(\vec{\beta}) = P\{t_1, \vec{x}_1, \ldots\ldots t_n, \vec{x}_n \mid \vec{\beta}\} \qquad (2.4)$$

Using the well-known theorem from probability theory

$$P(A,B) = P(A) \cdot P(B|A)$$

we may express $L_f(\vec{\beta})$ as a product of conditional probabilities.

$$L_f(\vec{\beta}) = \prod_{i=1}^{n} P\{t_i, \vec{x}_i | t_1, \vec{x}_1, \ldots, t_{i-1}, \vec{x}_{i-1}; \vec{\beta}\}$$

$$= \prod_{i=1}^{n} P\{t_i | t_1, \vec{x}_1, \ldots, t_{i-1}, \vec{x}_{i-1}; \vec{\beta}\} \cdot$$

$$\prod_{i=1}^{n} P\{\vec{x}_i | t_1, \vec{x}_1, \ldots, t_{i-1}, \vec{x}_{i-1}, t_i; \vec{\beta}\} \qquad (2.5)$$

The second term in the product, $L_p(\vec{\beta})$

$$L_p(\vec{\beta}) = \prod_{i=1}^{n} P\{\vec{x}_i | t_1, \vec{x}_1, \ldots, t_{i-1}, \vec{x}_{i-1}, t_i; \vec{\beta}\} \qquad (2.6)$$

Cox calls a conditional likelihood and this term is expected to carry most of the information about the regression coefficients. He continues by arguing if the baseline hazard function is truly arbitrary, only those instances at which a failure has has occurred can contribute to a knowledge of $\vec{\beta}$, so the product can be limited to the set of uncensored observations U. If there are no ties, the treatment is mathematically straight forward. As a practical matter, in most clinical studies where time intervals can be measured in days and patient numbers relatively small, the occurrence of ties is a rarity and will not be dealt with here. Thus,

$$L_p(\vec{\beta}) = \prod_{i \in U} P\{\vec{x}_i | t_i, n(t_i); \vec{\beta}\} \tag{2.7}$$

where $n(t_i)$ is, as above, the risk set of patients at t_i as defined in equation (2.6). Consider the conditional probability $P\{\vec{x}_i | t_i, n(t_i); \vec{\beta}\}$, given that a patient, i, of the risk set $n(t_i)$ fails at time t_i, what are the chances he will have the covariate values $\vec{x}_i$

$$P\{x_i | t_i, n(t_i); \vec{\beta}\} = EXP\{\vec{\beta} \cdot \vec{x}_i\} h'(t_i) \Delta t / \sum_{j \in n(t_i)} EXP\{\vec{\beta} \cdot \vec{x}_j\} h'(t_i) \Delta t$$

$$= EXP\{\vec{\beta} \cdot \vec{x}_i\} / \sum_{j \in n(t_i)} EXP\{\vec{\beta} \cdot \vec{x}_j\} \tag{2.8}$$

Cox then constructs the partial likelihood function $L_p(\vec{\beta})$ by taking the product of these conditional probabilities for all uncensored observations.

$$L_p(\vec{\beta}) = \prod_{i \in U} EXP\{\vec{\beta} \cdot \vec{x}_i\} / \sum_{j \in n(t_i)} EXP\{\vec{\beta} \cdot \vec{x}_j\} \tag{2.9}$$

He then treats L_p as an ordinary likelihood and proceeds to find the $\vec{\beta}$ which maximizes log $[L_p(\vec{\beta})]$, or $LL_p(\vec{\beta})$, by means of the Newton-Raphson method in the usual way. Efron[7] and Oakes[8] have found that the information carried in L_p is very high, usually greater than 90% of that carried in the full likelihood for a variety of models, in agreement with Cox's original assertion.

Following the Newton-Raphson method, we wish to solve the simultaneous equations

$$\vec{D}(\vec{\beta}) = \frac{\partial}{\partial \vec{\beta}} LL_p(\vec{\beta}) = 0 \tag{2.10}$$

This is done by an iterative method in which we assume an initial value for $\hat{\beta}^0 = \vec{0}$ and

$$\hat{\beta}^1 = \hat{\beta}^0 + \vec{I}^{-1}(\hat{\beta}^0) \cdot \vec{D}(\hat{\beta}^0) \tag{2.11}$$

where $\vec{I}(\vec{\beta}) = -\frac{\partial^2}{\partial \vec{\beta} \partial \vec{\beta}} LL_p(\vec{\beta})$ (2.12)

which is the familiar information matrix. At Stanford, we have performed these calculations first on a PDP-11/45 and more recently, on a

VAX-11/750. In our implementation, the iteration process runs until no component of $\vec{\beta}$ is changed by more than 1% of its current value or until nine iterations have been completed. Generally, four to six iterations are sufficient to find the solution. The total time required is under one minute per model on either computer.

As with the parametric methods, there are several procedures available for hypothesis testing or constructing confidence intervals. Cox asserts and Tsiatis[9] later proves that the estimates $\hat{\beta}_k$, are normally distributed about the true value with variance i^{-1}_{kk}. Thus, each component of $\vec{\beta}$ can be tested under the assumption that all other components are held fixed at their "best" values.

$$d\hat{\beta}_k = \text{SQRT}\{i^{-1}_{kk}(\vec{\beta})\} \tag{2.13}$$

where $d\hat{\beta}_k$ is the estimated standard error of $\hat{\beta}_k$. The normal variate, Z, is constructed as

$$Z = \hat{\beta}_k / d\hat{\beta}_k \tag{2.14}$$

and a P-value found in the usual way. Cox also suggests the use of a Rao statistic which he calls the "global null hypothesis" (HO: $\vec{\beta} = 0$). Under HO, this statistic

$$\text{GNH} = D(\vec{\beta}=0) \; \vec{i}^{-1}(\beta=0) \; \vec{D}(\vec{\beta}=0) \tag{2.15}$$

is asymptotically distributed as chi-squared with n degrees of freedom, where n is the dimensionality of $\vec{\beta}$. Notice that this method uses only the initial guess ($\vec{\beta} = 0$) and, hence, is relatively simple to calculate. In the special case where n = 1, this test reduces to the Mantel-Haentzel test, so we can expect the Cox method to give equal weight to all failures. In our computer program, we also calculate a Neyman-Pearson/Wilks[10] likelihood ratio statistic; under HO, this statistic

$$\text{WLR} = 2 \, [\, LL_p(\vec{\beta}) - LL_p(0) \,] \tag{2.16}$$

is also distributed as chi-squared with n degrees of freedom.

III. AN APPLICATION OF THE COX PROPORTIONAL HAZARDS MODEL

Recently, a retrospective study of stage I and II large cell non-Hodgkin's lymphoma patients who received radiation at Stanford University between 1968 and 1980 was carried out. A search of the tumor registry produced a total of 199 patients as candidates for this study, however, in 47 cases large cell histology was not confirmed and 4 others did not receive radiation. A summary of the selection of the final 148 patients is given in Table 3.1. For this study, both time to death and time to relapse were used as end points, however, only the study of relapse will be presented as an example here.

Eighteen covariates were selected for investigation. A listing of these covariates and the breakdown of the 148 patients among each is given in Table 3.2. Clearly, as more covariates are studied more data is required to support any conclusions drawn from the fitting procedure. For the proportional hazards model, there is no formula from which to calculate the number of failures, f, required to support a

Patients seen between 1968 and 1980		199
No path slides available	23	
Wrong histology on review	24	
Received no radiotherapy	4	
TOTAL	51	− 51
Patients remaining in study group		148
Relapses	80	
Non-relapses	68	

Table 3.1. Selection of patients for the Stage I and II large cell non-Hodgkin's lymphoma study.

model containing a given number of covariates, n. Two "rules of thumb" are in common use:

1) $n = f / 10$ 2) $n = \text{SQRT}\{ f \}$

In our experience with over 50 clinical studies, we have found these rules to be reasonably conservative. Thus, with 80 relapses, the data presented here should be able to support a model containing 8 or 9 covariates. This does not mean that all 18 covariates cannot be tested, but rather that a model in which more than 8 or 9 are adjusted at once may no longer fulfill the assumptions (asymptotic normality, etc.) on which the formulae of the procedure are based and that the resulting estimates for the regression coefficients cannot be trusted.

For each covariate x_j (j runs over the 18 components of x), the 148 patients were classified into one of two groups, arbitrarily termed "favorable" ($x_j = 0$) or "unfavorable" ($x_j = 1$), as may be seen in Table 3.2. This sort of dichotomizing is not required by the mathematics of the proportional hazards model. For example, the numerical age of the patient at referral (covariate 6, Table 3.2) could have been entered for x_6 and the corresponding regression coefficient would have represented the logarithm of the added relative risk per year of life. This may be appropriate and useful if the risk of failure depends upon the covariate in a more or less continuous monotonic fashion, however, in this particular instance, age did not appear to. When we studied survival, there was a sharp discontinuity at age 58 after which the patients did not do as well and we were anxious to see if this observation would hold for relapse as well. Moreover, physicians generally are concerned with identifying a subgroup of patients that are at higher or lower risk so that they may alter their therapy accordingly. To present them with information such as "the added risk is 2 per cent per year of life" may be mathematically elegant, but is not very helpful. What they need to know is, at what age does the added risk require more or less aggressive therapy. It is, therefore, our preference to dichotomize continuous and ranking variables even though there may be loss of information, introduction of bias and some arbitrariness in the process.

The problem of missing data came up in connection with covariate 3 in Table 3.2. This covariate intended to represent the maximum diameter of the tumor at the principal site of involvement and was dichotomized between 9 and 10 centimeters. In 96 cases, this dimension was actually recorded in the patient's chart, whereas in the other 52 cases, either cryptic notations or no information was present. The mathematically correct way to deal with this situation is to eliminate the 52 patients from the study group and proceed with the 96, however, this would have drastically reduced our patient numbers for the lack of data on a single covariate. Therefore, we decided to classify the 52 patients in the group with the most similar relapse rate which turned out to be the favorable group (size <= 9cm). This treatment

COVARIATE	FAVORABLE GROUP	UNFAVORABLE GROUP
1. X-ray Therapy (d-xRx)	extensive 62	limited 86
2. Sites of disease (sites)	<= 2 100	>= 3 48
3. Size of Prin Dis (size)	<= 9 cm 92	>= 10 cm 56
4. Age at Referral (age)	<= 57 years 97	>= 58 years 51
5. Ann Arbor Stage (stage)	I 57	II 91
6. Adj Chemotherapy (d-cRx)	received 27	not received 121
7. Protocol Participation (proto)	yes 51	no 97
8. Mediastinal Disease (m-dis)	absent 120	present 28
9. Extra Lymphatic Lesion (ex-les)	Absent 70	Present 78
10. Systemic Symptoms (symp)	absent 126	present 22
11. Reviewed Pathology (rpath)	Nodular 22	Diffuse 126
12. Staging Laparotomy (lap)	performed 70	omitted 78
13. Sclerosis (scler)	absent 90	present 58
14. Working Classif'n (wclas)	Non-immunoblastic 102	Immunoblastic 46
15. Gastro-intest Dis (gi-dis)	absent 117	present 31
16. Patient's Sex (sex)	female 63	male 85
17. Head and Neck Dis (hn-dis)	absent 19	present 129
18. Infradiaphragmic Dis (infra)	absent 63	present 85

Table 3.2. Eighteen covariates selected for investigation in the study. Column 1 gives a description and the abbreviation used for each covariate. Columns 2 and 3 give a description of the favorable and unfavorable groups for that covariate and the number of patients so classified.

should introduce the least bias and any inferences drawn can always be
checked using the smaller patient group.
   Having settled on the covariates, the proportional hazards model
which "best" explains the patient data is sought. Best in this sense
means the model containing only those covariates which have a significant and independent correlation with failure. The method we normally
follow is a step-up procedure in which the strongest 1 covariate model,
then 2 covariate model, and so on are found. In Table 3.3 the results
for all 18 one covariate models are shown. The type of delivered x-ray
therapy (d-xRx) is the most significant predictor of relapse with number of sites of disease (sites) a close second. Similar results would
have been found had we run 18 log-rank tests on the patient data dichotomized as shown in Table 3.2. If we had proceeded no further as was
commonly the practice only a few years ago, we might have concluded an
age at referral (age) and Ann Arbor stage (stage) are also important
prognostic indicators (P <= 0.01). Moreover, participation in Stanford
protocols (proto) and the absence of mediastinal disease (m-dis) appear
to confer a borderline benefit (P = 0.05. Note that undergoing a staging laparotomy (lap) does not (P = 0.34).

Covariate	Beta	dBeta	Z	P	GNH	WLR	I
d-xRx	0.8988	0.2443	3.6792	0.0002	14.36	14.57	4
sites	0.8272	0.2263	3.6546	0.0003	14.11	12.69	3
size	0.6307	0.2262	2.7881	0.0053	8.02	7.54	3
age	0.6440	0.2307	2.7916	0.0052	8.06	7.53	3
stage	0.6348	0.2477	2.5628	0.0104	6.79	7.08	4
d-cRx	0.8473	0.3545	2.3903	0.0168	6.06	7.06	4
proto	0.5031	0.2530	1.9883	0.0468	4.03	4.22	4
m-dis	0.5007	0.2684	1.8653	0.0621	3.55	3.16	3
ex-les	0.3333	0.2256	1.4776	0.1395	2.20	2.20	3
symp	0.4598	0.3036	1.5144	0.1299	2.33	2.06	3
rpath	0.3259	0.3171	1.0276	0.3041	1.06	1.14	4
lap	0.2178	0.2268	0.9605	0.3368	0.93	0.93	3
scler	0.2105	0.2291	0.9187	0.3583	0.85	0.83	3
wclas	0.1861	0.2406	0.7735	0.4392	0.60	0.59	3
gi-dis	0.0602	0.2798	0.2152	0.8296	0.05	0.05	3
sex	0.0431	0.2275	0.1894	0.8498	0.04	0.04	3
hn-dis	-0.0545	0.3256	0.1674	0.8671	0.03	0.03	3
infra	-0.0255	0.2304	0.1108	0.9118	0.01	0.01	3

Table 3.3. Proportional hazard models in which one covariate is adjusted. The first column gives the abbreviation for the covariate be-
ing adjusted. Columns 2 and 3 give $\hat{\beta}$ and $d\hat{\beta}$, respectively. Column 4
gives the normal variate $Z = \hat{\beta}/d\hat{\beta}$ and column 5 the corresponding P-
value. Columns 6 and 7 give the global statistics defined in section
II equations (2.15) and (2.16), respectively. The eighth column head-
ed I gives the number of iterations required to find the solution.

   Before continuing with two covariate models, we usually run a model
containing as many of the important covariates as can be supported by
the data (here, nine). The model for the nine major covariates is presented in Table 3.4 (note that only four iterations were required to
find this solution and that only four covariates retain P-values below

Covariate	Beta	dBeta	Z	P	GNH	WLR
d-xRx	1.1719	0.3118	3.7588	0.0002	47.24	51.10
sites	0.7646	0.2881	2.6537	0.0080		
size	0.5881	0.2511	2.3421	0.0192		
age	0.1871	0.2772	0.6751	0.4996		
stage	0.2112	0.3173	0.6657	0.5056		
d-cRx	1.4385	0.4074	3.5310	0.0004		
proto	-0.2661	0.3081	0.8636	0.3878		
m-dis	0.4599	0.3351	1.3725	0.1699		
ex-les	-0.1802	0.2447	0.7366	0.4614		

Table 3.4. Proportional hazards model in which nine covariates were simultaneously adjusted. The columns are labeled as described in the legend of Table 3.3. Four iterations were required to find this solution.

0.05). This model was run in order to obtain the values of the GNH (=47.24) and WLR (=51.10) statistics which then serve as goals; i.e., we shall seek a model involving fewer covariates which produce values of GNH and WLR nearly this big.

We then explore the various two covariate models obtained by adding to d-xRx each of the other covariates in turn. The results are shown in Table 3.5. Clearly, the best two covariate model is the one which combines d-xRx and sites (model 1). The P-values for both remain as low as in their one covariate models while the GNH and WLR statistics for the two covariate model (about 28) are close to the sum of these statistics (about 14) for each of the one covariate models. This indicates that both d-xRx and sites are independently good prognostic indicators for relapse. On the other hand, when d-xRx is combined with age (model 6) or proto (model 15), the P-value for d-xRx jumps up by an order of magnitude while age and proto lose the prognostic significance they had in their one covariate models. Moreover, neither GNH nor WLR (about 16.5 for age and 14.5 for proto) increase a great deal over the values in the one covariate d-xRx model (about 14). This is the most common indication of correlation among covariates and here is verified by the contingency tables shown in Fig. 3.1. Thus, we can

```
                age                              proto
         I-----|-----I                      I-----|-----I
         I unf | fav I                      I unf | fav I
         I     |     I                      I     |     I
         ====I=====|=====I=====|            ====I=====|=====I=====|
    d  unf I 42  |  9  I 51  |         d  unf I 72  | 14  I 86  |
    x    ----I------|------I------|         x    ----I------|------I------|
    R  fav I 44  | 53  I 97  |         R  fav I 25  | 37  I 62  |
    x    ====I=====|=====I=====|         x    ====I=====|=====I=====|
         I     |     I                      I     |     I
         I 86  | 62  I 148                  I 97  | 51  I 148
         I-----|-----I                      I-----|-----I
```

Figure 3.1. Contingency tables in which the covariates age and proto are correlated with d-xRx.

Covariate	Beta	dBeta	Z	P	GNH	WLR
1. d-xRx	0.9037	0.2445	3.6962	0.0002	28.57	27.40
sites	0.8333	0.2269	3.6729	0.0002		
2. d-xRx	0.9303	0.2459	3.7834	0.0002	23.07	23.00
size	0.6711	0.2278	2.9460	0.0032		
3. d-xRx	0.8875	0.2438	3.6405	0.0003	20.88	21.34
stage	0.6215	0.2479	2.5073	0.0122		
4. d-xRx	0.8870	0.2444	3.6290	0.0003	20.01	21.23
d-cRx	0.8253	0.3546	2.3276	0.0199		
5. d-xRx	0.9022	0.2443	3.6923	0.0002	18.01	17.84
m-dis	0.5100	0.2688	1.8970	0.0578		
6. d-xRx	0.7717	0.2629	2.9351	0.0033	16.56	16.50
age	0.3447	0.2480	1.3900	0.1645		
7. d-xRx	0.8882	0.2439	3.6412	0.0003	16.42	16.33
symp	0.4246	0.3054	1.3904	0.1644		
8. d-xRx	0.9332	0.2465	3.7861	0.0002	16.03	16.31
scler	0.3067	0.2307	1.3292	0.1838		
9. d-xRx	1.0037	0.2707	3.7078	0.0002	15.17	15.36
lap	-0.2255	0.2521	0.8944	0.3711		
10. d-xRx	0.9063	0.2444	3.7080	0.0002	15.19	15.29
wclas	0.2207	0.2410	0.9160	0.3597		
11. d-xRx	0.8866	0.2444	3.6268	0.0003	15.06	15.29
rpath	0.2589	0.3149	0.8223	0.4109		
12. d-xRx	0.9090	0.2451	3.7091	0.0002	14.63	14.84
sex	0.1179	0.2282	0.5168	0.6053		
13. d-xRx	0.8803	0.2575	3.4189	0.0006	14.41	14.62
ex-les	0.0541	0.2378	0.2276	0.8199		
14. d-xRx	0.9037	0.2455	3.6813	0.0002	14.40	14.61
gi-dis	-0.0560	0.2816	0.1989	0.8423		
15. d-xRx	0.8761	0.2818	3.1090	0.0019	14.38	14.60
proto	0.0470	0.2926	0.1606	0.8724		
16. d-xRx	0.8987	0.2443	3.6784	0.0002	14.37	14.58
infra	-0.0180	0.2309	0.0779	0.9379		
17. d-xRx	0.8985	0.2444	3.6768	0.0002	14.36	14.58
hn-dis	-0.0223	0.3260	0.0684	0.9455		

Table 3.5. Seventeen proportional hazards models in which a second covariate was added to d-xRx and the two adjusted simultaneously. The columns are labeled as described in Table 3.3. In each case, four iterations were required.

conclude that older people relapsed more often, not just because they were older, but because they often did not receive as aggressive therapy. When adjustment is made for therapy, age is unimportant. Likewise, those patients on protocol tended to receive more aggressive therapy and, therefore, did better. Finally, we note that lap adds nothing to d-xRx and that the two covariate model consisting of sites and size (the second and third strongest single covariates) produces a very much inferior model (GNH = 18, WLR = 16) to that of d-xRx and sites.

Twelve of the 16 possible three covariate models obtained by combining d-xRx, sites and then each of the other covariates are shown in Table 3.6. Again, the signals indicating correlation between d-xRx and age are present. Also present is an indication of correlation between sites and stage in model 8, which contains d-xRx and these two covariates. This correlation is to be expected since both are measures of the number of involved anatomic sites according to the Ann Arbor definition; stage is the customary dichotomization, one versus more than one site, whereas sites is an unconventional dichotomization, one or two versus more than two sites. In every model involving sites and/or stage constructed so far, sites has proven to be the stronger covariate. Thus, it would appear that, at least in our patients, the dichotomization between two and three anatomic sites would have made a better staging system than the conventional Ann Arbor system. This finding is further supported by the survival curves shown in Fig. 3.2. In both the favorable and unfavorable d-xRx groups, the two site subgroup has a relapse rate more similar to the one site subgroup. The large differences between these survival curves provide an independent verification of the independent prognostic significance of d-xRx and sites by the subgrouping method. Owing to correlation between sites and m-dis, m-dis is no longer significant near the P = 0.05 level. The best three covariate model (model 1) contains d-xRx, sites and d-cRx, delivered adjuvant chemotherapy, and produces GNH and WLR statistics near 40 which are nearly 75 per cent of those of the nine covariate model.

The best four covariate model is shown in Table 3.7 and is obtained by adding size to d-xRx, sites and c-cRx. The GNH and WLR statistics are about 90 per cent of those for the nine covariate model, so we may expect that essentially all the prognostic power for relapse is contained in these four covariates. In fact, no other significant covariate was found in any other model, except one. Surprisingly, when lap was added, it became significant at the 0.03 level, despite the fact that it had not shown any prognostic power in any model containing fewer covariates. This behavior can be explained as another example of correlation between the covariates which is somewhat more subtle than the examples we have seen so far. The presence of a three dimensional correlation between d-xRx, d-cRx and lap can be verified in the contingency tables presented in Fig. 3.3. This correlation can also be appreciated through the proportional hazards model. Comparing the best three covariate model with the four covariate model including lap, we note that a strange change has occurred in the β_k's. The β_k's for d-xRx and d-cRx have been increased significantly and the β_k for lap has assumed a large negative value. Owing to the correlation among these covariates, a negative value for lap will partially compensate for the larger positive values of d-xRx and d-cRx. A negative value for β_k indicates that our arbitrary designation of favorable-unfavorable is backwards for that covariate. Here, we assumed that undergoing a staging laparotomy would enable the physician to deliver more

	Covariate	Beta	dBeta	Z	P	GNH	WLR
1.	d-xRx	0.9315	0.2452	3.7994	0.0001	39.33	40.33
	sites	1.0481	0.2317	4.5246	0.0000		
	d-cRx	1.1462	0.3619	3.1670	0.0015		
2.	d-xRx	0.9266	0.2456	3.7723	0.0002	33.02	31.72
	sites	0.7072	0.2354	3.0036	0.0027		
	size	0.4956	0.2365	2.0953	0.0361		
3.	d-xRx	0.7881	0.2584	3.0500	0.0023	30.51	29.53
	sites	0.8405	0.2268	3.7050	0.0002		
	age	0.3570	0.2431	1.4684	0.1420		
4.	d-xRx	0.9371	0.2467	3.7978	0.0001	30.26	29.12
	sites	0.8339	0.2271	3.6714	0.0002		
	scler	0.3062	0.2308	1.3267	0.1846		
5.	d-xRx	0.9197	0.2446	3.7603	0.0002	29.71	29.05
	sites	0.8309	0.2273	3.6559	0.0003		
	symp	0.4106	0.3052	1.3450	0.1786		
6.	d-xRx	1.0220	0.2728	3.7457	0.0002	29.34	28.32
	sites	0.8383	0.2271	3.6918	0.0002		
	lap	-0.2464	0.2546	0.9676	0.3332		
7.	d-xRx	0.8925	0.2444	3.6523	0.0003	29.18	28.11
	sites	0.8342	0.2272	3.6723	0.0002		
	rpath	0.2604	0.3158	0.8245	0.4097		
8.	d-xRx	0.8987	0.2443	3.6785	0.0002	29.02	27.99
	sites	0.7022	0.2784	2.5224	0.0117		
	stage	0.2347	0.3041	0.7716	0.4403		
9.	d-xRx	0.9144	0.2453	3.7280	0.0002	28.83	27.68
	sites	0.8340	0.2270	3.6744	0.0002		
	sex	0.1208	0.2282	0.5293	0.5966		
10.	d-xRx	0.9035	0.2444	3.6964	0.0002	28.79	27.51
	sites	0.8204	0.2302	3.5631	0.0004		
	wclas	0.0813	0.2445	0.3324	0.7396		
11.	d-xRx	0.9030	0.2445	3.6933	0.0002	28.76	27.51
	sites	0.7993	0.2498	3.1997	0.0014		
	m-dis	0.0989	0.2962	0.3339	0.7384		
12.	d-xRx	0.9004	0.2448	3.6776	0.0002	28.71	27.50
	sites	0.8388	0.2276	3.6846	0.0002		
	hn-dis	-0.1043	0.3278	0.3181	0.7504		

Table 3.6. Twelve of 16 possible proportional hazards models in which a third covariate of our study is added to d-xRx and sites and all three adjusted simultaneously. The columns are labeled as described in Table 3.3. Again, four iterations were required to find the solutions. The other few possible models produced p-values > 0.8 for their added covariates (infra, proto, gi-dis and ex-les) and almost no change in GNH or WLR over the two parameter model.

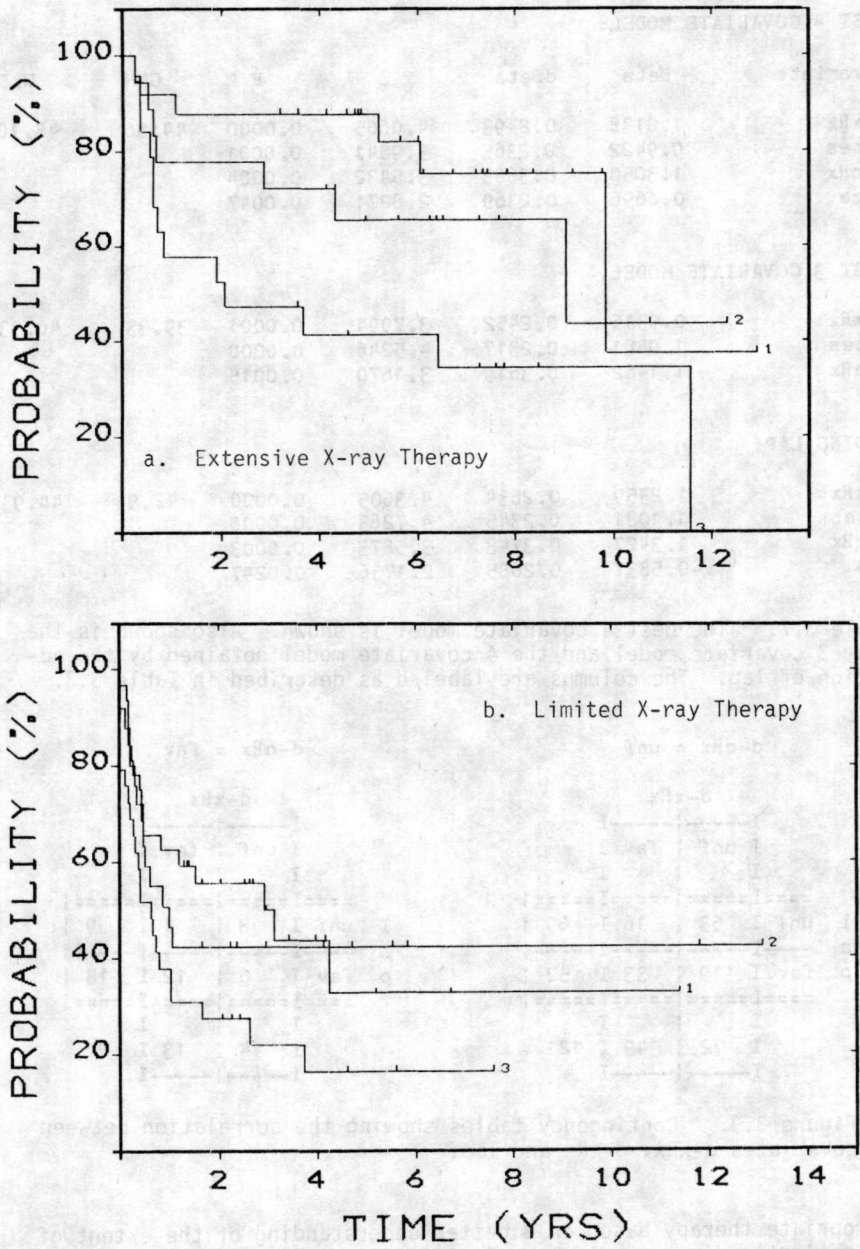

Figure 3.2. 6 $S_{pl}(t)$ curves for the study patients grouped by extent of radiotherapy (a & b) and further subgrouped by number of anatomic sites of involvement (1 = 1 site; 2 = 2 sites, 3 = 3 or more sites).

BEST 4 COVARIATE MODEL:

Covariate	Beta	dBeta	Z	P	GNH	WLR
d-xRx	1.0138	0.2493	4.0665	0.0000	44.86	48.10
sites	0.9422	0.2365	3.9841	0.0001		
d-cRx	1.3050	0.3683	3.5432	0.0004		
size	0.6696	0.2369	2.8271	0.0047		

BEST 3 COVARIATE MODEL:

d-xRx	0.9315	0.2452	3.7994	0.0001	39.33	40.33
sites	1.0481	0.2317	4.5246	0.0000		
d-cRx	1.1462	0.3619	3.1670	0.0015		

ADDING LAP:

d-xRx	1.2359	0.2834	4.3605	0.0000	42.93	44.93
sites	1.1081	0.2345	4.7263	0.0000		
d-cRx	1.3427	0.3743	3.5873	0.0003		
lap	-0.5837	0.2685	2.1736	0.0297		

Table 3.7. The best 4 covariate model is shown. Also shown is the best 3 covariate model and the 4 covariate model obtained by the addition of lap. The columns are labeled as described in Table 3.3.

```
                d-cRx = unf                              d-cRx = fav

                   d-xRx                                    d-xRx
               I-----|-----I                            I-----|-----I
               I unf | fav I                            I unf | fav I
               I     |     I                            I     |     I
               ====I=====|=====I=====|                  ====I=====|=====I=====|
          l  unf I  53 |  16 I  69 |                 l  unf I  8  |  1  I  9  |
          a    ----I-----|-----I-----|                a    ----I-----|-----I-----|
          p  fav I  19 |  33 I  52 |                 p  fav I  6  | 12  I  18 |
               ====I=====|=====I=====|                  ====I=====|=====I=====|
               I     |     I                            I     |     I
               I  72 |  49 I 121                        I  14 | 13  I  27
               I-----|-----I                            I-----|-----I
```

Figure 3.3. Contingency tables showing the correlation between covariates d-xRx, d-cRx and lap.

appropriate therapy based on a better understanding of the extent of the patient's disease. If the significance of lap were naively accepted, it might be concluded that a staging laparotomy actually spreads the disease, making aggressive therapy all the more important. Due to the correlation of these covariates, we should be suspicious that the sudden increase in the importance of lap is a mathematical artifact with no medical significance.

The results for the 14 five covariate models obtained by adding each of the other 14 covariates to those of the best four covariate model are shown in Table 3.8. Of the 14, only lap remains significant at the P = 0.05 level and adds more than two units to GNH and WLR. Like as fitting procedure, adding a parameter should increase GNH and WLR, but unless the increase is at least two units, the corresponding global P-value will actually decrease. Thus, the four covariate model involving d-xRx, sites, d-cRx and size provides the best explanation of the relapse observations in that each covariate has a strong correlation with relapse and that a knowledge of any of the other covariates does not add significant prognostic power.

BEST 4 COVARIATE MODEL:

Covariate	Beta	dBeta	Z	P	GNH	WLR
d-xRx	1.0138	0.2493	4.0665	0.0000	44.86	48.10
sites	0.9422	0.2365	3.9841	0.0001		
d-cRx	1.3050	0.3683	3.5432	0.0004		
size	0.6696	0.2369	2.8271	0.0047		

5 COVARIATE MODELS (added covariate only):

Covariate	Beta	dBeta	Z	P	GNH	WLR
lap	-0.5369	0.2700	1.9884	0.0468	48.70	51.96
sex	0.3548	0.2327	1.5244	0.1274	46.62	50.46
infra	-0.3630	0.2430	1.4940	0.1352	45.70	50.37
gi-dis*	-0.3933	0.2979	1.3202	0.1868	45.14	49.95
m-dis*	0.4055	0.3275	1.2380	0.2157	45.01	49.59
rpath	0.3753	0.3231	1.1617	0.2454	46.57	49.56
stage	0.2451	0.3081	0.7956	0.4263	45.42	48.73
hn-dis	-0.2811	0.3473	0.8094	0.4183	45.30	48.73
scler	0.1649	0.2367	0.6967	0.4860	46.21	48.58
ex-les	-0.1514	0.2406	0.6294	0.5291	45.27	48.50
symp	0.2032	0.3185	0.6380	0.5235	44.99	48.49
proto	-0.1783	0.2978	0.5987	0.5494	46.01	48.45
age	0.1275	0.2647	0.4815	0.6301	45.21	48.33
wclas	-0.0783	0.2473	0.3165	0.7516	45.46	48.20

Table 3.8. Results for the best 4 covariate model and 14 possible 5 covariate models are shown. Asterisks (*) indicate the models in which five iterations were required to find the solutions. Columns are labeled as in Table 3.3.

In testing n = 18 covariates, if any were accepted that were significant at the P = 0.05 level, there would be a $1 - (0.95)^{18} = 0.6$ probability of erroneously reporting a covariate as significant which happened to have a P-value this low by chance. (Perhaps lap is an example.) To avoid this error, the criterion for significance for a single covariate should be lowered to 0.01 or 0.005 which would reduce corresponding overall probability to 0.17 or 0.09, respectively, here. In fact a P-value of 0.05/n is the appropriate level for individual covariates if an overall confidence level of 0.95 is to be maintained.

In preparing this example analysis, over 80 proportional hazards models were constructed and examined (over 50 are actually presented). During this process, a great deal was learned about the interplay of covariates and their correlation with relapse for this group of patients. The effort required was less than a morning's work and could not have been carried out so quickly and effectively without a multivariate technique. Multivariate techniques are especially helpful when analyzing retrospective data in which randomization may not have been performed and unsuspected correlations between covariates might exist. We are indeed fortunate to have powerful tools such as the proportional hazards model to apply to these problems.

IV. DISCUSSION

A major advantage of the Cox method lies in the ease with which correlation among the covariates can be detected. The predictive power of each covariate can be assessed singly and in combination with others; thus, covariates with truly independent prognostic signficance can be identified. In section III, the proportional hazards model was applied to a case where the underlying assumptions are reasonably well met. Moreover, there are sufficient numbers of patients and failures to support an extensive analysis. This is not always the case and there are situations where the model should not be applied.

One of the clear signs of trouble with the model is failure to converge to a solution. In our case, this is detected when the limit of nine iterations is reached and usually occurs if the assumption of proportional hazards is not well fulfilled. If one subgroup of a particular covariate contains no failure, the corresponding regression coefficient will be infinite. A second more subtle example is provided by the data shown in Fig. 4.1.[11] Clearly, the hazard for the papillary subgroup is large for the first four years and zero thereafter, whereas the hazard for the non-papillary group is fairly constant out to 12 years. In this case, the maximum in $L_p(\vec{\beta})$ was so indistinct that the Newton-Raphson method failed to locate it. The solution was ultimately found, but only after extensive tinkering with the iteration step size. One way to check the appropriateness of the proportional hazards model is to make use of equations (2.3). Taking logarithms and dividing

$$g(t) = \frac{\log\{ S(t,1) \}}{\log\{ S(t,0) \}} \quad (4.1)$$

If the values of g at each failure do not remain approximately constant, the proportional hazards condition is not well fulfilled.

Although multivariate techniques can be extremely helpful in analyzing complex problems, they also present the difficulties associated with trying to comprehend n-dimensional relationships. When n becomes greater than two, even the mathematically sophisticated soon are overwhelmed. In the hands of the unsophisticated user, the proportional hazards model can be used to confirm preconceived notions rather than to increase one's understanding of the interplay of the variables. Moreover, new models and statistical tools are constantly being developed and the P-values they produce for the same data can vary a great deal. For example, between the generalized Wilcoxon and the corresponding proportional hazards model test, there is often a factor 3 difference in P-values and, in one instance, we have seen a factor 10

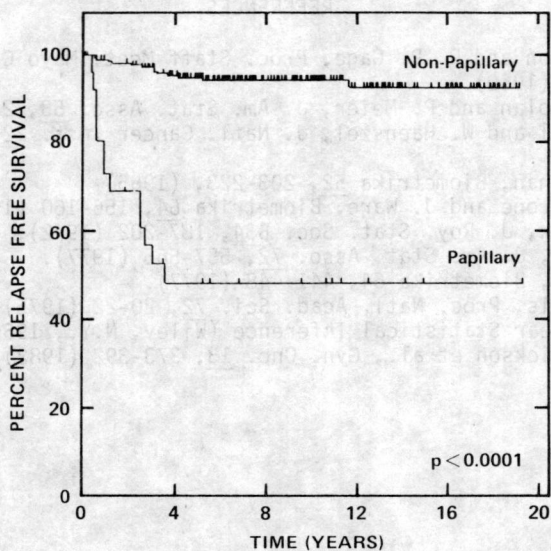

Figure 4.1. $\hat{S}_{p1}(t)$ for patients with papillary and non-papillary carcinoma of the endometrium. In this case, the assumption of proportional hazards is not fulfilled. P-value is GW test.

difference. In situations of borderline significance, P-values above or below 0.05 can be obtained just by judicious selection of testing procedures. It is up to the person providing statistical guidance to assure the fairness of the statistical procedures used.

In the example given in section III, many such "fairness" issues were touched upon. The importance ultimately assigned to a continuous or ranking covariate can be influenced by how it is entered initially; i.e., was it left with its actual value or was it dichotomized and, if dichotomized, how was the division point chosen (age). Likewise for categorical data, the way categories were defined and combined can influence the outcome. If a large number of covariates were involved, not all models can be constructed and examined; important ones might have been omitted in the initial selection process. If more than one covariate is used to represent the same attribute (sites and stage), the importance of both can be enhanced or diluted. All of these issues are matters of judgment and their manner of resolution can alter the results of any mathematical analysis. In fact, they are every bit as important as the critical examination of results and confirmation by other methods (such as subgrouping). Multivariate analyses and, in particular, the proportional hazards model are sophisticated tools and require sophistication on the part of the user to be properly applied.

REFERENCES

1. J. Berkson and R. P. Gage, Proc. Staff Meet. Mayo Clinic 25, 270-286 (1950).
2. E. L. Kaplan and P. Meier, J. Am. Stat. Asso. 53, 227-236 (1958).
3. N. Mantel and W. Haenszel, J. Natl. Cancer Inst. 22, 719-748 (1959).
4. E. A. Gehan, Biometrika 52, 203-223, (1965).
5. R. E. Tarone and J. Ware, Biometrika 64, 156-160 (1977).
6. D. R. Cox, J. Roy. Stat. Soc. B34, 187-202 (1972).
7. B. Efron, J. Am. Stat. Asso. 72, 557-565 (1977).
8. D. Oakes, Biometrika 64, 441-448 (1977).
9. A. Tsiatis, Proc. Natl. Acad. Sci. 72, 20-22 (1975).
10. Rao, Linear Statistical Inference (Wiley, N.Y., 1965) Section 6e.
11. M. Hendrickson et al., Gyn. Onc. 13, 373-392 (1982).

LEGAL QUESTIONS IN STATISTICS

Robert J. Shalek
Physics Department
The University of Texas System Cancer Center
Houston, Texas 77030

ABSTRACT

Statistical evidence in both civil and criminal cases is becoming increasingy available to courts. Some cases have depended in a major way upon such evidence. But as is indicated by decision reversals in higher courts and by a number of scholarly articles, there is uncertainty and a division of opinion upon the proper legal use of statistical information.

The problems in the use of statistics in court are many. A major issue is the understanding of the meaning of a statistical statement by the court, opposing lawyers and jury. A crisp statistical conclusion may be given greater weight than it deserves because of its apparent scientific validity while other "soft" unquantifiable evidence may be given less weight than it deserves. Even lawyers may be diffident because of their inability to challenge or even fully understand the steps in reaching a statistical conclusion. It is a rule in federal and state courts that expert testimony having a prejudicial effect significantly outweithing its probative value will not be admitted at a trial. However, in practice, courts are reluctant to apply that principle preferring to rely upon cross-examination and the testimony of opposing experts to counterbalance prejudical testimony. Judges are reluctant to exclude evidence because that can be a reason for ordering a new trial by a higher court.

A use of statistics in court is the description of conjunctive probability by the use of Bayes' theorem. The probability of the occurrence of causally independent events can be combined to yield an overall probability. It has been suggested how probabilities can be combined when they relate to pieces of evidence which confirm each other but are not independent. But the Bayes' theorem approach is criticized as overemphasizing the "hard" variables as against the "soft" variables. Further, the ability of laymen to assign probabilities from commonsense experience must be highly variable. As is the case with other types of evidence having a scientific base, the use of statistics is finding its way into the courtroom slowly and with suspicious regard by legal practitioners.

INTRODUCTION

The phrases employed in court to describe the decision making criteria seem superficially susceptible to statistical formulation. In civil trials, "more probable than not" or "by the greater weight and preponderance of the evidence" sound almost statistical. The criminal criteria of "beyond a reasonable doubt" or "to a moral certainty" are less statistical. Yet the legal process is moving to the use of statistics slowly and with uncertainty as to whether statistical methods will enhance or corrupt the existing truth-finding processes. In a heavily cited law review article, Tribe in 1971[1], concluded:

In an era when the power but not the wisdom of science is increasingly taken for granted, there has been a rapidly growing interest in the conjunction of mathematics and the trial processes. The literature of legal praise for the progeny of such a wedding has been little short of lyrical. Surely the time has come for someone to suggest that the union would be more dangerous than fruitful.

This negative view toward the use of statistics in court has been reflected in a number of subsequent review articles[2,3]. Yet the use of statistics in trials continues to increase, albeit hesitantly, as indicated by reversals in appellate courts.

Presenting statistical conclusions in court as an expert witness can be a challenging and even enjoyable experience; high stakes may be riding upon a performance that occurs over a relatively short period of time. On the other hand, appearing in court as a defendant is almost always unpleasant and may be traumatic. Statisticians and epidemiologists are much less likely to be subject to legal actions against them for flaws in their professional work than are physicians and medical physicists. However, mathematicians who participate in the creation of computer programs used in medical diagnosis or treatment may be opening avenues of legal exposure against themselves.

In this article the use of statistics as evidence, the role and functioning of the expert witness and risks of legal liability for mathematicians will be considered.

THE USE OF STATISTICS AS EVIDENCE

Problems in the use of statistics in court are not unlike questions arising in the use of other new scientific or medical evidence. There are the questions of understanding the evidence, evaluating its reliability and relating the evidence to other conventional evidence. A major problem is the understanding of the meaning of a statistical statement by the judge, opposing lawyers and a jury. A crisp statistical conclusion may be given greater weight than it deserves because of its apparent scientific validity while other "soft", unquantifiable evidence may be given less weight than it deserves. It is a rule in federal and state courts that expert testimony having a prejudicial effect significantly outweighing its probative value will not be admitted at a trial. However, in practice, courts are reluctant to apply that principle preferring to rely upon cross-examination and the testimony of opposing experts to counterbalance prejudicial testimony. Judges are reluctant to exclude evidence because that can be a reason for ordering a new trial by a higher court. Trial judges do not like being reversed.

The history of the use of statistics in two California cases illustrates the uncertainties of courts in admitting and evaluating statistical evidence. In People v. Collins (1963)[4], a witness described the person stealing a purse as a white woman with blond hair in a ponytail who was seen escaping in a partly yellow car driven by a black man with a moustache and beard. Suspects matching the description were arrested by the police. In the trial a mathematics instructor separated six characteristics of this event: a girl with blond hair; a girl with a ponytail; a partly yellow automobile; a man with a moustache; a black man with a beard; an interracial couple in a car. An independent probability was assigned to each of these characteristics. For example, a probability of one in ten was taken for seeing a partly yellow car and one in one thousand for seeing an interracial couple in a car. When the six probabilities were multiplied together it was concluded that the chance of seeing all six characteristics in one event was one in twelve million. The defendants were found guilty. The California Supreme Court reversed the decision. That court held that the six characteristics were not necessarily independent so that a product of the probabilities was not correct and that there was no empirical evidence supporting the choice of probabilities.

In a 1979 California paternity suit (Cramer v. Morrisson, 1979)[5], evidence by a medical school professor would have indicated a 98.3% probability that the defendant was the father of a child based upon human leukocyte antigen (HLA) tests. In this case, the trial court was cautious and did not allow this evidence admitted because there was a possibility that the statistical evidence would have a prejudicial effect upon the jury which would outweigh its probative value. A court of appeals reversed the trial court stating that all relevant evidence is admissible. Relevant evidence was defined as evidence having "any tendency in reason to prove or disprove any disputed fact that is of consequence to the determination of the action." The appeals court also stated "the more substantial the probative value, the greater must be the danger of prejudice to an adverse party, in order to justify a finding that the probative value is substantially outweighed by the danger of undue prejudice." In this case, the appeals court held that the HLA test interpretations were based upon objectively ascertainable data and not upon arbitrary probability values. The court concluded that the decision to bar the HLA test evidence constituted an abuse of discretion and a new trail was ordered.

A use of statistics in court is the decription of conjunctive probability by employing Bayes' Theorem. The probability of the occurrence of casually independent events can be combined to yield an overall probability. Numbers of legal writers have inquired into the utility of Bayes' Theorem as a means of seeking truth in trials[1,2,3]. If Y is the question in dispute and to be decided, it may have a certain probability of being true, P(Y), in the juror's mind based upon the general circumstances of the case. The impact of an additional piece of evidence, E, upon P(Y) can be evaluated by Bayes' Theorem. The theorem could be applied repeatedly for additional pieces of evidence. Two statements of Bayes' Theorem follow:

$$P(Y/E) = \frac{P(E/Y)}{P(E)} P(Y) \tag{1}$$

where P(Y) is the assessment of the probability of the truth of Y before the evidence E is considered and P(Y/E) is the probability of Y after considering evidence, E. The probability of E if Y occurs is described by P(E/Y) and P(E) is the probability of the occurrence of E whether or not Y occurs. A more explicit statement of Bayes' Theorem defines P(E) in the denominator:

$$P(Y/E) = \frac{P(E/Y)P(Y)}{P(E/Y)P(Y) + P(E/-Y)P(-Y)} \tag{2}$$

where P(E/-Y) is the probability of E occurring when Y has not occurred and P(-Y) is the asessed probability that Y did not occur prior to the receipt of evidence, E.

As an example of the use of Bayes' Theorem consider that a crime has been committed. Actor A was known to be in the area and witnesses identified a person of the actor's general appearance but were unable to identify the person. Supposing that this and other evidence would lead a reasonable person to believe that the probability was 0.75 that A committed the crime: $(P(Y) = 0.75; P(-Y) = 0.25$. Further, it was found that A did not return to his dwelling during the night of the crime (E). Suppose that neighbors gave evidence that it was A's custom to sleep at home and that he was, at most, away one night per month $(P(E/-Y) = 0.03)$. An estimate was made that if A committed a crime there was a 0.5 probability that he would not return $(P(E/Y) = 0.50)$. Substituting in equation (2) one obtains:

$$P(Y/E) = \frac{(0.50)(0.75)}{(0.50)(0.75) + (0.03)(0.25)} = 0.98$$

Thus, the probability that A committed the crime increases from 0.75 to 0.98 by considering additional evidence with Bayes' Theorem. But the Bayes' Theorem approach is criticized as overemphasizing the "hard" variables as against the "soft" variables. Further, the ability of laymen to assign probabilities from common sense experience must be highly variable and unreliable. In the example cited a 98% probability that A is guilty is impressive and perhaps conclusive, yet without the statistics the evidence remains circumstantial and likely insufficient for conviction in the light of centuries of criminal legal experience.

As is the case with other types of evidence having a scientific base, the use of statistics is finding its way into the courtroom slowly but consistently and with suspicious regard by legal practitioners. Lawyers are being advised to ask statistical experts to explain their model in simple terms indicating its likely correspondence to reality and to discuss other models which they considered and rejected. Continuing education courses in statistics for lawyers are beginning to be offered indicating an expected increase in the legal use of statistics.

THE EXPERT WITNESS

Medical physicists appear in court as expert witnesses particularly in mecial malpractice cases where an issue is radiation dose. The expert witness is asked to participate by either the plaintiff's or the defendant's lawyer. Usually the expert will communicte only with the lawyer and not with the parties (plaintiff or defendant), nor with other experts. Often the early communications are verbal to avoid discovery (forced production of documents by the other side). Late in the pretrial process there may be a deposition of the expert recorded under oath at which calculations, graphs or other documents prepared by the expert will be entered as exhibits. The opposing lawyers have full opportunity for questioning at a deposition. At trial the expert may be called to testify or his deposition may be read into the record. Increasingly, depositions are taken with the aid of a video recorder so that prior testimony can be seen and heard by the jury without the expert appearing in court.

What is the status and role of an expert witness? Expert testimony is permitted when some body of knowledge or unique experience is technically beyond the understanding of laymen. The judge will first hear the qualifications and the experience of the expert before qualifying him to testify as an expert. Lay witnesses may testify upon what they have heard, seen or otherwise observed, but the expert witness may go further to draw conclusions from evidence. Often an expert will be asked to draw conclusions from a hypothetical question which includes, as assumptions, facts in evidence for the case being tried. In a negligence case an expert for the plaintiff will usually be called upon to help establish what the national standard of practice was at the time of the patient's diagnosis or treatment. He will be asked whether that standard was adhered to in the present case. However, in many jurisdictions, he will not be able to say that deviations from the standard were negligent, since that would be answering the ultimate question and would invade the province of the jury. However, the testimony of medical physicists usually pertains to dose and does not include hypothetical questions. For the defense an expert physicist might testify as to standards of practice, dose and whether the defendants acted in a competent, prudent way.

The process of testifying can be an interesting and professionally satisfying experience for the expert witness. In a limited time the thoroughness of preparation, the skill of explanation and the credibility of the expert is tested in an atmosphere of high stakes riding on the outcome. The experience has a unique intensity.

POSSIBLE LEGAL LIABILITIES OF MATHEMATICIANS

While it appears that statistics will have a growing role is evidence, it seems to this writer that statisticians and epidemiologists are much less likely to be subject to legal actions against them for flaws in their professional work than are physicians and medical physicists. In radition medicine the possibility of a lawsuit for malpractice is never far away. A proven mistake that harms a patient is negligence. The plaintiff must prove that there is a standard of care that the defendant did not observe and that the patient suffered injury and damage as a result. This means that the defendant professionally rendered some service to the patient. Statisticians and epidemiologists seldom render service directly to patients in diagnosis or therapy and thus are not susceptible to negligence actions, the most common legal theory used against health professionals.

However, there is probably a fair number of mathematicians who have participated in the construction of computer programs used in diagnosis or therapy. The possibility exists here for liability even though no service is rendered and there is no negligence. One who sells a product unreasonably dangerous to the user is subject to strict liability for physical harm to the user if the product is expected to and does reach the user without modification. The general use to which the product is put must be foreseeable to the producer of the product. Whether a product is unreasonably dangerous is measured by an objective standard defined by the expectations of the ordinary consumer of that type of product. In weighing whether a product is unreasonably dangerous a court will consider the utility of the product against its danger and will decide whether adequate warnings of dangerous uses were made. As far as this writer knows, there have been no product liability cases where the product was a computer program that resulted in injury to a patient through its use. However, computer programs for patient related uses probably would be regarded as products subject to strict liability actions provided the criteria mentioned above were satisfied.

CONCLUSIONS

Physicians have long been familiar with the courtroom. In the last decade medical physicists have increasingly participated in the legal process as defendants and expert witnesses. It is predictable that mathematicians will play a larger role in litigation primarily as expert witnesses.

REFERENCES

1. L. H. Tribe. Harv. L. Rev. 84, 1329 (1971).
2. J. D. Jackson, N. IR. Legal Q. 31, 239 (1980).
3. R. L. Dickson, Forum, 17, 792 (1982).
4. People v. Collins, Cal. 2d, 68, 329 (1968); p. 2d 438, 33 (1968); Cal. Rptr. 66, 497 (1968).
5. Cramer v. Morrison, Cal. App. 3rd, 88, 873 (1979); App, Cal. Rptr. 153, 865 (1979).

SUMMARY OF REGRESSION METHODS

Raymond H. Myers
Virginia Polytechnic Institute and State University
Blacksburg, Virginia 24061

Health physicists should by inclination and training be good data analysts. There are three major reasons for this:
  (i) Mathematical training of a physicist is natural requisite.
  (ii) Training of a physicist allows the scientist to be analytical.
  (iii) The physicist knows the process that generates the data.
What is lacking that creates the need for a symposium such as this is formal training in statistics and data analysis. Professionals in other fields of application are receiving data analysis courses as a part of undergraduate and graduate training. Numbers of courses are increasing in several fields of engineering, the social sciences, biology and the physical sciences.

The participants at this symposium fall into two categories--those who have access to professional statistical support and those who are their own data analysts. In the latter case you should leave with definite ideas concerning what areas you need to pursue for your own edification. The following topics represent areas with which you are now familiar or have made note to pursue.

> maximum likelihood
> IRWLS
> PRESS residuals
> ordinary residuals
> data splitting
> partial residual plots
> logistic model
> discriminant analysis
> ridge analysis
> variance inflation factors
> hat diagonal
> Cook's D

In the following paragraphs we present some highlights of the statistical methods papers and mention a few areas for further exposition.

MODEL SELECTION CRITERIA

For quantifying prediction performance, the C_p statistic is a measure of "total error" in prediction, summed over the data locations. The PRESS statistic reflects model prediction performance by setting aside single data points one at a time and predicting them with the model constructed without their use.

OUTLIERS

Outliers in the y-direction can be diagnosed using the studentized residual, i.e.,

$$t_i = (y_i - \hat{y}_i)/(s_{-i}\sqrt{1-h_{ii}}).$$

Cutoff values are ±3 for situations where the data point in question is not under scrutiny a priori.

HAT DIAGONAL

Hat diagonal values given by

$$h_{ii} = x_i'(X'X)^{-1}x_i$$

represent diagnostics that measure the leverage associated with individual data points. The hat diagonal is a standardized distance from the point x_i to the vector of averages $\bar{x}$. It plays an important role in regression diagnostics.

REGRESSION DIAGNOSTICS

Regression diagnostics that are attentive to detection of influence are designed to identify data points that should be reexamined. There may be no necessity for eliminating the point from the data set. Indeed, the data point in question may be the most important and meaningful of all locations in the data set. On the other hand if errors are found in the data point, the analyst may wish to eliminate it. If the analyst conjectures in retrospect that perhaps the phenomenon is different at that location, it may lead to a new region of experimentation.

The diagnostics here involve the inspection of changes in statistical criteria (regression coefficients, fitted values, etc.) if observation i were to be "set aside".

MULTICOLLINEARITY

Symptoms and diagnosis of multicollinearity have been reviewed in previous papers. Large standard errors, poor prediction and hopeless tasks of variable screening are commonplace under severe multicollinearity. Perhaps the most telling formula indicating the effect of multicollinearity (at least one near zero eigenvalue) is given by

$$\sum_{i=1}^{m} \text{Var } \hat{\beta}_i/\sigma^2 = \sum_{i=1}^{m} 1/\lambda_i$$

where $\lambda_1, \lambda_2, \ldots, \lambda_m$ are the eigenvalues of the correlation matrix. Biased estimation techniques such as ridge regression are often used to combat multicollinearity.

LEAST SQUARES ASSUMPTIONS AND ROBUST REGRESSION

Robust regression is designed to produce an alternative to least squares as an estimation procedure in cases where the standard least squares assumptions do not hold. Under nonnormality of the error distribution, the M estimators discussed in previous papers tend to reduce the influence of data points where large residuals are produced.

REGRESSION PLOTS

Plots involving data sets often reveal many things. Plots of residuals or studentized residuals can reveal failures in the standard least squares assumptions. Model inadequacies may be detected. Failure of the homogeneous variance assumption may be a conclusion drawn from a plot of the residuals. One of the more interesting and instructive plots is the *partial* residual plot in which a set of *adjusted* y values is plotted against a regressor variable that has been adjusted for all other regressors. The plot reveals the true role of the regressor variable and the least squares slope of the plotted value is the regression coefficient in the multiple regression.

The future of data analysis in health physics or medical physics is indeed challenging. The closeness of the physicist to therapy and decision making by physicians makes it mandatory that good data analysis practices be used in the field. Participants should be able to help consulting statisticians use proper methods.

Regression Methods for Problems of Identification in the Health Sciences

G.T. Barnes
Department of Radiology, School of Medicine
University of Alabama at Birmingham, Birmingham, AL 35233

As was discussed in the preceeding papers, regression methods have proved to be extremely powerful in identifying the efficacy of different treatment modalities, assessing radiocarcinogenic risk and evaluating dose response models. To date, multivariate regression techniques have not been fully exploited to address the more general problem in the Health Sciences - the identification of disease. Based on the patient's current symptoms and past history and his experience, an attending physician orders a series of diagnostic exams and tests to identify the patient's problem. He in essence correlates the patient's symptoms and past history with the capabilities of the various diagnostic studies and systematically selects those that are needed to define the problem. That is, he applies an algorithm based on his experience and medical knowledge to identify the most likely cause of the patient's problem. As illustrated in Figure 1, the input information developed in the patient's workup can be thought of as the predictor variables in the familiar regression equation.

$$Y = X\beta + \varepsilon \qquad (1)$$

One can conceive of a situation where the number of regressor variables are initially only a few and are utilized to design optimal diagnostic exam and test strategies. The result of these exams and tests add to the number of regressor variables and the sequence is repeated until a diagnosis is reached.

Figure 1. The problem of identification in medicine involves synthesizing a wide range of information, i.e., symptoms, test results and exam results. The input information to this decision process is analogous to the predictor variables of multivariate regression.

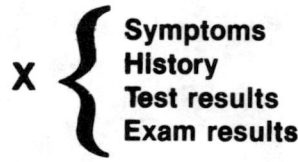

Although logical, such a senario doesn't exist at present. The reasons are many. The problem is extremely complex, and individuals with sufficient time and knowledge of medicine and statistical analysis to define and refine the decision making algorithms are virtually non-existent. Also, there is an understandable resistance by physicians to computer aided diagnosis. Motivating factors are the increased flexibility, power and cost effectiveness of computers - a factor which has positively influenced the growth of regression analysis during the past two decades - and the pressure for increased efficiency in the health care delivery system.

The latter motivating factor is a result of the excessive growth of the health care system during the past forty years and is evident in the recently introduced prospective payment system and diagnostically related groups (DRGs). In 1949, health care costs were 1.4% of the gross national product(1). In 1974 the percentage was 9%(1) and a conservative current estimate is 12%. The primary reason for this growth was the emergence of third party carriers and the majority of major U.S. employers offering medical insurance as a fringe benefit after World War II. With reasonably comprehensive medical coverage and no out-of-pocket expenses, individuals tended to overuse the available services. Health care insurance companies' profits were based on a percentage of their cash flow and thus, they had little concern about efficiency and rising cost. They also reimbursed hospitals on a cost plus basis which did not foster an efficient health care delivery system. Such is not the case at the present with prospective payment and methods of improving efficiency are highly desirable.

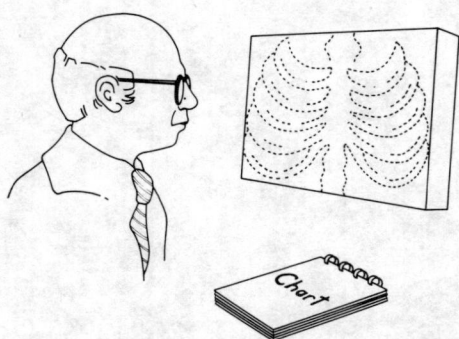

Figure 2. The reading of an x-ray film by radiologists is analogous to the regression problem - the predictor variables being the projected anatomical patterns and the patient's clinical history.

Although regression analysis has not been applied extensively to the general problem, it has applied with success to specific problems. An example in diagnostic imaging is the work by Partridge, Scott, Deverall and Macartney(2). Utilizing discriminate analysis they were able to objectively assess thoracic situs in congenital heart disease by measurement of the main bronchi. Their accuracy was 99.9%. Abnormal anatomical patterns are what radiologists are extensively trained to recognize. As illustrated in Figure 2 their input (regressor) information is the projected anatomy and the patient's clinical history, and it is of interest that a knowledge of the latter improves the accuracy of roentgenographic findings(3).

Initial attempts at computerized reading of chest films at the University of Missouri met with limited success(4). A problem with this initial study was the losses associated with converting from an analog to a digital image. Recently developed digital radiographic units do not have this limitation(5) and the inherent flexibility of the resultant high resolution digital image facilitates such analysis. Also, the time interval and energy subtraction capabilities of such systems significantly reduce structure noise and the complexity of the analysis.

REFERENCES

1. The National Income and Product Accounts of the United States, 1929-1974. United States Department of Commerce/Bureau of Economic Analysis, pp. 6-7, 88-89.
2. J.B. Partridge, O. Scott, P.B. Deverall, and F.J. Macartney, Circulation **15**, 188 (1975).
3. M.H. Schreiber, JAMA **185**, 137 (1963).
4. S.J. Dwyer, "Computer Applications in Diagnostic Radiology," Presented at the Annual Meeting of the American Association of Physicists in Medicine, Kansas City, Missouri, July 29 - August 1, 1974.
5. R.G. Fraser, E. Breatnach, G.T. Barnes, Radiology **148**, 1 (1983).

THE USE OF STATISTICS BY MEDICAL PHYSICISTS

G. Donald Frey, Ph.D., Medical University of South Carolina
Dept. of Radiology, Charleston, South Carolina 29425

The medical physicist's requirement for a knowledge of statistics has increased as medical physics has become more sophisticated. This is especially true in radiologic physics because the role of the radiologic physicist has expanded as the number and type of radiologic procedures has increased. The introduction of new modalities such as computerized tomography, digital radiography, magnetic resonance imaging, and positron emission tomography has altered the nature of radiologic practice and increased the need for medical physics participation within the radiology department.

The addition to the new modalities has increased the complexity of radiologic procedures. These technically more sophisticated procedures are more likely to produce quantitative rather than qualitative results, and in some cases such as quantitative thallium interpretation programs, the results that are presented bear little resemblance to direct imaging. As procedures become more quantitative and medical knowledge expands to utilize the increased information available, it is necessary to apply statistical principles to aid in analyzing results. For example, the output of some thallium quantitative programs use a color scale that relate uptake in the myocardium to a normal population data base. The scale units are standard deviations from the normal population. Information like this by its very nature requires a statistical approach to its collection, processing and interpretation. When quantitative procedures such as these are used, it is necessary to determine the range of normals at an individual institution and to be able to use statistical methods to calculated probabilities for abnormal outcomes. In most departments, the medical physicist is best equipped to supervise the conduct of quantitative procedures and to manage the quality control associated with the procedure.

Another role of the medical physicist that relates to statistics is in relation to medical research. While most large institutions have their own statistical staffs, many day-to-day statistical questions arise in the course of research that are best referred to someone within the radiology department, and the medical physicist is best equipped to function in this area. In addition, when a more advanced statistical analysis is required, the medical physicist can serve as the liaison between the research group and the statisticians. As research becomes more complex and requires more and more use of statistics in its evaluation, statistics becomes an integral part of the research process from the time the experiments are first planned until the completion of the evaluation of the data. In most radiologic research situations, the radiologic physicist is the member of the research team best suited to supply the needed statistical knowledge.

REGRESSION METHODS FOR PROBLEMS OF "YIELD" IN HEALTH SCIENCES

Jon H. Trueblood
Dept. of Radiology, Medical College of GA, Augusta, GA 30912

Frequently the health science data base has no clear apriori model and the "yield" from regression analysis is an arbitrary statistical model. The interpretation of this yield, or model, may give rise to controversy which can only be resolved by the collection of more definitive data which, in turn, rigorously supports a predictive model based on reasonable fundamental physical and socio-economic theory. Once established the predictive model bears its own fruit; the statistical analysis can be judged relative to the accepted model and thereby produces a more meaningful yield. An example of this problem of yield is the controversy which surrounds the models for the response variable as a function of low dose ionizing radiation. Indeed, these problems of "yield" using statistical regression analysis in the health sciences are very broad in scope. The present discussion is limited in scope and will address the current and future potential yield using regression analysis methods as applied to the problems associated with the patient care process. These thoughts are derived from the perspective of the author's many years of experience as a medical physicist.

The diagram below schematically represents the medical care system process for the sick patient. Regression analysis methods can be applied to this system but the yield from such analysis is dependent on a complex series of statistical methods, including regression analysis, applied to a very wide variety of data sets obtained in extremely diverse ways.

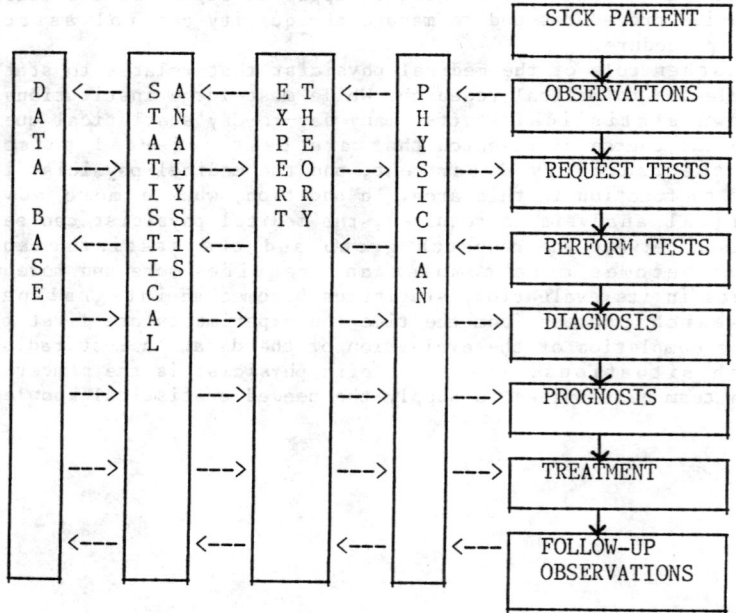

The flow depicted in the above diagram is basically familiar to any individual significantly involved in the care of patients. The manner in which the steps are carried out varies enormously. For example, the critical physician's role can range from an educated thought process wherein observations are judgementally correlated, to the structural process (scientific method) of a laboratory scientist gathering data with precisely modeled conclusions which are based on computer statistical analysis of a large population data base. This latter approach would be, for the statistician, the ideal solution to efficient care of sick patients. However, each step in this patient care process is beset by inherent problems which introduce systematic errors that can mask the random errors and make the yield from analysis by regression onto independent variables, which are difficult to define, of marginal value.

In the first step of the patient care process, the observations made by the physician can be in error in type and magnitude. For example, symptoms described by the patient may be mild rather than severe and the swelling observed by the physician may actually be a growth. In the second step, the expert theory (i.e. a "textbook case") may be based on statistical analysis of poorly defined data and may be biased by the personal opinion of the expert's unproven theory. As a result, more or fewer tests than necessary may then be conducted on the patient to obtain the appropriate data set on which to base the diagnostic decision, give a prognosis and prescribe a treatment. The follow-up observations suffer from the same physician subjective judgement process as the initial observations about the state of the patient's health. Hopefully these follow-up observations, which provide critical feedback to the data base and result in modification of expert theory, will establish an iterative self correcting process with a definitive set of variables and data collection techniques.

In this patient care process the application of multiple regression analysis can occur at several levels. At the basic level the approach can be used for analysis in specific diagnostic tests to determine the "best fit" curve for a multicompartment model which is established on the basis of a fundamental theory about a particular metabolic or physiological body function.[1] Typical independent variables are time, uptake and clearance constants, indicator concentrations, reaction time rates, pressures, PH levels, etc. Some specific examples include compartmental analysis of blood flow in renal, cardiac, brain and lung diagnostic studies using dye dilution, radionuclide concentration and other biochemical tracers.[2] A particularly successful situation is the example of applying the regression technique to find the "best-fit" standard curve for radioimmunoassay (RIA) in vitro laboratory tests.[3] In general the successful use of regression analysis (yield) in these diagnostic tests is dictated by the careful reduction of systematic error, precise analysis of the random error and a model which minifies the dependence between the chosen variables but magnifies the sensitivity of the response function to small changes in these variables.

The role of these diagnostic tests is to illuminate or establish a particular patient syndrome and, in turn, aid the physician in a logical sequential decision process, which is basically Bayesian in character[4] about the disease state of the patient. On the global level of the patient care process these syndromes, tests and decisions are factors which can be subjected to regression analysis with patient health as the response function while cost and time are parameters to be minimized. Obviously, there are no good apriori models at this global level at present. This deficiency should stimulate the medical science community to standardize a data collection registry in an effort to provide a large data base which could yield a rudimentary statistical model. An example of an approach to this problem is the national tumor registry program in radiation therapy.[5]

In general, finding the correct response function by the technique of regressing onto a set of data points derived from observations of the variation in an independent experimental variable should be a primary goal of the experimental scientist in medicine. Why then have medical physicists and other clinical academic scientists not enthusiastically endorsed the use of regression methods in the patient care process? Following is a list of some of the probable reasons:

1. These statistical techniques are not a traditional part of academic curriculum for (medical) physicists or physicians.
2. The cofactor dependence of many response variables is too complex for simple independent variable modeling.
3. Methods for data gathering are too subjective to apply precise data analysis successfully.
4. Improvements in the objective methods for obtaining data through sophisticated technology results in nonuniversal rapidly changing protocols.
5. Computer speed, memory and a limited access to a large data base have been a problem in the past.
6. A dynamic living environment may preclude establishing a reference data base.
7. Normative patient data is often dependent on unknown genetic population mix characteristics, region and protocol.
8. Physical scientists are reluctant to use purely statistical models, but prefer to postulate a theoretical model which is supported by statistical evidence.
9. Attempts to use the methods of regression analysis in the past have had a poor "yield" and consequently do not affect patient care significantly.

Although some of the above impediments to the wider use of regression analysis and other statistical methods will remain for years to come, interdisciplinary education (such as this symposium), new technologies and increasing availability of computer power will enhance the use of fundamental statistical techniques in the patient care process. Perhaps the best example of this evolutionary process will occur in the use of data collection technology of medical imaging. Within the next few years, a radiologist in most hospitals will have on-line access to millions of images obtained using a variety of different physical image data gathering processes, such as, ultrasound, computerized tomography (cross-sectional x-ray transmission), nuclear emission (chemical tracer) and nuclear magnetic resonance (proton density and biochemical content). The immediate access to these images, along with clinical laboratory findings and physician observations will provide a data base locally specific and demanding of statistical analysis. The three-dimensional character of computerized tomography, high resolution ultrasound and magnetic resonance images provides a new level of definition in the patient data set. Organ shape and volume, tissue characterization and distribution, etc. are potential variables for statistical analysis. The ability to determine the most efficient method to arrive at differential diagnosis will no longer be limited by the access to a large data base but only by the quality of the data obtained through the imaging device, the laboratory technician and the statistically oriented observant physician. Once the methods of statistics are properly implemented, expert systems and artificial intelligence techniques can be used to optimize the patient care delivery process.

The methods of multiple regression analysis have the potential for abetting each step in the patient care process to make the outcome better for the patient and generally more efficient as demanded by the provisions of the Federal DRG (diagnostic related groups) regulations. To have a significant yield from this analysis will require enhancing the objectivity and definitiveness of data obtained from the patient and provision for on-line access to a large patient data base. (Attainment of the goal of cost efficiency through the analysis of well defined data sets, ironically, may require for a time a greater expenditure of monies for highly sophisticated technology rather than less). Unfortunately, regardless of the quality of the current data set, the analysis will most likely proceed. The ultimate yield from this analysis, or perhaps more precisely the interpretation of the yield, holds the quality of life for individuals and for mankind in the balance. Controversy concerning interpretation of the yield will stem, not from the inexactness of the statistical science, but from the analysis of data derived from responses of the human organism to an infinitely complex laboratory, the natural environment.

REFERENCES

1. J. A. Jacquez (1968), "Tracer Kinetics," in Principles of Nuclear Medicine, ed. H.N. Wagner, Philadelphia. W.B. Sanders Co., 45-74.
2. C. M. Boyd and G. V. Dalrymple (1974), "Tracer Principles," in Basic Principles of Nuclear Medicine. eds. C. M. Boyd and G. V. Dalrymple, St. Louis: Mosby, 107-138.
3. Clin Lab Library Vol. 1: Program Part No. 09845-14251, (1978), Hewlett-Packard.
4. H. N. Wagner, Jr. and W. W. Walton, "The Diagnostic Process," in Principles of Nuclear Medicine, ed. H. N. Wagner, Philadelphia: W. B. Sanders Co., 15-22.
5. Cancer Program Manual: A Supplement on Tumor Registry (January, 1981), Chicago: Publication of the American College of Surgeons.

REGRESSION METHODS FOR PROBLEMS OF "YIELD" IN HEALTH SCIENCES: ONE PARTICIPANT'S REPORT FROM THE SYMPOSIUM

Larry D. Simpson, Ph.D.
University of Rochester Cancer Center
Rochester, New York 14642

Medical physicists, radiological physicists, and health physicists often find themselves at the interface between a clinical-biological world and the physical-mathematical world in their basic and applied interdisciplinary scientific efforts. Scientists from diverse areas are drawn to these interdisciplinary fields. With rare exceptions, all have one overriding parameter missing from their backgrounds, to be so involved at such an interface. Don Herbert is one such exception and you see before you the personal culmination of his efforts in guiding the majority of the rest of us towards a better appreciation and understanding of current, state-of-the-art multiple regression analyses as applied in the health sciences. I personally commend Don Herbert for his successful, well organized Symposium and, most importantly, his continued efforts at editing and bringing together for the reader's benefit, these important reports from that Symposium. He of course derived significant support and assistance from the Regional Southeast Chapter of the A.A.P.M.

Readers, you have before you, comprehensive expositions on multiple regression analysis, as well as examples of specific applications to the health sciences. Be aware, that in these written presentations, it has been difficult for individuals to convey the spirit and mode present during the give and take of the lecture-question format of the actual Symposium -- perhaps best expressed by a leader in this field, R. R. Hocking:

"The use of diagnostic procedures to reveal problems with the data and/or the model is highly recommended. We must, however, recognize that the fitting of equations to observational data (as opposed to data from carefully designed experiments) is at best a risky business. With enough manipulation of the data we can substantially modify the results. Perhaps a Bayesian approach might be more satisfying. Surely prior information on the problem should influence the analysis but, again, we must avoid automated procedures. There are many examples in which the most suspicious observation turned out to be the most interesting. I can only recommend extreme caution and common sense when scrutinizing the data and the form of the model."

What I would call, the "classical" regression analyses as well as "Bayesian" regression and sensitivity analyses have both been masterfully reviewed and taught. One hopes, as a minimum, that readers of these Proceedings will better cope, with less stress, with the fact that there may not be a unique order of variables in a model and that there may likely be no well-defined "best" equation. Alternatively, these Symposium Proceedings may permit you to initiate studies which result in quantitative "glimmers" or potentially verifiable or falsifiable interpretations of data, presented to you at some time in the future.

IMPRESSIONS AND PERSPECTIVES

Robert J. Shalek

The medical physicist is called upon to do many things in a hospital. In addition to the scientific and technical aspects of radiation as applied in therapy and diagnosis, he may have responsibilities for radiation, chemical, electrical and laser safety. Though he may not consider himself a mathematician or statistician he may well be the most knowledgeable person readily available to explain the mysteries of numbers. Most medical physicists understand standard deviation, standard deviation of the mean, linear regression, the student's t test and perhaps receiver operating characteristics (ROC) analysis, but robust regression, maximum likelihood, ridge regression and iterative reweighted least squares would sound like a foreign language for most. The clearly presented lectures at this symposium have allowed me to feel that I almost understand some of these things. I believe that the published proceedings will be a valuable reference for many a practicing medical physicist. It may help him preserve his reputation as the expert of everything difficult. More importantly, some physicists, and we have examples at this symposium will be stimulated to become competent statisticians and to make creative solutions to problems they encounter in their work.

CLOSING COMMENTS

Edward L. Chaney
University of North Carolina, Dept. of Radiation Oncology
Chapel Hill, North Carolina 27514

This symposium has broken new ground in at least two significant ways. First, it has been the beginning of what is hoped to be a continuing series of midyear topical symposia sponsored by the AAPM. It is no longer possible for the annual summer symposium to satisfy the broad based and expanding continuing education needs of the medical sciences community served by the AAPM, and this symposium represents a bold experiment in finding alterntive ways to satisfy those needs. Second, this symposium has addressed significantly for the first time by the AAPM a topic of vital but often overlooked importance to all fields of the health sciences. Much of the current literature, some of which has been discussed here, is testimony to the need for medically related scientists to more fully understand and correctly apply statistical analysis techniques. Hopefully, this symposium and the resulting proceedings will awaken awareness of those problems and focus attention on their resolution.

In closing, thanks are extended to all involved in making the symposium a success. Thanks to an outstanding faculty for well prepared lectures and manuscripts, and to institutional and commercial sponsors. Special acknowledgement is extended to the National Cancer Institute for grant support. Finally, the symposium would not have been possible without the foresight, perseverance and dedication of Don Herbert and the symposium committee members, Gary Barnes, Don Frey, Bob Shalek, Larry Simpson and Jon Trueblood.

INDEX

A

Actuarial analysis, 557, 559
Adequate regression model
  definition, 221, 256
Adjusted R^2, 493, 527
Akaike Information Criterion (AIC),
  265-267, 319, 320, 329, 395
Affine transformation, 489
Alias matrix, 271
Allocation rules in discriminant
  analysis
  minimum total error, 150, 494
  minimum total loss, 150, 494
  maximum "resemblance", 499
  hypothesis test of "membership",
    509
All possible regressions, 10
Analysis of variance for multiple
  regression model, 106, 107, 109,
  119
Antibody saturation curves, 546
A priori information
  in model building, 58
  on measurement errors in X, 285–
    287, 328–332
  on form of model, 267, 268,
    312, 317, 436
  on parameter vector, 58, 232,
    324
Assumptions for regression models
  linear, 5, 6, 218
  general linear, 86, 106, 115,
    222

B

Backward elimination, 10
Bayes' Factor, 268, 312, 436
Bayesian methods, 58, 225, 580–582
Bayesian-type techniques
  mixed-estimation, 226
  probability of a hypothesis,
    492
  probability of a model, 268,
    312, 436
  pure Bayes, 58, 225
  ridge regression, 274
BEIR III Report, assessment of, 307
  data of, 337–341
β/α ratio for LQ model, 193, 194,
  200, 206
Biased estimation, 35
  generalized ridge, 39–42
  principal components, 42–45,
    359, 447
  ridge regression, 35–42, 274–
    285, 289, 296–297
Bias in regression estimates
  due to identification errors in
    y, 241, 251, 256, 297
  due to measurement errors in X,
    284–289, 330–331
  due to mis-specification of X,
    11, 271
  due to perturbation of $X'X$, 13,
    35–42, 274–278
  due to small sample size, 223,
    289
Binomial regression, 84, 86, 103,
  115, 222
Bioassay, 162, 190
Bootstrap method, 147, 153
Bounded influence estimation, 52

C

Cancer (animal)
  liver, 107, 138
Cancer (human)
  breast, 128, 129–131, 148–150,
    308
  leukemia, 308
  lung, 96, 98, 99, 102
  malignant melanoma, 141
  non-Hodgkins lymphoma, 565
  other, 308
Cardiac output, 551

Case statistics, 238–241
Catenary pathways, 165
Censored data, 125, 136, 557
Centered variables, 423–426
    and Dystra's d, 426
Chi-squared statistic (goodness of fit)
    of model and sample, 222, 232, 311, 316, 326, 396
    of *a priori* and sample estimates of parameter vector, 227
Clinical trials
    historical controls vs randomization, 129, 133, 143–144
    Phase I, II, & III studies, 133
Coded levels, 212
Coefficient of determination, 6, 7, 119, 220, 221
Coefficient vector, 5
Collinearity, 12, 20, 34, 246, 269, 270, 315, 337, 586
    condition number, 233, 399
    diagnostics for, 233, 247, 399
    ill effects of, 13, 247, 284, 313
    and variance of model coefficients, 21, 220
Collinearity, remedies for
    Bayesian estimation, 58
    mixed estimation, 251, 272, 282, 285
    additional data, 314, 316, 337
    principal components regression 42
    ridge regression, 35–42, 274, 296, 297
Compartmental analysis, 546, 547
Computer
    small, 75
    graphics on, 80
Computer, regression software
    BMDP, 7, 147, 160
    GLIM-3, 107, 110
    MINITAB, 29
    P-STAT, 147, 160
    SAS, 29, 53, 161
    SPSS, 147, 160
    TROLL, 29
    graphics, 81
Condition number of matrix, 399
Confidence interval
    for model coefficient, 248
    for $E(x_I(P))$, 258–259, 360, 361 363
    for $E(y)$, 8, 9, 64, 65, 355, 360, 361, 363
    for spline knot, 362
    for misclassification rates, 505, 526
Confidence ellipsoid, 38, 251, 252, 296, 297, 364
    volume of, 220
Constraints
    linear, 226, 312, 321
    stochastic, 226, 272–274, 312–313, 321–325
Contour ellipsoid, 152, 234, 517, 537, 538
Cook's D, 19, 26, 52, 239, 241, 250, 251, 281, 296, 297
Correlation
    matrix, 247, 277
    and regression, 269
    between residuals, 425
Correlation coefficient, 269
    effect of measurement error on, 504
Covariance matrix, 6, 116
    trace$\{(X'X)^{-1}\}$ and collinearity, 39
    det$\{X'X\}$ and influential points, 27
    test of equality of, 497
Cox proportional hazards model, 124, 130, 557, 562–572, 576
C_p statistic, 11, 267
    (*see also* Akaike Information Criterion)
Criticism, 2
Cross-validation, 11, 310
    (*see also* PRESS, DUPLEX, Validation of model)
Cutoff values
    for condition numbers, 233
    for case statistics
        absolute, 23

size adjusted, 23
for VIF, 247

D

Data-augmentation, 37, 53, 58, 60, 224, 230, 324, 315, 333, 337
Data-splitting (*see* Cross-validation, PRESS, Validation of model)
Deletion diagnostics, 17, 24, 25, 26, 52, 239
  (*see also* Cook's *D*, PRESS, RSTUDENT)
  and masking, 322, 346
  single-row
    COVRATIO, 27, 311, 349
    DFBETA(S), 25, 311, 347, 348
    DFFIT(S), 26, 52, 311, 350
    hat matrix and, 22, 25–28
    residuals and, 23, 25–28
Dependent variable, 5, 18, 22, 84, 85, 88
Design of experiments, 212, 214, 218, 395, 444–446
Deviance, 93, 96, 106, 112, 118
Diagnostic radiology, 365, 487, 588, 591, 592
Diagnostically related groups (DRG), 589, 595
Digital radiographic unit, 589
Discriminant analysis
  probability of hypothesis vs test of hypothesis, 492, 495
  multiple groups, 160
  two groups, 146, 148, 589
Discriminant function
  linear (LDF), 146, 152, 233, 493
  variance of coefficients, 502
  quadratic (QDF) 147, 152, 496
Dose-response curve, 199, 256, 360–363
Dose-response models
  discriminant function, 233–236
  linear, 138
  logit, 138, 198
  multi-hit, 137, 138
  one-hit, 138

Poisson, 85, 309
probit, 88, 108, 138, 222, 233
Weibull, 138
Dose-response surface, 232, 411
Dual radiation action (DRA) model, 99
Dummy variable (*see* Indicator variable)
DUPLEX algorithm, 263

E

Effective renal plasma flow (ERPF), 547
Eigenvalues and eigenvectors
  symmetric matrix, 233, 243, 247, 248, 447
  asymmetric matrix, 498, 499, 517
Empirical model, 392, 547
Error mean square, 7
Errors
  Type I, 492, 495
  Type II, 492
  Type III, 509, 528
  False positive, 298, 492
  False negative, 298, 492
Errors-in-variables, 284–289, 330–331
Estimation of parameter vectors
  constrained, 226
  generalized least squares, 87, 222, 309, 393
  least squares, 5, 19, 35, 218
  matrix-weighted average, 58, 61, 224, 272
  maximum likelihood, 115
    small sample bias, 223
  mixed, 226, 272
  non-linear least squares, 218, 546
  point, 5, 19, 35, 60, 61, 251, 279, 296
  set (interval), 27, 38, 62, 251, 296, 297, 364
  weighted, 49, 87, 89, 275, 279
Estimation of response
  point, 5, 10, 18, 60, 86, 88, 256, 278, 360, 361, 363

set (interval), 8, 256, 355,
    360, 361, 363
Expert testimony
    prejudicial effect vs probative
        value, 580
Expert witness, 582
Extrapolation, 244, 245, 268
    hidden, 260, 268, 269
    low-dose, 132, 137
    maximum safe dose, 133, 137,
        138
Extreme levels of predictor
    variable, 22, 23, 101, 111,
        238, 386
Extreme levels of response
    variable, 403–404

F

"Fake data", 60, 292
False negative rate, 298, 366
False positive rate, 298, 366
F-distribution
    relation to Cook's D, 239
F-test
    partial, 7
    for significance of regression
        7
Feigl-Zelen model, 128
"Fictitious data", 58, 292
Fieller's theorem, 199, 258, 318
Filliben's probability plot
    correlation coefficient, 254,
        351, 419, 523
Forward selection, 10
Fractionated radiation, 192–194,
    454-460

G

Gehan-Wilcoxon test, 561
Generalized least squares, 222,
    309, 393
Generalized linear models, 28, 86,
    222, 223
"Goodness-of-fit statistics"
    (see Chi-squared, Deviance, R^2,
    PRESS)
Graphics (see Statistical graphics)

H

Hat matrix, 12, 22, 239
    "geometric" location of large h_i
        in sample space, 346, 443
    diagonal elements of, 12, 22,
        239, 345, 346, 586
    designed experiments, 445, 446
    against residuals, 240
    against predictor variables,
        346
    case (index), 241, 250, 280,
        345, 353, 354
Hazard function, 124, 126, 211, 558
    dependence on covariates, 128,
        210
    exponential, 127, 128, 210
    Gompertz, 216
    Weibull, 102, 125, 127
Heterogeneity of variance, 105, 120,
    223, 232, 238, 328, 411, 446
Heterogeneity factor (see Hetero-
    geneity of variance)
Hypothesis testing, 7, 492, 495

I

Identification of disease, 146,
    487, 588
Ill-conditioning of $X'X$ matrix, 13,
    20
Imaging, 587, 594
Independent variable, (see
    Predictor variable)
Indicator dilution model, 546
Indicator variables, 59, 210, 237,
    278, 309, 316, 356, 357, 360
Index (case) plots
    of Residuals (see Residuals)
    of h_i (see Hat matrix)
    of Cook's D, 250, 281, 422, 424,
        428, 439, 444, 512
    of DFBETAS, 347
    of COVRATIO, 349
    of DFFITS, 350
Influence function, 15, 25
Influential observations, 24, 238,
    239

Influential sub-sets, 28
    (*see also* Deletion diag.-masking)
Information theory, 266, 444
Intake, 162
    retention function, 164–165
Interactions
    of model and sample, 17, 311
    of predictor variables, 440
Internal radiation protection, 162
Interpolation, 256, 260
Inverse estimation, 256, 258, 355, 360, 361, 363
Isotonic regression, 135
Isoeffect curve, 194, 200, 441
Isorisk curve, 529, 530, 538, 539
Iteratively reweighted least squares (IRLS), 52, 84, 116

J

Jackknife validation (*see* PRESS)

K

Knot (spline function), 318, 357, 430
    as Estimated No Observed Effect Level (E-NOEL), 318

L

"Lack of fit" vs heterogeneity of variance, 18, 105, 120, 446
Law of Small Numbers, 236, 532
Latent root regression, 44
Least squares
    assumptions, 6, 46
    generalized, 222
    iteratively reweighted, 116, 222
    link, 222
    properties, 6, 35
    restricted, 226
    weighted, 71, 87, 89, 275, 279
Lachenbruch's "Leaving-one-out" method, 147, 153, 525, 527
Legal liabilities of medical scientists, 581–582
Legal questions, statistics in, 68, 351, 579

Leverage, 21, 23
Linear model, 5, 86, 218
    improper, 488, 533
    proper, 488, 533
Linear-Quadratic model
    high-dose, 193, 309, 392, 432, 441, 458, 459
    low-dose, 309, 316
L-L model, 309, 316
Link function, 86, 212, 222
Logarithmic transformation, 212, 358, 431
Logistic function, 198, 233, 234, 500, 529, 530, 538
Logistic regression, 105, 200, 392
Log-linear model, 86, 96
LQ-L model (*see* Linear-quadratic model)

M

Mahalanobis distance, 22, 153, 236, 493, 502
    relation to R^2, 233, 502
    "shrinkage" of in new sample, 493
    unbiased estimate of, 493
Malignancy Index, 487, 488
Management equation, 228
Mantel-Haenzel test, 561
Matrix-weighted average, 58, 225, 227, 315, 316
Mean square error of estimator, 35, 36
Mechanistic model, 392
Medical care system, model of, 491, 593
Method of scoring, 374, 375
Mixed-estimation, 226, 272–274, 312–313
Model selection criteria, 6–7, 10–11, 396 (*see also* AIC, C_p, posterior odds ratio, PRESS, R^2)
"Mouse-to-man" problem, 132, 138, 216, 385
Multicollinearity
    (*see* collinearity)
Multiple correlation coefficient, 7, 21, 319

"shrinkage" of in new sample, 262

N

Near-dependency of predictor variables (see Collinearity)
Newton–Raphson method, 374, 375
Nonlinear models, 99, 546
Nonlinear regression, 218, 444, 446, 546
Non-normal distribution, 34, 45
Normal distribution, tests for
  univariate, 252, 351, 524
  multivariate, 497, 514, 516
Normal equations, 219, 222
Nuclear medicine, 546

O

Orthogonal columns in X, 212, 397, 401, 446
Outliers, 13, 17, 238, 585
  "geometric" location of in sample space, 345, 443
  multivariate, 17
  and Normal probability plots, 524
  transformations and, 332
Over-dispersion factor (see Heterogeneity factor)
Overspecified model, 11

P

Parameter vector, 5, 18, 35
Parsimony in model building, 112, 220, 310
Partial F-test, 7
Partial regression coefficients, 19
  plots, 19, 20
Partial tolerance fraction (PTF), 455, 458, 459
Perturbation Index, 286, 287, 313, 328
Perturbations of assumptions, 58
Phase I, II, and III studies, 133
Piecewise regression, 448
Poisson regression, 84, 86, 89, 115, 307

"Polygon" models, 318, 319
Polynomial models, 64, 309, 395
  examples, 91, 316, 423–426
Pooling data (see Data augmentation, Matrix-weighted average)
Posterior odds ratio, 268, 312, 436
Predicted response
  as function of predictor variables, 5, 18
  as function of observed response, 22
Prediction interval, 8
Predictor variables, 5, 18
  random errors in, 284–289, 330–331
PRESS, 11, 24, 263, 264, 311, 352, 503, 527
  index plot, 352
Prior information, 58
Prior odds ratio, 268, 312, 436
Principal component regression, 42, 359, 447
Probability, concepts of
  frequency vs subjective, 59
Probability density function, 124, 127
Probability plot
  chi-squared (bivariate), 514–527
  Normal, 524
  correlation coefficient, 254, 351, 419, 523
  outliers in, 438, 524
Probit models, 108, 232, 395, 397, 411, 432
Prognostic stratum, 210, 356, 360
Projection matrix (see Hat matrix)
Proportions as responses
  arc-sine transformation, 142
  logit transformation, 554
  probit transformation, 214
Pseudo-observations, 313, 327, 353, 354, 358, 393
Pseudo-response vector, 222, 393

Q

Q-L model, 309

R

R^2, 6, 265, 275, 285, 493, 527
  adjusted, 493, 527
  $E(R^2)$ function of p, n, 7, 221
Radiation therapy, 192, 208, 385, 454
Radiocarcinogenesis, 307
Radioimmunoassay, 546, 554
Receiver operating charcteristic, 154, 158, 365–369, 525, 526, 527
  plot, 155, 158, 526
Regression diagnostics, 12, 14, 22, 25, 26, 52, 119, 238
Regressor variable hull (RVH), 260
Rejection of observations, 241–244
Replication, 103, 397, 446
Residual MS (mean square),
Residuals
  definition of, 13, 23
  mean square, 7
  ordinary, 6, 20, 23
  outliers, 238
  case (index), 240, 249, 280, 343, 344, 353, 354, 416, 417, 427, 437, 443, 511
  against expected response, 255, 420, 421
  against h_i, 240, 410, 416
  against X_j, 255, 256, 345, 418, 420, 437, 438, 587
  probability plots, 244, 251, 281, 351, 417, 418, 438
  PRESS, 11, 263, 264, 311, 503, 527
  index plot of, 352
  serial correlation, 416
  standardized, 14, 100, 120
  studentized, 14, 23, 24, 343–345
Residual SS (sum of squares), 5, 219, 222, 311, 312
Response surfaces, 232, 411
Response variable
  probit transformation of, 214
Restricted least squares (*see* Constrained estimation)

Ridge regression, 13, 35, 274–278
  Bayesian view, 73, 283
  choice of k, 36–39, 283, 297
  generalized, 39–42
  ridge trace, 37, 296, 297
Robust regression, 15, 45–53
  influence function, 15, 25
  L estimators, 51
  M estimators, 47
  R estimators, 51
Robust ridge regression, 53
RSTUDENT, 23, 24, 343–345

S

Sample size, 139, 140, 221, 236, 263, 397, 407–408, 446, 502
Sampled mean treatment plan, 454
Scaling and centering of predictor variables, 21, 423–425
Scatter diagram, 31–33, 155, 157, 200, 229, 235, 244, 245, 249, 337, 340, 346, 511, 513, 514, 538
Selection of variables procedures, 10, 153
  all possible regressions, 10
  backward elimination, 10
  forward selection, 10
  stepwise, 10, 58, 60
  variations,
Sensitivity analysis, 58, 59, 61, 72, 272, 273
Sensitivity index, 525
Sensitivity of diagnostic test, 366
Specificity of diagnostic test, 366
Signal detection theory, 367
Signal-to-noise and harmful collinearity, 284–289, 330–331
Significance levels, deletion diagnostics (*see* Cutoff values)
Similarity transformation, 548
Single-row diagnostics
  (*see* deletion diagnostics)
Splines, 317–318, 358, 430–434
Squared multiple correlation coefficient, 6, 265, 275, 285, 493, 527

adjusted, 493, 527
Standard error of estimated
 coefficient, 7
Standardized variables, 423–426
Statistical graphics, 75
Statistics in medicine,
 need for, 591-598
Stepwise regression, 10
Stochastic constraints, 227, 322
Surrogate
 biological system, 213–216,
  385, 387–391, 395
 model, 278–286, 332
Survival analysis
 actuarial, 559
 Kaplan–Meier, 559
 product limit, 559
 reduced sample, 559
Survival curves, 136, 203, 561, 573,
 577
Survival models
 Feigl–Zelen, 128
 Myers–Axtell–Zelen, 210, 211
Survival time, 124–127, 557–562
Survivorship function, 126

T

Target
 biological system, 213–216,
  387–391
Taylor series model (see Empirical
 model)
TDF, 455, 459
Theory of Proportionate Effects,
 452
"Threshold dose", 134, 317, 318,
 332, 357, 362, 363
Time
 as a dose-factor, 305, 306
 as a time to response, 305, 306
Time to failure, 125, 538
 "to tumor", 108
Toxicity models
 Multi-hit, 137
 One-hit, 137, 139
Transformations of variables, 17,
 154
 of response variable
  logit, 88, 105
  probit, 214

 of predictor variable
  log, 154, 400, 431
  spline, 431
t-test (Student's t-test)
 for regression coefficient, 7
Trans-scientific questions, 217
Two-alternative, forced choice
 (2-AFC), 367
Two-compartment model, 546

U

Underspecified model, 11
Uptake, 162
 retention function, 164

V

Validation of model, 261
 coefficients, 262
 data-splitting, 11, 263
 DUPLEX algorithm, 263
 new data, 262
 PRESS, 264
 training samples, 151
Variables
 dummy (response), 278, 501
 indicator (predictor), 85, 210,
  237
 predictor, 5
 response, 5
 transformation of
  coding, 212
  logarithm, 212, 358, 431
  logit, 88, 105
  probit, 88, 214
  spline, 358, 431
 selection procedures, 153
Variance-covariance matrix, 6
Variance inflation factor (VIF), 21
 247, 396
VIF (see Variance inflation factor)
Volumes of confidence ellipsoids
 (see Confidence ellipsoid)

W

Weighted least squares (see Least
 Squares, weighted)
Weighted residuals (see Chi-
 squared, PRESS, Robust
 regression, RSTUDENT)